Modeling, Analysis, and Control of 3D Printing Processes

Romdhane Ben Khalifa
University of Gabes, Tunisia

IGI Global
Scientific Publishing
Publishing Tomorrow's Research Today

Vice President of Editorial Melissa Wagner
Director of Acquisitions Mikaela Felty
Director of Book Development Jocelynn Hessler
Production Manager Mike Brehm
Cover Design Jose Rosado

Published in the United States of America by
IGI Global Scientific Publishing
701 East Chocolate Avenue
Hershey, PA, 17033, USA
Tel: 717-533-8845
Fax: 717-533-7115
Website: https://www.igi-global.com E-mail: cust@igi-global.com

Library of Congress Cataloging-in-Publication Data

LCCN: 2025003978 (CIP Data Pending)
ISBN13: 9798337305332
Isbn13Softcover: 9798337305349
EISBN13: 9798337305356

British Cataloguing in Publication Data
A Cataloguing in Publication record for this book is available from the British Library.

Table of Contents

Detailed Table of Contents

companies and specialists for B&C purposes are summarized in this study. Lastly, a brief discussion of possible research possibilities and the benchmarking qualities of these materials is provided.

M. Maniraj, Sardar Raja College of Engineering, Alangulam, India
S. M. Raj Kumar, Francis Xavier Engineering College, Tirunelveli, India

Additive Manufacturing (AM), commonly known as 3D printing, is emerging as a transformative technology capable of reshaping industrial production through material efficiency, design freedom, and decentralized fabrication. However, the long-term sustainability of AM is challenged by significant concerns related to energy consumption, material waste, and limited recyclability of polymers, metals, and composites. This chapter evaluates the environmental implications of AM within the framework of a circular economy, focusing on recycling strategies, energy efficiency, and material lifecycle management. Real-world examples from aerospace and automotive industries, such as Airbus's use of Ti6Al4V and BMW's recycled thermoplastics, illustrate both the opportunities and constraints of implementing sustainable AM at scale. The chapter explores mechanical and chemical recycling methods, biodegradable material alternatives, and AI-driven process optimizations to improve waste reduction and energy performance.

M. Maniraj, Sardar Raja College of Engineering, Alangulam, India
S. M. Raj Kumar, Francis Xavier Engineering College, Tirunelveli, India

The product development process is changing with Design for Additive Manufacturing (DfAM) systems which amplify additive manufacturing capabilities to create optimized designs that deliver high performance as well as sustainable design solutions. This section demonstrates crucial DfAM methods based on topology optimization models and generative design and lattice structures and multi-material implementation and support reduction strategies that boost product capabilities and reduce substance utilization. The research discusses material selection methods together with artificial intelligence-based design and simulation tools and sustainable operational approaches to enhance AM efficiency. The aerospace medical and automotive sectors conducted well-documented implementations of DfAM despite current issues regarding standards and industrial implementation and workforce preparedness. The combination of AI enablements and hybrid manufacturing with digital workflows makes DfAM ready to transform future manufacturing operations by providing sustainable design methods.

The chapter explores the transformative potential of integrating Computer-Aided Design (CAD), Computer-Aided Manufacturing (CAM), and Computer-Aided Engineering (CAE) in 3D printing and additive manufacturing. This amalgamation of digital technologies facilitates seamless design simulation and manufacturing integration, thereby revolutionizing product development across biomedical, aerospace, automotive, and manufacturing sectors. The chapter elucidates how modeling functions as a critical element in 3D printing, focusing on the importance of the CAD model's geometric modeling techniques and design conversion to a printable format. Using simulation techniques such as finite element analysis and computational fluid dynamics is crucial for optimizing part performance, preventing defects, and reducing material usage. The integration of machine learning serves as a solution to enhance process control and ensure quality.

This paper discusses the methodologies of topology optimization for 3D printed mechanical parts and how design, materials, and manufacturing processes interact. The computational topology optimization method, which optimizes material layout within a design space for given loads and boundary conditions, has gained significant importance in additive manufacturing for creating lightweight, high-performance components. Combining topology optimization with 3D printing allows for generating complex shapes and complex internal structures unattainable through conventional manufacturing techniques. The paper evaluates different topology optimization approaches, including density-based lev, el set methods, and evolutionary algorithms, to determine their benefits and drawbacks when applied to additive manufacturing limitations.

The evolution of additive manufacturing (AM), particularly 3D printing, from a prototyping tool to an industrial-scale production method has been heavily influenced by the development of advanced control systems. These systems are vital for achieving high precision, consistency, and efficiency across various AM technologies. This chapter presents a comprehensive review of control systems in AM, focusing on the theoretical underpinnings and practical implementations necessary for industrial adoption. It explores the integration of control engineering principles within the unique constraints of AM processes. The study synthesizes existing literature and technological developments concerning control strategies used in 3D printing. This includes the examination of open and closed-loop systems, sensor integration, real-time monitoring, and the application of advanced computational methods such as model predictive control, adaptive algorithms, and artificial intelligence.

Abstract: In modern manufacturing World, additive manufacturing (AM) often called as 3D Printing has become the dominant technique This Chapter represents study delver into the realm of AM with a specific focus on composite materials, starting with exploring various 3D printing processes tailored for composites. The study discusses the modeling, Analysis & Simulation of 3D printing processes, laying the foundation for a comprehensive understanding of this transformative technology. Numerical simulation of printed parts navigates through the intricacies of micro-, macro-, and meso-scale modeling, offering valuable insights into the complex world of 3D printing for composites. The study also scrutinizes the multifaceted factors that influence the characteristics of 3D-printed composites. Each facet contributes significantly to the overall performance and properties of printed composite materials. By exploring modeling, Analysis simulation, this study offers a roadmap of 3D printing in composites, fostering innovation and advancements in this burgeoning application

Over the past decade, silicon photonics has grown rapidly in the datacom industry, with applications in LIDAR, sensing, quantum photonics, and AI. This expansion has led commercial CMOS foundries to establish open-access silicon photonics process lines for large-scale production. Meanwhile, university research labs use e-beam lithography for small-scale prototyping, fostering innovation. However, transitioning from research to industry poses challenges due to the incompatibility of research tools with industry-standard fabrication. The CORNERSTONE rapid-prototyping platform addresses this by offering deep-UV lithography-based fabrication, enabling scalable production while maintaining design flexibility. It serves as a bridge between academia and industry, facilitating the commercialization of novel designs. CORNERSTONE's capabilities support the growing demand for photonic solutions and drive collaboration between stakeholders. This review explores its current strengths and future potential, highlighting its critical role in advancing silicon photonics.

Additive manufacturing (AM) has emerged as a transformative technology for producing complex, high-performance components across industries such as aerospace, healthcare, and automotive. Central to enhancing the precision, reliability, and scalability of AM is the integration of advanced sensors and real-time monitoring systems. These technologies enable in-situ process control, early defect detection, and adaptive feedback mechanisms that elevate both production quality and efficiency. This chapter delves into the diverse roles of sensor technologies—including thermal, optical, force, gas, and acoustic sensors—and how their deployment facilitates intelligent decision-making through machine learning and artificial intelligence. In addition, we discuss how real time data analytics and computational modeling are being used to predict and control process outcomes and move manufacturing away from reactive to predictive. The growing relevance of hybrid manufacturing, which combines additive and subtractive processes is also examined for its relevance to high precision applications.

Preface

INTRODUCTION

Additive manufacturing (AM), also known as 3D printing, has developed in the last few years from a tool for creating prototypes to a fully-fledged, transformative technology at the heart of Industry 4.0 and the fourth industrial revolution. The book Modeling, Analysis, and Control of 3D Printing Processes gives an in-depth look at scientific, engineering, and practical foundations supporting contemporary 3D printing processes, emphasizing modeling simulations and control methodologies for advanced manufacturing.

By its very nature, three-dimensional printing is an additive process- objects are made layer by layer from digital models rather than through a traditional subtractive process in which material is removed to create the final object. This revolution allows complex geometries, customized components, and structures optimized for performance to be made at low cost and with high efficiency compared to traditional methods. The seamless integration of CAD, CAM, and CAE tools enhances the workflow from design to finished product, reducing lead times and costs and increasing design flexibility.

The book discusses the complete 3D printing process chain from creating a digital model, preparing an STL file, and error checking to process planning, parameter selection, and G-code development for machine execution. A broad spectrum of additive manufacturing technologies is covered, including Fused Deposition Modeling (FDM), Stereolithography (SLA), and Powder Bed Fusion (PBF), each with different material compatibilities and application fields. The emphasis has been on composite plastics, innovative materials, and metals, which constitute the growing material palette for designers and engineers.

Great emphasis has been placed on modeling and simulating 3D printing processes. Only through proper simulation can one predict and control the thermal and mechanical behavior, microstructure development, and final part properties, which all play an essential role in quality assurance and reliability for demanding applications. The author tackles real-time monitoring challenges, sensor integration, and closed-loop control necessary for consistent results within industrial settings.

Themes of sustainability and circular economy are often addressed, with the specifics of materials recycling strategies, energy efficiency, and waste reduction within chapters. 3D printing makes it possible to enable localized, on-demand manufacturing and to help reuse and recycle polymers and Metals, making it the leading technology for a more sustainable future.

Emerging topics include 4D printing, wherein objects produced can change shape or function depending on external stimuli. It highlights the continued innovation where materials science, engineering, and digital fabrication converge.

By bridging theory and practice, this book becomes an introductory text and a resource for future researchers, engineers, and practitioners. It provides the reader with tools to deal with even very new and dynamic areas of additive manufacturing, from basic principles through advanced process control to trends that may develop in the future.

Objectives

This book attempts to formulate an elaborate framework to further develop the 3D printing process through a multidisciplinary perspective, incorporating models, analyses, and controls of additive manufacturing processes. The first objective is to create robust modeling and simulation methods to predict microstructural evolution's thermal and mechanical behavior and defects postulated to form in 3D printed materials. It can ultimately be integrated with advanced CAD/CAM/CAE tools to seamlessly transition from the digital design phase to actual realization, which manufacturers expect to be more accurate and reliable. Another essential objective of this study is to develop optimized process control strategies, real-time monitoring systems, and closed-loop feedback systems, which play a vital role in ensuring quality consistency while minimizing errors during the fabrication process. Another thing that this paper will focus on is the expansion of the material palette for 3D printing with composite plastics, innovative materials, and recyclable metals so that the printed parts will be functional yet sustainable and have a broad application range.

The book also aims to set the quality assurance process by providing quantifiable indicators for mechanical performance, dimensional accuracy, and surface finish under the framework of international standards to make the process reproducible and scalable. It emphasizes sustainability with goals aligned toward facilitating circular economy tenets through material reprocessing, waste diversion, and integrated energy-efficient manufacturing systems. Added to this are prospective new uses for 4D printing in edibles, medicines, and changeable constructions; all call for novel materials and innovative design strategies. The book will also close the gap between scientific inquiry and industrial application by reframing theoretical advances in topology optimization and AI-driven design as practical solutions for the aerospace, healthcare, and automotive production sectors. Therefore, the targeted interrelated goals would accelerate additive manufacturing technology adoption while mitigating technical, economic, and environmental hurdles.

Target Audience

The volume "Modeling, Analysis, and Control of 3D Printing Processes" is so designed that it would cater to a vast and multidisciplinary readership, considering the broad impact and fast-changing character of additive manufacturing (AM) technologies. The target audience for this book includes academic researchers and graduate students within disciplines such as mechanical engineering, materials science, industrial engineering, manufacturing technology, and applied physics. It will be an appropriate audience for advanced research or coursework involving modeling, simulation, and control of 3D printing processes; therefore, the authorships provide rigorous coverage of theoretical and practical aspects in this domain. For these scholars, some of the computational modeling methods detailed in the book will likely be helpful to them while conducting their independent investigations into finite element analysis, process simulation, and CAD/CAM/CAE integration, among other things, additive manufacturing. The case study's applicability, real-world examples, and up-to-date research results make this book valuable in conducting thesis work, laboratory projects, and joint research activities.

Beyond academia, the book remains relevant to contemporary engineers and technical professionals in sectors that currently use or are considering 3D printing technologies. The list is endless but includes aerospace, automotive, healthcare, dental, medical device manufacturing, consumer electronics, and even food and pharmaceutical industries. For these practitioners, this text provides practical insights on enhancing processes, assuring quality, and tackling real-life challenges when transitioning 3D printing from prototyping to mass production. The chapters covering process monitoring, sensor integration, closed-loop control, and parameter optimization grant engineers access to various tools and methodologies to help achieve repeatable, high-quality results in demanding industrial environments. Perplexity, bustiness, but mainly professionalism, must be observed while rewriting the following passage: The timely discussion of material selection, composite plastics, innovative materials, and recycling methods will be helpful for professionals responsible for material development, innovation, sustainability, or new product lines. The standards set by the book regarding consistency plus regulatory expectations will empower industry readers to untangle compliance puzzles plus certification within highly regulated spaces.

Entrepreneurs, innovators, and product designers also make up a significant part of the book's intended readership. As 3D printing is increasingly revolutionizing manufacturing by democratizing it and enabling fast and inexpensive prototyping, most start-ups and small to medium enterprises are considering how they can exploit additive manufacturing to gain a competitive edge. This book acts as a technical guide and a source of inspiration for these readers by showcasing the creative potential afforded through additive manufacturing, designing highly customized lightweight or functionally graded components. The chapters on topology optimization, generative design, and artificial intelligence applications in design and manufacturing processes are particularly relevant for those wishing to extend the limits of digital fabrication. This volume also considers the business-economic dimensions of additive manufacturing; among others, it includes a cost-benefit analysis of supply chain ramifications and three-dimensional printing in enabling distributed on-demand production systems.

The book will be helpful to educators and trainers in universities, technical colleges, and corporate training programs. It covers basic concepts, current trends, and hands-on applications, thus making it apt as the primary or secondary text in additive manufacturing classes, advanced manufacturing processes, digital fabrication, or innovative manufacturing systems. Illustrative diagrams, workflow illustrations, and problem sets facilitate effective teaching and learning, while interdisciplinary integration encourages students to understand the subject matter comprehensively. Materials are current regarding technological changes and industry best practices to keep those trained under continuing education and professional development programs adequately informed so they may hit the ground running as changes go with the AM revolution.

Policymakers, government officials, and standards organizations are the other audience segments often overlooked in this book. As additive manufacturing becomes more critical to the functioning of national and regional economies, it will increasingly require informed decision-making about investment in Research and Development, workforce training infrastructure, and regulatory frameworks. Thus, this writing provides valuable insight into these pressing issues by examining sustainability, circular economy principles, and material recycling strategies. With a clear and balanced exposition of the advantages, limitations, and social implications of 3D printing technology, the book leaves policymakers with the information needed to encourage AM growth while properly addressing concerns over safety protection, intellectual property rights, and environmental consequences.

Such a conclusion makes the book relevant to a broader readership of technology enthusiasts, makers, and hobbyists interested in digital fabrication who want to know more about 3D printing. The contents are scholarly and intricate, yet the authors deliberately set out to explain complicated ideas clearly and interestingly; thus, this work can be read by people with different levels of technical knowledge. Practical advice, troubleshooting help, and new applications such as 4D printing, food printing, and bioprinting will attract readers who like to play with the latest technologies and imagine what could be possible.

Indeed, "Modeling, Analysis, and Control of 3D Printing Processes" demands a diverse target audience ranging from academic researchers and graduate students to industry practitioners, entrepreneurs, educators, policymakers, and technology enthusiasts. Closing the gap between theoretical constructs and practical applications while simultaneously addressing both mature and new topics within additive manufacturing forms the basis upon which this book serves as an all-encompassing reference work and guide for those wishing to understand 3D printing technologies better or implement them innovatively. Its multidisciplinary character, combined with a focus on sustainability and quality, in addition to prevailing and emergent trends, will keep it more than relevant among writings even as the domain continues evolving. This book explores digital manufacturing's frontiers, whether readers want to advance scientific knowledge, enhance industrial processes, develop new products, educate the next generation, shape policy, or simply.

The Challenges

The book "Modeling, Analysis, and Control of 3D Printing Processes" elaborates on diverse multidimensional challenges that characterize the current landscape of additive manufacturing. Perhaps one of the most significant lingering issues is that parts produced via 3D printing exhibit much lower accuracy and repeatability than those made by traditional manufacturing processes, including injection molding and CNC machining. This limitation is particularly critical for aerospace, automotive, and biomedical industries because any variation in dimensional accuracy or mechanical properties makes a component unsuitable or unsafe for its intended application. A great variety exists in the causative factors for these inconsistencies. The foremost technological limitations inherent to printing technologies themselves are coupled with complexities in programming and calibrating three-dimensional printers. Inappropriate parameter selection- extrusion rates, layer height, etc.- will cause over- or under-extrusion with detrimental effects on surface finish, structure, integrity, and separation of layers, all compromising the final product's integrity and functionality.

Material selection and availability are another significant hurdle. Although traditional manufacturing has a vast catalog of certified materials, the options for 3D printing materials are significantly lower regarding performance and regulatory approval. This is especially problematic for industries requiring specialized or high-performance materials, such as biocompatible polymers for medical devices or high-strength alloys for structural components. On top of that, suitable materials are not easy to come by; several manufacturers have cited material issues as the top reason additive manufacturing isn't more widely adopted. The 3D printing materials available are less effective than those in traditional systems, discouraging companies from moving away from prototyping and into large-scale production.

Cost considerations are part and parcel of 3D printing. This promise of less waste and quick prototyping is negated by the fact that both pre-processing and post-processing can be very labor-intensive, not to mention costly. Pre-processing involves setting up digital files, machines, and calibrated parameters; all these are the work of skilled operators prone to mistakes. In addition, support removal, surface finishing, and sometimes additional curing or treatment must be done on the parts produced; these add time and cost to every part made. For instance, SLA parts must be washed to remove the excess resin; FDM parts often require sanding or vapor smoothing to achieve an acceptable surface quality. All these steps increase labor costs and overhead and create many opportunities for human error reintroduction, ultimately affecting process repeatability and reliability.

There are also significant challenges due to the technical limitations of existing 3D printing systems. Most 3D printers have limited build volume, speed, and capability to process several materials simultaneously. These technological limitations may prevent some elaborate designs from being realized or necessitate compromises in geometry and function for parts. In addition, the non-repeatability and unpredictability of the process compared with conventional manufacturing make it much more difficult to achieve close tolerances and high-quality finishes, especially in critical applications. Besides that, the workflow is complicated because it requires frequent calibration, maintenance, and troubleshooting, all without involving an expert in-house.

Quality control and standardization have always been concerns. Due to the variability in material properties, process parameters, and environmental conditions, part quality can vary significantly; thus, ensuring that every printed part meets the required specifications becomes challenging. Moreover, since there are no universally accepted standards for testing and certification of 3D-printed components, industries with strict regulatory requirements focus more on ensuring compliance by requiring companies to spend more on inspections and quality assurance resources.

Workforce and knowledge issues constitute yet another layer of challenge. A successful operation of an advanced 3D printing process requires a skilled workforce to deal with complex software, hardware, and material systems. A sharply reduced number of personnel possessing the required knowledge will slow the adoption rate and limit the effectiveness of additive manufacturing initiatives. Organizations attempting to scale their 3D printing operations continue to face ongoing challenges in training and retaining such talent.

Ecosystem and supply chain vulnerabilities have become more pronounced after recent global disruptions. The interplay between digital design, material availability, and machine availability means that any interruption in the supply chain will have knock-on effects on production timelines and costs. Integrating 3D printing into traditional manufacturing systems requires careful planning, strong data management, and interoperability between software and hardware platforms.

The challenges that "Modeling, Analysis, and Control of 3D Printing Processes" deal with are the complex interlinkage of diverse technical, economic, and organizational dimensions. These include limitations in dimensional accuracy and material options; high pre- and post-processing costs; technical constraints of existing printing systems; persistent quality control issues; labor shortages; and fragility in ecosystems. Only by addressing such challenges will additive manufacturing realize its full potential and move from being an often-used prototyping tool to becoming one of the hallmarks of contemporary production processes. The deep dive that the book takes on such matters gives valuable thought processes and workable approaches to researchers, engineers, and industrial executives who aim to find ways around or through the obstacles posed by massive 3D printing technology proliferation.

OVERVIEW OF THE BOOK

This book, " Modeling, Analysis, and Control of 3D Printing Processes," contains 17 chapters covering different themes related to the analysis and control of 3D Printing Processes. Here is a summary of the likely chapter topics:

Chapter 1 explores various materials used in 3D printing, including plastics, metals, ceramics, and composites. It highlights aerospace, healthcare, automotive, and consumer goods applications. The paper discusses the advantages of 3D printing, including enhanced design freedom, reduced material waste, and accelerated prototyping capabilities, while addressing its limitations and challenges. The research emphasizes the importance of matching specific 3D printing methods to suitable use cases by analyzing these technologies and their industrial applications. Additionally, it identifies future trends in the field, including advancements in multi-material printing, automation, and sustainability.

Chapter 2 presents the existing 3DP systems accessible from the literature. This study summarizes the corresponding materials employed by different companies and specialists up to this point for B&C purposes. Lastly, a brief discussion of possible research possibilities and the benchmarking qualities of these materials is provided.

Chapter 3 analyzes the environmental implications of AM based on a circular economy and compares them to material recycling strategies that make AM more sustainable. There are polymers, metals, and composite materials of AM, as well as an environmental impact analysis and related technical barriers analysis of their recycling. It discusses different recycling routes for thermoplastics, mechanical and chemical recycling, approaches to recover metal powders, and the possibility of biodegradable and bio–based alternatives. Moreover, the chapter discusses energy efficiency in AM and suggests process optimization, renewable energy integration, and waste reduction strategies.

Chapter 4 demonstrates crucial DfAM methods based on topology optimization models, generative design, lattice structures, and multi-material implementation and support reduction strategies that boost product capabilities and reduce substance utilization. The research discusses material selection methods, artificial intelligence-based design, simulation tools, and sustainable operational approaches to enhance AM efficiency. The aerospace, medical, and automotive sectors conducted well-documented implementations of DfAM despite current issues regarding standards, industrial implementation, and workforce preparedness. Combining AI enablements and hybrid manufacturing with digital workflows makes DfAM ready to transform future manufacturing operations by providing sustainable design methods.

Chapter 5 provides a detailed evaluation of existing methods and techniques for optimizing 3D printing parameters across different additive manufacturing technologies. The optimization of printing parameters is essential to obtain superior mechanical properties, dimensional accuracy, surface quality, and production efficiency in 3D printed components. The review systematically examines methodological approaches, including statistical design of experiments (DOE), artificial intelligence and machine learning (AI/ML) applications, simulation-based optimization, and in-process monitoring and feedback control. This chapter also demonstrates real-world applications of optimization methodologies through case studies from aerospace, automotive, medical, and consumer goods manufacturing industries, which show substantial improvements in product performance, production speed, and cost-effectiveness.

Chapter 6 discusses the methodologies of topology optimization for 3D printed mechanical parts and how design, materials, and manufacturing processes interact. The computational topology optimization method, which optimizes material layout within a design space for given loads and boundary conditions, has gained significant importance in additive manufacturing for creating lightweight, high-performance components. Combining topology optimization with 3D printing allows for generating complex shapes and complex internal structures unattainable through conventional manufacturing techniques. The paper evaluates different topology optimization approaches, including density-based lev, el set methods, and evolutionary algorithms, to determine their benefits and drawbacks when applied to additive manufacturing limitations. The research studies how material characteristics, printing settings, and post-treatment procedures affect topology optimization results to optimize structural performance and manufacturing capabilities.

Chapter 7 attempts to give an overall view of the main milestones in process planning and preparation for 3D printing while emphasizing the significance of each stage toward achieving the best results. The survey starts with the first step of developing a digital model, where several issues related to designing efficiency optimization, file types such as STL or OBJ, and the correction of mistakes identified to make the model printable are listed. From here, the workflow proceeds to model orientation and slicing, wherein slicing parameters like layer thickness, infill density, and support structures are chosen specifically for the printing technology and material used. A detailed analysis of how these parameters influence print quality, mechanical properties, and production efficiency is carried out.

Chapter 8 discusses various advanced modeling techniques, simulations, and real-time monitoring strategies that will improve predictive accuracy. Case studies for aerospace, automotive, and healthcare applications are presented as practical applications. The bridging of theory to the useful solution provided in this chapter is a real action for researchers and engineers to improve material performance, reduce defects, and optimize AM processes towards advancing high-quality, reliable, and scalable AM technologies.

Chapter 9 reviews the principles, difficulties, and the forefront of technology in control systems for 3D printing that achieves the precision, reliability, and efficiency of 3D printing processes. They examine the utilization of open-loop and closed-loop control mechanisms, integration of real-time monitoring with sensor networks, and advanced computational techniques, including model predictive control, adaptive algorithms, and artificial intelligence.

Chapter 10 represents a study delving into the realm of AM with a specific focus on composite materials, starting with exploring various 3D printing processes tailored for composites. The study discusses the modeling, Analysis, and Simulation of 3D printing processes, laying the foundation for a comprehensive understanding of this transformative technology. Numerical simulation of printed parts navigates through the intricacies of micro-, macro-, and meso-scale modeling, offering valuable insights into the complex world of 3D printing for composites.

Chapter 11 provides a detailed evaluation of existing methods and techniques for optimizing 3D printing parameters across different additive manufacturing technologies. The optimization of printing parameters is essential to obtain superior mechanical properties, dimensional accuracy, surface quality, and production efficiency in 3D printed components. The review systematically examines methodological approaches, including statistical design of experiments (DOE), artificial intelligence and machine learning (AI/ML) applications, simulation-based optimization, and in-process monitoring and feedback control. The review focuses on material-specific optimization methods for polymers, metals, composites, and ceramics by analyzing process-specific parameters and material-dependent variables. This chapter demonstrates real-world applications of optimization methodologies through case studies from aerospace, automotive, medical, and consumer goods manufacturing industries, which show substantial improvements in product performance, production speed, and cost-effectiveness.

Chapter 12 describes additive manufacturing as becoming more common in the contemporary industrial sector. This method uses the layer-by-layer deposition concept to create objects. As a result, the current study is critically important. It needs to focus on the fused-based modeling technique's elements, applications, and materials, such as nylon, metals, carbon fiber, polycaprolactone (PCL), polypropylene, polyethylene, and polylactic acid (PLA). Additionally, it has been determined that the FDM approach is frequently employed in developing items necessary in the current environment. Nozzle swabs and face mask supports are made using a variety of additive manufacturing processes. The application and development of fused deposition modeling (FDM), sometimes called three-dimensional (3D) printing, have grown dramatically due to the ongoing demand for increasingly complex designs and efficient material utilization. The fabrication of complicated designs is made possible by AM's ability to create near-net-shaped components with a lower buy-to-fly ratio.

Chapter 13 demonstrates that silicon photonics has grown rapidly in the Datacom industry over the past decade, with applications in LIDAR, sensing, quantum photonics, and AI. This expansion has led commercial CMOS foundries to establish open-access silicon photonics process lines for large-scale production. Meanwhile, university research labs use e-beam lithography for small-scale prototyping, fostering innovation. However, transitioning from research to industry poses challenges due to the incompatibility of research tools with industry-standard fabrication. The CORNERSTONE rapid-prototyping platform addresses this by offering deep-UV lithography-based fabrication, enabling scalable production while maintaining design flexibility. It serves as a bridge between academia and industry, facilitating the commercialization of novel designs. CORNERSTONE's capabilities support the growing demand for photonic solutions and drive stakeholder collaboration. This review explores its current strengths and future potential, highlighting its critical role in advancing silicon photonics.

Chapter 14 investigates how the use of advanced sensing technologies facilitates the continuous monitoring of critical parameters like temperature, material flow, stresses from the mechanical impacts, local environment details, and other parameters, which in turn brings the ability to continuously alter printing parameters and thus improve the process quality and productivity. A discussion of several sensor types, such as thermal, optical, force, gas, and acoustic sensors, in terms of defect prevention, material optimization, and closed-loop process control is presented. Moreover, the chapter examines the effect of Artificial intelligence and machine learning in data processing, predicting maintenance, and adaptive printing, and establishes such in autonomous manufacturing systems.

Chapter 15 discusses how different stimuli affect the color, flavor, texture, and form of food samples that are 4D printed. It also discusses the creation of 4D designs, food printing ink, and several techniques for 4D food printing. Thanks to recent, quick advancements in 3D printing technology and material research, smart materials or metamaterials with shape—or function-changing capabilities may now be printed.

Chapter 16 investigates the aging response of Maraging Steel 300 fabricated via Laser-Powder Bed Fusion (L-PBF). The as-built specimens undergo a post-heat treatment sequence comprising Solution Treatment (ST) at 840 °C for 2 hours, followed by water quenching and subsequent Aging Treatment (AT) at 450 °C, 490 °C, and 550 °C for 2 hours each. The effect of post-heat treatment on microstructure and micro hardness is analyzed, revealing a solidification sub-structure with semi-elliptical overlapped Melt Pool Boundaries (MPBs) and fine columnar dendritic/cellular grain morphology due to thermal gradient variations. Solute segregation occurs at grain boundaries, while intense undercooling during L-PBF results in a fine martensitic microstructure.

Chapter 17 investigates how Additive Manufacturing (AM), or 3D printing, enables material-efficient, customizable, and adaptable design. It examines the progress from prototyping pioneering AM, its current multifaceted status as an industrial technology, and its implications in multiple sectors such as aerospace, healthcare, automotive, and consumer goods. This chapter studies the principal AM techniques, including Direct Metal Laser Sintering (DMLS), Stereolithography (SLA), Fused Deposition Modeling (FDM), and Selective Laser Sintering (SLS), and materials comprising metals, polymers, and composites. End-to-end, the persistent challenges are covered, including cost, quality assurance, regulatory complexities, and intellectual property. In addition, the chapter imagines the future of AM, contemplating the incorporation of artificial intelligence (AI), the Internet of Things (IoT), bio printing, and its potential to democratize manufacturing, for instance, through localized production and disruptive practices.

The book is a reference for researchers, practitioners, and practitioner students who want to study advanced models, analysis, and simulation tools of the 3D printing process. It focuses on quality assurance, process optimization, and the fusion of innovative materials and technologies in additive manufacturing.

We want to thank the specialists who generously shared their knowledge and experience, as well as the reviewers who kindly provided feedback during the whole editorial process of this book. Modification of the content while maintaining quality.

Romdhane Ben Khalifa

National Engineering School of Gabes, University of Gabès, Tunisia

Chapter 1
An Overview of the Various 3D Printing

Romdhane Ben Khalifa
https://orcid.org/0000-0002-9171-9584
National Engineering School of Gabes (ENIG), Tunisia

Ghalia Nhassa
National Engineering School of Gabes (ENIG), Tunisia

Ali Snoussi
https://orcid.org/0000-0001-7211-2827
National Engineering School of Gabes (ENIG), Tunisia

ABSTRACT

This paper explores the integration of advanced digital manufacturing and production systems within the context of Industry 4.0, emphasizing the transformative potential of Additive Manufacturing (AM). The study examines how AM can revolutionize company operations and consumer interactions by introducing new business models that enhance competitiveness and profitability. It discusses how AM can address complex and unique customer needs through various technologies and techniques, optimizing production processes, and introducing sustainable business practices. The research also delves into the innovative solutions offered by AM, such as waste reduction, shortened supply chains, and extended product life cycles, which all contribute to material, energy, and cost savings. It acknowledges, however, that despite its benefits, AM is still in the early stages of adoption and faces challenges regarding capability, IT integration, and outcomes. This research is based on a quantitative approach using a structured online survey to fill gaps in existing literature.

DOI: 10.4018/979-8-3373-0533-2.ch001

INTRODUCTION

3D printing transformed manufacturing in wild ways. It shook up entire industries and still fires off bursts of unexpected innovation and unexpected efficiency boosts. I explore a mix of printing methods and zero in on how Material Extrusion (MEX) differs from styles like Stereolithography and Selective Laser Sintering, which skip using material extrusion. The tech is not just famous for creating mind-boggling, intricate designs; it also rattles the status quo in the face of modern production challenges. Think about items that mix lots of open porosity with rock-solid strength—they work great for water pollution cleanup, wedded with a dash of eco-friendly smarts (Scaffaro et al., 2024). Cloud Manufacturing (CM) jumps into the picture and shakes up old resource routines with a service-first twist, pushing additive methods deeper into our connected lives (Baumann & Roller, 2017). Knowing these shifting processes matters; it is not just about using them in various fields but also about blending them into the manufacturing setups of tomorrow. New techniques keep popping up, changing the game every day. Plain and straightforward, industry folks must stick with these twists to dodge pitfalls and grab the fresh opportunities ahead.

Definition of 3D printing

Three-dimensional printing is a blend of additive methods that turns digital designs into real things. It layers plastic, metal, or composites with clear orders, crafting funky shapes that old school makers cannot whip up. This cool tool in Industry 4.0 fires up speed, custom tweaks, and green vibes in aerospace and healthcare. Recent leaps in additive printing get wedded with cloud based setups that boost clarity and meld services, helping production pivot with market swings (Baumann & Roller, 2017). Manufacturing takes a wild turn as companies eagerly plunge into 3D printing's daring mix of fresh ideas and flexible operations (Savastano et al., 2019). Figure 1 visually represents the sequential steps of producing a part using additive manufacturing (AM) methods. The flowchart typically begins with the digital design phase, proceeds through preparation and printing, and ends with post processing and inspection.

Figure 1. Process Flowchart of Additive Manufacturing Techniques (Geiregat & De Vos, 2024)

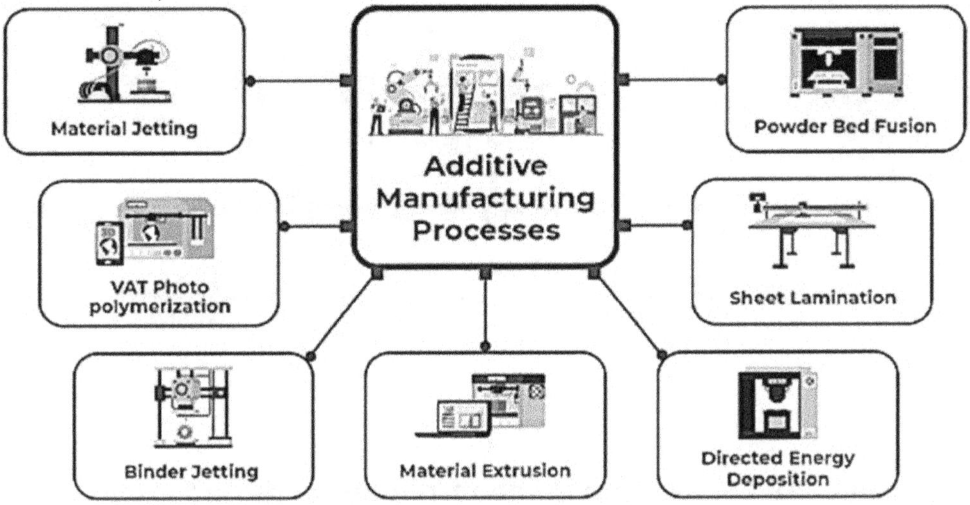

Brief history of 3D printing technology

3D printing started in the 1980s. It began as a neat trick to whip up prototypes and soon became a wild tool that shakes up how things get made. Early on, folks used it for quick prototypes, and then it took a turn into custom healthcare gadgets and delightfully odd craft projects. In 2002, cloud manufacturing jumped in when combining digital techniques with online services to offer on demand production resources (Baumann & Roller, 2017). Advances in 3D printing also helped rescue cultural treasures, letting people document and recreate ancient arts like Japanese lacquer work and Norwegian carvings, which now struggle under outdated methods and money troubles (Safonov et al., 2020). This shows that 3D printing still plays a surprisingly active role in making our world more unpredictable. The timeline in Figure 2 provides a clear, at-a-glance summary of the evolution of 3D printing, from its origins as a rapid prototyping tool to its current status as a versatile manufacturing technology used in diverse fields such as healthcare, aerospace, and consumer products (Ben Khalifa et al., 2024). It emphasizes the interplay between technological innovation, commercialization, and the broadening scope of applications over time.

Figure 2. Timeline of 3D Printing Technology Developments (1980- 2019) (Chahin et al., 2022)

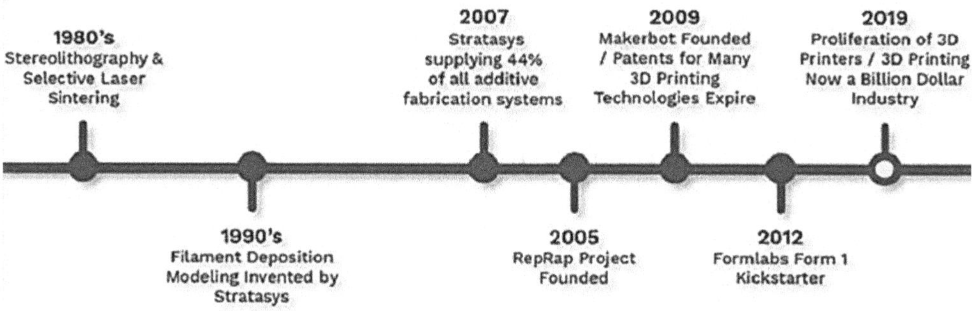

Importance of 3D printing in modern manufacturing

3D printing flips the manufacturing world on its head. It shatters the old production routines everyone once assumed were set in stone. This tool is at the heart of additive manufacturing and makes detailed designs and custom parts for super specific needs in an unexpected combo (Nhassa et al., 2024). When it teams up with Industry 4.0, 3D printing boosts efficiency and cuts material waste, saving companies time and money (Mehrpouya et al., 2019a). Hybrid techniques mixing additive with subtractive moves wedded together show promise in tackling tricky challenges, especially when handling messy geometries and nailing superb surface finishes (Ewart, 2019). These fresh ideas chop production cycles and trim lead times while nudging manufacturers to explore new, creative, and sustainable paths. 3D printing's influence keeps growing, paving the way for a future of personalized production and smart manufacturing that cements its spot as a key piece of the industry for years to come. Figure 3 visually distinguishes how each technology manipulates different material forms and energy sources to achieve precise, layer by layer construction of 3D objects (Ghalia Nhassa et al., 2025).

Figure 3. Mechanism of various 3D printing technologies: (a) Stereolithographic (SLA), (b1 2) Powder bed and powder jetting, (c) Selective laser sintering (SLS), (d) Semi solid extrusion (EXT) and (e) Fused deposition modelling (FDM) (Alhnan et al., 2016).

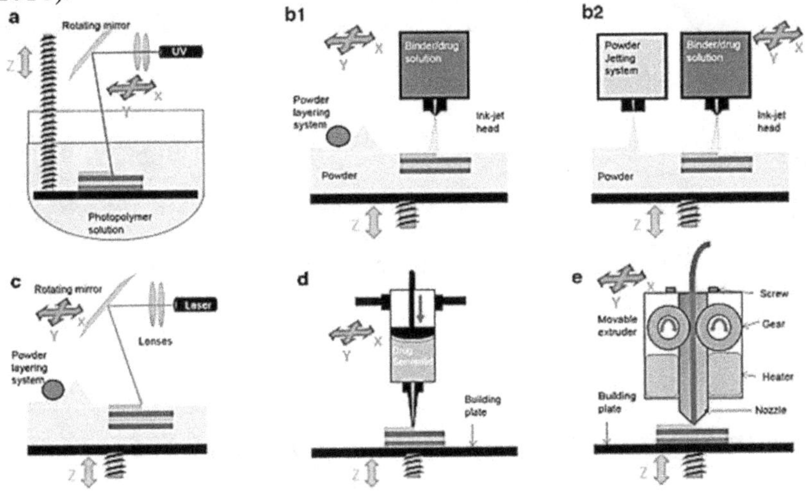

Overview of the essay structure

A strong essay on 3D printing methods kicks off with a clear point. You start by saying what 3D printing is and why it matters in today's design and manufacturing world (figure 4). Then you jump into each printing trick, discussing its cool traits, perks, and mess ups. Sometimes you toss in a real life story or mention a latest, quirky twist in 3D printing to show how these techniques pop up across various fields. One study notes that this mix wedges abstract ideas with down-to-earth results—additive manufacturing is busy fixing significant issues (Baumann & Roller, 2017). In short, a relaxed, freewheeling approach makes the confusing world of 3D printing feel friendlier and gives readers a neat peek into this game changing tech (Golubović et al., 2022).

Figure 4. Process of 3D Printing: Transforming Digital Designs into Physical Objects (Azlin et al., 2022)

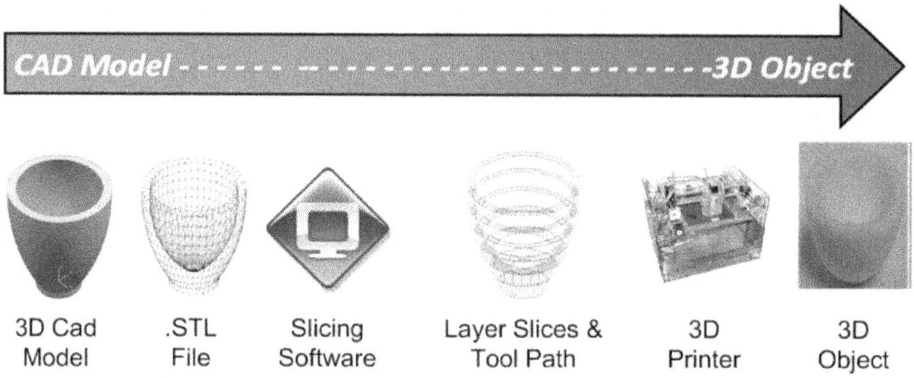

Purpose and significance of exploring various processes

3D printing methods matter. They boost our understanding of the tech's workings and shape its use in many fields. Techniques like Fused Deposition Modeling and Selective Laser Sintering come with quirks that mix up design choices, production speed, and what happens with materials (figure 5). Look at rapid prototyping: industries like fashion experiment on the fly, whipping up personalized, eco-friendly pieces (Corral et al., 2017a). Its many uses, like digitally preserving cultural heritage (Rodrigues Filho, 2023), help us see the tech's broad impact on society. Putting these techniques in place sparks fresh ideas in making and design while building an industry sometimes wedded to offbeat creativity.

Figure 5. Various 3D Printing Processes (Tyagi et al., 2024)

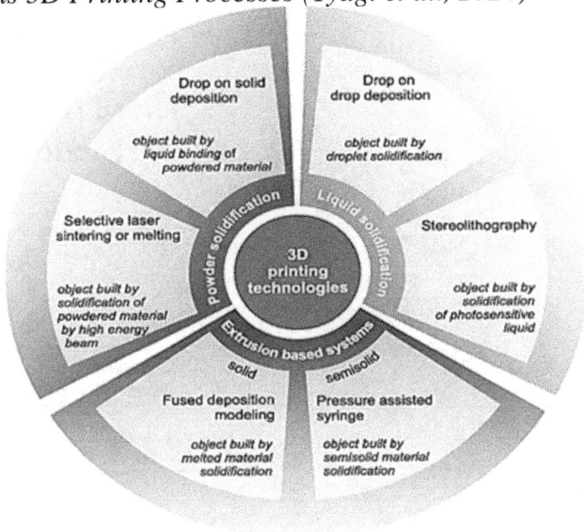

FUSED DEPOSITION MODELING (FDM)

Fused Deposition Modeling (FDM) is a 3D printing method that uses thermoplastic materials to build things one layer at a time. A heated filament streams out of a nozzle and gets wedged onto a build platform, slowly carving its form as guided by a CAD plan (figure 6). It works. FDM shines by crafting parts with funky curves and custom details on a budget, letting you easily whip up polymer and composite items (Festas et al., 2020). It is so user-friendly and flexible that many fields take notice—one can even find it in medicine for bio printing and custom prosthetics (Pereira Morais & Festas, 2022). Tech never stops, so FDM is geared to ramp production while dabbling with a broader mix of printing materials.

Figure 6. Process of Fused Deposition Modeling (FDM) (Przekop et al., 2023)

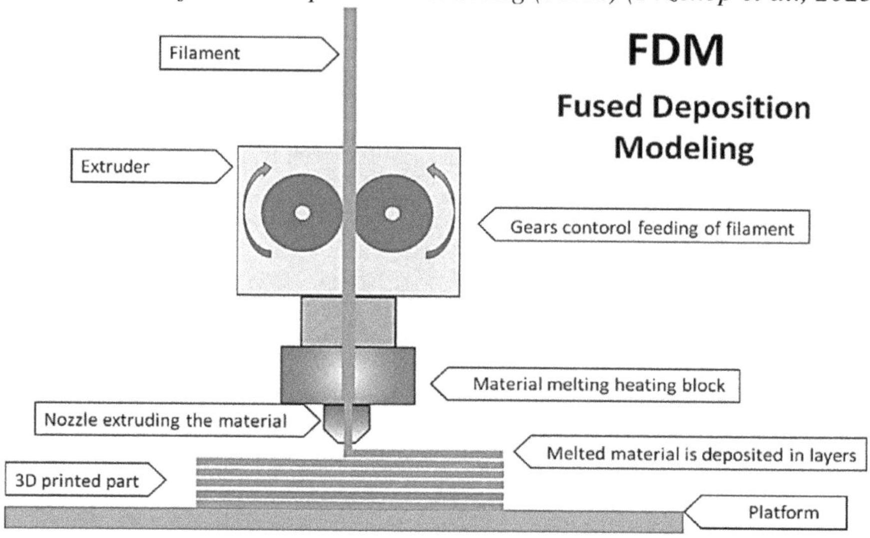

Fused Deposition Modeling is a big hit in 3D printing because it is cheap and easy enough for anyone. A hot nozzle melts strands of thermoplastic and drops them one by one onto a build platform. Layer by layer, it churns out detailed shapes and even quirky designs, making FDM a go to for rough prototypes and finished pieces. Its appeal pops up in the mix of materials it uses, and every day, ABS and PLA hang out with more offbeat plastics tuned for specific engineering twists. Sometimes designers even wed haptic tech with the design stage, boosting user feel and sharpening print accuracy (Armstrong et al., 2022). Ultimately, FDM has sparked leaps from quick test prints to practical components, proving it is a must-have in today's production game.

Materials commonly used in FDM

Fused deposition modeling (FDM) is a favorite in 3D printing because it works with various materials. People often use thermoplastics like PLA and ABS, each packing quirky perks. PLA gets picked for being eco-friendly and barely warping, making it a hit in school projects and prototypes. On the other hand, ABS shows off toughness and shoves off impacts, making it fit for parts that need to last. Material science lately has wedded fresh ideas with everyday twists, spawning composite filaments laced with odd additives that boost performance in surprising ways. Some hiccups stick around, like nailing a steady material supply and nailing the mix just right, so more research is needed to polish this manufacturing trick in various industrial arenas (Rajan et al., 2022).

Advantages of FDM technology

Fused Deposition Modeling, or FDM, shines in 3D printing (figure 7). It is prized for the benefits that keep it on top. It makes detailed shapes thanks to computer aided design systems wedded to the process, taking a rough idea and turning it into real stuff. FDM stays cool because it is cheap compared to other printing methods and works equally well for a handful of prototypes or a pile of finished pieces. It comes loaded with material choices—polymers, composites, and more—for projects ranging from early doodles to making medical gadgets. Recent studies have shown that its tight precision and flexible style keep pushing manufacturing ahead (Pereira Morais & Festas, 2022).

Figure 7. 3D Printing Process: The Creation of a Spiral-Shaped Object.

Limitations and challenges of FDM

Fused Deposition Modeling (FDM) is admired for its ease of use and flexibility in 3D printing, but it has many stumbling blocks. FDM parts are less challenging than traditional methods or other printing styles. FDM builds items layer by layer, and the layers get wedded together in quirky bonds that change strength depending on their direction, sometimes making the final product unreliable (figure 8). Nailing a smooth finish and perfect dimensions is tricky, too, so you do extra post-processing that wastes time and money. When you push for weird shapes or try hard materials, like in titanium blend tests that showed brittle quirks and design misfires (Baumann & Roller, 2017), FDM struggles with precision and overall quality. Ultimately, these issues mean you must think twice before picking FDM for your project.

Figure 8. Overview of the 3D Printing Process Utilizing Direct Material Filament

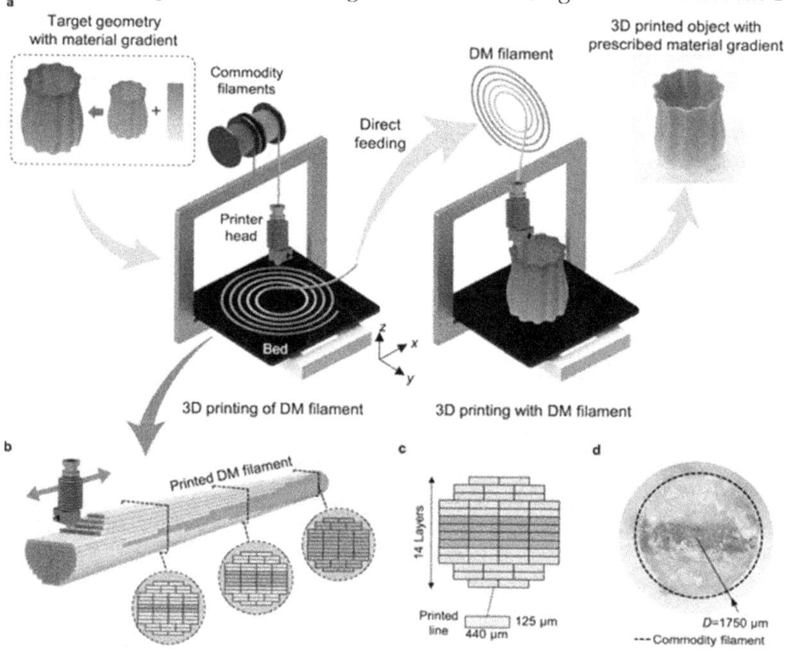

Applications of FDM in different industries

Fused Deposition Modeling (FDM) is shaking up many industries with its low price and oddly flexible style. Manufacturers use it to build prototype molds and tooling inserts, which cut startup costs for anyone diving into particulate injection molding (PIM) processes (Figure 9). Research shows you can make PIM cavity inserts with old school machining and FDM, and each way brings a mix of perks and quirks (Ewart, 2019). Engineering education digs it too—tech folks mix classroom ideas with real projects, letting them tackle everyday problems without sweating giant spending leaps (Ewart, 2019). FDM proves its broad usefulness and weds advanced ideas with down to earth practicality, boosting how things run in areas like airplanes, cars, and everyday products.

Figure 9. Overview of 3D Printing Applications (Tyagi et al., 2024)

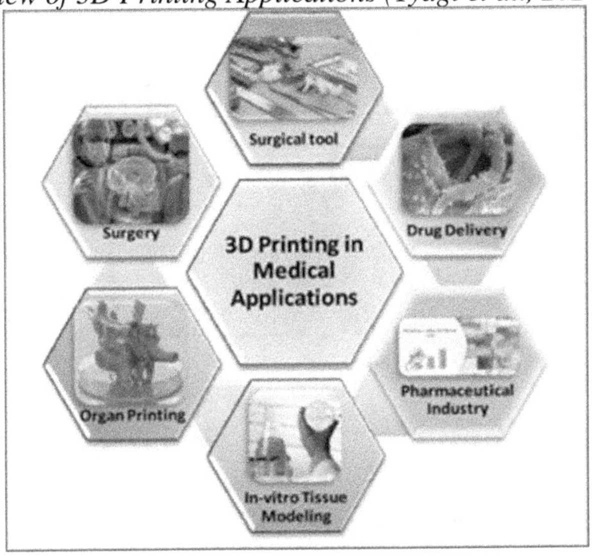

STEREOLITHOGRAPHY (SLA)

Stereolithography (SLA) is one of the first game-changing 3D printing methods. It uses photo polymerization, where ultraviolet light locks liquid resin into complex layers, giving you models with wild detail and spot on accuracy (figure 10). You can create bizarre, complex shapes that old school manufacturing sometimes cannot nail. SLA does not only dress up prototypes; it dives into real world gigs like printing tissues and organs, whipping up custom implants, and molding anatomical models for surgery. Its tricks include killer resolution and a surface as smooth as silk, which matters in medicine. The process does have a catch, though, slow print times and extra post-processing drag out production (Pereira Morais & Festas, 2022) (Festas et al., 2020). Even if SLA fuels bold innovation, merging it into big production means you have to tackle these snags head-on.

Figure 10. Schematic of UV Curing Process in Resin-Based 3D Printing (Yehia et al., 2024)

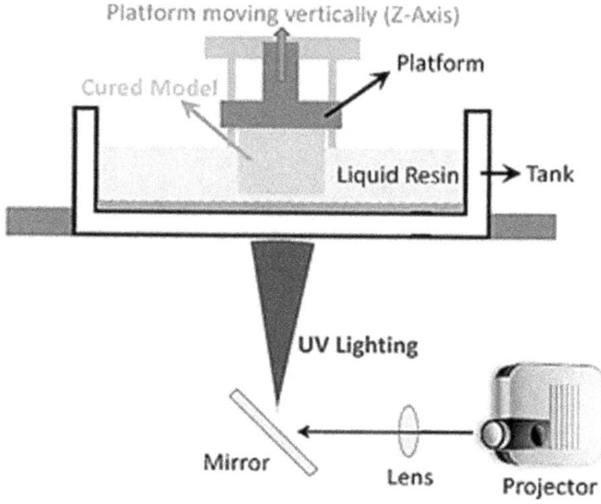

Overview of the SLA process

Stereolithography (SLA) is one of the earliest and most popular 3D printing methods for making polymer parts. A laser slowly cures liquid resin, wedded to a burst of chemical magic that sparks polymerization and builds each layer in wild detail (figure 11). It has a knack for crafting parts with sharp precision and smooth, almost glassy surfaces, a trait that makes it a favorite for both prototypes and final products. SLA dishes out a smorgasbord of material options, letting users play with the toughness and look of their prints. Recent buzz in the 3D printing world claims SLA is necessary for hitting the right design quirks and function needs (Kafle et al., 2021) (Livesu et al., 2017). Knowing how SLA works is key when picking the right tricks and materials for each off beat project.

Figure 11. Cause and effect diagram of SLA process parameters (Gowda et al., 2014).

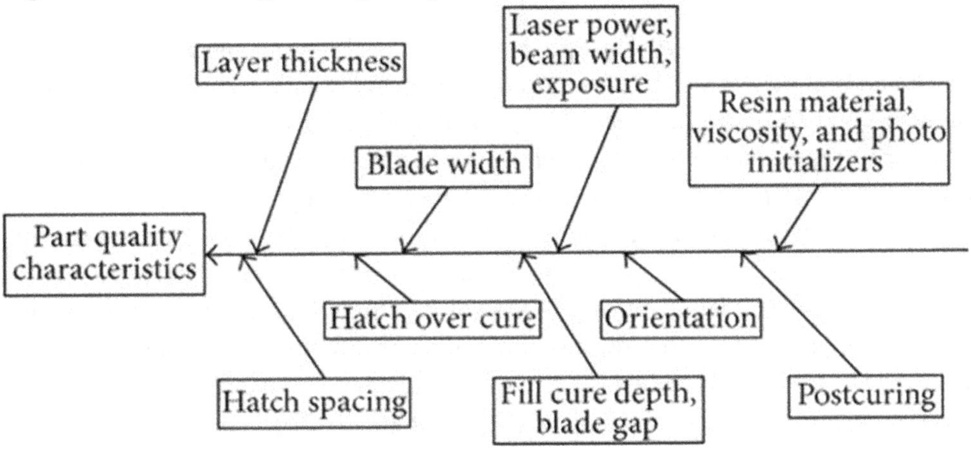

Types of materials used in SLA

3D printing pushes Stereolithography (SLA) into a broad spectrum of materials. Back then, plain resins ran the show for quick prototypes. Now, a melange of tough, squishy, and heat-hardy resins appears, each wedded to its own offbeat need (figure 12). Their main traits—strength and a small quantity of stretch play a big part in everything from rough industrial models to hospital parts. Studies show that every resin reacts quirkily when facing challenges like radiation (Michiels et al., 2016). Sorting and tweaking these materials cranks up printed parts' detail and overall punch, riding on new geometric tweaks in Additive Manufacturing (Livesu et al., 2017).

Figure 12. Process of 3D printing a microneedle array (Razzaghi et al., 2024).

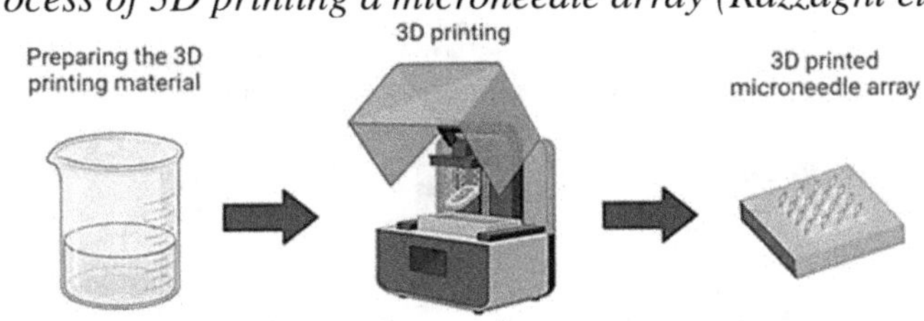

Benefits of using SLA for prototyping

Stereolithography (SLA) flips prototyping by packing in benefits that crank up speed and nail design details. It churns out insanely detailed, high res models so engineers and designers can get a clear peek at product ideas before big production. That early glimpse fuels fast tweaks, letting teams refine concepts with early feedback while dodging long waits seen in old school manufacturing. Also, SLA works with a mix of photopolymers that mimic all sorts of material vibes—like flexibility and toughness, which is key for checking a product's performance. Sometimes, SLA weds with Additive Manufacturing (AM), cutting through development hassles and boosting agile practices that react quickly to market swings (Baumann & Roller, 2017a). All these tricks show how SLA stays a heavy hitter in prototyping today, sparking fresh ideas all over the place (Corral et al., 2017b).

Rawbacks and limitations of SLA

Stereolithography (SLA) kicked open fresh doors in 3D printing, but it isn't all smooth sailing. The steep cost for gear and materials keeps small operations and weekend makers on the sidelines. Curing, cleaning, and the extra post processing steps drag things out and muddle the workflow. Compared with other printing tricks, SLA outputs tend to be more fragile and break more easily (Karkun & Dharmalingam, 2022). Many resins pack a toxic punch, so you must watch your steps when handling them (figure 13). SLA weds razor sharp detail and the ability to craft intricate designs with hiccups, so you have to weigh its perks and pitfalls before picking this route (Sahafnejad Mohammadi et al., 2022).

Figure 13. Polymer based SLS Challenges (Singh et al., 2024)

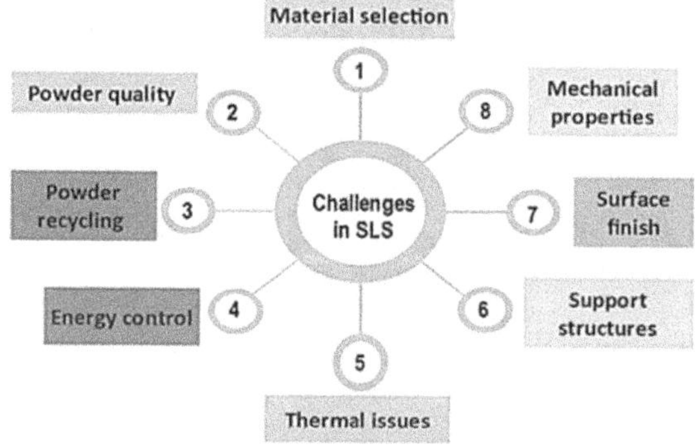

Real world applications of SLA technology

Stereolithography (SLA) pushes boundaries in countless ways. It shapes dental models and quirky surgical guides that fit just right, boosting results and quickening surgeries. Its pinpoint precision weds careful planning with fast production, giving both doctors and patients a leg up. Designers and engineers use SLA to whip up prototypes quickly, letting them test wild ideas that spark fresh product twists. Sometimes the method feels a bit rough and ready. SLA also chimes into Cloud Manufacturing (CM), supplying on-demand resources that match old-school approaches (Baumann & Roller, 2017). A mishmash of geometric algorithms tweaks its process into custom, effective fixes for a grab bag of industrial gigs (Livesu et al., 2017). Ultimately, this tech is set to remix the future of additive manufacturing.

Description of the SLS process

Selective Laser Sintering (SLS) uses a laser to melt powder into solid shapes, one layer after another. A thin layer of thermoplastic dust spreads over the build plate (figure 14). Then, a high-energy laser zaps chosen spots from a digital 3D plan, melting that powder into a rough pattern. The build platform sinks just a bit, fresh powder lands, and the process repeats until the object is fully in shape. One incredible perk is that SLS can whip up tricky designs without extra supports because the loose powder naturally weds everything together during the build. It works with loads of materials from polymers to composites and pops up in many areas, as seen in recent studies (Kafle et al., 2021) (Sabbatini et al., 2021).

Indeed, Selective Laser Sintering (SLS) leaps forward, shaking up 3D printing for polymer projects. It fires a powerful laser that welds with powder, melting and fusing it layer by layer to whip up shapes that old methods cannot manage. The big hitch is its small pool of materials, a problem researchers are busy trying to patch by tweaking processing and sharpening model precision (Mearns, 2017). Stack it next to techniques like fused deposition modeling or stereolithography and SLS pops, especially since it can tailor both the feel and the form of its final pieces (Kafle et al., 2021). Fixing these material wrinkles is key if SLS is to branch out into more industries and show off its mojo in the world of 3D printing.

Figure 14. Schematic Diagram of Selective Laser Sintering (SLS) 3D Printing (Kafle et al., 2021)

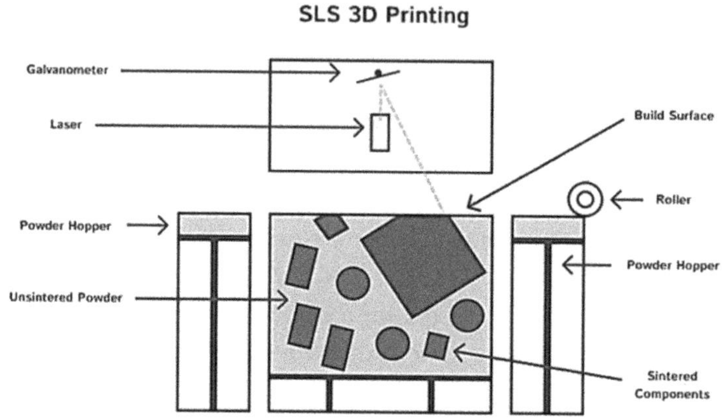

Materials suitable for SLS printing

Selective laser sintering (SLS) rocks with a wide mix of materials. It uses quirky thermoplastics and special metal powders made just for 3D printing. Nylon, polystyrene, and polycarbonate add their punch, boosting strength, flexibility, and heat endurance. The process carves out crazy, intricate shapes that old school methods cannot dream up, opening up fresh chances in fields like aerospace and healthcare. Breakthroughs in material science have tossed extra options into the mix, letting composites deliver unique, job-specific perks, and figuring out how material vibes mash with processing tricks is key to wedded SLS crafting high-performance parts (Livesu et al., 2017) (Kafle et al., 2021).

Advantages of SLS over other methods

Selective Laser Sintering (SLS) forges tough, twisty parts without extra support structures—a real pain with Fused Deposition Modeling (FDM) and Stereolithography (SLA). SLS builds layer by layer, melting powder and juggling material quirks to craft pieces that blend rock solid strength with heat smarts—a neat trick for industry (figure 15). It dances with a range of thermoplastics, letting a prototype and a finished part pop out in one go, bumping up production (Kafle et al., 2021). High-Speed Sintering (HSS) banks on infrared and sometimes trips over powder hardness, while SLS is wedded to consistent quality all over its build bed (Thomas et al., 2006).

Figure 15. Advantages of the SLS method

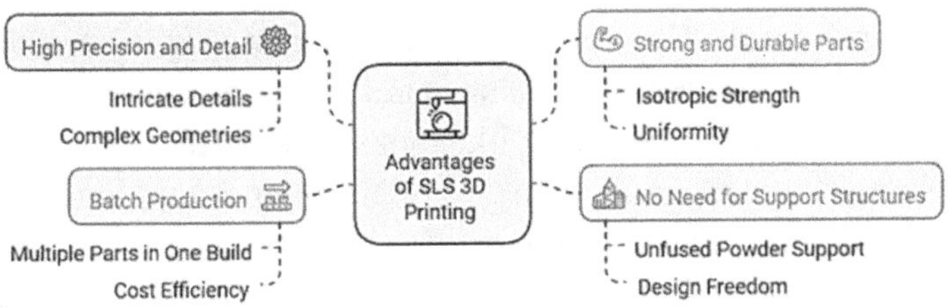

Challenges faced in SLS technology

Selective Laser Sintering (SLS) has flipped the whole scene in 3D printing, but it still bumps into some significant snags. The machines and special powders cost a fortune and can easily shut out small outfits (Ogoh & Fairweather, 2019). Running SLS rigs demands wedded, skilled techs a rough deal when the industrial support is scattered (Table 3). You also face issues keeping material qualities steady while slogging through tedious post-processing (Tareq et al., 2021). Custom orders and fast turnarounds rule these days, so sorting out these hiccups is key for SLS to shine in the wild world of 3D printing.

Table 3. Challenges Faced in SLS Technology

Challenge	Description
High Cost of Equipment	Selective Laser Sintering machines are expensive, limiting accessibility for small businesses.
Material Limitations	The range of materials suitable for SLS is still limited compared to other 3D printing technologies.
Post Processing Requirements	Parts often require extensive post-processing to achieve the desired surface finish and accuracy.
Thermal Issues	SLS can suffer from issues related to thermal distortion and uneven cooling.
Limited Build Size	The build volume of SLS machines can restrict the maximum size of parts produced.

Industries that utilize SLS for production

Selective laser sintering (SLS) weds advanced laser tech with everyday ingenuity to craft custom parts that impress in insane detail. Aerospace jumps on SLS to whip up feather-light components and build like tanks, perfect for slashing fuel use and firing up performance. Auto makers use it for speedy prototyping and for shaping weird, twisty forms that still clear tough safety checks. Medical teams cash in on SLS by crafting one-of-a-kind implants and prosthetics tailored to each patient, easing compatibility puzzles and boosting treatment gains. It also plays nicely with materials like polyamides and polycarbonates, stretching its playground in odd ways. Yet, bumps like lower strength and minor shape hiccups leave plenty of room for tweaking (Pinkerton, 2016).

DIGITAL LIGHT PROCESSING (DLP)

Digital Light Processing (DLP) stands out in 3D printing because it churns detailed parts with pinpoint light (figure 16). It cures a whole layer simultaneously with a digital projector instead of zapping one dot at a time like old-school SLA. DLP started for plain resins, but smart tech tweaks now let it tackle mixes filled with ceramics and metals. That twist has unlocked chances to make items for dental implants and cutting-edge biomaterials (Chaudhary, 2023). DLP also weds with biocompatible acrylate hydrogels, and tossing in nanoparticles boosts both toughness and compatibility, giving tissue engineering a real kick. Things run fast. DLP is set to flip the script further in both industry and medicine.

Figure 16. Schematic representation of DLP 3D printing process (Sekmen et al., 2024).

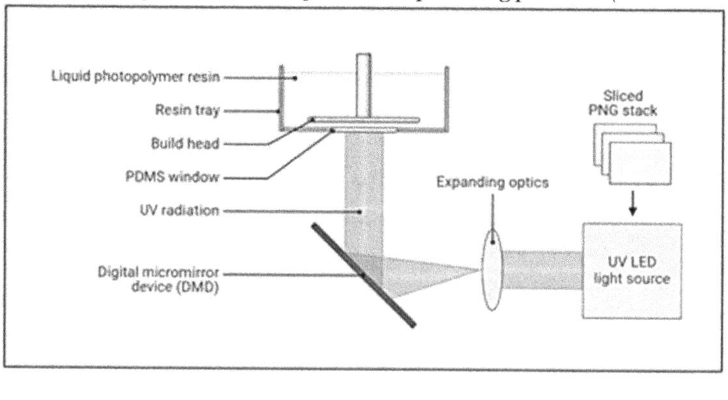

Introduction to the DLP process

Digital Light Processing (DLP) rocks 3D printing by quickly whipping up prototypes with crazy fine detail. It opts to flash its digital light onto photopolymer resin, curing it one layer at a time, which lets you craft wild shapes that regular techniques struggle with. This method rides the rapid prototyping wave; some research (cite52) hints that it chops down manufacturing costs while squeezing production times. DLP tech gets cooler as it evolves, now wedded with slick geometric algorithms that ease model prep and juggle machine commands. In a nutshell, DLP does more than deliver intricate designs—it fires up industrial innovation and stakes its claim as a major player in 3D printing.

Comparison of DLP with SLA

Digital Light Processing (DLP) and Stereolithography (SLA) are vat photopolymerization techniques that each jam to their beat. DLP wedges a clever digital projector with the art of zapping an entire resin layer in one go. It speeds printing up a ton and works excellently in busy, high volume settings. Sometimes it trips when you need different resins in one build. The risk of one resin mingling with another, plus the extra time to switch things out, can mess up project schedules (Matte et al., 2019). SLA packs a punch with details that pop. It beams out prints so crisp they capture every little microstructure, making it a favorite in pharmaceutical and dental work (Awad et al., 2023). It cures resin one layer at a time using a focused UV beam. That laser focus means it prints more slowly, especially when the object is large. At the end of the day, both methods churn out parts with smooth, quality finishes. The right pick depends on what your project craves razor-sharp precision, funky material mixes, fast production, or keeping costs low. Look over these bits carefully; sometimes the choice is not so obvious.

3D printing, especially Digital Light Processing (DLP), thrives on picking the right materials. DLP blasts light onto photopolymer resins until they harden, and you get prints loaded with crazy detail. The resins can be tweaked for many tasks, whether for dental models or crafting intricate jewelry. New composite blends feature ceramics wedded with offbeat additives to boost toughness and usability. Researchers even built 3D printed zeolite monoliths that filter out water contaminants (Chen et al.). Cool, right? Geometric algorithms for additive manufacturing race ahead, showing how vital it is to dial in material properties with the printing process. Manufacturers keep shaking things up, so expect a wild mix of DLP materials that meet business and lab needs (Livesu et al., 2017).

Figure 17 visually encapsulates the promise of DLP 3D printing as a versatile, high resolution, and rapid prototyping platform for flexible devices. It balances this optimism with a realistic appraisal of current material, technical, and standardization challenges, and it points toward ongoing innovations, particularly in materials science and hybrid fabrication, that are expected to drive the next wave of innovative, high-performance flexible devices.

Figure 17. Perspective on DLP printing of flexible devices (Yi et al., 2025)

Pros and cons of DLP technology

Digital Light Processing (DLP) technology packs both serious wins and a few face palms in 3D printing. It churns out prints with crazy high resolution, super smooth surfaces, and nails the dimensions perfectly for making stuff like dental prosthetics and quirky models (Rezaie et al., 2023). Fast curing of each layer pushes production speed way beyond old school methods and fires up output. Trouble pops up, too. Photopolymerization means that when ultraviolet light gets wedded to polymer materials, it sometimes warps details and introduces distortions. This same light can break down the material over time, slowly weakening its strength (Gomez, 2023). Plus, DLP gear and materials often come with a hefty price tag that scares off some users. DLP technology rocks in many ways, but also stumbles here and there. You must balance the cool perks against the bumps in the road to decide if it fits your needs.

Applications of DLP in various fields

Digital Light Processing (DLP) is a groundbreaker that mixes tech smarts with everyday magic. It makes parts fast with pinpoint precision and shakes up our old ideas. People need to check the side effects and ethical quirks that pop up along the way. DLP jumps into bio work by whipping up wild, life like structures that mimic nature. This move has pushed big leaps in tissue making and healing science. Instead of just setting up cell scaffolds, the tech sparks questions about what happens when these creations settle inside our bodies and if odd surprises might crop up. It even spins detailed mini models for surgery practice and custom care, making you wonder how accurate each little replica is to someone's blueprint. Studies show DLP can mimic nature's quirky intricacies. Sometimes it dresses up designs to echo the unusual circular buckling seen in pumpkins by wedded core shell barrel architectures with responsive hydrogels (Wu et al., 2019). That kind of detail sparks wild chances for mimicry in meds and materials science, yet forces us to think about how these fancy builds might play with our environment. DLP's nimbleness has rocked fields like aerospace and automotive, where exact shapes trim weight and kick performance into gear, saving fuel. The tech whips up lightweight, tough parts in no time, though you still have to size up if these new ideas play nice with nature. New tricks in geometric algorithms for 3D crafting are stirring up fresh opportunities, ensuring the meld between planning, building, and actual function gets smoother and meets tough industry yardsticks (Livesu et al., 2017). Overall, DLP is shifting from a basic build tool to a wild game changer that reworks industry playbooks on many fronts. As it stretches its influence, folks must chew over how it might lead to a greener, leaner industrial future while facing big moral and social puzzles.

CONCLUSION

3D printing is a wild, ever changing craft that jumps into many industries. Engineers build water cleanup systems using 3D fabrication methods that tackle big problems like water pollution with solid fixes (Scaffaro et al., 2024). Cloud Manufacturing throws a twist, making printing operations nimble and fresh and sparking creative energy in old school factories (Baumann & Roller, 2017). When these methods get wedded, you can see clear signs that 3D printing has a bright future—research even hints at cooler, more eco-smart ways to build things. 3D printing's broad range shakes up industry and everyday life, so we must keep digging and trying new stuff.

Summary of key 3D printing processes

Manufacturing is shifting fast with a dash of quirky 3D printing magic. FDM (Fused Deposition Modeling) is excellent for cooking prototypes and wallet friendly items by stacking plastic layers. SLA (Stereolithography) spins out super fine details—a real gem when crafting snazzy jewelry pieces or dental devices. SLS (Selective Laser Sintering) zaps powdered material with a laser that weds it into tough forms, working great for rough test runs and cranking out large batches. Precision and custom builds grab more attention nowadays, so intelligent geometric algorithms pump up performance and stretch these methods to new limits (Livesu et al., 2017). 3D printing fits with Industry 4.0's offbeat, bright vibe, slicing time and materials while backing a more scattered production setup (Mehrpouya et al., 2019). These twists flip the usual playbook on design and delivery, sparking fresh ideas across many industries.

Importance of selecting the proper process

3D printing significantly affects a project's success since it directly hits on product quality, how things work, and cost. Every printing method has perks and bumps, so you need to be smart when pairing available tech with what your project demands. Additive manufacturing, for instance, varies a ton in material fit and detail—not using the proper process can mess things up. Online printing gigs are booming, wedded with a creative buzz that gets users innovating and teaming up, showing that a good pick can boost design partnerships (Rayna et al., 2015). New interactive visualization tools let you look at and fiddle with complex 3D models, adding another angle to consider when choosing the best approach for healthcare and astronomy (Punzo et al., 2017).

Future trends in 3D printing technology

3D printing keeps pushing past old limits, and many quirky trends hint at huge breakthroughs. Digital twin tech jumps in by making virtual doppelgangers of real structures to smooth out design and build work. This trick is set to ramp up speed and precision on projects as Industry 5.0 weds human spark with digital smarts. Smart manufacturing, boasting that classic Industry 4.0 vibe, backs tough plans to boost eco-friendliness and custom-made outcomes. As tech evolves, chances to mess around with snazzy new materials and offbeat processes keep growing, setting the stage for radical shifts and opening fresh doors for research.

Impact of 3D printing on manufacturing and design

3D printing flipped manufacturing and design on its head. It now gives access to wild efficiency boosts and custom tweaks that once seemed out of reach. It churns out bizarre shapes that old methods could not manage, shaking up lands like aerospace and healthcare. This tech whips together lightweight parts with intricate insides, ramping up performance, cutting waste, and slashing costs. It also uses flexible stuff like Polyvinyl Alcohol (PVA) to dodge those usual support headaches when printing tricky designs, nudging creative minds toward wild ideas without dropping quality. 3D printing revamps production and dares designers to chase bold new paths; it even weds old tricks with fresh spins that might reshape design's future.

Final thoughts on the evolution of 3D printing processes

Manufacturing is shifting fast, and breakthroughs in 3D printing scream that change. Engineers mix machine learning with 3D and 4D printing, wedging these tools to whip up products that tweak themselves for any odd requirement, ramping up performance and efficiency. This tech mashup fires up fresh ideas while handling the urgent push for eco-friendly moves and quick market flips, proving that digital change beats at production's core (Savastano et al., 2019). Sure, bumps still crop up—especially when trying to sync common standards and scale these tricks—but the chance to churn out detailed, custom items remains massive. Researchers and industry folks keep chasing these quirky possibilities. As 3D printing keeps evolving, it opens wild paths for creative design and making, locking in its spot as a must have in modern manufacturing.

REFERENCES

Alhnan, M. A., Okwuosa, T. C., Sadia, M., Wan, K.-W., Ahmed, W., & Arafat, B. (2016). Emergence of 3D Printed Dosage Forms : Opportunities and Challenges. *Pharmaceutical Research, 33*(8), 1817–1832. DOI: 10.1007/s11095-016-1933-1 PMID: 27194002

Armstrong, M., Mehrabi, H., & Naveed, N. (2022). An overview of modern metal additive manufacturing technology. *Journal of Manufacturing Processes, 84*, 1001–1029. DOI: 10.1016/j.jmapro.2022.10.060

Awad, A., Xu, X., Ong, J. J., Goyanes, A., & Basit, A. W. (2023). Vat Photopolymerisation additive manufacturing for pharmaceutical applications. In *Nano-and Microfabrication Techniques in Drug Delivery : Recent Developments and Future Prospects* (p. 99-124). Springer. DOI: 10.1007/978-3-031-26908-0_5

Azlin, M. N. M., Ilyas, R. A., Zuhri, M. Y. M., Sapuan, S. M., Harussani, M. M., Sharma, S., Nordin, A. H., Nurazzi, N. M., & Afiqah, A. N. (2022). 3D Printing and Shaping Polymers, Composites, and Nanocomposites : A Review. *Polymers, 14*(1), 180. Advance online publication. DOI: 10.3390/polym14010180 PMID: 35012202

Baumann, F. W., & Roller, D. (2017). Survey on additive manufacturing, cloud 3D printing and services. arXiv preprint arXiv:1708.04875.

Chahin, S., Afify, A., Mohsen, H., & Youssef, M. (2022). ROLE OF 3D PRINTED GREEN WALLS IN HEALING ARCHITECTURE. *BAU Journal - Health and Wellbeing, 5*(1, Article 1).

Chaudhary, R. (2023). *Additive Manufacturing by Digital Light Processing*. DLP. DOI: 10.1007/s40964-022-00336-0

Corral, L. C., Walker, K. J., Hubert, S. K., Smith, K. R., & Cheramie, L. M. (2017). *Exploring the abilities of 3D printing and its viability for consumption in the fashion industry. 74*(1).

Ewart, P. (2019). *A comparison of processing techniques for producing prototype injection moulding inserts.*

Festas, A., Ramos, A., & Davim, J. (2020). Medical devices biomaterials–A review. *Proceedings of the institution of mechanical engineers, part L. Journal of Materials: Design and Applications, 234*(1), 218–228.

Geiregat, M., & De Vos, S. (2024). *Leerpad over FFF 3D-printing voor FabLab-gebruikers.*

Golubović, Z., Milovanović, A., & Mitrović, A. (2022). *Characterization of 3D printed parts*. 56-56.

Gomez, S. (2023). *An Experimental Study on the Mechanical Properties and Chemical Composition of LCD 3D Printed Specimens.*

Gowda, R. B. S., Udayagiri, C. S., & Narendra, D. D. (2014). Studies on the Process Parameters of Rapid Prototyping Technique (Stereolithography) for the Betterment of Part Quality. *International Journal of Manufacturing Engineering, 2014*(1), 804705. DOI: 10.1155/2014/804705

Kafle, A., Luis, E., Silwal, R., Pan, H. M., Shrestha, P. L., & Bastola, A. K. (2021). 3D/4D printing of polymers : Fused deposition modelling (FDM), selective laser sintering (SLS), and stereolithography (SLA). *Polymers, 13*(18), 3101. DOI: 10.3390/polym13183101 PMID: 34578002

Karkun, M. S., & Dharmalingam, S. (2022). 3D printing technology in aerospace industry–a review. *International Journal of Aviation, Aeronautics, and Aerospace, 9*(2), 4. DOI: 10.15394/ijaaa.2022.1708

Khalifa, R. B., Nhassa, G., & Snoussi, A. (2024). An Overview of Additive Manufacturing's Role in industry 4.0. Modeling, Analysis, and Control of Smart Energy Systems, 137-152.

Livesu, M., Ellero, S., Martínez, J., Lefebvre, S., & Attene, M. (2017). *From 3D models to 3D prints : An overview of the processing pipeline. 36*(2), 537-564.

Matte, C.-D., Pearson, M., Trottier-Cournoyer, F., Dafoe, A., & Kwok, T. H. (2019). Automated storage and active cleaning for multi-material digital-light-processing printer. *Rapid Prototyping Journal, 25*(5), 864–874. DOI: 10.1108/RPJ-08-2018-0211

Mearns, C. D. (2017). *Analysis of Selective Laser Sintering print parameter modelling methodologies for energy input minimisation : A thesis presented in partial fulfilment of the requirements for the degree of Master of Engineering in Mechatronics at Massey University, Albany, New Zealand.*

Mehrpouya, M., Dehghanghadikolaei, A., Fotovvati, B., Vosooghnia, A., Emamian, S. S., & Gisario, A. (2019). The potential of additive manufacturing in the smart factory industrial 4.0 : A review. *Applied Sciences (Basel, Switzerland), 9*(18), 3865. DOI: 10.3390/app9183865

Michiels, S., D'Hollander, A., Lammens, N., Kersemans, M., Zhang, G., Denis, J., Poels, K., Sterpin, E., Nuyts, S., Haustermans, K., & Depuydt, T. (2016). Towards 3D printed multifunctional immobilization for proton therapy : Initial materials characterization. *Medical Physics*, *43*(10), 5392–5402. DOI: 10.1118/1.4962033 PMID: 27782703

Nhassa, G., Ben khalifa, R., & Snoussi, A. (2023, December). Numerical Analysis of the Mechanical Behavior of 3D-Printed Parts. In International Conference Design and Modeling of Mechanical Systems (pp. 474-486). Cham: Springer Nature Switzerland.

Nhassa, G., Ben Khalifa, R., & Snoussi, A. (2025). Impact of 3D Printing Settings on Polylactic Acid Filament Mechanical Behaviors Based on the Taguchi Method. *Recent Patents on Mechanical Engineering*, *18*, 1–11. DOI: 10.2174/0122127976328835241016094600

Ogoh, G. I., & Fairweather, N. B. (2019). The state of the responsible research and innovation programme : A case for its application in additive manufacturing. *Journal of Information. Communication and Ethics in Society*, *17*(2), 145–166. DOI: 10.1108/JICES-12-2018-0093

Pereira Morais, A., & Festas, J. (2022). *Medical Applications of Materials Manufactured by the AM Process.*

Pinkerton, A. J. (2016). Lasers in additive manufacturing. *Optics & Laser Technology*, *78*, 25–32. DOI: 10.1016/j.optlastec.2015.09.025

Przekop, R. E., Gabriel, E., Pakuła, D., & Sztorch, B. (2023). Liquid for Fused Deposition Modeling Technique (L-FDM)—A Revolution in Application Chemicals to 3D Printing Technology : Color and Elements. *Applied Sciences (Basel, Switzerland)*, *13*(13), 7393. Advance online publication. DOI: 10.3390/app13137393

Punzo, D., van der Hulst, J., Roerdink, J. B., Fillion-Robin, J.-C., & Yu, L. (2017). SlicerAstro : A 3-D interactive visual analytics tool for HI data. *Astronomy and Computing*, *19*, 45–59. DOI: 10.1016/j.ascom.2017.03.004

Rajan, K., Samykano, M., Kadirgama, K., Harun, W. S. W., & Rahman, M. M. (2022). Fused deposition modeling : Process, materials, parameters, properties, and applications. *International Journal of Advanced Manufacturing Technology*, *120*(3), 1531–1570. DOI: 10.1007/s00170-022-08860-7

Rayna, T., Striukova, L., & Darlington, J. (2015). Co-creation and user innovation : The role of online 3D printing platforms. *Journal of Engineering and Technology Management*, *37*, 90–102. DOI: 10.1016/j.jengtecman.2015.07.002

Razzaghi, M., Alexander Ninan, J., & Akbari, M. (2024). Advancements in Materials for 3D-Printed Microneedle Arrays : Enhancing Performance and Biocompatibility. *Micromachines, 15*(12), 1433. Advance online publication. DOI: 10.3390/mi15121433 PMID: 39770187

Rezaie, F., Farshbaf, M., Dahri, M., Masjedi, M., Maleki, R., Amini, F., Wirth, J., Moharamzadeh, K., Weber, F. E., & Tayebi, L. (2023). 3D printing of dental prostheses : Current and emerging applications. *Journal of Composites Science, 7*(2), 80. DOI: 10.3390/jcs7020080 PMID: 38645939

Rodrigues Filho, C. R. (2023). *Analysing the Impact of Data Quality and Processing Techniques on the Accuracy of Virtual Reconstruction-A Case Study of a Deactivated Grain Milling Factory.*

Sabbatini, B., Cambriani, A., Cespi, M., Palmieri, G. F., Perinelli, D. R., & Bonacucina, G. (2021). An overview of natural polymers as reinforcing agents for 3D printing. *ChemEngineering, 5*(4), 78. DOI: 10.3390/chemengineering5040078

Safonov, A., Maltsev, E., Chugunov, S., Tikhonov, A., Konev, S., Evlashin, S., Popov, D., Pasko, A., & Akhatov, I. (2020). Design and fabrication of complex-shaped ceramic bone implants via 3d printing based on laser stereolithography. *Applied Sciences (Basel, Switzerland), 10*(20), 7138. DOI: 10.3390/app10207138

Sahafnejad-Mohammadi, I., Karamimoghadam, M., Zolfagharian, A., Akrami, M., & Bodaghi, M. (2022). 4D printing technology in medical engineering : A narrative review. *Journal of the Brazilian Society of Mechanical Sciences and Engineering, 44*(6), 233. DOI: 10.1007/s40430-022-03514-x

Savastano, M., Amendola, C., Bellini, F., & D'Ascenzo, F. (2019). Contextual impacts on industrial processes brought by the digital transformation of manufacturing : A systematic review. *Sustainability (Basel), 11*(3), 891. DOI: 10.3390/su11030891

Scaffaro, R., Mistretta, M. C., & Balsamo, M. (2024). Innovative 3D-printed devices for water pollutant removal : Comprehensive review on printing parameters, composition, properties and performances of the latest 3D-systems. *Polymer Testing, 140*, 108627. DOI: 10.1016/j.polymertesting.2024.108627

Sekmen, K., Rehbein, T., Johlitz, M., Lion, A., & Constantinescu, A. (2024). Thermal analysis and shrinkage characterization of the photopolymers for DLP additive manufacturing processes. *Continuum Mechanics and Thermodynamics, 36*(2), 351–368. DOI: 10.1007/s00161-022-01137-0

Singh, S., Kaur, D., Singh, M., Balu, R., Mehta, A., & Vasudev, H. (2024). Challenges and issues in manufacturing of components using polymer-based selective laser sintering (SLS) : A review. [IJIDeM]. *International Journal on Interactive Design and Manufacturing*. Advance online publication. DOI: 10.1007/s12008-024-02049-w

Tareq, M. S., Rahman, T., Hossain, M., & Dorrington, P. (2021). Additive manufacturing and the COVID-19 challenges : An in-depth study. *Journal of Manufacturing Systems*, *60*, 787–798. DOI: 10.1016/j.jmsy.2020.12.021 PMID: 33897085

Thomas, H. R., Hopkinson, N., & Erasenthiran, P. (2006). *High speed sintering–continuing research into a new rapid manufacturing process.*

Tyagi, N., Bhardwaj, V., Sharma, D., Tomar, R., Chaudhary, V., Khanuja, M., Singh, M. K., & Sharma, G. (2024). 3D Printing Technology in the Pharmaceutical and Biomedical Applications : A Critical Review. *Biomedical Materials & Devices*, *2*(1), 178–190. DOI: 10.1007/s44174-023-00105-9

Wu, D., Song, J., Zhai, Z., Hua, M., Kim, C., Frenkel, I., Jiang, H., & He, X. (2019). Visualizing morphogenesis through instability formation in 4-D printing. *ACS Applied Materials & Interfaces*, *11*(50), 47468–47475. DOI: 10.1021/acsami.9b19730 PMID: 31741387

Yehia, H. M., Hamada, A., Sebaey, T. A., & Abd-Elaziem, W. (2024). Selective laser sintering of polymers : Process parameters, machine learning approaches, and future directions. *Journal of Manufacturing and Materials Processing*, *8*(5), 197. DOI: 10.3390/jmmp8050197

Yi, J., Yang, S., Yue, L., & Lei, I. M. (2025). Digital light processing 3D printing of flexible devices : Actuators, sensors and energy devices. *Microsystems & Nanoengineering*, *11*(1), 51. DOI: 10.1038/s41378-025-00885-8 PMID: 40108126

Chapter 2
Materials for 3D Printing:
State of the Art

Ruchira Srivastava
https://orcid.org/0000-0002-5957-9534
Amity University, Greater Noida, India

Ayushi Thakur
https://orcid.org/0000-0002-4172-6751
Amity University, Greater Noida, India

ABSTRACT

Three-dimensional printing is a rapidly developing technology. The globe uses 3D printing a lot these days. 3D printing technology is being utilized more and more in the automobile, locomotive, aviation, healthcare, and agricultural sectors to produce and customize large quantities of open-source designs. An item may be created directly from a computer-aided design (CAD) model using 3D printing technology by depositing material layer by layer. The technique and materials utilized for 3D printing in the manufacturing sector are summarized in this study. It is difficult to choose the right materials and create their blend, but it is essential for successful execution. The existing 3DP systems that are accessible from the literature and the corresponding materials that have been employed up to this point by different companies and specialists for B&C purposes are summarized in this study. Lastly, a brief discussion of possible research possibilities and the benchmarking qualities of these materials is provided.

DOI: 10.4018/979-8-3373-0533-2.ch002

INTRODUCTION

A promising technique, 3D printing (3DP), sometimes referred to as additive manufacturing (AM), can create three-dimensional complicated form prototypes straight from computer-aided design (CAD) models without the need for any equipments or human involvement. AM is widely utilized in various sectors to help with product design, manufacturing, and commercialization because of its unique features (Shahrubudin et al., 2019). The literature begins in 1984, when the third Industrial Revolution began with the creation of stereolithography and, therefore, the introduction of commercial additive manufacturing (AM) (Tay et al., 2016). When Charles ("Chuck") Hull applied for a patent on the method in 1984, he came up with the name "stereolithography" (Zhou et al., 2020). Chuck Hull established 3D Systems, Following the discovery of stereolithography, a technique that allows the creation of single three-dimensional objects from successive solidified layers of ultraviolet light-sensitive resin, the world's first 3D printing company was established in 1986. The process began with digital data processed by Computer-Aided Drafting (CAD) software (Dey et al., 2022). Scott Crump established Stratasys, the top 3D printing firm in the world, in 1988 after discovering fused deposition modelling, or FDM. FDM's single-step process speeds up the manufacturing cycle time by enabling the quick manufacture of 3D objects from CAD blueprints. Melted material is pushed out of the print head's nozzle and applied to the developing workpiece in FDM (Jandyal et al., 2022). One horizontal plane is deposited at a time by moving the head in two dimensions under computer control, then starting a new slice by moving slightly upward. The first Laminated Object Manufacturing (LOM) product was marketed by Helisys in 1991. Using pieces of paper that have been sliced and adhered downstairs, the sculpture is realized. The University of Texas created the selective laser sintering (SLS) technology a year later, and DTM Corporation, a start-up that 3D Systems eventually purchased, brought it to market (Su et al., 2022). The SLS method sinters powdered material using a laser as the power source. Adrian Bowyer's concept gave rise to the Replicating Rapid Prototype Project at the University of Bath in 2005, which gained notoriety under the abbreviation RepRap. The British researcher's project, created under open-source licenses, sought to create a 3D printer that manufactures most of its own parts. 2010 saw the unveiling of Urbee, the first 3D-printed prototype automobile in history (Blanco, 2020). Meanwhile, Organovo Inc., a bioprinting-focused regenerative medicine company, announced the release of the first fully bio-printed blood vessels at the Tissue Engineering and Regenerative Medicine International Society (TERMIS) meeting in Orlando, Florida (Gopinathan & Noh, 2018). This likely marked the beginning of the fourth industrial revolution. Significant advancements in metal 3D printing technology and materials have been made over the last 20 years. As a result, the technologies have been widely investigated and used by R&D organizations, industry, and academia (Zastrow, 2020). An outline of metal 3D printing techniques and the metallic material systems utilized in these techniques is presented after a brief history of the technology. Current problems and certification concerns are covered next, followed by microstructure and characteristics and how they relate to process parameters (Srivastava & Srivastava, 2020). A basic process has been depicted in figure 1 that how 3D printing process takes place.

Even while 3D metal printing has many benefits, there are also major technological and scientific obstacles that must be overcome before these technologies can be widely used and used (Bahadur et al., 2023). First off, layer wise additive consolidation produces very unique microstructures and crystal patterns that are seldom similar to those made using traditional manufacturing techniques like casting or deformation processing (Iftekar et al., 2023). Second, a variety of "as produced" microstructures and properties result from the intrinsic complexity of the physical processes involved in the gradual consolidation of metals, which are accompanied by wide variations in feedstock materials, processing parameters, processing protocols, and equipment architectures (Srivastava, Thakur, & Rana, 2024). Third, The structural and microstructural flaws in metals processed with 3D printing/AM techniques and post-processing procedures like heat treatment and hot-isostatic pressing cycles to reduce these flaws and alter the microstructures to ones appropriate for service conditions are not well understood, despite a great deal of research conducted by R&D organizations (Wang et al., 2020).

Figure 1. 3D Printing Process

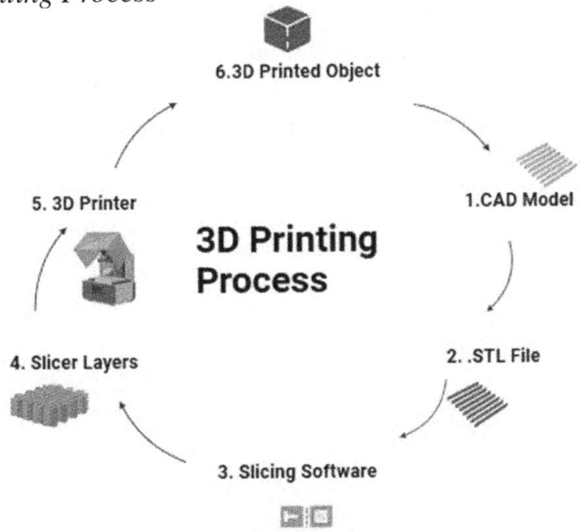

Building and construction (B&C) applications have more recently adopted this technology to address some of the major problems, including a lack of trained labour, high production costs and construction duration, worker health and safety difficulties in hazardous environments, etc (Srivastava, Thakur, Bahadur, & Rana, 2024). An increasing number of novel applications are being made possible by 3D printing. From a basic perspective, each particular application requires consideration of the available materials, production speed, and resolution of 3D printing technologies (Jain et al., 2021). An overview of the principles of 3D printing procedures and the latest advancements in innovative 3D printing materials, including composites, smart materials, ceramic materials, electronic materials, and biomaterials, is given in this article (Thakur et al., 2024) In conclusion, 3D printing technology has become a versatile and potent equipment in the advanced manufacturing sector in recent years (Chen et al., 2019). Many nations have made extensive use of this technology, particularly in the manufacturing sector. As a result, this article provides an overview of the many types of 3D printing technologies, their applications, and, finally, the materials that are utilized in the manufacturing sector (Thakar et al., 2022).

2. VARIOUS 3D PRINTING PROCESS

Many 3D printing technologies have been created, each serving a distinct purpose. Material extrusion, binding jetting, material jetting, directed energy deposition, powder bed fusion, sheet lamination, and vat photopolymerization are the seven classes into which ASTM categorized 3D printing technologies in accordance with ASTM Standard F2792 (Ford & Minshall, 2019). Since every equipment and technology has specific uses, there is no argument about which one works better. These days, 3D printing technology is being utilized more and more to create a wide range of items rather than just prototypes (Tian et al., 2021). A detailed categorization of 3D printing process is shown in Figure 2.

Figure 2. Types of 3D Printing processes

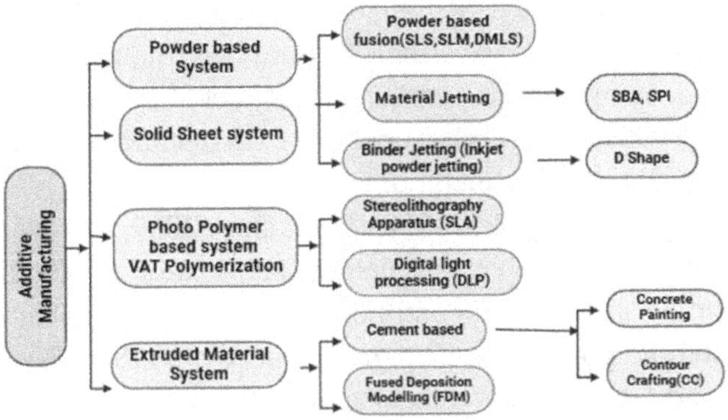

Most people agree that the first 3D printing method was stereolithography (SL), which was also the first to be made available for purchase. SL uses photopolymer resins and is a laser-based procedure (Palmara et al., 2021). In order to create extremely exact components, they must react with the laser and cure to form a solid in a precise manner. The photopolymer resin is kept in a vat with a moveable platform within, albeit this is a complicated procedure (Regehly et al., 2020). The resin hardens exactly where the laser strikes it, and a laser beam is aimed in the X-Y axes over the resin's surface in accordance with the 3D data that is sent into the machine (the.stl file) (Quan et al., 2020). The platform within the vat descends slightly (in the Z axis) once the layer is finished, and the laser traces out the next layer. The platform may then be lifted out of the vat for removal after the entire sculpture is finished (El-Sayegh et al., 2020).

Stereolithography is widely acknowledged as one of the most precise 3D printing techniques with superior surface quality. The necessary post-processing procedures and the materials' stability over time—which might cause them to become more brittle—are, however, limiting issues (Bozkurt & Karayel, 2021).

As a 3D printing method that uses photopolymers, DLP, or digital light processing, is comparable to stereolithography. The light source is the primary distinction (Rao et al., 2022). DLP is often quicker than SL since it applies light to the whole surface of the vat of photopolymer resin in a single pass using a more traditional light source, like an arc lamp, a liquid crystal display panel, or a deformable mirror device (DMD) (Vidal et al., 2020).

When discussing a laser-based 3D printing method that uses powdered materials, the words "laser sintering" and "laser melting" are synonymous. Based on the 3D data that is input into the system, the laser is tracked in the X-Y axes across a powder bed of tightly packed powdered material (Alabi, 2020).

The most popular and well-known 3D printing method is probably extrusion of thermoplastic material. Due to its length, the method is most often known as Fused Deposition Modelling (FDM); nevertheless, Stratasys, the firm that created it, has registered this trade name as a trademark (Cornejo et al., 2022).

Since the early 1990s, Stratasys has operated FDM technology, which is currently an industrial-grade 3D printing method.

The widespread use of entry-level 3D printers since 2009, however, mostly makes use of a similar technique known as Freeform Fabrication (FFF), albeit in a more rudimentary form because Stratasys still owns the patents (Prashar et al., 2023).

Though substantial post-processing may be necessary, Stratasys' FDM process is a dependable and precise method that is quite office/studio friendly (Nesic et al., 2020). As anticipated, the FFF approach generates far less accurate models at the entry level, although improvements are ongoing.

The jetting method is used in two different 3D printing processes. Binder jetting is the process of selectively spraying a binder onto a powder bed of component material to fuse it one layer at a time, producing or printing the necessary part (Amores et al., 2022). When a layer is finished, the powder bed drops gradually, and a roller or blade smoothes the powder across the bed's surface before the jet heads move on to the next pass, allowing the binder for the next layer to form and fuse with the one before it. This process is similar to other powder bed systems (Javaid et al., 2022).

Material jetting is a 3D printing technique in which the actual building materials—in a liquid or molten state—are jetted selectively through a number of jet heads, while other jet heads concurrently jett support components (Mayfield et al., 2022). Nonetheless, the materials are typically liquid photopolymers that undergo curing with a UV light pass during the deposition of each layer.

This product's nature permits the simultaneous deposition of several components, enabling the production of a single part from several materials with various traits and attributes (Placone et al., 2020). A particularly accurate 3D printing technique that creates pieces with a flawless finish is material jetting. Mcor Technologies created and produces SDL, a proprietary 3D printing method. Because of the similarities in how the final item is formed by stacking and molding paper, there is a desire to link this technique to Helisys' Laminated Object Manufacturing (LOM) technology from the 1990s. But that's the extent of any resemblance (Bijalwan et al., 2024).

Material of 3D Printing

The materials that are now suitable for 3D printing have advanced significantly since the inception of the technology as shown in figure 3. These days, there are many different kinds of materials available in various states (powder, filament, pellets, granules, resin, etc.) (Karyappa et al., 2024). With material qualities that are more specifically suited to the application, unique materials are now often designed for specific platforms carrying out devoted applications (the dentistry industry is one example). However, it would be impossible to include all of the proprietary materials from the several 3D printer companies here (Roudný & Syrový, 2022).

Rather, this essay will take a more general look at the most common kinds of content. There are also a few materials that are particularly noteworthy.

Figure 3. Materials of 3D printing

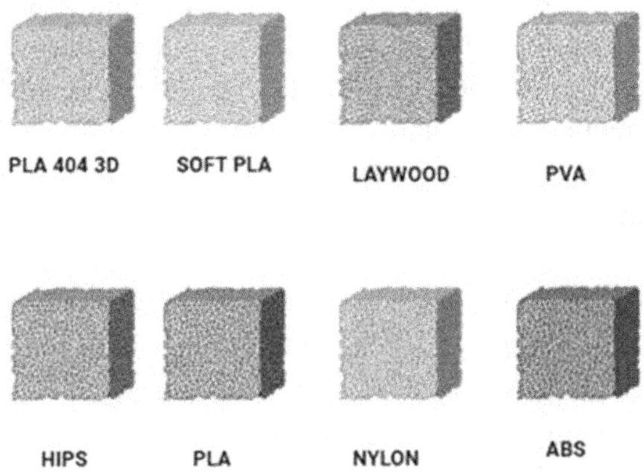

Polyamide, often known as nylon, is frequently utilized in filament form with the FDM process or in powder form with the sintering process. It is a dependable plastic substance that is robust, flexible, and long-lasting for 3D printing (Somers et al., 2024). Although it is white by nature, it can be coloured either before or after printing. Alumide, another popular 3D printing material for sintering, may be created by combining this substance (in powder form) with powdered aluminium. Another popular material for 3D printing is ABS, which is frequently found in filament form on low-end FDM 3D printers (Shao et al., 2022). It is a very durable plastic that is available in a variety of colours. Another factor contributing to ABS's widespread use is the fact that it can be purchased in filament form from a variety of non-proprietary suppliers (Kumar et al., 2022). For precisely this reason, PLA, a biodegradable plastic, has become popular with 3D printing. It may be used in filament form for the FDM process or in resin form for DLP/SL operations. It is available in a range of colours, including translucent, which has worked well for some 3D printing applications. Nevertheless, it lacks ABS's flexibility and durability. For entry-level extrusion 3D printers, Lay Wood is a specially designed 3D printing material (Ashby & Jones, 2012). It is also known as WPC and is a wood/polymer composite that is available in filament form. For industrial-grade 3D printing, an increasing variety of metals and metal composites are utilized. Derivatives of cobalt and aluminium are two of the most prevalent (Budinski & Budinski, 1983).

Some prevailing categories of 3D materials are discussed below

Metals: The manufacturing, automotive, medical, and aerospace industries are paying close attention to metal 3D printing technology due to its many benefits (El-Eskandarany, 2013). Metal materials are used in complex manufacturing processes, such as printing human organs and aerospace components, because of their exceptional physical qualities. Stainless steel, titanium alloys, cobalt-based alloys, nickel-based alloys, and aluminium alloys are a few examples of these materials (Stevens, 2010). The use of cobalt-based alloys in 3D printed dental applications is appropriate. This is due to its high specific stiffness, resilience, elongation, recovery capacity, and heat-treated environments. Additionally, 3D printing technology can use nickel base alloys to create aerospace parts (Czaja et al., 2009). 3D printed products created from nickel base alloys can be used in dangerous environments. This is because of its exceptional corrosion resistance and capacity to tolerate temperatures of up to 1200 °C (Shaheen et al., 2023). Finally, titanium alloys may be used to build the item utilizing 3D printing technology. The alloy titanium has highly special qualities, including low density, strong corrosion resistance, oxidation resistance, and ductility. It is utilized in high-stress environments with high operating temperatures, such as the biomedical and aerospace industries (Iordachescu & Quintino, 2008).

Ceramics: A relatively recent class of materials, ceramics may be 3D printed with varying degrees of success. The important thing to keep in mind with these materials is that, after printing, the ceramic components must go through the same fire and glazing procedures as any ceramic component manufactured using conventional manufacturing techniques (Fleck et al., 1994).

Polymers: Polymer components, ranging from prototypes to functional constructions with challenging geometries, are produced using 3D printing technology. Polylactic acid (PLA), acrylonitrile butadiene styrene (ABS), polypropylene (PP), or polyethylene (PE) are examples of extruded thermoplastic filaments that may be deposited in consecutive layers to create a 3D printed object using fused deposition modeling (FDM) (Mueller et al., 2006). Lately, PEEK and PMMA, two thermoplastic filaments with greater melting temperatures, have become suitable materials for 3D printing. The 3D printing business makes extensive use of liquid polymer materials or those with low melting points because of their affordability, portability, and ease of processing (Xu et al., 2015). Most typically used as inert materials, polymer materials are crucial to biomaterials and medical device goods because they support mechanically numerous orthopaedic implants and help the devices operate efficiently (Oropallo & Piegl, 2016).

Composites: Composite materials' exceptional flexibility, low weight, and adjustable properties have been revolutionizing high-performance industries. Composite materials come in two varieties: carbon fibre reinforced polymer composites and glass fibre reinforced polymer composites (Murtezani et al., 2022). The aerospace industry makes extensive use of carbon fibre reinforced polymer composite structures due to their high specific stiffness, strength, superior fatigue performance, and resistance to corrosion (Dabin et al., 2016). Glass fiber reinforced polymer composites are also frequently used for a range of 3D printing applications due to their remarkable performance and low cost. High thermal conductivity and a low coefficient of thermal expansion are characteristics of fiberglass (Patpatiya et al., 2022). Furthermore, fiberglass is ideal for use in 3D printing applications since it is non-flammable and unaffected by the curing temperatures used in manufacturing procedures (Gu, 2016).

Bio inspired material: The potential of 3D printing biomaterials for a variety of medical (and other) uses is the subject of much investigation. Several top universities are researching living tissue in order to create applications such as printing external tissues for body replacements and human organs for transplantation (Awad et al., 2020). Additional study in this field focuses on creating food products, with meat serving as the main example.

Smart Material: The term "smart materials" refers to materials that may change an object's geometry and shape in response to environmental factors like heat and water (Kristiawan et al., 2021). Two examples of 3D printed items using intelligent materials are the soft robotics system and the self-evolving structure. Smart materials also include materials for 4D printing. Group smart materials include, for example, form memory polymers and form memory alloys. Some shape-memory alloys, including nickel-titanium, are utilized in biomedical implants and microelectromechanical systems (Tan et al., 2022). Shape memory polymer (SMP) is a type of functional material that reacts to stimuli like light, electricity, heat, certain chemicals, and so on. By using 3D printing technology, the complex shape of shape memory polymer could be easily and conveniently produced (Ribeiro et al., 2021). This material's quality is assessed based on its part density, surface roughness, and dimensional precision. Meanwhile, while producing nickel-titanium 3D printed goods, transformation temperatures, microstructure repeatability, and density are crucial considerations (Mukai et al., 2021).

Paper: Mcor Technologies provides the patented SDL technique, which uses standard A4 copier paper as a 3D printing medium. The firm has a completely different business plan than other 3D printing providers; while the machine's capital expenditure is in the middle range, a readily available, reasonably priced supply of materials that can be purchased locally is the main focus (Meng et al., 2023).

Paper-based 3D printed models are safe, eco-friendly, readily recyclable, and don't need any post-processing.

Food: Extruder experiments for 3D printing food items have significantly increased in the past few years. Chocolate is the most popular (and coveted) (Chung et al., 2020). Additionally, there are printers that use sugar and some trials that use meat and pasta. Research is being done to use 3D printing technology to create perfectly balanced complete meals in the future (Baldi et al., 2022).

Lunar Dust: Future moon colonization may benefit from the direct production of multi-layered components from lunar dust using the 3D printing technology (Wei et al., 2022).

Textile: The advancement of 3D textile printing technology will make the jewellery and apparel industries shine. Short processing times for products, lower packaging costs, and lower supply chain expenses are some benefits of 3D printing technology in the fashion business (Zub et al., 2022).

Finally, with its digital materials for the Object Connex 3D printing platform, Stratasys is one business that does provide a distinctive (proprietary) material offering (Humnabad et al., 2022). This product enables the creation of unique materials with the necessary qualities by combining common Object 3D printing components in different and predetermined concentrations during the printing process. By mixing the current core elements in various ways, up to 140 distinct digital materials can be created (Pradhan et al., 2023).

When it comes to the usefulness, robustness, and performance of 3D printed items, materials are vital. Metals (such as titanium, stainless steel, and aluminum), ceramics (such as zirconia and alumina), polymers (such PLA, ABS, and nylon), and composites (combining fibers with metals or polymers) are examples of common materials. Every material has certain qualities that are appropriate for a given use (Dvorak et al., 2021). For example, ceramics are used in dentistry and electronics because of their resilience to heat and chemicals, whereas titanium alloys are used extensively in aircraft and biomedical implants because of their strength and biocompatibility. Because of their variety and simplicity of printing, polymers are preferred in consumer goods, automotive parts, and prototypes (Biswas et al., 2021). Composites are becoming more and more common in sectors like sports and automotive that need lightweight yet durable components.

The material choice is crucial and frequently influenced by the printed object's intended mechanical qualities, cost-effectiveness, and end-use environment (Ghazal et al., 2023).

Global effects of 3D Printing in different domains:

Manufacturing Sector: The manner that things are made is already being impacted by 3D printing; the technology's nature allows for innovative approaches to considering the manufacturing process's effects on society, the economy, the environment, and security, with generally positive outcomes (Patel et al., 2022).

One of the main arguments in favour of this claim is that 3D printing may enable production to move closer to the consumer or end user, easing the constraints on the present supply chain (Truxova et al., 2020). Similar to how Amazon runs its company, 3D printing's customization value and capacity to manufacture small production batches on demand are certain ways to engage customers AND cut down on or eliminate inventories and stock piling (Maines et al., 2021).

World economy: If 3D printing technology is widely utilized, it might have an impact on the international economy. Inequality between exporting and importing nations may be lessened by switching manufacturing and distribution from the existing model to a localized production based on demand, on site, customized production model (Pereira et al., 2019). New industries and even new professions, including those involved in the manufacture of 3D printers, might be spawned by 3D printing (Ntousia & Fudos, 2019). Professional services ranging from new product designers, printer operators, material suppliers, and intellectual property legal challenges and settlements are available in the context of 3D printing. For many IP holders, the current issue with 3D printing is piracy (Ranjan et al., 2022). 3D printing has both positive and negative effects on the developing world. Recycled and other locally produced materials can reduce production costs, for instance. However, many poor nations may suffer greatly from the loss of manufacturing employment, which would take decades to recover from (Chun, 2021).

Application: The foundation of 3D printing in "Rapid Prototyping" was based on the ideas of industrial prototyping, which aims to expedite the initial phases of product development by providing a rapid and simple method of creating prototypes that enable several iterations of a product to arrive more quickly and efficiently at an ideal solution as detailed in figure 4 (Eisenberg, 2013). These guarantees trust prior to production tooling and saves time and money at the beginning of the whole product development process (Ukobitz, 2021). Even if it is occasionally disregarded, prototyping is still most likely the most common use of 3D printing in modern times (Chaudhuri et al., 2017).

Since the advent of 3D printing for prototyping, advancements and improvements in both the technique and the materials have led to the adoption of the processes for use lower down the product development process chain (Tuli, Khatun, & Rashid, 2024). Applications for casting and tooling were created by leveraging the benefits of the various techniques. Once more, a growing number of industrial sectors are adopting and using these applications (Zhang, Yang, & Sheng, 2018).

The advancements in final production procedures are also continuing to make adoption easier (Al Jassmi, Al Najjar, & Mourad, 2018).

Here is a basic summary of the industrial vertical markets that stand to gain the most from industrial

Figure 4. Some applications of 3D printing

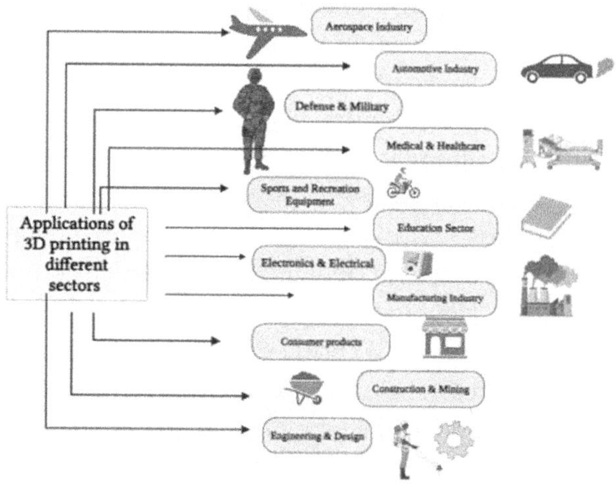

3D printing across all of these wide-ranging applications:

Medical & Health care: Although the medical industry was among the first to use 3D printing, it is also thought to have enormous growth potential because of the technologies' capacity for customization and personalization as well as their potential to enhance people's lives as procedures advance and materials that satisfy medical grade standards are created (Tissen, 2022). Through direct use of the technologies, both stock items, such as hip and knee implants, and custom patient-specific products, such as hearing aids, shoe orthotic insoles, customized prosthetics, and one-off implants for patients with conditions like cancer, osteoarthritis, and osteoporosis, as well as accident and trauma victims, are manufactured (Kalaskar, 2022). The technologies are used to produce tools that are vacuum-formed over plastic to create dental aligners and to create patterns for the downstream metal casting of dental crowns, in addition to developing prototypes to support the development of new products for the medical and dental industries (Dimitrov, Schreve, & de Beer, 2006). Another new use that is helping surgeons in their job and patients in their recuperation are 3D printed surgical guides for particular procedures. Additionally, technology for 3D printing human organs, medications, skin, bone, and tissue is being developed. Commercialization of these technologies is still decades away, though (Fu et al., 2017).

Automotive Sector: These days, our industry is quickly changing to create, develop, and manufacture new items thanks to 3D printing technology. 3D printing technology has revolutionized the car industry by enabling the rapid creation of lighter and more intricate structures (Espalin et al., 2014). For instance, Local Motor created the first 3D-printed electric car in 2014. By creating the OLLI, a 3D-printed bus, Local Motors broadened the application of 3D printing technology beyond only cars. The 3D-printed, self-governing, electric, recyclable, and highly intelligent bus is called OLLI (Shahzad, Umair, & Waqar, 2022). Additionally, Ford is a pioneer in the usage of 3D printing technology, using it to create engine components and prototypes. Additionally, BMW manufactures hand tools using 3D printing technology (Rael & San Fratello, 2017).

Food Industry: Both the culinary and aerospace industries have benefited from 3D printing technology. The need for creating specialized foods for people with specific nutritional needs—athletes, kids, pregnant women, patients, and so forth—is currently increasing (Zhong & Zhao, 2017). This can be achieved by lowering the quantity of unneeded ingredients and increasing the number of nutritious ones. However, creating bespoke foods requires a very creative and thorough approach, which is where 3D food printing comes into play (Besklubova, Skibniewski, & Zhang, 2021). Food layer manufacturing, sometimes referred to as 3D-food printing, is created by directly using computer-aided design data to deposit consecutive layers one after the other (Esene et al., 2023). Certain materials may be combined and processed into a variety of intricate shapes and structures utilizing 3D printing technology. New food products with intricate and intriguing shapes and designs may be made using sugar, chocolate, pureed food, and flat foods like spaghetti, pizza, and crackers (Alqutaibi et al., 2024).

Construction Industry: In addition to offering countless opportunities for the fulfilment of geometric complexity, 3D printing technology may be regarded as an ecologically friendly derivative. Modelling may convey information and knowledge about 3D buildings and is a digital representation of functional and physical qualities (Park et al., 2022). It may serve as a trustworthy resource for decisions at every stage of the building's life cycle, from conception to destruction in order to design or construct the structure. This cutting-edge, cooperative technology will provide more effective approaches to the planning, development, and upkeep of the built environment (Miryousefi Ata, Kazemian, & Jafari, 2022).

Textile Industry: 3D printed shoes, jewellery, consumer products, and apparel are starting to appear on the market as 3D printing technology penetrates the retail sector. While it may not seem like the most obvious mix, 3D printing and fashion are beginning to become commonplace worldwide (Subramani et al., 2024). For example, major corporations like Adidas, Nike, and New Balance are working to develop the ability to produce 3D printed shoes in large quantities. These days, sneakers, personalized shoes, and athlete's shoes are constructed using 3D printing (Antezana et al., 2023).

Electronics Industry: Manufacturers are beginning to see 3D printing's promise realized in a variety of intriguing ways as it becomes more widely available to the scientific, technological, and manufacturing domains (Kaliaraj et al., 2023). Currently, a range of 3D printing techniques are extensively used for structural electronic devices, including active electronic materials, electrodes, and devices with mass customization and flexible design, by integrating the conductors into the devices (Robayo-Salazar et al., 2023).

Future of 3D printing: The requirements of the modern day and the technologies employed to satisfy them are evolving along with the globe, which is changing and developing daily. With several successful applications across various industries, 3D printing technology has exciting prospects for the future (Kurian & Thankachan, 2023). It is intended to be useful in resolving some of the issues we have been facing lately. The masks made using this technique to ward against the coronavirus epidemic, which has recently been a global concern, are without a question among the greatest illustrations of this circumstance. It is intended to be useful in resolving some of the issues we have been facing lately (Wang et al., 2023). The masks made using this technique to ward against the coronavirus epidemic, which has recently been a global concern, are without a question among the greatest illustrations of this circumstance. One of the biggest issues during pandemic emergencies is the absence of protective face masks, particularly for medical professionals like physicians, nurses, and paramedics (Tabassum & Mir, 2023). However, it is anticipated that things will change in the future and that many lives will be saved by the production and usage of organs that are capable of fulfilling their intended roles in organ transplantation. These are only a few instances of what the 3D printer will be able to achieve in the future. A few years ago, its capabilities appeared like something out of science fiction films, but that has since changed (Baigarina, Shehab, & Ali, 2023). Consequently, it is thought that this technology will lead to new opportunities in the future.

CONCLUSION

The manufacturing sector is currently embracing 3D printing technology, which has several advantages for individuals, businesses, and the government. In order to develop strategies for boosting the use of 3D printing technology, more data is needed. Businesses and the government may modernize and improve the infrastructure for 3D printing technology with the help of more understanding about the technology. Thus, this article's goal is to give a general overview of the many types of 3D printing technologies, the materials used in the manufacturing industry, and, lastly, the uses of 3D printing technology. Researchers can look at the many kinds of 3D printers and the materials that work well for each kind of machine in the future. Due to these benefits, 3D printer technology is becoming more and more popular across a variety of industries. It can generate intricately formed components that are challenging to produce using conventional methods, and it can do it more quickly and easily at a lower cost. Its benefits make it particularly useful in the biomedical industry, where it is chosen for a variety of applications on a daily basis. Although this miraculous technology is still being researched and refined, its usage in the bioprinting and pharmaceutical industries is not yet mainstream but is expected to lead to revolutionary advancements once it does. It has the potential to save countless lives and will revolutionize the medical industry when it is produced soon. In every field of health, from functional prostheses to the treatment of cancer and birth abnormalities, 3D printing-powered innovations are being developed to give patients a higher quality of life and a longer lifespan.

REFERENCES

Al Jassmi, H., Al Najjar, F., & Mourad, A. H. I. (2018, March). Large-Scale 3D printing: The way forward. [). IOP Publishing.]. *IOP Conference Series. Materials Science and Engineering*, *324*, 012088. DOI: 10.1088/1757-899X/324/1/012088

Alabi, M. O. (2020). Big data, 3D printing technology, and industry of the future. In *Additive Manufacturing: Breakthroughs in Research and Practice* (pp. 503-520). IGI Global.

Alqutaibi, A. Y., Alghauli, M. A., Aljohani, M. H. A., & Zafar, M. S. (2024). Advanced additive manufacturing in implant dentistry: 3D printing technologies, printable materials, current applications and future requirements. *Bioprinting (Amsterdam, Netherlands)*, *42*, 00356. DOI: 10.1016/j.bprint.2024.e00356

Amores, I. D., González-Gutiérrez, J., García, I. M., Franco, J. M., & Gallegos, C. (2022). 3D printing–Present and future–A Chemical Engineering perspective. *Chemical Engineering Research & Design*, *187*, 598–610. DOI: 10.1016/j.cherd.2022.08.049

Antezana, P. E., Municoy, S., Ostapchuk, G., Catalano, P. N., Hardy, J. G., Evelson, P. A., Orive, G., & Desimone, M. F. (2023). 4D printing: The development of responsive materials using 3D-printing technology. *Pharmaceutics*, *15*(12), 2743. DOI: 10.3390/pharmaceutics15122743 PMID: 38140084

Ashby, M. F., & Jones, D. R. (2012). *Engineering materials 1: an introduction to properties, applications and design* (Vol. 1). Elsevier.

Awad, A., Fina, F., Goyanes, A., Gaisford, S., & Basit, A. W. (2020). 3D printing: Principles and pharmaceutical applications of selective laser sintering. *International Journal of Pharmaceutics*, *586*, 119594. DOI: 10.1016/j.ijpharm.2020.119594 PMID: 32622811

Bahadur, P. S., Srivastava, R., & Srivastava, V. (2023). Review on the Development of Smart Materials, Including Shape Memory Alloys, Characterization, and Recent Applications. *Modeling, Characterization, and Processing of Smart Materials*, 107-139.

Baigarina, A., Shehab, E., & Ali, M. H. (2023). Construction 3D printing: A critical review and future research directions. *Progress in Additive Manufacturing*, *8*(6), 1393–1421. DOI: 10.1007/s40964-023-00409-8

Baldi, J., Grò, A., Orsini, U., & Favale, L. (2022). *3D-Printed Custom-Made Instruments. In 3D Printing in Bone Surgery*. Springer International Publishing.

Besklubova, S., Skibniewski, M. J., & Zhang, X. (2021). Factors affecting 3D printing technology adaptation in construction. *Journal of Construction Engineering and Management, 147*(5), 04021026. DOI: 10.1061/(ASCE)CO.1943-7862.0002034

Bijalwan, V., Rana, S., Yun, G. J., Singh, K. P., Jamil, M., & Schlögl, S. (2024). 3D Printing of Covalent Adaptable Networks: Overview, Applications and Future Prospects. *Polymer Reviews (Philadelphia, Pa.), 64*(1), 36–79. DOI: 10.1080/15583724.2023.2227692

Biswas, M. C., Chakraborty, S., Bhattacharjee, A., & Mohammed, Z. (2021). 4D printing of shape memory materials for textiles: Mechanism, mathematical modeling, and challenges. *Advanced Functional Materials, 31*(19), 2100257. DOI: 10.1002/adfm.202100257

Blanco, I. (2020). The Use of Composite Materials in 3D Printing. *Journal of Composites Science, 4*(2), 42. DOI: 10.3390/jcs4020042

Bozkurt, Y., & Karayel, E. (2021). 3D printing technology; methods, biomedical applications, future opportunities and trends. *Journal of Materials Research and Technology, 14*, 1430–1450. DOI: 10.1016/j.jmrt.2021.07.050

Budinski, K. G., & Budinski, M. K. (1983). *Engineering materials*. Pearson Education India.

Chaudhuri, A., Rogers, H., Søberg, P., Baricz, N., & Pawar, K. (2017, July). Identifying future 3D printing related services: insights from Denmark and Germany. In *22nd International Symposium on Logistics* (p. 73). Centre for Concurrent Enterprise, Nottingham University Business School.

Chen, Z., Li, Z., Li, J., Liu, C., Lao, C., Fu, Y., Liu, C., Li, Y., Wang, P., & He, Y. (2019). 3D printing of ceramics: A review. *Journal of the European Ceramic Society, 39*(4), 661–687. DOI: 10.1016/j.jeurceramsoc.2018.11.013

Chun, H. (2021). A study on the impact of 3D printing and artificial intelligence on education and learning process. *Scientific Programming, 2021*(1), 2247346. DOI: 10.1155/2021/2247346

Chung, J. J., Im, H., Kim, S. H., Park, J. W., & Jung, Y. (2020). Toward biomimetic scaffolds for tissue engineering: 3D printing techniques in regenerative medicine. *Frontiers in Bioengineering and Biotechnology, 8*, 586406. DOI: 10.3389/fbioe.2020.586406 PMID: 33251199

Cornejo, J., Cornejo-Aguilar, J. A., Vargas, M., Helguero, C. G., Milanezi de Andrade, R., Torres-Montoya, S., Asensio-Salazar, J., Rivero Calle, A., Martínez Santos, J., Damon, A., Quiñones-Hinojosa, A., Quintero-Consuegra, M. D., Umaña, J. P., Gallo-Bernal, S., Briceño, M., Tripodi, P., Sebastian, R., Perales-Villarroel, P., De la Cruz-Ku, G., & Russomano, T. (2022). Anatomical Engineering and 3D printing for surgery and medical devices: International review and future exponential innovations. *BioMed Research International*, *2022*(1), 6797745. DOI: 10.1155/2022/6797745 PMID: 35372574

Czaja, A. U., Trukhan, N., & Müller, U. (2009). Industrial applications of metal–organic frameworks. *Chemical Society Reviews*, *38*(5), 1284–1293. DOI: 10.1039/b804680h PMID: 19384438

. Dabin, M., Narushima, T., Beirne, S. T., Ritz, C. H., & Grady, K. (2016). 3d modelling and printing of microtonal flutes.

Dey, D., Srinivas, D., Panda, B., Suraneni, P., & Sitharam, T. G. (2022). Use of industrial waste materials for 3D printing of sustainable concrete: A review. *Journal of Cleaner Production*, *340*, 130749. DOI: 10.1016/j.jclepro.2022.130749

Dimitrov, D., Schreve, K., & de Beer, N. (2006). Advances in three dimensional printing–state of the art and future perspectives. *Rapid Prototyping Journal*, *12*(3), 136–147. DOI: 10.1108/13552540610670717

Dvorak, K., Dvorakova, J., Zarybnicka, L., & Horak, Z. (2021). Influence of 3D printing topology by DMLS method on crack propagation. *Materials (Basel)*, *14*(23), 7483. DOI: 10.3390/ma14237483 PMID: 34885639

Eisenberg, M. (2013). 3D printing for children: What to build next? *International Journal of Child-Computer Interaction*, *1*(1), 7–13. DOI: 10.1016/j.ijcci.2012.08.004

El-Eskandarany, M. S. (2013). *Mechanical alloying: For fabrication of advanced engineering materials*. Univ. Press of Mississippi.

El-Sayegh, S., Romdhane, L., & Manjikian, S. (2020). A critical review of 3D printing in construction: Benefits, challenges, and risks. *Archives of Civil and Mechanical Engineering*, *20*(2), 1–25. DOI: 10.1007/s43452-020-00038-w

Esene, J. E., Nasman, P. R., Akuoko, Y., Tahir, A., & Woolley, A. T. (2023). Past, current, and future roles of 3D printing in the development of capillary electrophoresis systems. *Trends in Analytical Chemistry*, *162*, 117032. DOI: 10.1016/j.trac.2023.117032 PMID: 37008739

Espalin, D., Muse, D. W., MacDonald, E., & Wicker, R. B. (2014). 3D Printing multifunctionality: Structures with electronics. *International Journal of Advanced Manufacturing Technology*, *72*(5-8), 963–978. DOI: 10.1007/s00170-014-5717-7

Fleck, N. A., Kang, K. J., & Ashby, M. F. (1994). Overview no. 112: The cyclic properties of engineering materials. *Acta Metallurgica et Materialia*, *42*(2), 365–381. DOI: 10.1016/0956-7151(94)90493-6

Ford, S., & Minshall, T. (2019). Invited review article: Where and how 3D printing is used in teaching and education. *Additive Manufacturing*, *25*, 131–150. DOI: 10.1016/j.addma.2018.10.028

Fu, K., Yao, Y., Dai, J., & Hu, L. (2017). Progress in 3D printing of carbon materials for energy-related applications. *Advanced Materials*, *29*(9), 1603486. DOI: 10.1002/adma.201603486 PMID: 27982475

Ghazal, A. F., Zhang, M., Mujumdar, A. S., & Ghamry, M. (2023). Progress in 4D/5D/6D printing of foods: Applications and R&D opportunities. *Critical Reviews in Food Science and Nutrition*, *63*(25), 7399–7422. DOI: 10.1080/10408398.2022.2045896 PMID: 35225117

Gopinathan, J., & Noh, I. (2018). Recent trends in bioinks for 3D printing. *Biomaterials Research*, *22*(1), 11. DOI: 10.1186/s40824-018-0122-1 PMID: 29636985

Gu, D. (2016). Materials creation adds new dimensions to 3D printing. *Science Bulletin*, *61*(22), 1718–1722. DOI: 10.1007/s11434-016-1191-y

Humnabad, P. S., Tarun, R., & Das, I. (2022). An overview of direct metal laser sintering (DMLS) technology for metal 3D printing. *Journal of Mines, Metals and Fuels*, 127-133.

Iftekar, S. F., Aabid, A., Amir, A., & Baig, M. (2023). Advancements and limitations in 3D printing materials and technologies: A critical review. *Polymers*, *15*(11), 2519. DOI: 10.3390/polym15112519 PMID: 37299318

Iordachescu, D., & Quintino, L. (2008). Steps toward a new classification of metal transfer in gas metal arc welding. *Journal of Materials Processing Technology*, *202*(1-3), 391–397. DOI: 10.1016/j.jmatprotec.2007.08.081

Jain, K., Shukla, R., Yadav, A., Ujjwal, R. R., & Flora, S. J. S. (2021). 3D printing in development of nanomedicines. *Nanomaterials (Basel, Switzerland)*, *11*(2), 420. DOI: 10.3390/nano11020420 PMID: 33562310

Jandyal, A., Chaturvedi, I., Wazir, I., Raina, A., & Haq, M. I. U. (2022). 3D printing–A review of processes, materials and applications in industry 4.0. *Sustainable Operations and Computers*, *3*, 33–42. DOI: 10.1016/j.susoc.2021.09.004

Javaid, M., Haleem, A., Singh, R. P., & Suman, R. (2022). 3D printing applications for healthcare research and development. *Global Health Journal (Amsterdam, Netherlands)*, *6*(4), 217–226. DOI: 10.1016/j.glohj.2022.11.001

Kalaskar, D. M. (Ed.). (2022). *3D printing in medicine*. Woodhead Publishing.

Kaliaraj, G. S., Shanmugam, D. K., Dasan, A., & Mosas, K. K. A. (2023). Hydrogels—A promising materials for 3D printing technology. *Gels (Basel, Switzerland)*, *9*(3), 260. DOI: 10.3390/gels9030260 PMID: 36975708

Karyappa, R., Zhang, D., Zhu, Q., Ji, R., Suwardi, A., & Liu, H. (2024). Newtonian liquid-assisted material extrusion 3D printing: Progress, challenges and future perspectives. *Additive Manufacturing*, *79*, 103903. DOI: 10.1016/j.addma.2023.103903

Kristiawan, R. B., Imaduddin, F., Ariawan, D., Ubaidillah, , & Arifin, Z. (2021). A review on the fused deposition modeling (FDM) 3D printing: Filament processing, materials, and printing parameters. *Open Engineering*, *11*(1), 639–649. DOI: 10.1515/eng-2021-0063

Kumar, N., Ramakrishnan, S. A., Lopez, K. G., Chin, B. Z., Devyapriya, S., Kumar, L., & Anantharajan, S. K. (2022). Current trends and future scope in 3D printing for surgical management of spine pathologies. *Bioprinting (Amsterdam, Netherlands)*, *26*, e00197. DOI: 10.1016/j.bprint.2022.e00197

Kurian, M., & Thankachan, S. (2023). Introduction: Ceramics classification and applications. In *Ceramic catalysts* (pp. 1-17). Elsevier.

Maines, E. M., Porwal, M. K., Ellison, C. J., & Reineke, T. M. (2021). Sustainable advances in SLA/DLP 3D printing materials and processes. *Green Chemistry*, *23*(18), 6863–6897. DOI: 10.1039/D1GC01489G

Mayfield, C. K., Ayad, M., Lechtholz-Zey, E., Chen, Y., & Lieberman, J. R. (2022). 3D-printing for critical sized bone defects: Current concepts and future directions. *Bioengineering (Basel, Switzerland)*, *9*(11), 680. DOI: 10.3390/bioengineering9110680 PMID: 36421080

Meng, M., Wang, J., Huang, H., Liu, X., Zhang, J., & Li, Z. (2023). 3D printing metal implants in orthopedic surgery: Methods, applications and future prospects. *Journal of Orthopaedic Translation*, *42*, 94–112. DOI: 10.1016/j.jot.2023.08.004 PMID: 37675040

Miryousefi Ata, S., Kazemian, A., & Jafari, A. (2022). Application of concrete 3D printing for bridge construction: Current challenges and future directions. In *Construction Research Congress 2022* (pp. 869-879). DOI: 10.1061/9780784483961.091

Mueller, U., Schubert, M., Teich, F., Puetter, H., Schierle-Arndt, K., & Pastre, J. (2006). Metal–organic frameworks—Prospective industrial applications. *Journal of Materials Chemistry, 16*(7), 626–636. DOI: 10.1039/B511962F

Mukai, S., Mukai, E., Santos-Junior, J. A., Shibli, J. A., Faveri, M., & Giro, G. (2021). Assessment of the reproducibility and precision of milling and 3D printing surgical guides. *BMC Oral Health, 21*(1), 1–7. DOI: 10.1186/s12903-020-01362-6 PMID: 33388028

Murtezani, I., Sharma, N., & Thieringer, F. M. (2022). Medical 3D printing with a focus on Point-of-Care in Cranio-and Maxillofacial Surgery. A systematic review of literature. *Annals of 3D Printed Medicine, 6*, 100059.

Nesic, D., Durual, S., Marger, L., Mekki, M., Sailer, I., & Scherrer, S. S. (2020). Could 3D printing be the future for oral soft tissue regeneration? *Bioprinting (Amsterdam, Netherlands), 20*, e00100. DOI: 10.1016/j.bprint.2020.e00100 PMID: 32679657

Ntousia, M., & Fudos, I. (2019, June). 3D printing technologies & applications: an overview. In *Proceedings of the CAD 2020 Conference,* Singapore (Vol. 243, p. 248).

Oropallo, W., & Piegl, L. A. (2016). Ten challenges in 3D printing. *Engineering with Computers, 32*(1), 135–148. DOI: 10.1007/s00366-015-0407-0

Palmara, G., Frascella, F., Roppolo, I., Chiappone, A., & Chiadò, A. (2021). Functional 3D printing: Approaches and bio applications. *Biosensors & Bioelectronics, 175*, 112849. DOI: 10.1016/j.bios.2020.112849 PMID: 33250333

Park, Y. G., Yun, I., Chung, W. G., Park, W., Lee, D. H., & Park, J. U. (2022). High-resolution 3D printing for electronics. *Advancement of Science, 9*(8), 2104623. PMID: 35038249

Patel, R., Desai, C., Kushwah, S., & Mangrola, M. H. (2022). A review article on FDM process parameters in 3D printing for composite materials. *Materials Today: Proceedings, 60*, 2162–2166. DOI: 10.1016/j.matpr.2022.02.385

Patpatiya, P., Chaudhary, K., Shastri, A., & Sharma, S. (2022). A review on polyjet 3D printing of polymers and multi-material structures. *Proceedings of the Institution of Mechanical Engineers. Part C, Journal of Mechanical Engineering Science, 236*(14), 7899–7926. DOI: 10.1177/09544062221079506

Pereira, G. R., Gasi, F., & Lourenço, S. R. (2019). Review, analysis, and classification of 3D printing literature: Types of research and technology benefits. *IJAERS*, *6*(6), 167–187. DOI: 10.22161/ijaers.6.6.18

Placone, J. K., Mahadik, B., & Fisher, J. P. (2020). Addressing present pitfalls in 3D printing for tissue engineering to enhance future potential. *APL Bioengineering*, *4*(1), 010901. DOI: 10.1063/1.5127860 PMID: 32072121

Pradhan, S. R., Singh, R., & Banwait, S. S. (2023). 3D printing assisted investment casting of dental crowns for recycling of DMLS waste. *Arabian Journal for Science and Engineering*, *48*(3), 3289–3299. DOI: 10.1007/s13369-022-07155-z

Prashar, G., Vasudev, H., & Bhuddhi, D. (2023). Additive manufacturing: Expanding 3D printing horizon in industry 4.0. [IJIDeM]. *International Journal on Interactive Design and Manufacturing*, *17*(5), 2221–2235. DOI: 10.1007/s12008-022-00956-4

Quan, H., Zhang, T., Xu, H., Luo, S., Nie, J., & Zhu, X. (2020). Photo-curing 3D printing technique and its challenges. *Bioactive Materials*, *5*(1), 110–115. DOI: 10.1016/j.bioactmat.2019.12.003 PMID: 32021945

Rael, R., & San Fratello, V. (2017). Clay bodies: Crafting the future with 3D printing. *Architectural Design*, *87*(6), 92–97. DOI: 10.1002/ad.2243

Ranjan, R., Kumar, D., Kundu, M., & Moi, S. C. (2022). A critical review on Classification of materials used in 3D printing process. *Materials Today: Proceedings*, *61*, 43–49. DOI: 10.1016/j.matpr.2022.03.308

Rao, C. H., Avinash, K., Varaprasad, B. K. S. V. L., & Goel, S. (2022). A review on printed electronics with digital 3D printing: Fabrication techniques, materials, challenges and future opportunities. *Journal of Electronic Materials*, *51*(6), 2747–2765. DOI: 10.1007/s11664-022-09579-7

Regehly, M., Garmshausen, Y., Reuter, M., König, N. F., Israel, E., Kelly, D. P., Chou, C.-Y., Koch, K., Asfari, B., & Hecht, S. (2020). Xolography for linear volumetric 3D printing. *Nature*, *588*(7839), 620–624. DOI: 10.1038/s41586-020-3029-7 PMID: 33361791

Ribeiro, D., Cimino, S. R., Mayo, A. L., Ratto, M., & Hitzig, S. L. (2021). 3D printing and amputation: A scoping review. *Disability and Rehabilitation. Assistive Technology*, *16*(2), 221–240. DOI: 10.1080/17483107.2019.1646825 PMID: 31418306

Robayo-Salazar, R., de Gutiérrez, R. M., Villaquirán-Caicedo, M. A., & Arjona, S. D. (2023). 3D printing with cementitious materials: Challenges and opportunities for the construction sector. *Automation in Construction*, *146*, 104693. DOI: 10.1016/j.autcon.2022.104693

Roudný, P., & Syrový, T. (2022). Thermal conductive composites for FDM 3D printing: A review, opportunities and obstacles, future directions. *Journal of Manufacturing Processes*, *83*, 667–677. DOI: 10.1016/j.jmapro.2022.09.026

Shaheen, I., Hussain, I., Zahra, T., Javed, M. S., Shah, S. S. A., Khan, K., Hanif, M. B., Assiri, M. A., Said, Z., Arifeen, W. U., Akkinepally, B., & Zhang, K. (2023). Recent advancements in metal oxides for energy storage materials: Design, classification, and electrodes configuration of supercapacitor. *Journal of Energy Storage*, *72*, 108719. DOI: 10.1016/j.est.2023.108719

Shahrubudin, N., Lee, T. C., & Ramlan, R. J. P. M. (2019). An overview on 3D printing technology: Technological, materials, and applications. *Procedia Manufacturing*, *35*, 1286–1296. DOI: 10.1016/j.promfg.2019.06.089

Shahzad, Q., Umair, M., & Waqar, S. (2022). Bibliographic analysis on 3D printing in the building and construction industry: Printing systems, material properties, challenges, and future trends. *Journal of Sustainable Construction Materials and Technologies*, *7*(3), 198–220. DOI: 10.47481/jscmt.1143239

Shao, L., Chang, Y. C., Hao, C., Fei, M. E., Zhao, B., Bliss, B. J., & Zhang, J. (2022). A chemical approach for the future of PLA upcycling: From plastic wastes to new 3D printing materials. *Green Chemistry*, *24*(22), 8716–8724. DOI: 10.1039/D2GC01745H

Somers, P., Münchinger, A., Maruo, S., Moser, C., Xu, X., & Wegener, M. (2024). The physics of 3D printing with light. *Nature Reviews. Physics*, *6*(2), 99–113. DOI: 10.1038/s42254-023-00671-3

Srivastava, R., & Srivastava, V. K., (2020). Changing Scenario of Manufacturing Industries Due to Covid 19. *International Journal for Modern Trends in Science and Technology*, *6*(10), 143–146. DOI: 10.46501/IJMTST061026

Srivastava, R., Thakur, A., Bahadur, P. S., & Rana, A. (2024). A State-of-the-Art Review of Recent Developments and Challenges for 2D Implantable Materials. *Engineering Materials for Efficient Energy Storage and Conversion*, 1-24.

Srivastava, R., Thakur, A., & Rana, A. (2024). The Basic Understanding of Nano Materials: A Review. *Innovations in Nanomaterials-Based Corrosion Inhibitors*, 1-20.

Stevens, C. V. (2010). *Industrial applications of natural fibres: structure, properties and technical applications*. John Wiley & Sons.

Su, C., Chen, Y., Tian, S., Lu, C., & Lv, Q. (2022). Natural materials for 3D printing and their applications. *Gels (Basel, Switzerland)*, *8*(11), 748. DOI: 10.3390/gels8110748 PMID: 36421570

Subramani, R., Mustafa, M. A., Ghadir, G. K., Al-Tmimi, H. M., Alani, Z. K., Haridas, D., Rusho, M. A., Rajeswari, N., Rajan, A. J., & Kumar, A. P. (2024). Advancements in 3D printing materials: A comparative analysis of performance and applications. *Applied Chemical Engineering*, *7*(2), ACE-3867. DOI: 10.59429/ace.v7i2.3867

Tabassum, T., & Mir, A. A. (2023). A review of 3d printing technology-the future of sustainable construction. *Materials Today: Proceedings*, *93*, 408–414. DOI: 10.1016/j.matpr.2023.08.013

Tan, L., Wang, Z., Jiang, H., Han, B., Tang, J., Kang, C., Zhang, N., & Xu, Y. (2022). Full color 3D printing of anatomical models. *Clinical Anatomy (New York, N.Y.)*, *35*(5), 598–608. DOI: 10.1002/ca.23875 PMID: 35384062

Tay, Y. W. D., Panda, B., Paul, S. C., Tan, M. J., Qian, S. Z., Leong, K. F., & Chua, C. K. (2016, August). Processing and properties of construction materials for 3D printing. []. Trans Tech Publications Ltd.]. *Materials Science Forum*, *861*, 177–181. DOI: 10.4028/www.scientific.net/MSF.861.177

Thakar, C. M., Parkhe, S. S., Jain, A., Phasinam, K., Murugesan, G., & Ventayen, R. J. M. (2022). 3d Printing: Basic principles and applications. *Materials Today: Proceedings*, *51*, 842–849. DOI: 10.1016/j.matpr.2021.06.272

Thakur, A., Srivastava, R., Bahadur, P. S., & Rana, A. (2024). Exploring the Various Biomedical Applications of Smart Materials. In *Discovery, Disruption, and Future Implications of Nanomaterials* (pp. 249-274). IGI Global. DOI: 10.4018/979-8-3693-4397-5.ch008

Tian, Y., Chen, C., Xu, X., Wang, J., Hou, X., Li, K., Lu, X., Shi, H. Y., Lee, E.-S., & Jiang, H. B. (2021). A review of 3D printing in dentistry: Technologies, affecting factors, and applications. *Scanning*, *2021*(1), 9950131. DOI: 10.1155/2021/9950131 PMID: 34367410

Tissen, L. N. (2022). 3D printing and the art world: current developments and future perspectives. In *Advances in 3D printing*. IntechOpen.

Truxova, V., Safka, J. I. R. I., Seidl, M., Kovalenko, I., Volesky, L. U. K. A. S., & Ackermann, M. I. C. H. A. L. (2020). Ceramic 3D printing: Comparison of SLA and DLP technologies. *MM Sci. J*, *2020*(2), 3905–3911. DOI: 10.17973/MMSJ.2020_06_2020006

Tuli, N. T., Khatun, S., & Rashid, A. B. (2024). Unlocking the future of precision manufacturing: A comprehensive exploration of 3D printing with fiber-reinforced composites in aerospace, automotive, medical, and consumer industries. *Heliyon*, *10*(5), e27328. DOI: 10.1016/j.heliyon.2024.e27328 PMID: 38495162

Ukobitz, D. V. (2021). Organizational adoption of 3D printing technology: A semi-systematic literature review. *Journal of Manufacturing Technology Management*, *32*(9), 48–74. DOI: 10.1108/JMTM-03-2020-0087

Vidal, L., Kampleitner, C., Brennan, M. Á., Hoornaert, A., & Layrolle, P. (2020). Reconstruction of large skeletal defects: Current clinical therapeutic strategies and future directions using 3D printing. *Frontiers in Bioengineering and Biotechnology*, *8*, 61. DOI: 10.3389/fbioe.2020.00061 PMID: 32117940

Wang, B., Zhang, Z., Pei, Z., Qiu, J., & Wang, S. (2020). Current progress on the 3D printing of thermosets. *Advanced Composites and Hybrid Materials*, *3*(4), 462–472. DOI: 10.1007/s42114-020-00183-z

Wang, S., Chen, X., Han, X., Hong, X., Li, X., Zhang, H., Li, M., Wang, Z., & Zheng, A. (2023). A review of 3D printing technology in pharmaceutics: Technology and applications, now and future. *Pharmaceutics*, *15*(2), 416. DOI: 10.3390/pharmaceutics15020416 PMID: 36839738

Wei, X., Zou, N., Zeng, L., & Pei, Z. (2022). PolyJet 3D printing: Predicting color by multilayer perceptron neural network. *Annals of 3D Printed Medicine*, *5*, 100049.

Xu, K., Liu, S., & Ai, Y. (2015). Application of Shearlet transform to classification of surface defects for metals. *Image and Vision Computing*, *35*, 23–30. DOI: 10.1016/j.imavis.2015.01.001

Zastrow, M. (2020). The new 3D printing. *Nature*, *578*(7793), 20–23. DOI: 10.1038/d41586-020-00271-6 PMID: 32025009

Zhang, H., Yang, D., & Sheng, Y. (2018). Performance-driven 3D printing of continuous curved carbon fibre reinforced polymer composites: A preliminary numerical study. *Composites. Part B, Engineering*, *151*, 256–264. DOI: 10.1016/j.compositesb.2018.06.017

Zhong, N., & Zhao, X. (2017). 3D printing for clinical application in otorhinolaryngology. *European Archives of Oto-Rhino-Laryngology*, *274*(12), 4079–4089. DOI: 10.1007/s00405-017-4743-0 PMID: 28929219

Zhou, L. Y., Fu, J., & He, Y. (2020). A review of 3D printing technologies for soft polymer materials. *Advanced Functional Materials*, *30*(28), 2000187. DOI: 10.1002/adfm.202000187

Zub, K., Hoeppener, S., & Schubert, U. S. (2022). Inkjet printing and 3D printing strategies for biosensing, analytical, and diagnostic applications. *Advanced Materials*, *34*(31), 2105015. DOI: 10.1002/adma.202105015 PMID: 35338719

Chapter 3
Additive Manufacturing for a Circular Economy: Material Recycling Strategies

M. Maniraj
https://orcid.org/0009-0000-5343-3895
Sardar Raja College of Engineering, Alangulam, India

S. M. Raj Kumar
https://orcid.org/0000-0002-8978-0108
Francis Xavier Engineering College, Tirunelveli, India

ABSTRACT

Additive Manufacturing (AM), commonly known as 3D printing, is emerging as a transformative technology capable of reshaping industrial production through material efficiency, design freedom, and decentralized fabrication. However, the long-term sustainability of AM is challenged by significant concerns related to energy consumption, material waste, and limited recyclability of polymers, metals, and composites. This chapter evaluates the environmental implications of AM within the framework of a circular economy, focusing on recycling strategies, energy efficiency, and material lifecycle management. Real-world examples from aerospace and automotive industries, such as Airbus's use of Ti6Al4V and BMW's recycled thermoplastics, illustrate both the opportunities and constraints of implementing sustainable AM at scale. The chapter explores mechanical and chemical recycling methods, biodegradable material alternatives, and AI-driven process optimizations to improve waste reduction and energy performance.

DOI: 10.4018/979-8-3373-0533-2.ch003

1. INTRODUCTION

Additive manufacturing (AM) or 3D printing has become a disruptive technology that is being used in areas like aerospace, automotive and consumer goods sector, to name a few. Unlike traditional subtractive manufacturing processes that involve cutting away material to remove excess material from larger block, AM builds objects by adding layer layer from digital models, taking out wasting of material to create highly complex and customized design (Zhou et al. 2024). AM has become a key enabler of innovation with its capability to quickly prototype, produce in a decentralized manner and effectively use materials. While the environmental impact of AM technologies have been sparingly reported earlier, as they scale, mounting questions on its impact on material consumption, material waste generation, and energy consumption emerges. As a manufacturing method, AM is often promoted as less wasteful than traditional manufacturing, but in reality its material lifecycle (and the lifecycle of the products it produces) remain a major environmental challenge to overcome in order to strengthen AM's sustainability (Rahman et al. 2024).

A circular economy concept can also be used for mitigating the negative environmental effects of AM. The circular economy is a departure from the conventional linear model of production and consumption (extract–use–dispose) and presents a more sustainable way of doing business with the focus of maximizing resource efficiency, reusing materials and recycling. Through designing AM processes that enable circular economy principles, materials can be recovered, put back into the production cycle, and their dependency on virgin resources decreased, as well as their waste minimized (Zhao et al. 2024). In this context, this approach is highly relevant as AM uses a wide variety of materials including polymers, metals and composites, most of which lack end-of-life management possibilities. The way to meet these conditions for AM would be to achieve progress in material development, improved recycling techniques and changing the design principles to promote reusability and recyclability.

The objective of this chapter is to investigate the sustainability challenges of AM and to analyze potential methods of increasing material recycling in this area. This will evaluate the environmental impact of AM materials commonly used and look at existing AM material recycling processes and their restrictions, and proposing new directions in recovering AM materials. In addition, the chapter discusses the wider facets of sustainability in AM from the perspectives of energy efficiency, regulatory framework, and ethical implications. This chapter aims at contributing to the information flows already in existence concerning sustainable manufacturing and related implications and opportunities in particular for integrating AM into a circular economy by providing a comprehensive analysis of the factors listed above.

Figure 1. Concept of additive manufacturing in a circular economy (Hendrixson, 2020)

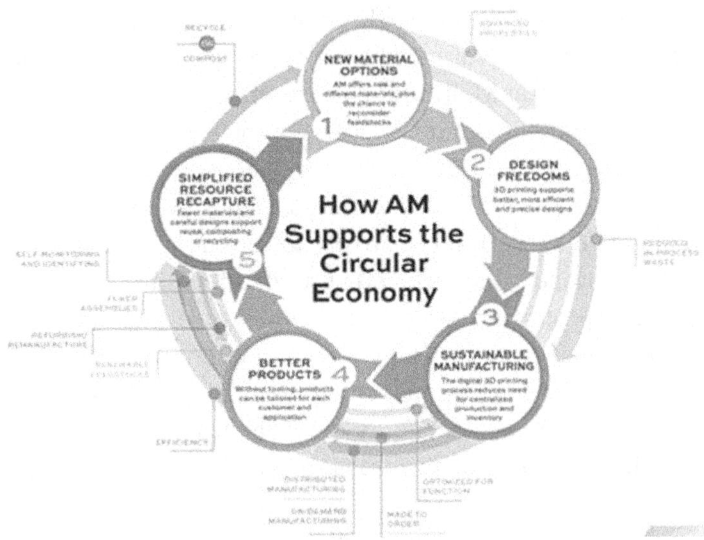

2. ENVIRONMENTAL IMPACT OF 3D PRINTING MATERIALS

Additive manufacturing inherently depends on the lifecycle of its materials, from raw material extraction, processing, usage, to disposal, and the environmental impact of AM is associated with it. Although AM is commonly seen as a greener alternative to conventional manufacturing, as it reduces waste of excess material, the environmental footprint of AM is highly complex when the entire material lifecycle is considered (Mohamed & Mohamed, 2025). This means significant amount of resource extraction and energy input is required to produce raw material for AM to be produced in the form of polymer filaments, metal powders, or composite resins. In the processing of thermoplastics including polylactic acid (PLA) and ABS, multiple processing steps including refining raw feedstocks into filament are involved, each with its associated energy consumption and carbon emissions. Titanium and aluminum powders, which are used to make metal AM materials, are extremely energy intensive because atomization processes are used to produce fine, printable particles. However, further environmental challenges occur during post processing and end of life disposal when these materials have limited recycling options and therefore lead to increased material waste (Zaid & El Ouni, 2024).

Another critical factor affecting the sustainability of AM is the energy consumption. The energy demand of different 3D printing technologies is different due to their required material and operating mechanism. Based on such powder metal AM techniques as Selective Laser Melting and Electron Beam Melting, these techniques demand extremely high temperature of almost a thousand degrees, powerful laser or electron beam sources, and quite high energy consumption (Chen et al. 2024). On the other hand, polymer based like Fused Deposition Modeling (FDM) and Stereolithography tend to work on relatively lower energy but still large amounts of power are needed to heat, extrude and cure. In addition to the energy demands of AM, the pre-and post-processing steps, namely material preparation, surface finishing, heat treatments, can prove to be even more indirect energy demanding and contribute to significant additional environmental costs. AM can be more precise in material usage, but its total energy footprint has to be evaluated to decide if its sustainability benefits over conventional techniques of manufacturing are real (Olawumi et al. 2024).

In addition to its reputation for reducing waste, AM still produces a lot of material waste throughout the production cycle. Material inefficiencies and disposal challenges occur with failed prints which are caused by design flaws, mechanical errors or machine calibration issues. In addition to these processes, many AM processes also require support structures to hold parts during printing, especially in metal and resin based systems, which are then discarded as waste after production (Zhao et al. 2024). Furthermore, additional material byproducts are generated in post processing, i.e. grinding, sanding and chemical treatments, which can be hard to recycle. When composite materials or multi material prints are made, the task of separating and reclaiming the constituent materials becomes more difficult. However, if there are no efficient strategies for waste management and material recovery, the environmental benefits of AM can be greatly undermined. To address these challenges we need to improve process efficiency, develop novel recycling solutions, and extend the principles of a circular economy into the AM industry (Ma et al. 2024).

3. COMMONLY USED MATERIALS IN ADDITIVE MANUFACTURING AND THEIR CHALLENGES

Polymer, metal and composite materials all are the principal categories of materials used in this additive manufacturing. None of these material classes is sustainable, recyclable or free from environmental impact. On the one hand, AM is a waste reducing technology compared to other manufacturing processes, yet the inherent properties of these materials normally become hurdles in realizing a completely closed loop economy. A key challenge in advancing AM to be more sustainable continues to be the ability to recycle and reuse the materials in an efficient manner without loss of performance (Ng et al. 2024).

FDM printing thermoplastics such as pla, ABS, and PETG in particular is the most widely used material in AM because of their affordability, ease of processability and compatibility to print techniques such as FDM. Because PLA is biodegradable under industrial composting conditions, it is often viewed as an environmentally friendly alternative to traditional plastics. Yet, the large scale production of PLA is associated with land use, water consumption, and environmental impact of agricultural inputs (Sapkota et al. 2024). However, ABS is a petroleum based plastic, which combines high strength and durability with greater environmental concerns as it is non biodegradable and its carbon footprint is higher. Alternatively, PETG is more recyclable than ABS and it does try to provide better toughness and better resistance to chemicals, but most manufactories do not have a good way to make sure that the material maintains its quality after multiple recycling cycles. One of the major issues with polymer recycling in AM is degradation since thermoplastics degrade in their molecular structure when melted and re extruded repeatedly. As a result, they have diminished mechanical properties and are not reusable in high performance applications. The difficulty in recycling hinges on the contamination from dyes, fillers and additives that prevents the consistency of the material properties across recycled batches (Shanmugam et al. 2024).

The metals used in AM (i.e. titanium, aluminum, and stainless steel) present significant sustainability challenges in that these materials provide many advantages for high performance applications such as aerospace and medical applications. Production of metal powders for AM is energy intensive, with complex atomization or plasma based methods that consume a lot of electricity and produce greenhouse gas emissions. For example, titanium is valued for its strength-to weight ratio and corrosion resistance, but is one of the most energy demanding metals to produce because of the amounts of processing that is required. Although aluminum is more recyclable than most other metals, reprocessing still faces problems related to its oxidation and impurity management (Behseresht & Park, 2024). Due to its versatility, stainless steel is widely used, but due to contamination risks and difficulty in maintaining consistent powder quality, stainless steel recycling within AM is limited. The reusability of metal powders in AM is very challenging due to oxidation, moisture absorption, and particle size variation, which may adversely affect the print quality and mechanical property. Metal recycling in AM suffers from frequent replenishment with virgin material due to the accumulation of defects due to powder aging. One key area of improvement to make metal based AM more sustainable is to improve powder recovery methods and optimize reusability (Erdaş et al. 2024).

The complexity associated with the sustainability of AM is increased by composites and multi material structures. Many times, composite materials (i.e., containing two or more different materials, such as carbon fiber reinforced polymers or metal–ceramic hybrids) are used to impart strength, thermal resistance and lightweight properties in their use. But because they're multi phase, they are very difficult to recycle. Technically, it is challenging and economically not viable to separate the individual components of composites and this contributes to a higher likelihood of disposal rather than reuse (Jamal et al. 2024). Recycling is further complicated by multi material 3D printing, which allows for integration of different substances within a single print. In the case where polymers, metals, or ceramics are fused together in one structure, the mechanical or chemical separation of material incompatibilities is very difficult. The challenges are increased upon the lack of standard recycling procedures and the extra cost for the material recovery. As the AM technology continues to progress, the development of sustainable solutions for recycling composite and multi material is essential to reducing waste and closing the loop in using AM materials (Ahmad et al. 2024).

4. RECYCLING STRATEGIES IN ADDITIVE MANUFACTURING

An important role for improving the AM sustainability lies in the recycling strategies. Especially, the problem of material waste and the related environmental issues, are continuing to grow. However, AM is inherently more material efficient than traditional manufacturing, but there are still challenges in guaranteeing that materials are effectively recoverable and recyclable for future uses. In order to increase circularity in AM, a number of recycling strategies have been explored, such as mechanical and chemical recycling, metal powder recovery, and development of biodegradable materials. But each method has its technical and implementation challenges that need to be addressed in order to have widespread implementation (Romani et al. 2024).

One of the most common methods of reprocessing thermoplastics used in AM is mechanical recycling, which is especially applicable to materials such as PLA, ABS, and PETG. Traditionally, failed prints, support and other plastic waste are shredded into small granules and melted down to be extruded as filament to be reused. On the one hand, this method is quite straightforward and economically sound, on the other hand, it is restricted by the degradation in the polymer properties in the course of repeated processing. With each recycling cycle, the molecular weight of the material is reduced, which reduces the mechanical strength, print quality, and increases brittleness. This process is further complicated by contaminating additives, dyes, and mixed polymer types, greatly hampering efforts in obtaining consistent material performance. In spite of the challenges surrounding vertical recycling, emerging advancements in filament recycling technologies including precision filtering and advanced extrusion techniques continue to make it possible to use mechanically recycled polymers within AM (Di et al. 2024).

Alternatively, chemical recycling breaks polymers down into their monomer components, which can be repolymerized into new materials with properties comparable to those of virgin plastics. Depolymerization, as this method is also called, allows for more material to be recovered of higher quality, thus limiting the deleterious effect of contaminants, promoting a truly closed loop style recycling. For example, PLA can be chemically converted back to lactic acid that can be reused to make high quality PLA filament. Chemical recycling has obvious advantages in material purity and sustainability, but is only at early stages of development for AM, and much more research will be required to improve its cost efficiency and scalability. Furthermore, the energy requirements and the chemical inputs needed for depolymerization must be controlled in order to justify the environmental benefits over the resource costs (Li et al. 2024).

Recycling of metals in AM poses a very different problem set than that of plastics, due to the nature of how metal powders are made and reused. Unlike polymers, metal powders can not be melted and reformed, however, they can degrade over time because of processes including: oxidation, moisture absorption and changed in particle size distribution all affecting print quality and mechanical properties (Yakubov et al. 2024). Recycling and reusing powders is difficult without significant quality loss, as many of these metal AM processes such as Selective Laser Melting and Electron Beam Melting require highly controlled powder characteristics. In order to tackle this issue, methods of purifying powder have been developed, including sieving, heat treatments, and chemical refinement to remove contaminants and improve the quality of powder. Other researchers are studying atomization driven recycling methods that involve remelting and reprocessing of metal waste to produce new powder that can be utilized in AM. However, these approaches have been demonstrated to be promising, however the high energy requirement for recycling of metal powder and the economic feasibility of these are major frustration to widespread uptake (Golvaskar et al. 2024).

Another potential means to sustainably AM is via biodegradable and bio-based alternatives, including materials like polylactic acid and polyhydroxyalkanoates (PHA), which can be used as green alternatives to petroleum based plastics. PLA is made from renewable resources such as corn starch and (safely) composted under industrial conditions and has a lower carbon footprint compared to conventional thermoplastics. It is even more promising since PHA is a microbial derived biopolymer that is biodegradable in various environments such as marine and soil conditions (Peng et al. 2024). Although these materials have environmental advantages, their costs make them commercially unviable, and they offer lower mechanical properties than traditional plastics, and also require dedicated recycling and disposal infrastructure for their adoption on a large scale. To bridge this gap, there lies potential in the development of enhanced bio based materials that possess better mechanical strength, heat resistance and recyclability than the ones currently available, which in turn could lead towards making them more viable option for AM applications tomorrow (Bian et al. 2024).

Recycling strategies in AM are still progressing but there is more to do in order to optimize these processes for efficiency, cost and scalability. In order to achieve a truly circular approach to AM, however, a combination of improved mechanical and chemical recycling techniques, enhanced metal powder recovery methods, as well as development of biodegradable materials will be required. Through technological innovation, industry collaboration, and regulatory support, these challenges can be addressed to move AM toward a more sustainable future with more waste materials eliminated and greater resource efficiency. Figure 2 shows the recycling strategy in additive manufacturing (Al Rashid & Koç, 2023).

Figure 2. Recycling strategy in additive manufacturing (Al Rashid & Koç, 2023)

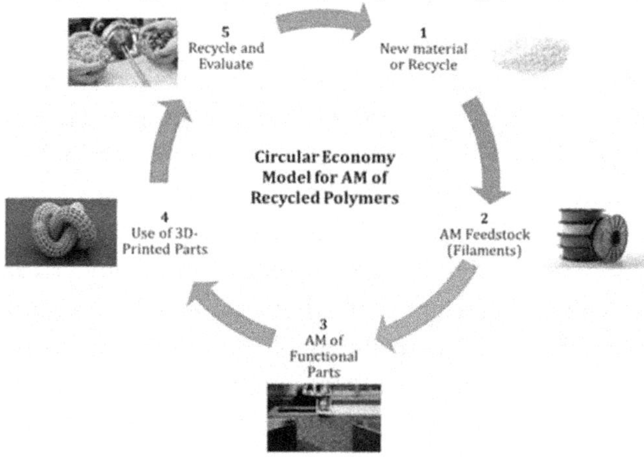

5. ENERGY EFFICIENCY AND SUSTAINABILITY IN ADDITIVE MANUFACTURING

Additive manufacturing is an energy intensive process and therefore energy efficiency is a critical factor in determining the total sustainability of AM. Despite the fact that AM is regarded as a resource efficient alternative compared to conventional manufacturing processes, its energy consumption is largely determined by the particular technology used. One of these laser based additive manufacturing such as Selective Laser Melting and Direct Metal Laser Sintering uses extremely high temperature to fuse metal powders and made solid component. On the other hand, these processes depend on the use of very powerful lasers or electron beams whose needs are also high in electricity and therefore also in terms of their carbon footprint (Fidan et al. 2024). Furthermore, since metal AM is conducted in inert gas environments, powder handling systems, and even post processing processes, like heat treatment and surface finishing, add more energy burden in this technology. Similarly, even in polymer based AM methods like the Stereolithography and FDM, consistent heating for curing, heating and extrusion of material and so forth enhance energy consumption. As AM adoption continues to grow, it is essential to tackle these energy intensive processes in order to make sure that its sustainability benefits outweigh its environmental costs (Mattera et al. 2025).

Incorporating renewable energy sources into AM facilities offers a means of decreasing the carbon footprint of the technology. Energy for AM production can be supplied by solar, wind and hydroelectric power to reduce dependence on fossil fuels. AM companies and some research institutions are already beginning to explore possibilities of using renewable energy within them, either by loading AM equipment with solar panels or electric power drawn from green energy grids (Abazari et al. 2024). Especially for industries such as aerospace and automotive manufacturing, where the goals for sustainability become more stringent, this approach is promising. Moreover, AM production is decentralized, as on demand and local manufacturing decreases the necessity for long distances of transportation and thus, supposes more energy impact reduction. Despite this, challenges exist in scaling up renewable energy integration, including the ability to operationalize energy storage and reliability of the grid to continuously operate AM systems.

Energy efficiency in AM is also achieved through process optimization. Researchers and engineers are currently looking for different strategies in reducing energy consumption while producing parts with the same level of quality and accuracy. The first approach is to optimize laser scanning paths and exposure times in metal AM to minimize energy wastage while retaining structural integrity (Nyamekye et al. 2024). Thermal management in polymer based AM is improved with better heating and cooling cycle, reducing unnecessary power usage during printing. Moreover, hybrid manufacturing techniques combining AM with traditional machining or casting can increase productivity by relying on the advantages of one of these methods against energy consuming steps. Another emerging avenue is AI driven process optimization where machine learning algorithms monitor real time printing data and adjust printing parameters in real time such that energy usage is optimized (Goh et al. 2024).

However, many of the environmental benefits claimed for AM as compared to traditional manufacturing may only be realized if energy efficiency can be greatly improved. The AM industry can make meaningful strides towards a sustainable future through a recognition of laser based process energy demand and integrating renewable energy sources as well as optimizing production techniques. As the technical and economic barriers of energy efficient AM are overcome, research and industry collaboration continued will be essential to ensure that the adoption of the technology is geared towards a more sustainable and a resource conscious manufacturing landscape (Selvam et al. 2024).

6. WASTE MANAGEMENT AND CIRCULAR ECONOMY IN ADDITIVE MANUFACTURING

With great promise to drastically cut material usage, additive manufacturing is equally reliant on good waste management to demonstrate its sustainability. AM with a circular economy approach will minimize waste generation, maximize material recovery, and ensure end of life component will be reintegrated into a production cycle. The most effective of the three is design for recyclability, in which products are designed to be repurposed from the point of departure with thoughts of material reuse and minimal waste. Since designers have the options to optimize the parts geometry, reduce the excessive amount of support structures and choose more recyclable materials, this all can mean that waste from printing and post processing can be considerably decreased. Innovations like lattice structures, topology optimization and generative design algorithms allows manufacturers to use only the required material that will provide the same strength and functionality, leading to more efficient and sustainable AM process (Villa et al. 2024).

One of the other significant challenges in AM waste management generally stems from a high turnover rate of failed prints, due to errors in part design, inconsistent material properties, incorrectly calibrated machines, or environmental conditions like inconsistent temperature. The failures also result in material wastage, increase energy consumption and operational costs. Failed prints can be mitigated with better simulation software, i.e. process reliability, real time monitoring systems and adaptive printing technologies. Predictive maintenance and automated error detection are becoming more and more of a role in minimizing failed prints by finding potential defects before they ruin an entire build. Furthermore, material reprocessing techniques such as filament extruder systems that enable waste plastic to be turned to the usable feedstock, have practicality for polymer based AM waste reduction. As powder recovery and refining processes are suitable for metal AM, they are used to reclaim unused and degraded material for reuse, although achieving consistent powder quality over multiple reuse cycles remains challenging (Ononiwu et al. 2024).

There are a few examples of circular economy initiatives in AM that have proven successful in the sense that strategic waste management can result in more sustainable manufacturing. New closed loop recycling systems have also been pioneered by both companies and research institutions where used or defective prints can be shredded, reprocessed, and turned into new feedstock (Zhao et al. 2024). The repurpose of industrial waste is also looked at, for example converting discarded plastic found in consumer goods as printable filament or using metal scraps from aerospace manufacturing to make AM grade powders. To reduce dependence on virgin resources, researchers in construction and in large scale AM have researched the use of recycled concrete and bio-based materials. Furthermore, various industry, academia and policy education activities contribute toward developing standardization and guidelines promoting sustainable AM practices including take-back programs for used, 3D printed products and incentives for companies using eco-friendly materials (Priyadarshini et al. 2024).

Therefore, integration of waste management strategies into AM is necessary for making AM a truly circular technology. AM can be a sustainable manufacturing approach through focusing on design for recyclability, improving print success rates and implementing large scale recycling initiatives. Innovations in material science, process optimization and digital manufacturing will continue to evolve and the industry can bring about a change in how waste is dealt with, and maintaining the AM in the forefront of environmentally responsible production (Aslam et al. 2025).

7. POLICY, REGULATION, AND INDUSTRY INITIATIVES

Additive manufacturing's rapid rise has raised an important question of its environmental impact when it comes to material recycling and waste management. Nevertheless, regulatory frameworks for recycling of the AM material are still in their infancy compared to the traditional manufacturing sectors. Although AM is subject to the general environmental regulations on waste disposal, emissions, and material handling, specific policies addressing the specific issues of this technology are needed (Klenam et al. 2025). Currently, regions have different policies, for instance, the European Union has devoted that through directives like Waste Framework Directive and Circular Economy Action Plan on recycling of plastics and metals in industrial processes. In the United States, for example, agencies like the Environmental Protection Agency (EPA) and Occupational Safety and Health Administration (OSHA) are responsible for some parts of material handling and emissions, but AM guidelines are still evolving for the industry. The absence of standard regulations for the AM material reuse and recycling creates challenges for AM manufacturers to recycle or reuse AM materials in their processes within the circular economy context, indicating the necessity of clearer policies related to the whole lifecycle of the AM material (Mathur et al. 2025).

Existing sustainability initiatives lead by industry have provided a means to fill the void left in the absence of environmental regulations to increase the quality of the environmental footprint of AM process. Closed loop recycling programs have been introduced by many organizations allowing users to return used or defective prints for reprocessing back into new feedstock. There are some AM material suppliers who now provide recycled filaments and metal powders that deliver same high performance as virgin resources, while reducing reliance on virgin resources (Sheikh et al. 2025). Moreover, partnerships between industry leaders and academic institutions have enabled the development of bio-based AM materials and biodegradable materials with bio-based or petroleum-derived sources to be used as alternatives to petroleum derived polymers. The International Organization for Standardization (ISO) and the American Society for Testing and Materials (ASTM) are among the certification programs developing sustainability benchmarks for AM materials and processes with which manufacturers can follow through adoption of more eco-friendly practices. However, widespread sustainable AM solution adoption is still hindered by cost barriers, technological limitations, and a lack of universal standards pertaining to material quality and recyclability (Vadivelu et al. 2025).

The establishment of regulatory frameworks, offering of incentives towards green manufacturing along with support for AM sustainability research initiatives falls to the government and policy makers as an important step in promoting sustainable AM practices. Implementing policies like tax incentives for companies investing in recycled materials, funding research to identify low impact AM technologies, and mandates for fostering responsible waste management can expedite this transition. In addition, public-private partnerships can be found to the advantage of innovation in the form of governmental research initiatives that lead to more efficient recycling methods and materials that are sustainable. Additionally, the education and workforce development programs on sustainable manufacturing can provide engineers and designers the knowledge to integrate the principles of a circular economy in AM workflows. As the AM industry develops, it will be crucial for legislation to be clear, for the industry to work together, and governments to support as we want the technology to advance the move to a more sustainable, resource efficient, manufacturing future (Sheikh et al. 2025).

8. FUTURE DIRECTIONS AND INNOVATIONS IN SUSTAINABLE ADDITIVE MANUFACTURING

With research and industry efforts, additive manufacturing is fast becoming centric of sustainability with the desire to develop fully recyclable and environmentally friendly materials in the future. In the past decade, advances in material science have enabled the invention of new polymers, metals and composites increasingly developed to be recycled more easily without degradation of performance. An area where the potential for development is promising is the expansion of the available biodegradable and bio-based materials, including improved formulations of polylactic acid and polyhydroxyalkanoates with improved mechanical properties, and maintained compostability. There are also efforts to develop recyclable high performance engineering plastics: modified polyether ether ketone and polyethylene terephthalate glycol (PETG) for the same reason – materials degradation during the recycling process. But in metal AM, researchers are looking at alloys that are better recyclable and developing new ways to recover powder for effective reuse of metal feedstocks. Further, the integration of recycled waste materials from the metalworking or automotive industries (repurposed carbon fiber, for instance, or recovered metal scraps through reuse, remanufacturing, and recycling) indicates the ability of AM to achieve some level of a circular economy (Krishna et al. 2024).

In the field of AM, artificial intelligence (AI) and machine learning are starting to have an important role on optimizing on material usage and waste. Materials can be analyzed for material behavior, predict print failures and be optimally designed for part designs that minimize materials consumption without sacrificing performance standards. We are developing machine leaning models that can fine incrementing parameters in real fly while achieving higher success rates and less reprints. These advances result in a dramatic decrease of waste due to failed prints and overhang support structures, thus creating a more extreme stabilizing dialog of engineering AM. Additionally, with the use of AI generated quality control systems, the efficiency of post processing and recycling can be improved as they can identify defects in printed parts and indicate their suitability for remanufacturing. As a result, AM facilities can use AI to predict maintenance as well as optimize processes to lower energy use, reduce scrap rates, and extend material usability for a more resource efficient manufacturing ecosystem (Mohammed et al. 2025).

In an attempt to move towards more sustainable AM processes, the development of closed loop recycling systems is critical. The systems are designed as closed loop systems that can continuously recover and reuse material without virgin feedstock while maintaining high quality production outputs. These systems include both efficient mechanical and chemical recycling processes as well as improved material tracking and sorting technologies capable of making high purity materials out of the recycled material. Decentralized recycling units are already being tested by companies that allow users to convert failed prints and waste material into new filaments or powders directly in their own facilities. This approach cuts transportation emissions related to external recycling processes and offers on demand access to sustainable materials. Besides, some other researchers are also trying to develop blockchain based AM material traceability systems monitoring the life cycle of the AM materials for a transparent and accountable sustainable production. Over time, as closed loop recycling systems mature, they will be an important lever for reducing the environmental impact of AM and promoting a more circular, self supporting manufacturing method (Bigliardi et al. 2024).

Advanced materials, AI driven optimization, and closed loop recycling technologies have the potential to converge to redefine the sustainability of AM. To overcome the current limitations and make AM fully compliant with the lines of the circular economy, continued investment in research and innovation and industry collaboration will be required. This opens up the way for AM to become much not only more efficient in production, but also to pave the way in sustainable manufacturing, thereby cutting waste and resource consumption worldwide.

9. ETHICAL AND SOCIAL CONSIDERATIONS

Innovations in additive manufacturing are often seen as a source of ethical and social implications beyond only new technologies and their economic benefits such as raising environmental justice, enabling equitable access to sustainable practices, and consumer responsibility. An important ethical question in AM is that regarding sourcing of raw materials and the environmental justice problems of their extraction and disposal. The metals and polymers used in AM are based on resource intensive processes that pollute, deforest, and destroy habitat (Graziosi et al. 2024). Ras-metal and petroleum derived feedstocks are typically extracted in areas of weak environmental protection resulting in ecological and health degradation of local communities. The same is true for the disposal of AM waste, which, most notably, features non recyclable thermoplastics and metal powder, and the consequences of environmental damage over the long term especially if this waste is sent to countries that do not have adequate recycling infrastructure. To face these challenges, greater supply chain transparency, the innovation of sustainable material alternatives, and more stringent waste policies that do not burden marginalized communities with the environmental costs inherent in our production processes, have to be implemented (Agarwal & Ojha, 2024).

Although the potential benefits of the technology are considerable, small and medium sized enterprises (SMEs) are facing considerable barriers in adopting sustainable AM practices. Large corporations can afford to support these methods through eco friendly material, energy efficient equipment and closed loop recycling system however small businesses cannot due to high cost of sustainable options. However, many recycled filaments, bio based polymers and energy efficient AM machines are more expensive than the conventional alternatives, thus limiting their accessibility to SMEs who operate with limited budget (Soori et al. 2024). Moreover, the absence of clear regulatory frameworks and universal sustainability standards for the industry poses uncertainty for the businesses trying to incorporate green practices. In addition, many SMEs themselves do not have the technical and logistical means to properly recycle AM materials, while increasing the level of accompanying waste generation. Thus, to address these disparities, policymakers along with industry leaders have to come up with incentive, subsidies, and educational programs to support SMEs in transitioning to more sustainable AM workflows. The gap can also be bridged through co operation in the form of knowledge sharing between larger corporations, the research institutions and small scale businesses (Gift et al. 2024).

By making material choices, how they print, and how they dispose of it, consumers influence the sustainability of AM. With the increase of consumer grade 3D printing, there is more plastic waste as hobbyists and small scale manufacturers are printing disposable or short lived items that will be thrown away without any thought to environmental impact. However, many users are not aware of the recyclability limitations of popular AM materials and thus dispose of them improperly, contributing to the landfill problem. Greater awareness and education on sustainable AM practices, such as the benefits of using biodegradable filaments, recycling failed prints, and minimizing unnecessary production, together with encouraging responsible consumer behavior, is needed to encourage more responsible behavior in the consumer. As far as AM companies that manufacture and sell AM materials and printers go, they also have a moral duty to promote sustainability by providing information about recycling, introducing takeback programs for used materials and introducing eco-friendly alternatives. When the use of AM grows, it will be important to encourage a culture of sustainability among consumers to guarantee that AM will help in creating a more environmentally and socially conscious manufacturing ecosystem (Bendoly et al. 2024).

Through consideration of these ethical and social dimensions, AM can evolve as a technology that improves production efficiency and also meets other principles of environmental justice and equity in sustainability. To realize this vision, we need a coordinated effort from the industry leaders, policymakers, SMEs and the consumers together to build a balanced ecosystem, for AM to be deployed without exacerbating the environmental or social injustices.

10. CONCLUSION AND RECOMMENDATIONS

The sustainability of additive manufacturing is increasingly considered a critical issue that must be addressed to ensure its long-term viability as a manufacturing technology. While AM holds the potential to significantly reduce material waste, enable mass customization, and support decentralized manufacturing, its environmental implications pose a cause of concern. In this chapter, the environmental implications of AM material, material recycling issues, energy consumption, and waste management, as well as the roles of regulations, industry action, and emerging technologies in promoting sustainability, have been addressed. A central lesson is that while AM holds the potential to be more sustainable than traditional manufacturing, circularity requires advances in material recyclability, energy efficiency, and the widespread implementation of waste reduction. The discussion of the inadequacies of polymer, metal, and composite recycling highlights the need for cutting-edge strategies to maintain material quality and extend the life of AM feedstocks. The application of AI, closed-loop recycling, and bio-based material was also presented as promising directions to reduce the environmental impact of AM.

Future research must focus on the development of high-performance, fully recyclable AM material that will not degrade with multiple cycles of use. More research must be done to optimize mechanical and chemical recycling processes for polymers and metals to make reprocessed material as strong and long-lasting as necessary for industrial applications. Further research into energy-efficient AM processes, including laser-based and powder-bed fusion, must continue to reduce the overall carbon footprint of AM production. Policy proposals must include the implementation of clear regulatory frameworks that mandate responsible material disposal, provide incentives for sustainable AM practice, and facilitate the development of industry-standard recycling protocols. Governments must encourage the use of AM in a way that is sustainable by investing in research programs, subsidizing companies that transition to environmentally friendly material, and encouraging responsible waste management practices on the industrial and consumer level.

Cooperation among industry stakeholders, policymakers, and researchers will be key to accelerating the transition to a sustainable AM ecosystem. Material suppliers and AM technology companies must partner with research institutions and universities to develop and commercialize sustainable alternatives. Companies must also take a proactive role in integrating sustainability into production processes, either by improving material sourcing, recycling, or energy-efficient print technology.

Policymakers, meanwhile, must create regulatory environments that support circular economy strategies and facilitate sustainable AM development. The collaboration of these efforts will be key to making AM a truly sustainable manufacturing technology. Through an atmosphere of innovation, accountability, and collaboration, AM can not only remain competitive but also enable a more resource-efficient and environmentally friendly future.

REFERENCES

Abazari, R., Sanati, S., Bajaber, M. A., Javed, M. S., Junk, P. C., Nanjundan, A. K., Qian, J., & Dubal, D. P. (2024). Design and advanced manufacturing of NU-1000 metal–organic frameworks with future perspectives for environmental and renewable energy applications. *Small, 20*(15), 2306353. DOI: 10.1002/smll.202306353 PMID: 37997226

Agarwal, A., & Ojha, R. (2024). Prioritizing implications of Industry-4.0 on the sustainable development goals: A perspective from the analytic hierarchy process in manufacturing operations. *Journal of Cleaner Production, 444*, 141189. DOI: 10.1016/j.jclepro.2024.141189

Ahmad, M., Javaid, M., & Haleem, A. (2024). A study on fused deposition modeling (FDM) and laser-based additive manufacturing (LBAM) in the medical field. *Intelligent Pharmacy, 2*(3), 381–391. DOI: 10.1016/j.ipha.2024.02.010

Al Rashid, A., & Koç, M. (2023). Additive manufacturing for sustainability and circular economy: Needs, challenges, and opportunities for 3D printing of recycled polymeric waste. *Materials Today Sustainability, 24*, 100529. DOI: 10.1016/j.mtsust.2023.100529

Aslam, J., Saleem, A., Lai, K. H., & Kim, Y. B. (2025). Critical successes factors for the adoption of additive manufacturing: Integrated impact for circular economy model. *Technological Forecasting and Social Change, 213*, 124041. DOI: 10.1016/j.techfore.2025.124041

Behseresht, S., & Park, Y. H. (2024). Additive manufacturing of composite polymers: Thermomechanical FEA and experimental study. *Materials (Basel), 17*(8), 1912. DOI: 10.3390/ma17081912 PMID: 38673269

Bendoly, E., Chandrasekaran, A., Lima, M. D. R. F., Handfield, R., Khajavi, S. H., & Roscoe, S. (2024). The role of generative design and additive manufacturing capabilities in developing human–AI symbiosis: Evidence from multiple case studies. *Decision Sciences, 55*(4), 325–345. DOI: 10.1111/deci.12619

Bian, D., Tong, Z., Gong, G., Huang, H., Fang, L., Yang, H., Gu, W., Yu, H., & Zheng, Y. (2024). Additive manufacturing of biodegradable molybdenum–from powder to vascular stent. *Advanced Materials, 36*(32), 2401614. DOI: 10.1002/adma.202401614 PMID: 38837830

Bigliardi, B., Bottani, E., Gianatti, E., Monferdini, L., Pini, B., & Petroni, A. (2024). Sustainable Additive Manufacturing in the context of Industry 4.0: A literature review. *Procedia Computer Science, 232*, 766–774. DOI: 10.1016/j.procs.2024.01.076

Chen, K., Liu, Q., Chen, B., Zhang, S., Ferrara, L., & Li, W. (2024). Effect of raw materials on the performance of 3D printing geopolymer: A review. *Journal of Building Engineering*, *84*, 108501. DOI: 10.1016/j.jobe.2024.108501

Di, L., Yang, Y., & Wang, S. (2024). Additive manufacturing thermoplastic recycling: Profit-driven planning and optimization. *Journal of Cleaner Production*, *436*, 140598. DOI: 10.1016/j.jclepro.2024.140598

Erdaş, M. U., Yıldız, B. S., & Yıldız, A. R. (2024). Experimental analysis of the effects of different production directions on the mechanical characteristics of ABS, PLA, and PETG materials produced by FDM. *Materialprüfung (Munchen)*, *66*(2), 198–206. DOI: 10.1515/mt-2023-0206

Fidan, I., Naikwadi, V., Alkunte, S., Mishra, R., & Tantawi, K. (2024). Energy efficiency in additive manufacturing: Condensed review. *Technologies*, *12*(2), 21. DOI: 10.3390/technologies12020021

Gift, M. M., Senthil, T. S., Hasan, D. S., Alagarraja, K., Jayaseelan, P., & Boopathi, S. (2024). Additive Manufacturing and 3D Printing Innovations: Revolutionizing Industry 5.0. In *Technological Advancements in Data Processing for Next Generation Intelligent Systems* (pp. 255-287). IGI Global.

Goh, G. D., Wong, K. K., Tan, N., Seet, H. L., & Nai, M. L. S. (2024). Large-format additive manufacturing of polymers: A review of fabrication processes, materials, and design. *Virtual and Physical Prototyping*, *19*(1), e2336160. DOI: 10.1080/17452759.2024.2336160

Golvaskar, M., Ojo, S. A., & Kannan, M. (2024). Recent Advancements in Material Waste Recycling: Conventional, Direct Conversion, and Additive Manufacturing Techniques. *Recycling*, *9*(3), 43. DOI: 10.3390/recycling9030043

Graziosi, S., Faludi, J., Stanković, T., Borgianni, Y., Meisel, N., Hallstedt, S. I., & Rosen, D. W. (2024). A vision for sustainable additive manufacturing. *Nature Sustainability*, *7*(6), 698–705. DOI: 10.1038/s41893-024-01313-x

Hendrixson, S. (2020). Infographic: How additive manufacturing supports the circular economy. *Additive Manufacturing*.

Jamal, M. A., Shah, O. R., Ghafoor, U., Qureshi, Y., & Bhutta, M. R. (2024). Additive manufacturing of continuous fiber-reinforced polymer composites via fused deposition modelling: A comprehensive review. *Polymers*, *16*(12), 1622. DOI: 10.3390/polym16121622 PMID: 38931971

Klenam, D., Asumadu, T., Bodunrin, M., Obiko, J., Genga, R., Maape, S., McBagonluri, F., & Soboyejo, W. (2025). Toward sustainable industrialization in Africa: The potential of additive manufacturing–an overview. *Frontiers in Manufacturing Technology*, *4*, 1410653. DOI: 10.3389/fmtec.2024.1410653

Krishna, R. S., Rehman, A. U., Mishra, J., Saha, S., Korniejenko, K., Rehman, R. U., & Qureshi, T. S. (2024). Additive manufacturing of geopolymer composites for sustainable construction: Critical factors, advancements, challenges, and future directions. *Progress in Additive Manufacturing*, ●●●, 1–59.

Li, Y., Wang, S., Qian, S., Liu, Z., Weng, Y., & Zhang, Y. (2024). Depolymerization and Re/upcycling of biodegradable PLA plastics. *ACS Omega*, *9*(12), 13509–13521. DOI: 10.1021/acsomega.3c08674 PMID: 38559974

Ma, S., Fu, S., Yang, H., He, P., Sun, Z., Duan, X., Jia, D., Colombo, P., & Zhou, Y. (2024). Exploiting bifunctional 3D-Printed geopolymers for efficient cesium removal and immobilization: An approach for hazardous waste management. *Journal of Cleaner Production*, *437*, 140599. DOI: 10.1016/j.jclepro.2024.140599

Mathur, V., Dsouza, V., Srinivasan, V., & Vasanthan, K. S. (2025). Volumetric Additive Manufacturing for Cell Printing: Bridging Industry Adaptation and Regulatory Frontiers. *ACS Biomaterials Science & Engineering*, *11*(1), 156–181. DOI: 10.1021/acsbiomaterials.4c01837 PMID: 39746181

Mattera, G., Caggiano, A., & Nele, L. (2025). Energy Efficiency Optimisation in Wire arc Additive Manufacturing of Invar 36 Alloy via Intelligent Data-Driven Techniques. *International Journal of Precision Engineering and Manufacturing-Green Technology*, 1-13.

Mohamed, R. A., & Mohamed, A. F. A. (2025). Exploring the environmental benefits of 3D printing technology in concrete construction; a review. *Progress in Additive Manufacturing*, *10*(1), 279–289. DOI: 10.1007/s40964-024-00622-z

Mohammed, R., Shaik, A. S., Mohammed, S., Bunga, K. K., Aggala, C., Babu, B. P., & Badruddin, I. A. (2025). Advancements and Challenges in Additive Manufacturing: Future Directions and Implications for Sustainable Engineering. *Advance Sustainable Science Engineering and Technology*, *7*(1), 0250108–0250108. DOI: 10.26877/asset.v7i1.1079

Ng, W. L., Goh, G. L., Goh, G. D., Ten, J. S. J., & Yeong, W. Y. (2024). Progress and opportunities for machine learning in materials and processes of additive manufacturing. *Advanced Materials*, *36*(34), 2310006. DOI: 10.1002/adma.202310006 PMID: 38456831

Nyamekye, P., Lakshmanan, R., Tepponen, V., & Westman, S. (2024). Sustainability aspects of additive manufacturing: Leveraging resource efficiency via product design optimization and laser powder bed fusion. *Heliyon, 10*(1), e23152. DOI: 10.1016/j.heliyon.2023.e23152 PMID: 38163216

Olawumi, M. A., Oladapo, B. I., & Olugbade, T. O. (2024). Evaluating the impact of recycling on polymer of 3D printing for energy and material sustainability. *Resources, Conservation and Recycling, 209*, 107769. DOI: 10.1016/j.resconrec.2024.107769

Ononiwu, N., Ako, P., Anyaoha, C., Ikwuagwu, C., & Jacobs, I. (2024). Sustainable considerations in additive manufacturing processes: A review. *Journal of Mechanical Engineering Science*, •••, 9853–9871. DOI: 10.15282/jmes.18.1.2024.5.0780

Peng, B., Xu, H., Song, F., Wen, P., Tian, Y., & Zheng, Y. (2024). Additive manufacturing of porous magnesium alloys for biodegradable orthopedic implants: Process, design, and modification. *Journal of Materials Science and Technology, 182*, 79–110. DOI: 10.1016/j.jmst.2023.08.072

Priyadarshini, J., Singh, R. K., Mishra, R., He, Q., & Braganza, A. (2024). Implementation of additive manufacturing in the healthcare supply chain for circular economy goals: Paradoxical tensions and solutions from an Industry 5.0 perspective. *Information Systems Frontiers*, •••, 1–23. DOI: 10.1007/s10796-024-10482-1

Rahman, M., Islam, K. S., Dip, T. M., Chowdhury, M. F. M., Debnath, S. R., Hasan, S. M. M., Sakib, M. S., Saha, T., Padhye, R., & Houshyar, S. (2024). A review on nanomaterial-based additive manufacturing: Dynamics in properties, prospects, and challenges. *Progress in Additive Manufacturing, 9*(4), 1197–1224. DOI: 10.1007/s40964-023-00514-8

Romani, A., Perusin, L., Ciurnelli, M., & Levi, M. (2024). Characterization of PLA feedstock after multiple recycling processes for large-format material extrusion additive manufacturing. *Materials Today Sustainability, 25*, 100636. DOI: 10.1016/j.mtsust.2023.100636

Sapkota, A., Ghimire, S. K., & Adanur, S. (2024). A review on fused deposition modeling (FDM)-based additive manufacturing (AM) methods, materials and applications for flexible fabric structures. *Journal of Industrial Textiles, 54*, 15280837241282110. DOI: 10.1177/15280837241282110

Selvam, P. P., Prabhakaran, S., Vinod, B., & Jishnu, T. (2024). A review of energy efficiency and Machine learning analysis for additive manufacturing of direct laser metal deposition. *Materials Today: Proceedings*. Advance online publication. DOI: 10.1016/j.matpr.2024.03.034

Shanmugam, V., Babu, K., Kannan, G., Mensah, R. A., Samantaray, S. K., & Das, O. (2024). The thermal properties of FDM printed polymeric materials: A review. *Polymer Degradation & Stability*, *228*, 110902. DOI: 10.1016/j.polymdegradstab.2024.110902

Sheikh, M. F., Padmanabhan, R., Hadid, M., & Kerbache, L. (2025). Towards Additive Manufacturing Enabled Sustainable Spare Parts Supply Chain: A Multi-Objective MILP Model for Cost and Carbon Emissions Minimization. *Procedia Computer Science*, *253*, 952–963. DOI: 10.1016/j.procs.2025.01.157

Soori, M., Dastres, R., Arezoo, B., & Jough, F. K. G. (2024). Intelligent robotic systems in Industry 4.0: A review. *Journal of Advanced Manufacturing Science and Technology*, 2024007-0.

Vadivelu, V., Nachimuthu, P., & Karthikeyan, P. (2025). Transforming Supply Chains (SCs) with Additive Manufacturing (AM) A Paradigm Shift in Production. *Metal Additive Manufacturing: Principles, Techniques and Applications*, 359-406.

Villa, A., Gianchandani, P. K., & Baino, F. (2024). Sustainable approaches for the additive manufacturing of ceramic materials. *Ceramics*, *7*(1), 291–309. DOI: 10.3390/ceramics7010019

Yakubov, V., Ostergaard, H., Bhagavath, S., Leung, C. L. A., Hughes, J., Yasa, E., Khezri, M., Löschke, S. K., Li, Q., & Paradowska, A. M. (2024). Recycled aluminium feedstock in metal additive manufacturing: A state of the art review. *Heliyon*, *10*(5), e27243. DOI: 10.1016/j.heliyon.2024.e27243 PMID: 38463898

Zaid, O., & El Ouni, M. H. (2024). Advancements in 3D printing of cementitious materials: A review of mineral additives, properties, and systematic developments. *Construction & Building Materials*, *427*, 136254. DOI: 10.1016/j.conbuildmat.2024.136254

Zhao, H., Wang, Y., Liu, X., Wang, X., Chen, Z., Lei, Z., ... & Singh, A. (2024). Review on solid wastes incorporated cementitious material using 3D concrete printing technology. *Case Studies in Construction Materials*, e03676.

Zhao, J., Yang, Y., Kobir, M. H., Faludi, J., & Zhao, F. (2024). Driving additive manufacturing towards circular economy: State-of-the-art and future research directions. *Journal of Manufacturing Processes*, *124*, 621–637. DOI: 10.1016/j.jmapro.2024.06.018

Zhou, L., Miller, J., Vezza, J., Mayster, M., Raffay, M., Justice, Q., Al Tamimi, Z., Hansotte, G., Sunkara, L. D., & Bernat, J. (2024). Additive manufacturing: A comprehensive review. *Sensors (Basel)*, *24*(9), 2668. DOI: 10.3390/s24092668 PMID: 38732776

Chapter 4
Design for Additive Manufacturing: Bridging Creativity and Precision in the 3D Printing Era

M. Maniraj
https://orcid.org/0009-0000-5343-3895
Sardar Raja College of Engineering, Alangulam, India

S. M. Raj Kumar
https://orcid.org/0000-0002-8978-0108
Francis Xavier Engineering College, Tirunelveli, India

ABSTRACT

The product development process is changing with Design for Additive Manufacturing (DfAM) systems which amplify additive manufacturing capabilities to create optimized designs that deliver high performance as well as sustainable design solutions. This section demonstrates crucial DfAM methods based on topology optimization models and generative design and lattice structures and multi-material implementation and support reduction strategies that boost product capabilities and reduce substance utilization. The research discusses material selection methods together with artificial intelligence-based design and simulation tools and sustainable operational approaches to enhance AM efficiency. The aerospace medical and automotive sectors conducted well-documented implementations of DfAM despite current issues regarding standards and industrial implementation and workforce preparedness. The combination of AI enablements and hybrid manufacturing with digital workflows makes DfAM ready to transform future manufacturing operations by providing sustainable design methods.

DOI: 10.4018/979-8-3373-0533-2.ch004

1. INTRODUCTION

The emergence of additive manufacturing (AM), commonly referred to as 3D printing, has fundamentally reshaped the landscape of product design and manufacturing. Unlike traditional subtractive and formative manufacturing methods, which often impose significant geometric and material constraints, AM builds objects layer by layer, enabling unprecedented design flexibility and material efficiency (Gibson et al. 2021). This shift has not only revolutionized industries such as aerospace, healthcare, automotive, and consumer goods but has also redefined how engineers and designers approach product development. With the ability to fabricate complex geometries, lightweight structures, and customized components directly from digital models, AM has unlocked a new era of innovation where creativity is no longer restricted by conventional manufacturing limitations. However, while AM offers these advantages, leveraging its full potential requires a fundamental rethinking of design methodologies, leading to the emergence of Design for Additive Manufacturing (DfAM) as a critical discipline (Gibson et al. 2021b).

DfAM encompasses a set of principles, strategies, and best practices specifically tailored to optimize designs for AM technologies. Unlike traditional design for manufacturing, which primarily aims to simplify geometries to accommodate conventional processes, DfAM embraces the unique capabilities of AM to create complex, high-performance, and functionally optimized structures. It addresses key factors such as topology optimization, material distribution, support structure minimization, and part consolidation, ensuring that designs are not only manufacturable but also take full advantage of AM's ability to reduce weight, improve performance, and enhance material efficiency. Furthermore, DfAM considers the specific constraints of AM processes, such as layer-based anisotropy, surface finish limitations, and post-processing requirements, ensuring that products meet both functional and manufacturing requirements (Diegel et al. 2020).

This chapter aims to provide a comprehensive exploration of DfAM, bridging the gap between creativity and precision in additive manufacturing. It will examine the fundamental principles that guide DfAM, the computational tools that enable optimized design, and the challenges that must be navigated to achieve successful implementation. In addition, it will address the role of materials in AM, the sustainability implications of DfAM, and the integration of these principles into industrial workflows. By investigating case studies and real-world applications, this chapter will offer valuable insights for researchers, engineers, designers, and educators looking to harness the transformative power of AM. Ultimately, the goal is to establish DfAM as not just a design methodology but a paradigm shift that empowers the creation of more efficient, innovative, and sustainable products in the 3D printing era.

2. FUNDAMENTALS OF DFAM

DfAM represents a fundamental departure from traditional Design for Manufacturing principles, which have historically been constrained by the limitations of subtractive and formative manufacturing processes. Conventional DFM focuses on simplifying geometries, minimizing tooling complexity, and optimizing manufacturing efficiency by adhering to constraints such as machining accessibility, mold parting lines, and material waste reduction. In contrast, DfAM embraces the design freedom enabled by layer-by-layer fabrication, allowing for the creation of complex, highly intricate geometries that were previously impossible or prohibitively expensive to manufacture. This shift redefines how designers approach product development, shifting the emphasis from manufacturability simplification to functionality enhancement, lightweighting, and multi-functional integration (Alfaify et al. 2020).

One of the defining characteristics of DfAM is unprecedented design freedom, enabling features such as intricate lattice structures, topology-optimized forms, and part consolidation. Unlike traditional methods where increased geometric complexity often results in higher manufacturing costs and production challenges, AM thrives on complexity without adding significant fabrication constraints. This capability allows for biomimetic structures, customized medical implants, and aerospace components that optimize material use while maintaining strength and performance. However, with this freedom comes a new set of challenges that designers must carefully consider.

The challenges and constraints of AM primarily stem from process-specific limitations, including anisotropic material properties, build orientation effects, and the need for support structures in certain printing techniques. Anisotropy arises due to the layer-by-layer deposition of material, which often results in weaker mechanical properties along the z-axis. This requires strategic design decisions to optimize load-bearing orientations and ensure structural integrity. Build orientation also affects surface quality, residual stresses, and print success rates, making orientation planning an integral part of DfAM. Additionally, many AM processes require support structures to stabilize overhanging features during fabrication. These supports increase material consumption, post-processing time, and overall production cost, making their minimization or elimination a key design goal in DfAM (Moreno Nieto & Moreno Sánchez, 2021).

Understanding the classification of AM processes is essential for effective DfAM implementation, as different technologies impose unique design considerations. Fused Deposition Modeling (FDM), one of the most widely used AM methods, builds parts using thermoplastic filament extrusion. While FDM is cost-effective and accessible, it typically exhibits anisotropic strength, requiring careful consideration of layer orientation and support removal strategies. Selective Laser Sintering (SLS) utilizes a high-powered laser to sinter powdered materials, enabling the production of complex geometries without the need for support structures. However, powder handling, porosity control, and post-processing are critical considerations for SLS-designed components. Stereolithography (SLA), which uses a UV laser to cure photopolymer resins, offers high resolution and smooth surface finishes, making it ideal for detailed prototypes and medical applications, though its material selection is more limited than other AM processes. Metal Additive Manufacturing, which includes techniques such as Direct Metal Laser Sintering and Electron Beam Melting, allows for the fabrication of high-strength, lightweight metal components. These methods require careful thermal management to mitigate residual stresses, warping, and microstructural defects that can compromise mechanical performance (Lopez Taborda et al. 2021).

Incorporating DfAM principles into the design workflow requires a deep understanding of these AM processes and their implications on geometry, material properties, and post-processing requirements. By leveraging AM's unique strengths while addressing its inherent limitations, designers can create optimized components that push the boundaries of conventional engineering, paving the way for innovative and efficient manufacturing solutions (Ma et al. 2025).

3. KEY DESIGN STRATEGIES FOR ADDITIVE MANUFACTURING

Designing for additive manufacturing requires a shift in mindset that embraces complexity, efficiency, and functionality while accounting for the unique constraints of layer-based fabrication. Unlike conventional manufacturing, where intricate geometries often lead to increased production costs, additive manufacturing thrives on complexity. Geometric complexity and topology optimization play a fundamental role in AM-driven design strategies. Topology optimization involves using computational algorithms to determine the most efficient material distribution for a given set of constraints, such as load-bearing requirements or weight reduction goals. By leveraging these techniques, designers can create organic, high-performance structures that are impossible to manufacture through traditional means (Jihong et al. 2021). This capability is particularly valuable in industries such as aerospace and automotive, where reducing mass while maintaining strength is critical for improving efficiency and performance.

One of the most transformative aspects of additive manufacturing is lightweighting through lattice and cellular structures. By replacing solid material with finely tuned internal lattices, components can achieve significant weight reductions without sacrificing mechanical integrity. These structures, inspired by naturally occurring patterns like bone or honeycombs, not only improve weight-to-strength ratios but also enable the customization of mechanical properties such as flexibility, shock absorption, and heat dissipation. The ability to engineer such micro-architectures is particularly useful in biomedical implants, where porosity can be tailored for bone integration, or in high-performance automotive components that require a balance of rigidity and impact resistance (Garner et al. 2022).

The introduction of multi-material and functionally graded structures further expands the design potential of AM. Unlike traditional processes that often require separate manufacturing steps to integrate multiple materials, AM allows for the gradual transition between different materials within a single print. This capability enables the creation of functionally graded materials (FGMs), where properties such as stiffness, conductivity, or thermal resistance can be varied within a single component (Teacher & Velu, 2024). Such advancements open up new possibilities in applications ranging from wear-resistant coatings and heat exchangers to biomedical implants that mimic the natural transition between bone and cartilage. However, achieving these designs requires advanced multi-material printing technologies and careful consideration of interfacial bonding and material compatibility.

While AM enables unparalleled design freedom, certain processes still impose limitations, particularly in the form of support structures required to print overhanging features. Support structures consume additional material, increase post-processing time, and can sometimes degrade surface quality upon removal. As a result, minimizing or eliminating support structures is a key objective in DfAM. Techniques such as reorienting parts to reduce unsupported overhangs, incorporating self-supporting angles, or using advanced support-free AM processes like SLS or powder-bed fusion help designers optimize manufacturability. Additionally, computational tools are increasingly being used to generate dissolvable or breakaway support structures that reduce the impact on post-processing and material waste.

Another significant advantage of DfAM is the ability to consolidate multiple components into a single multi-functional part. Traditional assembly-based manufacturing often requires multiple pieces to be fabricated separately and then assembled, leading to added complexity, increased weight, and potential points of failure. Additive manufacturing removes these constraints by enabling part consolidation, where a single printed component integrates multiple functions. This approach simplifies supply chains, reduces assembly costs, and improves mechanical performance. A notable example is the aerospace sector, where traditionally multi-part assemblies, such as fuel nozzles and engine brackets, have been successfully redesigned as single-piece AM-optimized components, reducing weight and increasing durability (Vaneker et al. 2020).

Despite these benefits, designers must also consider surface finish requirements, as AM parts often exhibit rougher surfaces due to the layer-by-layer deposition process. Surface roughness can impact mechanical properties, fluid dynamics, and overall part aesthetics, making post-processing an essential consideration. Processes such as sanding, chemical smoothing, bead blasting, or laser polishing can enhance surface quality, but these steps introduce additional time and cost. Therefore, DfAM strategies often include designing with as-printed surface textures in mind, optimizing layer orientation to minimize roughness, and integrating hybrid AM-subtractive techniques to achieve precise surface finishes where necessary (Obilanade et al. 2022).

By applying these key design strategies, engineers and designers can harness the full potential of additive manufacturing to create innovative, high-performance, and optimized components. The ability to merge complexity with efficiency, minimize material usage while maximizing strength, and integrate multiple functionalities into a single part positions AM as a transformative force in modern engineering and product development.

4. MATERIAL CONSIDERATIONS IN DFAM

Material selection is a fundamental aspect of DfAM, as the success of an additively manufactured component depends not only on its geometry but also on the properties and behavior of the chosen material. Unlike conventional manufacturing, where material properties are well understood and standardized for specific processes, additive manufacturing introduces new complexities due to its unique fabrication methods. AM-compatible materials include polymers, metals, ceramics, and composites, each offering distinct advantages and limitations that influence the design and performance of printed components. Polymers such as PLA, ABS, and nylon are commonly used in FDM due to their ease of processing and cost-effectiveness, while high-performance thermoplastics like PEEK and ULTEM provide superior mechanical and thermal properties for demanding applications. Metal additive manufacturing processes, such as direct metal laser sintering and electron beam melting, enable the production of high-strength, lightweight components from titanium, aluminum, stainless steel, and nickel-based superalloys, making them ideal for aerospace and medical applications. Ceramics are also gaining traction in AM for applications requiring extreme heat resistance, biocompatibility, or electrical insulation, while composite materials, including fiber-reinforced polymers and metal-matrix composites, are expanding the possibilities of AM by offering enhanced strength-to-weight ratios and multi-functional properties (Goala & Sarkar, 2024).

While AM provides access to a broad range of materials, each category presents specific constraints that must be carefully managed during the design process. One of the most significant challenges in AM materials is anisotropy, where mechanical properties vary based on build orientation due to the layer-by-layer nature of fabrication. This anisotropic behavior can lead to weaknesses in the z-direction, requiring strategic design decisions such as reorienting parts for optimal strength distribution or incorporating reinforcing geometries. Fatigue performance is another critical concern, especially for load-bearing applications, as the presence of internal voids or microstructural defects can lead to premature failure under cyclic loading. Porosity, which results from incomplete fusion of powder particles in metal AM or void formation in polymer printing, affects mechanical integrity, density, and thermal properties. To mitigate these issues, post-processing techniques such as heat treatment, hot isostatic pressing, and surface finishing are often necessary to improve material performance and consistency (Zohdi & Yang, 2021).

Selecting the right material for a given AM application requires a thorough understanding of its functional requirements, processing compatibility, and cost implications. In aerospace applications, for instance, the emphasis is on lightweight, high-strength materials such as titanium and aluminum alloys that offer excellent strength-to-weight ratios while being able to withstand extreme environmental conditions. Additive manufacturing enables the creation of topology-optimized, hollow, or lattice-filled structures that reduce weight without compromising performance, significantly benefiting aircraft and spacecraft components. In biomedical applications, material selection is guided by biocompatibility, corrosion resistance, and the ability to customize patient-specific geometries. Titanium and cobalt-chrome alloys are widely used for orthopedic implants due to their excellent mechanical properties and compatibility with human bone, while polymer-based AM materials like PEEK are employed for medical implants requiring radiolucency and high strength (Agrawal, 2021).

Several case studies highlight the transformative role of material-specific DfAM strategies in industry. In aerospace, companies such as GE Aviation have successfully leveraged metal AM to produce complex fuel nozzles that consolidate multiple components into a single, lightweight part with superior performance and durability. In the biomedical field, patient-specific implants manufactured using titanium AM have demonstrated improved osseointegration and reduced recovery times, showcasing the power of customization in medical treatments. These advancements underscore the importance of material considerations in DfAM, where understanding the interaction between material properties, AM processes, and application-specific requirements is key to unlocking the full potential of additive manufacturing. As material research continues to evolve, new alloys, polymers, and composite formulations will further expand the design possibilities, pushing the boundaries of what can be achieved through AM-driven innovation (Djokikj & Kandikjan, 2023).

5. COMPUTATIONAL TOOLS AND SIMULATION IN DFAM

The integration of computational tools and simulation techniques is essential in DfAM, enabling designers to fully exploit the capabilities of AM while mitigating potential risks and inefficiencies. One of the most transformative advancements in this domain is generative design and topology optimization. Generative design leverages algorithmic approaches to explore vast design possibilities based on predefined parameters such as material constraints, load conditions, and performance requirements. By using iterative computational processes, generative design algorithms can propose highly optimized structures that maximize strength while minimizing material usage, often producing organic, biomimetic geometries that would be impractical or impossible to manufacture using traditional methods. Topology optimization, closely related to generative design, refines existing structures by redistributing material where it is most needed for mechanical performance, ensuring lightweight and high-strength designs that fully leverage the layer-by-layer fabrication of AM. These tools not only enhance design efficiency but also reduce waste, making AM an inherently more sustainable manufacturing approach (Sbrugnera Sotomayor et al. 2021).

In addition to design optimization, computational simulation plays a crucial role in predicting and mitigating potential manufacturing challenges. Finite Element Analysis (FEA) and build simulation allow engineers to assess the structural integrity and manufacturability of AM components before fabrication, minimizing costly trial-and-error iterations. FEA evaluates stress distribution, deformation, and thermal properties under operational loads, ensuring that printed components meet performance and safety standards. Build simulation, on the other hand, focuses on the manufacturing process itself, identifying issues such as residual stresses, thermal distortions, and warping, which are particularly problematic in metal additive manufacturing. By simulating the layer-wise deposition process, designers can optimize build orientation, support structure placement, and heat management strategies to improve the reliability of printed parts. These predictive models are increasingly used to fine-tune AM workflows, reducing the risk of defects and improving first-time-right manufacturing success rates (Ismail & Ibrahim, 2021).

Beyond optimization and simulation, design automation for AM is gaining traction as a means of streamlining the design-to-production workflow. Parametric modeling and rule-based design automation allow for rapid design iterations, particularly useful in applications requiring mass customization, such as personalized medical implants or tailored automotive components. These automation tools integrate with AM-specific constraints, ensuring that generated designs adhere to process limitations, material properties, and performance requirements. The ability to automate repetitive design tasks significantly reduces development time while enabling greater exploration of innovative geometries that capitalize on AM's capabilities (Wiberg et al. 2023).

Artificial intelligence (AI) and machine learning are further revolutionizing AM design optimization by analyzing vast datasets to identify patterns and correlations that may be imperceptible through traditional engineering methods. AI-driven design approaches can enhance generative design by incorporating past manufacturing data, refining material distribution strategies, and predicting optimal print parameters based on historical performance. Machine learning algorithms are also being used for real-time defect detection and adaptive process control, improving the reliability and repeatability of AM processes. By continuously learning from print outcomes, AI-based systems can refine design and production strategies, pushing AM towards greater efficiency, accuracy, and quality (Ladani, 2021).

Despite these advancements, adopting computational tools in DfAM presents several challenges. The complexity of generative algorithms and simulation models requires specialized expertise, often creating a steep learning curve for engineers transitioning from conventional design methods. Validation remains a critical concern, as algorithmically generated designs and simulated performance predictions must be rigorously tested to ensure real-world feasibility. Integrating these advanced computational tools into existing design workflows can also be challenging, as compatibility between software platforms, AM hardware, and material databases is not always seamless. Additionally, the computational power required for high-resolution simulations and AI-driven optimizations can be resource-intensive, limiting accessibility for smaller enterprises. Overcoming these challenges requires continued advancements in user-friendly software interfaces, improved interoperability between design and manufacturing platforms, and the development of standardized validation methodologies (Dasappa Ashoka & Sidnekoppa, 2024).

As computational tools continue to evolve, their integration into DfAM will become increasingly indispensable, enabling more efficient, innovative, and reliable additive manufacturing processes. The synergy between generative design, simulation, automation, and AI-driven optimization represents a transformative shift in how products are conceived, tested, and fabricated. These advancements are not only redefining what is possible in AM but also positioning it as a foundational technology for next-generation engineering and manufacturing.

6. DFAM FOR SUSTAINABILITY AND THE CIRCULAR ECONOMY

DfAM plays a crucial role in advancing sustainability and fostering a circular economy by optimizing material usage, reducing waste, and enabling the efficient production of high-performance, long-lasting products. One of the most significant advantages of AM over traditional manufacturing is its ability to achieve material efficiency through precise deposition, ensuring that only the required material is used to build a part. Unlike subtractive manufacturing methods such as milling or machining, where large portions of material are cut away and often discarded as scrap, AM significantly reduces material waste, making it an inherently more sustainable process. In industries where material costs and environmental impact are major concerns, such as aerospace and medical device manufacturing, this efficiency translates into lower costs, reduced raw material extraction, and minimized landfill contributions. By leveraging DfAM principles, designers can further enhance material efficiency through strategies such as topology optimization and lattice structures, ensuring that products maintain their strength and functionality while using the least amount of material possible (Su et al. 2024).

Although AM offers significant material savings, its energy consumption remains a critical factor in evaluating its overall sustainability. Some AM technologies, particularly laser-based metal printing processes such as Selective Laser Melting and Electron Beam Melting, require high energy inputs to achieve precise thermal control and material fusion. This energy demand, if sourced from non-renewable sources, can contribute to the environmental footprint of AM despite its material efficiency advantages. However, the shift toward more energy-efficient AM processes, renewable energy integration in production facilities, and the development of lower-energy printing techniques is helping mitigate these concerns. Additionally, the lightweighting potential of AM-designed parts contributes to long-term sustainability, particularly in applications such as aerospace and automotive manufacturing, where reduced component weight leads to improved fuel efficiency and lower emissions over a product's lifecycle (Almeida et al. 2021).

A key consideration in the sustainability of AM products is lifecycle management, which includes assessing their durability, repairability, and end-of-life strategies. DfAM facilitates the design of parts that can be easily repaired or refurbished, extending their usable life and reducing the need for full replacements. In contrast to traditional manufacturing, where parts often need to be re-machined or replaced entirely when damaged, AM allows for on-demand repairs and the production of replacement components, particularly beneficial in remote locations or industries requiring long-term part availability, such as aerospace and military applications.

Additionally, AM promotes localized production, reducing the carbon footprint associated with transportation and supply chain logistics.

Recycling and the reuse of AM materials are emerging as important factors in ensuring the long-term sustainability of the technology. While thermoplastic-based AM processes such as FDM can incorporate recycled polymers, challenges remain in recycling metal powders used in powder bed fusion processes, as repeated exposure to high temperatures can alter particle size distribution and material properties. However, advances in material recovery techniques, such as closed-loop powder recycling systems and solvent-based polymer reclamation, are improving the feasibility of AM material reuse. Research into bio-based and biodegradable materials is also paving the way for more environmentally friendly AM materials that reduce dependence on petroleum-based polymers and minimize environmental impact (Romani & Levi, 2024).

Several case studies highlight the impact of DfAM-driven sustainable solutions across various industries. In the aerospace sector, lightweight, topology-optimized AM components have led to significant fuel savings and emissions reductions, demonstrating how DfAM contributes to sustainability over a product's lifetime. In the medical field, patient-specific implants and prosthetics produced via AM have reduced material waste by eliminating the need for standardized stock sizes, while also enhancing patient outcomes through precise customization. In consumer goods, companies are increasingly turning to AM for on-demand production and localized manufacturing, reducing the need for mass production and minimizing overproduction waste.

By integrating sustainability principles into DfAM, additive manufacturing can evolve into a more environmentally responsible technology that aligns with circular economy goals. Future advancements in recyclable AM materials, energy-efficient printing technologies, and lifecycle-conscious design methodologies will further enhance the sustainability of AM, making it a key driver in the transition toward a more resource-efficient and environmentally friendly manufacturing ecosystem. As research and industry practices continue to evolve, the role of DfAM in sustainability will become increasingly essential, helping balance the benefits of advanced manufacturing with the pressing need for ecological responsibility.

7. INDUSTRIAL ADOPTION OF DFAM

DfAM in industrial settings is rapidly transforming manufacturing workflows, product development, and supply chain dynamics. Across industries, companies are recognizing the potential of AM to produce lightweight, high-performance components while reducing material waste and production lead times. However, despite its advantages, widespread industrial adoption remains a gradual process due to technical, economic, and organizational barriers. Many industries still operate within traditional manufacturing paradigms that prioritize established design methodologies, supply chain logistics, and large-scale production efficiencies. The transition to DfAM requires a shift in mindset, engineering expertise, and manufacturing infrastructure, all of which can present significant hurdles. Companies must navigate challenges such as the high initial costs of AM equipment, the complexity of AM-specific design tools, and the need for standardized qualification processes to ensure product reliability. Furthermore, the lack of universally accepted certification protocols for AM-produced parts, especially in high-stakes applications such as aerospace and medical devices, continues to slow adoption (Obi, 2022).

Successfully integrating DfAM into industrial workflows requires a combination of technical innovation, workforce training, and strategic investment. One of the key strategies is the implementation of hybrid manufacturing approaches, where AM is used in conjunction with conventional methods to optimize performance and cost-effectiveness. By focusing on low-volume, high-complexity applications, companies can leverage AM's strengths while maintaining traditional manufacturing processes for mass production. Another crucial factor is investing in computational design tools that support generative design, topology optimization, and build simulations, ensuring that AM components are not only manufacturable but also optimized for performance and sustainability. Workforce training is equally important, as engineers and designers accustomed to subtractive or formative manufacturing must develop new skills in AM-specific design principles, material science, and process optimization. Industry leaders are increasingly collaborating with academic institutions and research organizations to create specialized training programs that equip engineers with the expertise needed for successful DfAM implementation (Prabhu et al. 2020).

Cross-disciplinary collaboration is essential for the widespread adoption of DfAM. Unlike traditional manufacturing, where design and production are often separate stages, AM requires a holistic, integrated approach that brings together designers, material scientists, computational engineers, and manufacturing specialists from the outset. This level of collaboration ensures that design decisions align with material properties, process constraints, and performance expectations. Companies that have successfully adopted DfAM often foster multi-disciplinary teams that work together to iterate on designs, test material behavior, and refine production techniques. Additionally, partnerships between industry and academia have proven instrumental in advancing DfAM knowledge and accelerating the development of AM-specific standards and best practices.

Several industries have demonstrated the tangible benefits of DfAM, particularly in aerospace, medical devices, and automotive manufacturing. In aerospace, companies like GE Aviation and Airbus have integrated AM to lightweight engine components and structural parts, resulting in significant fuel savings and improved performance. GE's famous fuel nozzle, redesigned for AM, consolidated multiple components into a single lightweight part with enhanced durability and efficiency, reducing weight and assembly time while improving combustion performance. In the medical field, DfAM has enabled the production of patient-specific implants and prosthetics, using biocompatible materials and lattice structures that enhance osseointegration. Customizable cranial implants, spinal cages, and dental prosthetics have demonstrated superior performance compared to conventionally manufactured alternatives, highlighting AM's potential for personalized healthcare solutions. In the automotive sector, companies such as Bugatti and Ford have utilized DfAM to create high-strength, lightweight components such as brake calipers, suspension parts, and intake manifolds, enhancing vehicle efficiency and performance while reducing production complexity (Sæterbø & Solvang, 2024).

The growing adoption of DfAM across industries signals a shift toward more efficient, flexible, and sustainable manufacturing practices. As companies continue to overcome adoption barriers through technological advancements, process standardization, and workforce development, the role of AM in industrial production will only expand. The future of DfAM lies in its ability to seamlessly integrate into existing workflows while pushing the boundaries of design innovation, ultimately transforming how products are conceptualized, manufactured, and utilized across diverse applications.

8. EDUCATION AND WORKFORCE DEVELOPMENT FOR DFAM

As additive manufacturing (AM) continues to revolutionize product design and production, integrating DfAM into engineering education has become essential. The growing demand for expertise in AM-driven design highlights the need for curricula that equip students with the necessary skills to leverage AM's full potential. Traditional engineering programs have long been rooted in conventional manufacturing methods such as machining, casting, and injection molding, but AM introduces an entirely new paradigm that requires a fundamental rethinking of design approaches. DfAM education must bridge the gap between theoretical knowledge and practical application, ensuring that future engineers and designers can create optimized, functional, and manufacturable components using AM technologies. Understanding material behavior, process limitations, support structure optimization, and computational design tools is crucial for preparing students to navigate the unique challenges of AM. As industries increasingly adopt AM for production, the ability to design specifically for AM will be a defining skill for engineers entering the workforce (Borgianni et al. 2022).

A key element of effective DfAM education is hands-on learning, where students gain direct experience with AM processes, materials, and design principles. Exposure to real-world AM challenges helps students understand the intricacies of printing methods such as FDM, Selective Laser Sintering, Stereolithography (SLA), and metal additive processes like Direct Metal Laser Sintering. Industry partnerships play a vital role in this experiential learning, enabling students to work on collaborative projects, internships, and research initiatives that align with industrial applications. By engaging with companies that specialize in AM, students develop problem-solving skills and gain insights into real-world constraints such as cost considerations, material selection, and post-processing requirements. Universities and technical institutes are increasingly forming partnerships with AM industry leaders to integrate DfAM coursework with professional applications, ensuring that graduates are prepared to contribute immediately upon entering the workforce (Prabhu et al. 2023).

The use of AM simulation software has also become an integral part of DfAM education, enabling students to design, optimize, and validate their concepts before fabrication. Simulation tools allow for topology optimization, generative design, finite element analysis (FEA), and build process simulation, ensuring that students learn how to optimize AM-specific designs while minimizing manufacturing defects. These software tools provide a virtual environment where students can analyze thermal stresses, mechanical performance, and material distribution, helping them refine their designs before committing to physical production. The ability to work with computational tools such as Autodesk Netfabb, nTopology, ANSYS Additive Suite, and Siemens NX for AM gives students a competitive edge as industries increasingly rely on digital design workflows.

Despite these advancements, training the next generation of AM designers comes with significant challenges. One of the primary obstacles is the lack of standardized educational frameworks for DfAM, as curricula vary widely between institutions and often fail to keep pace with the rapid evolution of AM technologies. Many engineering programs still focus predominantly on traditional manufacturing, leaving students with limited exposure to AM-specific design principles. Additionally, the interdisciplinary nature of DfAM requires knowledge spanning materials science, computational modeling, mechanical engineering, and process optimization, making it difficult to integrate all necessary topics within a single curriculum. Access to high-quality AM equipment and materials is another challenge, as industrial-grade 3D printers and advanced simulation software can be costly, limiting opportunities for hands-on learning in some educational institutions (Havenga et al. 2023).

Overcoming these barriers requires a concerted effort from academic institutions, industry leaders, and policymakers to ensure that DfAM education evolves alongside technological advancements. Expanding online learning resources, offering AM-focused certification programs, and fostering greater collaboration between universities and AM companies will help build a workforce equipped with the skills necessary to drive AM innovation. As the demand for AM expertise continues to grow, the ability to design efficiently for additive processes will be a defining competency for engineers, designers, and manufacturing professionals, shaping the future of industrial production and technological development.

9. FUTURE TRENDS IN DFAM

As additive manufacturing (AM) continues to evolve, DfAM is undergoing significant advancements that are shaping the future of product development and industrial production. One of the most promising trends is the development of multi-material and hybrid AM technologies, which enable the fabrication of components with varying material properties within a single build process. Unlike conventional AM, where single-material printing is the standard, multi-material AM allows for functionally graded structures, integrating materials with distinct mechanical, thermal, or electrical properties in a seamless manner. This capability is particularly transformative in applications such as aerospace, where components must exhibit high strength in certain areas while remaining lightweight elsewhere, or in healthcare, where implants can mimic the natural transition between soft and hard tissues. Hybrid AM, which combines additive and subtractive manufacturing processes, further enhances design possibilities by integrating precise machining within the AM workflow. This approach ensures superior surface finishes and tight tolerances while maintaining the geometric complexity enabled by AM (Dilberoglu et al. 2021).

Another major shift in DfAM is the integration of artificial intelligence (AI)-driven generative design, which is redefining the way products are conceptualized and optimized. AI-powered generative design leverages machine learning algorithms to explore vast design spaces, rapidly generating and refining complex geometries based on predefined performance criteria. Unlike traditional design approaches that rely on human intuition and incremental improvements, AI-driven DfAM autonomously identifies optimal configurations, often resulting in organic, highly efficient structures that would be unintuitive or impossible to conceive manually. These AI-generated designs are particularly valuable in industries where performance optimization is critical, such as automotive and aerospace, where reducing weight while maintaining strength is a primary objective. The ability of AI to analyze vast amounts of manufacturing data also facilitates process optimization, improving the reliability and repeatability of AM by predicting and mitigating defects before fabrication (Aljabali et al. 2025).

In parallel with AI-driven design, the development of automated AM design validation is addressing one of the most significant challenges in additive manufacturing: ensuring that designs are not only optimized for performance but also manufacturable with minimal risk of failure. Traditionally, validating AM designs has been a time-consuming process that involves extensive simulation, prototyping, and iterative testing. Emerging validation tools now leverage computational models to automatically assess printability, predict residual stresses, optimize support structures, and simulate thermal distortions. These automated validation systems significantly reduce development time by identifying potential issues before physical fabrication, ensuring that AM components meet required performance and quality standards from the outset. As these tools become more sophisticated, they will enable real-time design feedback, allowing engineers to refine DfAM strategies dynamically and reducing dependency on physical prototyping.

The impact of AM on the evolution of design paradigms is perhaps the most profound transformation driven by DfAM. Traditional design methodologies have long been constrained by the limitations of subtractive and formative manufacturing, necessitating compromises in geometry, material use, and assembly complexity. AM has broken these constraints, allowing designers to shift from manufacturability-driven design to function-driven design, where components are created based on optimal performance rather than process limitations. This paradigm shift is fostering innovation across multiple disciplines, from bioinspired structures that mimic nature's efficiency to highly integrated, multi-functional components that consolidate assemblies into single, optimized parts. As AM technologies continue to mature, the boundary between design and manufacturing will continue to blur, leading to a future where fully digital, AI-assisted, and simulation-driven design processes redefine how products are conceptualized, engineered, and produced (Raut et al. 2025).

With advancements in multi-material printing, AI-powered design, automated validation, and new design paradigms, the future of DfAM is set to push the boundaries of engineering innovation. These developments will not only enhance the efficiency and performance of AM components but also accelerate the adoption of additive manufacturing across industries, paving the way for a new era of design freedom, sustainability, and technological progress.

10. CONCLUSION

The evolution of DfAM has redefined the way products are conceptualized, engineered, and manufactured, offering unparalleled design freedom while introducing new challenges that require innovative solutions. Throughout this chapter, key insights into DfAM principles, material considerations, computational tools, sustainability, industrial adoption, and future trends have demonstrated that additive manufacturing is more than just an alternative production method—it is a transformative force that demands a rethinking of conventional design approaches. Unlike traditional manufacturing methods, which often impose geometric constraints and require extensive assembly, AM enables highly complex, lightweight, and functionally optimized structures that were previously impossible to produce. The integration of generative design, topology optimization, and AI-driven methodologies has further expanded the potential of AM, allowing engineers to push the boundaries of efficiency and performance. However, leveraging these advantages requires a deep understanding of process-specific constraints, including material anisotropy, support structure minimization, and post-processing considerations, all of which influence the final quality and reliability of AM-produced parts.

Despite the remarkable progress in DfAM, significant challenges remain that must be addressed for broader industrial adoption. One of the key hurdles is the standardization of AM processes and materials, particularly in highly regulated industries such as aerospace, medical devices, and automotive manufacturing. The lack of universally accepted certification protocols creates uncertainty in part validation and quality assurance, necessitating further research and collaboration between academia, industry, and regulatory bodies. Additionally, while computational tools have greatly enhanced design optimization and simulation capabilities, their complexity and integration challenges present barriers for widespread use. Many organizations struggle with the steep learning curve associated with advanced simulation software and AI-driven design tools, underscoring the need for more intuitive, user-friendly platforms that streamline the DfAM workflow. Another critical challenge lies in workforce development, as many engineering curricula still focus primarily on traditional manufacturing, leaving a skills gap that must be addressed through specialized education and training programs.

Moving forward, researchers, designers, and industries must actively engage in advancing DfAM methodologies to unlock the full potential of additive manufacturing. This requires fostering a culture of innovation that encourages interdisciplinary collaboration, bringing together experts in materials science, computational modeling, mechanical engineering, and industrial design to refine and expand the capabilities of AM. Continued investment in research and development will be essential to overcoming existing limitations, from enhancing multi-material printing technologies to improving energy efficiency and recyclability in AM processes. The integration of AI and automation into AM workflows will also play a pivotal role in streamlining design validation, predictive maintenance, and defect mitigation, ultimately driving AM toward greater reliability and scalability. Industries must embrace DfAM not only as a means of improving efficiency but as a strategic tool for reshaping supply chains, enabling localized manufacturing, and reducing environmental impact.

As additive manufacturing continues to evolve, the role of DfAM will become increasingly critical in shaping the next generation of engineering solutions. By bridging the gap between creativity and precision, DfAM empowers designers to create optimized, high-performance products that align with modern demands for efficiency, sustainability, and innovation. The challenge now is to accelerate its adoption, ensuring that the knowledge, tools, and infrastructure necessary for successful AM implementation are widely accessible. The future of manufacturing lies in the ability to rethink traditional constraints, embrace digital transformation, and leverage AM-driven design to create solutions that are not only functional but revolutionary.

REFERENCES

Agrawal, R. (2021). Sustainable material selection for additive manufacturing technologies: A critical analysis of rank reversal approach. *Journal of Cleaner Production, 296*, 126500. DOI: 10.1016/j.jclepro.2021.126500

Alfaify, A., Saleh, M., Abdullah, F. M., & Al-Ahmari, A. M. (2020). Design for additive manufacturing: A systematic review. *Sustainability (Basel), 12*(19), 7936. DOI: 10.3390/su12197936

Aljabali, B. A., Parupelli, S. K., & Desai, S. (2025). Generalized Design for Additive Manufacturing (DfAM) Expert System Using Compliance and Design Rules. *Machines (Basel), 13*(1), 29. DOI: 10.3390/machines13010029

Almeida, H., Pei, E., & Vitorino, L. (2021). Sustainablity for 3D Printing. In *Sustainability for 3D Printing* (pp. 1–13). Springer International Publishing.

Borgianni, Y., Pradel, P., Berni, A., Obi, M., & Bibb, R. (2022). An investigation into the current state of education in Design for Additive Manufacturing. *Journal of Engineering Design, 33*(7), 461–490. DOI: 10.1080/09544828.2022.2102893

Dasappa Ashoka, V. K., & Sidnekoppa, L. (2024). Challenges and Opportunities within Design for Additive Manufacturing.

Diegel, O., Nordin, A., & Motte, D. (2020). *A practical guide to design for additive manufacturing*. Springer.

Dilberoglu, U. M., Gharehpapagh, B., Yaman, U., & Dolen, M. (2021). Current trends and research opportunities in hybrid additive manufacturing. *International Journal of Advanced Manufacturing Technology, 113*(3-4), 623–648. DOI: 10.1007/s00170-021-06688-1

Djokikj, J., & Kandikjan, T. (2023). DfAM: Application of the design rules in the early design stages. *Procedia CIRP, 118*, 659–663. DOI: 10.1016/j.procir.2023.06.113

Garner, E., Wu, J., & Zadpoor, A. A. (2022). Multi-objective design optimization of 3D micro-architected implants. *Computer Methods in Applied Mechanics and Engineering, 396*, 115102. DOI: 10.1016/j.cma.2022.115102

Gibson, I., Rosen, D., Stucker, B., Khorasani, M., Gibson, I., Rosen, D., ... & Khorasani, M. (2021b). Design for additive manufacturing. Additive manufacturing technologies, 555-607.

Gibson, I., Rosen, D. W., Stucker, B., Khorasani, M., Rosen, D., Stucker, B., & Khorasani, M. (2021a). *Additive manufacturing technologies* (Vol. 17). Springer. DOI: 10.1007/978-3-030-56127-7

Goala, S., & Sarkar, P. (2024). Selection methodology for additive manufacturing processes considering design, material, and manufacturability. *Rapid Prototyping Journal*, *30*(1), 15–32. DOI: 10.1108/RPJ-05-2023-0155

Havenga, S., van Zyl, I., Snaddon, B., & Chisin, A. (2023). Towards the effective development of Design for Additive Manufacturing (DFAM) curricula: an exploration of strategies and solutions in education. In MATEC Web of Conferences (Vol. 388, p. 05009). EDP Sciences.

Ismail, M. I., & Ibrahim, M. (2021). Analysis on Lighweighting Approach Topology Optimization Design for Additive Manufacturing (DfAM). *Research Progress in Mechanical and Manufacturing Engineering*, *2*(2), 860–868.

Jihong, Z. H. U., Han, Z. H. O. U., Chuang, W. A. N. G., Lu, Z. H. O. U., Shangqin, Y. U. A. N., & Zhang, W. (2021). A review of topology optimization for additive manufacturing: Status and challenges. *Chinese Journal of Aeronautics*, *34*(1), 91–110. DOI: 10.1016/j.cja.2020.09.020

Ladani, L. J. (2021). Applications of artificial intelligence and machine learning in metal additive manufacturing. *JPhys Materials*, *4*(4), 042009. DOI: 10.1088/2515-7639/ac2791

Lopez Taborda, L. L., Maury, H., & Pacheco, J. (2021). Design for additive manufacturing: A comprehensive review of the tendencies and limitations of methodologies. *Rapid Prototyping Journal*, *27*(5), 918–966. DOI: 10.1108/RPJ-11-2019-0296

Ma, Q., Dong, K., Li, F., Jia, Q., Tian, J., Yu, M., & Xiong, Y. (2025). Additive manufacturing of polymer composite millimeter-wave components: Recent progress, novel applications, and challenges. *Polymer Composites*, *46*(1), 14–37. DOI: 10.1002/pc.28985

Moreno Nieto, D., & Moreno Sánchez, D. (2021). Design for additive manufacturing: Tool review and a case study. *Applied Sciences (Basel, Switzerland)*, *11*(4), 1571. DOI: 10.3390/app11041571

Obi, M. (2022). Investigating an approach for the dissemination of formalised Design for Additive Manufacturing knowledge (Doctoral dissertation, Loughborough University).

Obilanade, D., Törlind, P., & Dordlofva, C. (2022). Surface roughness and design for additive manufacturing: A design artefact investigation. *Proceedings of the Design Society, 2*, 1421–1430. DOI: 10.1017/pds.2022.144

Prabhu, R., Bracken, J., Armstrong, C. B., Jablokow, K., Simpson, T. W., & Meisel, N. A. (2020). Additive creativity: Investigating the use of design for additive manufacturing to encourage creativity in the engineering design industry. *International Journal of Design Creativity and Innovation, 8*(4), 198–222. DOI: 10.1080/21650349.2020.1813633

Prabhu, R., Simpson, T. W., Miller, S. R., Cutler, S. L., & Meisel, N. A. (2023). Teaching designing for additive manufacturing: formulating educational interventions that encourage design creativity. 3D printing and additive manufacturing, 10(2), 356-372.

Raut, J. M., Pimpalshende, A., Chandak, M. A., Patil, T. R., Pinjarkar, L., & Nair, S. (2025). A novel multi-method framework for 3D printed fiber-reinforced polymer concrete utilizing advance additive manufacturing techniques. Asian Journal of Civil Engineering, 1-14.

Romani, A., & Levi, M. (2024). Large-Format Material Extrusion Additive Manufacturing for Circular Economy Practices: A Focus on Product Applications with Materials from Recycled Plastics and Biomass Waste. *Sustainability (Basel), 16*(18), 7966. DOI: 10.3390/su16187966

Sæterbø, M., & Solvang, W. D. (2024). Metal additive manufacturing adoption in SMEs: Technical attributes, challenges, and opportunities. *Journal of Manufacturing Processes, 128*, 175–189. DOI: 10.1016/j.jmapro.2024.07.074

Sbrugnera Sotomayor, N. A., Caiazzo, F., & Alfieri, V. (2021). Enhancing design for additive manufacturing workflow: Optimization, design and simulation tools. *Applied Sciences (Basel, Switzerland), 11*(14), 6628. DOI: 10.3390/app11146628

Su, J., Ng, W. L., An, J., Yeong, W. Y., Chua, C. K., & Sing, S. L. (2024). Achieving sustainability by additive manufacturing: A state-of-the-art review and perspectives. *Virtual and Physical Prototyping, 19*(1), e2438899. DOI: 10.1080/17452759.2024.2438899

Teacher, M., & Velu, R. (2024). Additive manufacturing of functionally graded materials: A comprehensive review. *International Journal of Precision Engineering and Manufacturing, 25*(1), 165–197. DOI: 10.1007/s12541-023-00864-x

Vaneker, T., Bernard, A., Moroni, G., Gibson, I., & Zhang, Y. (2020). Design for additive manufacturing: Framework and methodology. *CIRP Annals*, *69*(2), 578–599. DOI: 10.1016/j.cirp.2020.05.006

Wiberg, A., Persson, J. A., & Ölvander, J. (2023). A Design Automation Framework Supporting Design for Additive Manufacturing. In International Design Engineering Technical Conferences and Computers and Information in Engineering Conference (Vol. 87295, p. V002T02A083). American Society of Mechanical Engineers. DOI: 10.1115/DETC2023-116415

Zohdi, N., & Yang, R. (2021). Material anisotropy in additively manufactured polymers and polymer composites: A review. *Polymers*, *13*(19), 3368. DOI: 10.3390/polym13193368 PMID: 34641184

Chapter 5
Modelling, Simulation, and Design for 3D Printing: A CAD/CAM/CAE Approach

Romdhane Ben Khalifa
https://orcid.org/0000-0002-9171-9584
National Engineering School of Gabes (ENIG), Tunisia

Ghalia Nhassa
National Engineering School of Gabes (ENIG), Tunisia

Ali Snoussi
National Engineering School of Gabes (ENIG), Tunisia

ABSTRACT

The chapter explores the transformative potential of integrating Computer-Aided Design (CAD), Computer-Aided Manufacturing (CAM), and Computer-Aided Engineering (CAE) in 3D printing and additive manufacturing. This amalgamation of digital technologies facilitates seamless design simulation and manufacturing integration, thereby revolutionizing product development across biomedical, aerospace, automotive, and manufacturing sectors. The chapter elucidates how modeling functions as a critical element in 3D printing, focusing on the importance of the CAD model's geometric modeling techniques and design conversion to a printable format. Using simulation techniques such as finite element analysis and computational fluid dynamics is crucial for optimizing part performance, preventing defects, and reducing material usage. The integration of machine learning serves as a solution to enhance process control and ensure quality.

DOI: 10.4018/979-8-3373-0533-2.ch005

INTRODUCTION

The technology of additive manufacturing, or 3D printing, has transformed multiple industries by enabling the creation of complex designs and customized products with simplified procedures (Wu et al., 2017).

The technology stands apart from conventional subtractive manufacturing processes because it constructs objects through sequential layer addition, which provides designers with extensive design flexibility and material options (Spencer, 2018). The growing requirement for complex components has driven manufacturers to adopt new methods, and additive manufacturing has proven to be a promising solution for creating multi-functional and multi-material designs (Jin et al., 2020).

3D printing demonstrates strong potential for materials, mechanical, and biomedical engineering through its precise material deposition capabilities (Rojek et al., 2025). The ability to precisely position materials in three dimensions has established 3D printing as a revolutionary manufacturing technology (Garcia et al., 2013). The rapid prototyping features of 3D printing technology enable fast design and model production. In contrast, its multiple printing techniques and materials support functional items and scale models, making it suitable for scientific and industrial applications (Królikowski et al., 2019).

The customization capabilities of 3D printing prove essential in healthcare, aerospace, and automotive industries because they enable the creation of customized solutions and optimized designs. The aerospace and defense sectors started using 3D printing for quick prototyping and design verification. This technology creates complex designs at reduced weights, enabling it to make lightweight components such as hinges, brackets, and air ducts (Pathak et al., 2023). The development of new materials and manufacturing techniques enables the 3D printing of essential turbine blades and engine parts, leading to better performance and fuel efficiency, shorter production times, and reduced costs and material waste for sustainable operations.

Over the last ten years, 3D printing has transformed our manufacturing processes through the simplified creation of unusual and complex shapes. The aerospace industry and healthcare practices witnessed significant growth in market demand for complex, detailed designs. Researchers currently explore various product parameters, including temperature, layer thickness, speed, and material properties, to define product strength, appearance, and price points. Researchers combine physical testing with digital modeling techniques to discover optimal printer parameters for various materials and printing equipment (Natarajan et al., 2024). Researchers use machine learning algorithms to explore automatic printing operations while simultaneously minimizing waste output (Olsson et al., 2019). When scientists modify support scaffolds and infill patterns, they discover unexpected results as variables develop connections that create intellectual puzzles (Cristea et al., 2019). Maintaining production efficiency while enhancing product quality remains a delicate challenge, according to Du et al. (. The development is substantial, yet particular regions continue to demand further exploration. Most studies examine one experimental factor while neglecting multiple-factor combinations (Oberoi et al., 2018). The behavior of polymers and metals during 3D printing remains a complex problem that researchers have not yet fully solved. Research into product toughness modifications caused by printing behaviors represents an attractive field for upcoming scientific investigation (Morettini et al., 2022). People usually study individual machines or printing methods, yet fail to analyze how different techniques might work together to advance production outcomes (Ahmed et al., 2020). The optimization of print settings creates benefits for both creating attractive prints and sustainable, eco-friendly production at reduced costs. The world's growing environmental concerns make it essential to establish precise parameters that minimize waste production while maximizing operational efficiency (Rojek et al., 2025). The research is essential for contemporary makers beyond theoretical concepts (Mladenovska et al., 2021). The review explores widely used methods while discussing present findings and highlighting significant knowledge gaps in the field. The paper examines present methods alongside fresh knowledge and proposes a plan to close knowledge gaps to stimulate innovative additive manufacturing (Zhou et al., 2024).

KEY 3D PRINTING PARAMETERS AND THEIR INFLUENCE

The quality and functionality of 3D-printed objects are highly dependent on a complex interplay of printing parameters, including layer thickness, printing speed, infill density, nozzle temperature, and build orientation (Yin et al., 2022). The resolution and surface finish of the printed part are significantly influenced by layer thickness, with thinner layers resulting in smoother surfaces but requiring longer printing times (Tian et al., 2021). The rate of material deposition and the overall build time are affected by printing speed. However, excessive speeds can compromise print quality due to inadequate layer adhesion and increased material deformation (Zhou et al., 2024). The printed object's internal structure and mechanical properties are determined by infill density, which offers a trade-off between weight, strength, and material consumption (Maurya et al., 2021). The nozzle temperature is crucial for controlling the viscosity and flow behavior of the printing material, ensuring proper melting and bonding of layers. At the same time, the build orientation affects support structure requirements, surface finish, and mechanical strength, necessitating careful consideration to minimize material usage and maximize structural integrity (Kishore et al., 2021) (Chua & Leong, 2013; Wang et al., 2023).

Figure 1. Process of optimization of 3d printing parameters (Ranjan & Saha, 2025)

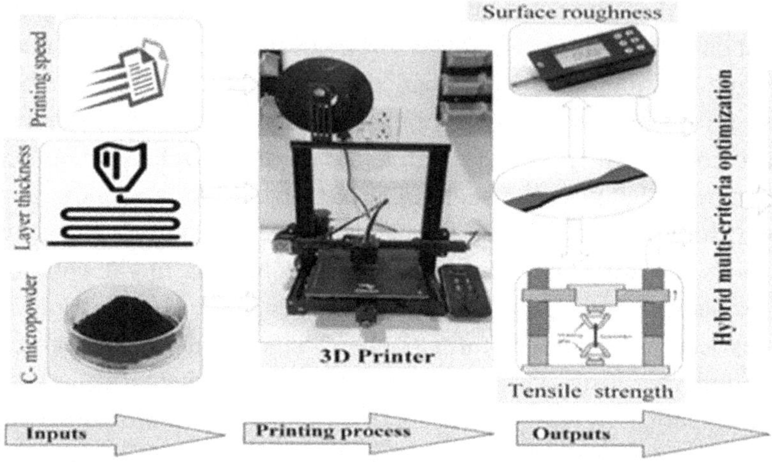

Material Properties and Selection

The selection of materials is a fundamental element in 3D printing because the properties of selected materials must match the requirements of the final printed object (Zhou et al., 2024). The mechanical strength, thermal stability, chemical resistance, and biocompatibility of polymers, metals, ceramics, and composites differ (Mladenovska et al., 2021). The optimization of printing parameters needs to consider material-specific characteristics because temperature, pressure, and printing speed variations produce substantial effects on the final product properties. The production of food products demands precise attention to material attributes and process variables, as well as post-treatment protocols, to reach exact printing outcomes (Liu et al., 2017). The additive manufacturing process produces parts with strength properties that differ from the original material, so designers must evaluate anisotropic material properties when creating structural designs. Researchers continue to develop new materials for 3D printing through research that aims to improve mechanical performance, decrease material costs, and increase printable material diversity (Ahmed et al., 2020). The printing process becomes complex because of multiple factors, including dispensing orifice dimension, solvent and fluid viscosity, surface tension, fluid-surface interactions, polymer concentration, humidity, and temperature (Chang et al., 2011). All printing-related aspects require precise control to generate parts with accurate and precise results. Figure 2 presents a cause-and-effect (fishbone) diagram that categorizes the key process parameters affecting the outcomes of Fused Filament Fabrication (FFF) — a widely used 3D printing technique.

Process Parameters

Build time can vary a lot (Almy & Tontowi, 2018). Going too fast can mess up the print quality because the layers will not stick together well, and the material can change shape. The two most significant factors that affect print time are layer height and print speed. Other important things include how much fill you use, how many walls there are, and the layer height, since these affect how much material you use. The surface's roughness and the accuracy of the dimensions mainly depend on layer height (Mendřický & Fris, 2020). It is also key to consider how process settings impact cell accuracy, with printing speed being the most significant influence on accuracy in the printing plane. For height accuracy, the thickness of the layer and what is known as 'bridging' are critical (Filippidis et al., 2020). The quality and performance of parts produced by ME are highly dependent on the build parameters used. These parameters can be grouped into temperature-related parameters, infill-related parameters, and speed-related parameters (Loh et al., 2020). The extrusion temperature and speed determine the morphological characteristics of printed parts, such as the degree of crystallinity and the coalescence between deposited lines (Cabreira & Santana, 2020). The infill percentage affects material consumption, printing time, and mechanical properties, which enables the production of lightweight yet structurally sound parts (Buj-Corral et al., 2019). The selection of process parameters is essential for the success of printing, especially if the aim is to obtain components that possess load capacity (Morettini et al., 2022). The mechanical characteristics of the specimen decrease when the raster angle increases (Cristea et al., 2019).

The figure systematically illustrates how various factors influence print quality, mechanical properties, and process reliability.

Figure 2. Ishikawa diagram of FFF parameters (Alzyod, H., & Ficzere, P., 2023)

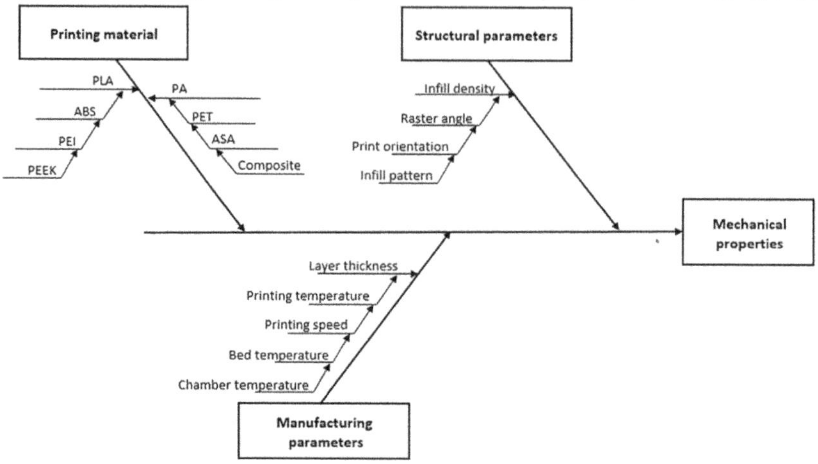

OPTIMIZATION TECHNIQUES FOR 3D PRINTING PARAMETERS

Several methods of optimizing your printing include experimental design, simulation, and feedback control. Experimental design techniques, such as Taguchi and response surface methodology, can systematically explore the parameter space and determine optimal parameters for a given material and printing process (Babagowda et al., 2018). Numerical simulation provides information on material behavior during printing and can be helpful in virtually optimizing parameters prior to physical printing, e.g., finite element analysis and computational fluid dynamics. Feedback control systems take into account real-time feedback input on temperature, pressure, and material flow, and such systems can change printing parameters continuously to stabilize quality and consider environmental conditions and material specifications. An increasing number of machine learning techniques using artificial neural networks are being used to study how the printing parameters can be used to control part properties, thus paving the way for machine learning techniques to be suggested for parameter-optimized printing quality prediction, quality checking, and printing method (Mahmood et al., 2020).

The optimization methods are particularly relevant where accuracy, print duration, and material usage must be carefully balanced, as in producing removable partial dentures using digital light processing (Hussein & Hussein, 2022). Furthermore, part orientation can be optimized to adjust printing parameters such as support structures, printing time, layering, shrinkage, and surface roughness. It shows the importance of optimizing printing processes to improve mechanical properties and print accuracy.

Different methodological approaches can be employed to optimize 3D printing parameters, each with different strengths and limitations regarding efficiency, accuracy, and flexibility.

Experimental design approaches, such as Taguchi methods, response surface methodology, and factorial designs, enable systematic parameter space exploration to identify optimal settings for specific materials and printing processes (Kam et al., 2020). Some of these methods involve experiments wherein printing parameters are varied per a predefined schedule, and printed part characteristics are measured and examined to determine optimal parameter settings (Ren et al., 2017). Numerical simulation techniques, such as finite element analysis and computational fluid dynamics, can predict material behavior during printing, allowing preprinting virtual optimization of parameters (Lee et al., 2019). Heat transfer, fluid flow, and stress distribution in the printed part can be simulated by these techniques and utilized to understand the effect of different printing parameters on part functionality and quality. Machine learning techniques can predict optimal printing parameters by learning from experimental data and iteratively optimizing the parameter values to achieve target part properties (Equival et al., 2024). Machine learning algorithms can be trained on a database of printing parameters and corresponding part properties. They can then predict the optimal parameter values for new materials or parts (Chari et al., 2018). Real-time monitoring and control systems can change printing parameters during the printing process to balance environmental or material characteristics variations. Such systems usually consist of sensors that quantify temperature, pressure, and flow of material and control algorithms that modify the printing parameters accordingly. Process parameter optimization is critical to realizing desired results in additive manufacturing, often including incremental tuning of laser power or feed rate for each new layer (Li et al., 2021). As a case in point, process parameter optimization, such as incremental tuning of laser power or feed rate for each new layer, is crucial to desired results in additive manufacturing (Kim et al., 2015).

Figure 1 typically represents a system architecture integrating artificial intelligence (AI) to enhance the precision, adaptability, and autonomy of 3D printing processes through real-time monitoring and feedback control.

Figure 3. Block diagram of AI-enabled closed-loop 3D printing (Adarsh, S.H. et al.2025)

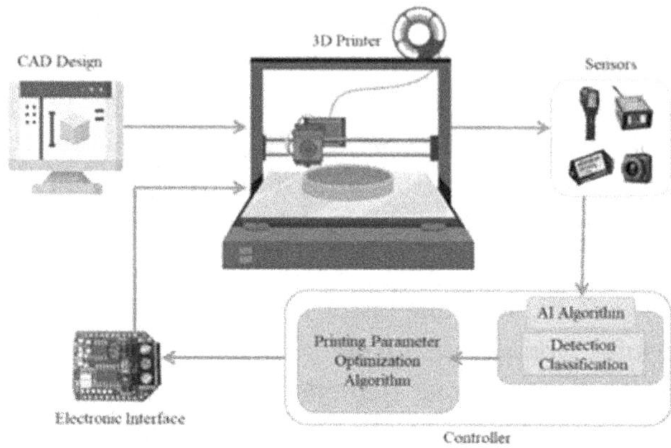

Statistical Design of Experiments (DOE)

The systematic approach of statistical experimental design optimizes 3D printing parameters by conducting experiments to determine how input parameters affect output responses, according to Wang et al. (2020). Response surface methodology evaluates how parameters affect dimensional accuracy (Almy & Tontowi, 2018). The analysis uses statistical models and regression analysis to identify important parameters and their best values (Fernández & Martín, 2023). Machine learning is a computer program for artificial intelligence to predict optimal process parameters that optimize production processes, according to Hooda et al. (2021). The process has been applied to model the printing process and forecast optimal parameters. Machine learning algorithms in manufacturing operations decrease the time required to configure machines and process parameters for new products. Process engineers perform this task through their experience with product similarity assessments, physical testing, and parameter correlation analysis of desired product quality outputs. The artificial intelligence system uses input characteristics to predict new product process parameters, which it receives as output data. Machine learning models acquire the ability to detect variations in machine parameters that result from different product categories. The system uses learned patterns to forecast new parameter values for upcoming products with specific characteristics (Khdoudi & Masrour, 2018). Physics-based modeling applies mathematical calculations and computational models to predict how processes and materials will act during the 3D printing operation so that parameters can be optimized virtually. These models apply fundamental physics principles such as heat transport, fluid dynamics, and solid mechanics to simulate complex interactions while printing. Multi-scale process modeling, improved in-situ monitoring, and post-fabrication processes are crucial in maintaining consistency for defect avoidance and quality assurance, especially in mission-critical applications (Panwisawas et al., 2020). Physics-based modeling can help optimize the printing parameters by predicting the effect of different parameter values on part performance and quality. For example, finite element analysis can simulate the thermal stresses and deformations during printing, allowing engineers to identify parameter settings that minimize these effects.

Artificial Intelligence and Machine Learning (AI/ML)

3D printing parameter optimization will experience a revolutionary transformation through emerging technologies of machine learning and artificial intelligence, which regulate and optimize print operations. These technologies utilize algorithms and data analysis techniques to extract knowledge from experimental results, leading to continuous parameter setting adjustments to achieve desired part characteristics (Omairi & Ismail, 2021). New additive manufacturing algorithms are being developed, and future uses incorporating state-of-the-art artificial intelligence methods are being explored (Jin et al., 2020). Machine learning integration with 3D printing enables better process control and optimization (Goh et al., 2020). The role of machine learning in additive manufacturing will become increasingly vital because it enables predictive maintenance while improving material characterization and automating process control (Ghilan et al., 2020). The most significant potential of AI and machine learning in 3D printing exists in creating predictive models that use given parameters to forecast printing outcomes.

Machine learning models acquire complex input-output relationships through training with extensive data collections of printing parameters and their associated part characteristics (Rojek et al., 2025). Machine learning in additive manufacturing allows real-time compensation analysis of extensive datasets to produce enhanced results while creating necessary data environments for effective learning and optimization (Anantrasirichai & Bull, 2021). Unsupervised machine learning enables automatic material phase detection through data observation, providing superior material characterization methods, including topological superconductors (Dwivedi et al., 2019). AI techniques for real-time error correction in additive manufacturing eliminate prolonged processing times and real-time loss of control, resulting in unintended goals (Omairi & Ismail, 2021). AI software monitors process parameters and sensor readings online, enabling it to detect printing path deviations and modify parameters for mistake correction. The system architecture or flowchart in Figure 3 shows how fuzzy logic and artificial neural networks (ANNs) are integrated to optimize the key parameters in Fused Deposition Modeling (FDM). This hybrid model aims to combine the interpretability of fuzzy systems with the learning capacity of neural networks.

Figure 4. Optimizing FDM 3D printing parameters using fuzzy neural network (Wei et al., 2024)

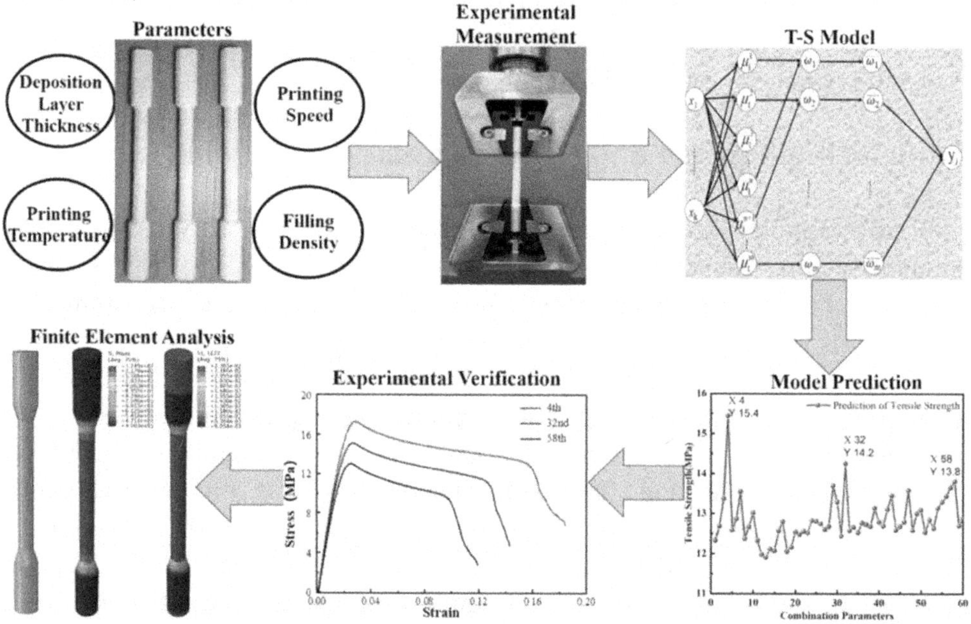

AI is vital in automation because it performs real-time data analysis and control functions, maximizing manufacturing process optimization. The real-time monitoring capabilities of AI systems provide better performance than human operators because they can handle processes automatically (Dwivedi et al., 2019). Laser processing operations become more precise and efficient when artificial intelligence streamlines them for applications in the automotive, aerospace, and biomedical industries (Murzin, 2024). Machine learning techniques added to metal-based additive manufacturing optimization methods will decrease human intuition requirements and the effort needed (Silbernagel et al., 2019). Modern machine-learning techniques enable algorithms to develop experience-based learning abilities, leading specialists to obtain answers that improve quality (Grünberger, 2020). Implementing machine learning systems in 3D printing operations enables businesses to automate their processes while decreasing human involvement to achieve higher printing efficiency and dependability (Plathottam et al., 2023). Through sensor data analysis, AI optimizes equipment maintenance for laser machining by identifying failure precursors, which reduces downtime and maintenance expenses (Murzin, 2024). AI systems display various advantages, yet researchers must detect biases that develop during training since these biases affect decision-making processes (Anantrasirichai & Bull, 2021). AI modeling depends heavily on data quality because this factor ensures both reliable and precise results (Cascón et al., 2024).

Figure 3 is probably a schematic or infographic showing how machine learning is applied to different stages and domains of additive manufacturing. However, I cannot access the original figure, based on typical content in similar scholarly works.

Figure 5. Emerging applications of ML in 3D printing (Rojek, I. et al., 2025)

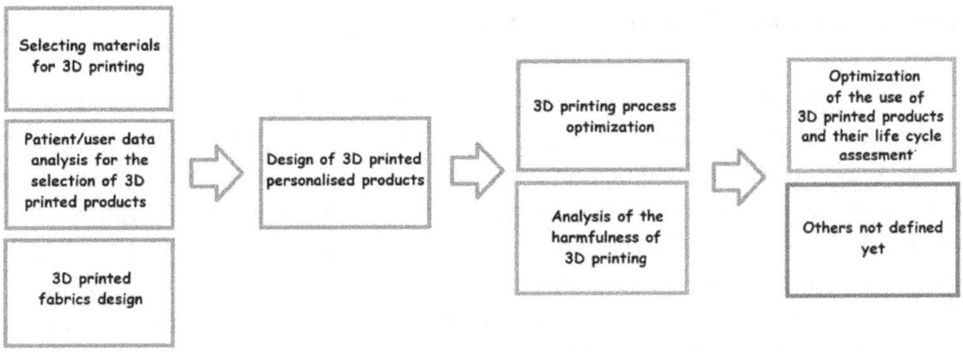

Simulation-Based Optimization

Simulation-based optimization uses computer simulations to model 3D printing processes to predict how parameter configurations affect component quality and performance (Murzin, 2024). The simulations optimize printing parameters by identifying configurations that reduce defects, maximize strength, and realize other target objectives. Simulation methods are increasingly used in manufacturing industries to evaluate and optimize performance metrics of advanced manufacturing systems. The simulation of additive manufacturing processes accurately predicts temperature distribution, residual stress, and deformation of the part being manufactured (Shin et al., 2020). We can observe the effects of modifications or interventions on finished products through simulation, which helps optimize the manufacturing process (D'Emilia et al., 2020). The physics-based models in additive manufacturing use heat transfer, fluid flow, and structural mechanics to predict the final characteristics of the part (Kumar et al., 2022). The 3D printing process requires numerical models, including finite element analysis, computational fluid dynamics, and discrete element modeling, to simulate its various components. The simulation models analyze material movement, thermal heat distribution, and stress development (Druzgalski et al., 2020). Simulation optimization is vital for material property prediction, defect minimization, design, and process parameter enhancement. The determination of end-product quality and performance depends heavily on these factors. The method enables a thorough investigation of design possibilities while making it easier to detect important factors that affect product quality. Implementing multi-physics simulation in production aims to improve product quality and process efficiency while decreasing time requirements and expenses. A complete understanding of the process becomes possible through simulation, which enables continuous improvement by predicting virtual outcomes (Zhang et al., 2024).

Simulation and Modeling Techniques

Simulation and modeling methods play a key role in 3D printing parameter optimization, where manufacturers can predict and improve the printing process before implementation. These methods leverage computational platforms to simulate the printing process, thus making it possible to find the best parametric settings without the need for time-consuming trial-and-error experiments (Alharbi et al., 2020). Finite element analysis is a valuable tool for predicting the mechanical behavior of printed parts under different loading conditions. Computational fluid dynamics simulations can model the material flow through the printer nozzle and predict the resulting layer formation.

Topology optimization serves to create designs that maximize stiffness while, at the same time, reducing weight (Scapin & Peroni, 2021). Models like these may include material behavior, printing conditions, and ambient conditions to project the result of the printing operation.

Predictive modeling is the development of mathematical models forecasting the 3D printing process's behavior under different conditions. The models employ machine learning algorithms to process historical data and determine patterns to forecast future outcomes (Balani et al., 2018). The model incorporates input variables like layer thickness, print speed, and material properties to predict the ensuing part quality and mechanical properties. With continuous monitoring and optimization of print parameters, real-time optimization can translate to substantial improvement in the quality and consistency of the parts produced. Finite Element Analysis (FEA) is a computational method applied to simulate the behavior of a physical system when subjected to different conditions. FEA is utilized to forecast stress patterns, thermal behavior, and deformation of 3D printed parts. FEA recognizes potential failure points and optimizes printing and design parameters, thereby improving structural integrity (Ficzere et al., 2021).

Computational Fluid Dynamics (CFD) simulates material flow in 3D printing. CFD simulations can be utilized to optimize printing parameters and nozzle design, providing even material distribution and avoiding defects (Du et al., 2019).

In-Process Monitoring and Feedback Control

In-process monitoring combined with feedback control methods is used to supervise the 3D printing process, making adjustments to the printing parameters where necessary. This ensures the printing process is maintained within specified parameters, guaranteeing that the final component conforms to the specifications. The real-time monitoring provision is achieved by incorporating sensors within the 3D printing machine to take measurements of temperature, pressure, and layer thickness. Feedback control systems interpret sensor information and automatically regulate print parameters like laser power, print speed, and material feed rate to preserve ideal operating conditions.

Enabled by in-process monitoring, this method is a powerful tool for detecting anomalies, reducing defects, and ensuring consistent quality in the parts produced. Real-time feedback loops allow instant adjustment while printing, ensuring accurate control of material deposition and minimizing the chance of error or inconsistency. In-situ monitoring systems with advanced capabilities offer extensive observation of the additive manufacturing process, allowing real-time adjustment and improved process control. This level of regulation significantly enhances the quality of the parts produced and optimizes resource usage efficiency by minimizing waste and reducing the need for subsequent processing. By observing material properties and printing conditions in real-time, manufacturers can institute immediate adjustments to prevent defects and ensure consistent quality. In-situ monitoring is crucial in process control improvement, material loss minimization, and additive manufacturing process optimization. It provides excellent visibility into the additive manufacturing process, allowing real-time anomaly detection and correction to ensure optimum print conditions. Manufacturers can achieve greater precision and efficiency by continuously monitoring critical parameters and providing feedback for real-time correction.

Gathering information about manufacturing operations allows for real-time quality control and creates a platform for analyzing relationships between quality characteristics and assets.

This approach overlaps process temperature data to determine anomalies (Bernhard et al., 2020). Machine learning software can be trained to recognize patterns and anomalies in the data, enabling the prediction and prevention of defects. This is a crucial aspect of the real-time quality control system. Integrating in-process monitoring and control systems improves printed parts' quality, minimizes waste, and optimizes resource usage (Ompusunggu et al., 2020; Wolter et al., 2011). Industries that require assured component quality and reproducible processes, such as aerospace and healthcare sectors, stand to gain significantly from adopting in-situ process monitoring and closed-loop control systems (Scime & Beuth, 2017). Advances in sensor technology and the Internet of Things have facilitated real-time system monitoring, which is necessary to maintain the integrity of the manufacturing environment and prevent unplanned repairs and operational disruptions (Sundaram & Zeid, 2023; Vernezi et al., 2021). Smart manufacturing implements real-time production process monitoring, from material selection to the final product, via Internet of Things technology (Sari et al., 2020). Random failures may result in production interruption and produce waste of raw materials or system malfunctioning (Yuan et al., 2019). Therefore, the ability to monitor in real-time is essential to ensuring safety in the realm of smart manufacturing (Ding et al., 2020).

Material-Specific Optimization Strategies

Material-specific optimization strategies are critical in 3D printing to attain the optimum level of results. Polymers, metals, ceramics, and composites have distinct properties that influence their performance in the printing process (Budzik et al., 2021). Optimization strategies guarantee that the printing process is closely adjusted to get the most out of the material, leading to improved part quality and function. Knowing the thermal behavior of the material is important in managing heat distribution while printing to avert warping or cracking. The mechanical behavior of the printed part can be significantly affected by changing printing settings, such as layer height, printing speed, and infill density. Controlling the printing environment temperature and the cooling rate for polymers is essential in preventing warping and facilitating adequate layer adhesion (Abdulhameed et al., 2019). Also, the infill's configuration and density greatly influence the final component's mechanical strength and weight, providing a flexible means of satisfying individual performance needs. In the case of metals, optimizing laser energy and scanning speed is crucial to attaining the desired density and microstructural characteristics (Yuan et al., 2020). The mechanical properties of metal components can also be improved through post-processing procedures like heat treatment. The support structures aim to give enough support while printing and minimizing the material use and post-processing effort. In the case of ceramics, it is essential to manage the slurry characteristics and binder content to obtain high density and dimensional accuracy. Printing with composite materials involves optimizing the print parameters by carefully balancing the matrix and reinforcement phase characteristics.

It is vital to tune printing parameters to specific material properties to attain the desired mechanical and thermal behavior for the final product (Picard et al., 2020). Acrylonitrile Butadiene Styrene is one of the most common polymers used for Fused Deposition Modeling (Zárybnická et al., 2019). Printing with ABS needs cautious temperature control to evade warping and facilitate layer adhesion (Dul et al., 2021).

CASE STUDIES AND INDUSTRIAL APPLICATIONS

3D printing parameter optimization is essential across multiple industries because case studies demonstrate how it enhances quality, productivity, and financial performance. Aerospace engineering benefits from optimized print parameters for titanium and nickel-based superalloys to create lightweight aircraft and engine assembly components with enhanced corrosion resistance (Tetsuka & Shin, 2020). The optimization of 3D printing parameters for biocompatible matrices leads to customized medical implants and prosthetics that demonstrate better functionality and biocompatibility. The automobile industry depends on optimized 3D printing parameters for prototyping tooling and component tailoring to produce lightweight components with shorter production times and greater design flexibility (Jankovics & Barari, 2019). The three-dimensional printing technology now operates across multiple sectors, including aircraft manufacturing, geoscience, education, fashion, medical, and pharmaceutical industries (Fan et al., 2020). The extensive utilization of 3D printing across different industries demonstrates its transformative power to transform production methods (Trifol et al., 2021). The technology of 3D printing provides cost-effective solutions and facilitates better collaboration while making design and production more accessible to everyone (Nadagouda et al., 2020; Trenfield et al., 2019). Research and science have experienced breakthroughs through the revolutionary production techniques enabled by 3D printing technology (Coakley & Hurt, 2016). The manufacturing process generates customized structures with special arrangements focusing on individual anatomical needs (Mathews et al., 2020).

Figure 6 likely presents a structured overview or classification chart highlighting the broad and diverse real-world applications of 3D printing across various sectors. Such figures often take the form of a radial chart, pyramid, or tree diagram to show categories and subcategories of use cases.

Figure 6. Applications of 3D Printing (Deepak Kumar et al., 2022)

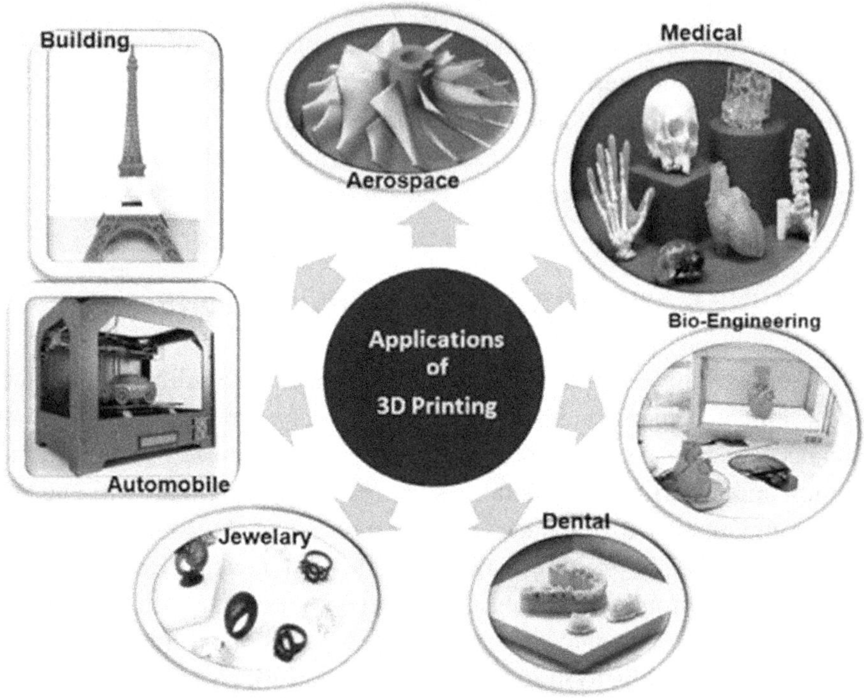

CHALLENGES AND FUTURE DIRECTIONS

Notwithstanding the great strides that have been made in optimizing 3D printing parameters, there are still a few challenges that remain, thwarting the realization of the full potential of this technology. Among them is the narrow range of materials that we have available at our disposal, especially when we want to move beyond classical polymers (Oberoi et al., 2018). The internal complexities of such materials during printing make it challenging to predict suitable printing conditions for novel substances. More research is needed to develop better models that more effectively predict material behavior over a range of printing parameters.

The pharmaceutical industry will benefit significantly from how 3D printing is used in personalized medicine, effectively addressing the unique needs of heterogeneous populations by creating customized products with ease and excellent quality (Dhyani et al., 2020) (Wang et al., 2023). The growing need for innovative assembly and production technologies opens a new era for unmatched accuracy, improved product quality, high-speed production rates, and lower costs, all in efforts to maximize efficiency and independence in production, eventually leading to shorter times to market (Spencer, 2018). By successfully combining 3D printing with other Industry 4.0 technologies, production operations get considerable improvements, with consequent economic advantages like lower warehouse and repair expenses, improved quality due to better personalization, and increased employee safety (Molino et al., 2020). Moreover, integrating 3D printing technology into the food supply chain makes food processing personalized to clients' precise requirements, optimizing operations where possible (Zhang et al., 2024). Besides, 3D printing technology provides quick prototyping, low costs, and the ability to create complicated geometries (Ali et al., 2020; Rojek et al., 2025; Tang, 2018). This innovative method has become a revolutionary material, mechanical, and biomedical engineering frontier. There is no question that 3D printing is best suited for producing low-cost customized goods.

While its development has been remarkable, 3D printing is still hampered by limitations like low speeds, a narrow range of materials available, and the need for considerable post-processing to achieve standards within industries (Zhou et al., 2024). These limitations will have to be addressed for broad application across industries. The long-term vision for 3D printing is to create complicated and fully functional parts with unmatched levels of personalization while unlocking opportunities for SMEs to participate in the manufacturing industry (Garcia et al., 2013). Machine learning and artificial intelligence integration can improve real-time printing parameters and part quality without generating waste (Królikowski et al., 2019). As production relies on automation more and more, future factories will most likely use a lot of AI tech (Dwivedi et al., 2019).

With the broader acceptance of AI technologies, its application in the construction industry is expected to expand (Na et al., 2022). AI can potentially revolutionize operations and processes in various industries (Dwivedi et al., 2019). The construction industry has been behind other industries in adopting new technologies (Na et al., 2022). In addition, computer-based systems have the potential to understand human biofield technology, monitor health parameters, and provide remote diagnostic services, eventually making healthcare service delivery better (Dwivedi et al., 2019).

CONCLUSION

Parameter optimization in 3D printing requires a comprehensive approach that involves material science, process management, and advanced monitoring. The ability of researchers and engineers to effectively control these parameters is vital in realizing the full potential of 3D printing for high-value-added personalized parts that are useful in diverse sectors. Technology-based construction will continue to evolve as research and development move forward to tackle existing problems and expand its application scope. Addressing these challenges with a simultaneous opening of new opportunities will lead to widespread adoption of 3D printing across various industries, making production more efficient, sustainable, and innovative (Jin et al., 2020; Mahmood et al., 2020). New materials and processes with unique design versatility will enhance the design freedom when creating parts with the best functional performance; such materials are particularly suitable for additive manufacturing. Moreover, it would provide real-time control over complex production systems through combined intelligence ANS (Anantrasirichai & Bull, 2021). However, construction offers huge potential areas where 3D printing can be used more extensively due to increased automation within processes, greater efficiency at reduced waste rates, and improved geometrical accuracy (Filip et al., 2023; Tahmasebinia et al.,2018). In the future, his application shall revolutionize reduced costs and time savings while reducing environmental pollution, onsite accidents, and injuries. Research and development will continue playing a vital role in addressing existing constraints and enlarging applications of three-dimensional printing under innovative frameworks. Furthermore, machine learning would contribute significantly by streamlining processes for better efficiency and reliability in three-dimensional printing.

REFERENCES

Abdulhameed, O., Al-Ahmari, A., Ameen, W., & Mian, S. H. (2019). Additive manufacturing: Challenges, trends, and applications. *Advances in Mechanical Engineering, 11*(2), 1687814018822880. Advance online publication. DOI: 10.1177/1687814018822880

Adarsh, S. H., & Nagamadhu, M. (2025). Effect of Printing Parameters on Mechanical Properties and Warpage of 3D-Printed PEEK/CF-PEEK Composites Using Multi-Objective Optimization Technique. *Journal of Composites Science, 9*(5), 208. DOI: 10.3390/jcs9050208

Ahmed, W., Alnajjar, F., Zaneldin, E., Al-Marzouqi, A. H., Gochoo, M., & Khalid, S. (2020). Implementing FDM 3D Printing Strategies Using Natural Fibers to Produce Biomass Composite [Review of Implementing FDM 3D Printing Strategies Using Natural Fibers to Produce Biomass Composite]. Materials, 13(18), 4065. Multidisciplinary Digital Publishing Institute. DOI: 10.3390/ma13184065

Alharbi, M., Kong, I., & Patel, V. I. (2020). Simulation of uniaxial stress–strain response of 3D-printed polylactic acid by nonlinear finite element analysis. *Applied Adhesion Science, 8*(1), 5. Advance online publication. DOI: 10.1186/s40563-020-00128-1

Ali, A., Ahmad, U., & Akhtar, J. (2020). 3D Printing in Pharmaceutical Sector: An Overview. In *IntechOpen eBooks*. IntechOpen., DOI: 10.5772/intechopen.90738

Almy, R. D., & Tontowi, A. E. (2018). The effect of 3d printing machine parameters in the extrusion process of biocomposite materials (PMMA and HA) on dimensional accuracy. *Sinergi, 22*(2), 83. DOI: 10.22441/sinergi.2018.2.003

Alzyod, H., & Ficzere, P. (2023). Material-Dependent Effect of Common Printing Parameters on Residual Stress and Warpage Deformation in 3D Printing: A Comprehensive Finite Element Analysis Study. *Polymers, 15*(13), 2893. DOI: 10.3390/polym15132893 PMID: 37447538

Anantrasirichai, N., & Bull, D. (2021). Artificial intelligence in the creative industries: a review [Review of Artificial intelligence in the creative industries: a review]. Artificial Intelligence Review, 55(1), 589. Springer Science+Business Media. DOI: 10.1007/s10462-021-10039-7

Babagowda, M., Kadadevara Math, R. S., Goutham, R., & Srinivas Prasad, K. R. (2018). Study of Effects on Mechanical Properties of PLA Filament, which is blended with Recycled PLA Materials. *IOP Conference Series. Materials Science and Engineering, 310*, 12103. DOI: 10.1088/1757-899X/310/1/012103

Balani, S. B., Cantarel, A., Chabert, F., & Nassiet, V. (2018). Influence of parameters controlling the extrusion step in fused filament fabrication (FFF) process applied to polymers using numerical simulation. *AIP Conference Proceedings*, *1960*, 140003. DOI: 10.1063/1.5034995

Bernhard, R. J., Neef, P., Wiche, H., Hoff, C., Hermsdorf, J., Kaierle, S., & Wesling, V. (2020). Defect detection in additive manufacturing via a toolpath overlaid melt-pool-temperature tomography. *Journal of Laser Applications*, *32*(2), 022055. Advance online publication. DOI: 10.2351/7.0000055

Brüggemann, H., Stempin, S., & Meier, J.-M. (2020). Consideration of digitalization for the purpose of resource efficiency in a learning factory. *Procedia Manufacturing*, *45*, 140–145. DOI: 10.1016/j.promfg.2020.04.085

Budzik, G., Woźniak, J., Paszkiewicz, A., Przeszłowski, Ł., Dziubek, T., & Dębski, M. (2021). Methodology for the Quality Control Process of Additive Manufacturing Products Made of Polymer Materials. *Materials (Basel)*, *14*(9), 2202. DOI: 10.3390/ma14092202 PMID: 33923048

Buj-Corral, I., Domínguez-Fernández, A., & Durán-Llucià, R. (2019). Influence of Print Orientation on Surface Roughness in Fused Deposition Modeling (FDM) Processes. *Materials (Basel)*, *12*(23), 3834. DOI: 10.3390/ma12233834 PMID: 31766409

Cabreira, V., & Santana, R. M. C. (2020). Effect of infill pattern in Fused Filament Fabrication (FFF) 3D Printing on materials performance. *Matéria (Rio de Janeiro)*, *25*(3), e-12826. Advance online publication. DOI: 10.1590/s1517-707620200003.1126

Cascón, I., Gómez-Omella, M., Fernández, D., Gil, A., Alberdi, N., & González, H. (2024). Towards Zero-Defect Manufacturing Based on Artificial Intelligence through the Correlation of Forces in 5-Axis Milling Process. *Machines (Basel)*, *12*(4), 226. DOI: 10.3390/machines12040226

Chang, C. C., Boland, E. D., Williams, S. K., & Hoying, J. B. (2011). Direct-write bioprinting three-dimensional biohybrid systems for future regenerative therapies [Review of Direct-write bioprinting three-dimensional biohybrid systems for future regenerative therapies]. Journal of Biomedical Materials Research Part B Applied Biomaterials, 1, 160. Wiley. DOI: 10.1002/jbm.b.31831

Chari, V. S., Venkatesh, P. R., Krupashankar, , & Dinesh, V. (1943). Krupashankar, & Dinesh, V. (2018). Effect of processing parameters on the FDM process. *AIP Conference Proceedings*, *20061*, 020061. Advance online publication. DOI: 10.1063/1.5029637

Chua, C. K., & Leong, K. F. (2013). 3D Printing and Additive Manufacturing. DOI: 10.1142/9008

Coakley, M. F., & Hurt, D. E. (2016). 3D Printing in the Laboratory: Maximize Time and Funds with Customized and Open-Source Labware. *SLAS Technology*, *21*(4), 489–495. DOI: 10.1177/2211068216649578 PMID: 27197798

Cristea, D., Pop, M. A., Faraian, C., & Munteanu, D. (2019). The influence of additive manufacturing parameters on the structural and mechanical properties of acrylonitrile butadiene styrene (ABS) parts produced by fused filament fabrication. *IOP Conference Series. Materials Science and Engineering*, *682*(1), 12013. DOI: 10.1088/1757-899X/682/1/012013

D'Emilia, G., Ilio, A. D., Gaspari, A., Natale, E., & Stamopoulos, A. G. (2020). Uncertainty assessment for measurement and simulation in selective laser melting: A case study of an aerospace part. *Acta IMEKO*, *9*(4), 96. DOI: 10.21014/acta_imeko.v9i4.720

da Silva, C. E. S., Salgado, E. G., Mello, C. H. P., Oliveira, E., & Leal, F. (2013). Integration of computer simulation in design for manufacturing and assembly. *International Journal of Production Research*, *52*(10), 2851–2866. DOI: 10.1080/00207543.2013.853887

Deepak Kumar, S., Arun Manohar, G., & Surya Teja, R. (2022). The state of art 3D printing: A case study of Ganesh Idol. *Materials Today: Proceedings*, *56*, 455–461. Advance online publication. DOI: 10.1016/j.matpr.2022.01.418

Dhyani, A., Singh, N., Kumar, V., & Dhyani, A. (2020). Applications of 3-Dimensional (3D) Printing in the Biomedical Field. *International Journal of Current Research and Review*, *12*(19), 71–75. DOI: 10.31782/IJCRR.2020.121918

Ding, H., Gao, R. X., Isaksson, A., Landers, R. G., Parisini, T., & Yuan, Y. (2020). State of AI-Based Monitoring in Smart Manufacturing and Introduction to Focused Section. *IEEE/ASME Transactions on Mechatronics*, *25*(5), 2143–2154. DOI: 10.1109/TMECH.2020.3022983

Druzgalski, C., Ashby, A. S., Guss, G., King, W. E., Roehling, T. T., & Matthews, M. J. (2020). Process optimization of complex geometries using feed-forward control for laser powder bed fusion additive manufacturing. *Additive Manufacturing*, *34*, 101169. DOI: 10.1016/j.addma.2020.101169

Du, Z., Liu, J., & He, Y. (2019). Intelligent Analysis Method of Assembly Process Based on Semantic Elements. *IOP Conference Series. Materials Science and Engineering*, *563*(4), 42055. DOI: 10.1088/1757-899X/563/4/042055

Duan, G., & Yan, X. (2020). A Real-Time Quality Control System Based on Manufacturing Process Data. *IEEE Access : Practical Innovations, Open Solutions, 8*, 208506–208517. DOI: 10.1109/ACCESS.2020.3038394

Dul, S., Fambri, L., & Pegoretti, A. (2021). High-Performance Polyamide/Carbon Fiber Composites for Fused Filament Fabrication: Mechanical and Functional Performances. *Journal of Materials Engineering and Performance, 30*(7), 5066–5085. DOI: 10.1007/s11665-021-05635-1

Dwivedi, Y. K., Hughes, L., Ismagilova, E., Aarts, G., Coombs, C., Crick, T., Duan, Y., Dwivedi, R., Edwards, J. S., Eirug, A., Galanos, V., Ilavarasan, P. V., Janssen, M., Jones, P., Kar, A. K., Kizgin, H., Kronemann, B., Lal, B., Lucini, B., & Williams, M. D. (2019). Artificial Intelligence (AI): Multidisciplinary perspectives on emerging challenges, opportunities, and agenda for research, practice, and policy. *International Journal of Information Management, 57*, 101994. DOI: 10.1016/j.ijinfomgt.2019.08.002

Equbal, M. A., Equbal, A., Khan, Z. A., & Badruddin, I. A. (2025). Machine learning in additive manufacturing: A comprehensive insight. *International Journal of Lightweight Materials and Manufacture, 8*(2), 264–284.

Fan, D., Li, Y., Wang, X., Zhu, T., Wang, Q., Cai, H., Li, W., Tian, Y., & Liu, Z. (2020). Progressive 3D Printing Technology and Its Application in Medical Materials [Review of Progressive 3D Printing Technology and Its Application in Medical Materials]. Frontiers in Pharmacology, 11. Frontiers Media. DOI: 10.3389/fphar.2020.00122

Fernández, F. M., & Martín, M. J. (2023). Analysis of the Effect of the Surface Inclination Angle on the Roughness of Polymeric Parts Obtained with Fused Filament Fabrication Technology. *Polymers, 15*(3), 585. DOI: 10.3390/polym15030585 PMID: 36771886

Ficzere, P., Lukács, N. L., & Borbás, L. (2021). The Investigation of Interlaminar Failures Caused by Production Parameters in Case of Additive Manufactured Polymers. *Polymers, 13*(4), 556. DOI: 10.3390/polym13040556 PMID: 33668572

Filip, G. A., Abanda, F. H., & Fru, F. A. (2023). Construction 3D-printing reduces the incidence of long latency respiratory diseases among construction workers in the UK. *Safety in Extreme Environments, 5*(3), 177–197. DOI: 10.1007/s42797-023-00078-4

Filippidis, P., Papazetis, G., & Vosniakos, G.-C. (2020). Process parameter investigation for 3D printing of cellular structured parts. *Procedia Manufacturing, 51*, 717–724. DOI: 10.1016/j.promfg.2020.10.101

Garcia, C. R., Rumpf, R. C., Tsang, H., & Barton, J. H. (2013). Effects of extreme surface roughness on 3D printed horn antenna. *Electronics Letters, 49*(12), 734–736. DOI: 10.1049/el.2013.1528

Garcia, F. L., Moris, V. A. da S., Nunes, A. O., & Silva, D. A. L. (2018). Environmental performance of additive manufacturing process – an overview. *Rapid Prototyping Journal, 24*(7), 1166–1177. DOI: 10.1108/RPJ-05-2017-0108

Gargalis, L., Madonna, V., Giangrande, P., Rocca, R., Ashcroft, I., Hague, R., & Galea, M. (2020).. . *Development and Testing of Soft Magnetic Rotor for a Switched Reluctance Motor Built Through Additive Manufacturing Technology., 263*, 263–268. Advance online publication. DOI: 10.23919/ICEMS50442.2020.9290893

Ghilan, A., Chiriac, A. P., Niţă, L. E., Rusu, A. G., Neamţu, I., & Chiriac, V. M. (2020). Trends in 3D Printing Processes for Biomedical Field: Opportunities and Challenges [Review of Trends in 3D Printing Processes for Biomedical Field: Opportunities and Challenges]. Journal of Polymers and the Environment, 28(5), 1345. Springer Science+Business Media. DOI: 10.1007/s10924-020-01722-x

Ginestra, P. S., Ceretti, E., & Fiorentino, A. (2016). Potential of modeling and simulations of bioengineered devices: Endoprostheses, prostheses and orthoses [Review of Potential of modeling and simulations of bioengineered devices: Endoprostheses, prostheses and orthoses]. Proceedings of the Institution of Mechanical Engineers Part H Journal of Engineering in Medicine, 230(7), 607. SAGE Publishing. DOI: 10.1177/0954411916643343

Goh, G. D., Sing, S. L., & Yeong, W. Y. (2020). A review on machine learning in 3D printing: applications, potential, and challenges [Review of A review on machine learning in 3D printing: applications, potential, and challenges]. Artificial Intelligence Review, 54(1), 63. Springer Science+Business Media. DOI: 10.1007/s10462-020-09876-9

Grünberger, T. (2020). Machine Learning for Process Monitoring Systems. *PhotonicsViews, 17*(3), 56–59. DOI: 10.1002/phvs.202000026

Hooda, N., Chohan, J. S., Gupta, R., & Kumar, R. (2021). Deposition angle prediction of Fused Deposition Modeling process using ensemble machine learning. *ISA Transactions, 116*, 121–128. DOI: 10.1016/j.isatra.2021.01.035 PMID: 33518221

Huang, Z., Shi, T., & Yue, X. (2019). The Optimization of FDM 3D Printer's Structure Based on Finite Element Analysis. MATEC Web of Conferences, 257, 2004. DOI: 10.1051/matecconf/201925702004

Hussein, M. O., & Hussein, L. A. (2022). Optimization of Digital Light Processing Three-Dimensional Printing of the Removable Partial Denture Frameworks; The Role of Build Angle and Support Structure Diameter. *Materials (Basel), 15*(6), 2316. DOI: 10.3390/ma15062316 PMID: 35329767

Jankovics, D., & Barari, A. (2019). Customization of Automotive Structural Components using Additive Manufacturing and Topology Optimization. *IFAC-PapersOnLine, 52*(10), 212–217. DOI: 10.1016/j.ifacol.2019.10.066

Jin, Z., Zhang, Z., Demir, K., & Gu, G. X. (2020). Machine Learning for Advanced Additive Manufacturing. *Matter, 3*(5), 1541–1556. DOI: 10.1016/j.matt.2020.08.023

Kam, M., İpekçi, A., & Şengül, Ö. (2020). Effect of FDM Process Parameters on the Mechanical Properties and Production Costs of 3D Printed PowerABS Samples. [IJAEFEA]. *International Journal of Analytical Experimental and Finite Element Analysis, 7*(3). Advance online publication. DOI: 10.26706/ijaefea.3.7.20200806

Khdoudi, A., & Masrour, T. (2018). Prediction of Industrial Process Parameters using Artificial Intelligence Algorithms. arXiv (Cornell University). https://doi .org//arxiv.1812.10537DOI: 10.48550

Kim, D. B., Denno, P., & Jones, A. (2015). A model-based approach to refining process parameters in smart manufacturing. *Concurrent Engineering, Research and Applications, 23*(4), 365–376. DOI: 10.1177/1063293X15591038

Kiran, A. R., Li, Y., Hodek, J., Brázda, M., Urbánek, M., & Džugan, J. (2022). Heat Source Modeling and Residual Stress Analysis for Metal Directed Energy Deposition Additive Manufacturing. *Materials (Basel), 15*(7), 2545. DOI: 10.3390/ma15072545 PMID: 35407876

Kishore, R. A., Sivaraj, S., Dhinakaran, V., Prabhagar, V. S. P. P., & Srinivas, S. (2021). Strength of 3D prints with variable print orientation. *Journal of Physics: Conference Series, 2027*(1), 12021. DOI: 10.1088/1742-6596/2027/1/012021

Królikowski, T., Knitter, R., & Stachnik, M. (2019). Thermo-mechanical tests using 3d printed elements. *Procedia Computer Science, 159*, 2551–2559. DOI: 10.1016/j.procs.2019.09.430

Kumar, S., Gopi, T., Harikeerthana, N., Gupta, M. K., Gaur, V., Królczyk, G., & Wu, C. (2022). Machine learning techniques in additive manufacturing: A state of the art review on design, processes, and production control. *Journal of Intelligent Manufacturing, 34*(1), 21–55. DOI: 10.1007/s10845-022-02029-5

Lee, J., Walker, J., Natarajan, S., & Yi, S. (2019). Prediction of geometric characteristics in polycaprolactone (PCL) scaffolds produced by extrusion-based additive manufacturing technique for tissue engineering. *Rapid Prototyping Journal, 26*(2), 238–248. DOI: 10.1108/RPJ-08-2018-0219

Li, G., Odum, K., Yau, C., Soshi, M., & Yamazaki, K. (2021). High productivity fluence-based control of Directed Energy Deposition (DED) part geometry. *Journal of Manufacturing Processes, 65*, 407–417. DOI: 10.1016/j.jmapro.2021.03.028

Liu, Z., Zhang, M., Bhandari, B., & Wang, Y. (2017). 3D printing: Printing precision and application in the food sector. *Trends in Food Science & Technology, 69*, 83–94. DOI: 10.1016/j.tifs.2017.08.018

Loh, G. H., Pei, E., González-Gutiérrez, J., & Monzón, M. (2020). An Overview of Material Extrusion Troubleshooting. *Applied Sciences (Basel, Switzerland), 10*(14), 4776. DOI: 10.3390/app10144776

Mahmood, M. A., Visan, A. I., Ristoscu, C., & Mihăilescu, I. N. (2020). Artificial Neural Network Algorithms for 3D Printing [Review of Artificial Neural Network Algorithms for 3D Printing]. Materials, 14(1), 163. Multidisciplinary Digital Publishing Institute. DOI: 10.3390/ma14010163

Mathews, D. A. P., Baird, A. D., & Lucky, M. (2020). Innovation in Urology: Three-Dimensional Printing and Its Clinical Application [Review of Innovation in Urology: Three-Dimensional Printing and Its Clinical Application]. Frontiers in Surgery, 7. Frontiers Media. DOI: 10.3389/fsurg.2020.00029

Maurya, N. K., Maurya, M., Saxena, A., Tyagi, A., Kiran, G. K., Singh, H., Tripathi, H., Kumar, P., & Yadav, S. (2021). Investigation of mechanical properties and form error of the components fabricated by rapid prototyping: A review [Review of Investigation of mechanical properties and form error of the components fabricated by rapid prototyping: A review]. Materials Today Proceedings, 47, 3901. Elsevier BV. DOI: 10.1016/j.matpr.2021.03.553

Mendřický, R., & Fris, D. (2020). Analysis of the Accuracy and the Surface Roughness of FDM/FFF Technology and Optimisation of Process Parameters. Tehnicki Vjesnik - Technical Gazette, 27(4). DOI: 10.17559/TV-20190320142210

Mladenovska, D., Lazarevska, A. M., & Krstanoski, M. (2021). Attributes relevant for sustainable additive manufacturing – material-driven approach. *IOP Conference Series. Materials Science and Engineering, 1190*(1), 12001. DOI: 10.1088/1757-899X/1190/1/012001

Molino, M., Cortese, C. G., & Ghislieri, C. (2020). The Promotion of Technology Acceptance and Work Engagement in Industry 4.0: From Personal Resources to Information and Training.

Morettini, G., Palmieri, M., Capponi, L., & Landi, L. (2022). Comprehensive characterization of the mechanical and physical properties of PLA structures printed by the FFF-3D printing process in different directions. *Progress in Additive Manufacturing*, *7*(5), 1111–1122. DOI: 10.1007/s40964-022-00285-8

Murzin, S. P. (2024). Artificial Intelligence-Driven Innovations in Laser Processing of Metallic Materials. *Metals*, *14*(12), 1458. DOI: 10.3390/met14121458

Na, S., Heo, S., Han, S., Shin, Y.-S., & Roh, Y.-S. (2022). Acceptance Model of Artificial Intelligence (AI)-Based Technologies in Construction Firms: Applying the Technology Acceptance Model (TAM) in Combination with the Technology–Organisation–Environment (TOE) Framework. *Buildings (Basel, Switzerland)*, *12*(2), 90. DOI: 10.3390/buildings12020090

Nadagouda, M. N., Ginn, M., & Rastogi, V. (2020). A review of 3D printing techniques for environmental applications [Review of A review of 3D printing techniques for environmental applications]. Current Opinion in Chemical Engineering, 28, 173. Elsevier BV. DOI: 10.1016/j.coche.2020.08.002

Natarajan, M., Pasupuleti, T., & Kiruthika, J. D. P. (2024). "Optimization and Regression Modeling of Additive Manufacturing (Fused Deposition Modeling) of PETG Material for Automobile Applications," SAE Technical Paper 2024-28-0234, DOI: 10.4271/2024-28-0234

Navarro, J., Din, M., Janes, M. E., Swayambunathan, J., Fisher, J. P., & Dreher, M. L. (2019). Effect of print orientation on microstructural features and mechanical properties of 3D porous structures printed with continuous digital light processing. *Rapid Prototyping Journal*, *25*(6), 1017–1029. DOI: 10.1108/RPJ-10-2018-0276

Oberoi, G., Nitsch, S., Edelmayer, M., Janjić, K., Müller, A. S., & Agis, H. (2018). 3D Printing—Encompassing the Facets of Dentistry [Review of 3D Printing—Encompassing the Facets of Dentistry]. Frontiers in Bioengineering and Biotechnology, 6. Frontiers Media. DOI: 10.3389/fbioe.2018.00172

Olsson, N. O. E., Shafqat, A., Arica, E., & Økland, A. (2019). "3D-Printing Technology in Construction: Results from a Survey", Lill, I. and Witt, E. (Ed.) 10th Nordic Conference on Construction Economics and Organization (Emerald Reach Proceedings Series, Vol. 2), Emerald Publishing Limited, Leeds, pp. 349-356. DOI: 10.1108/S2516-285320190000002044

Omairi, A., & Ismail, Z. H. (2021). Towards Machine Learning for Error Compensation in Additive Manufacturing. *Applied Sciences (Basel, Switzerland), 11*(5), 2375. DOI: 10.3390/app11052375

Ompusunggu, A. P., Vonderscher, Y., & Motl, D. (2020). Physics-based vibration feature for detecting eccentric workpiece/runout faults during continuous generating gear grinding processes. *Mechanical Systems and Signal Processing, 153*, 107536. DOI: 10.1016/j.ymssp.2020.107536

Panwisawas, C., Tang, Y. T., & Reed, R. C. (2020). Metal 3D printing as a disruptive technology for superalloys. *Nature Communications, 11*(1), 2327. Advance online publication. DOI: 10.1038/s41467-020-16188-7 PMID: 32393778

Pathak, K., Saikia, R., Das, A., Das, D., Islam, M. A., Pramanik, P., Parasar, A., Borthakur, P. P., Sarmah, P., Saikia, M., & Borthakur, B. (2023). 3D printing in biomedicine: Advancing personalized care through additive manufacturing. *Exploration of Medicine, 1135*, 1135–1167. Advance online publication. DOI: 10.37349/emed.2023.00200

Picard, M., Mohanty, A. K., & Misra, M. (2020). Recent advances in additive manufacturing of engineering thermoplastics: challenges and opportunities [Review of Recent advances in additive manufacturing of engineering thermoplastics: challenges and opportunities]. RSC Advances, 10(59), 36058. Royal Society of Chemistry. DOI: 10.1039/D0RA04857G

Plathottam, S. J., Rzonca, A., Lakhnori, R., & Iloeje, C. O. (2023). A review of artificial intelligence applications in manufacturing operations [Review of A review of artificial intelligence applications in manufacturing operations]. Journal of Advanced Manufacturing and Processing, 5(3). Wiley. DOI: 10.1002/amp2.10159

Ranjan, R., & Saha, A. (2025). Selection of best 3d printing parameters for fused deposition modeling part using a novel hybrid multi-criteria optimization technique. *Surface Review and Letters, 32*(03), 2450121. DOI: 10.1142/S0218625X2450121X

Ren, L., Zhou, X., Song, Z., Zhao, C., Liu, Q., Xue, J., & Li, X. (2017). Process Parameter Optimization of Extrusion-Based 3D Metal Printing Utilizing PW–LDPE–SA Binder System. *Materials (Basel), 10*(3), 305. DOI: 10.3390/ma10030305 PMID: 28772665

Rojek, I., Mikołajewski, D., Kempiński, M., Galas, K., & Piszcz, A. (2025). Emerging Applications of Machine Learning in 3D Printing. *Applied Sciences (Basel, Switzerland), 15*(4), 1781. DOI: 10.3390/app15041781

Sari, M. W., Herianto, H., Dharma, I. G. B. B., & Tontowi, A. E. (2020). Design of Product Monitoring System Using Internet of Things Technology for Smart Manufacturing. *IOP Conference Series. Materials Science and Engineering*, *835*(1), 12048. DOI: 10.1088/1757-899X/835/1/012048

Scapin, M., & Peroni, L. (2021). Numerical Simulations of Components Produced by Fused Deposition 3D Printing. *Materials (Basel)*, *14*(16), 4625. DOI: 10.3390/ma14164625 PMID: 34443146

Scime, L., & Beuth, J. (2017). Anomaly detection and classification in a laser powder bed additive manufacturing process using a trained computer vision algorithm. *Additive Manufacturing*, *19*, 114–126. DOI: 10.1016/j.addma.2017.11.009

Shin, D. S., Lee, C. H., Kühn, U., Lee, S., Park, S. J., Schwab, H., Scudino, S., & Kosiba, K. (2020). Optimizing laser powder bed fusion of Ti-5Al-5V-5Mo-3Cr by artificial intelligence. *Journal of Alloys and Compounds*, *862*, 158018. DOI: 10.1016/j.jallcom.2020.158018

Silbernagel, C., Aremu, A., & Ashcroft, I. (2019). Using machine learning to aid in the parameter optimisation process for metal-based additive manufacturing. *Rapid Prototyping Journal*, *26*(4), 625–637. DOI: 10.1108/RPJ-08-2019-0213

Spencer, O. O. (2018). Additive Manufacturing Technology Development: A Trajectory Towards Industrial Revolution. *American Journal of Mechanical and Industrial Engineering*, *3*(5), 80. DOI: 10.11648/j.ajmie.20180305.12

Sundaram, S., & Zeid, A. (2023). Artificial Intelligence-Based Smart Quality Inspection for Manufacturing. *Micromachines*, *14*(3), 570. DOI: 10.3390/mi14030570 PMID: 36984977

Tahmasebinia, F., Niemelä, M., Sepasgozar, S. M. E., Lai, T., Su, W., Reddy, K. R., Shirowzhan, S., Sepasgozar, S. M. E., & Marroquín, F. A. (2018). Three-Dimensional Printing Using Recycled High-Density Polyethylene: Technological Challenges and Future Directions for Construction. *Buildings (Basel, Switzerland)*, *8*(11), 165. DOI: 10.3390/buildings8110165

Tang, Y. (2018). Development Trend and Application Prospect of 3D Printing Technology in Intelligent Manufacturing. *International Robotics & Automation Journal*, *4*(1). Advance online publication. DOI: 10.15406/iratj.2018.04.00089

Tetsuka, H., & Shin, S. R. (2020). Materials and technical innovations in 3D printing in biomedical applications [Review of Materials and technical innovations in 3D printing in biomedical applications]. Journal of Materials Chemistry B, 8(15), 2930. Royal Society of Chemistry. DOI: 10.1039/D0TB00034E

Tian, Y., Chen, C., Xu, X., Wang, J., Hou, X., Li, K., Lu, X., Shi, H., Lee, E., & Jiang, H. B. (2021). A Review of 3D Printing in Dentistry: Technologies, Affecting Factors, and Applications [Review of A Review of 3D Printing in Dentistry: Technologies, Affecting Factors, and Applications]. Scanning, 2021, 1. Wiley. DOI: 10.1155/2021/9950131

Trenfield, S. J., Awad, A., Madla, C. M., Hatton, G. B., Firth, J., Goyanes, Á., Gaisford, S., & Basit, A. W. (2019). Shaping the future: recent advances of 3D printing in drug delivery and healthcare [Review of Shaping the future: recent advances of 3D printing in drug delivery and healthcare]. Expert Opinion on Drug Delivery, 16(10), 1081. Taylor & Francis. DOI: 10.1080/17425247.2019.1660318

Trifol, J., Jayaprakash, S., Baniasadi, H., Ajdary, R., Kretzschmar, N., Rojas, O. J., Partanen, J., & Seppälä, J. (2021). 3D-Printed Thermoset Biocomposites Based on Forest Residues by Delayed Extrusion of Cold Masterbatch (DECMA). *ACS Sustainable Chemistry & Engineering*, *9*(41), 13979–13987. DOI: 10.1021/acssuschemeng.1c05587

Tyflopoulos, E., Hofset, T. A., Olsen, A., & Steinert, M. (2021). Simulation-based design: A case study combining optimization methodologies for angle-ply composite laminates. *Procedia CIRP*, *100*, 607–612. DOI: 10.1016/j.procir.2021.05.131

Vernezi, M., Lukyanov, A., Dolgov, V., & Zotov, V. V. (2021). About diagnostics of the technical condition of metal-cutting machines. *IOP Conference Series. Materials Science and Engineering*, *1029*, 12124. DOI: 10.1088/1757-899X/1029/1/012124

Wang, C., Tan, X., Tor, S. B., & Lim, C. S. (2020). Machine learning in additive manufacturing: State-of-the-art and perspectives. *Additive Manufacturing*, *36*, 101538. DOI: 10.1016/j.addma.2020.101538

Wang, S., Chen, X., Han, X., Hong, X., Li, X., Zhang, H., Li, M., Wang, Z., & Zheng, A. (2023). A Review of 3D Printing Technology in Pharmaceutics: Technology and Applications, Now and Future [Review of A Review of 3D Printing Technology in Pharmaceutics: Technology and Applications, Now and Future]. Pharmaceutics, 15(2), 416. Multidisciplinary Digital Publishing Institute. DOI: 10.3390/pharmaceutics15020416

Wang, T., Kaur, L., Beniwal, A. S., Furuhata, Y., Aoyama, H., & Singh, J. (2023). Physico-chemical and Textural Properties of 3D Printed Plant-based and Hybrid Soft Meat Analogs. *Plant Foods for Human Nutrition (Dordrecht, Netherlands)*, *78*(2), 375–382. DOI: 10.1007/s11130-023-01068-4 PMID: 37199825

Wei, H., Tang, L., Qin, H., Wang, H., Chen, C., Li, Y., & Wang, C. (2024). Optimizing FDM 3D printing parameters for improved tensile strength using the Takagi–Sugeno fuzzy neural network. *Materials Today. Communications, 38*, 108268. DOI: 10.1016/j.mtcomm.2024.108268

Wolter, B., Dobmann, G., & Boller, C. (2011). NDT-Based Process Monitoring and Control. Strojniški Vestnik –. *Jixie Gongcheng Xuebao, 2011*(3), 218–226. DOI: 10.5545/sv-jme.2010.172

Wu, J., Aage, N., Lefèbvre, S., & Wang, C. C. L. (2017). Topology Optimization for Computational Fabrication. Eurographics. DOI: 10.2312/egt.20171030

Yin, J., Suo, Y., Lv, T., Ma, K., Wang, X., & Zhang, Z. (2022). *Application of 3D Printing Technology in the Construction Industry and Its Development Prospects.* Atlantis Highlights in Intelligent Systems/Atlantis Highlights in Intelligent Systems., DOI: 10.2991/ahis.k.220601.002

Yuan, S., Li, J., Yao, X., Zhu, J., Gu, X., Gao, T., Xu, Y., & Zhang, W. (2020). Intelligent optimization system for powder bed fusion of processable thermoplastics. *Additive Manufacturing, 34*, 101182. DOI: 10.1016/j.addma.2020.101182

Yuan, Y., Ma, G., Cheng, C., Zhou, B., Zhao, H., Zhang, H., & Ding, H. (2019). *Artificial Intelligence Diagnosis and Monitoring in Manufacturing. arXiv*. Cornell University., DOI: 10.1093/nsr/nwz190

Zárybnická, L., Dvořák, K., Dostalova, Z., & Vojáčková, H. (2019). Study of Different Printing Design Type Polymer Samples Prepared by Additive Manufacturing. *Periodica Polytechnica. Chemical Engineering, 64*(2), 255–264. DOI: 10.3311/PPch.13991

Zhang, H., Zhou, X., Fang, Y., Xiong, Z., & Zhang, T. (2024). AI-driven 3D bioprinting for regenerative medicine: From bench to bedside [Review of AI-driven 3D bioprinting for regenerative medicine: From bench to bedside]. Bioactive Materials, 45, 201. Elsevier BV. DOI: 10.1016/j.bioactmat.2024.11.021

Zhang, S., Yu, Q., Wan, S., Cao, H., & Huang, Y. (2024). Digital supply chain: Literature review of seven related technologies. *Manufacturing Review, 11*, 8. DOI: 10.1051/mfreview/2024006

Zhou, L., Miller, J., Vezza, J., Mayster, M., Raffay, M., Justice, Q., Tamimi, Z. A., Hansotte, G., Sunkara, L. D., & Bernat, J. (2024). Additive Manufacturing: A Comprehensive Review [Review of Additive Manufacturing: A Comprehensive Review]. Sensors, 24(9), 2668. Multidisciplinary Digital Publishing Institute. DOI: 10.3390/s24092668

Chapter 6
Methodologies for Topology Optimization in 3D-Printed Mechanical Parts

Romdhane Ben Khalifa
https://orcid.org/0000-0002-9171-9584
National Engineering School of Gabes (ENIG), Tunisia

ABSTRACT

This paper discusses the methodologies of topology optimization for 3D printed mechanical parts and how design, materials, and manufacturing processes interact. The computational topology optimization method, which optimizes material layout within a design space for given loads and boundary conditions, has gained significant importance in additive manufacturing for creating lightweight, high-performance components. Combining topology optimization with 3D printing allows for generating complex shapes and complex internal structures unattainable through conventional manufacturing techniques. The paper evaluates different topology optimization approaches, including density-based lev, el set methods, and evolutionary algorithms, to determine their benefits and drawbacks when applied to additive manufacturing limitations.

INTRODUCTION

3D printing techniques, as demonstrated by (Wu et al.,2017), are a game-changer in the field of manufacturing. By combining topology optimization and additive manufacturing, these techniques can produce complex geometries with superior

DOI: 10.4018/979-8-3373-0533-2.ch006

performance characteristics. The construction process of 3D printing, relying on additive manufacturing methods that add layers one at a time (Hoang et al., 2016), offers many benefits. Complex designs optimized for performance become achievable through this method, surpassing traditional manufacturing restrictions to create engineered materials with previously impossible properties (Babbar et al., 2021; Jared et al., 2017). Additive manufacturing provides a competitive benefit for metal components through its ability to produce complex geometries and high-volume manufacturing (Eriksson et al., 2018). The mathematical technique of topology optimization finds the most efficient material distribution in design space through constraint application and objective maximization to reach optimal performance (Brackett et al., 2011). The design principles used in conventional manufacturing face limitations, while additive manufacturing benefits from its vast design freedom, according to (Wang et al., 2020).

Implementing topology optimization for mechanical parts created through 3D printing introduces specific manufacturing challenges alongside potential opportunities. The design freedom of 3D printing comes with manufacturing restrictions that need to be evaluated during optimization processes (Brackett et al., 2011). Building orientation, support requirements, material characteristics, and printing resolution represent the main manufacturing constraints that need consideration (Figure 1). In recent decades, significant interest has been shown in multi-functional material systems, which help decrease the minimum number of components needed for manufacturing products (Daminabo et al., 2020). Successful fabrication and functionality of resulting designs require optimization methodologies to manage performance targets alongside manufacturing limitations while developing 3D-printed mechanical parts. The requirement for specialized computational tools and algorithms to handle additive manufacturing complexities has become essential.

Lately, additive manufacturing and design have shaken up how we build and refine mechanical parts. New methods let us make shapes that traditional processes cannot handle. Companies chasing sustainability now demand light and tough parts, pushing folks to explore topology optimization and figure out the best way to spread materials. Hence, everything works just right under set limits. Designers are blending these ideas with 3D printing to produce intricate geometries. However, with rising pressures in fields like aerospace and automotive, we need to dig deeper into these methods, both for present projects and tomorrow's challenges (Arora et al.). It seems the literature buzzes with overlapping themes. People talk about algorithm advances—genetic algorithms and machine learning, for instance, that boost both speed and accuracy in design. Many studies celebrate these wins (Li, H et al., 2024), yet an undercurrent urges us to weigh their upsides and downsides. Then there are multiple reports showing that finite element methods (FEM) work pretty well for checking how optimized designs perform under real-world stress.

Figure 1. Design is going through topology optimization

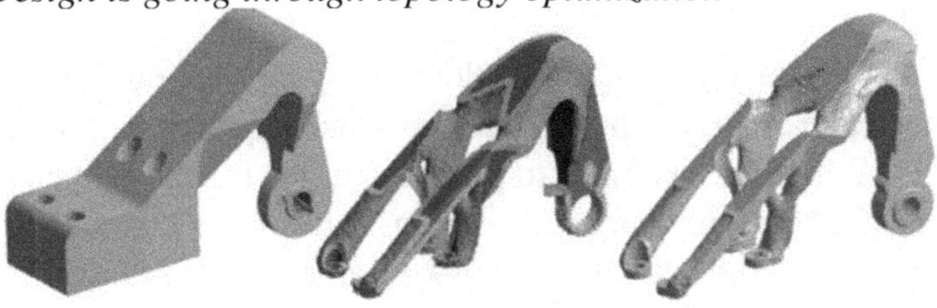

Recently, research has turned toward multi-objective frameworks that consider the environment and cost, a sign that sustainability is finally getting its due in engineering. Design methods are constantly being tested across various industries, and the results show that these techniques are as flexible as they are innovative. With experts starting to mix in insights from other fields, topology optimization in 3D printing will only get more interesting. There is also a recurring focus on how computational tools can fine-tune designs. Tools like genetic algorithms and particle swarm optimization have been credited with helping to nail down the best material layouts while keeping parts sturdy (Li, H et al., 2024). Matching these methods with the right materials is also key. Different substances, like polymers versus metals, have quirks that can significantly affect outcomes. Plus, the rise of machine learning means our optimization processes can learn from previous tries, tweaking parameters on the fly for even better designs. On the practical side, several analyses confirm that these optimized designs cut down on material waste and boost performance in the aerospace and automotive industries (Sandeep et al., 2021).

BACKGROUND

Additive manufacturing technologies have shown fast development in the last decades, transforming product design, manufacturing, and delivery to end-users (Alfaify et al., 2020). The transformative power of additive manufacturing is evident in its ability to enable the development of components that weigh less yet demonstrate greater strength (Sandeep et al., 2021). This technology has proven to be a game-changer in multiple industrial sectors, creating complex geometries while bypassing traditional manufacturing process restrictions (Carrola et al., 2021; Peles et al., 2023). The manufacturing process through additive methods allows functional part construction during a single operation, reducing raw material waste (Vayre et al., 2012).

Additive manufacturing technology allows designers and engineers to develop innovative mechanical parts with high-performance capabilities (Guenther et al., 2021; Pfaff et al., 2018). The ability to place materials in three dimensions shows promise to transform manufacturing operations, offering unmatched product customization capabilities and market entry opportunities for small businesses (Garcia et al., 2013). The remarkable progress of 3D printing technology remains limited by its slow printing speeds, restricted materials, and the need for detailed post-processing to fulfill industrial requirements (Zhou et al., 2024). Figure 2 visually encapsulates the progression from manual, intuition-driven structural optimization to advanced, computationally enabled topology optimization and its integration with metal additive manufacturing. The figure highlights how these developments have enabled the creation of highly efficient, lightweight, and complex structures while pointing to the field's challenges and future research directions.

Figure 2. Background of topology optimization (Ibhadode et al., 2023)

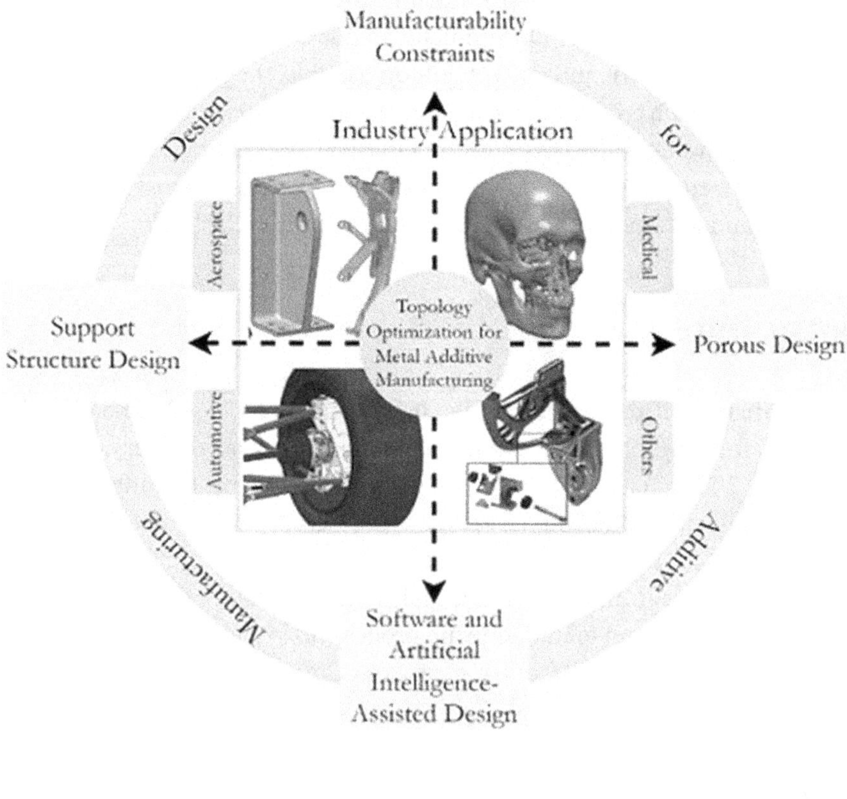

Topology Optimization

The fundamental operation of topology optimization relies on mathematical programming techniques. This powerful optimization method arranges material distribution in specified design areas to achieve performance targets under defined constraints (Dani et al., 2020). The benefits of topology optimization, particularly when combined with additive manufacturing, are significant. It allows the production of complex mechanical components, which was impossible through conventional manufacturing techniques (Tebianian et al., 2023). Through topology optimization, structural engineers gain access to possibilities to discover required performance-maximizing solutions. The essential structure for load support during product life cycles can be determined through topology optimization methods (Burkhalter, 2020). The technique benefits industries requiring weight reduction and performance enhancement, particularly in aerospace, automotive, and biomedical engineering applications (Andronov et al., 2021). Figure 3 emphasizes the importance of engineering objectives (like dynamic performance) and practical constraints (such as manufacturability), making it highly relevant for modern engineering design and advanced manufacturing contexts.

Figure 3. Flowchart of Topology optimization solution (Tian et al., 2019).

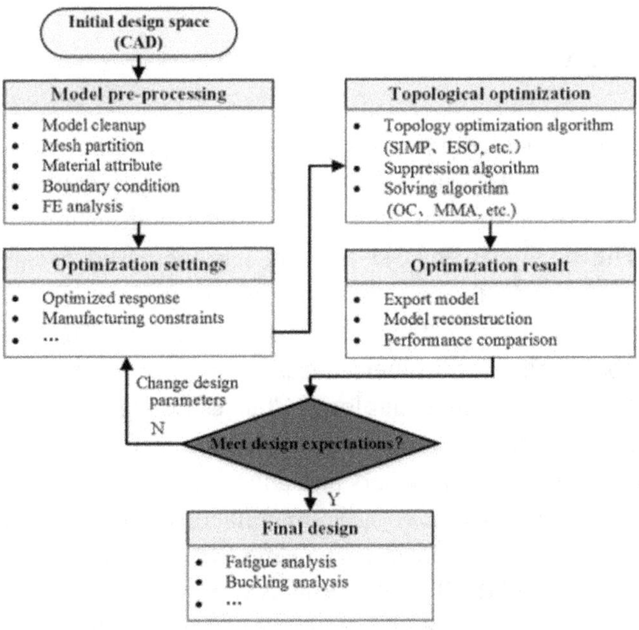

3D Printing of Mechanical Parts

The technology of 3D printing has transformed mechanical part design, manufacturing, and deployment across various sectors. The evolution of additive manufacturing technologies has significantly changed the design optimization approaches used by structural engineers and architects (Stefańska et al., 2021). The additive manufacturing process enables the creation of complex shapes and intricate designs that conventional manufacturing methods cannot replicate (Jayakrishna et al., 2023). Additive manufacturing is built on 3D data, serving as its foundation (Yin et al., 2022). The rapid prototyping applications of this technology have evolved into a manufacturing process that produces functional parts with customized material properties and complex geometries. This manufacturing technique has become one of the standard production techniques (Srinivasan et al., 2021). 3D printing facilitates personalized, immediate part production while significantly reducing waste materials and shortening product development timelines (Despeisse & Ford, 2015). The ability of AM technologies to convert digital three-dimensional models into functional components of any geometry makes them highly useful (Raheem et al., 2021). The current additive manufacturing processes have reached the point where they fulfill all required performance standards, while also contributing to a more environmentally conscious and responsible manufacturing process.

METHODOLOGY

The additive manufacturing process known as 3D printing enables sustainable manufacturing through its waste-reducing layer-by-layer construction method, which differs from traditional subtractive manufacturing techniques (Nadagouda et al., 2020). The popularity of 3D printing has increased because it provides ready access to fast production at low costs for pre-production models compared to traditional manufacturing methods (Garcia et al., 2018). The capability to create intricate shapes with built-in internal truss structures for weight reduction is a primary benefit of 3D printing technology (Chua & Leong, 2013). The production method relies on three-dimensional models to create materials through sequential layering of plastics and metal powders and adhesive materials (Hao & Lin, 2020; He & Khan, 2021). The popular additive manufacturing method known as Fused Deposition Modeling enables customers to modify their models frequently with ease (Vorkapić et al., 2020). The different AM techniques operate with unique capabilities, material compatibility, and distinct surface quality characteristics. 3D printing is a revolutionary manufacturing system that generates three-dimensional objects by employing multiple production methods and raw materials, including filaments, resins, and powder grains, through sequential layer construction (Anwajler et al., 2023). The different methods present distinct benefits, drawbacks, and specific implementation areas (Wilson et al., 2021). Figure 4 provides a synthesized overview of topology optimization (TO) methodologies, workflows, and challenges specific to static structural design. It integrates theoretical frameworks (linear/nonlinear) and practical considerations (e.g., additive manufacturing). The figure likely combines visual elements such as flowcharts, comparative tables, and case studies to illustrate key concepts.

Figure 4. Topology Optimization: A Review for Structural Designs Under Statics Problems (Tang et al., 2024)

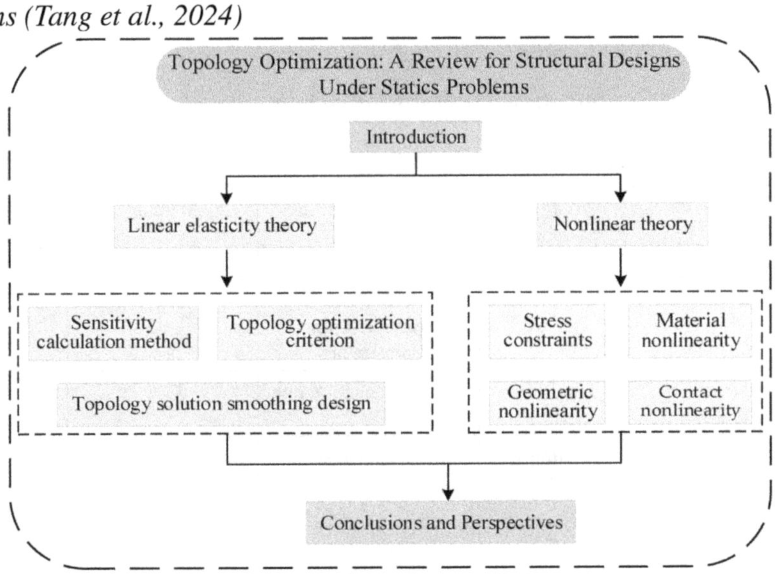

Design Optimization Process

Additive manufacturing and topology optimization enable designers to create mechanical parts with improved functionality through complex geometric structures (Niesler & Hermatschweiler, 2015). The integrated methodology enables advanced prediction of Fused Deposition Modeling product surface quality through specified process parameters (Boschetto et al., 2016). Engineers use the combined strength of computational optimization with additive manufacturing to develop innovative designs that extend structural performance and functional capabilities (Table 1). The mathematical method of topology optimization determines the most efficient material arrangement inside a defined design space by applying performance targets and boundary conditions (Pandey et al., 2020). Stress distribution analysis and additive manufacturing create a strong system for designing mechanical parts that achieve optimized performance characteristics through fabrication. The production of small quantities or basic fused deposition modeling methods through 3D printing leads to cost savings, yet large-scale manufacturing with intricate printing techniques remains expensive (Nadagouda et al., 2020; Zhou et al., 2024). The main application of 3D printing involves creating molds and patterns for conventional manufacturing processes, including injection molding. Research and development activities continue to enhance 3D printing capabilities and manufacturing applications despite ongoing challenges with scalability, material properties, and surface finish (Equbal et al., 2018).

Table 1. Topology Optimization Methods for 3D-Printed Parts

Method	Objective	Constraints	Computational Complexity	Suitability for 3D Printing
SIMP (Solid Isotropic Material with Penalization)	Minimize compliance	Volume fraction	Medium	High
Level Set Method	Minimize stress	Volume fraction, manufacturing constraints	High	Medium
Evolutionary Structural Optimization (ESO)	Maximize stiffness	Volume reduction rate	Low	Medium
Bidirectional Evolutionary Structural Optimization (BESO)	Minimize mass	Stress, displacement	Medium	High
Moving Morphable Components (MMC)	Multi-objective	Geometric, manufacturing	High	Very High

Computational Techniques

The evaluation of topology-optimized designs under different loading conditions heavily depends on advanced computational techniques, which include finite element analysis and computational fluid dynamics. Through simulation techniques, engineers discover critical stress areas, formation behaviors, and failure mechanisms, which enable them to enhance design elements for maintaining structural stability. The simulation software, with its ability to produce an exact digital model that reproduces all printing process complexities, instills confidence in the accuracy of the process. Researchers employ data-driven methods to handle these difficulties by analyzing the metal powder characteristics and the extensive freedom of printing strategies (Panwisawas et al., 2020).

The optimization process uses computational algorithms to enhance material distribution throughout the design space continuously. It considers the manufacturing constraints of additive manufacturing, including but not limited to minimum feature size, overhang angles, and support structures (Cucinotta et al., 2019). These constraints are crucial in ensuring the manufacturability of the design. The optimization process produces designs that achieve structural efficiency while maintaining manufacturability through the chosen additive manufacturing method (Graziosi et al., 2017). Machine learning algorithms use extensive datasets of process parameters and material properties to forecast the mechanical behavior of 3D-printed parts (Mahmood et al., 2020).

Simulation tools, especially finite element analysis, come in handy to check that these parts can stand up to real use (Ficzere et al., 2021). These pieces show how computational strategies, material science, and practical design combine to define the future of 3D-printed mechanical parts. Meanwhile, earlier studies leaned on classical methods based on FEA to boost structural performance while slashing material use (Li,H et al., 2024). That groundwork paved the way for newer, more sophisticated tactics like generative design, where machine learning helps evolve parts based on performance. More recent literature has highlighted the benefits of multi-objective frameworks, which allow for a nuanced view of the trade-offs between performance and material efficiency.

Using Pareto optimization, designers can juggle many constraints simultaneously, ultimately enhancing what 3D-printed components can do. Simulation tools also get much love. The iterative nature of these design methods, bolstered by ongoing computational feedback, lets engineers tweak designs in real-time based on what they observe on the ground (Mahmood et al., 2020). This kind of flexibility fits with today's drive for sustainable manufacturing: fewer resources wasted and a smaller environmental footprint are high on the agenda(Ahmed & Zaneldin, 2020). As these methods keep improving, there is a clear call for approaches that mix old-school techniques with new computational wizardry. Such a balanced mix could unlock even better 3D printing solutions and push the envelope on what is possible (Ahmed & Zaneldin, 2020).

Design, material properties, and actual manufacturing processes might be the key to fully harnessing the promise of 3D printing. When you blend diverse theoretical frameworks, you get a complex patchwork of approaches for topology optimization in 3D-printed parts. Researchers have long championed methods that zero in on efficiency and precision, with algorithms like genetic and particle swarm optimization leading to notable advances in strength and weight reduction (Li, H et al., 2024). Meanwhile, parametric design techniques have made it possible to rapidly experiment with countless configurations (Ahmed & Zaneldin, 2020). This speed helps prototype faster and meshes well with the industry's demand for quick iterations. Still, some experts caution that not every theoretical model works perfectly in the real world; there is a strong call for empirical validation to ensure these neat models hold water against manufacturing challenges, especially considering the material limits imposed by current 3D printing technologies.

Figure 5. Summary of significant topology optimization and metal additive manufacturing techniques (Kok et al., 2018)

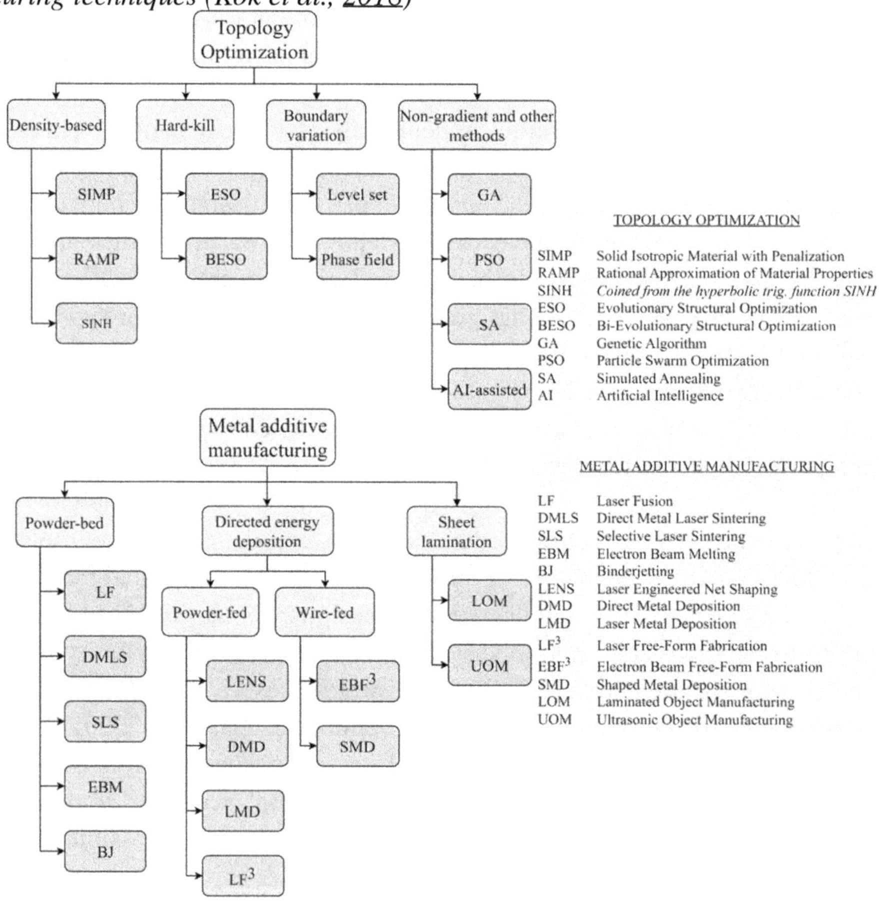

Material Considerations

3D-printed part performance characteristics and mechanical properties depend heavily on the selection of materials used. The selected material must fulfill application functional needs while also being suitable for the chosen additive manufacturing process. Researchers continue to develop new materials and material combinations that enable new additive manufacturing possibilities (Daminabo et al., 2020; Ghaffar et al., 2018; Tetsuka & Shin, 2020). The selection of materials requires evaluation alongside printing methods, sterilization requirements, production expenses, and duration (Hoang et al., 2016; Yeole et al., 2016). The theory of elasticity demands precise material property knowledge for proper structural integrity calculations through model-based elasticity theory (Ficzere et al., 2021).

The growing integration of advanced material science with topology optimization shows real promise for addressing these issues, hinting that a cross-disciplinary approach might be exactly what the field needs (Figure 6). In mechanical design, these optimization techniques have become potent tools against today's environmental and economic hurdles. This review has highlighted how blending algorithmic innovations with material science has transformed additive manufacturing. For example, methods leveraging genetic algorithms and machine learning have boosted design precision and efficiency, enabling creators to produce highly intricate, previously unattainable structures (Li, H et al., 2024).

Figure 6. A simple academic example of topology optimization

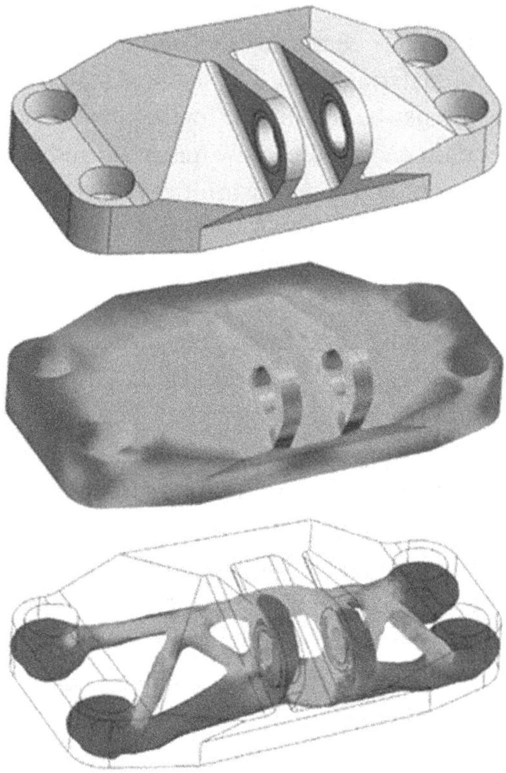

Guidelines for Choosing an Optimization Algorithm

The choice of optimization algorithm depends heavily on the design problem's complexity and specific requirements (Table 2). Different algorithms may suit different objectives (weight reduction, increased stiffness). The resolution required from the print object, the medical properties of the material, available materials, and how long it will take (Oberoi et al., 2018) should also be considered. When choosing an optimization algorithm, it is important to consider the computational cost per iteration (Goh et al., 2020). Gradient-based algorithms may be suitable for smooth and well-behaved objective functions, while evolutionary algorithms may be more appropriate for non-convex and discontinuous problems.

Table 2. Key Factors for Algorithm Selection

Factor	SIMP	Level Set	BESO
Complexity	Low (density-based)	High (boundary tracking)	Moderate (evolutionary)
3D Suitability	Excellent (mesh flexibility)	Limited (2D/simple 3D)	Good (supports 3D)
Multi-Material	Yes (graded densities)	No	No
Convergence Speed	Fast (penalty-driven)	Slow (contour recalculation)	Moderate (iterative addition/ removal)

RESULTS

Experimental testing is essential to validate the effectiveness of topology optimization in 3D-printed mechanical parts. By comparing the experimental results with the simulation predictions, engineers can gain confidence in the computational models' accuracy and the optimization process's reliability. A printed specimen's tensile and flexural characteristics are determined by tensile and bending tests (Babagowda et al., 2018). Experimental modal analysis can be performed to correlate the natural frequencies with the elastic modulus of parts derived from flexural bending tests; This optimization level is important to industries requiring lightweight parts (Vigliotti & Pasini, 2011).

Assessment of Performance and Future Directions

While 3D printing offers numerous advantages, it also presents several challenges that must be addressed to ensure the technology's sustainability (Liu et al., 2016).

By addressing these challenges and capitalizing on the opportunities presented by topology optimization and additive manufacturing, engineers can unlock new possibilities for designing and fabricating high-performance mechanical parts with enhanced functionality and reduced material consumption. The capacity to produce intricate shapes using 3D printing can lead to lighter parts with improved performance (Aman, 2020; Дородов et al., 2021). Before printing complex shapes, the machine was tested in advance with a more straightforward specimen design, a specimen design for a flexural strength test (Almy & Tontowi, 2018). Topology optimization is used with 3D printing in the construction industry, where speed, direct data conversion, handling of complex geometry, accuracy, and cost-effectiveness are critical (Yin et al., 2018). Additive manufacturing techniques make creating materials with controlled surface roughness possible, which is helpful in applications needing particular surface qualities (Garcia et al., 2013). 3D printing technology substantially contributes to numerous industries due to its remarkable advantages (Ahmed & Zaneldin, 2020). For the manufacture of biofuel component equipment, a robotic 3D printing system can offer the most freedom when creating mock-ups and test samples (Bashmur et al., 2022). Topology optimization also allows for the mass customization of designs, waste reduction, and using recycled materials (Tahmasebinia et al., 2018).

Optimized Designs

Selecting appropriate materials remains highly sensitive to the amount of design freedom, according to Hahn et al. (2018). The literature presents three main structural optimization approaches: topology and analytic methods based on Michell-type theories and shape derivative methods (Ureta et al., 2019). The optimized designs produce complex geometries that traditional manufacturing methods cannot produce. The complex shapes produced by these designs result in parts that achieve both weight reduction and enhanced performance. Topology optimization enables the development of multimaterial elements through its ability to integrate different materials for specified properties. The mechanical properties of 3D-printed components need additional research to determine the most effective infill settings for optimal performance. The research shows that 3D printing enables the production of distinctive topographies, including gecko-like protuberances, which are unique structures that mimic the gripping ability of gecko feet. However, creating them with conducting polymers remains an unexplored application (Dominguez-Alfaro et al., 2024).

DISCUSSION

3D-printed mechanical parts benefit from topology optimization techniques, a crucial process for reducing material waste in modern manufacturing. According to (Yang & Li, 2020), designers must fully understand the constraints of topology optimization. The primary benefit of this technique is its ability to produce designs that combine lightweight construction with maximum structural efficiency. Topology optimization algorithms achieve this by removing material from stress-free areas, leading to strength-to-weight ratio optimization. The optimization process, by reducing material waste to a minimum, has become vital for sustainable manufacturing operations. The development of additive manufacturing technology allows designers to create functional materials through structural designs instead of chemical compositions (Li et al., 2023). The integration of topology optimization with additive manufacturing enables the production of architected materials with customized properties (Plancher et al., 2020). Materials design is the quantitative process of reversing material relationships between process variables, structure, and material properties while treating materials as connected subsystems (Arróyave & McDowell, 2019). Material and binder jetting additive manufacturing techniques have advanced their printing resolution capabilities while expanding their material selection range (Chavez et al., 2020; Zhou et al., 2024). The techniques face ongoing challenges regarding achieving scalability, dependability, and cost-effectiveness for large-scale production requirements.

Design Implications

3D-printed parts experience substantial changes in their performance and functionality by implementing topology optimization in design processes. Optimization techniques enable the development of parts that achieve better stiffness along with enhanced strength and longer fatigue life (Wang et al., 2016). Manufacturers achieve sustainable design through topology optimization by optimizing material distribution and minimizing material waste. Through topology optimization, designers can generate intricate shapes that traditional manufacturing techniques cannot produce (Sakin & Kiroglu, 2017). Additive manufacturing enables the optimization of components through manufacturing-related constraints, resulting in improved performance. The topology optimization needs to be adjusted according to the angular position of the cylinder to achieve maximum drag reduction in specific flow conditions (Pelacci et al., 2021). The main objective of generative design methodologies is weight reduction and performance enhancement for manufacturing applications (Aman, 2020). Integrating additive manufacturing with topology optimization enables designers to develop lightweight aircraft components that maintain their original design characteristics (Levashkin et al., 2019). The application of additive manufacturing extends beyond creating complex shapes because it enables superior structural performance through microstructure regulation (Palanivel et al., 2014). The development requires topology optimization methods that consider specific requirements and manufacturing constraints. (Bałon et al., 2020).

Limitations and Future Work

Despite their advantages, future research should tackle multiple limitations in topology optimization for 3D-printed parts. The main drawback of topology optimization algorithms is their high computational expense. The technology faces barriers because of its limited large-scale deployment and building information modeling capabilities and requirements for mass customization and printed project life cycle costs (Wu et al., 2016). The solution to these challenges requires advanced algorithms together with robust computing systems. The precision of topology optimization outcomes heavily depends on the material data used during simulation. The requirement of support structures presents a significant challenge in specific 3D-printing operations. The total material consumption, printing duration, and expenses rise because of support structures. Future research must create topology optimization methods that account for support structure requirements and reduce their necessity. Unconstrained sintering produces parts that maintain permanent shape deviations, while constrained deformation develops residual stress (Ritchie et al., 2023). Complex component geometries, elevated component aspect ratios, and internal constraint features are the most susceptible to sintering distortion. Additive manufacturing technologies need additional research to become widely adopted because not every standard manufacturing material can be processed (Wong & Hernandez, 2012). Future research should expand additive technologies to include architecture and design, according to Drebezgova et al. (2019). The effectiveness of this technology remains uncertain for real-world applications because all demonstrated construction examples have occurred within laboratory control environments (Sakin & Kiroglu, 2017; Perkins & Skitmore, 2015).

CONCLUSION

Topology optimization in 3D-printed mechanical parts offers significant advantages, including improved performance, reduced material waste, and the ability to create complex geometries (Srinivas & Babu, 2017). Despite these advantages, additive manufacturing has unique drawbacks in manufacturing lattice structures (Seharing et al., 2014). The technology is still limited by the expense and time needed to create devices using current 3D technology (Martelli et al., 2016). Nevertheless, ongoing research and development efforts address these limitations and pave the way for the widespread adoption of topology optimization in various industries (Hager et al., 2016; Martelli et al., 2016). Further research is needed to address topology optimization's limitations and fully realize the potential of additive manufacturing (Camacho et al., 2018). It is crucial to remember that additive manufacturing is still a developing technology with substantial obstacles to overcome at every stage of its life cycle (Ford & Despeisse, 2016). The capability of AM to produce intricate shapes without the requirement for conventional processing steps is one of its main advantages over other production techniques (Dass & Moridi, 2019). The capacity to create parts with complex 3D molded components, which would be challenging or impossible to achieve using conventional production methods, is an exceptional accomplishment (Dhinakaran et al., 2020). The design freedom afforded by AM enables the creation of customized and optimized designs for particular applications. The technology is still relatively young, and its consistency and dependability are still being developed. Furthermore, problems like repeatability, scalability, and material qualities still need to be resolved before they can be widely used in industrial applications (Faes et al., 2014; Ransikarbum et al., 2020; Rouf et al., 2022; Výtisk et al., 2019).

In conclusion, topology optimization is a powerful tool for designing high-performance 3D-printed parts. As additive manufacturing technology continues to evolve, the combination of topology optimization and 3D printing will play an increasingly important role in shaping the future of manufacturing. This approach seeks to optimize material constructs, design topologies, and stochastic processes concurrently to realize a comprehensive design paradigm (Jared et al., 2017). AM offers a significant advantage by reducing the number of manufacturing steps between virtual design and ready-to-use parts (Guessasma et al., 2015). A designer, manufacturer, or software provider can either develop original software or enhance an existing one (Brackett et al., 2011). The integration of data security, product quality control, manufacturing planning, in-process monitoring, process optimization, and design highlights the logical flow of the AM process (Wang et al., 2020).

REFERENCES

Дородов, П. В., Kasatkin, V., Lekomcev, P. L., Petrov, V. A., & Niyazov, A. M. (2021). Rigidity, creep and dynamic strength of plastics for three-dimensional printing of machine parts. In IOP Conference Series Earth and Environmental Science (Vol. 677, Issue 5, p. 52045). IOP Publishing. DOI: 10.1088/1755-1315/677/5/052045

Ahmed, W., & Zaneldin, E. (2020). The Innovation of Using 3D Printing Technology in Mechanical and Manufacturing Engineering. *IOP Conference Series. Materials Science and Engineering*, *938*(1), 12002. DOI: 10.1088/1757-899X/938/1/012002

Alfaify, A., Saleh, M., Abdullah, F. M., & Al-Ahmari, A. (2020). Design for Additive Manufacturing: A Systematic Review [Review of Design for Additive Manufacturing: A Systematic Review]. Sustainability, 12(19), 7936. Multidisciplinary Digital Publishing Institute. DOI: 10.3390/su12197936

Almy, R. D., & Tontowi, A. E. (2018). The effect of 3d printing machine parameters in the extrusion process of biocomposite materials (PMMA and HA) on dimensional accuracy. In SINERGI (Vol. 22, Issue 2, p. 83). Mercu Buana University. DOI: 10.22441/sinergi.2018.2.003

Aman, B. (2020). Generative Design for Performance Enhancement, Weight Reduction, and Its Industrial Implications. In arXiv (Cornell University). Cornell University. https://doi.org//arxiv.2007.14138DOI: 10.48550

Andronov, V. N., Šimota, J., Pelikán, L., Beránek, L., & Novák, V. (2021). Analysis of strength properties of 3D printed lightweight structures made of AlSi10Mg material. In IOP Conference Series Materials Science and Engineering (Vol. 1178, Issue 1, p. 12002). IOP Publishing. DOI: 10.1088/1757-899X/1178/1/012002

Anwajler, B., Zdybel, E., & Tomaszewska-Ciosk, E. (2023). Innovative Polymer Composites with Natural Fillers Produced by Additive Manufacturing (3D Printing)—A Literature Review [Review of Innovative Polymer Composites with Natural Fillers Produced by Additive Manufacturing (3D Printing)—A Literature Review]. Polymers, 15(17), 3534. Multidisciplinary Digital Publishing Institute. DOI: 10.3390/polym15173534

Arróyave, R., & McDowell, D. L. (2019). Systems Approaches to Materials Design: Past, Present, and Future. In Annual Review of Materials Research (Vol. 49, Issue 1, p. 103). Annual Reviews. DOI: 10.1146/annurev-matsci-070218-125955

Babagowda, K. M. R. S., Goutham, R., & Srinivas Prasad, K. R. (2018, February). Study of Effects on Mechanical Properties of PLA Filament which is blended with Recycled PLA Materials. In IOP conference series: materials science and engineering (Vol. 310, p. 012103). IOP Publishing.

Babbar, A., Rai, A., & Sharma, A. (2021). The latest trend in building construction: Three-dimensional printing. *Journal of Physics: Conference Series*, *1950*(1), 12007. DOI: 10.1088/1742-6596/1950/1/012007

Bałon, P., Rejman, E., Świątoniowski, A., Kiełbasa, B., Smusz, R., Szostak, J., Cieślik, J., & Kowalski, Ł. (2020). Thin-walled Integral Constructions in the Aircraft Industry. *Procedia Manufacturing*, *47*, 498–504. DOI: 10.1016/j.promfg.2020.04.153

Bashmur, K. A., Тынченко, В. С., Bukhtoyarov, V. V., & Saramud, M. V. (2022). Choice of technology and evaluation of the production capabilities of a 3d printer robot for creating elements of experimental equipment for producing biofuel components. In arXiv (Cornell University). Cornell University. https://doi.org//arxiv .2202.00889DOI: 10.48550

Boschetto, A., Bottini, L., & Veniali, F. (2016). Integration of FDM surface quality modeling with process design. In *Additive manufacturing* (Vol. 12, p. 334). Elsevier BV., DOI: 10.1016/j.addma.2016.05.008

Brackett, D., Ashcroft, I., & Hague, R. (2011). Topology Optimization for Additive Manufacturing. DOI: 10.26153/tsw/15300

Burkhalter, D. (2020). *Simulation-Driven Design of a Portable Basketball Hoop System.*, DOI: 10.3390/proceedings2020049131

Camacho, D. D., Clayton, P., O'Brien, W. J., Seepersad, C. C., Juenger, M. C. G., Ferron, R. D., & Salamone, S. (2018). Applications of additive manufacturing in the construction industry – A forward-looking review [Review of Applications of additive manufacturing in the construction industry – A forward-looking review]. Automation in Construction, 89, 110. Elsevier BV. DOI: 10.1016/j.autcon.2017.12.031

Cannella, F., Garinei, A., Marsili, R., & Speranzini, E. (2017). Dynamic mechanical analysis and thermoelasticity investigate composite structural elements made with additive manufacturing. In *Composite Structures* (Vol. 185, p. 466). Elsevier BV., DOI: 10.1016/j.compstruct.2017.11.029

Carrola, M., Asadi, A., Zhang, H., Papageorgiou, D. G., Bilotti, E., & Koerner, H. (2021). Best of Both Worlds: Synergistically Derived Material Properties via Additive Manufacturing of Nanocomposites. In Advanced Functional Materials (Vol. 31, Issue 46). Wiley. DOI: 10.1002/adfm.202103334

Chavez, L. A., Ibave, P., Wilburn, B. R., Alexander, D., Stewart, C. M., Wicker, R. B., & Lin, Y. (2020). The Influence of Printing Parameters, Post-Processing, and Testing Conditions on the Properties of Binder Jetting Additive Manufactured Functional Ceramics. In Ceramics (Vol. 3, Issue 1, p. 65). Multidisciplinary Digital Publishing Institute. DOI: 10.3390/ceramics3010008

Chua, C. K., & Leong, K. F. (2013). 3D Printing and Additive Manufacturing. DOI: 10.1142/9008

Cucinotta, F., Raffaele, M., & Salmeri, F. (2019). A stress-based topology optimization method by a Voronoi tessellation, Additive Manufacturing oriented. In The International Journal of Advanced Manufacturing Technology (Vol. 103, p. 1965). Springer Science+Business Media. DOI: 10.1007/s00170-019-03676-4

Daminabo, S. I. C., Goel, S., Grammatikos, S., Nezhad, H. Y., & Thakur, V. K. (2020). Fused deposition modeling-based additive manufacturing (3D printing): Techniques for polymer material systems. *Materials Today. Chemistry*, *16*, 100248. DOI: 10.1016/j.mtchem.2020.100248

Dani, I., Drossel, W., Milaev, N., Korn, H., Hannemann, C., Hohlfeld, J., & Wertheim, R. (2020). Sustainability of industrial components using additive manufacturing and foam materials. *Procedia Manufacturing*, *43*, 10–17. DOI: 10.1016/j.promfg.2020.02.102

Dass, A., & Moridi, A. (2019). State of the Art in Directed Energy Deposition: From Additive Manufacturing to Materials Design. *Coatings*, *9*(7), 418. DOI: 10.3390/coatings9070418

Despeisse, M., & Ford, S. (2015). The Role of Additive Manufacturing in Improving Resource Efficiency and Sustainability. In IFIP, information and communication technology advances are made p. 129). Springer Science+Business Media. DOI: 10.1007/978-3-319-22759-7_15

Dhinakaran, V., Stalin, B., Ravichandran, M., Balasubramanian, M., Chairman, C. A., & Pritima, D. (2020). Wire Arc Additive Manufacturing Perspectives and Recent Developments. In IOP Conference Series Materials Science and Engineering (Vol. 988, Issue 1, p. 12102). IOP Publishing. DOI: 10.1088/1757-899X/988/1/012102

Dominguez-Alfaro, A., Mitoudi-Vagourdi, E., Dimov, I. B., Picchio, M. L., Lopez-Larrea, N., de Lacalle, J. L., Tao, X., Serrano, R. R., Gallastegui, A., Vassardanis, N., Mecerreyes, D., & Malliaras, G. G. (2024). Light-Based 3D Multi-Material Printing of Micro-Structured Bio-Shaped, Conducting and Dry Adhesive Electrodes for Bioelectronics. *Advancement of Science*, *11*(27), 2306424. Advance online publication. DOI: 10.1002/advs.202306424 PMID: 38251224

Drebezgova, M. Yu., Чернышева, Н., Klimenko, V. A., & Olisov, A. V. (2019). Architectural design of residential modules using additive technologies. In IOP Conference Series Materials Science and Engineering (Vol. 698, Issue 3, p. 33007). IOP Publishing. DOI: 10.1088/1757-899X/698/3/033007

Equbal, A., Equbal, M. A., Sood, A. K., Pranav, R., & Equbal, M. I. (2018, December). A review and reflection on part quality improvement of fused deposition modelled parts. [). IOP Publishing.]. *IOP Conference Series. Materials Science and Engineering*, *455*(1), 012072.

Eriksson, M., Lervåg, M., Sørensen, C., Robertstad, A., Brønstad, B. M., Nyhus, B., Aune, R. E., Ren, X., & Akselsen, O. M. (2018). Additive manufacture of superduplex stainless steel using WAAM. In MATEC Web of Conferences (Vol. 188, p. 3014). EDP Sciences. DOI: 10.1051/matecconf/201818803014

Faes, M. G. R., Abbeloos, W., Vogeler, F., Valkenaers, H., Coppens, K., Goedemé, T., & Ferraris, E. (2014). Process Monitoring of Extrusion-Based 3D Printing via Laser Scanning. In arXiv (Cornell University) (Vol. 6, p. 363). Cornell University. https://arxiv.org/abs/1612.02219

Ficzere, P., Lukács, N. L., & Borbás, L. (2021). The Investigation of Interlaminar Failures Caused by Production Parameters in the Case of Additive Manufactured Polymers. In Polymers (Vol. 13, Issue 4, p. 556). Multidisciplinary Digital Publishing Institute. DOI: 10.3390/polym13040556

Ford, S., & Despeisse, M. (2016). Additive manufacturing and sustainability: an exploratory study of the advantages and challenges. In *Journal of Cleaner Production* (Vol. 137, p. 1573). Elsevier BV., DOI: 10.1016/j.jclepro.2016.04.150

Garcia, C. R., Rumpf, R. C., Tsang, H., & Barton, J. H. (2013). Effects of extreme surface roughness on 3D printed horn antenna. In Electronics Letters (Vol. 49, Issue 12, p. 734). Institution of Engineering and Technology. DOI: 10.1049/el.2013.1528

Garcia, F. L., Moris, V. A. da S., Nunes, A. O., & Silva, D. A. L. (2018). Environmental performance of additive manufacturing process – an overview. In Rapid Prototyping Journal (Vol. 24, Issue 7, p. 1166). Emerald Publishing Limited. DOI: 10.1108/RPJ-05-2017-0108

Ghaffar, S. H., Corker, J., & Fan, M. (2018). Additive manufacturing technology and its implementation in construction as an eco-innovative solution. *Automation in Construction*, *93*, 1–11. DOI: 10.1016/j.autcon.2018.05.005

Goh, G. D., Sing, S. L., & Yeong, W. Y. (2020). A review on machine learning in 3D printing: applications, potential, and challenges [Review of A review on machine learning in 3D printing: applications, potential, and challenges]. Artificial Intelligence Review, 54(1), 63. Springer Science+Business Media. DOI: 10.1007/s10462-020-09876-9

Graziosi, S., Rosa, F., Casati, R., Solarino, P., Vedani, M., & Bordegoni, M. (2017). Designing for Metal Additive Manufacturing: A Case Study in the Professional Sports Equipment Field. In Procedia Manufacturing (Vol. 11, p. 1544). Elsevier BV. DOI: 10.1016/j.promfg.2017.07.288

Guenther, E., Kahlert, M., Vollmer, M., Niendorf, T., & Greiner, C. (2021). Tribological Performance of Additively Manufactured AISI H13 Steel in Different Surface Conditions. In Materials (Vol. 14, Issue 4, p. 928). Multidisciplinary Digital Publishing Institute. DOI: 10.3390/ma14040928

Guessasma, S., Zhang, W., Zhu, J., Belhabib, S., & Nouri, H. (2015). Challenges of additive manufacturing technologies from an optimisation perspective. In *International Journal for Simulation and Multidisciplinary Design Optimization* (Vol. 6). EDP Sciences., DOI: 10.1051/smdo/2016001

Hager, I., Golonka, A., & Putanowicz, R. (2016). 3D Printing of Buildings and Building Components as the Future of Sustainable Construction? In *Procedia Engineering* (Vol. 151, p. 292). Elsevier BV., DOI: 10.1016/j.proeng.2016.07.357

Hahn, M., Gies, S., & Tekkaya, A. E. (2018). Light enough or go lighter? In *Materials & Design* (Vol. 163, p. 107545). Elsevier BV., DOI: 10.1016/j.matdes.2018.107545

Hao, B., & Lin, G. (2020). 3D Printing Technology and Its Application in Industrial Manufacturing. *IOP Conference Series. Materials Science and Engineering*, *782*(2), 22065. DOI: 10.1088/1757-899X/782/2/022065

He, F., & Khan, M. (2021). Effects of Printing Parameters on the Fatigue Behaviour of 3D-Printed ABS under Dynamic Thermo-Mechanical Loads. In Polymers (Vol. 13, Issue 14, p. 2362). Multidisciplinary Digital Publishing Institute. DOI: 10.3390/polym13142362

Hoang, D., Perrault, D., Stevanović, M., & Ghiassi, A. (2016). Surgical applications of three-dimensional printing: a review of the current literature & how to get started [Review of Surgical applications of three-dimensional printing: a review of the current literature & how to get started]. Annals of Translational Medicine, 4(23), 456. AME Publishing Company. DOI: 10.21037/atm.2016.12.18

Ibhadode, O., Zhang, Z., Sixt, J., Nsiempba, K. M., Orakwe, J., Martinez-Marchese, A., Ero, O., Shahabad, S. I., Bonakdar, A., & Toyserkani, E. (2023). Topology optimization for metal additive manufacturing: Current trends, challenges, and future outlook. *Virtual and Physical Prototyping*, *18*(1), e2181192. Advance online publication. DOI: 10.1080/17452759.2023.2181192

Jared, B. H., Aguiló, M. A., Beghini, L. L., Boyce, B., Clark, B. W., Cook, A., Kaehr, B., & Robbins, J. (2017). Additive manufacturing: Toward holistic design. In *Scripta Materialia* (Vol. 135, p. 141). Elsevier BV., DOI: 10.1016/j.scriptamat.2017.02.029

Javaid, M., Haleem, A., Singh, R. P., Suman, R., & Rab, S. (2021). Role of additive manufacturing applications towards environmental sustainability. In Advanced Industrial and Engineering Polymer Research (Vol. 4, Issue 4, p. 312). Elsevier BV. DOI: 10.1016/j.aiepr.2021.07.005

Jayakrishna, M., Vijay, M., & Khan, B. (2023). An Overview of Extensive Analysis of 3D Printing Applications in the Manufacturing Sector. *Journal of Engineering*, *1*, 1–23. Advance online publication. DOI: 10.1155/2023/7465737

Kok, Y., Tan, X. P., Wang, P., Nai, M. L. S., Loh, N. H., Liu, E., & Tor, S. B. (2018). Anisotropy and Heterogeneity of Microstructure and Mechanical Properties in Metal Additive Manufacturing: A Critical Review. *Materials & Design*, *139*, 565–586. DOI: 10.1016/j.matdes.2017.11.021

Levashkin, D., Ogin, P. A., & Vasilyev, F. V. (2019). Efficiency of Hybrid Cyclic Processing with the Use of Additive Technologies on CNC Machines for the Manufacture of Composite Aviation Parts due to the Reduction of Processing Errors. In *Materials Science Forum* (Vol. 946, p. 959). Trans Tech Publications. https://doi.org/DOI: 10.4028/www.scientific.net/MSF.946.959

Li, H., Shi, X., Wu, B., Corradi, D. R., Pan, Z., & Li, H. (2024). Wire arc additive manufacturing: A review on digital twinning and visualization process. *Journal of Manufacturing Processes*, *116*, 293–305. DOI: 10.1016/j.jmapro.2024.03.001

Li, X., Chua, J. W., Yu, X., Li, Z., Zhao, M., Wang, Z., & Zhai, W. (2023). 3D-Printed Lattice Structures for Sound Absorption: Current Progress, Mechanisms and Models, Structural-Property Relationships, and Future Outlook [Review of 3D-Printed Lattice Structures for Sound Absorption: Current Progress, Mechanisms and Models, Structural-Property Relationships, and Future Outlook]. Advanced Science, 11(4). Wiley. DOI: 10.1002/advs.202305232

Liu, Z., Jiang, Q., Zhang, Y., Li, T., & Zhang, H. (2016). Sustainability of 3D Printing: A Critical Review and Recommendations [Review of Sustainability of 3D Printing: A Critical Review and Recommendations]. Volume 2: Materials; Biomanufacturing; Properties, Applications and Systems; Sustainable Manufacturing. DOI: 10.1115/MSEC2016-8618

Mahmood, M. A., Visan, A. I., Ristoscu, C., & Mihăilescu, I. N. (2020). Artificial Neural Network Algorithms for 3D Printing [Review of Artificial Neural Network Algorithms for 3D Printing]. Materials, 14(1), 163. Multidisciplinary Digital Publishing Institute. DOI: 10.3390/ma14010163

Martelli, N., Serrano, C., Brink, H. van den, Pineau, J., Prognon, P., Borget, I., & Batti, S. E. (2016). Advantages and disadvantages of 3-dimensional printing in surgery: A systematic review [Review of Advantages and disadvantages of 3-dimensional printing in surgery: A systematic review]. Surgery, 159(6), 1485. Elsevier BV. DOI: 10.1016/j.surg.2015.12.017

Nadagouda, M. N., Ginn, M., & Rastogi, V. (2020). A review of 3D printing techniques for environmental applications [Review of A review of 3D printing techniques for environmental applications]. Current Opinion in Chemical Engineering, 28, 173. Elsevier BV. DOI: 10.1016/j.coche.2020.08.002

Niesler, F., & Hermatschweiler, M. (2015). Two-Photon Polymerization — A Versatile Microfabrication Tool. *Laser-Technik-Journal, 12*(3), 44–47. DOI: 10.1002/latj.201500019

Oberoi, G., Nitsch, S., Edelmayer, M., Janjić, K., Müller, A. S., & Agis, H. (2018). 3D Printing—Encompassing the Facets of Dentistry [Review of 3D Printing—Encompassing the Facets of Dentistry]. Frontiers in Bioengineering and Biotechnology, 6. Frontiers Media. DOI: 10.3389/fbioe.2018.00172

Palanivel, S., Nelaturu, P., Glass, B., & Mishra, R. S. (2014). Friction stir additive manufacturing for high structural performance through microstructural control in an Mg-based WE43 alloy. In *Materials & Design (1980-2015)* (Vol. 65, p. 934). Elsevier BV., DOI: 10.1016/j.matdes.2014.09.082

Pandey, S., Srivastava, R., & Narain, R. (2020). Study on the effect of powder feed rate on direct metal deposition layer of precipitation hardened steel. In *Materials Today Proceedings* (Vol. 26, p. 2272). Elsevier BV., DOI: 10.1016/j.matpr.2020.02.492

Panwisawas, C., Tang, Y. T., & Reed, R. C. (2020). Metal 3D printing as a disruptive technology for superalloys. In Nature Communications (Vol. 11, Issue 1). Nature Portfolio. DOI: 10.1038/s41467-020-16188-7

Pelacci, M., Robins, A., & Szyniszewski, S. (2021). Drag Reduction of a Circular Cylinder Through the Use of an Architectured Lattice Material. In arXiv (Cornell University). Cornell University. https://doi.org//arxiv.2102.08938DOI: 10.48550

Peles, A., Paquit, V., & Dehoff, R. (2023). Deep-Learning Quantitative Structural Characterization in Additive Manufacturing. In arXiv (Cornell University). Cornell University. https://doi.org//arxiv.2302.06389DOI: 10.48550

Perkins, I., & Skitmore, M. (2015). Three-dimensional printing in the construction industry: A review [Review of Three-dimensional printing in the construction industry: A review]. International Journal of Construction Management, 15(1), 1. Taylor & Francis. DOI: 10.1080/15623599.2015.1012136

Pfaff, A., Jäcklein, M., Hoschke, K., & Wickert, M. (2018). Designed Materials by Additive Manufacturing—Impact of Exposure Strategies and Parameters on Material Characteristics of AlSi10Mg Processed by Laser Beam Melting. *Metals*, *8*(7), 491. DOI: 10.3390/met8070491

Plancher, E., Suard, M., Dendievel, R., Blandin, J., Martin, G., & Lhuissier, P. (2020). Behavior by design made possible by additive manufacturing: The case of a whistle-blower mechanical response. In *Materials Letters* (Vol. 282, p. 128669). Elsevier BV., DOI: 10.1016/j.matlet.2020.128669

Raheem, A., Hameed, P., Whenish, R., Elsen, R., Aswin, G., Jaiswal, A. K., Prashanth, K. G., & Manivasagam, G. (2021). A Review on Development of Bio-Inspired Implants Using 3D Printing [Review of A Review on Development of Bio-Inspired Implants Using 3D Printing]. Biomimetics, 6(4), 65. Multidisciplinary Digital Publishing Institute. DOI: 10.3390/biomimetics6040065

Rajkumar, S. M. (2022). Effect of infill pattern and build orientation on mechanical properties of FDM printed parts: An experimental modal analysis approach. In arXiv (Cornell University). Cornell University. https://doi.org//arxiv.2202.05692DOI: 10.48550

Ransikarbum, K., Pitakaso, R., & Kim, N. (2020). A Decision-Support Model for Additive Manufacturing Scheduling Using an Integrative Analytic Hierarchy Process and Multi-Objective Optimization. In Applied Sciences (Vol. 10, Issue 15, p. 5159). Multidisciplinary Digital Publishing Institute. DOI: 10.3390/app10155159

Ritchie, S. M., Kovačević, S., Deshmukh, P., Christodoulides, A. D., Malen, J. A., Mesarovic, S. Dj., & Panat, R. (2023). Shape distortion in sintering results from nonhomogeneous temperature activating a long-range mass transport. In Nature Communications (Vol. 14, Issue 1). Nature Portfolio. DOI: 10.1038/s41467-023-38142-z

Rouf, S., Malik, A., Singh, N., Raina, A., Naveed, N., Siddiqui, M. I. H., & Haq, M. I. U. (2022). Additive manufacturing technologies: Industrial and medical applications. *Sustainable Operations and Computers, 3*, 258–274. DOI: 10.1016/j.susoc.2022.05.001

Sakin, M., & Kiroglu, Y. C. (2017). 3D Printing of Buildings: Construction of the Sustainable Houses of the Future by BIM. In *Energy Procedia* (Vol. 134, p. 702). Elsevier BV., DOI: 10.1016/j.egypro.2017.09.562

Sandeep, B., Kannan, T. T. M., Chandradass, J., Ganesan, M., & Rajan, A. J. (2021). Scope of 3D printing in manufacturing industries-A review. *Materials Today: Proceedings, 45*, 6941–6945. DOI: 10.1016/j.matpr.2021.01.394

Seharing, A., Azman, A. H., & Abdullah, S. (2014). A review on integration of lightweight gradient lattice structures in additive manufacturing parts [Review of A review on integration of lightweight gradient lattice structures in additive manufacturing parts]. Advances in Mechanical Engineering, 12(6). SAGE Publishing. DOI: 10.1177/1687814020916951

Srinivas, M., & Babu, B. S. (2017). A Critical Review on Recent Research Methodologies in Additive Manufacturing [Review of A Critical Review on Recent Research Methodologies in Additive Manufacturing]. Materials Today Proceedings, 4(8), 9049. Elsevier BV. DOI: 10.1016/j.matpr.2017.07.258

Srinivasan, D., Meignanamoorthy, M., Ravichandran, M., Mohanavel, V., Alagarsamy, S. V., Chanakyan, C., Sakthivelu, S., Karthick, A., Prabhu, T. R., & Rajkumar, S. (2021). 3D Printing Manufacturing Techniques, Materials, and Applications: An Overview. Advances in Materials Science and Engineering, 2021(1). DOI: 10.1155/2021/5756563

Stefańska, A., Cygan, M., Batte, K., & Pietrzak, J. (2021). Application of Timber and Wood-based Materials in Architectural Design using Multi-objective Optimisation Tools. *Construction Economics and Building, 21*(3). Advance online publication. DOI: 10.5130/AJCEB.v21i3.7642

Tahmasebinia, F., Niemelä, M., Sepasgozar, S. M. E., Lai, T., Su, W., Reddy, K. R., Shirowzhan, S., Sepasgozar, S. M. E., & Marroquín, F. A. (2018). Three-Dimensional Printing Using Recycled High-Density Polyethylene: Technological Challenges and Future Directions for Construction. *Buildings (Basel, Switzerland), 8*(11), 165. DOI: 10.3390/buildings8110165

Tang, T., Wang, L., Zhu, M., Zhang, H., Dong, J., Yue, W., & Xia, H. (2024). Topology Optimization: A Review for Structural Designs Under Statics Problems. *Materials (Basel), 17*(23), 5970. DOI: 10.3390/ma17235970 PMID: 39685406

Tebianian, M., Aghaie, S., Jafari, N. R., Hosseini, S. R. E., Pereira, A. B., Fernandes, F. A. O., Farbakhti, M., Chen, C., & Huo, Y. (2023). A Review of the Metal Additive Manufacturing Processes [Review of A Review of the Metal Additive Manufacturing Processes]. Materials, 16(24), 7514. Multidisciplinary Digital Publishing Institute. DOI: 10.3390/ma16247514

Tetsuka, H., & Shin, S. R. (2020). Materials and technical innovations in 3D printing in biomedical applications [Review of Materials and technical innovations in 3D printing in biomedical applications]. Journal of Materials Chemistry B, 8(15), 2930. Royal Society of Chemistry. DOI: 10.1039/D0TB00034E

Tian, X., Wang, Q., Liu, G., Liu, Y., Xie, Y., & Deng, W. (2019). Topology optimization design for offshore platform jacket structure. *Applied Ocean Research, 84*, 38–50. DOI: 10.1016/j.apor.2019.01.003

Ureta, F. G., Pietroni, N., & Zorin, D. (2019). Structurally optimized shells. In arXiv (Cornell University). Cornell University. https://doi.org//arxiv.1904.12240DOI: 10.48550

Vayre, B., Vignat, F., & Villeneuve, F. (2012). Metallic additive manufacturing: state-of-the-art review and prospects. In Mechanics & Industry (Vol. 13, Issue 2, p. 89). EDP Sciences. DOI: 10.1051/meca/2012003

Vigliotti, A., & Pasini, D. (2011). *Structural Optimization of Lattice Materials.*, DOI: 10.1115/DETC2011-47390

Vorkapić, M., Hasan, M. A., Tanović, D., Baltić, M., & Tomić, B. (2020). Implementation of 6R strategy in FDM printing process: Case: Small electronic enclosure box. In Journal of Engineering Management and Competitiveness (Vol. 10, Issue 2, p. 141). University of Novi Sad, Technical faculty Mihajlo Pupin, Zrenjanin. DOI: 10.5937/jemc2002141V

Výtisk, J., Kočí, V., Honus, S., & Vrtek, M. (2019). Current options in the life cycle assessment of additive manufacturing products. In Open Engineering (Vol. 9, Issue 1, p. 674). De Gruyter. DOI: 10.1515/eng-2019-0073

Wang, C., Tan, X., Tor, S. B., & Lim, C. S. (2020). Machine learning in additive manufacturing: State-of-the-art and perspectives. In *Additive manufacturing* (Vol. 36, p. 101538). Elsevier BV., DOI: 10.1016/j.addma.2020.101538

Wang, X., Jiang, M., Zhou, Z., Gou, J., & Hui, D. (2016). 3D printing of polymer matrix composites: A review and prospective [Review of 3D printing of polymer matrix composites: A review and prospective]. Composites Part B Engineering, 110, 442. Elsevier BV. DOI: 10.1016/j.compositesb.2016.11.034

Wilson, S., Thomas, R. J., Mary, N., Bosco, E. T., & Gopinath, A. (2021). Development and fabrication of fused deposition modelling 3D printer. *IOP Conference Series. Materials Science and Engineering*, *1132*(1), 12019. DOI: 10.1088/1757-899X/1132/1/012019

Wong, K. V., & Hernandez, A. (2012). A Review of Additive Manufacturing [Review of A Review of Additive Manufacturing]. ISRN Mechanical Engineering, 2012, 1. Hindawi Publishing Corporation. DOI: 10.5402/2012/208760

Wu, J., Aage, N., Lefèbvre, S., & Wang, C. C. L. (2017). Topology Optimization for Computational Fabrication. In Eurographics. DOI: 10.2312/egt.20171030

Wu, P., Wang, J., & Wang, X. (2016). A critical review of the use of 3-D printing in the construction industry [Review of A critical review of the use of 3-D printing in the construction industry]. Automation in Construction, 68, 21. Elsevier BV. DOI: 10.1016/j.autcon.2016.04.005

Yang, C., & Li, Q. M. (2020). Advanced lattice material with high energy absorption based on topology optimisation. In *Mechanics of Materials* (Vol. 148, p. 103536). Elsevier BV., DOI: 10.1016/j.mechmat.2020.103536

Yeole, S. N., Madhav, C. H. V., & Kesav, R. S. N. H. (2016). Importance and utilization of 3D printing in various applications. https://www.researchgate.net/profile/Shivraj_Yeole/publication/293487409_Importance_and_utilization_of_3D_printing_in_various_applications/links/5720d8aa08aeaced78907019.pdf

Yin, H., Qu, M., Zhang, H., & Lim, Y. (2018). 3D Printing and Buildings: A Technology Review and Future Outlook [Review of 3D Printing and Buildings: A Technology Review and Future Outlook]. Technology|Architecture + Design, 2(1), 94. Taylor & Francis. DOI: 10.1080/24751448.2018.1420968

Yin, J., Suo, Y., Lv, T., Ma, K., Wang, X., & Zhang, Z. (2022). Application of 3D Printing Technology in the Construction Industry and Its Development Prospects. In Atlantis Highlights in Intelligent Systems/Atlantis highlights in intelligent systems. DOI: 10.2991/ahis.k.220601.002

Zhou, L., Miller, J., Vezza, J., Mayster, M., Raffay, M., Justice, Q., Tamimi, Z. A., Hansotte, G., Sunkara, L. D., & Bernat, J. (2024). Additive Manufacturing: A Comprehensive Review [Review of Additive Manufacturing: A Comprehensive Review]. Sensors, 24(9), 2668. Multidisciplinary Digital Publishing Institute. DOI: 10.3390/s24092668

Chapter 7 Process Planning and Preparation for 3D Printing From Design to Execution

Romdhane Ben Khalifa
https://orcid.org/0000-0002-9171-9584
National Engineering School of Gabes (ENIG), Tunisia

ABSTRACT

The pace of innovation in 3D printing, or additive manufacturing, has changed the product design and fabrication process in many sectors, including healthcare, aerospace, and general manufacturing. However, turning a digital design into a physical object even high-quality requires a comprehensive workflow that covers all activities from planning to production. This paper attempts to give an overall view of the main milestones in process planning and preparation for 3D printing while emphasizing the significance of each stage toward achieving the best results. The survey starts with the first step of developing a digital model where several issues are listed that relate to designing efficiency optimization, file type such as STL or OBJ and correction of mistakes identified so as to make the model printable. From here, the workflow proceeds to model orientation and slicing wherein for slicing parameters like layer thickness, infill density and support structures are chosen parameter specifically for the printing technology and material used.

DOI: 10.4018/979-8-3373-0533-2.ch007

INTRODUCTION

The ability to fully control process planning and preparation emerges because 3D printing grants users complete freedom to move from design to physical output (Egan, 2019). The easy accessibility of 3D printers combined with their simple operation has created a false impression that object creation can be achieved through a single button click after designing a 3D digital model (Jaksic, 2015). The successful execution of 3D printing depends on correct thinking about material properties, printer capabilities, and part performance needs (Chen, 2016). The specific applications need advanced expertise and sophisticated parameter optimization for biological additive manufacturing and deep learning stress prediction (Goh et al.,2020). All these complexities are involved in a detailed process plan: material selection, pre-processing of 3D models, printing parameters definition, and post-processing techniques. The process blueprint provides step-by-step guidance for maintaining consistency and preventing errors during three-dimensional printing operations. The combination of precise scaffold design with three-dimensional printing of direct functional scaffolds and process modeling enables this complex activity (An et al.,2015). Process evolution simulation and outcome prediction within artificial environments generate insights that drive continuous process enhancement (Zhang et al.,2024). The success of three-dimensional printing projects depends heavily on proper planning procedures.

The technology of 3D printing brings revolutionary changes to society while its effects become visible across manufacturing, education, and healthcare sectors (Arvanitidi et al., 2019; Huleihil, 2017; Noorani et al., 2019). The additive manufacturing principles enable 3D printing to create solid items by adding material layers according to computer-aided design models (Chen et al., 2021; Marew & Birhanu, 2021). Unlike traditional subtractive manufacturing processes, 3D printing produces complex geometries with intricate design features because it operates without molds or special tooling (Yin et al., 2022). The process begins with digital input of the intended product design. The conversion process produces separate horizontal sections for each layer.

The 3D printer receives instructions from these cross-sections, which show the material deposition sequence for building the final three-dimensional structure (James et al., 2017). 3D printing enables the creation of products through various materials, including polymers, metals, ceramics, and composites, which possess unique mechanical, thermal, and chemical properties for specific uses (Garcia et al., 2013). The third printing in different sectors prompted significant research and development efforts to improve materials, processes, and uses. Medical is one of the fields that will benefit enormously from 3D printing; it promises to make personalized medical products, drugs, and devices more easily available (Marew

& Birhanu, 2021). Academics have investigated laboratory equipment customized with 3D printing, possibly saving time and money (Coakley & Hurt, 2016). In dentistry, surgical guides are produced using 3D printing technology, but with the continuous improvement of the technique, more applications and higher accuracy are anticipated (Jeong et al., 2023).

The inherent promise of 3D printing for personalized medicine is again marred by its intrinsic limitations, such as speed, material availability, and complex post-processing (Zhou et al., 2024). Device efficacy or effectiveness, most importantly, cannot be substituted with inconclusive research to demonstrate clinical success; Diment et al., 2017. In addition, quality control issues became a significant hurdle to the widespread adoption of 3D printing in regulated industries like the pharmaceutical sector or aerospace. So, although there is vast potential in personalized medicine, the journey toward that goal will require technology, which limits regulatory uncertainty, and ethical considerations do not come to pass.

This paper's methodology follows a comprehensive and systematic process for 3D printing planning and preparation, which includes everything from digital design to the final product. First, digital models are designed through CAD software or 3D scanning, with special attention given to file preparation and error correction so that the models are printable. In medical applications or modeling, the anatomical data of patients is segmented and modified to obtain suitable personalized designs. The procedure then continues to slicing, where specific software slices the model into layers, and parameters like layer thickness, infill density, and support structures are defined. The best parameters are obtained using optimization algorithms such as group search optimization and orthogonal experimentation that illustrate a trade-off between mechanical strength and surface quality. Production planning incorporates advanced scheduling and 3D bin packing algorithms to exploit printer utilization effectively while decreasing lead time, especially in multi-printer settings. Specific machines will address such issues as material selection, calibration of printers, and cloud-based monitoring systems so that printing remains consistent and reliable over time. Integrating support removal, curing, and surface finishing enhances printed parts' functional and aesthetic properties. Support quality definition identifies critical quality attributes throughout the process based on Quality by Design (QbD) principles, while risk assessment strategies focus on identifying potential failures. Real-time detection of errors with automatic workflow adjustments enhances reliability further. This methodology thus combines technical accuracy with algorithmic optimization along with operational efficiency in the context of producing high-quality complex 3D printed parts across a variety of applications. The Figure 1 begins with modeling, where a digital design is created using CAD software; followed by printing, where the 3D printer builds the object layer by layer based on the digital model; and ending with finishing, which involves post-processing steps like surface smoothing and removal of support structures to achieve the final desired quality and functionality of the printed object.

Figure 1. Different steps of 3D printing (Mohapatra et al., 2022)

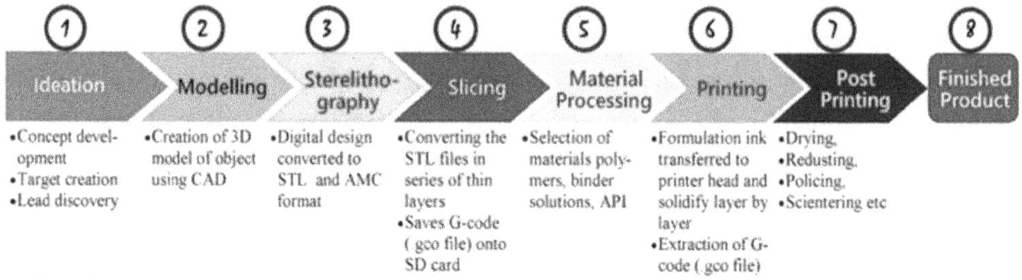

DESIGN PHASE

Before starting the printing operation, a detailed process plan is required to determine the appropriate printing parameters and sequence of operations (Lolla et al., 2023). This plan is formulated by considering several factors, such as layer thickness, speed of printing, temperature settings, and orientation in the build, which subsequently affect the end product's quality, accuracy, and mechanical properties (Bahraminasab, 2020). The characteristics of the materials, the particular 3D printing method employed, and the design's complexity govern these parameters.

The preliminary design stage is critical for 3D printing because it demands intensive knowledge concerning materials, the printing process, and the interaction of the final product in its intended use (Zhou et al., 2024). This includes the process during which a design is developed, requirements are defined, and production planning is prepared (Carrier, 1987). The design process starts with digitizing an initial shape but still involves considering human body proportions when designing wearable artifacts (Minaoglou et al., 2024). The industrial applications of 3D printing offer such freedom that designers can seriously consider and bring to life design complexities that were once impossible and only theoretically exploitable (Niesler & Hermatschweiler, 2015). Design with 3D printing allows the quick and practical realization of concepts so that designers can rapidly build and test their ideas (Zhao, 2020). Thus, it will speed up the design phase and facilitate innovation among various sectors because it can easily translate digital designs into physical prototypes. Additive manufacturing as a new approach influences environmental impacts throughout the product lifecycle and enhances achievable sustainability-

Panmirzaei et al. Therefore, a proper review of 3D molding and multiobjective optimization techniques will be necessary. Life Cycle Assessment and Computer-Aided Design will be integrated to enhance sustainability over 3D printing processes, etc. Furthermore, these techniques help reduce adverse environmental effects while promoting greener production methods alongside better product design.

CAD Modeling

Making a digital model is an important step in 3D printing, usually performed through Computer-Aided Design software that allows an accurate and detailed image of the intended object (Jaksic, 2015). Such a digital model acts as the blueprint for the process of 3D printing by guiding the printer to make the tangible object layer by layer. The CAD program permits intricate design personalization, making it possible to create complex shapes and interiors that are challenging or impossible to achieve through conventional manufacturing methods (Yin et al., 2022). CAD modeling offers freedom in optimizing designs for functional performance, visual appeal, and ease of manufacture. Changes in design parameters enable improved material efficiency and product performance at increased structural integrity (Gardan, 2020). Additionally, CAD systems often include simulators that allow designers to test and verify their designs virtually before production. This helps pinpoint potential problems and fine-tune the design for accuracy and efficiency in replication.

Figure 2. Format of STL file for one example (Živanović, S et al.,2019)

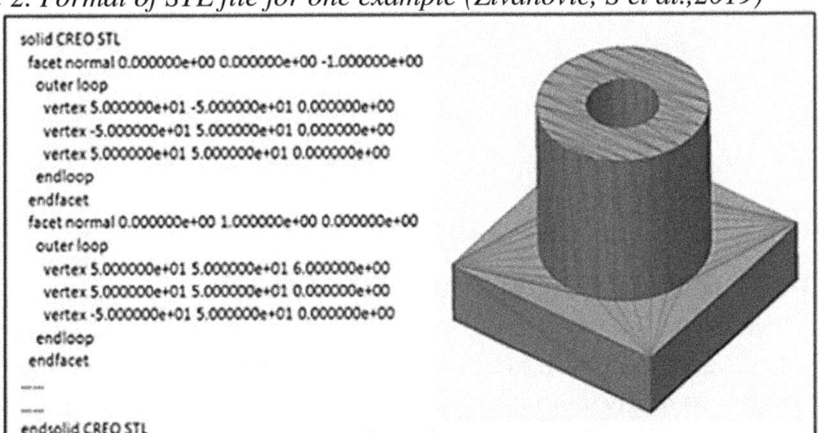

A design for 3D printing has to be optimized considering what the technology can and cannot do so that a successful production will be realized (Jayakrishna et al., 2023; Olasek & Wiklak, 2014). This also covers the optimization of designs relative to structural integrity, material usage, and functionality enhancement. Several important factors in designing parts influence whether or not 3D printing will succeed (figure 3)

. These include part orientation within a given building volume, generation of support structures, and selection of materials (Garcia et al., 2013). Good part orientation minimizes support structures and, hence, material wastage and time spent on post-processing. Supporting structures must be strategically designed so as not to allow deformation or collapse during the printing process – especially for complex geometries with overhangs. The choice of materials determines the mechanical properties, visual appeal, and longevity of the printed object; thus, a complete knowledge of what is available and suitable for the intended use is determined. Topology optimization is used in parameterization using elements for the design space. It provides a solution for optimum material distribution based on performance measures alongside constraints imposed by (Wu et al.,2017).

Figure 3. General steps involved in 3D printing (Kumar et al., 2021)

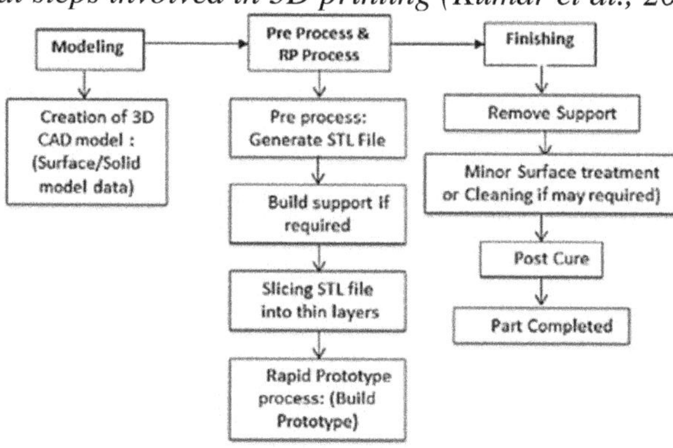

DESIGN FOR MANUFACTURABILITY (DFM)

DFAM is significantly different from the conventional manufacturing design approach due to its nearly unlimited design freedom and new opportunities in material and topology design (Wang et al., 2020). Successful three-dimensional printing requires Design for Additive Manufacturing principles, which are optimized designs that take full advantage of additive manufacturing's unique capabilities while somewhat alleviating potential challenges (Alfaify et al., 2020). The key tenet of DFAM is that parts are designed with self-supporting geometries, so there is hardly any need for support structures, thus saving materials and reducing post-processing time and costs. Another central concept in DFAM is part consolidation, which refers to assembling many parts into one integrated component, thereby minimizing assembly time and enhancing the performance of the assembled product. Moreover, DFAM advocates lightweight yet robust designs that use materials better while reducing weight without compromising structural integrity. The strategic application of attributes like lattice structures, infill patterns, and functionally graded materials will enhance the designers' performance and functionality.

DFAM requires a thorough understanding of material properties, process parameters, and equipment capabilities. In creating or redesigning existing components, it employs a 6R approach that allows quick and easy changes to be made to models in response to client requirements before production begins (Vorkapić et al., 2020). The 6R approach involves remanufacturing, repair, redesign, repurposing, recycling, and reduction (Anwajler et al., 2023). Engineers and designers must work closely to optimize the designs for the chosen 3D printing process, build orientation support structures, and material properties (Păcurar,2019). Manufacturers who adhere to DFAM principles will then be able to fully exploit all the benefits additive manufacturing can offer, with substantial improvements in product performance, cost-effectiveness, and sustainability.

Additive manufacturing technologies arguably mark a revolution in the industrial world, providing unmatched freedom of design and personalization (Jiménez et al., 2019). In AM, the design freedom is much greater than in the traditional processes where parts and functions are integrated (Schneberger et al., 2019). This enables intricate geometries and complex internal structures that can seldom be realized with conventional manufacturing techniques (Dezaki & Ariffin, 2020). For production methods where material placement is crucial, AM allows designers control over manufactured items' outer morphology and inner geometry (Wooldridge, 2020).

PREPARATION PHASE

The preparation phase is crucial for any successful 3D printing outcome; it encompasses data, build setup, and material selection so that the digital design matches the physical printing process (figure 4). Several important steps are involved in this process, namely the turning of the CAD model into an appropriate format, for instance, STL; slicing the model into thin layers; and creating toolpaths for the printer, as well as choosing suitable printing parameters (Garcia et al., 2018). The STL file format has become and is still the standard file format for 3D printing. This format represents the surface geometry of an object as a collection of triangles (Giannopoulos et al., 2015). Slicing software will further slice the STL file into horizontal layers. Each layer corresponds to a cross-section of the object at a particular height. It must be noted that the quality of the STL file and the slicing parameters directly influence the accuracy, surface finish, and mechanical properties of any printed part (Budinoff, H.D et al., 2021).

Figure 4. Workflow of the 3D printing process (Shen, Y et al., 2022)

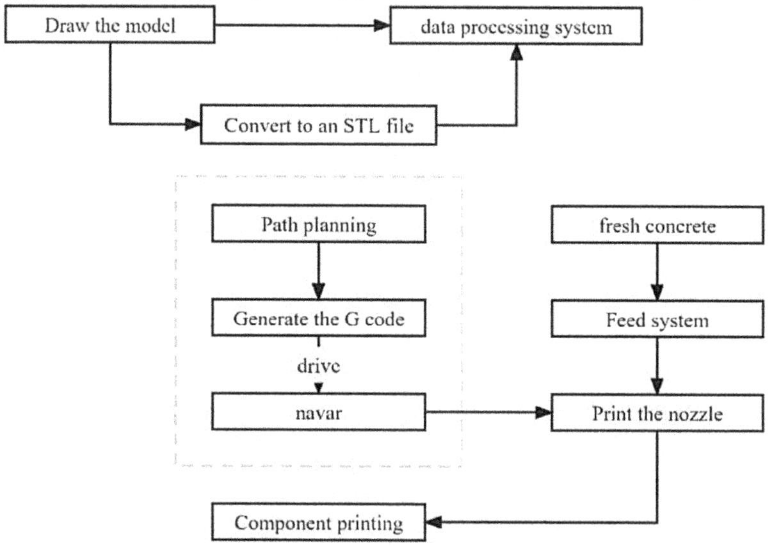

Material Selection and Handling

The selection of materials in 3D printing represents a crucial decision point because it determines the final product's mechanical properties, appearance, and functional capabilities (He & Khan, 2021). 3D printing enables users to work with polymers, metals, ceramics, and composite materials. Each category has its strengths and weaknesses. The selection of materials depends on particular desired properties, which include strength, stiffness, impact resistance to temperature resistance, and electrical compatibility. Polymers lead the 3D printing market because these materials offer versatile applications, simple processing methods, and affordable alternatives (Alghamdi et al., 2021). Metals are preferred for aerospace, automotive, and medical applications because they provide strength and heat stability (Panwisawas et al., 2020). Ceramics possess outstanding hardness properties, wear resistance, and thermal stability, making them appropriate for tooling mold applications and high-temperature environments.

Slicing

The 3D model is sliced into thin horizontal layers during slicing, which dictates the movement of the 3D printer. The coordinates in the layered structure determine each layer's size and shape (Morampudi et al., 2021). Layer thickness is a function of the time used for printing and materials consumed during the process; more delicate layers are created with finer details but take longer to print. The final STL file is sliced into horizontal layers (Majko et al., 2020). Such large and complex STL models can be sliced efficiently (Choi & Kwok, 2002). The slicing software improves speed, accuracy, and material usage in printing by generating toolpaths that dictate how exactly the print head or laser moves during manufacturing. One example of modern cutting-edge slicing software is Cura, which takes a three-dimensional model as input and outputs numerous horizontal layers (Frizziero et al., 2019). Slicing sets parameters such as layer height, infill density, and support structures so that the process is optimal for what is desired. The figure 5 illustrates the general three-dimensional (3D) printing process, which begins with creating a 3D geometric model, typically using CAD software-followed by process planning steps such as orienting and positioning the object, slicing the model into layers, adding support structures if needed, and generating the toolpath for the printer

Figure 5. (a) General three-dimensional (3D) Printing Process; (b) 3D Triangle Mesh Slicing in OYZ space and in OXYZ space (Pham, G. N et al., 2018)

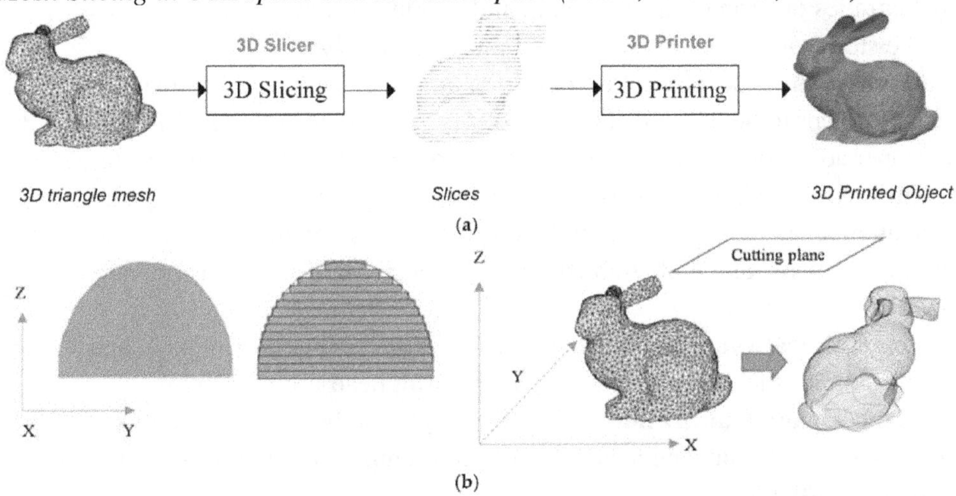

Printer Setup and Calibration

Only proper setup and calibration of the printer will enable it to print 3D objects accurately and reliably. Calibration is adjusting the settings for specific parameters, such as nozzle temperature, bed temperature, speed of printing, and extrusion rate, to achieve the best flow of the material and adhesion. If not correctly calibrated, several defects in printing are seen, such as warping, delamination, and dimensional error. Good bed adhesion is required so that the first layer sticks well to the build platform without warping or delamination. Optimal bed adhesion is achieved through proper adjustment of bed temperature, velocity of printing, and thickness of the first layer. The initial layer adherence at the printing plate, object orientation, inappropriate geometry, geometric dimensions' errors, misalignment, and insufficient support structures are among the causes of failures in printing (Jaksic, 2015).

The printer's mechanical parts, including its print head, build platform, and motion system, require careful alignment and calibration to ensure accurate and steady printing.

The setup and calibration of the printer play an important role in determining print quality (Birtchnell & Hoyle, 2014; Hsieh, 2017). It includes correct assembly, leveling of the build platform so that first-layer adhesion is not compromised, and calibration regarding dimensions and firmware of the printer (9 Steps on How To Set Up a 3D Printer, 2024; Hsieh, 2017). Nozzle height is explicitly set to improve the first layer (Satterfield, 2020). Printing accuracy must be checked, especially after any upgrades or changes in environmental conditions have been made.

The steps for printer setup and calibration generally include:

Assembly and Leveling: Proper printer assembly with all components level and square (9 Steps to Set Up a 3D Printer, 2024). The leveled build platform is requisite for proper adhesion of the first layer (9 Steps on How to Set Up a 3D Printer, 2024).

Calibration: Calibration is done so that the printer produces quality prints (Birtchnell & Hoyle, 2014). It entails setting the physical dimensions of the printer, configuring firmware, and establishing parameters for the built-in software (Hsieh, 2017).

Axis Calibration: During the calibration, you must calibrate the X and Y axes using a simple ruler to measure how far the print head travels (2024).

Nozzle Height Calibration: Adjust the height for the best first layer (Satterfield, 2020). The nozzle must be heated to printing temperature before leveling the bed (Satterfield, 2020).

Accuracy Assessment: The accuracy of the 3D printer needs to be assessed, especially after any upgrades or changes in environmental conditions (Davydov et al., 2021). New printers may also require some calibration (Issal, 2023).

The 3D printing methodology appears in Figure 5 as a flowchart that requires professional involvement from engineering or design experts. The process starts with idea generation, followed by 3D CAD design that creates an STL file before the printer receives G-code instructions. The printing operation builds material through successive layers, which depend on adjustable parameters such as layer thickness, nozzle temperature, and extrusion rate. The printing process includes controlled cooling, curing, and finishing stages. The flowchart confirms all stages from design creation through final validation to maintain consistent quality and precision in 3D printing operations.

Figure 6. Methodology flowchart of printing 3D (Sovetova & Calautit, 2024)

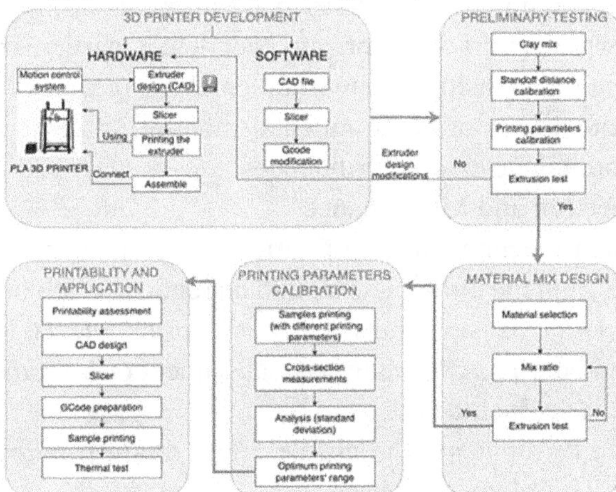

Printer Setup

A good printer setup is necessary to achieve dimensional accuracy and quality results in 3D printing. This includes print parameters, hardware readiness, and a controlled environment.

Print Parameter Configuration

The critical parameters are edge print speed, inner print speed, and inner slope. All three directly influence dimensional accuracy. Therefore, by changing these parameters, one can very effectively control the amount of material placed. Control is very much needed to obtain the desired geometry of the printed parts.

- The operator must ensure that the software settings (temperature, speed, and material profiles) match the specific filament or resin used. These settings may need to be adjusted for the best print quality when using different materials or part geometries.

- Build Platform Preparation

  The build platform should be cleaned before every print to ensure that dust, particles, or leftover glue do not interfere with first-layer adhesion and cause failures in the print process.
- Still, the bed must be leveled so that the distance between the nozzle and the build plate is constant throughout the surface, which is suitable for even extrusion and good first-layer adhesion.
- Environmental Controls

  The environment where printing takes place must be controlled regarding temperature, humidity, and airflow. Variations in these parameters can lead to warping, cracking, or any other defect, particularly when printing with materials sensitive to environmental changes, such as ABS or nylon.
- The use of enclosures or environmental monitoring devices helps stabilize conditions and reduce external influences.
- Routine Inspection and Maintenance

  Inspection and maintenance of critical components such as nozzles, belts, bearings, and the build plate should be regular. Worn or clogged nozzles lead to poor extrusion and subsequently print defects; loose or damaged belts and bearings cause layer misalignment or even mechanical failure
- Maintenance tasks include:

  Cleaning or replacing the nozzle after certain printing hours or when extrusion problems are observed.
- Lubricating moving components (lead screws, rails, bearings) to minimize friction and wear.
- Regular checks and tightening of belts, screws, and frame components ensure the structure holds together and the print remains accurate.
- Checking and Cleaning of the fans as calendrical connections to prevent overheating and electrical faults.
- Material Handling

  Store filament dry and airtight to avoid absorbing moisture, which can cause problems with extrusion and diminish print quality.

  Appropriate procedures should be followed when changing materials to ensure the old material is disposed of and to prevent cross-contamination or clogs.
- Controlling printer parameters, preparing build platforms, managing environments, and performing regular maintenance checks can enhance the reliability, dimensional accuracy, and overall quality of 3D-printed parts manifolds.

Support Structure Generation

Support structures are temporary constructions designed to hold features that overhang in place so they do not fall during printing. It gives a solid base for the layers that follow, allowing accurate and reliable printing of complex geometries.

The geometry of the component, material characteristics, and printing parameters are critical in deciding the support structure design. The type, density, and orientation of supports can all affect printing time, the quantity of material used, and even the surface finish on the final printed part.

Complex geometries require a support structure to provide these overhanging parts with a foundation; however, these structures were designed to be easily removable without damaging the final product (Kumar et al., 2020). Support structures provide stability during the printing process, allowing proper adhesion of each layer to the one beneath.

Figure 7. Example of Support Structure

Parameter Settings

The nozzle temperature affects the quality of the parts produced. At an optimal temperature, the printed parts are of good quality with a smoother finish and better detail resolution (Loh et al., 2020). An optimized temperature would also obtain overhangs and better bridging (Loh et al., 2020). The inappropriate setting of the nozzle temperature would cause under-extrusion and poor adhesion between layers, resulting in a weak structure.

Printing speed refers to the rate at which the print head moves along the X and Y axes, parameters that influence both the time taken to print and the quality of the produced part (Almy & Tontowi, 2018). The height of layers determines the printed part's resolution and surface finish; thinner layers yield better detail but take longer to print (Mendřický & Fris, 2020). The two important factors determining printed specimens' morphological features are extrusion temperature and speed, which affect the cooling process.

Infill density determines the weight and strength of the printed object; increasing the infill density results in stronger but heavier parts, which also affects printing time (Kędziora et al., 2023). Build orientation affects mechanical properties, surface texture, and printing time (Dezaki et al., 2021).

Figure 8. 3D printing parameters (Manoj Prabhakar et al., 2021)

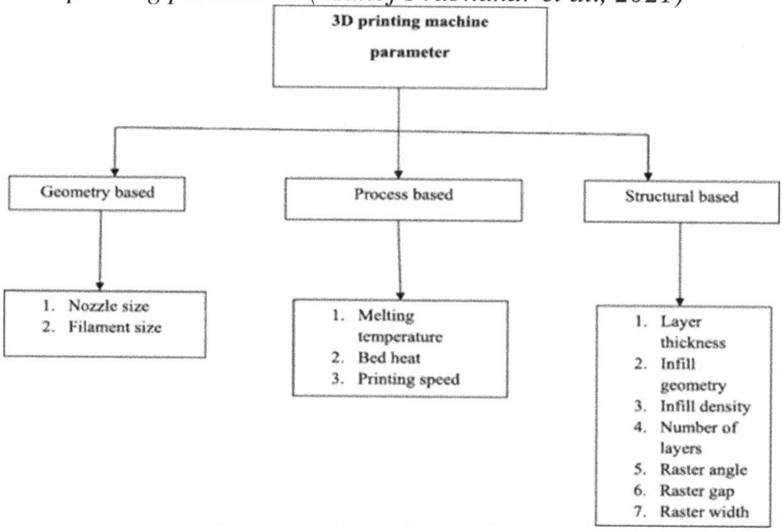

EXECUTION PHASE

The execution phase is the real 3D printing where, according to the prepared design, the object is formed by depositing materials layer by layer. Continuous monitoring and control are needed for this process to maintain quality and accuracy. This phase covers a wide range of activities, including monitoring the printing process, controlling material flow, and solving any problems that may occur during printing. 3D printing makes an object three-dimensionally, layer by layer, based on toolpaths prepared in the preparation phase. The 3D printer makes each layer of the final object by accurately depositing materials through extrusion, powder bed fusion, or stereolithography (Ghilan et al., 2020).

Fused Deposition Modeling (FDM) is a process in which a thermoplastic filament is extruded through a heated nozzle, which deposits the material layer by layer on the build platform (Mwema & Akinlabi, 2020). Monitoring in real-time is required to ensure that any problems arising, such as clogging the nozzle, warping of the material, or failure of the printer itself, are promptly detected and corrected.

Printing Process

While printing, the 3D printer follows the toolpaths created by the slicing software and puts down material, layer upon layer, to build the final object (Dul et al., 2021).

Printing parameters, including nozzle temperature, bed temperature, printing speed, and extrusion rate, are also kept within certain limits to ensure satisfactory material flow, adhesion, and part quality.

The printer extrudes the material through a diameter nozzle of 0.3 to 0.8 mm. The machine places the paths side by side to form one layer of material with a thickness ranging between 0.05 and 0.4 mm (Paprocka & Kempa, 2021). Infill density is an important parameter that determines the strength, weight, and duration of printing for the part (Martín et al., 2021).

Monitoring and Control

Real-time monitoring of the 3D printing process is crucial for detecting and solving any issues that may arise, such as nozzle clogging, printing material warping, or even printer malfunction (Christensen & Rybicki, 2017). This data, along with other sensor data, visual inspection, and camera feeds, provides beneficial information about the state of the printing process so that interventions can be made in time.

Post-Processing

Support material removal is one of the simplest to most complex operations in finishing surfaces, painting, and assembly. The material determines the post-processing methods, the surface finish required, and the application intended for the printed part. In addition to support structures being removed from parts printed by a printer, processing can also improve surface finishes or make the parts more straightforward to handle. According to Sağbaş (2020), part processing after printing involves changes to improve mechanical properties.

Part Consolidation

Once printed, 3D parts often need to be joined together with other parts to form usable products or systems. Part consolidation, enabled by 3D printing, simplifies the assembly process and reduces the number of separate components required (Chua & Leong,2013). Three-dimensional printing allows multiple pieces to be integrated into a single design, thus decreasing overall complexity and time of assembly.

Challenges During Printing

Notwithstanding the development in 3D printing technologies, numerous problems are encountered during the printing operation that subsequently affect the printed components' quality, accuracy, and dependability (Zhou et al., 2024). These include material-related issues such as warping, cracking, and delamination caused by thermal stresses, poor bed adhesion, or inappropriate material properties. Specific printer-related problems include nozzle clogs, extruder malfunctions, and axis misalignment that interrupt operations and result in defective parts. Some continuing challenges in FDM are surface quality issues like volumetric error and shape deviation from finish (Taufik & Jain, 2020). Because pre-processing techniques play a significant role in controlling surface finish and accuracy requirements when handling different geometries under high time constraints, Chohan and Singh (2017) cited that material properties, process parameters, and post-processing treatments must be considered for effective three-dimensional food printing. According to Liu et al. (2017), parameter optimization concerning processing conditions and post-treatment processes is crucial in improving mechanical property characteristics and surface quality for parts manufactured via three-dimensional printing. For good print results under three-dimensional additive manufacturing, processing parameters, material properties, and environmental conditions must be carefully considered (Oberoi et al., 2018).

ADVANTAGES OF 3D PRINTING

3D printing is a disruptive technology with applications in various sectors, including biomedicine, aerospace, and manufacturing (Babbar et al., 2021). The key advantages of 3D printing over conventional manufacturing processes are creating intricate geometries, product customization, and reduced material waste (Nadagouda et al., 2020). Through 3D printing, the healthcare sector is taking the production of tailored prosthetics, implants, and surgical guides to a new level (Liu et al., 2020). With such an application of 3D printing, on-demand personalized medicine could be produced to enhance patient adherence and outcomes during treatment (Zafar et al., 2019). These factors directly translate into data handling complex geometries with high precision, ecological benefits, and cost-effectiveness, which have been the reasons for its increasing popularity in diverse industries(Yin et al.,2018). Medical interventions can be crafted that are more specific to individual patients' needs, thereby enhancing the outcomes of treatments (Mathews et al.,2020).

LIMITATIONS OF 3D PRINTING

3D printing is spectacular because of its numerous advantages, but it also has some limitations that must be addressed to utilize the technology entirely. The cost and time of producing devices through 3D printing are still hindering its access in hospitals (Martelli et al., 2016). One of the significant limitations is that only a few materials are available under 3D printing compared to the traditional manufacturing process (Martelli et al., 2016). Thirdly, another limitation is mass production; generally, 3D printing is best suited for small-batch or customized production runs, not for scaling up. However affordable 3D printing may be, there is one such limitation regarding the availability of 3D printable biomaterials for just about every application (Choonara et al., 2016). For most medical fields, the safety and efficacy of devices manufactured using 3D printing technology remain unknown Diment et al., 2017}. Moreover, clinical research and regulatory frameworks constitute hurdles toward broader adoption of medical applications using 3D printing technology {Liaw & Guvendiren,2017; Zhang et al.,2019}.

APPLICATIONS

Post-Processing

Final processing is an important step in 3D printing, including many activities done after the printing to improve the printed part's quality, functionality, and appearance. All these activities are aimed at surface finish, removal of support structures, improvement in mechanical properties, and preparation for final use. The nature of post-processing depends on materials, the technology used in printing, and goals. However, in almost all cases, there is a need to remove supports from parts with overhangs or complex geometries (Wong, 2016). Support structures are sacrificial elements that provide temporary support during printing but must be removed without damaging the main structure (Jeong et al., 2023). The surface finishing technique involves sanding, polishing, and coating to enhance smoothness, texture, and aesthetic value. Such techniques help eliminate layer lines and reduce surface roughness, improving the piece's aesthetics (Dawood et al., 2015). Hot isostatic pressing and other processes eliminate internal voids, thus enhancing mechanical properties such as fatigue life (Dawood et al., 2015). Biocompatibility must always be ensured for bone scaffolds made from 3D printing and implants (Li et al., 2022). A post-processing step involving sterilization surface modification and coatings with biocompatible materials meant to enhance cell adhesion proliferation and differentiation Oleksy et al.(2023). Proper post-processing can significantly enhance the performance and longevity of 3D-printed parts.

The biomedical field is one of the most prominent fields where 3D printing has shown remarkable promise and may soon become significant in solving some major human health challenges (Aimar et al., 2019). Furthermore, it could significantly enhance research and medical practice while changing production in almost every sector(Amaya et al., 2024; Coakley & Hurt, 2016). One of the most optimistic aspects of this technology is the production of tissue scaffolds, where 3D printing enables scaffold manufacturing to be standardized, mold-free, and capable of virtually any design with a wide range of materials and techniques(Kantaros & Piromalis, 2021). The scaffolds are designed to provide only a temporary support structure while promoting the appropriate regeneration behavior by guiding cell proliferation. The design geometry plays a critical role in scaffold selection for particular criteria regarding material and geometric features to provide an adequate environment for cells (Liu & Yan, 2018). With 3D printing technology, tailored designs for patients can be made from data obtained through computed tomography scans, ensuring perfect accuracy in integrating the patient's anatomy (Kumar et al., 2016).

Support Removal

Support removal is one of the crucial stages in post-processing 3D-printed items, especially those containing complex geometries or overhangs that require support during the printing process. Support structures are sacrificial elements placed intentionally to prevent the part from collapsing or deforming during printing due to lack of support. These can be homogeneous or heterogeneous with the part, but they must always be easy to detach. The technique for removing supports depends on materials, geometry, and manufacturing processes. Manual removal is breaking away or cutting off the supports physically using hand tools like pliers, knives, or sandpaper. This is best for simple support structures and brittler materials like glass or ceramics. In chemical removal processes, the support material is dissolved in a solvent wherein part materials are unaffected. It is thus best for pretty complex support structures and fragile parts that may get damaged if removed manually. Breakaway supports can easily be snapped off with minimal effort from the printed component.

On the other hand, soluble supports will dissolve in a specified solution without harming the part material itself. Heat treatment can also weaken these supporting structures so that they can be removed more easily without causing any damage to the parts themselves. The aim here is clean removals without affecting surface finish or structural integrity.

Surface Finishing

Surface finishing techniques are essential for the aesthetic, functionality, and performance of 3D printed components. These techniques address issues of surface smoothness while eliminating roughness and imperfections such as layer lines, voids, and protruding features. A broadly utilized surface finishing technique is sanding, which uses abrasive materials to eliminate layer lines and hone rough surfaces, thus improving the part's overall finish. Another surface finishing technique used on 3D-printed parts is polishing, which gives a smooth, glossy finish. Coating refers to applying a thin layer of material on the surface of the 3D-printed component to improve aesthetics, protect it from environmental factors, or enhance functionality. The machining processes include grinding, polishing, and grid-blasting to remove or shape material surfaces (Zhang et al., 2021).

Heat Treatment and Assembly

Heat treatment is an important post-process that can significantly improve the mechanical properties and dimensional stability of 3D-printed parts. In this process, the printed part is subjected to controlled heating and cooling cycles so that microstructural changes occur, internal stresses are relieved, and the part improves performance. Assembly makes the final product by joining several 3D-printed parts or integrating them with other components. Press fitting is characterized by a tight interference fit where one part is inserted into another. Heat staking is a process in which localized heating causes a plastic boss or protrusion to melt and flow to secure another component. Bonding 3D-printed parts can be accomplished using epoxy, cyanoacrylate, or structural adhesives. These provide quite a solid and durable connection.

The properties of 3D-printed parts are affected by printing parameters, material characteristics, and post-processing treatments. A part produced by additive manufacturing must have some predetermined dimensional accuracy and surface quality (Kent et al., 2017). Requirements for above-normal can only be met through hybrid manufacturing with subtractive finishing operations (Ituarte et al., 2015).

Structured Patterns and Sintering

Structured patterns are now considered a potential approach to enhancing the performance of 3D-printed components (Zou et al., 2023). Since such patterns can be conceived and applied to the geometry of the component, they will affect not only the mechanical and thermal properties but also tribological characteristics; hence, options are broadened for functionality customization in totally three-dimensional printed parts. Sintering is an essential post-process for powder-based 3D printing, and it plays a crucial role in transforming powder particles into a solidified cohesive mass (Ritchie et al., 2023). The grain size and porosity of ceramics are directly influenced by the sintering temperature, casting technique, and pressure; these factors determine the mechanical strength, corrosion resistance, and optical properties of ceramic samples.

Lightweighting via Infill

Lightweighting is a design approach that minimizes the weight of a component or structure while preserving its structural integrity and functionality (Pernet et al., 2022). it is most applicable in the aerospace, automotive, and sports equipment industries, where reduction of weight leads to enhanced performance, fuel efficiency, and maneuverability (Pernet et al., 2022). Infill patterns and densities are among the key parameters that influence the mechanical properties of 3D-printed components (Cojocaru et al., 2022). Infill density compromises strength and lightness (Khaliq et al., 2023). The printed specimens' strength and weight rely on various infill patterns such as linear, wiggle, honeycomb, and triangle shapes(Cojocaru et al., 2022; Dezaki et al., 2021). The optimization of mechanical properties along with material usage can be achieved through combination infill patterns(Dezaki & Ariffin, 2020). Such an innovative heterogeneous infill strategy enhances the strength-to-weight ratio and material efficiency and reduces build time(Sajjad et al., 2019). A decrease in mass and reduced material consumption are associated with less infill(Cojocaru et al.,2022). Infill structures possess stiffness, strength, and printability latticed shells (Kędziora et al.,2023). Processing parameters affect the quality and cost of products made by three-dimensional printing; one such parameter is infill patterns(Khan et al.,2018). Different mechanical properties for particular investigations may result from each infill pattern (Abdullahi & Sanusi,2019; Yankin et al.,2023). No studies discuss which infill pattern has better mechanics or sustainability(Pernet et al.,.2022).

Quality Assurance

One of the important aspects of 3D printing is quality assurance, which ensures the final product conforms to the stipulated requirements and performance standards. Geometrical accuracy and dimensional stability are maintained while ensuring that the 3D printed parts are functional and interchangeable; this involves measuring actual dimensions against those planned and specified in the design model for a particular part. Internal defects, voids, or cracks in 3D-printed components can be scanned using non-destructive testing techniques such as ultrasonic testing, radiography, and computed tomography, which do not cause any damage to the components being tested. The mechanical properties of 3D-printed specimens are determined using destructive test methods, including tensile, flexural, and impact tests. The surface quality of parts made by 3D printing is found by determining surface roughness.

CONCLUSION

3D printing changed manufacturing by allowing rapid prototyping and personalized production (Rajkumar, 2022). While modeling plastic products for 3D printing, it is necessary to consider the mechanical properties tested regarding the performance characteristics of those materials under mechanical testing (Дородов et al., 2021). The effect of the parameters of 3D printing on tensile tests has been conducted using plastic materials (Bowo et al., 2021). Through process planning and preparation phase optimizations, designers and engineers can tap into the full spectrum of design possibilities that three-dimensional printing offers for high-quality functional parts across many domains. This research used matrix material PLA filament, as per literature reviews (Ficzere et al., 2021). Such material properties exhibited anisotropic behavior; interlayer bonding, deposition filaments, and configuration might have contributed to such characteristics exhibiting dependency on the direction of 3D-printed samples (Ficzere et al., 2021.). Three-dimensional printing introduces defects within tolerances, such as missing materials leading to voids between layers(Fayazbakhsh et al.,2018). The two most important parameters in enhancing the strength of three-dimensional printed components are layer orientation and infill percentage during printing processes. (Raney et al.,2017)

REFERENCES

Дородов, П. В., Kasatkin, V., Kasatkina, N. Y., Petrov, V. A., & Litvinyuk, A. A. (2021). Investigation of the main mechanical characteristics of plastics for three-dimensional printing of machine parts models. *IOP Conference Series. Earth and Environmental Science*, *677*(5), 52041. DOI: 10.1088/1755-1315/677/5/052041

Abdulhameed, O., Al-Ahmari, A., Ameen, W., & Mian, S. H. (2019). Additive manufacturing: Challenges, trends, and applications. *Advances in Mechanical Engineering*, *11*(2), 1687814018822880. Advance online publication. DOI: 10.1177/1687814018822880

Aimar, A., Palermo, A., & Innocenti, B. (2019). The Role of 3D Printing in Medical Applications: A State of the Art [Review of The Role of 3D Printing in Medical Applications: A State of the Art]. Journal of Healthcare Engineering, 2019, 1. Hindawi Publishing Corporation. DOI: 10.1155/2019/5340616

Alfaify, A., Saleh, M., Abdullah, F. M., & Al-Ahmari, A. (2020). Design for Additive Manufacturing: A Systematic Review [Review of Design for Additive Manufacturing: A Systematic Review]. Sustainability, 12(19), 7936. Multidisciplinary Digital Publishing Institute. DOI: 10.3390/su12197936

Alghamdi, S. S., John, S., Choudhury, N. R., & Dutta, N. K. (2021). Additive Manufacturing of Polymer Materials: Progress, Promise and Challenges [Review of Additive Manufacturing of Polymer Materials: Progress, Promise and Challenges]. Polymers, 13(5), 753. Multidisciplinary Digital Publishing Institute. DOI: 10.3390/polym13050753

Almy, R. D., & Tontowi, A. E. (2018). THE EFFECT OF 3D PRINTING MACHINE PARAMETERS IN EXTRUSION PROCESS OF BIOCOMPOSITE MATERIALS (PMMA AND HA) ON DIMENSIONAL ACCURACY. *SINERGI*, *22*(2), 83. DOI: 10.22441/sinergi.2018.2.003

Amaya, J., Perero, B. S., Helguero, C. G., Hurel, J., Peralta, J., Flores, F. A., & Alvarado, J. D. (2024). Future trends of additive manufacturing in medical applications: An overview [Review of Future trends of additive manufacturing in medical applications: An overview]. Heliyon, 10(5). Elsevier BV. DOI: 10.1016/j.heliyon.2024.e26641

An, J., Teoh, J. E. M., Suntornnond, R., & Chua, C. K. (2015). Design and 3D Printing of Scaffolds and Tissues. *Engineering (Beijing)*, *1*(2), 261–268. DOI: 10.15302/J-ENG-2015061

Anwajler, B., Zdybel, E., & Tomaszewska-Ciosk, E. (2023). Innovative Polymer Composites with Natural Fillers Produced by Additive Manufacturing (3D Printing)—A Literature Review [Review of Innovative Polymer Composites with Natural Fillers Produced by Additive Manufacturing (3D Printing)—A Literature Review]. Polymers, 15(17), 3534. Multidisciplinary Digital Publishing Institute. DOI: 10.3390/polym15173534

Babbar, A., Rai, A., & Sharma, A. (2021). Latest trend in building construction: Three-dimensional printing. *Journal of Physics: Conference Series, 1950*(1), 12007. DOI: 10.1088/1742-6596/1950/1/012007

Bahraminasab, M. (2020). Challenges on optimization of 3D-printed bone scaffolds [Review of Challenges on optimization of 3D-printed bone scaffolds]. BioMedical Engineering OnLine, 19(1). BioMed Central. DOI: 10.1186/s12938-020-00810-2

Birtchnell, T., & Hoyle, W. (2014). What is 3D Printing? In *Palgrave Macmillan UK eBooks* (p. 36). Palgrave Macmillan., DOI: 10.1057/9781137365668_3

Bowo, F. P., Suzen, Z. S., & Oktriadi, Y. (2021). Pengaruh Infill Geometry, Printing Speed dan Nozzle Temperature terhadap Kekuatan Impak Menggunakan Filamen ST PLA. *Jurnal Syntax Admiration, 2*(7), 1257–1268. DOI: 10.46799/jsa.v2i7.265

Budinoff, H. D., & McMains, S. (2021). Will it print: A manufacturability toolbox for 3D printing. *International Journal on Interactive Design and Manufacturing, 15*(4), 613–630. DOI: 10.1007/s12008-021-00786-w

Cabreira, V., & Santana, R. M. C. (2020). Effect of infill pattern in Fused Filament Fabrication (FFF) 3D Printing on materials performance. *Matéria (Rio de Janeiro), 25*(3), e-12826. Advance online publication. DOI: 10.1590/s1517-707620200003.1126

Carrier, A. (1987). Sequential steps in project execution. *International Journal of Project Management, 5*(1), 25–28. DOI: 10.1016/0263-7863(87)90006-8

Chen, X. (2016).. . *Making Fabrication Real., 17*, 17–20. Advance online publication. DOI: 10.1145/2984751.2984785

Chohan, J. S., & Singh, R. (2017). Pre and post processing techniques to improve surface characteristics of FDM parts: A state of art review and future applications. *Rapid Prototyping Journal, 23*(3), 495–513. DOI: 10.1108/RPJ-05-2015-0059

Choi, S.-H., & Kwok, K. T. (2002). Hierarchical slice contours for layered-manufacturing. *Computers in Industry, 48*(3), 219–239. DOI: 10.1016/S0166-3615(02)00040-4

Choonara, Y. E., Toit, L. C. du, Kumar, P., Kondiah, P. P. D., & Pillay, V. (2016). 3D-printing and the effect on medical costs: a new era? [Review of 3D-printing and the effect on medical costs: a new era?]. Expert Review of Pharmacoeconomics & Outcomes Research, 16(1), 23. Taylor & Francis. DOI: 10.1586/14737167.2016.1138860

Christensen, A., & Rybicki, F. J. (2017). Maintaining safety and efficacy for 3D printing in medicine [Review of Maintaining safety and efficacy for 3D printing in medicine]. 3D Printing in Medicine, 3(1). BioMed Central. DOI: 10.1186/s41205-016-0009-5

Chua, C. K., & Leong, K. F. (2013). 3D Printing and Additive Manufacturing. DOI: 10.1142/9008

Coakley, M. F., & Hurt, D. E. (2016). 3D Printing in the Laboratory: Maximize Time and Funds with Customized and Open-Source Labware. *SLAS Technology*, *21*(4), 489–495. DOI: 10.1177/2211068216649578 PMID: 27197798

Cojocaru, V., Frunzăverde, D., & Micloşină, C.-O. (2022). On the Behavior of Honeycomb, Grid and Triangular PLA Structures under Symmetric and Asymmetric Bending. *Micromachines*, *14*(1), 120. DOI: 10.3390/mi14010120 PMID: 36677181

Davydov, R., Polyakov, N., Korenchenkova, D. E., Чулкин, С. Г., Rud', V. Yu., & Krasnoscshekov, V. (2021). Testing the methodology for evaluating the accuracy of the 3d printing. *Journal of Physics: Conference Series*, *1942*(1), 12057. DOI: 10.1088/1742-6596/1942/1/012057

Dawood, A., Marti, B. M., Sauret-Jackson, V., & Darwood, A. (2015). 3D printing in dentistry [Review of 3D printing in dentistry]. BDJ, 219(11), 521. Springer Nature. DOI: 10.1038/sj.bdj.2015.914

Despeisse, M., & Ford, S. (2015). The Role of Additive Manufacturing in Improving Resource Efficiency and Sustainability. In IFIP advances in information and communication technology (p. 129). Springer Science+Business Media. DOI: 10.1007/978-3-319-22759-7_15

Dezaki, M. L., & Ariffin, M. K. A. M. (2020). The Effects of Combined Infill Patterns on Mechanical Properties in FDM Process. *Polymers*, *12*(12), 2792. DOI: 10.3390/polym12122792 PMID: 33255897

Dezaki, M. L., Ariffin, M. K. A. M., Serjouei, A., Zolfagharian, A., Hatami, S., & Bodaghi, M. (2021). Influence of Infill Patterns Generated by CAD and FDM 3D Printer on Surface Roughness and Tensile Strength Properties. *Applied Sciences (Basel, Switzerland)*, *11*(16), 7272. DOI: 10.3390/app11167272

Diment, L., Thompson, M. S., & Bergmann, J. (2017). Clinical efficacy and effectiveness of 3D printing: a systematic review [Review of Clinical efficacy and effectiveness of 3D printing: a systematic review]. BMJ Open, 7(12). *BMJ (Clinical Research Ed.)*, *7*(12), e016891. Advance online publication. DOI: 10.1136/bmjopen-2017-016891 PMID: 29273650

Dul, S., Fambri, L., & Pegoretti, A. (2021). High-Performance Polyamide/Carbon Fiber Composites for Fused Filament Fabrication: Mechanical and Functional Performances. *Journal of Materials Engineering and Performance*, *30*(7), 5066–5085. DOI: 10.1007/s11665-021-05635-1

Egan, P. (2019). Integrated Design Approaches for 3D Printed Tissue Scaffolds: Review and Outlook [Review of Integrated Design Approaches for 3D Printed Tissue Scaffolds: Review and Outlook]. Materials, 12(15), 2355. Multidisciplinary Digital Publishing Institute. DOI: 10.3390/ma12152355

Equbal, A., Equbal, M. A., Sood, A. K., Pranav, R., & Equbal, M. I. (2018, December). A review and reflection on part quality improvement of fused deposition modelled parts. [). IOP Publishing.]. *IOP Conference Series. Materials Science and Engineering*, *455*(1), 012072.

Fayazbakhsh, K., Movahedi, M., & Kalman, J. (2018). The impact of defects on tensile properties of 3D printed parts manufactured by fused filament fabrication. *Materials Today. Communications*, *18*, 140–148. DOI: 10.1016/j.mtcomm.2018.12.003

Ficzere, P., Lukács, N. L., & Borbás, L. (2021). The Investigation of Interlaminar Failures Caused by Production Parameters in Case of Additive Manufactured Polymers. *Polymers*, *13*(4), 556. DOI: 10.3390/polym13040556 PMID: 33668572

Frizziero, L., Liverani, A., Donnici, G., Osti, F., Neri, M., Maredi, E., Trisolino, G., & Stilli, S. (2019). New Methodology for Diagnosis of Orthopedic Diseases through Additive Manufacturing Models. *Symmetry*, *11*(4), 542. DOI: 10.3390/sym11040542

Garcia, C. R., Rumpf, R. C., Tsang, H., & Barton, J. H. (2013). Effects of extreme surface roughness on 3D printed horn antenna. *Electronics Letters*, *49*(12), 734–736. DOI: 10.1049/el.2013.1528

Garcia, F. L., Moris, V. A. da S., Nunes, A. O., & Silva, D. A. L. (2018). Environmental performance of additive manufacturing process – an overview. *Rapid Prototyping Journal*, *24*(7), 1166–1177. DOI: 10.1108/RPJ-05-2017-0108

Gardan, J. (2020). Additive manufacturing technologies for polymer composites: State-of-the-art and future trends. In *Elsevier eBooks* (p. 3). Elsevier BV., DOI: 10.1016/B978-0-12-819535-2.00001-6

Ghilan, A., Chiriac, A. P., Niţă, L. E., Rusu, A. G., Neamţu, I., & Chiriac, V. M. (2020). Trends in 3D Printing Processes for Biomedical Field: Opportunities and Challenges [Review of Trends in 3D Printing Processes for Biomedical Field: Opportunities and Challenges]. Journal of Polymers and the Environment, 28(5), 1345. Springer Science+Business Media. DOI: 10.1007/s10924-020-01722-x

Giannopoulos, A. A., Chepelev, L., Sheikh, A., Wang, A., Dang, W., Akyuz, E., Hong, C. J., Wake, N., Pietila, T., Dydynski, P., Mitsouras, D., & Rybicki, F. J. (2015). 3D printed ventricular septal defect patch: a primer for the 2015 Radiological Society of North America (RSNA) hands-on course in 3D printing. 3D Printing in Medicine, 1(1). DOI: 10.1186/s41205-015-0002-4

Goh, G. D., Sing, S. L., & Yeong, W. Y. (2020). A review on machine learning in 3D printing: applications, potential, and challenges [Review of A review on machine learning in 3D printing: applications, potential, and challenges]. Artificial Intelligence Review, 54(1), 63. Springer Science+Business Media. DOI: 10.1007/s10462-020-09876-9

Guessasma, S., Zhang, W., Zhu, J., Belhabib, S., & Nouri, H. (2015). Challenges of additive manufacturing technologies from an optimisation perspective. *International Journal for Simulation and Multidisciplinary Design Optimization*, *6*, A9. Advance online publication. DOI: 10.1051/smdo/2016001

He, F., & Khan, M. (2021). Effects of Printing Parameters on the Fatigue Behaviour of 3D-Printed ABS under Dynamic Thermo-Mechanical Loads. *Polymers*, *13*(14), 2362. DOI: 10.3390/polym13142362 PMID: 34301117

Hoang, D., Perrault, D., Stevanović, M., & Ghiassi, A. (2016). Surgical applications of three-dimensional printing: a review of the current literature & how to get started [Review of Surgical applications of three-dimensional printing: a review of the current literature & how to get started]. Annals of Translational Medicine, 4(23), 456. AME Publishing Company. DOI: 10.21037/atm.2016.12.18

Hsieh, C. T. (2017). Investigation of Delta Robot 3D Printer for a Good Quality of Printing. *Applied Mechanics and Materials*, *870*, 164–169. . DOI: 10.4028/www.scientific.net/AMM.870.164

Issal, A. (2023). Ender 3 (V2/Pro) Calibration: 5 Simple Steps. https://all3dp.com/2/ender-3-calibration-how-to-calibrate-your-ender-3/

Ituarte, I. F., Chekurov, S., Salmi, M., Tuomi, J., & Partanen, J. (2015). Post-processing opportunities of professional and consumer grade 3D printing equipment: A comparative study. *International Journal of Rapid Manufacturing*, *5*(1), 58. DOI: 10.1504/IJRAPIDM.2015.073548

Jaksic, N. (2015). What to Do When 3D Printers Go Wrong: Laboratory Experiences. DOI: 10.18260/p.25066

Jayakrishna, M., Vijay, M., & Khan, B. (2023). An Overview of Extensive Analysis of 3D Printing Applications in the Manufacturing Sector. *Journal of Engineering*, *1*, 1–23. Advance online publication. DOI: 10.1155/2023/7465737

Jeong, M., Radomski, K., López, D., Liu, J. T., Lee, J. D., & Lee, S. J. (2023). Materials and Applications of 3D Printing Technology in Dentistry: An Overview [Review of Materials and Applications of 3D Printing Technology in Dentistry: An Overview]. Dentistry Journal, 12(1), 1. Multidisciplinary Digital Publishing Institute. DOI: 10.3390/dj12010001

Jiménez, M., Romero, L., Domínguez, I. A., & Escudero, M. (2019). del M. E., & Somonte, M. D. (2019). Additive Manufacturing Technologies: An Overview about 3D Printing Methods and Future Prospects. *Complexity*, *2019*(1), 9656938. Advance online publication. DOI: 10.1155/2019/9656938

Kantaros, A., & Piromalis, D. (2021). Fabricating Lattice Structures via 3D Printing: The Case of Porous Bio-Engineered Scaffolds. *Applied Mechanics*, *2*(2), 289–302. DOI: 10.3390/applmech2020018

Kędziora, S., Decker, T., & Museyibov, E. (2023). Application of Functionally Graded Shell Lattice as Infill in Additive Manufacturing. *Materials (Basel)*, *16*(12), 4401. DOI: 10.3390/ma16124401 PMID: 37374591

Kent, N., Jolivet, L., O'Neill, P., & Brabazon, D. (2017). An evaluation of components manufactured from a range of materials, fabricated using PolyJet technology. *Advances in Materials and Processing Technologies*, *3*(3), 318–329. DOI: 10.1080/2374068X.2017.1330856

Khaliq, J., Gurrapu, D. R., & Elfakhri, F. (2023). Effects of Infill Line Multiplier and Patterns on Mechanical Properties of Lightweight and Resilient Hollow Section Products Manufactured Using Fused Filament Fabrication. *Polymers*, *15*(12), 2585. DOI: 10.3390/polym15122585 PMID: 37376230

Khan, S. F., Zakaria, H., Chong, Y. L., Saad, M. A. M., & Basaruddin, K. S. (2018). Effect of infill on tensile and flexural strength of 3D printed PLA parts. *IOP Conference Series. Materials Science and Engineering*, *429*, 12101. DOI: 10.1088/1757-899X/429/1/012101

Kumar, A., Nune, K. C., Murr, L. E., & Misra, R. D. K. (2016). Biocompatibility and mechanical behaviour of three-dimensional scaffolds for biomedical devices: Process–structure–property paradigm. *International Materials Reviews*, *61*(1), 20–45. DOI: 10.1080/09506608.2015.1128310

Kumar, A., Verma, S., & Jeng, J. (2020). Supportless Lattice Structures for Energy Absorption Fabricated by Fused Deposition Modeling. 3D Printing and Additive Manufacturing, 7(2), 85. DOI: 10.1089/3dp.2019.0089

Kumar, R., Kumar, M., & Chohan, J. S. (2021). Material-specific properties and applications of additive manufacturing techniques: A comprehensive review. *Bulletin of Materials Science*, *44*(3), 181. DOI: 10.1007/s12034-021-02364-y

Li, Z., Wang, Q., & Liu, G. (2022). A Review of 3D Printed Bone Implants [Review of A Review of 3D Printed Bone Implants]. Micromachines, 13(4), 528. Multidisciplinary Digital Publishing Institute. DOI: 10.3390/mi13040528

Liaw, C.-Y., & Güvendiren, M. (2017). Current and emerging applications of 3D printing in medicine [Review of Current and emerging applications of 3D printing in medicine]. Biofabrication, 9(2), 24102. IOP Publishing. DOI: 10.1088/1758-5090/aa7279

Liu, J., Su, Z., Wang, C., & Xu, Z. (2022). Effect of an Adaptive-Density Filling Structure on the Mechanical Properties of FDM Parts with a Variable Cross-Section. *Materials (Basel)*, *15*(24), 8746. DOI: 10.3390/ma15248746 PMID: 36556552

Liu, J.-Y., & Yan, C. (2018). 3D Printing of Scaffolds for Tissue Engineering. In InTech eBooks. DOI: 10.5772/intechopen.78145

Liu, L., Yang, X., Bhandari, B., Meng, Y., & Prakash, S. (2020). Optimization of the Formulation and Properties of 3D-Printed Complex Egg White Protein Objects. *Foods*, *9*(2), 164. DOI: 10.3390/foods9020164 PMID: 32046351

Liu, Z., Jiang, Q., Zhang, Y., Li, T., & Zhang, H. (2016). Sustainability of 3D Printing: A Critical Review and Recommendations [Review of Sustainability of 3D Printing: A Critical Review and Recommendations]. Volume 2: Materials; Biomanufacturing; Properties, Applications and Systems; Sustainable Manufacturing. DOI: 10.1115/MSEC2016-8618

Liu, Z., Zhang, M., Bhandari, B., & Wang, Y. (2017). 3D printing: Printing precision and application in food sector. *Trends in Food Science & Technology*, *69*, 83–94. DOI: 10.1016/j.tifs.2017.08.018

Loh, G. H., Pei, E., González-Gutiérrez, J., & Monzón, M. (2020). An Overview of Material Extrusion Troubleshooting. *Applied Sciences (Basel, Switzerland)*, *10*(14), 4776. DOI: 10.3390/app10144776

Lolla, R., Srinath, A., Murali, G., Murugan, M., Venkatesan, E. P., & Hasan, N. (2023). Effect of Infill Patterns with Machine Learning Techniques on the Tensile Properties of Polylactic Acid-Based Ceramic Materials with Fused Filament Fabrication. *ACS Omega*, *8*(28), 24786–24796. DOI: 10.1021/acsomega.2c08164 PMID: 37483243

Lubombo, C., & Huneault, M. A. (2018). Effect of infill patterns on the mechanical performance of lightweight 3D-printed cellular PLA parts. *Materials Today. Communications*, *17*, 214–228. DOI: 10.1016/j.mtcomm.2018.09.017

Majko, J., Sága, M., Vaško, M., Handrik, M., & Ságová, Z. (2020). FE modeling of continuous fiber reinforced thermoplastic composite structures produced by additive manufacturing. *IOP Conference Series. Materials Science and Engineering*, *776*(1), 12080. DOI: 10.1088/1757-899X/776/1/012080

Prabhakar, M. M., Saravanan, A. K., Lenin, A. H., Mayandi, K., & Ramalingam, P. S. (2021). A short review on 3D printing methods, process parameters and materials. *Materials Today: Proceedings*, *45*, 6108–6114.

Martelli, N., Serrano, C., Brink, H. van den, Pineau, J., Prognon, P., Borget, I., & Batti, S. E. (2016). Advantages and disadvantages of 3-dimensional printing in surgery: A systematic review [Review of Advantages and disadvantages of 3-dimensional printing in surgery: A systematic review]. Surgery, 159(6), 1485. Elsevier BV. https://doi.org/DOI: 10.1016/j.surg.2015.12.017

Martín, M. J., Auñón, J. A., & Fernández, F. M. (2021). Influence of Infill Pattern on Mechanical Behavior of Polymeric and Composites Specimens Manufactured Using Fused Filament Fabrication Technology. *Polymers*, *13*(17), 2934. DOI: 10.3390/polym13172934 PMID: 34502974

Martin. (2024). 3D Printer Calibration: Step-by-Step to Perfect Results. https://the3dprinterbee.com/3d-printer-calibration/

Martinez-Marquez, D., Mirnajafizadeh, A., Carty, C. P., & Stewart, R. A. (2018). Application of quality by design for 3D printed bone prostheses and scaffolds. *PLoS One*, *13*(4), e0195291. Advance online publication. DOI: 10.1371/journal.pone.0195291 PMID: 29649231

Mathews, D. A. P., Baird, A. D., & Lucky, M. (2020). Innovation in Urology: Three Dimensional Printing and Its Clinical Application [Review of Innovation in Urology: Three Dimensional Printing and Its Clinical Application]. Frontiers in Surgery, 7. Frontiers Media. DOI: 10.3389/fsurg.2020.00029

Mendřický, R., & Fris, D. (2020). Analysis of the Accuracy and the Surface Roughness of FDM/FFF Technology and Optimisation of Process Parameters. Tehnicki Vjesnik - Technical Gazette, 27(4). DOI: 10.17559/TV-20190320142210

Minaoglou, P., Efkolidis, N., Manavis, A., & Kyratsis, P. (2024). A Review on Wearable Product Design and Applications [Review of A Review on Wearable Product Design and Applications]. Machines, 12(1), 62. Multidisciplinary Digital Publishing Institute. DOI: 10.3390/machines12010062

Mohapatra, S., Kar, R. K., Biswal, P. K., & Bindhani, S. (2022). Approaches of 3D printing in current drug delivery. *Sensors International*, *3*, 100146. DOI: 10.1016/j.sintl.2021.100146

Morampudi, P., Ramana, V. S. N. V., Prabha, K. A., Swetha, S., & Rao, A. (2021). 3D-printing analysis of surface finish. *Materials Today: Proceedings*, *43*, 587–592. DOI: 10.1016/j.matpr.2020.12.085

Mwema, F. M., & Akinlabi, E. T. (2020). Basics of Fused Deposition Modelling (FDM). In SpringerBriefs in applied sciences and technology (p. 1). Springer Nature. DOI: 10.1007/978-3-030-48259-6_1

Nadagouda, M. N., Ginn, M., & Rastogi, V. (2020). A review of 3D printing techniques for environmental applications [Review of A review of 3D printing techniques for environmental applications]. Current Opinion in Chemical Engineering, 28, 173. Elsevier BV. DOI: 10.1016/j.coche.2020.08.002

Niesler, F., & Hermatschweiler, M. (2015). Two-Photon Polymerization — A Versatile Microfabrication Tool. *Laser-Technik-Journal*, *12*(3), 44–47. DOI: 10.1002/latj.201500019

Oberoi, G., Nitsch, S., Edelmayer, M., Janjić, K., Müller, A. S., & Agis, H. (2018). 3D Printing—Encompassing the Facets of Dentistry [Review of 3D Printing—Encompassing the Facets of Dentistry]. Frontiers in Bioengineering and Biotechnology, 6. Frontiers Media. DOI: 10.3389/fbioe.2018.00172

Olasek, K., & Wiklak, P. (2014). Application of 3D printing technology in aerodynamic study. *Journal of Physics: Conference Series*, *530*, 12009. DOI: 10.1088/1742-6596/530/1/012009

Oleksy, M., Dynarowicz, K., & Aebisher, D. (2023). Advances in Biodegradable Polymers and Biomaterials for Medical Applications—A Review [Review of Advances in Biodegradable Polymers and Biomaterials for Medical Applications—A Review]. Molecules, 28(17), 6213. Multidisciplinary Digital Publishing Institute. DOI: 10.3390/molecules28176213

Otitoju, T. A., Okoye, P. U., Chen, G., Li, Y., Okoye, M. O., & Li, S. (2020). Advanced ceramic components: Materials, fabrication, and applications. *Journal of Industrial and Engineering Chemistry*, *85*, 34–65. DOI: 10.1016/j.jiec.2020.02.002

Păcurar, A. (2019). Introductory Chapter: Applications of Design for Manufacturing and Assembly. In IntechOpen eBooks. IntechOpen. DOI: 10.5772/intechopen.82549

Panwisawas, C., Tang, Y. T., & Reed, R. C. (2020). Metal 3D printing as a disruptive technology for superalloys. *Nature Communications*, *11*(1), 2327. Advance online publication. DOI: 10.1038/s41467-020-16188-7 PMID: 32393778

Paprocka, I., & Kempa, W. M. (2021). Model of Production System Evaluation with the Influence of FDM Machine Reliability and Process-Dependent Product Quality. *Materials (Basel)*, *14*(19), 5806. DOI: 10.3390/ma14195806 PMID: 34640202

Perkins, I., & Skitmore, M. (2015). Three-dimensional printing in the construction industry: A review [Review of Three-dimensional printing in the construction industry: A review]. International Journal of Construction Management, 15(1), 1. Taylor & Francis. DOI: 10.1080/15623599.2015.1012136

Pernet, B., Nagel, J., & Zhang, H. (2022). Compressive Strength Assessment of 3D Printing Infill Patterns. *Procedia CIRP*, *105*, 682–687. DOI: 10.1016/j.procir.2022.02.114

Pham, G. N., Lee, S.-H., Kwon, O.-H., & Kwon, K.-R. (2018). A 3D Printing Model Watermarking Algorithm Based on 3D Slicing and Feature Points. *Electronics (Basel)*, *7*(2), 23. DOI: 10.3390/electronics7020023

Rajkumar, S. M. (2022). Effect of infill pattern and build orientation on mechanical properties of FDM printed parts: An experimental modal analysis approach. arXiv (Cornell University). https://doi.org//arxiv.2202.05692DOI: 10.48550

Raney, K., Lani, E., & Kalla, D. K. (2017). Experimental characterization of the tensile strength of ABS parts manufactured by fused deposition modeling process. *Materials Today: Proceedings*, *4*(8), 7956–7961. DOI: 10.1016/j.matpr.2017.07.132

Ritchie, S. M., Kovačević, S., Deshmukh, P., Christodoulides, A. D., Malen, J. A., Mesarovic, S. Dj., & Panat, R. (2023). Shape distortion in sintering results from nonhomogeneous temperature activating a long-range mass transport. *Nature Communications*, *14*(1), 2667. Advance online publication. DOI: 10.1038/s41467-023-38142-z PMID: 37160902

Sağbaş, B. (2020). Effect of Orientation Angle on Surface Quality and Dimensional Accuracy of Functional Parts Manufactured by Multi Jet Fusion Technology. *European Mechanical Science*, *4*(2), 47–52. DOI: 10.26701/ems.678901

Sajjad, R., Butt, S. U., Ali, A., & Saeed, H. A. (2019). Investigating the Impacts of Heterogeneous Infills on Structural Strength of 3D Printed Parts. *Key Engineering Materials*, *799*, 276–281. . DOI: 10.4028/www.scientific.net/KEM.799.276

Satterfield, C. (2020). How to Calibrate your FDM 3D Printer: The Ultimate Guide. https://3dprintknowledge.com/how-to-calibrate-your-fdm-3d-printer-the-ultimate-guide/

Schneberger, J.-H., Häfele, T., Kaspar, J., & Vielhaber, M. (2019). Hybrid Additive Manufacturing - Requirements Engineering Framework for Process Chain Considerations. Proceedings of the ... International Conference on Engineering Design, 1(1), 3651. DOI: 10.1017/dsi.2019.372

Shen, Y., Lin, L., Wei, S., Yan, J., & Xu, T. (2022). Research on the Preparation and Mechanical Properties of Solidified 3D Printed Concrete Materials. *Buildings (Basel, Switzerland)*, *12*(12), 2264. DOI: 10.3390/buildings12122264

Sovetova, M., & Calautit, J. K. (2024). Design, calibration and performance evaluation of a small-scale 3D printer for accelerating research in additive manufacturing in construction. *Cleaner Engineering and Technology*, *22*, 100786. DOI: 10.1016/j.clet.2024.100786

9Steps on How To Set Up a 3D Printer. (2024). https://www.xometry.com/resources/3d-printing/set-up-a-3d-printer/

Taufik, M., & Jain, P. K. (2020). Part surface quality improvement studies in fused deposition modelling process: a review [Review of Part surface quality improvement studies in fused deposition modelling process: a review]. Australian Journal of Mechanical Engineering, 20(2), 527. Taylor & Francis. DOI: 10.1080/14484846.2020.1723342

Vayre, B., Vignat, F., & Villeneuve, F. (2012). Metallic additive manufacturing: State-of-the-art review and prospects. *Mechanics & Industry*, *13*(2), 89–96. DOI: 10.1051/meca/2012003

Vorkapić, M., Hasan, M. A., Tanović, D., Baltić, M., & Tomić, B. (2020). Implementation of 6R strategy in FDM printing process: Case: Small electronic enclosure box. *Journal of Engineering Management and Competitiveness, 10*(2), 141–150. DOI: 10.5937/jemc2002141V

Wang, C., Tan, X., Tor, S. B., & Lim, C. S. (2020). Machine learning in additive manufacturing: State-of-the-art and perspectives. *Additive Manufacturing, 36*, 101538. DOI: 10.1016/j.addma.2020.101538

Wong, K. C. (2016). 3D-printed patient-specific applications in orthopedics [Review of 3D-printed patient-specific applications in orthopedics]. Orthopedic Research and Reviews, 57. Dove Medical Press. DOI: 10.2147/ORR.S99614

Wooldridge, E. (2020). Phantom holes: Optimised internal structural design for use with additive manufacturing, typical fused filament fabrication systems. *International Journal of Rapid Manufacturing, 9*(2/3), 137. DOI: 10.1504/IJRAPIDM.2020.107732

Wu, J., Aage, N., Lefèbvre, S., & Wang, C. C. L. (2017). Topology Optimization for Computational Fabrication. Eurographics. DOI: 10.2312/egt.20171030

Yankin, A., Alipov, Y., Temirgali, A., Serik, G., Danenova, S., Talamona, D., & Perveen, A. (2023). Optimization of Printing Parameters to Enhance Tensile Properties of ABS and Nylon Produced by Fused Filament Fabrication. *Polymers, 15*(14), 3043. DOI: 10.3390/polym15143043 PMID: 37514431

Yin, H., Qu, M., Zhang, H., & Lim, Y. (2018). 3D Printing and Buildings: A Technology Review and Future Outlook [Review of 3D Printing and Buildings: A Technology Review and Future Outlook]. Technology|Architecture + Design, 2(1), 94. Taylor & Francis. DOI: 10.1080/24751448.2018.1420968

Yin, J., Suo, Y., Lv, T., Ma, K., Wang, X., & Zhang, Z. (2022). *Application of 3D Printing Technology in the Construction Industry and Its Development Prospects*. Atlantis Highlights in Intelligent Systems/Atlantis Highlights in Intelligent Systems., DOI: 10.2991/ahis.k.220601.002

Zafar, A., Alruwaili, N. K., Rizwanullah, Md., Bukhari, S. N. A., Amir, M., Ahmed, M. M., & Fazil, M. (2019). 3D Printing Technology in Design of Pharmaceutical Products [Review of 3D Printing Technology in Design of Pharmaceutical Products]. Current Pharmaceutical Design, 24(42), 5009. Bentham Science Publishers. DOI: 10.2174/1381612825666190116104620

Zhang, H., Zhou, X., Fang, Y., Xiong, Z., & Zhang, T. (2024). AI-driven 3D bioprinting for regenerative medicine: From bench to bedside [Review of AI-driven 3D bioprinting for regenerative medicine: From bench to bedside]. Bioactive Materials, 45, 201. Elsevier BV. DOI: 10.1016/j.bioactmat.2024.11.021

Zhang, J., Hu, Q., Wang, S., Tao, J., & Gou, M. (2019). Digital Light Processing Based Three-dimensional Printing for Medical Applications [Review of Digital Light Processing Based Three-dimensional Printing for Medical Applications]. *International Journal of Bioprinting*, *6*(1), 242. DOI: 10.18063/ijb.v6i1.242 PMID: 32782984

Zhang, Y., Gulati, K., Li, Z., Di, P., & Liu, Y. (2021). Dental Implant Nano-Engineering: Advances, Limitations and Future Directions [Review of Dental Implant Nano-Engineering: Advances, Limitations and Future Directions]. Nanomaterials, 11(10), 2489. Multidisciplinary Digital Publishing Institute. DOI: 10.3390/nano11102489

Zhao, C. (2020). Research on Cultural and Creative Product Design System Based on 3DP Technology. *Journal of Physics: Conference Series*, *1648*(4), 42097. DOI: 10.1088/1742-6596/1648/4/042097

Zhou, L., Miller, J., Vezza, J., Mayster, M., Raffay, M., Justice, Q., Tamimi, Z. A., Hansotte, G., Sunkara, L. D., & Bernat, J. (2024). Additive Manufacturing: A Comprehensive Review [Review of Additive Manufacturing: A Comprehensive Review]. Sensors, 24(9), 2668. Multidisciplinary Digital Publishing Institute. DOI: 10.3390/s24092668

Živanović, S., Tabaković, S., & Randjelović, S. (2019). Rapid prototyping of art sculptural shapes according to the sample. Advanced Technologies and Materials, 44(1), 27-32. DOI: <ALIGNMENT.qj></ALIGNMENT>10.24867/ATM-2019-1-005

Zou, Y., Song, Y., & Liu, X. (2023). Designing for Strength: Enhancing Mechanical Performance through Structured Patterns in 3D Printed Elastomer. *Materials Research Express*, *10*(9), 95302. DOI: 10.1088/2053-1591/acf64f

Chapter 8
Thermal and Mechanical Analysis of Materials in Additive Manufacturing

G. Boopathy
https://orcid.org/0000-0003-2515-7277
Vel Tech Rangarajan Dr. Sagunthala R&D Institute of Science and Technology, Chennai, India

V. Srinivasan
https://orcid.org/0009-0006-2988-6476
Vel Tech Rangarajan Dr. Sagunthala R&D Institute of Science and Technology, Chennai, India

Andrejs Krasnikovs
Riga Technical University, Latvia

Balaji Ganesan
https://orcid.org/0000-0001-5799-9423
Hindustan Institute of Technology and Science, India

ABSTRACT

Additive manufacturing (AM) has seen very little advancement in terms of its constraints, particularly with respect to thermal manageability and mechanical durability. Accordingly, the accuracy, reliability, and part performance of AM are significantly influenced by the mechanical and thermal analysis of materials. Therefore, the work investigates a heat transfer mechanism, residual stress formation and microstructural evolution and mechanical behavior of printed materials such as tensile strength, fatigue resistance and defect mitigation. This chapter discusses various techniques of advanced modeling, simulation and real time monitoring

DOI: 10.4018/979-8-3373-0533-2.ch008

strategies that will improve the predictive accuracy. Case studies for the applications in aerospace, automotive and healthcare industry are presented as practical applications. The bridging of theory to the practical solution provided in this chapter is a real action for researchers and engineers to improve material performance, reduce defects and optimize AM process towards advancing high quality, reliable and scalable AM technologies.

INTRODUCTION

Additive manufacturing (AM), commonly referred to as 3D printing, is a revolutionary technology that uses layer-by-layer methods to directly fabricate products from digital models. By utilizing this method, designers can create more complex geometry than what is typically obtainable using traditional subtractive manufacturing techniques which in turn increases design flexibility, reduces the amount of material used, and reduces scrap. Its growing application across sectors such as aerospace, automotive, healthcare, and consumer products, where light weight structures and customization is expected, underline the relevance of AM.

Thermal and mechanical analysis of AM processes is of utmost importance due to its direct impact on performance and reliability of the manufactured components. The knowledge of the heat transfer mechanisms is important for understanding the layer deposition for thermal analysis. It addresses related problems of thermal gradients and residual stresses, which can result in warping and dimensional inaccuracies. For the mechanical side, material properties like tensile strength and any elasticity must be known in order to predict how parts will be behaving under operational conditions. Thermal and mechanical properties interplay is essential to achieve optimum AM processes with structural integrity of the final products.

This chapter will provide a comprehensive overview of fundamentals of additive manufacturing, analyse their importance in terms of thermal and mechanical analysis and integrate them to improve manufacturing outcomes. The work presents a discussion on AM technologies and material properties as well as latest analytical developments into the field.

FUNDAMENTALS OF ADDITIVE MANUFACTURING

Because of the evolution of AM technologies, there is a variety of methods with their own capabilities and applications. AM processes occur in the context of dominated by the material properties, in turn determining the nature of the success and performance of the manufactured components. Looking forward as AM continues to mature, there will be on-going research into materials and their properties should this emerging manufacturing technology unlock new applications and expand capabilities.

Brief History and Evolution of AM Technologies

AM has gone through a great deal of evolution since its inception in the 1980s. In 1986 Chuck Hull developed the first AM technology, Stereolithography (SLA) which cures photopolymer resin layer by layer using at ultraviolet light (Chmelko et al., 2023). From this pioneering method, subsequent developments on AM technologies were laid. Stratasys introduced early in the 1990s Fused Deposition Modelling (FDM), which has been one of the most accepted AM processes, partly because of its simplicity and low cost (Puerta et al., 2023). In FDM thermoplastic material is extruded out of a heated nozzle to build complex geometries easily. Further advances in technology gave rise to a variety of other AM techniques like Selective Laser Sintering (SLS), Direct Metal Laser Sintering (DMS) and so on. In the 1980s develop sintered by layer (SLS) (Zhang et al., 2024), whereby powder material is sintered in solid structures strata by strata. Moreover, this made our AM process more flexible, in the sense that we could also have polymers as well as metals. In the market where high strength components such as in aerospace and automotive are required (Mirzaali et al., 2022), DMLS has proven more popular as a metal powder specific technique similar to DMLS.

Advancements in materials science, computer aided design (CAD), and manufacturing process have served as the driver of AM technology evolution, furthering precision, speed, and versatility of the AM applications. Reemergence of AM as an enabling technology over the last decade has been closely linked to Industry 4.0, where digital technologies and automation are redefining manufacturing landscapes (Miao, 2024). Such innovations have extended the capabilities of AM, including multi material printing, bioprinting and the use of advanced materials such as composites and biomaterials to further include it as an integral part of modern manufacturing strategies (Yusof et al., 2024). Its continuously evolving AM technologies represent a wider shift towards customisation, sustainability and efficiency in production processes. Figure 1 represents the various properties that affect the FDM based 3D printing.

Figure 1. Various properties that affect FDM based 3D printing (Image by Krajangsawasdi et al., (2021) from "Fused Deposition Modelling of Fibre Reinforced Polymer Composites: A Parametric Review (https://www.mdpi.com/2504-477X/5/1/29)" is licensed under CC BY 4.0)

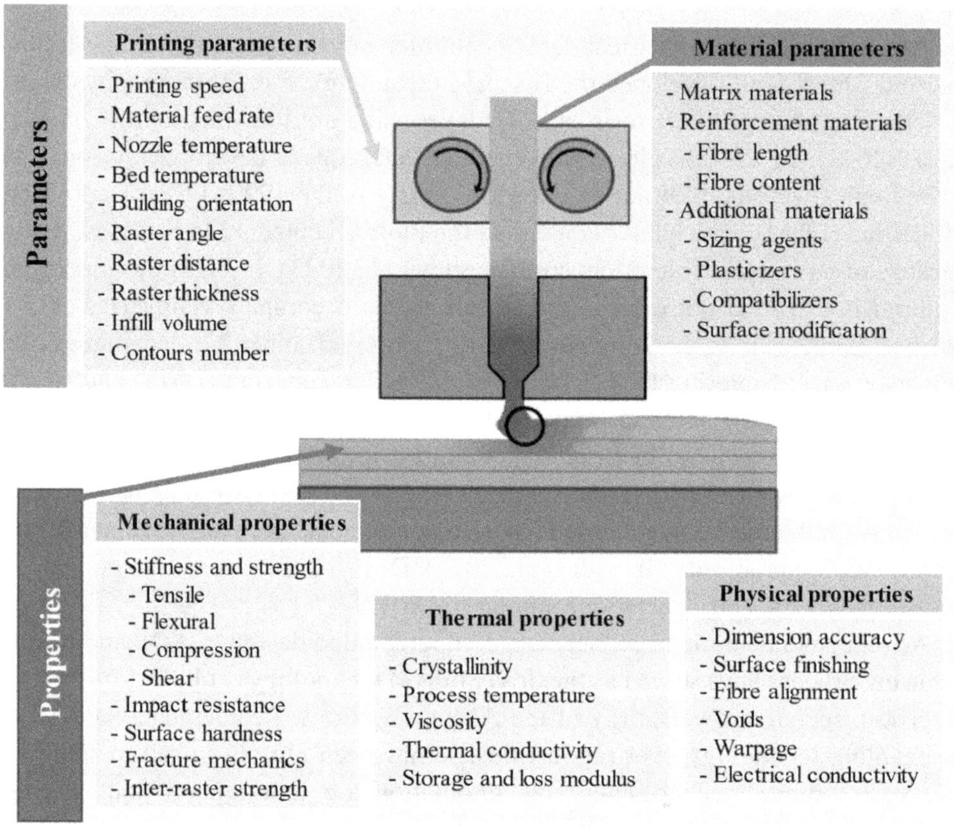

Overview of Key AM Methods

As several key AM methods have evolved, each of which has their advantages and limitations and therefore they are valuable for different applications, and materials. The general rule of thumb that most AM methods are regarded as additive has created challenges with production part cost since this way of manufacturing often results in a significant material and support material waste. Table 1 gives the classification of AM techniques and Figure 2 represents the Wire Arc based Additive Manufacturing (WAAM) technique.

Table 1. Additive Manufacturing (AM) Techniques

AM Method	Description	Applications	Advantages
Fused Deposition Modelling (FDM)	Extrusion of thermoplastic filaments through a heated nozzle, deposited layer by layer onto a build platform.	Prototyping and low-volume production.	Affordable, easy to use, capable of complex geometries.
Stereolithography (SLA)	Laser curing of liquid photopolymer resin into solid layers.	Jewelry, dental, and medical devices.	High precision, smooth surface finish.
Selective Laser Sintering (SLS)	Laser selectively fuses powdered materials (polymers/metals) into solid structures without support structures.	Functional prototypes and end-use parts.	Complex geometries, no need for support structures.
Direct Metal Laser Sintering (DMLS)	Specialized SLS focusing on metal powders.	Aerospace and automotive industries.	Excellent mechanical properties, suitable for high-performance components.
Binder Jetting	Liquid binder selectively deposited onto a powder bed, cured to form solid parts.	Ceramics and metals.	High speed, suitable for large parts.
Material Jetting	Deposition of material droplets, cured to form solid layers.	Prototypes and models.	High resolution, smooth surface finish.
Wire Arc Additive Manufacturing (WAAM)	Uses an electric arc to melt metal wire, depositing it layer by layer.	Large components with complex geometries.	Suitable for large components.

Figure 2. The schematic diagram of WAAM-CMT equipment. (Image by Yang et al., (2022) from "Microstructure and Mechanical Properties of TiB_2/AlSi7Mg0.6 Composites Fabricated by Wire and Arc Additive Manufacturing Based on Cold Metal Transfer (WAAM-CMT) (https://www.mdpi.com/1996-1944/15/7/2440)" is licensed under CC BY 4.0)

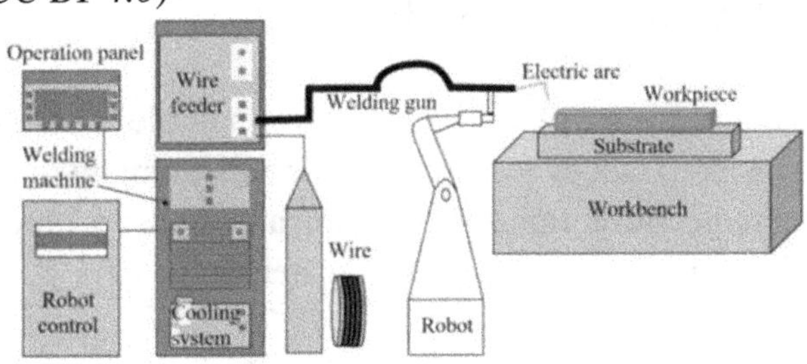

Role of Material Properties in Determining Process Outcomes

Additive manufacturing processes are sensitive to material properties. However, the choice of material has an effect not only on the mechanical properties of the final product, but also on the processing parameters and/or AM success. For example, thermoplastic materials used in FDM differ thermally and mechanically from metals used in DMLS. Achieving dimensional accuracy and part integrity requires heat transfer and layer adhesion, which are sensitive to thermal conductivity, melting point, and viscosity of the material (Arıkan, 2023).

Furthermore, the microstructure of the materials can vary widely depending on the AM process used. For example, because DMLS metal production involves rapid cooling and solidification processes, the grains that form may be different and be associated with different mechanical properties than metals produced conventionally. The result can be anisotropic mechanical properties, i.e. properties that depend on how the load is applied (Chafik et al., 2024). These material characteristics are important to allow for the notion of best process parameters and the definition of final products that meet those specifications for their intended use application.

THERMAL ANALYSIS IN ADDITIVE MANUFACTURING

Thermal analysis of additive manufacturing is highly dependent on mechanisms of heat transfer, thermal gradients and residual stresses. Researchers use mathematical models and simulation tools to predict thermal behaviour and optimize AM processes so as to attain better part quality and performance. Practical usage of the thermal analysis is shown by further case studies in such spreading of the capabilities of the additive manufacturing in various industries.

Heat Transfer Mechanisms in AM

Heat transfer mechanisms play a very important role in producing the final product in AM. The main means for heat transfer happening in the layer deposition process are conduction, convection and radiation and there are unique thermal roles in AM at all the three mechanisms of heat transfer.

Conduction

This is a major mode of heat transfer within the printed material. When a layer is deposited, heat that is produced from the deposited layer by the gas from the nozzle or the laser source is conducted through the added material to the underlying layers. Also, the thermal conductivity of material being used in AM changes vastly between polymers and metals. Taking for instance materials like metals that exhibit a high thermal conductivity can dissipate heat more quickly than polymers, which results in distinct solidification behaviours (Mantecón et al. 2022). Proper layer adhesion and minimization of warping and delamination are both dependent on this conduction process.

Convection

Heat transfer from the material surface to the surrounding environment: air or cooling medium is called heat conduction. Convective behaviour can be affected in AM processes by factors including the speed of the print head, by the ambient temperature, and by the presence of any cooling systems. If the thermal gradient is effective, then convection can manage the thermal gradients and prevent over heating of the material which would lead to defects (Gudur et al., 2023). For instance, Mantecón et al. have shown that the motion of the extruder alters temperature profiles, attenuating thermal fluctuations in upper layers because of the heat sink behaviour of the substrate material (Mantecón et al., 2022).

Radiation

However, in processes which need high energy sources such as lasers it becomes significant at higher temperatures. When the material is heated it emits thermal radiation and can influence the temperature of nearby layers and also the overall thermal profile of the build. Radiation is typically not a major contributing factor to the thermal environment of AM but can still contribute, during rapid heating and cooling process cycles (Tian et al., 2022). The knowledge of these heat transfer mechanisms is critical to maximizing the quality of AM processing, as these directly determine the thermal gradients and residual stresses that develop during layer deposition.

Full understanding of these heat transfer mechanisms is critical to the optimization of AM processes, as such heat transfers directly influence the thermal gradients and residual stresses developed within the layer deposition cycle.

Thermal Gradients and Residual Stresses

AM of parts is a critical factor that significantly influences the quality and dimensional accuracy of the parts produced with it. AM processes exhibit inherent rapid heating and cooling cycles that can produce large temperature differentials in the material and cause issues such as mechanical and structural properties.

Thermal Gradients in Additive Manufacturing

In both the AM and SM processes, steep thermal gradients are created during the AM process, more particularly in processes like SLM and Direct Energy Deposition (DED) using laser or energy source. They can be on the order of ~10^6K/m (Cho et al., 2023). With rapid solidification of molten material, cooling rates across the part become nonuniform resulting in differential thermal expansion and contraction. (He et al. 2023) point out that this phenomenon is particularly pronounced in metals, since complete melting and then quenching can cause large thermal stresses. Also thermal gradients may influence the microstructure of the printed part. For one, He et al. explained that it's the ordered arrangement of temperature gradients that's critical for the competitive growth of columnar grains in high-entropy alloys made by laser powder bed fusion (He et al., 2022). Disruption of these gradients can result in defect formation such as porosity and lack of fusion which degrade the mechanical properties of the part.

Residual Stresses and Their Implications

Internal stresses, that is residuals stresses, remain in the material after the stress causing them are removed. On the extreme micro scale, these stresses are due to thermal gradients and phase transformations during and after the AM melting and solidification processes. Residual stress and distortion are significant challenges to achieve dimensional accuracy that people can use (Bastola et al., 2023). These high thermal gradients can locally cause areas of tension and compression, leading to warping or cracking, most particularly in thin walled structures.

The consequences of residual stresses are significant. Such behaviour can result in reductions in fatigue resistance and the component as a whole. For instance, parts with high residual stress can experience failure from cyclic loading conditions due to the residual stresses as an initiation site of cracks (Alibe et al., 2023). Furthermore, such dimensional accuracy of the final product can be defeated resulting in a deviation from the intended specifications and a resulting part that is unusable in its intended application.

Strategies for Mitigating Thermal Gradients and Residual Stresses

Various approaches are possible to counteract thermal gradients and residual stresses. The other approach is the optimization of process parameters (such as laser power, scanning speed and layer thickness) in order to control the thermal profile during the AM process (Luo et al., 2024). For instance, by following Turk, adjustments are suggested to the scanning strategy to reduce overheating, to the thermal gradients, minimizing residual stresses (Turk et al., 2024).

Also, residual stresses can be relieved post processing using heat treatment. Due to these treatments, stresses are redistributed to the material and are allowed to improve the overall mechanical properties and dimensional stability of the part (Mahmood et al., 2022). The use of advanced monitoring techniques during AM of components also enables in-situ monitoring and detection of thermal gradients and residual stresses, and associated issues such as on the fly correction (Khanafer et al., 2024).

Modelling Thermal Behaviour

AM is an essential link to facilitate integration of mathematical modelling and simulation tools into thermal behaviour and quality of final products. Overcoming rheological limitations of AM requires continued improvement in modelling techniques and computational tools so that the capability of producing complex geometries with better mechanical properties also increases.

Mathematical models of heat distribution

Calculation of heat in AM requires mathematical modelling of heat for the description of the thermal behaviour of materials for the manufacturing process. Typically, the transient thermal dynamics which occur in depositing layers are simulated using FEA along with these models. By using FEA, its thermal profile gets annotated with details like heat source, material properties, environmental conditions.

Researchers also applied this on a simplified modelling approach to analyze the thermal behaviour during hybrid plasma arc laser additive manufacturing (Wang et al., 2022). By using their finite element model, they have found how substrate preheating affects thermal cycles and in turn reduces the cooling rate in the melt pool. They also used a study to develop a finite element that obtained great correlation between the simulation and experimental results by demonstrating the model's accuracy with wire arc additive manufactured parts (Di et al., 2022). Such predictions of microstructure and mechanical properties of a final product from these phenomena that of melting, solidification and heat accumulation directly impact are so important that these models are needed.

Further advanced then is thermal modelling coupled with advanced modelling techniques such as ML in order to enhance predictive ability. For example, a researcher offers a simplified example by using machine learning with numerical simulation to quickly find thermal histories of wire and arc additive manufacturing. First, we explore an ML approach to accelerate the modelling process and optimize AM parameters (Le et al., 2023)., using experimental datasets to improve the accuracy of thermal prediction.

Simulation Tools for Predicting Thermal Performance

Tools to simulate thermal performance in additive manufacturing are numerous, and have different capabilities for different applications. They enable the engineer to have some thermal analysis pre-production, enabling process parameter optimization and material selection.

For example, the commercial finite element software is being used for thermal simulations in AM. In (Poblete et al., 2023), for instance, they used finite element-based methods to describe the thermal history of 3D printed parts, including temperature dependent material properties and rapid cooling events. These tools facilitate modelling of heat transfer at the time scale small enough to capture the thermal dynamics associated with the AM process.

The necessity for choosing appropriate simulation tools so as to predict the behaviour of the material and how this can optimize the part functionality is important. As demands on part geometry and material properties increase, complexity of the simulation tools should also rise to produce accurate predictions (Kaščák et al., 2023; Boopathy et al., 2024). As a consequence, it is essential to be adaptable in addressing problems resulting from the employment of advanced AM techniques and materials.

Additionally, graph theory has been applied as an innovative approach in the development of thermal modelling for directed energy deposition (DED). A new graph theory based mesh free computational thermal modelling approach for fast and accurate prediction of thermal history critical for the understanding and preventing of flaw formation during the AM process was presented by (Riensche et al., 2022). Figure 3 gives the principle of DED system.

Figure 3. Diagram of directed energy deposition (DED) (Image by Cho et al., (2023) from "Investigation of Effect of Processing Parameters for Direct Energy Deposition Additive Manufacturing Technologies (https://www.mdpi.com/2504-4494/7/3/105)" is licensed under CC BY 4.0)

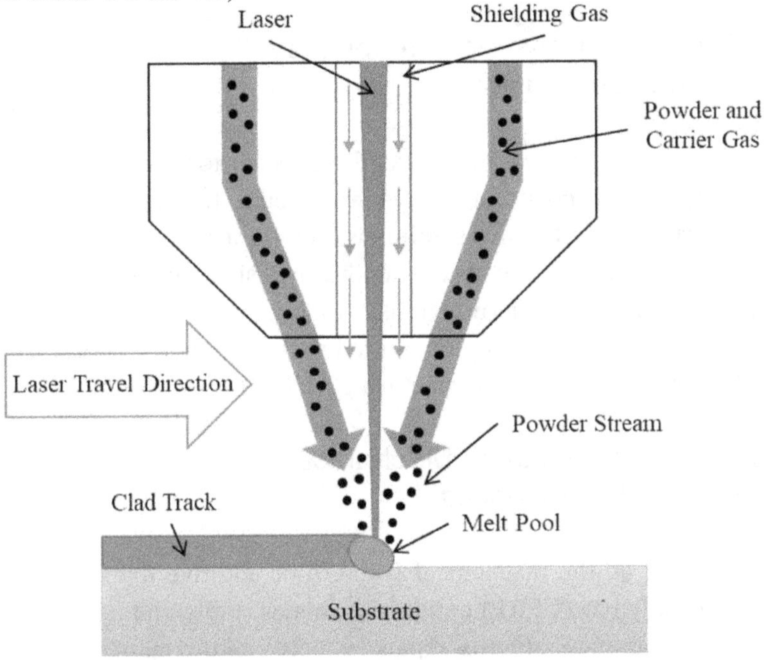

Case Studies

AM thermal analysis is crucial for the optimization of the process and production of good quality final products; further, it provides insight into the changes during the manufacturing, and the solidification of the metal. Thermal analysis is applied to several case studies and its practical applications in different AM techniques are discussed, with an emphasis on thermal management and material performance.

Thermal Management in Cold Plates Using Artificial Neural Networks

Cold plate heat transfer coefficients and pressure drops with surface roughness were predicted by artificial neural network (ANN) models. These models demonstrate the significance of finding optimal thermal management systems for additively manufactured components, particularly for applications that require effective heat dissipation. With such approach, the ANN enhances the heat transfer performance without losing the operational efficiency to the limits of allowable pressure drop, which offers a robust framework for designing thermal management system for AM. (Randriambololona et al., 2022)

Thermal Stability of Polymer Composites in Laser Powder Bed Fusion

Laser powder bed fusion was investigated as a means to process activated carbon/ UHMWPE polymer composites and TGA was used to test the thermal stability of the material. Furthermore, thermal stability is important which allows the composite to be resistant to high thermal loads during the laser sintering process. Thus, it emphasizes the need to analyse thermal transformations in order to avoid degradation and ensure control of the integrity of printed parts and their importance in phase material selection for additive manufacturing. (Khalil et al., 2022)

Impact of Thermal Effects on Mechanical Strength of CFRP Composites

Thermal effects on the mechanical strength of additive manufactured carbon fiber reinforced polymer (CFRP) composites under stable and cyclic temperature conditions were examined, and discrepancies between the matrix CTE and fibers CTE resulted in thermal stress. Fiber matrix debonding and subsequent mechanical deterioration resulted from this stress. These results underscore the role of thermal analysis in delivering real value on the mechanical usability of composite parts made using additive manufacturing. (Muna et al., 2022)

Advanced Thermal Management Structures in AM

An advanced design and manufacturing approach for structures with improved thermal management performance was reviewed through a comprehensive review with special emphasis on unique capabilities of AM for producing complex heat sinks and heat exchangers. It reviewed AM's ability to create intricate thermally enhanced geometries and the need for effective thermal analysis to realize the potential within AM technologies in thermal management applications. (Sun et al., 2024)

Multi-Criteria Thermal Analysis Framework

An industrial part scale thermal analysis framework for additive manufacturing processes using a multi criteria h adaptive finite element approach was developed to evaluate thermal performance in the complex geometries achieved via shape optimization analysis. It was demonstrated as a means of addressing thermal conduction and heat distribution challenges to advance thermal modelling and inform design and also improve the functionality of additively manufactured component. (Moreira et al., 2022)

Thermal Conductivity of Lattice Structures

A modified temperature profile method was used to measure the effective thermal conductivity of additively manufactured lattice structures, in particular the effects of heat treatment on the thermal conductivity of Ti6Al4V structures. The key the findings showed was that these structures can be optimized to improve cooling capabilities - a benefit provided by additive manufacturing of advanced thermal metamaterials. (Nagano et al., 2022)

Thermal Effects on Mechanical Behaviour of CFRP Samples

Numerical modelling was performed to study thermal effects on mechanical behaviour of additive manufactured CFRP samples and it was found how thermal gradients affect the mechanical performance of these composites. The research focused around designing additive manufacturing processes that will prevent and mitigate likely defects through the use of thermal analysis. (Mieloszyk & Muna, 2023)

Thermal Insulation Applications of 3D Printed Materials

A review was made of the thermal conductivity of emerging 3D printed thermal insulating materials in fabrication. The use of these materials for sustainable complementary use was highlighted and demonstrated their potential for use in thermal analysis to bring the development of an efficient thermal insulator to additive manufacturing. (Dhangar et al., 2023)

MECHANICAL BEHAVIOUR OF PRINTED MATERIALS

Determine the effect of mechanical behaviour printed materials during additive manufacturing from varieties of factors like material type, process parameters and post processing techniques. Thus, these influences, and related failure mechanism, need to be understood in order to allow AM process optimization and realize part reliability. Using smart make strategies to decrease instance and limit residual strictness and microstructure change, lenders can improve the effectiveness of additively fabricated segments in a large scope of uses.

Mechanical Properties of AM Materials

AM research is producing materials that possess customizability of their mechanical properties that are required for their use in a variety of applications. This section reviews main mechanical properties of these AM materials along with tensile and compressive strength and elasticity depending on materials, and other parameters.

Tensile Strength

The tensile strength is a critical mechanical property that denotes the maximum force than can be applied at tensile or pulling load at which the material breaks (Gebrehiwot et al. 2023; Bhoominathan et al., 2024). For many of the materials and print parameters additive manufacturing, the changes in tensile strength are quite large. One example of this is that in their research to optimize this infill density and nozzle temperature for the tensile strength of printed parts, infill density and nozzle temperature could be improved using recycled polylactic acid (rPLA). In addition, tensile strength of dual-wire electron beam AM manufactured intermetallic Ni-Al-Cr materials was studied and it was found that tensile strength correlates with grain size heterogeneity, as finer grains yield higher tensile strength. (Astafurova et al., 2024)

Compressive Strength

Structured Compressive strength measures in Axial loads a material's ability not to fail. Compressive strength of CoCrFeMoNiV complex concentrated alloys is highly influenced in wire and arc additive manufacturing by alloy composition and microstructure. The results have shown that carbides improve strength and wear resistance, and material optimization is important for improving performance. (Treutler et al., 2022)

Elasticity

The type of material and the additive manufacturing process it was subject to will determine how elastic the material is. However, polymers generally have less elasticity than metals and combining reinforcing fiber with composites can greatly enhance composite elastic properties. For example, optimization of the resin content in glass fiber reinforced composites was shown to increase interlaminar strength, thus improving the elasticity of the resulting product. (Perez-De-Eulate et al., 2023)

Influence of Material Type

The mechanical properties of AM materials are significantly influenced by the type of material used:

Polymers

AM is well used with polymers thanks to their ease in processing and versatility, although their mechanical properties can vary significantly. One such example is that PLA has some good tensile strength, however, the material is brittle and prone to failing under impact (Gebrehiwot, 2023). This study demonstrates studies on hybrid PLA/basalt bio composites showing improved properties through material optimization such as incorporation of fillers or reinforcements (Faizaan, 2024).

Metals

Whereas polymers are added to metal parts using additive manufacturing, such as SLM, metal additive manufacturing processes produce parts with superior mechanical properties. Regarding the microstructure, AM can be used to allow better control during the AM process and have better strength and ductility. For example a significant influence of the mechanical properties of CoCrFeMoNiV alloys is alloy composition, and formation of carbides with exceptions decreasing strength and performance. (Treutler et al., 2022).

Composites

Composite materials of polymer and reinforcing fibers have properties that are greater than the properties of either one of the component materials. Research has shown that inclusion of continuous fiber reinforcement in additive manufacturing significantly improves the tensile strength and stiffness and these materials are very promising for demanding applications (Grozav, 2023; kumaar et al., 2024).

Factors Affecting Mechanical Properties

Various factors such as process parameters and post processing techniques have an effect on mechanical properties of AM materials. Knowledge of these factors is of the driving force for AM components performance optimization in a wider range of applications.

Process Parameters

Layer Thickness: Mechanical properties of additive manufacturing parts are dependent on the layer thickness. As the layers become thinner, mechanical properties improve due to improved interlayer bonding, and the surface finish becomes smoother. Once layer thickness in the FDM process is reduced tensile strength and elasticity of printed parts improves. On the other hand, the final outcome of thicker layers is porosity and weak performance in mechanical properties, especially due to lack of bonding among layers (Kim et al., 2023). Figure 4 depicts the direction of the observation plane and scanning strategy in the additively manufactured specimen.

Figure 4. Illustration of the direction of the observation plane and scanning strategy in the additively manufactured specimen. (Image by Kim et al., (2023) from "Correlation between Cellular Structure Morphology and Anisotropic Yield in Additively Manufactured Stainless Steel 316L (https://www.mdpi.com/1996-1944/16/4/1666)" is licensed under CC BY 4.0)

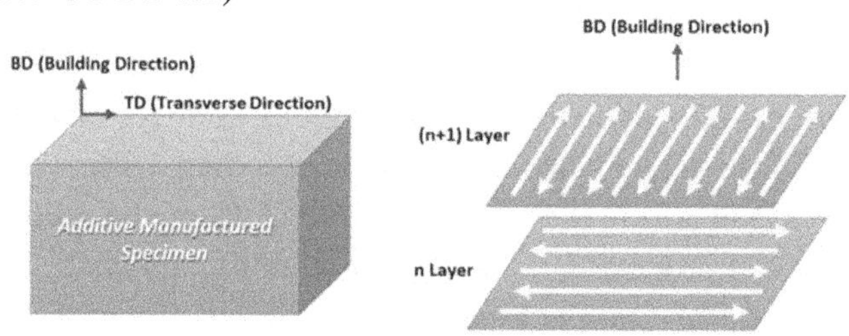

Print Speed: Cooling rate and thermal history of materials during additive manufacturing process are largely controlled by print speed. The deflection in time that the material spends under the jet may decrease, thereby reducing the time for proper material bonding, thus causing defects such as delamination or weak interlayer adhesion. But print speed and other parameters have been shown to greatly improve the tensile behaviour of materials such as acrylonitrile styrene acrylate (ASA) and, in the process, improve the quality of printed parts (Magri et al., 2022).

Infill Density: Strength and weights of printed parts are influenced directly by the infill density. On the basis, higher infill densities generally result in a more favourable tensile strength through greater availability of material volume, which results in a more favourable distribution of load and improved structural integrity. Although, excessive infill densities can cause higher material usage and longer print time, therefore, it is imperative to find the appropriate balance between efficiency and strength to satisfy the best performance (Daly et al., 2024).

Infill Pattern: In fact, the infill pattern selection has a very prominent role in deciding the mechanical properties of additively manufactured parts. Different kinds of pattern like honeycomb or grid provide different strength and flexibility in accord with application requirements. It has been shown in the past, through research, that the variation in infill pattern has a strong influence on the tensile strength of materials such as PLA and could have a biased result for tensile performance under load, wherein specific patterns outperform others, and therefore pattern selection is an important design consideration (Jasmin et al., 2022).

Extrusion Temperature: The filament viscosity and interlayer bonding in additive manufacturing are dependent on extrusion temperature. Adhesion of layers can be improved with higher temperatures but, at the same time, higher temperatures may bring challenges such as warping and stringing. Studies on carbon fiber reinforced PLA composites show that this parameter is critical to optimize extrusion temperature and that fine tuning of this parameter significantly affects the mechanical performance of the material (Kargar & Ghasemi-Ghalebahman, 2023).

Post-Processing Techniques

Annealing: Residual stresses and the improvement of mechanical properties in additive manufactured parts can be achieved by annealing, a thermal treatment process. It is well known that post processing heat treatments, such as annealing, improve the strength and ductility of titanium alloys manufactured through AM, making them perform and be reliable.

Surface Finishing: Surface finishing plays an important role in improving surface quality and fatigue resistance of additive manufacturing parts by surface finishing techniques such as sanding, polishing and chemical treatment. These processes also reduce surface defects in the parts, which enhance the overall mechanical performance of the parts. In the studies, surface treatments are found to play a major role on the mechanical properties of 3D printed fibreglass reinforced thermoplastics which makes them a necessary step towards high quality and reliable components (Hermosilla et al., 2023).

Heat Treatments: It is necessary to perform specific heat treatments to optimize the microstructure of additively manufactured materials (especially metals). Mechanical properties as strong as strength and ductility have been enhanced in controlled thermal processes for components manufactured utilizing laser powder bed fusion. Such treatments can modify microstructures and therefore are critical to enhancing AM part performance and reliability (Zhao et al., 2023). Figure 5 depicts the flowchart on post processing techniques in additive manufacturing

Figure 5. Illustration on post processing techniques in additive manufacturing

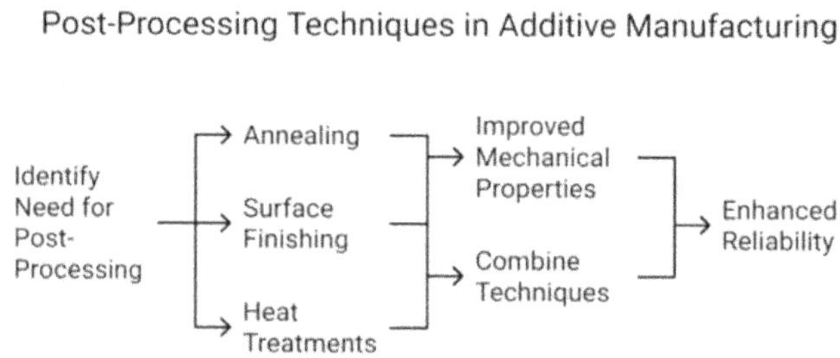

Post-Processing Optimization: The synergistic combination of post-processing techniques can result in improvements of mechanical properties. For example, annealing can be used in conjunction to surface finishing to generate parts that have improved strength as well as surface quality, suitable for use in functional applications.

Failure Mechanisms in Additive Manufacturing

Although AM methods have several benefits, such as increased material efficiency and design flexibility, they are also have some common defects. Understanding these mechanisms is crucial for improving the reliability and performance of AM components. This section discusses about the common defects encountered in AM, including warping, delamination, and microcracks, along with methods for their prevention and mitigation.

Warping: Thermal stresses that develop during cooling are common in the additive manufacturing defect known as warping. Polymers in particular have very high thermal expansion coefficients, so this distortion is present. However, as the transformer of the material flows is proportional to the material thickness, warping can produce significant dimensional inaccuracies on large or thin-walled components. However, thermal gradients which are common to material layer by layer deposition do exacerbate the issue leading to curled or lifted edges (Morales et al., 2021) which may render printed parts of poor quality and failure to function.

Delamination: Delamination, the lack of bonding between layers within a part is a defect responsible for causing separation of parts and compromising structural integrity of additive manufacturing parts. These causes are usually inadequate extrusion temperature and incompatibility between the layer and material when it is printed, causing a poor adhesion. High residual stresses make components fabricated by the directed energy deposition techniques especially susceptible to delamination. However, the presence of this defect can greatly damage the mechanical properties of a part and it is very likely that under load, the part will fail, making it imperative to optimize the process to reduce its appearance (Wu et al., 2023;).

Microcracks: Thus, small fractures, also known as microcracks, which form during the additive manufacturing process cooling phase can grow under stress causing catastrophic failure. Since micro cracks are very critical material defects, they usually act as the initiation point for crack propagation. Research shows that the impact of these surface defects is much worse than internal ones and they compromise fatigue resistance and overall durability of AM components, thus demanding process control and post-processing to eliminate their influence.

Methods for Prevention and Mitigation

Optimizing Process Parameters: Process parameter optimization is one of the most effective means to control defects like warping and delamination. Detrimental factors, e.g. layer thickness, printing speed and extrusion temperature can be adjusted with respect to layer adhesion and the amount of thermal stress. For instance, if the bed temperature is increased then it improves the amounts of warping by synthesizing better adhesion between the first layer and the built platform.

Post-Processing Techniques: Residual stresses and poor mechanical properties of AM parts can be alleviated through the use of postprocessing such as annealing. Annealing enables redistributing the stresses in the material to increase ductility and diminishes the chances of crack formation. The surface quality can also be improved by surface finishing techniques, i.e., sanding or coating, which would further reduce the defect impact on mechanical performance.

Material Selection: Prevention of defects is crucial and determines the material to use in this particular application. Put otherwise, a lower warping risk is associated with a lower thermal expansion coefficient. Moreover, reinforcement materials like fibers or fillers can improve mechanical properties and make the samples sensitive to defects.

Real-Time Monitoring: The use of real time monitoring systems during the AM process can detect the defects as are occurring. In situ imaging, machine learning algorithms can be used to identify anomalies in the printing process, and be used for instant corrective actions. Such proactive approach can greatly lower defects occurrence and significantly enhance overall part printed quality.

Design Considerations: The AM process can also be considered when designing part: this will limit defects (Buffa et al., 2023). For example, the introduction of such features like a more heat dissipative or avoiding sharp corners features can help reduce thermal stresses in the graphene and improve separation of layers while being introduced. Furthermore, the mechanical properties will increase and the occurrences of defects will be decreased by the orientations of the part while getting printed. (Hyer, 2022).

INTEGRATION OF THERMAL AND MECHANICAL ANALYSES

For the optimization of the performance and reliability of printed components, it is necessary to integrate thermal and mechanical analysis in AM. Thermal and the mechanical properties interplay, thermal and mechanical performance balancing strategies, as well as the role of real-time monitoring and control in improving the AM processes are this section's focus.

Interplay between Thermal and Mechanical Properties

Thus, AM thermal and mechanical properties need to be interplayed in order to optimize the performance and reliability of printed components. The effect of thermal conditions on mechanical strength and durability are discussed in this section with a view on the factors that help in balancing the requirements of thermal management with mechanical performance.

Effects of Thermal Conditions on Mechanical Strength and Durability

The relation of the thermal states at AM to the final mechanical properties in the knifed product is critical. Due to such factors as thermal history, cooling rate and temperature gradient, the mechanical performance also varies accordingly, from being uniform in small grains to columnar structure in large grains. As an example, compatibilizers are used to enhance thermal protection by increasing mechanical strength through decreasing the polymer decomposition during processing in polypropylene wollastonite composites. It illustrates that the mechanical properties are conditioned by the thermal conditions. (Leontiadis et al., 2022)

In addition, polymeric materials are severely damaged by thermal cycling as it induce water sorption, and that of polymer chains, while reducing the mechanical performance. Studies of denture base materials have shown that thermal cycling decreases flexural strength, but that failure due to the interaction of different materials with thermal cycling is of little importance, indicating that the effect occurs in similar material groups. As these findings show, thermal management must be effective throughout the AM process to achieve the final product with desired performance and durability (Çakmak et al., 2022).

Strategies for Balancing Thermal Management and Mechanical Performance

Strategies such as optimal material selection, design enhancements, process parameter adjustments, and post-processing techniques help in balancing thermal and mechanical performance, ensuring high-quality, reliable AM components. Table 2 gives the strategies for balancing thermal management and mechanical performance.

Table 2. Strategies for balancing thermal management and mechanical performance

Strategy	Description
Material Selection	• Choosing materials with favorable thermal and mechanical properties ensures optimal performance in AM applications. • High thermal conductivity materials enhance heat dissipation. • Materials with lower thermal expansion coefficients reduce the risk of warping and residual stresses.
Design Optimization	• Advanced methodologies such as topology optimization and biomimicry help fabricate structures that reject thermal load while maintaining mechanical integrity. • Incorporating innovative design principles enhances thermal management and mechanical performance.
Process Parameter Optimization	• Adjusting parameters like layer thickness, print speed, and infill density can balance thermal and mechanical properties. • Reducing thermal gradients and improving layer adhesion enhances mechanical strength and durability.
Post-Processing Techniques	• AM parts are prone to thermal stresses, which can be mitigated using post-processing techniques. • Annealing and surface finishing help relieve residual stresses and enhance interlayer bonding. • These techniques improve the mechanical properties, making components stronger and more durable.

Real-Time Monitoring and Control

For the benefit of printed components, the integration of real-time monitoring and control systems in AM is required. It discusses the use of sensors in thermal and mechanical data measurements, implementation of adaptive systems for the optimization of the process.

Role of Sensors in Capturing Thermal and Mechanical Data

A critical part of the AM process requiring real time monitoring of the processes is where thermal and mechanical data is captured. The day they change vehicles, tents or any other pieces of our equipment undergo continuous sensor reading by various sensors, including thermal imaging cameras, acoustic sensors, etc. to gauge temperature gradients, layer adhesion, dimensional accuracy and surface defects among others.

This demands heavy use of condition monitoring tools including of real time sensors for maintenance of such additive manufactured components. The tools can detect both anomalies and defects in the manufacturing operations for support to correction measures on-the-spot. Thermal imaging on build platform enables users to track hotspots that could give rise to warping and thermal related defect.

In-situ process inspections of temperature and strain are achievable through the embedded Fiber Bragg Grating (FBG) sensor technology. Studies prove FBG sensors effectively monitor deforming and residual strain behaviour in polymer components made through AM technology thus proving their ability for real-time surveillance. Manufacturers who utilize these technologies can make data-based choices during production which enhances both product reliability and total product quality (Mieloszyk et al., 2022).

Adaptive Systems for Process Optimization

Such machine learning algorithms based adaptive systems can be used to dynamically analyze the real time data to optimize process parameters for minimizing defects and improving performance during the AM process. For instance, machine learning based systems had been used for process monitoring in automated manufacturing of composites where in situ data such as thermal history is crucial in optimizing the manufacturing conditions. Through this approach, there are adaptive responses in a changing environment so that outcomes are of higher quality (Mujtaba et al., 2022).

This concept is further increased by the ability to integrate physical and digital representations of the AM process and brings about the concept of a digital twin. Digital twins can combine sensor data with predictive models to provide real time simulation and optimization of process dynamics. Multisensor fusion based digital twin has demonstrated great potential to support in situ quality monitoring and defect correction to enhance the process efficiency and product quality in the AM processes (Chen et al., 2023).

It is also possible to use adaptive control systems to adjust process parameters in real-time, i.e., changing the value of process parameters to the value that will decrease the cost function from the sensor feedback. For instance, a thermal sensor in which temperature increases would cause an adjustment of the laser power or the print speed to minimize failures. Additively manufactured component secured by a reliable and a high production quality using a proactive approach to process control.

ADVANCES IN MATERIAL ANALYSIS FOR AM

Those include material science, innovation tools and techniques, and a growing sustainability focus that make AM field into a rapidly evolving field. In this section, the material analysis for AM in the context of novel materials, emerging tools and techniques as well as sustainability considerations is discussed.

Novel Materials for AM

The advancement of AM has led to the development of novel materials that enhance the performance and applicability of 3D-printed components. This section explores the significance of nanomaterials, composites, and bio-compatible materials in AM, highlighting their unique properties and potential applications.

Nanomaterials

AM is partially leveraging nanomaterial capabilities by virtue of high-quality mechanical, thermal and electrical properties of major nanomaterials. Perhaps among these brief examples is the example in which amphiphilic chitin has been incorporated into plasticized polylactic acid (PLA)/starch biocomposites to improve material toughness and the quality of interfacial interactions within the matrix. Improvements in modulus and mechanical performance are demonstrated, demonstrating the potential of nanomaterials to create high on mechanical performance components created through AM (Olaiya et al., 2022)

In addition, the enhanced thermal stability and mechanical strength can be achieved by integrating nanoparticles into the polymer matrices, such as carbon nanotubes or graphene. Such nanocomposites have demonstrated their superior properties of higher tensile strength and thermal conductivity in these application industries, including aerospace and automotive. In this, nanomaterials find themselves growing integrated into more and more AM technologies, demonstrating what can be done with the capabilities of AM enhanced by their integration (Dwivedi et al., 2024).

Composites

AM is greatly relying on composite materials based on polymer materials and reinforcements like fiber or nanoparticles to provide tailored mechanical properties required for a particular application. For example, it is demonstrated that advanced composites improve the performance and durability of available 3D printable removable partial denture base materials by addressing microstructural deficiencies. Moreover, continuous fiber reinforcement further enhances tensile strength and stiffness of these composites by incorporating it into AM, making them suitable for high structural demanding applications.(Boopathy et al., 2024).

Furthermore, polyaryletherketone (PAEK) has been designed for AM with high mechanical properties and high thermal stability. Due to their material performance characteristics, particularly in aerospace, these are positively suited to PAEK. Composite materials used in AM demonstrates its adaptive ability for the optimization of both its mechanical and thermal properties to match certain engineering and industrial requirements, which further demonstrates its versatility in developing the future of AM technologies (Chen et al., 2022).

Bio-Compatible Materials

Very recently, biocompatible materials have attracted high demand in AM, especially in the biomedical field. Such materials as polycaprolactone (PCL), polylactic acid (PLA) are used widely to produce scaffolds and implants due to their biocompatibility and biodegradability. Equal to recent research, using biodegradable materials not only promotes sustainable and environmentally friendly 3D printing streams but also cuts down the environmental footprint of the manufacturing process (Subramani et al., 2024).

Further advancements in hydrogel materials for bioprinting have advanced possibilities for creating soft, flexible structures that resemble biological tissues. One of the advantages of 3D printed hydrogels is its potential applications in smart windows and other applications that do not require biomedical application of the material and there are 3D printed hydrogels that could serve as examples for this. Thus, ability to fabricate such complex geometries with biocompatible materials is important, since it is a cornerstone in AM technologies in general; and is essential to develop functional implants and scaffolds in tissue engineering.

Emerging Tools and Techniques

AM has made its emergence in the last two decades due to many breakthroughs in materials, processes, and technologies. The aim of this section is to describe the emerging tools and methodologies that increase the proficiency of AM, specifically, through the means of analysis methods driven by AI and new simulation and modelling tools.

AI-driven analysis methods

AI is transforming the field of additive manufacturing by enabling more efficient and accurate analysis of processes and materials. AI-driven methods can optimize various aspects of AM, from design to production, by leveraging data analytics and machine learning algorithms.

- ***Surface Roughness Prediction:*** For example, different AI techniques have been applied to predict surface roughness of additively manufactured components in a variety of manufacturing processes including FDM and Wire Electrical Discharge Machining (WEDM). Predictions of surface roughness are both accurate and helpful to the manufacturer for improving the quality of the final product and minimizing the amount of work for its post processing. The advances described here show the applications of using AI models to maximize manufacturing processes and guaranteeing good product quality (Batu et al., 2023)
- ***Real-Time Quality Monitoring:*** AI systems function as essential components for real-time quality control in AM procedures which improve managerial decisions while decreasing product flaws and enhancing manufacturing operation effectiveness. Such continuous monitoring of thermal and mechanical properties through these systems allows manufacturers to find and solve problems during production thereby achieving reliable process consistency. The application of AI demonstrates its power to enhance AM workflows by bringing changes to both quality standards and operational speed (Okuyelu & Adaji, 2024).
- ***Predictive Maintenance:*** The functionality of predictive maintenance in AM systems depends on AI technologies. Every AM system benefits from AI because it uses historical information and pattern identification to predict equipment breakdowns in advance and support prompt maintenance. The systemic application fetches two advantages for operations while lowering unexpected breakdown-related expenses which is essential for contemporary industrial production (Hao, 2024).

Advanced Simulation and Modelling Platforms

The development of advanced simulation and modelling platforms is crucial for optimizing the AM process and predicting material behaviour. These tools enable engineers to simulate various scenarios and assess the impact of different parameters on the final product.

- ***Finite Element Analysis (FEA):*** FEA stands as a primary simulation technique that manufacturers extensively use in additive manufacturing to model thermal and mechanical processes throughout printing. Managers use simulation tools to understand material performance and exercise process control which helps manufacturers maximize their part functions. FEA enables users to see both stress distributions and thermal gradients so they can detect possible failure locations before actual production thus improving the dependability of their end products.
- ***Multiscale Modelling:*** Currently scientists use multiscale modelling procedures which combine data starting at atomic scales and advancing through to macroscopic measurements. The technique proves extremely beneficial for researchers to study how microstructural changes impact the mechanical properties of AM materials. Through multiscale modelling scientists obtain comprehension about processing-to-performance relationships which guides their design decisions. The article lacks evidence to confirm the stated claim although the reference list included no references to support it.
- ***Digital Twin Technology:*** In AM, the idea of a digital twin or digital replica of a physical system has taken off. Real time monitoring and simulation of the AM process is made possible by this technology allowing the manufacture to dynamically optimize the AM process operation. Digital twins use sensor data to integrate with predictive models to deliver process performance insights and to facilitate adaptive control.

Sustainability Considerations

The AM industry is paying increasing attention to sustainability since industries tend to use more resources more effectively while leaving less of an environmental impact. The second consideration on this section discusses on two critical aspects of sustainability in AM, which are recycling and reuse of materials and energy efficient processing methods.

Recycling and Reuse of Materials

- ***Recycling of Thermoplastic Materials:*** The potential for recycling of 3D-printable thermoplastic cellulose composites has been investigated and found to implicate the melt flow index (MFI) of recycled material keeping relatively similar to virgin material. Thus, it shows that recycled thermoplastics do not lose their processing characteristics and mechanical performance. This is indicative of recycling of thermoplastic materials in additive manufacturing to support the advent of a circular economy with less waste and usage of less resources (Immonen et al., 2022).
- ***Recycled Concrete in AM:*** This study focuses on the use of sustainable recycled materials used in 3D printed concrete blocks for their thermal insulation in traditional buildings. It makes use of waste materials and promotes their reuse, which helps in surrounding the construction practice with sustainability. In addition, by introducing recycled materials it is also economical as materials costs are decreased resulting in practical as well as ecologically friendly approach in the contemporary construction (Volpe et al., 2023).
- ***Mechanical Recycling of Thermosets:*** However, thermoset materials used for additive manufacturing have both challenges and opportunities in terms of mechanical recycling. Thermosets traditionally have been very difficult to recover valuable materials for reuse in new applications due to their traditional difficult recycling. The development of sustainable strategies for recycling thermoset composites in AM represents a current gap, which has been shown to be a viable opportunity for recycling strategies, and can support reducing waste and increasing resource efficiency (Armstrong et al., 2023).
- ***Dynamic Bonds in 3D Printing:*** We introduce the concept of dynamic bonds in light-based 3D printing that allows recycling of the printed structures. Such materials possess dynamic bonds that permit reprocessing and reuse with minimal loss of mechanical property for an additive manufacturing method that is sustainable in the context of materials legacy. Plastics in AM have become a challenge plaguing the industry, and this innovative approach contributes to supporting the development of materials that are fully recyclable, providing an environmentally friendly approach in the industry (Zhu et al., 2023).

Energy-Efficient Processing Methods

- ***Energy Consumption Assessment:*** Here we introduce the concept of dynamic bonds in light-based 3D printing that allows recycling of the printed structures. Such materials possess dynamic bonds that permit reprocessing and reuse with minimal loss of mechanical property for an additive manufacturing method that is sustainable in the context of materials legacy. Plastics in AM have become a challenge plaguing the industry, and this innovative approach contributes to supporting the development of materials that are fully recyclable, providing an environmentally friendly approach in the industry (Cozzolino et al., 2023).

- ***Green Technology Adoption:*** A significant role of green technology and renewable energy in manufacturing process is the integration of green technology and renewable energy into manufacturing processes which helps enhance energy efficiency and decreases the carbon emission. The adoption of energy efficient process and use of renewable energy sources is possible in AM to enhance sustainability goals. Manufacturers can achieve a more sustainable production by optimizing energy usage by printing process and powering AM operations with renewable energy (Rusmayadi, 2023).
- ***Sustainable Manufacturing Practices:*** The author highlighted the significant role of sustainable manufacturing in review of polymer materials in AM development and challenges. AM technologies are an alternative to the traditional manufacturing due to advantages like less material waste and lesser energy consumption. This results in a paradigm of more sustainable and efficient production. They imply the potential of AM to enable environmentally friendly manufacturing processes (Miao, 2024).
- ***Circular Economy in AM:*** In the AM sector, the term circular economy has gained momentum and is directed towards resource efficiency and long-term product lifetime. Circular Economy principles proved to have a great potential in terms of production of personalized products with low environmental impact taking advantage of the synergy on 3D printing. This method puts the emphasis on the recovery, reuse and recycling of materials, in the interest of sustainable practices and in line with other goals for sustainable manufacturing (Firtikiadis et al., 2022)

APPLICATIONS AND CASE STUDIES

AM has found applications across various industries, including aerospace, automotive, healthcare, and electronics. This section highlights specific case studies that illustrate the use of AM in producing high-strength lightweight components, custom implants and prosthetics, and heat-resistant and mechanically robust components.

Aerospace and Automotive

- ***Selective Laser Melting of Nickel Superalloys:*** Research on SLM of non-weldable nickel superalloy has been performed and proven that high performance components for aerospace applications can be manufactured. In terms of microstructure, crack formation and texture, the superalloys with their superlative high temperature resistance and mechanical strength were analyzed. The result shows that AM can make very lightweight, high strength components that can endure extreme operating conditions and meet the specificities of the aerospace industry (Starikov et al., 2023).
- ***Lattice Structures in Automotive Applications:*** In the automotive industry, it is becoming more common to use lattice structure because they are light in weight and can absorb much impact energy. Research of these structures for automotive application has demonstrated that these structures can reduce component weight by a substantial amount with no sacrifice in handling capability. Since the mechanical properties depend on geometry, lattice geometry needs to be optimized, which is an advantage of AM to produce lattice geometries that cannot be made using the conventional manufacturing techniques (Gao et al., 2022).

Healthcare

- ***3D-Printed Custom Implants:*** AM has greatly advanced healthcare with most being used to manufacture custom implants. 3D printing has been compared with normal abutments in terms of advantages of patient specific solutions (e.g. custom CAD / CAM implant abutments). Tailored implants to individual anatomical requirements can provide the fit, the function, and ultimately the patient care, while minimizing complications. The fact that this personalization is being done shows how AM can be transformative in modern healthcare (Țârtea et al., 2023).
- ***Biocompatible Materials for Prosthetics:*** AM has been advancing significantly, where development of biocompatible materials for prosthetics is one of the great achievements. Biodgrade polymers use is highlight as a creative and green alternative of 3D printing prosthetics. Limitations of customized prosthetics can be overcome with AM technologies to thereby customize prosthetics to better meet individual patient needs for comfort and functionality. This capability shows the strong promise of AM in a revolutionary role towards prosthetic design and manufacture in the healthcare realm.

Electronics

- ***Thermal Management in Electronics:*** Advanced materials have been integrated into AM and together with-it, heat-resistant components have been developed for use in electronic applications. For example, atomic layer deposition applied to make high quality aluminum nitride (AlN) thin films represents potential for efficient thermal management in electronic devices. AM materials can help achieve high thermal conductivity to enhance performance and reliability of electronic components, as a critical role in the performance and reliability of electronic components that high thermal conductivity can be played (Zhang, 2023).
- ***Nanofluid-Filled Heat Pipes:*** It has been explored what the thermal performance of the grooved cylindrical heat pipes filled with nanofluids are for electronics cooling application and they show much higher heat transfer efficiency. It is well suited for high power electronic components where cooling is enhanced by these heat pipes. The challenge of this application is that it demonstrates the potential of AM for fabricating novel, efficient, and reliable cooling solutions for electronic devices (Saad et al., 2022).
- ***Microchannel Heat Sinks:*** AM technologies have greatly improved multiscale design and optimization of microchannel heat sinks for electronic cooling. Excellent thermal management capability of microchannel heat sinks has been demonstrated for compact electronic systems to address the challenges of power density increasing. With the capability of architectural engineering for intricate geometries that AM can create, these heat sinks gain advantages in comprehensive reviews of fluid flow and heat transfer. It adds an extra capability, superior heat dissipation, which helps high power electronic components perform and last better. Figure 6 depicts the advanced thermal management in electronics.

Figure 6. Thermal management flowchart in electronics.

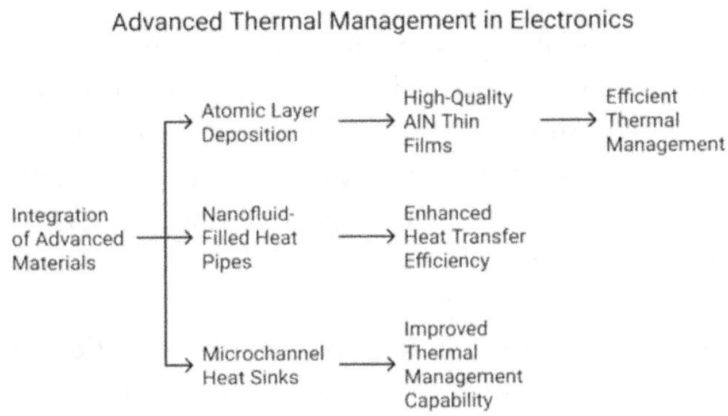

CHALLENGES AND FUTURE DIRECTIONS

As AM continues to evolve, it faces several challenges that must be addressed to fully realize its potential. This section discusses current limitations in thermal and mechanical analyses, the need for standardization in material testing and analysis, and the prospects for integrating advanced computational tools and real-time systems.

Current Limitations in Thermal and Mechanical Analyses

Complexity of Thermal and Mechanical Interactions

AM is a key challenge to be tackled due to the complexity of thermal and mechanical interactions in this process. Heating and cooling cycles incur residual stresses and defects that are difficult to predict and model with high accuracy, so there are inevitable residual stresses and defects. Furthermore, the additional complexity is that there is no generally accepted control and gauge values for the assembling mechanical properties of materials created utilizing SLM. This work presents some of the efforts towards developing a greater understanding of the thermal behaviour of soft materials, given that, by the virtue of this challenge, more research is needed to develop robust analytical models to accurately describe the thermal response of AM materials under various thermal conditions to achieve process reliability as well as improve material performance. (Vysochinskiy & Akhtar, 2022).

Material Behaviour Variability

There is significant variability of the mechanical properties of additive manufacturing materials due to influence of the ecological and the printing parameters. Likewise, variations of the mechanical performance are also seen with modifications of microstructure and bonding mechanisms using various AM techniques. However, such variability presents significant challenge to defining consistent AM material performance metrics and standards. These tensile properties studies of materials produced in wire arc illustrate the great difficulty in achieving the homogeneous mechanical properties, i.e. standardised testing and optimisation to ensure the reliance, reliability, and reliability of optimum performance from AM finished SME components (Leonetti et al., 2023).

Need for Standardization in Material Testing and Analysis

Reliability and comparability of AM materials by different uses depends on standardization. There are currently no standardized methods of testing AM materials and this makes it difficult for manufacturers and researchers alike.

Development of Testing Standards

Unfortunately, it is necessary to commence with the development of general testing standards applicable to AM materials possessing special properties. While the ASTM F3122–14 standard does cover some of the general guidelines of AM process technology it does not cover all the minutia associated with the AM process technology. Improved benchmarking and comparison of material properties through use of standardized testing protocols would help stabilize the AM technologies in the material process.

Material Database Creation

Furthermore, it is critical to develop open material database which includes tensile test properties and other mechanical properties of AM materials. These databases are available to researchers and manufacturers for the access of standardised data in material selection and the optimization of the process as an aid. For example, need for transparency that along with collaborative effort to raise reliability and performance of AM processes through investigation of variability is required for creation of open material databases of tensile properties of AM materials. And these are such precious resources that are important for driving innovation and in general for keeping quality level of AM applications at the top (Garcia-Granada et al., 2023).

Prospects for Integrating Advanced Computational Tools and Real-Time Systems

Advanced Computational Tools

To further improve the process' predicting capabilities of AM processes, it's important for integration of advanced computational tools (such as FEA and CFD). These tools allow for the detailed simulation of thermal and mechanical behaviour for the purpose of manufacturer process parameter and material selection optimization in order to produce superior product quality and performance. Studies on system integration, which are focussed on application of computational tools in AM as demonstrated, show that computational tools can be used to increase the efficiency and streamline the operations in such a way that reliable and high performing AM outcomes can be achieved (Heilala et al., 2023).

Real-Time Monitoring and Adaptive Control

Real-time monitoring systems equipped with sensors can be successfully implemented for quality control of AM processes. We use these systems to capture continuous thermal and mechanical data that enables as nearly immediate as possible adjustments to optimize the systems' performance. It is further possible to improve process optimization by using adaptive control systems based on machine learning algorithms that process real time data and take decisions. By integrating the real-time system, it will be critical to address the defects and reliability of AM components (Khanafer et al., 2024).

SUMMARY

In this discussion, the exploration of AM is fired periodically via an examination of the transformative potential of AM across different industries as materials, process and technology improve. Thermal and mechanical analyses should be integrated in the optimization of the performance and reliability of AM components. For printed part quality enhancement, it is necessary to understand what the interplay between the thermal conditions and the mechanical properties results in, namely, the thermal gradients and residual stresses can cause significant damages to the mechanical strength and the durability of the printed part. The discussion on material selection on novel materials such as nanomaterials, composites and biocompatible materials show the need of material selection for best performance in AM application. Advanced materials allow for the incorporation that improves mechanical properties and expands the AM technology application range in the area of aerospace, healthcare and electronics.

The field of AM is being revolutionized by emerging tools and techniques, specifically by AI driven analysis methods and advanced simulation platforms. These innovations offer a real-time means of monitoring and adaptive control to allow manufacturers to dynamically optimize processes to achieve better quality product. In addition, the sustainability objectives of reuse and recycling of materials and energy efficient processing methods are emerging in industries as they attempt to reduce its environmental footprint. Despite the remarkable progress in AM neither the material testing or the analysis nor the thermal and mechanical complexities have been fully addressed to date. These areas will continue to be areas of research and development into which we will need investment to increase the capabilities of additive manufacturing technologies and their broader adoption in to other sectors.

REFERENCES

Alibe, A. M., Janga, A., Girgiri, A., & Alibe, I. (2023). Residual stress characterization of an additively manufactured titanium alloy. *Interantional Journal of Scientific Research in Engineering and Management, 07*(09), 1–11. DOI: 10.55041/IJSREM25885

Arıkan, V. (2023). Temperature-dependent debonding resistance of 316 stainless steel, inconel 625, and ti-6al-4v alloys. *Black Sea Journal of Engineering and Science, 6*(3), 287–295. DOI: 10.34248/bsengineering.1312954

Armstrong, M., Mehrabi, H., Naveed, N., & Gregg, C. (2023). Evaluating energy consumption in atomic diffusion additive manufacturing versus sand casting. Journal of Engineering &Amp. *Technological Advances, 8*(1), 46–65. DOI: 10.35934/segi.v8i1.81

Astafurova, E., Reunova, K., Zagibalova, E., Astapov, D., Astafurov, S., & Kolubaev, E. (2024). Microstructure, phase composition, and mechanical properties of intermetallic ni-al-cr material produced by dual-wire electron-beam additive manufacturing. *Metals, 14*(1), 75. DOI: 10.3390/met14010075

Bastola, N., Jahan, M. P., Rangasamy, N., & Rakurty, C. S. (2023). A review of the residual stress generation in metal additive manufacturing: Analysis of cause, measurement, effects, and prevention. *Micromachines, 14*(7), 1480. DOI: 10.3390/mi14071480 PMID: 37512791

Batu, T., Lemu, H. G., & Shimels, H. (2023). Application of artificial intelligence for surface roughness prediction of additively manufactured components. *Materials (Basel), 16*(18), 6266. DOI: 10.3390/ma16186266 PMID: 37763543

Bhoominathan, R., Boopathy, G., Jacob, S., Pakiyaraj, A., & Karthik, M. K. (2024). Analysis of failure modes and energy absorption in glass fiber and chopped basalt sandwich panels under low-velocity impact. *Materials Today: Proceedings*. Advance online publication. DOI: 10.1016/j.matpr.2024.05.054

Boopathy, G., Palani, S., Ganesan, B., & Kumaar, K. (2024). Advancements in design, fabrication, and applications of functionally graded composite materials. In *Engineering materials for efficient energy storage and conversion* (pp. 25–50). IGI Global., DOI: 10.4018/979-8-3693-2798-2.ch002

Boopathy, G., Sathish, S., & Kumar, M. H. (2024). Introduction to precision manufacturing for micro-and nanofabrication. *Microfabrication and Nanofabrication: Precision Manufacturing, 11*, 1–24. DOI: 10.1515/9783111335537-001

Buffa, G., Costa, A., Palmeri, D., Pollara, G., Barcellona, A., & Fratini, L. (2023). A new control parameter to predict micro-warping-induced job failure in lpbf of ti6al4v titanium alloy. *International Journal of Advanced Manufacturing Technology, 126*(3-4), 1143–1157. DOI: 10.1007/s00170-023-11179-6

Çakmak, G., Dönmez, M. B., Akay, C., Abou-Ayash, S., Schimmel, M., & Yılmaz, B. (2022). Effect of thermal cycling on the flexural strength and hardness of new-generation denture base materials. *Journal of Prosthodontics, 32*(S1), 81–86. DOI: 10.1111/jopr.13615 PMID: 36266244

Chafik, R., Ebeid, S., & Sayed, M. (2024). Printing parameters and material characteristics affecting mechanical properties of fdm printed parts. *Engineering Research Journal (Shoubra), 53*(2), 220–225. DOI: 10.21608/erjsh.2023.251818.1252

Chen, L., Duan, G., Zhang, C., Cheng, P., & Wang, Z. (2022). 3d printed hydrogel for soft thermo-responsive smart window. *International Journal of Extreme Manufacturing, 4*(2), 025302. DOI: 10.1088/2631-7990/ac5ae3

Chen, L., Yao, X., Liu, K., Tan, C., & Moon, S. K. (2023). Multisensor fusion-based digital twin in additive manufacturing for in-situ quality monitoring and defect correction. *Proceedings of the Design Society, 3*, 2755–2764. DOI: 10.1017/pds.2023.276

Chen, P., Wang, H., Su, J., Tian, Y., Wen, S., Su, B., Yang, C., Chen, B., Zhou, K., Yan, C., & Shi, Y. (2022). Recent advances on high-performance polyaryletherketone materials for additive manufacturing. *Advanced Materials, 34*(52), 2200750. Advance online publication. DOI: 10.1002/adma.202200750 PMID: 35385149

Chmelko, V., Šulko, M., Škriniarová, J., Margetin, M., Gašparík, M., Koščo, T., & Semeš, M. (2023). Strength and cyclic properties of additive vs. conventionally produced material alsi10mg. *Materials (Basel), 16*(7), 2598. DOI: 10.3390/ma16072598 PMID: 37048892

Cho, K. T., Nuñez, L., Shelton, J., & Sciammarella, F. M. (2023). Investigation of effect of processing parameters for direct energy deposition additive manufacturing technologies. *Journal of Manufacturing and Materials Processing, 7*(3), 105. DOI: 10.3390/jmmp7030105

Cozzolino, E., Franchitti, S., Borrelli, R., Pirozzi, C., & Astarita, A. (2023). Energy consumption assessment in manufacturing ti6al4v electron beam melted parts post-processed by machining. *International Journal of Advanced Manufacturing Technology, 125*(3-4), 1289–1303. DOI: 10.1007/s00170-022-10794-z PMID: 36644782

Daly, M., Tarfaoui, M., Bouali, M., & Bendarma, A. (2024). Effects of infill density and pattern on the tensile mechanical behavior of 3d-printed glycolyzed polyethylene terephthalate reinforced with carbon-fiber composites by the fdm process. *Journal of Composites Science*, *8*(4), 115. DOI: 10.3390/jcs8040115

Dhangar, M., Chaturvedi, K., Mili, M., Patel, S. S., Khan, M. A., Bhargaw, H. N., Srivastava, A. K., & Verma, S. (2023). Emerging 3d printed thermal insulating materials for sustainable approach: A review and a way forward. *Polymers for Advanced Technologies*, *34*(5), 1425–1434. DOI: 10.1002/pat.5989

Di, Y., Zheng, Z., Pang, S., & Li, J. J. (2022). Numerical and experimental investigation of the thermal behavior and microstructure of wire arc additive manufactured 316l stainless steel straight wall part. *IOP Conference Series. Materials Science and Engineering*, *1270*(1), 012084. DOI: 10.1088/1757-899X/1270/1/012084

Dwivedi, H., Tushar, , Singh, S., Rani, P., Ananya, , & Kumar, S. (2024). Analysis of the microstructural and mechanical properties of 3d-printed removable partial denture base materials. *Journal of Pharmacy & Bioallied Sciences*, *16*(Suppl 1), S681–S683. DOI: 10.4103/jpbs.jpbs_939_23 PMID: 38595564

Faizaan, M., Shenoy, S., & Kini, C. R. (2024). Tensile and flexural performance of hybrid fdm and compression moulded pla/basalt biocomposite. *Materials Science Forum*, *1120*, 77–84. DOI: 10.4028/p-duYO7M

Firtikiadis, L., Minaoglou, P., Manavis, A., Aidinli, K., & Efkolidis, N. (2022). Circular economy through customised 3d printed products: a case of souvenir. Proceedings - The Eleventh International Symposium GRID 2022, 467-471. DOI: 10.24867/GRID-2022-p51

Gao, J., Hu, Z., Yang, Q., Liang, X., & Wu, H. (2022). Fluid flow and heat transfer in microchannel heat sinks: Modelling review and recent progress. *Thermal Science and Engineering Progress*, *29*, 101203. DOI: 10.1016/j.tsep.2022.101203

Garcia-Granada, A. A., Rostro-Gonzàlez, H., Puigoriol-Forcada, J. M., & Reyes-Pozo, G. (2023). Open material database for tensile test properties of additive manufacturing materials. *IOP Conference Series. Materials Science and Engineering*, *1294*(1), 012043. DOI: 10.1088/1757-899X/1294/1/012043

Gebrehiwot, S., Gebrehiwot, A. S. Z., Espinosa-Leal, L., Linderbäck, P., & Remes, H. (2023). Optimising the mechanical properties of additive-manufactured recycled polylactic acid (rpla) using single and multi-response analyses methods... DOI: 10.21203/rs.3.rs-3115087/v1

Gebrehiwot, S., Gebrehiwot, A. S. Z., Espinosa-Leal, L., Linderbäck, P., & Remes, H. (2023). Optimising the mechanical properties of additive-manufactured recycled polylactic acid (rpla) using single and multi-response analyses methods... DOI: 10.21203/rs.3.rs-3115087/v1

Grozav, S. D., Sterca, A. D., Kočiško, M., Pollák, M., & Ceclan, V. (2023). Artificial neural network-based predictive model for finite element analysis of additive-manufactured components. *Machines (Basel), 11*(5), 547. DOI: 10.3390/machines11050547

Hao, C. (2024). Artificial intelligence and operational efficiency of chinese manufacturing firms. *International Cognitive Journal, 1*(1), 28–39. DOI: 10.69659/qcqt7937

He, X., Wang, H., Zhu, Z., Wang, L. Z., Liu, J. Q., Haghdadi, N., Nai, S. M. L., Huang, J., Primig, S., Ringer, S. P., & Liao, X. (2022). Texture evolution in a crmnfeconi high-entropy alloy manufactured by laser powder bed fusion. *Journal of Materials Science, 57*(21), 9714–9725. DOI: 10.1007/s10853-022-07025-x

He, Y., Wei, J., Peng, Y., Wang, F., Wang, Y., & Liu, J. (2023). Deformation prediction and experimental investigation on alternating additive-subtractive hybrid manufacturing of 316l stainless steel thin-walled parts. *International Journal of Advanced Manufacturing Technology, 129*(11-12), 5271–5285. DOI: 10.1007/s00170-023-12592-7

Heilala, J., Parchegani, S., & Piili, H. (2023). Additive manufacturing systems integration. *IOP Conference Series. Materials Science and Engineering, 1296*(1), 012024. DOI: 10.1088/1757-899X/1296/1/012024

Hermosilla, R., Oñate, A., Castillo, R., Fuente, A. D., Sepúlveda, J., Escudero, B., & Medina, C. (2023). Influence stacking sequence and heat treatments on the out-of-plane mechanical properties of 3d-printed fiberglass-reinforced thermoplastics. *International Journal of Advanced Manufacturing Technology, 125*(9-10), 4753–4764. DOI: 10.1007/s00170-023-11050-8

Hyer, H. (2022). Status of additive manufacturing capabilities for processing refractory alloys under the mo-99 program.. DOI: 10.2172/1909108

Immonen, K., Metsä-Kortelainen, S., Nurmio, J., Tribot, A., Turpeinen, T., Mikkelson, A., Kalpio, T., Kaukoniemi, O.-V., & Kangas, H. (2022). Recycling of 3d printable thermoplastic cellulose-composite. *Sustainability (Basel), 14*(5), 2734. DOI: 10.3390/su14052734

Jasim, M., Abbas, T., & Huayier, A. (2022). The effect of infill pattern on tensile strength of pla material in fused deposition modeling (fdm) process. *Engineering and Technology Journal, 40*(12), 1–8. DOI: 10.30684/etj.2021.131733.1054

Kargar, E., & Ghasemi-Ghalebahman, A. (2023). Experimental investigation on fatigue life and tensile strength of carbon fiber-reinforced pla composites based on fused deposition modeling. *Scientific Reports, 13*(1), 18194. Advance online publication. DOI: 10.1038/s41598-023-45046-x PMID: 37875509

Kaščák, Ľ., Varga, J., Bidulská, J., Bidulský, R., & Grande, M. A. (2023). Simulation tool for material behaviour prediction in additive manufacturing. *Acta Metallurgica Slovaca, 29*(2), 113–118. DOI: 10.36547/ams.29.2.1846

Khalil, Y., Hopkinson, N., Kowalski, A., & Fairclough, J. P. A. (2022). Investigating the feasibility of processing activated carbon/uhmwpe polymer composite using laser powder bed fusion. *Polymers, 14*(16), 3320. DOI: 10.3390/polym14163320 PMID: 36015577

Khanafer, K., Cao, J., & Kokash, H. (2024). Condition monitoring in additive manufacturing: A critical review of different approaches. *Journal of Manufacturing and Materials Processing, 8*(3), 95. DOI: 10.3390/jmmp8030095

Khanafer, K., Cao, J., & Kokash, H. (2024). Condition monitoring in additive manufacturing: A critical review of different approaches. *Journal of Manufacturing and Materials Processing, 8*(3), 95. DOI: 10.3390/jmmp8030095

Kim, D. W., Han, S. B., Lee, Y. S., Park, D. Y., Lee, H., Park, S. H., & Song, H. (2023). Correlation between cellular structure morphology and anisotropic yield in additively manufactured stainless steel 316l. *Materials (Basel), 16*(4), 1666. DOI: 10.3390/ma16041666 PMID: 36837296

Kumaar, K. R., Boopathy, G., Bejaxhin, B. H., & Ramanan, N. (2024). Considerations on fractured edge dislocations of mechanical tested hand-laid hemp flax GFRP hybrid polymer composites with SEM verifications. *Journal of Polymer & Composites, 11*(13), 1–19. https://journals.stmjournals.com/jopc/article=2024/view=133571/

Le, V. T., Bui, M. C., Pham, T. Q. D., Tran, H. S., & Tran, X. V. (2023). Efficient prediction of thermal history in wire and arc additive manufacturing combining machine learning and numerical simulation. *International Journal of Advanced Manufacturing Technology, 126*(9-10), 4651–4663. DOI: 10.1007/s00170-023-11473-3

Leonetti, D., de Munnik, M., Kassing, L., Khan, D., Moulin, J., & Snijder, B. (2023). Rotating bending fatigue behaviour and quasi-static tensile properties of wire arc additively manufactured 3081 stainless steel. *ce/papers*, *6*(3-4), 732–738. DOI: 10.1002/cepa.2441

Leontiadis, K., Tsioptsias, C., Messaritakis, S., Terzaki, A., Xidas, P., Mystikos, K., Tzimpilis, E., & Tsivintzelis, I. (2022). Optimization of thermal and mechanical properties of polypropylene-wollastonite composite drawn fibers based on surface response analysis. *Polymers*, *14*(5), 924. DOI: 10.3390/polym14050924 PMID: 35267749

Luo, W., Hui, J., Wang, J., & Yan, Z. (2024). Study on the influence of selective laser melting temperature on the melting bath. *Journal of Physics: Conference Series*, *2791*(1), 012024. DOI: 10.1088/1742-6596/2791/1/012024

Magri, A. E., Ouassil, S., & Vaudreuil, S. (2022). Effects of printing parameters on the tensile behavior of 3d-printed acrylonitrile styrene acrylate (asa) material in z direction. Polymer Engineering &Amp. *Polymer Engineering and Science*, *62*(3), 848–860. DOI: 10.1002/pen.25891

Mahmood, M. A., Chioibasu, D., Rehman, A. U., Mihai, S., & Popescu, A. C. (2022). Post-processing techniques to enhance the quality of metallic parts produced by additive manufacturing. *Metals*, *12*(1), 77. DOI: 10.3390/met12010077

Mantecón, R., Rufo-Martín, C., Castellanos, R., & Díaz-Álvarez, J. (2022). Experimental assessment of thermal gradients and layout effects on the mechanical performance of components manufactured by fused deposition modeling. *Rapid Prototyping Journal*, *28*(8), 1598–1608. DOI: 10.1108/RPJ-12-2021-0329

Miao, R. (2024). Development, prospect and challenge for additive manufacturing of polymer materials. Transactions on Environment. *Energy and Earth Sciences*, *1*, 90–96. DOI: 10.62051/s182f353

Mieloszyk, M., Majewska, K., & Andrearczyk, A. (2022). Embedded optical fibre with fibre bragg grating influence on additive manufactured polymeric structure durability. *Materials (Basel)*, *15*(7), 2653. DOI: 10.3390/ma15072653 PMID: 35407984

Mieloszyk, M., & Muna, I. I. (2023). Numerical modeling of thermal effects on mechanical behavior of additive manufactured carbon fiber reinforced polymer. *Health Monitoring of Structural and Biological Systems*, *XVII*, 35. Advance online publication. DOI: 10.1117/12.2657985

Mirzaali, M. J., Moosabeiki, V., Rajaai, S., Zhou, J., & Zadpoor, A. A. (2022). Additive manufacturing of biomaterials—Design principles and their implementation. *Materials (Basel), 15*(15), 5457. DOI: 10.3390/ma15155457 PMID: 35955393

Mohd Yusof, A. A. B., Md Yusop, A. H. B., & Abu Kasim, H. B. (2024). Analysis of crack propagation of pla fabricated by the additive manufacturing technique. *Engineering Research Express, 6*(3), 035518. DOI: 10.1088/2631-8695/ad62ae

Morales, M. A., Maranon, A., Hernández, C., & Porras, A. (2021). Development and characterization of a 3d printed cocoa bean shell filled recycled polypropylene for sustainable composites. *Polymers, 13*(18), 3162. DOI: 10.3390/polym13183162 PMID: 34578062

Moreira, C. A., Caicedo, M., Cervera, M., Chiumenti, M., & Baiges, J. (2022). A multi-criteria h-adaptive finite-element framework for industrial part-scale thermal analysis in additive manufacturing processes. *Engineering with Computers, 38*(6), 4791–4813. DOI: 10.1007/s00366-022-01655-0

Mujtaba, A., Islam, F., Kaeding, P., Lindemann, T., & Prusty, B. G. (2022). Machine-learning based process monitoring of automated composites manufacturing to enable digital twin development. DOI: 10.21203/rs.3.rs-2220331/v1

Muna, I. I., Mieloszyk, M., Rimašauskienė, R., Maqsood, N., & Rimašauskas, M. (2022). Thermal effects on mechanical strength of additive manufactured cfrp composites at stable and cyclic temperature. *Polymers, 14*(21), 4680. DOI: 10.3390/polym14214680 PMID: 36365673

Nagano, H. Umemoto, R., Ueno, A., & Ogushi, T. (2022). Effective thermal conductivity measurement of additively manufactured lattice structures by application of modified temperature profile method. SSRN Electronic Journal. DOI: 10.2139/ssrn.4313766

Okuyelu, O., & Adaji, O. (2024). Ai-driven real-time quality monitoring and process optimization for enhanced manufacturing performance. *Journal of Advances in Mathematics and Computer Science, 39*(4), 81–89. DOI: 10.9734/jamcs/2024/v39i41883

Olaiya, N. G., Maraveas, C., Salem, M., Raja, S. P., Rashedi, A., El-Bahy, Z. M., & Olaiya, F. G. (2022). Viscoelastic and properties of amphiphilic chitin in plasticised polylactic acid/starch biocomposite. *Polymers, 14*(11), 2268. DOI: 10.3390/polym14112268 PMID: 35683940

Perez-de-Eulate, N. G., Ares-Elejoste, P., Goenaga, G., Urrutxua, M., Vallejo, F. J. L., Ballestero, J., Allue, A., & Gómez-Alonso, J. L. (2023). On the influence of flame-retardant additives on uv-curable thermosetting glass fiber-reinforced composites. *Polymers*, *15*(1), 240. DOI: 10.3390/polym15010240 PMID: 36616588

Puerta, A. P. V., Fernández-Vidal, S., & Girot, F. (2023). Impact of in-soil ageing effect on pla printed parts tensile properties. *Polymers*, *15*(4), 862. DOI: 10.3390/polym15040862 PMID: 36850145

Randriambololona, A. M., Shaeri, M. R., & Sarabi, S. (2022). Artificial neural network models to predict heat transfer coefficients and pressure drops in cold plates with surface roughness. *Proceedings of the 8th World Congress on Mechanical, Chemical, and Material Engineering*. DOI: 10.11159/htff22.167

Riensche, A., Severson, J., Yavari, R., Piercy, N., Cole, K. D., & Rao, P. (2022). Thermal modeling of directed energy deposition additive manufacturing using graph theory. *Rapid Prototyping Journal*, *29*(2), 324–343. DOI: 10.1108/RPJ-07-2021-0184

Robles Poblete, F., Ireland, M., Slattery, L., Davids, W. G., & Lopez-Anido, R. A. (2023). In situ, real-time temperature mapping and thermal fe simulations of large-format 3d printed petg/cf vertical wall. *Materials (Basel)*, *16*(19), 6486. DOI: 10.3390/ma16196486 PMID: 37834624

Rusmayadi, G., Salawati, U., Haslinah, A., & Judijanto, L. (2023). The effect of investment in green technology and renewable technology adoption on energy efficiency and carbon emissions reduction in indonesian manufacturing companies. *West Science Interdisciplinary Studies*, *1*(11), 1175–1183. DOI: 10.58812/wsis.v1i11.347

Saad, I., Maalej, S., & Zaghdoudi, M. C. (2022). Investigation of the thermal performance of a nanofluid-filled grooved cylindrical heat pipe for electronics cooling. *Journal of Advanced Research in Fluid Mechanics and Thermal Sciences*, *99*(2), 135–154. DOI: 10.37934/arfmts.99.2.135154

Starikov, K., Polozov, I., Borisov, E., Kim, A., Voevodenko, D., Gracheva, A., Shamshurin, A., & Popovich, A. (2023). Selective laser melting of non-weldable nickel superalloy: Microstructure, cracks and texture. *Metals*, *13*(11), 1886. DOI: 10.3390/met13111886

Subramani, R., Mustafa, M. A., Ghadir, G. K., Al-Tmimi, H. M., Alani, Z. K., Rusho, M. A., & Kumar, A. P. (2024). Exploring the use of biodegradable polymer materials in sustainable 3D printing. *Applied Chemical Engineering*, *7*(2), 3870–3870.

Ţârtea, D. A., Ionescu, M., Manolea, H. O., Mercu , V., Obădan, E., Amărăscu, M. O., Mără escu, P. C., Dăguci, L., & Popescu, S. M. (2023). Comparative study of dental custom cad-cam implant abutments and dental implant stock abutments. *Journal of Clinical Medicine, 12*(6), 2128. DOI: 10.3390/jcm12062128 PMID: 36983129

Tian, H., Yi, X., Lu, L., Chen, X., Zhang, G., Cai, Y., & Han, J. (2022). Heat propagation simulation of thin-walled ti6al4v alloy components fabricated by double-wire gtaw additive manufacturing. *Science and Technology of Welding and Joining, 27*(6), 446–454. DOI: 10.1080/13621718.2022.2063566

Treutler, K., Lorenz, S., Hamje, J., & Wesling, V. (2022). Wire and arc additive manufacturing of a cocrfemoniv complex concentrated alloy using metal-cored wire—Process, properties, and wear resistance. *Applied Sciences (Basel, Switzerland), 12*(13), 6308. DOI: 10.3390/app12136308

Treutler, K., Lorenz, S., Hamje, J., & Wesling, V. (2022). Wire and arc additive manufacturing of a cocrfemoniv complex concentrated alloy using metal-cored wire—Process, properties, and wear resistance. *Applied Sciences (Basel, Switzerland), 12*(13), 6308. DOI: 10.3390/app12136308

Turk, T., Dominguez, C. E., Sutton, A. T., Bernardin, J. D., Park, J., & Leu, M. C. (2024). Spot pattern welding scanning strategy for sensor embedding and residual stress reduction in laser-foil-printing additive manufacturing. *Rapid Prototyping Journal, 31*(1), 83–99. DOI: 10.1108/RPJ-01-2024-0042

Volpe, S., Sangiorgio, V., Petrella, A., Notarnicola, M., Varum, H., & Fiorito, F. (2023). 3d printed concrete blocks made with sustainable recycled material. VITRUVIO - International Journal of Architectural Technology and Sustainability, 8, 70-83. DOI: 10.4995/vitruvioijats.2023.18832

Vysochinskiy, D., & Akhtar, N. (2022). On the influence of cross-section size on measured strength of slm-produced alsi10mg-alloy. *Key Engineering Materials, 926*, 132–140. DOI: 10.4028/p-g078sc

Wang, C., Sun, Y., Chen, G., Chen, X., Ding, J., Suder, W., Diao, C., & Williams, S. (2022). A simplified modelling approach for thermal behaviour analysis in hybrid plasma arc-laser additive manufacturing. *International Journal of Heat and Mass Transfer, 195*, 123157. DOI: 10.1016/j.ijheatmasstransfer.2022.123157

Wu, Q., Tang, W., Yang, J., & Shi, X. (2023). Comparison of thermomechanical responses of single-arc and dual-arc parallel additive manufacturing. *Science and Technology of Welding and Joining, 28*(7), 580–588. DOI: 10.1080/13621718.2023.2219909

Zhang, M., & Shang, J. (2024). Molecular dynamics simulation of ultrasound-assisted 3d printing of aluminum matrix composites. *Fourth International Conference on Applied Mathematics, Modelling, and Intelligent Computing (CAMMIC 2024)*, 66. DOI: 10.1117/12.3035661

Zhang, W., Li, J., Fang, J., Hui, L., Qin, L., Gong, T., Sun, F., & Feng, H. (2023). Atomic layer deposited high quality aln thin films for efficient thermal management. *Journal of Materials Chemistry. A, Materials for Energy and Sustainability*, *11*(40), 21846–21856. DOI: 10.1039/D3TA04618D

Zhao, R., Shmatok, A., Fischer, R., & Prorok, B. C. (2023). Investigation of causal relationships between printing parameters, pore properties and porosity in laser powder bed fusion. *Metals*, *13*(2), 330. DOI: 10.3390/met13020330

Zhu, G., Houck, H. A., Spiegel, C. A., Selhuber-Unkel, C., Hou, Y., & Blasco, E. (2023). Introducing dynamic bonds in light-based 3d printing. *Advanced Functional Materials*, *34*(20), 2300456. Advance online publication. DOI: 10.1002/adfm.202300456

Chapter 9
From Theory to Practice: Control System Challenges in Additive Manufacturing

M. Maniraj
https://orcid.org/0009-0000-5343-3895
Sardar Raja College of Engineering, Alangulam, India

S. M. Raj Kumar
https://orcid.org/0000-0002-8978-0108
Francis Xavier Engineering College, Tirunelveli, India

ABSTRACT

The evolution of additive manufacturing (AM), particularly 3D printing, from a prototyping tool to an industrial-scale production method has been heavily influenced by the development of advanced control systems. These systems are vital for achieving high precision, consistency, and efficiency across various AM technologies. This chapter presents a comprehensive review of control systems in AM, focusing on the theoretical underpinnings and practical implementations necessary for industrial adoption. It explores the integration of control engineering principles within the unique constraints of AM processes. The study synthesizes existing literature and technological developments concerning control strategies used in 3D printing. This includes the examination of open and closed-loop systems, sensor integration, real-time monitoring, and the application of advanced computational methods such as model predictive control, adaptive algorithms, and artificial intelligence.

DOI: 10.4018/979-8-3373-0533-2.ch009

1. INTRODUCTION

Additive manufacturing (AKA 3D printing) has changed the way how product are designed, prototyped and made in a wide range of industries. This transformative technology has been applied for the production of complex geometries, scrap material reduction and accelerated innovation cycles in aerospace and automotive, healthcare and consumer goods. Unlike the traditional subtractive manufacturing method which remove the material to get the final product, additive manufacturing creates the object by adding materials in layer by layer, providing us great design flexibility and even fabricating the complex internal structure that was impossible or very costly in the tradition manufacturing. With technological development, its applications are continually widening and hence affecting industries through improved material science, computational design, and process automation (Zhou et al. 2024).

However, at the core of which additive manufacturing's success relies on is the powerful contribution of control systems that guarantee the accuracy, efficiency and reliability of the printing procedure. The effectiveness of control mechanisms is very important for a 3D printer to be able to deposit material accurately, regulate thermal conditions and maintain quality in the print. Open loop controls were the early 3D printing systems, wherein parameters were predetermined and did not have feedback with the print in progress often causing the prints to fail due to changes in material behaviour and other environment factors. However, modern 3D printers integrate with sophisticated closed loop feedback system that dynamically adjusts parameters in real time and optimizing material flow, temperature and movement which leads to highest quality prints with as little errors as possible. The job of the control systems is to synchronize multiple sub-systems including actuators, sensors, and computational algorithms to make the process go smoothly for each job. An absence of these precise and adaptive control mechanisms would make it virtually impossible to reach repeatability and scalability in the additive manufacturing process (Rahman et al. 2024).

The scope of this chapter is to bridge between theoretical control principles and their realization for additive manufacturing. Although a theory providing mathematical foundation for system stability, error correction and optimization is given by control theory, its role in real world implementation is full of complexities as encountered in material variations, environmental conditions and process requirements. With the integration of advanced control technique like model predictive control, adaptive algorithm and artificial intelligence, it is possible to improve 3D printing system's performance and efficiency. Yet, solutions to these challenges need to be implemented and they involve issues of computational power, sensor integration, and real time data processing. This chapter offers a comprehensive treatment of the technological advancements, limitations, and future opportunities via an exploration of the interplay between control engineering and additive manufacturing. By reading this discussion, readers will understand how the control system of 3D printing affects the precision and scalability of 3D printing, which will lead to the future of digital manufacturing.

2. FUNDAMENTALS OF CONTROL SYSTEMS IN ADDITIVE MANUFACTURING

Additive manufacturing processes are precision, consistent and reliable thanks to control systems. In its essence, a control system in 3D printing controls the movement of mechanical components, the deposition of material, and the optimal thermal condition in the printing process. Whether these systems are effective or not will determine how accurate the printed parts can be created, how efficient they are at using the available material, and the quality of the final products. Two types of system, namely open loop and closed loop control, constitute the basis of control in additive manufacturing (Mattera et al. 2024). Figure 1 shows the open loop and closed loop control systems.

Figure 1. Open loop vs. closed loop control systems.

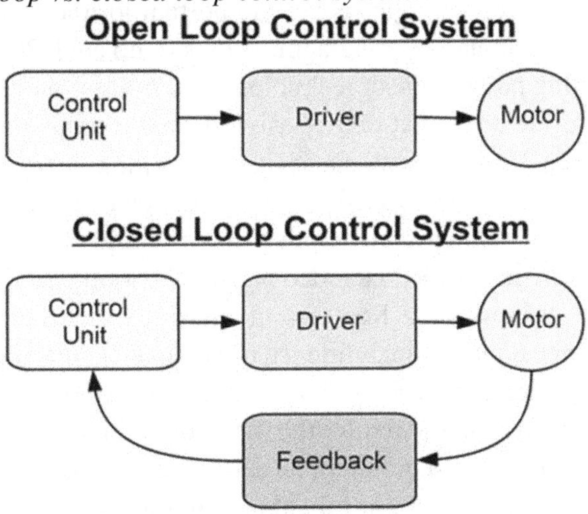

In open loop control systems, feedback is not required and the instructions programmed in the system are tested the way they are. Open-loop control was the main control method used in early 3D printing technologies where the printers followed pre defined tool paths and material flow rates that were not equipped with any mechanism to detect deviations or correct errors that may occur during the printing process. Open loop systems are simple and inexpensive, yet sensitive to property, environmental and mechanical variations, resulting in numerous failed prints or dimensional inaccuracies. On the other hand, closed loop control systems have feedback mechanisms implemented which constantly observe the printing process and adjust on the fly. The data (such as print head position, extrusion rate and temperatures) is collected by sensors and used as feedbacks to a control algorithm that dynamically changes the settings to optimize performance. Print accuracy is significantly improved along with reduced material waste and increase in process stability for closed loop systems, which compensate for the change in system behavior (Yang et al. 2024).

Further refinement of the precision of control systems in additive manufacturing is given by the integration of feedback and feedforward mechanisms. The feedback control for each layer of material is based on real time data from the sensor to mitigate discrepancies as per the desired parameters in order to get the accuracy of deposition. On the other hand, feedforward control predicts errors that might happen before they occur by studying the system inputs and adjusting predictively. By offering modern 3D printers the opportunity to proactively set parameters like nozzle temperature, laser intensity and material deposition speed, it can optimize the process for high quality results, in the same way. In particular, the implementation of these advanced control strategies is very important for the consistency in large scale or high precision manufacturing applications (Guo et al. 2024).

From a computing power, sensor technology and process modeling perspective additive manufacturing control systems have followed roughly on the heels of the state of the art. When 3D printing was in its early days, simple proportional integral derivative (PID) controllers were used to control motion and temperature in a basic form of automation. With increasing applications of additive manufacturing, more complex control strategies, which included machine learning, adaptive control, and predictive model were introduced. Real time monitoring system have been developed to further enhance the control capabilities to in-situ defect detection and corrective actions during the printing process. As 3D printing technologies become more complex, the control system changes to solve the issues of multi material printing, hybrid manufacturing, and high speed production (Parisi et al. 2024). As such, this further highlights the enormous role of control engineering in defining the future path of additive manufacturing; to enable the technology to satisfy increasing demands for accuracy, efficiency and scale of additive manufacturing in industrial applications.

3. CHALLENGES IN CONTROLLING 3D PRINTING PROCESSES

The challenge in controlling the 3D printing process is that the process is dynamic, material deposition is variable, and the environmental factors are variable, and we have very tight requirements on the precision of the parts we produce. Real time control of temperature, material flow and print speed are therefore important to the success of an additive manufacturing system. These parameters are interdependent, so it is difficult to achieve consistency in them, for instance, extrusion temperature fluctuations may change the material viscosity, which in turn will affect flow rate and print head movement. Similarly, variations in layer adhesion and structural integrity can be dependent upon printing speed, thus requiring sophisticated control algorithms to continually adjust settings according to sensor feedback. Maintaining uniform material deposition on a range of complex geometries requires both thermal regulation, motion control, material feed precision that are dynamically managed to avoid defects (Sierra-Vega et al. 2025).

The other major problem is that 3D printing materials have an inherent variability. The printing process behavior is different for different polymers, metals, ceramics and composite materials, the viscosity of material changes during the process and the shrinkage and bonding behavior also changes. In predictability of print quality, there is inconsistency, even within one material type; for instance, composition, moisture content, and batch variations introduce unpredictable deviations. In addition, environmental factors such as ambient temperature, humidity, and airflow make the situation even more difficult because these factors can affect the material properties and result in phenomena, for example, uneven cooling or premature solidification. The stability of these variables relies on adaptive control strategies, which can respond to fluctuating changes in real time, and thus making the process stabilization a complex and ever lingering problem in additive manufacturing (Vemuganti et al. 2024).

The thermal behavior is crucial to the success of a 3D print in processes that involve melting and solidification, such as FDM and SLM. Care must be taken in the heat distribution to avoid warping, residual stress and layer delamination. In metal additive manufacturing at high temperatures, undesirable microstructural changes or porosity due to excessive heat accumulation will weaken the final part. On the contrary, weak layer adhesion or incomplete fusion of particles due to insufficient heat may cause structural defects. Real time monitoring and adaptive control mechanism for modulation of heat input, cooling rates and sequencing is necessary in controlling thermal gradients. However, it remains a challenge to design the control systems based on complex realworld thermal interactions, as advanced thermal models and machine learning algorithms are increasingly included in the control systems to predict and mitigate issues such as power thermal management and module thermal runaway in the solar farm (Wen et al. 2024).

Besides measurements, which are a material and thermal issue, the precision requirements of complex geometries present another great challenge for control of 3D printing. A number of additive manufacturing applications, particularly in aerospace and biomedical fields, involve need for intricate designs having details of small dimensions and tight tolerances. To achieve such precision sub micron accuracy in print head positioning, motion control and layer alignment is required. Geometric distortions induced by any minor deviation of the motion trajectories or the extrusion rates also impacts functional consequences in critical components. Controlling support material strategies and deposition planning becomes the challenge when printing overhanging structure and internal cavities is concerned, to avoid collapse or deformation. To solve these problems, advanced kinematic control algorithms, multi-axis coordination, and error compensation techniques are needed but demanding in computational resource and highly responsive actuation system (Fei et al. 2024).

These present challenges shed light on how 3D printing processes are in fact a complex set to control and are therefore best controlled through a multidisciplinary approach that embeds various controls, mechanical, thermal, and computational. With the expansion of additive manufacturing, further research on intelligent control systems, real time monitoring, and predictive analytics will play a critical role in realizing advancements in the precision, reliability, and scalability in that field.

4. CONTROL SYSTEM REQUIREMENTS FOR DIFFERENT 3D PRINTING TECHNOLOGIES

Additive manufacturing control requirements depend greatly on the individual technology employed. Control of each 3D printing method is required because the fabrication of objects from these processes is done in a layer by layer fashion using distinct physical and chemical processes. These requirements are complex due to differences in material behavior, energy input, and environmental interactions, and control system design is a key factor in the optimization of performance of each of these techniques (Liu et al. 2025).

In FDM control systems must control extrusion flow, layer adhesion and warping prevention to maintain part dimensional accuracy. To avoid over or under extrusion, the movement of the print head and the feeding of thermoplastic filament have to be precisely synchronized during the extrusion process. The temperature control is of equal importance: the viscosity of the material and the strength of the bonding between layers depend on it. Poor layer adhesion and excessive sagging can result from the variations in the temperature. This would eventually degrade the mechanical integrity of the final object. Another common problem with FDM is warping, which is typically due to uneven cooling during cooling that creates internal stresses and layer separation. These effects are mitigated in control systems where heated build plates are regulated, cooling fan speeds are altered and print speeds are adjusted dynamically in accordance to thermal conditions. Real time extrusion monitoring with pressure sensors and machine learning algorithms is included in advanced FDM printers to guarantee consistent material flow and eliminate defects (Behseresht et al. 2024).

In contrast, the problem of controlling stereolithography (SLA) that involves curing liquid resin with ultraviolet (UV) light is different. It needs to be cured in such a way that each layer cures uniformly without being overexposed or not fully polymerized. UV exposure duration and intensity need to be finely controlled on the basis of the resin composition and layer thickness. Besides exposure control, flow of resin must be managed, especially in systems which have recoating mechanisms or limited material resin vats. Viscosity changes due to temperature variations and polymerization rates must be accounted in control algorithms that also guarantee a constant supply of uncured resin for subsequent layers. Optical sensors for monitoring of curing uniformity and application of corrective actions in real time are often integrated into SLA printers, which contributes to overall part accuracy and mitigates risk of defects (Pilch et al. 2024).

Selective Laser Sintering (SLS) and Selective Laser Melting (SLM) make the control requirements even more complex because high power lasers and fine metal or polymer powder are used. In order to sinter or melt uniformly without introducing defects such as porosity, cracking, or incomplete fusion, it is necessary to have precise control over laser intensity and scan speed. Localized heating causes thermal stress that must be monitored so as to prevent part warpage and accumulation of residual stresses leading to part failure. To achieve better quality of printed parts, powder bed uniformity and particle distribution are critical in the metal additive manufacturing process, which necessitates the monitoring and adjustment of powder spreading mechanism in real time. High speed infrared cameras with a closed loop feedback detect the deviation in the heat distribution and adjust the laser parameters in realtime optimizing the energy input to achieve consistent fusion of layers. In addition, computational models to predict thermal expansion and stress formation are integrated into the process design for greater process stability when producing high performance metal components (Yehia et al. 2024).

With the addition of multi material and hybrid printing, materials may be required to be deposited simultaneously, along with different thermal, mechanical, and chemical properties associated with the printed models. However, these materials require sophisticated control strategies that consider the differences in melting point, nature of the bonding and curing time. Hybrid systems from additive and subtractive manufacturing processes require the precise synchronization between material deposition and the following subtractive processes, like milling or laser finishing. These technologies use several control systems based on a real time material property sensing, adaptive toolpath generation and advanced motion planning for smooth transitions between different materials and fabrication techniques. Multi material control systems are more and more often complemented with artificial intelligence and machine learning to predict potential defects and optimize blending of the materials in order to provide greater performance and durability (Chatzipapas et al., 2024).

Different 3D printing technologies have diverse requirements, which necessitate that their control approaches should be specialized to each process. With advances in additive manufacturing (3D printing) more and more, the future of This will include real time monitoring, adaptive algorithms and predictive modeling are crucial to overcome technical limitations to open the door of precision and functionality in all forms of 3D printing.

5. ROLE OF SENSORS AND REAL-TIME MONITORING IN 3D PRINTING

In this line, integration of sensors and real-time monitoring has become essential to the improvement of precision, reliability, and flexibility in 3D printing processes. Using closed loop control, which is heavily dependent on feedback from continuous sensors such as sensors, printers can make changes within material flow, temperature and positioning that can recalibrate parameters to ensure ideal conditions along the duration of print. Without such monitoring, defects, dimensional inaccuracy, or simply print failures can occur as a result of deviations from ideal printing conditions that have gone unnoticed. In industrial and high precision applications, where material costs and production time are important, consistency is achieved by being able to detect anomalies in real time and corrects them immediately (Jhun et al. 2024).

Additive manufacturing utilizes various types of sensors to monitor different aspects of printing process. Temperature sensors aid in regulating the temperature levels for processes that includes fused deposition modeling (FDM), selective laser sintering (SLS) and selective laser melting (SLM). A stable thermal environment during the print prevents warping, and keeps the layers from separating, helping to keep the internal stresses within the print as low as possible. Commonly optical sensors (cameras and laser systems) are used for monitoring the layer deposition, for detection of surface irregularities, for ensuring of alignment accuracy. These sensors supply visible responses to algorithms that are utilized to take care of speed price, appropriate misalignments, and make up for adjustments in material circulation.

In particular, pressure sensors are useful in extrusion based printing techniques as they provide real time monitoring of flow of filament or resin so that clogging or under extrusion can be prevented. Another emerging approach is acoustic monitoring where sound signatures created during printing are analyzed for irregularities such as nozzle blockages, layer separation, or mechanical failures (Mandal et al. 2024).

While sensor integration results in an improvement in the information available about the process, the acquisition, processing and real time control and decision-making poses a significant challenge. Multiple sensors generate a vast amount of data that needs to be efficiently processed for the extraction of meaningful insights without much latency. Predictive algorithms must run on high speed data acquisition systems which, in turn, must execute in the presence of relevant data messages to ascertain beforehand potential failures and address them before defects occur. Techniques of smart processing of the sensor data utilizing machine learning and artificial intelligence have been experimented to identify the patterns and to optimize the print parameters dynamically. However, sensor calibration, noise interference, and data synchronization are issues to deal with, in which case sophisticated filtering techniques and robust algorithms are necessary to guarantee, to a certain extent, reliable performance (Luo et al. 2024).

Studies using case studies have shown that sensor based feedback systems are effective in improving print quality and reduce amount of waste material. For example, in infrared cameras have been utilized in metal additive manufacturing to observe thermal distributions in laser sintering, to allow adaptive control of laser power so that the sintered part is not over heated and hence contains porosity. Pressure sensors were integrated to the extruder systems in FDM printing to detect the filament feed rate fluctuation and make an automatic adjustment to keep the material feed rate consistent. Real time correction of layer misalignment was achieved with optical inspection systems which were also successfully deployed to detect and correct layer misalignment, improving the accuracy of the large scale, of the large scale prints. The potential of real time monitoring and the integration of sensors into additively manufactured devices shows the great impact of this process on additive manufacturing, and announces more autonomous and intelligent 3D printing systems that can correct themselves, control performance and produce with greater accuracy, almost requiring no human intervention (Tao et al. 2024).

6. ADVANCED CONTROL TECHNIQUES IN ADDITIVE MANUFACTURING

With additive manufacturing growing to meet the speed, precision and material efficiency requirements of today, traditional methods of control are being enhanced with and supplemented by advanced techniques which allow real time adaptability and process optimisation. Predictive modeling, adaptive learning as well as artificial intelligence can be leveraged in modern control strategies to improve print quality, reduce error and maintain process steadiness under changing conditions (Aravind Raj, 2025). In addition to improving the accuracy in deposition of layers and usage of materials, these sophisticated approaches pave the way for fully autonomous and intelligent 3D printing systems.

Model Predictive Control (MPC) is one of the most powerful techniques in these modern control system and this technique can be used to achieve real time optimization of process parameters with the anticipation of future states of the system. Traditional proportional integral derivative (PID) controllers react to apply current errors, but MPC uses a mathematical model of the print process to predict how variables like temperature, feedrate, and motion trajectory will change over time. Because the control system is capable of simulating continuous future outcomes, this system can be used to make proactive adjustments to minimize deviations as well as to maximize energy efficiency. The ability to predict this feature is especially important for high speed and high precision printing as even small changes in the feature can result in large changes in the final product. For example, in metal additive manufacturing, MPC has been used for regulating laser power and scanning pattern to avoid thermal distortion and ensure uniform layer fusion (Mu et al. 2024), and more reliable stronger printed structures were produced. Figure 2 shows the model predictive control in additive manufacturing (van Blitterswijk et al. 2024).

Second, adaptive and selftuning algorithms are another important development in control systems that enable 3D printers to tune process parameters online due to realt w changes. Whereas static control settings rely upon the static conditions (material properties, environmental condition) the adaptive algorithms adaptively adjust their control action based on the real time sensor feedback. For instance, in the fused deposition modeling (FDM), the flow rate of the extruder can be dynamically modified based on the changes of filament diameter, environmental temperature or nozzle wear. For instance, in stereolithography (SLA) based printing technologies which uses the resin-based method of printing allows UV curing time to be modified according to resin viscosity and layer thickness to prevent under curing, and over curing. These self tuning capabilities help to increase the reliability of the printing process, especially for applications where long duration or multi material printing require, and behavior of the material can introduce cumulative errors (Gerdes et al. 2024).

Figure 2. Model predictive control in additive manufacturing (van Blitterswijk et al. 2024)

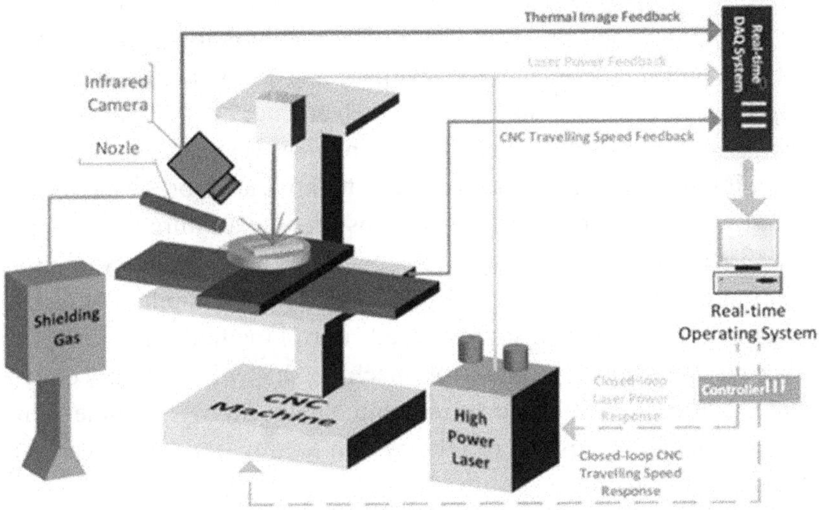

With additive manufacturing, machine learning and artificial intelligence are becoming equally important in improving the control systems of the additive manufacturing processes. With AI backed predictive quality assurance, printing companies can determine potential defecting issues before they manifest by integrating sensor data of the real time and historical print performance. The system is able to detect patterns related to poor adhesion, thermal stresses, or geometric distortions and automatically correct or notify operators of impending failures before they occur (Ejjami et al. 2024). It also gives printers enhanced error detection and self optimization abilities where printers can learn from previous mistakes and perfect their operations through time. One such example is convolutional neural networks (CNNs) trained with thousands of print samples can examine layer by layer images to find irregularities with deposition, and then revise print parameters to keep the quality up. In addition to optimizing individual prints however, this AI integration goes as high as optimizing over a fleet of machines and synthesizing data from multiple machines to improve global performance across a manufacturing facility.

However, the solution for the most complex 3D printing applications is emerging to be hybrid control systems which leverage the power of classical control methods along with AI driven approaches. Hybrid systems considered here are a blend of PID, MPC and the machine learning and the adaptive algorithms which can offer a reconciliation among stability, on the other hand, real time responsiveness and lastly, process optimization over long time period. As a practical example, these hybrid controllers are necessary in multi material and hybrid printing, and in some cases, multiple control strategies must run in parallel. For example, by integrating metal printing with subtractive post processing, classical motion control provides accurate tool motion whereby, AI based defect prediction is used to make an adaptive changes in laser power and/or depth of cut. The combination of determinism and data rigidity in control architecture for AM processes provides a more reliable and flexible approach to control design (Taghizadeh & Zhu, 2024).

However, with these advanced control techniques, 3D printers are beginning to become more intelligent and autonomous, able to fabricate more complex materials and geometries, piquing the interest of startups that see a future using 3D printers in small and local factories making a variety of electronic devices or in fabrication labs supporting the bottom of the value chain in developing countries. With increased computational power, sophisticated control methods can be deployed for the task of providing surgical motions to a machine, this will be simply one of many applications in which computational power and AI algorithms continue to improve, pushing boundaries of precision, efficiency and automation in how products are designed and manufactured.

7. COMPUTATIONAL CHALLENGES AND SOFTWARE CONSIDERATIONS

Therefore the control systems used in additive manufacturing are critically dependent on the computational frameworks that support process optimization, realtime control and predictive analytics. In high precision and multi materials applications, 3D printing processes are very complex, therefore the necessary software must support designs with complex geometries, the description of material behaviors and adequacy for environment fluctuations. These computational frameworks are hosted at their core by toolpath generation algorithms that describe the movement of print heads and the deposition of material, and are most influential in determining the efficiency of their printing. Print time must be minimized, material waste must be minimized, and structural integrity must be maintained with strict accuracy requirements, and these algorithms must optimize the trajectories to achieve this. Real time sensor feedback and adaptive corrections are integrated in the system of advanced toolpath generation techniques which enable the system to modify the deposition pattern of layers in a dynamic manner. Mid print toolpath adjustment is advantageous to process stability especially when process input parameters like material properties and thermal behavior can cause defects (Baktheer & Classen, 2024).

With the advancement of the manufacturing technology, the control systems have moved from traditional proportional integral derivative (PID) controllers to more advanced ones driven by AI. Although PID controllers are still widely used because they are simple and effective at regulating temperature, motion and flow rate, they are limited in their ability to adapt to dynamic conditions in processes that require adaptive response. Machine learning and predictive modeling is used to predict deviations before they happen in order to proactively adjust the control system to improve print reliability within the AI-driven control systems. For example, model predictive control (MPC) takes the traditional feedback control and predicts the future system states to optimally select the process parameters. Other deep learning models have enabled more accurate control by mining for complex sensor data patterns and automatically adjusting printer parameters to achieve the desired material properties and geometric tolerances. These prints have greatly improved the print accuracy and repeatability but have added considerable computation that needs powerful processing and data handling architectures (Yu and Dunn, 2024).

There exists one of the key tradeoffs in implementing advanced computational control systems to date – the computational efficiency versus the real-time responsiveness. AI algorithms, predictive modeling, demand massive data processing that encompasses uses such as large scale simulation, integration of data analytical and simulation techniques and real time analytics. Although, the high computational load results in delays in processing that are detrimental to fast printing operations. strikes a balance between sophisticated control capabilities and low latency, especially in the case of additive manufacturing at an industrial scale where printing rates need to be high and coordination between multiple axes is needed. Since the latency and computational robustness can no longer be achieved through cloud computing, edge computing solutions, that process the data at the data source rather than remotely, are emerging as the potential solution. Furthermore, parallel computing along with GPU acceleration has been taken advantage of, to tackle subtle control tasks in real time (Liu et al. 2024).

Although we have explored here only a small subset of additive manufacturing processes suitable for its applications, future trends in software driven control systems will continue moving forward the bounders of automation, efficiency and scalability in additive manufacturing. Digital twins (virtual replicas of physical printing systems) will be integrated to enable the realtime simulation and optimization of the printing process before execution and reduce trial and error iterations as well as increase firstprint success rates. Distributed manufacturing is also expected to benefit from cloud based control platforms where printers in different locations can be optimally and remotely monitored for optimizing the overall use of these machines. During the time that artificial intelligence becomes more and more integral to the software architecture of any software built with it, such features will continue to democratize the ability to adaptively control printing parameters over time to create smarter and more autonomous production environments. Combining elements of sensor technology, adaptive control, and computational advancement, the next generation of high performance 3D printing systems will depend on these three advancements together working in concert (Wagner & Higgs III, 2024).

8. SCALABILITY AND INDUSTRIAL ADAPTATION OF 3D PRINTING CONTROL SYSTEMS

Due to the transition from prototyping to full scale industrial production of additive manufacturing, the scalability of the control system becomes the most important factor to ensure productivity, consistence, and reliability. There are many challenges associated with scaling up 3D printing, which are different compared to those seen in small scale or desktop printing. However, one of the main challenges associated with depositing materials onto large print areas is the difficulty of having complete control over deposition rates, thermals, and mechanical motion, all of which tend to add up and result in major defects. Additive manufacturing processes at large scale necessitate highly coordinated control mechanisms to control movements of the print head, material flow, environmental factors over extended build volumes, and avoid errors such as warping, layer misalignment and thermal distortion (Yang et al. 2024).

In order to tackle these challenges, distributed control architectures have been developed to synchronize a multi head printing system. Industrial scale additive manufacturing more commonly employs several print heads or laser sources working in parallel not similarly as traditional single extruder or single laser systems. Special mechanisms are necessary for advanced motion planning and feedback assemblies to coordinate these elements without collisions, uniform layer deposition, and efficient toolpath. In these distributed control systems, the control of a print head's operation is informed in real time, with each print head updated based on changes in material behavior and the print conditions. However, such architectures are highly useful in high throughput manufacturing environments where manufacturing speed has to be maintained without sacrificing quality (Santhoshkumar et al. 2024).

It is also challenging to ensure consistency in high volume production such as even tiny differences in process parameters lead to batch to batch variations. In contrast to additive manufacturing, repeatability is traditionally achieved in traditional manufacturing through the control of machining, but no such controls exist in additive manufacturing, where extruded parts are made from polymers of varying quality under thermal expansion and machine wear. The role of continuous monitoring of process variables and adjustment of control parameters is maintained with the help of advanced sensing and feedback mechanisms. The frequency of automated in line quality assurance systems integrated with artificial intelligence driven analytics is increasing as they are used to detect and correct in bounds in errors before they can propagate in through the entire print job. The machine learning algorithms can then utilize these deviations from the printed parts in past prints to realize the patterns of deviations and optimize control strategies to achieve reliability in successive production runs (Ukwaththa et al. 2024).

There also exists a need for robust control solutions in production logistics such as print scheduling, material replenishment and subsequent integration of post processing. Additive manufacturing systems are becoming an interconnected component in the smart factory environments where a cloud-based control, process monitoring platforms and real time connection to the digital manufacturing ecosystems are established. These systems can also provide remote diagnostics, predictive maintenance and adaptive process performance optimization, thus improving operational efficiency overall. With the technique growing into a mass production technology, the control architectures will be crucial to the ongoing and growing transfer of laboratory scale innovation into industrial scale manufacturing, with seamless, efficent and quality requirements that are suitable for commercial applications (Paxton, Jefferey, et al. 2024).

9. ENERGY EFFICIENCY AND SUSTAINABILITY CONSIDERATIONS

Since additive manufacturing is more and more used in the industry, energy efficiency and sustainability are major issues. Although 3D printing is hailed for minimizing waste in comparison to conventional subtractive manufacturing, it is not energy effective per se. Overwhelmingly, most 3D printing processes, often utilizing high powered lasers, heated build chambers, and require post processing, consume considerable energy. The control systems managing this energy consumption have the challenge of balancing power usage with structural integrity and precision of printed parts. Additive manufacturing techniques can require different amounts of energy, as for example selective laser melting (SLM) and electron beam melting (EBM), which generally need a large amount of heat to be input, whereas fused deposition modeling (FDM) is, in comparison, less energy demanding. Nevertheless, inefficiencies can occur in lower-energy processes as well due to long print times, excessive rework, and poor process planning (Fidan et al. 2024).

Both material waste and energy consumption are greatly reduced through control strategies. A dynamic energy management system approach is to adjust power input based on real time process requirements. For instance, in laser based printing, adaptive laser intensity control can be used to save unnecessary energy expenditure while keeping the melting conditions consistent. The same holds true for intelligent temperature regulation in extrusion based printing, thus the heating elements work only when required and avoid excessive loss of energy in the form of heat loss due to prolonged idle state. Another effective strategy is optimizing the toolpath planning to avoid unnecessary movement and deposition, not only speeding up the printing, but the whole power consumption as well. Predictive control algorithms can be incorporated into printers to determine energy behavior as a function of print parameters, and estimate the energetic cost of employed parameters to strike a balance between efficiency and part quality (Graziosi et al. 2024).

Yet, achieving energy savings in additive manufacturing typically comes at the expense of contradictions among these three factors: process speed, quality, and sustainability. However, high speed printing usually needs more energy input to keep material flow consistency and to avoid defects while slower and more controlled processes are often more energy efficient but may not be adequate for high volume manufacturing. Also they may decrease the mechanical strength or surface finish quality, even significantly, when compounded with energy efficient strategies such as lower power settings or reduced layer overlap. Therefore, the control systems need to be designed to dynamically evaluate these trade-offs and make real time adjustments that reflect the target of a particular application. Nowadays, machine learning and artificial intelligence are being employed to optimize these decisions based on historical data and predict the most energy efficient print setting without sacrificing part performance (Chen et al. 2024).

In addition to direct energy consumption, sustainability in additive manufacturing encompasses material efficiency and lifecycle. In order to make the process more sustainable, it is possible to minimize material waste by optimizing the deposition, reducing support structures and implementing recycling mechanisms for unused powder or filament. Even some of the control systems are now equipped with real time defect detection and automatic correction features that prevent products from being sold without successful print, further reducing wasted materials and energy. Moreover, the trend to employ biodegradable or recycled materials in 3D printing also needs control algorithms to account for the special properties of these sustainable alternatives. With the industry trending towards greener manufacturing solutions, energy aware control strategies will gradually evolve to provide additive manufacturing with technical excellence and environmental responsibility (Nyamekye et al. 2024).

10. SECURITY, ETHICS, AND RELIABILITY OF CONTROL SYSTEMS IN 3D PRINTING

With advancements in additive manufacturing systems as well as their networking, security, ethics and reliability have become key points of attention necessary for a responsible and efficient deployment of 3D printing technologies. However, cyber security risks are the primary challenges in automated additive manufacturing and this risk increase when industrial scale printers are coupled with cloud based platforms, remote monitoring solutions and AI induced control frameworks. Vulnerabilities exist as a result of being able to modify an organizations design files, intercept sensor data, or manipulate print parameters remotely and can lead to defective products, compromised proprietary designs, or even malicious tampering. Since most 3D printers can be operated from their own control system, unauthorized access and control, or even tampering with the software on the control system, could allow the attacker to introduce microstructural weaknesses that are undetectable from immediate inspection but would result in catastrophic failure for aerospace, medical or defense applications. In order to mitigate these risks the control architectures must incorporate robust encryption protocols, secure authentication mechanisms and real-time intrusion detection systems. The security and integrity of printed parts has also been proposed to be enhanced via implementation of blockchain (Rautio & Broms, 2024) based verification of design files and process logs.

Another big concern in digital manufacturing is that design files and process parameters are easily copied, modified and redistributed without authorization, and thus intellectual property protection is always required. While conventional manufacturing is tied to physical tooling and controlled supply chains, additive manufacturing enables the rapid digital transfer of complex geometries and functional designs. An accessibility such as this causes concerns about counterfeiting, reverse engineering and production without permission. Present strategies for manufacturers and designers to put into place, for implementing digital rights management (DRM) for 3D printing files, watermarking techniques that would be available in the printed structures, and access controlled platforms to prevent the file from freely distributed. Additionally, cloud-based licensing models are explored by companies in which users are able to print parts without owning the source files thus protecting them from theft of design. Despite that, it is a great legal and logistical challenge to enforce the intellectual property rights in decentralized manufacturing (Venkatalakshmi, 2024).

The landscape then becomes even more complicated, especially when thinking about the ethical... However, AI powered control systems can be used to optimize print quality, minimize material waste, and increase overall efficiency, but the lack of transparency and accountability in the decision making process is not a good aspect. Consequently, machine learning models have to be relied on to make alterations to the operational process, bringing with it the risk of bias in training data and low interpretability of the modifications generated by the AI, especially in the case of safety critical applications. As an example, a deep learning based AI controlled 3D printer is part of medical implant production and it is critical that parameter adjustments that the AI control makes do not have a negative impact on biocompatibility or mechanical integrity, however, the degree of complexity in some deep learning models may including making it difficult to tell how specific decisions made were arrived at. Standards for transparency, explainability, and human oversight need to be established as ethical guidelines for AI driven manufacturing to ensure trust on the technology (Kanthimathi et al. 2024).

Recurring challenge in making control systems work reliably all of the time, under all conditions, without failures. Instability may be introduced into the printing process due to the randomness of a real world manufacturing environments, resulting from considerations such as otherwise predictable fluctuations in material properties or power supply disruptions, or sensor malfunctions. These risks can be mitigated in control systems by incorporating redundancy in control systems, fail safe mechanisms and self diagnostic capabilities. On another hand, as the machine learning algorithms analyze the operation data to predict the failure before it occur, the predictive maintenance made with AI, start to become more valuable for system reliability improvement. Unfortunately, any adoption of such advanced control strategies needs to be in accordance with the need for verifiability and regulatory compliance, depending on the industry in which additive manufacturing is applied, especially if the purpose is mission critical (Ma et al. 2024).

In an increasingly digital and automated world, it will be essential that cybecrime experts, engineers, policymakers, and ethicists work together as the industry continues to evolve to ensure that the control systems on additive manufacturing can still retain the ability to be secure, fair, and reliable.

11. FUTURE DIRECTIONS AND INNOVATIONS IN CONTROL SYSTEMS FOR ADDITIVE MANUFACTURING

Advances in artificial intelligence, computational power, as well as real time process optimization, are a growing determinant on the future of control systems in the additive manufacturing course. According to the Futuristc society, the use of AI enhanced control systems will be a key part in making 3D printing autonomous and self correcting as well as to make it handle complex geometries with higher precision. They will continue to refine machine learning algorithms which can learn from past prints, predict potential defects, and dynamically adjust parameters such as extrusion rate, laser intensity, and thermal distribution. It will use AI to detect anomalies — even real-time anomalies — and correct them — allowing printers to disable or alter a process in real time, as the machine is operating. This will lead to less material waste, increased reliability. In addition, printers are being encouraged to adapt by conducting adaptive control using reinforcement learning models, which refine a printer's operating strategy over time by taking feedback from completed builds (Di Salvo, 2024).

Another promising area of the innovation is the application of quantum computing and edge computing to real time control. Using the power of quantum computing to perform complex optimizations at speeds that are beyond the power of classical computers, quantum computing can allow for instant near instantaneous simulation of the process parameters, heat dissipation models and stress distributions that occur in an additive manufacturing system. While quantum enhanced optimization is still relatively in its infancy, it has shown to drastically increase multi-variable control in applications with high precision accuracy such as aerospace and biomedical engineering. On the other hand, edge computing has already begun to impact by not linking data to central cloud servers to process. It offers latency in feedback control loops, where printers are able to update frequently based on sensor inputs, with no delay due to the network. With edge AI models embedded in 3D printers, control decisions can be continuously refined and thereby 3D printers can be made more autonomous and efficient for large scale manufacturing operations (Duarte et al. 2024).

This is set to redefine how digital twins are integrated in additive manufacturing and also in optimizing process and predictive maintenance. A digital twin is a virtual copy of a physical system updated in real time using real time data. Within the realm of 3D printing, digital twins can run the whole printing process and have operators test different control strategies, identify points of failure and optimize toolpaths prior to start of 3D printing. Using data from the physical printer's sensors, compared to those of its digital counterpart, control systems can identify the first signs of wear, material inconsistency, or mechanical failure, and can correct the problem 'in real time'. Besides, this approach makes printing systems more reliable and also allows remote monitoring and diagnostics that save the downtime and the maintenance costs for industrial facilities (Okoli & Kabaso, 2024).

To develop control systems for additive manufacturing, interdisciplinary research will be required, making use of knowledge from mechanical engineering, artificial intelligence, material science, and computational modeling. Combination of these fields will lead to new control strategies for the emerging technologies of multi material printing, biofabrication and large scale additive manufacturing. Coupling of academia with industry will help with the adoption of next generation control algorithms for additive manufacturing, making it continue to outperform at the precision, efficiency and scalability frontrunners. With the expansion of research in the future, the 3D printing control systems will, however, be governed by a more smart, faster and adaptive framework that can smoothly amalgamate the concepts of AI, real time computing and predictive analytics, transforming digital manufacturing (Wagmare et al. 2024).

12. CONCLUSION

Control systems in additive manufacturing have driven the development from 3D printing to a prototyping tool to a reliable and scalable production technology. Feedback mechanisms have improved in their precision, the algorithms are adaptive, and frameworks control AI driven with vast improvements in the use of precision, efficiency, and material optimization of various 3D printing processes. With an integration of real time monitoring, forecasting and clever automation present in the current printer, they are able to realign parameters dynamically, decreased the error and increase repeatability making them recommended for high performance applications in businesses for instance aerospace, medical and automobile. However, challenges like material variability, its computational complexity, and scalability are still major problems that need more refinement and innovation. This trajectory, which has seen additive manufacturing to continue to shape ability to control deposition processes with high accuracy, to regulate thermal behaviour and to optimize energy consumption to meet the ever growing demand for quality, energy efficient and cost effective process.

The progress in this field depends on collaboration among engineers, data scientist, material researcher and industry leader in the future. New techniques for more effective refinement of additive manufacturing processes are made possible by the convergence of control theory, artificial intelligence, sensor technology, and high performance computing. Further pushing the boundaries of what is possible, digital twins will continue to develop, edge compute for real time feedback will continue, and AI enhanced decision making will continue so that 3D printing systems can operate more autonomously and reliably. To accomplish these advances will need to be coordinated across many domains, involving robotics, software engineering and manufacturing science, and produce robust control solutions that can account for the particulars of various printing technologies. Standardized control protocols, security of next generation additive manufacturing systems ensures reliability and cybersecurity, as well as standardized AI ethics will play a role as well.

However, bridging this gap between theoretical models and practical implementations in the pursuit system of 3D printing is still a first order challenge in further advancement 3D printing control systems. Control theories give a base to keep stable, optimize, and make ad hoc control, though its practical application is subject to unpredictable real parameters, in particular due to limited computing potential and inconsistency of materials. Continuous refinement of the predictive models and adaptive control algorithms as well as machine learning that can work with more and more data of sensors is the key to this gap. Through blending theoretical insights in theory with practical experimentation and experimentation in practice, the industry is able to produce enhanced smarter, smarter, more sensitive 3D printing systems that can capture levels of precision and efficiency that have never before been achieved. With the continued advancement of additive manufacturing, the role of advanced control systems will be crucial to push the next step of innovation so that higher complexity, higher quality, and more sustainable products can be produced in a digital fabrication manner and introduce new possibilities for digital fabrication.

REFERENCES

Aravind Raj, S. (2025). Quality control in additive manufacturing: A review of traditional and advanced techniques. *Engineering Research Express.*

Baktheer, A., & Classen, M. (2024). A review of recent trends and challenges in numerical modeling of the anisotropic behavior of hardened 3D printed concrete. *Additive Manufacturing*, *89*, 104309. DOI: 10.1016/j.addma.2024.104309

Behseresht, S., Love, A., Valdez Pastrana, O. A., & Park, Y. H. (2024). Enhancing fused deposition modeling precision with serial communication-driven closed-loop control and image analysis for fault diagnosis-correction. *Materials (Basel)*, *17*(7), 1459. DOI: 10.3390/ma17071459 PMID: 38611974

Chatzipapas, K., Nika, A., & Krimpenis, A. A. (2024). Introduction of Hybrid Additive Manufacturing for Producing Multi-Material Artificial Organs for Education and In Vitro Testing. *Designs*, *8*(3), 51. DOI: 10.3390/designs8030051

Chen, M. T., Gong, Z., Zhang, T., Zuo, W., Zhao, Y., Zhao, O., Zhang, G., & Wang, Z. (2024). Mechanical behavior of austenitic stainless steels produced by wire arc additive manufacturing. *Thin-walled Structures*, *196*, 111455. DOI: 10.1016/j.tws.2023.111455

Di Salvo, S. (2024). Insights into new trends and contemporary challenges in 3D printing in architectural sector. *Engineering Innovations*, *8*, 43–61. DOI: 10.4028/p-VPN0Le

Duarte, L. C., Figueredo, F., Chagas, C. L., Cortón, E., & Coltro, W. K. (2024). A review of the recent achievements and future trends on 3D printed microfluidic devices for bioanalytical applications. *Analytica Chimica Acta*, *1299*, 342429. DOI: 10.1016/j.aca.2024.342429 PMID: 38499426

Ejjami, R., & Boussalham, K. (2024). Industry 5.0 in Manufacturing: Enhancing Resilience and Responsibility through AI-Driven Predictive Maintenance, Quality Control, and Supply Chain Optimization. International Journal For Multidisciplinary Research, 6(4).

Fei, G., Parra-Cabrera, C., Zhong, K., Clays, K., & Ameloot, R. (2024). From grayscale photopolymerization 3D printing to functionally graded materials. *Advanced Functional Materials*, *34*(32), 2314635. DOI: 10.1002/adfm.202314635

Fidan, I., Naikwadi, V., Alkunte, S., Mishra, R., & Tantawi, K. (2024). Energy efficiency in additive manufacturing: Condensed review. *Technologies*, *12*(2), 21. DOI: 10.3390/technologies12020021

Gerdes, S., Gaikwad, A., Ramesh, S., Rivero, I. V., Tamayol, A., & Rao, P. (2024). Monitoring and control of biological additive manufacturing using machine learning. *Journal of Intelligent Manufacturing*, *35*(3), 1055–1077. DOI: 10.1007/s10845-023-02092-6

Graziosi, S., Faludi, J., Stanković, T., Borgianni, Y., Meisel, N., Hallstedt, S. I., & Rosen, D. W. (2024). A vision for sustainable additive manufacturing. *Nature Sustainability*, *7*(6), 698–705. DOI: 10.1038/s41893-024-01313-x

Guo, Y., Zhang, Y., Pan, Z., & Zhou, W. (2024). Recent progress of sensing and machine learning technologies for process monitoring and defects detection in wire arc additive manufacturing. *Journal of Manufacturing Processes*, *125*, 489–511. DOI: 10.1016/j.jmapro.2024.07.060

Jhun, J., Lee, D. H., Rehman, A. U., Kang, S., & Kim, J. H. (2024). Development of a real-time geometric quality monitoring system for extruded filaments of 3D concrete printing construction. *IEEE Access : Practical Innovations, Open Solutions*, *12*, 69981–69999. DOI: 10.1109/ACCESS.2024.3401472

Kanthimathi, T., Rathika, N., Fathima, A. J., Rajesh, K. S., Srinivasan, S., & Thamizhamuthu, R. (2024, January). Robotic 3D Printing for Customized Industrial Components: IoT and AI-Enabled Innovation. In 2024 14th International Conference on Cloud Computing, Data Science & Engineering (Confluence) (pp. 509-513). IEEE.

Liu, B., Qian, B., Liang, Y., Dai, P., Li, R., & Wei, Q. (2025). Fast and accurate auto-disturbances-rejection temperature control system based on particle swarm optimized fuzzy control: Applied for particle three-dimensional (3D) printing prosthetic orthotic plate. *Additive Manufacturing*, *101*, 104704. DOI: 10.1016/j.addma.2025.104704

Liu, Y., Hu, X., & Chen, S. (2024). Multi-Material 3D Printing and Computational Design in Pharmaceutical Tablet Manufacturing. *Journal of Computer Science and Artificial Intelligence*, *1*(1), 34–38. DOI: 10.54097/40ftw337

Luo, X., Wu, H., Wang, C., Jin, Q., Luo, C., Ma, G., Guo, W., & Long, Y. (2024). 3D printing of self-healing and degradable conductive ionoelastomers for customized flexible sensors. *Chemical Engineering Journal*, *483*, 149330. DOI: 10.1016/j.cej.2024.149330

Ma, T., Zhang, Y., Ruan, K., Guo, H., He, M., Shi, X., Guo, Y., Kong, J., & Gu, J. (2024). Advances in 3D printing for polymer composites: A review. *InfoMat*, *6*(6), e12568. DOI: 10.1002/inf2.12568

Mandal, A., Morali, A., Skorobogatiy, M., & Bodkhe, S. (2024). 3D Printing of Polyvinylidene Fluoride-Based Piezoelectric Sensors for Noninvasive Continuous Blood Pressure Monitoring. *Advanced Engineering Materials*, *26*(1), 2301292. DOI: 10.1002/adem.202301292

Mattera, G., Nele, L., & Paolella, D. (2024). Monitoring and control the Wire Arc Additive Manufacturing process using artificial intelligence techniques: A review. *Journal of Intelligent Manufacturing*, *35*(2), 467–497. DOI: 10.1007/s10845-023-02085-5

Mu, H., He, F., Yuan, L., Hatamian, H., Commins, P., & Pan, Z. (2024). Online distortion simulation using generative machine learning models: A step toward digital twin of metallic additive manufacturing. *Journal of Industrial Information Integration*, *38*, 100563. DOI: 10.1016/j.jii.2024.100563

Nyamekye, P., Lakshmanan, R., Tepponen, V., & Westman, S. (2024). Sustainability aspects of additive manufacturing: Leveraging resource efficiency via product design optimization and laser powder bed fusion. *Heliyon*, *10*(1), e23152. DOI: 10.1016/j.heliyon.2023.e23152 PMID: 38163216

Okoli, N. J., & Kabaso, B. (2024). Building a smart water city: Iot smart water technologies, applications, and future directions. *Water (Basel)*, *16*(4), 557. DOI: 10.3390/w16040557

Parisi, F., Sangiorgio, V., Parisi, N., Mangini, A. M., Fanti, M. P., & Adam, J. M. (2024). A new concept for large additive manufacturing in construction: Tower crane-based 3D printing controlled by deep reinforcement learning. *Construction Innovation*, *24*(1), 8–32. DOI: 10.1108/CI-10-2022-0278

Paxton, N. C., Zhao, J., & Sauret, E. (2024). Polymer 3D printing in perspective: Assessing challenges and opportunities in industrial translation against the metal benchmark. *International Journal of Advanced Manufacturing Technology*, *133*(1), 59–80. DOI: 10.1007/s00170-024-13744-z

Pilch, M., Topa-Skwarczyńska, M., Chachaj-Brekiesz, A., Jamróz, P., Kiesiewicz, D., Noworyta, M., & Ortyl, J. (2024). Luminescence labeled surfaces mapping system dedicated for use in quality control of 3D prints produced by stereolithography 3D printing (SLA) and laser engraving. *Sensors and Actuators. A, Physical*, *365*, 114828. DOI: 10.1016/j.sna.2023.114828

Rahman, M., Islam, K. S., Dip, T. M., Chowdhury, M. F. M., Debnath, S. R., Hasan, S. M. M., & Houshyar, S. (2024). A review on nanomaterial-based additive manufacturing: Dynamics in properties, prospects, and challenges. *Progress in Additive Manufacturing*, *9*(4), 1197–1224. DOI: 10.1007/s40964-023-00514-8

Rautio, S., & Broms, M. (2024). Security scenarios: 3D printed firearms. *Security and Defence Quarterly*, *48*(4), 59–79.

Santhoshkumar, P., Negi, A., & Moses, J. A. (2024). 3D printing for space food applications: Advancements, challenges, and prospects. *Life Sciences in Space Research*, *40*, 158–165. DOI: 10.1016/j.lssr.2023.08.002 PMID: 38245341

Sierra-Vega, N. O., Ashraf, M., O'Connor, T., Kopcha, M., Di Prima, M., Coburn, J., & Zidan, A. (2025). Emerging 3D printing technologies for solid oral dosage forms: Processes, materials and analytical tools for real-time assessment. *International Journal of Pharmaceutics*, *673*, 125415. DOI: 10.1016/j.ijpharm.2025.125415 PMID: 40023346

Taghizadeh, M., & Zhu, Z. H. (2024). A comprehensive review on metal laser additive manufacturing in space: Modeling and perspectives. *Acta Astronautica*, *222*, 403–421. DOI: 10.1016/j.actaastro.2024.06.027

Tao, Y., Yu, P., Zhang, X., Li, M., Liu, T., & Zhao, J. (2024). Displacement-reconstruction-realized components by structure-sensing integration via a hybrid 3D printing strategy. *Composites. Part A, Applied Science and Manufacturing*, *177*, 107896. DOI: 10.1016/j.compositesa.2023.107896

Ukwaththa, J., Herath, S., & Meddage, D. P. P. (2024). A review of machine learning (ML) and explainable artificial intelligence (XAI) methods in additive manufacturing (3D printing). *Materials Today. Communications*, *41*, 110294. DOI: 10.1016/j.mtcomm.2024.110294

van Blitterswijk, R. H., Farshidianfar, M. H., & Khajepour, A. (2024). Model predictive control for real-time microstructure control in laser heat treatment. *Journal of Manufacturing Processes*, *118*, 116–127. DOI: 10.1016/j.jmapro.2024.03.038

Vemuganti, S., Soliman, E., & Taha, M. M. R. (2024). Manipulating interfacial bond for controlling load transfer in 3D printed fiber reinforced polymer composites. *Journal of Reinforced Plastics and Composites*, •••, 07316844241238502. DOI: 10.1177/07316844241238502

Venkatalakshmi, S. (2024). Role of Cybersecurity to Safeguard 3D Bioprinting in Healthcare: Challenges and Opportunities. Computational Intelligence in Bioprinting, 93-126.

Wagmare, R., Harshe, R., Pednekar, J., & Patro, T. U. (2024). Additive manufacturing of continuous fiber-reinforced polymer composites: Current trend and future directions. *Progress in Additive Manufacturing*, •••, 1–28.

Wagner, J. J., & Higgs, C. F. III. (2024). Coupled CFD-DEM simulation of interfacial fluid–particle interaction during binder jet 3D printing. *Computer Methods in Applied Mechanics and Engineering*, *421*, 116747. DOI: 10.1016/j.cma.2024.116747

Wen, Y., Che, Q. T., Wang, S., Park, H. J., & Kim, H. W. (2024). Elaboration of dimensional quality in 3D-printed food: Key factors in process steps. *Comprehensive Reviews in Food Science and Food Safety*, *23*(1), e13267. DOI: 10.1111/1541-4337.13267 PMID: 38284586

Yang, Y., Jiang, R., Han, C., Chen, J., Li, H., Wang, Y., ... & Wang, D. (2024). Frontiers in laser additive manufacturing technology. Additive Manufacturing Frontiers, 200160.

Yehia, H. M., Hamada, A., Sebaey, T. A., & Abd-Elaziem, W. (2024). Selective laser sintering of polymers: Process parameters, machine learning approaches, and future directions. *Journal of Manufacturing and Materials Processing*, *8*(5), 197. DOI: 10.3390/jmmp8050197

Yu, K., & Dunn, M. L. (2024). Design and extrusion-based 3D printing of continuous fiber composites: Status, challenges, and opportunities. *Langmuir*, *40*(17), 8751–8759. DOI: 10.1021/acs.langmuir.4c00139 PMID: 38644644

Zhou, L., Miller, J., Vezza, J., Mayster, M., Raffay, M., Justice, Q., Al Tamimi, Z., Hansotte, G., Sunkara, L. D., & Bernat, J. (2024). Additive manufacturing: A comprehensive review. *Sensors (Basel)*, *24*(9), 2668. DOI: 10.3390/s24092668 PMID: 38732776

Chapter 10
Modeling, Analysis, and Simulation of 3D Printing of Composite Plastics

Amit Subhash Vaidya
https://orcid.org/0000-0003-0067-0445
MIT, India

ABSTRACT

Abstract: In modern manufacturing World, additive manufacturing (AM) often called as 3D Printing has become the dominant technique This Chapter represents study delver into the realm of AM with a specific focus on composite materials, starting with exploring various 3D printing processes tailored for composites. The study discusses the modeling, Analysis & Simulation of 3D printing processes, laying the foundation for a comprehensive understanding of this transformative technology. Numerical simulation of printed parts navigates through the intricacies of micro-, macro-, and meso-scale modeling, offering valuable insights into the complex world of 3D printing for composites. The study also scrutinizes the multifaceted factors that influence the characteristics of 3D-printed composites. Each facet contributes significantly to the overall performance and properties of printed composite materials. By exploring modeling, Analysis simulation, this study offers a roadmap of 3D printing in composites, fostering innovation and advancements in this burgeoning application

DOI: 10.4018/979-8-3373-0533-2.ch010

INTRODUCTION

Additive manufacturing (AM), often called 3D printing or rapid prototyping, is a group of processes that produce three-dimensional parts by depositing, fusing, joining, or solidifying materials layer by layer. It was initially developed since 1980s. However, not until the 2010s did it gain tremendous interests from scientist and engineers. Since then, the technology in additive manufacturing has evolved dramatically from a mere method of prototyping to sophisticated and versatile solutions in many fields. Novel AM materials, innovative AM methods, and new AM related software and hardware have been extensively and continuously developed. With the capability of combining with more innovative materials developed from the material science and engineering field, AM has become a process that can print high performance components along with other benefit such as light-weighting and low investment in limited equipment and space. It has been considered to be the preferred prototyping method in new product development for geometry verification, concept demonstration, and so on. AM becomes so versatile that it is applied in almost every science, technology, engineering, medical, and other fields for various purposes, including part prototyping, design verification, tooling, customized part production, fully functional production components, etc.

The versatility of AM technology has allowed it to use various materials for the printing process. Those materials include various Composites, thermoplastic composites, thermosets, thermoset composites, concrete, metals, ceramics, food, tissues, medicines, etc. Among all the materials, neat composite thermoplastics, such as ABS and PLA, are most commonly used because of the low cost of printers required, the wide variety of commercially available feedstock materials, low cost of the feedstock material, ease of printability, etc. However, the AM parts printed using neat thermoplastics generally have low range properties. Those parts are often conceptual prototypes rather than functional components. One main approach to improve the properties of the neat thermoplastics is to add reinforcements, especially fibers. The addition of discontinuous fibers offers improved specific modulus and strength without considerable compromise of printability. With the development of continuous fiber reinforced plastic composites for AM in 2014, it is made feasible to print components with significantly enhanced properties. Additive manufacturing of plastic composites is a unique process with the characteristics that other manufacturing processes of plastic composites do not have. It differs from the conventional processing methods in several aspects, including manufacturing approach, flexibility, production rate, cost, etc. Therefore, AM of composites is described in a separate chapter as a unique process. Additive manufacturing possesses the following advantages that differentiate it from the conventional processing methods.

In modern manufacturing, additive manufacturing (AM) has become the dominant technique, surpassing traditional machining. Traditional machining poses challenges such as warping and internal stress, while rapid prototyping through AM offers cost-effective, precise, and efficient solutions. This approach is now widely used for complex design production, revolutionizing the manufacturing industry. This study delves into the realm of AM with a specific focus on composite materials, starting with exploring various 3D printing processes tailored for composites. The study discusses the modeling of 3D printing processes, laying the foundation for a comprehensive understanding of this transformative technology. Numerical simulation of printed parts navigates through the intricacies of micro-, macro-, and memo-scale modeling, offering valuable insights into the complex world of 3D printing. The study also scrutinizes the multifaceted factors that influence the characteristics of 3D-printed composites. These factors include interfacial bonding, the influence of loads and environmental conditions, and the pivotal role played by fiber content. Each facet contributes significantly to the overall performance and properties of printed composite materials. By exploring the modeling, simulation, and influencing factors, this study offers a roadmap for researchers and industry professionals to navigate the intricate landscape of 3D printing in composites, fostering innovation and advancements in this burgeoning field.

Types of Additive manufacturing techniques of composites

AM, also known as 3D printing, revolutionizes how physical objects are created. It operates on the layering principle, where successive layers of material are deposited or fused together to form a complete 3D structure based on a digital design. This technology has found its place in various industries, from aerospace and automotive to healthcare and consumer goods.

Among the various 3D printing techniques, several have gained prominence. The FDM involves melting thermoplastic material and extruding it onto a build platform through a nozzle, whereas the SLS utilizes a laser to sinter powdered material, allowing for the creation of intricate, functional parts. Then PBF (powder bed fusion) involves various heat sources such as thermal, laser, or electron beams employed to melt and fuse the powder. The SLA employs a laser to solidify the liquid resin, producing highly detailed, smooth-surfaced models. In addition, 3D plotting uses controlled extrusion to create objects, offering a different approach to AM.

As the 3D printing landscape continues to evolve, researchers are exploring new techniques and materials to push the boundaries of what can be achieved. The choice of which technique to employ depends on various factors, from the type of materials available to the desired production speed and the level of detail required. Moreover, cost considerations and the performance specifications of the final products play a crucial role in determining the most suitable approach. This dynamic field of technology promises exciting innovations in the future as it continues to transform how one designs and manufactures objects in diverse industries.

Stereolithography (SLA)

The SLA, the earliest form of 3D printing, was invented and patented by Charles Hull in 1984, despite initial research in 3D printing being conducted as early as the 1970s (Bagaria et al., 2018). The SLA is a 3D printing technique that transforms liquid photopolymers into solid objects layer-by-layer using a specialized SLA machine. To perform SLA, a computer-aided design (CAD) file is first created and converted to an STL file, which provides the geometric information needed for 3D printing. The SLA setup consists of four main components: a UV-curable photopolymer liquid, a perforated table, a UV laser, and a computer for process control.

During the printing operation, the perforated table is submerged in photopolymer liquid. As the table descends, the UV laser hardens the liquid polymer that comes into contact, forming a solid layer. The table continues to move downward, building the object layer by layer. Each new layer fuses with the previous one. After the final layer is completed, the 3D-printed object is separated from the remaining liquid polymer by submerging it in another resin (Fig1). The object is then cured in a UV oven to strengthen the layer bonds and improve the surface finish.

Nonwoven mats composed of E-glass, carbon fiber, and Para-aramid fibers have been utilized in the fabrication of composite components through SLA using photo curable resins, including both acrylic-based and epoxy-based variants (Karalekas, 2003). Among these combinations, the maximum strength and modulus were observed in the composite featuring E-glass fiber reinforcement with an acrylic-based resin. Incorporating glass fibers into this composite significantly enhances strength (48.8%) and stiffness (50.0%). When the fiber mat has an areal density of 17 gsm, this composite's tensile strength and modulus reach 55.2 MPa and 2.85 GPa, respectively. However, the introduction of glass fibers has a negative impact on epoxy-based composites, reducing strength and modulus. Furthermore, higher fiber mat densities corresponded to more pronounced reductions in mechanical characteristics (Karalekas, 2003). Investigation on SLA carbon fiber polymers has been relatively limited, primarily due to the challenges associated with printing components using resin infused with fibers. The SLA offers numerous advantages in the field of AM. It achieves a superior surface finish, resulting in finely detailed and visually appealing parts. It can easily and accurately produce complex components that may be challenging with other manufacturing methods. Another advantage is the high thermal resistance of parts created through SLA, ensuring their stability at elevated temperatures. Additionally, 3D-printed objects made using this process can serve as excellent patterns for casting, making it a versatile and valuable choice for various applications (Jandyal et al., 2022). None the less, SLA also presents certain disadvantages. One significant limitation is its relatively slow processing speed because it exposes a smaller surface area (approximately 0.15 mm) to the laser at a time, leading to longer production times for intricate parts. Another challenge is the substantial initial investment cost associated with acquiring and setting up SLA equipment. Additionally, manufacturing overhanging parts can be complex and demanding within this process, necessitating careful consideration and the use of support structures to achieve the desired outcomes (Jandyal et al., 2022).

Figure 1. SLA (van de Werken et al., 2020).

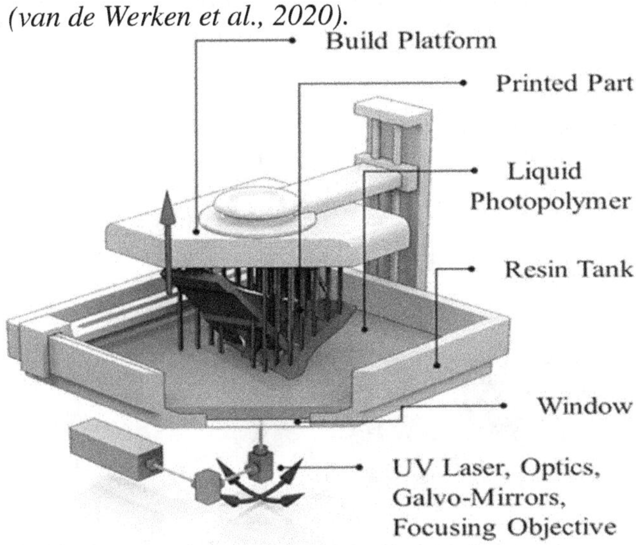

Powder bed fusion (PBF)

The PBF is one of the most prevalent techniques in the spectrum of AM, drawing significant attention for its cost-effectiveness. PBF essentially eliminates the need for external support structures, as the powder itself serves that function. The process is versatile, accommodating a range of materials such as plastics, metals, glass, and their alloys. Moreover, it offers the advantage of recycling unused powder for future projects (Fig..2).

In PBF, various heat sources such as thermal, laser, or electron beams are employed to melt and fuse the powder. Specific sub-techniques include SLS for plastics and SLM (selective laser melting) for metals. The process involves layering new powder over the preceding layer using mechanisms (e.g., a blade or roller) (Murr, 2015).

The powder and build chambers, often part of the same machine, can adjust along a z-axis to set the height of the final product (Murr et al., 2012). The workflow in PBF usually involves nine steps, starting with converting a 3D CAD model into an STL file. The STL file is then loaded onto the printer's bed using software. Multiple copies or different parts can be printed simultaneously to enhance productivity. The powder chamber is filled manually or automatically, and the heat source fuses the powder layer by layer. After the final part is created, any remaining powder can be recycled for other projects (Singh et al., 2020). Although PBF offers numerous benefits, such as cost-efficiency and a wide material range, it also has drawbacks. The process is time-intensive, requiring pre-heating and possibly vacuum generation. Additionally, the structural and surface qualities of the finished products are generally inferior to those produced by other manufacturing methods.

Figure 2. PBF (Singh et al., 2020).

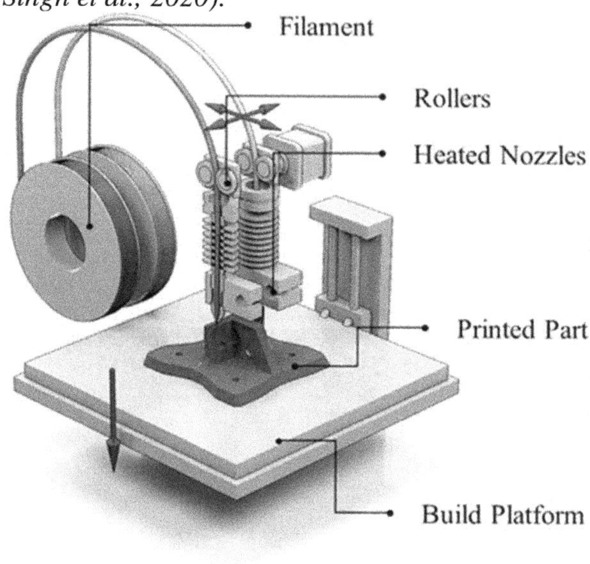

Fused deposition modeling (FDM)

The FDM is a widely used 3D printing method for polymers. This method (Fig. 3) entails melting a thermoplastic polymer filament within a print head and extruding the liquefied substance onto a printing surface. The procedure is executed incrementally, one layer at a time: the print head dispenses the required material for a specific XY plane, ascends vertically along the Z-axis by the thickness of one layer, and repeats this sequence. Incorporating support structures, particularly those crafted from soluble substances, can broaden the spectrum of attainable shapes and improve the surface finish of the printed components. Typical materials are used in FDM printing, primarily thermoplastics, such as acrylonitrile butadiene styrene (ABS), polycarbonate (PC), polyamide (nylon), acrylonitrile styrene acrylate, glycol-modified polyethylene terephthalate (PETG), polylactic acid (PLA), and polystyrene. Additionally, high-performance thermoplastics such as polyetherimide (ULTEM), polyphenylene sulfide (PPS), polyphenylsulfone (PPSU), and the polyaryletherketone (PAEK) family can be 3D printed using FDM, although they are typically more overpriced and challenging to work with (3DXTECH, 2019).

Lately, there has been a growing interest among researchers in the use of fiber reinforcement within FDM. Much of this research has centered on developing filaments containing short fiber additives. Incorporating fibers into the filament provides several benefits, such as minimizing printing head expansion during deposition and improving rigidity (Zhong et al., 2001).

For example, in a study by Carneiro et al. (Carneiro et al., 2015), they examined glass fiber-reinforced polypropylene (PP) and found significant improvements in modulus and strength. These enhancements amounted to 30% and 40%, respectively, compared to pure PP. Another approach involved incorporating vapor-grown carbon fibers (VGCFs) and single-wall carbon nanotubes (SWNTs) into acrylonitrile butadiene styrene (ABS) for FDM. VGCFs, in particular, could be well-aligned during extrusion, resulting in an 18% increase in tensile strength for parts containing 5 wt.% VGCFs and a 31% increase for SWNT-filled parts. Notably, the strain-to-failure of components containing VGCFs and SWNTs exhibits a significant reduction (Shofner et al., 2002). ABS filled with aligned VGCFs and SWNTs demonstrates remarkable developments in modulus, reaching as high as 93% (Shofner et al., 2003).

Another avenue explored was the use of thermotropic liquid crystalline polymers (TLCPs), such as ABS and PP, in fiber-reinforced FDM components, aimed at overcoming the limitations associated with the relatively short length compared to the diameter of the reinforcing fibers (Gray IV et al., 1998). The temperature during the processing significantly impacts the surface characteristics and mechanical properties of TLCPs. In this regard, a 100% increase in tensile modulus was observed for ABS containing 40 wt.% TLCP, while PP shows a 150% increase. A higher carbon fiber ratio offers greater thermal stability due to a higher maximum decomposition temperature. In a study by Ning et al. (Ning et al., 2016), the impact of weight ratio and fiber length on the physical properties of FDM samples with an ABS matrix was investigated. They found that 5% and 7.5% carbon fiber volume yield the most substantial improvements in tensile strength and Young's modulus, respectively. The elongated carbon fibers also lead to higher tensile strength and Young's modulus but reduce toughness and ductility.

FDM technology is worth considering because of its efficient printing speed and cost-effectiveness. It is particularly suitable for creating proof-of-concept models and basic prototypes. Nevertheless, FDM has several drawbacks, including its relatively weak mechanical properties, noticeable surface roughness, and constraints when choosing thermoplastic materials (Chohan et al., 2017). FDM typically operates within a resolution range spanning from 50 to 200 mm (Wang et al., 2017).

Figure 3. FDM (van de Werken et al., 2020).

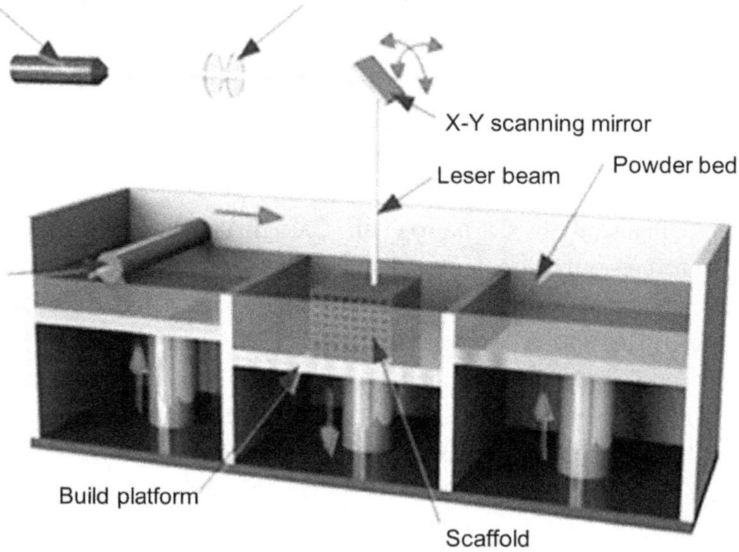

Selective laser sintering (SLS)

The SLS is an AM technique that operates with powdered materials. In this technique, a laser scans and solidifies the powder layer by layer to construct a 3D structure, as depicted in Fig. 4. SLS is commonly employed with materials such as wax and ceramics (Bertrand et al., 2007; Liu et al., 2007; Shishkovsky et al., 2007), metals (Agarwala et al., 1995; Chatterjee et al., 2003), and polymers (Childs et al., 1999; Tan et al., 2003). Among the polymers used in SLS, notable examples include nylon, particularly polyamide (PA) (Good ridge et al., 2011), and semi-crystalline thermoplastics (e.g., PE) (Hao et al., 2006; Salmoria et al., 2007), PEEK (Schmidt et al., 2007), and PCL (Kruth et al., 2007; Williams et al., 2005).

SLS can be classified into various methods: solid-state sintering (SSS), partial melting with liquid-phase sintering, complete melting, and chemically induced binding. SSS is a thermal process that occurs in the temperature range between half of T_{melt} and T_{melt}, where T_{melt} denotes the material's melting point. In the liquid-phase sintering with a partial melting approach, the binder material typically becomes liquid, whereas the structural material prevails in a solid state. Complete melting involves completely liquefying the powder, resulting in characteristics similar to those of bulk materials (Kruth et al., 2005). SLS is versatile and can be used with a wide range of materials, although it often requires longer processing times and the pre-heating of powders.

The SLS offers several distinct advantages. It excels at producing complex parts that are challenging to manufacture using other methods. One key benefit is that it does not necessitate external support structures during the printing process, simplifying production. SLS is well-suited for mass production due to its efficiency and ability to create multiple parts simultaneously. Additionally, it delivers good accuracy and precision in the final products, making it a valuable choice for various manufacturing applications (Jandyal et al., 2022).

However, SLS has several disadvantages. One of the primary drawbacks is its high manufacturing cost, which can be a significant factor in the production process. Additionally, SLS often requires extensive post-processing to achieve the desired finish and quality, adding time and cost to the overall manufacturing process. Another challenge is that when dealing with large surfaces or tiny holes, achieving precise and accurate results can be difficult, posing limitations in certain applications (Jandyal et al., 2022).

Figure 4. SLS, with red arrows indicating the relative mechanical movement of the components (van de Werken et al., 2020).

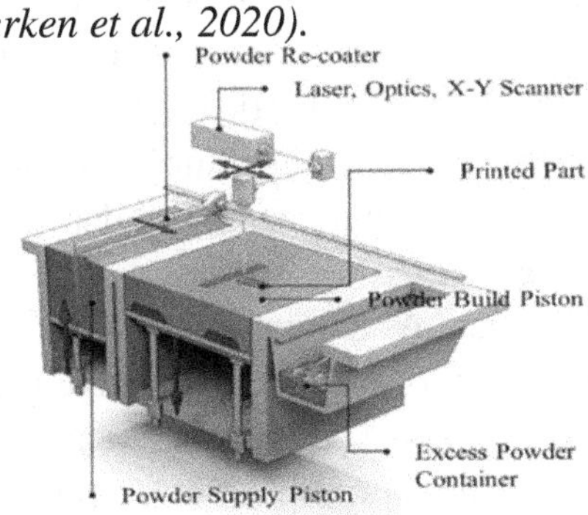

Fused Pellet Fabrication

Fused filament fabrication (FFF), also known as fused deposition modeling (with the trademarked acronym FDM), or filament freeform fabrication, is a 3D printing process that uses a continuous filament of a thermoplastic material (3DXTECH, 2019). Filament is fed from a large spool through a moving, heated printer extruder head, and is deposited on the growing work. The print head is moved under computer control to define the printed shape. Usually the head moves in two dimensions to deposit one horizontal plane, or layer, at a time; the work or the print head is then moved vertically by a small amount to begin a new layer. The speed of the extruder head may also be controlled to stop and start deposition and form an interrupted plane without stringing or dribbling between sections. "Fused filament fabrication" was coined by the members of the RepRap project to give an acronym (FFF) that would be legally unconstrained in its use. Fused filament printing is now the most popular process (by number of machines) for hobbyist-grade 3D printing (Agarwala et al., 1995). Other techniques such as photopolymerisation and powder sintering may offer better results, but they are much costlier. Filaments are commonly used as the feedstock in additive manufacturing of thermoplastics and thermoplastic composites. The filament is a material form for both discontinuous and continuous fiber thermoplastic composite. The feed rate of the filament based AM process is largely limited to the size of the filament diameter and the filament feed rate. The filament is also required to have a tight tolerance for proper feeding and printing. The need for a higher feed rate and a lower raw material cost has led to the development of fused pellet fabrication (FPF), another extrusion-based AM process especially for printing large scale parts. Fused pellet fabrication process is also called fusion pellet fabrication or fused particle fabrication process. The working principle of the FPF process is similar to that of the FDM process, namely, extruding thermoplastic composite melt through a nozzle and depositing it for additive manufacturing. However, the FPF process uses pellets as its feedstock material instead of filaments in the FDM process. As shown in Fig.5; the FPF printer is equipped with an extruder that converts thermoplastic composite pellets into viscous melts for printing. Thermoplastic composite pellets are fed in the extruder and heated to the processing temperature of the thermoplastic composite before being extruded out of the nozzle for depositing. IN contrast to the FDM process or other AM processes that only allow printing small build volumes at low deposition rates, the FPF process is capable of printing components with significantly larger built volumes at a relatively high deposition rate. The process has been successfully used to print large structures such as tooling in a relatively short time (Error! Reference source not found.). Additionally, the FPF process has other advantages compared to the FDM process. The FPF process uses pelletized feedstock material which is easier to produce and handle. The issues caused by low filament quality such as inconsistency of the filament diameter and inadequate rigidity of the filament, etc. encountered in the FDM process, can be circumvented in the FPF process. However, the FPF process is suited for printing discontinuous fiber thermoplastic composites, which results in limited mechanical properties, while the FDM process is able to print continuous fiber thermoplastic composites.

Figure 5. FPF, with labelling indicating the relative relation of the components (Wikipedia et al., 2020).

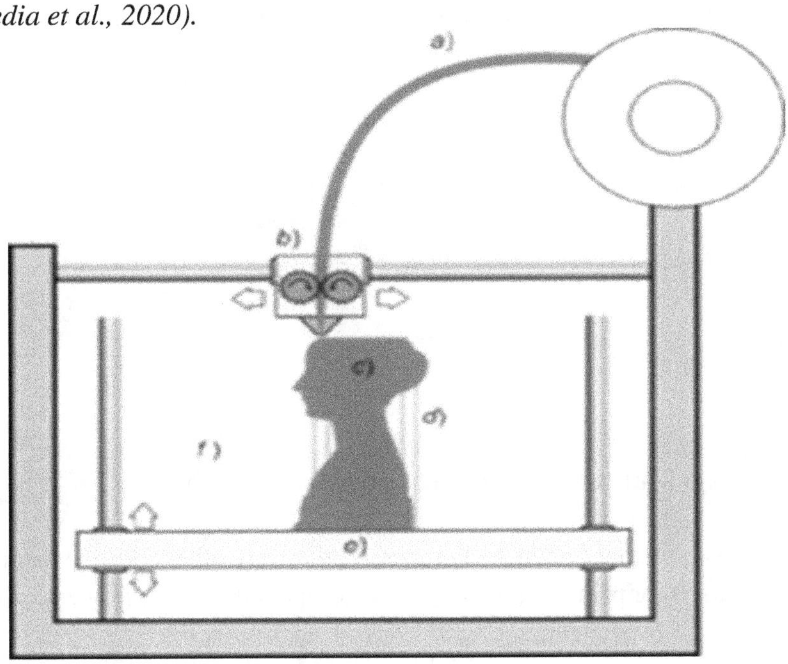

Laminated Object Manufacturing

Laminated object manufacturing (LOM) is another AM process that can be used to print continuous fiber reinforced thermoplastic composites. It was initially developed for thermoset based continuous fiber prepregs. The same concept is adopted for printing thermoplastic composites after several modifications are made, for example, increasing the capacity of the heating unit from curing thermosets to melting thermoplastics. The LOM process has a similar principle with the automated tape layup process in which thermoplastic composite preforms are stacked and consolidated layer by layer. Fig. 6 illustrates the setup of the LOM process. Continuous fiber thermoplastic composite preforms, for example, woven fabric reinforced thermoplastic composite preforms, move through a forming station where heating, consolidation, and cutting of the composite preform take place. The preform is firstly heated by a laser unit and consolidated by a heated roller. A laser or a blade cuts along the periphery of the consolidated preform area. The platform is moved down with the consolidated layers and fresh preform is rolled into position for consolidation. The preform width is normally greater than the part width in order for the take-up roll to pull the new preform to the printing station.

Figure 6. LOM, with labelling indicating the relative relation of the components in LOM

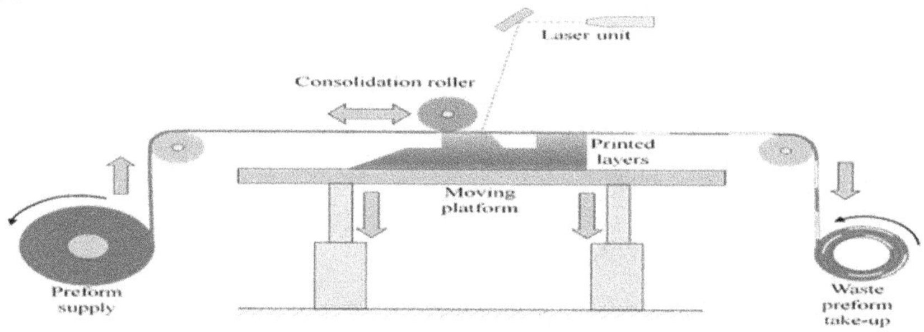

Other Additive Manufacturing Processes

Besides the abovementioned AM processes, there are other additive manufacturing processes used for printing thermoplastic composites. Those processes include;

1. Liquid deposition modeling, or LDM, is an AM process that utilizes a solution as the printing material to produce parts. It is also referred to as direct ink writing. The solution normally has a high viscosity and is in a gel or paste state. After the solution is extruded from the printer nozzle, its solvent vaporizes rapidly and the material solidifies; the geometrical shape can therefore be retained after printing. It is mainly used for printing thermosets and thermoset composites. It allows for a high loading of reinforcements because of the low viscosity of the solution. The process has been sometimes used in printing thermoplastic composites, mainly PLA matrix composites. PLA is dissolved in a highly volatile solvent, such as dichloromethane, and reinforcements are mixed and dispersed in the solution for printing.
2. Selective laser melting (SLM) is sometimes referred to as laser beam melting. It is a process that is similar to selective laser sintering. Thermoplastic or thermoplastic composite powders are fused together by laser. Instead of sintering

the powders in the selective laser sintering process that maintains the material in a solid form, the laser beam melting process uses high-power laser to melt and fuses thermoplastic powders or thermoplastic composite powders together. After one powder layer is melted and fused together, the powder bed lowers down and a roller or rake levels the powders for the next melting cycle. Since complete melting of the thermoplastic composite powder is involved in the SLM process, the printed part possesses better fusion and enhanced mechanical properties compared to the SLS part. However, geometry control is much more challenging in SLM because of the geometry instability induced by its higher processing temperature (more thermal shrinkage) and the transitions from solid to liquid and back to solid.

Figure 7. SLM & LDM Setup, with red arrows indicating the relative mechanical movement of the components (van de Werken et al., 2020).

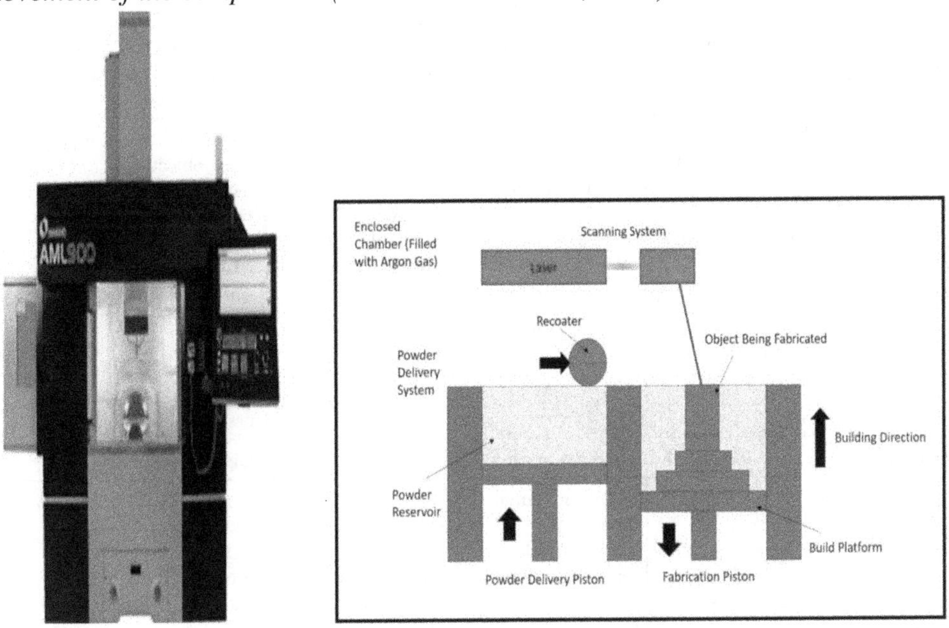

COMPARISON OF ADDITIVE MANUFACTURING PROCESSES

Table 1.

AM method	Feedstock	Reinforcements	Advantages	Disadvantages
Fusion deposition modeling (FDM)	Filament	-Discontinuous fiber - Particulate - Nano-material - Continuous fiber	- Great mechanical properties -Able to integrate different materials	- Printing of continuous fiber composite is still under development - High filament quality required
Fused pellet fabrication (FPF)	Pellet	-Discontinuous fiber - Particulate - Nano-material	- Cost-effective - High printing rate	- Relatively low mechanical property
Selective laser sintering (SLS)	Powder	-Discontinuous fiber - Particulate - Nano-material	- No support required	- Limited feedstock options - Powder form required - Low mechanical property
Laminated object manufacturing (LOM)	Fabric	- Continuous fiber	- Great mechanical properties	- Waste material generated - Relatively high cost
Stereo lithography (SLA)	Powder Liquid	- Continuous Fiber - - Wood, Silica, metal, Carbon	- Is Fast. - Is Accurate. - Works with Different Materials. - Sustainability - Multi-Part Assemblies Are Possible. ... - Texturing Is Possible.	- Limited in building large-sized objects. - Post-processing may be required to remove excess resin and achieve desired aesthetics. - Material choices are often limited compared to other 3D printing methods. - Handling liquid resins can be messy and require safety precautions
AM method	**Feedstock**	**Reinforcements**	**Advantages**	**Disadvantages**
Liquid Deposition Modeling (LDM)	Liquid	- -Discontinuous fiber -Clay, porcelain, - alumina, zirconia	-low cost of systems - -Wide range of mtl available - Large size part capabilities - Very versatile Applications	- Limited layer resolution - Limited part geometry - Difficult to operate properly - Require Mtl. Expertise

continued on following page

Table 1. Continued

AM method	Feedstock	Reinforcements	Advantages	Disadvantages
Selective Laser Melting (SLM)	Filament	Aluminum Oxide Hybrid fiber Composites Metal Matrix Composites (C,Al,Ti, NA etc.)	- No design limitations - Reduced part weight - R Reduced R&D cycle & cost - Wider Application Range	- Large Post-processing required - Heat Related Errors in Parts (Porosity, Residual Stress, Blistering etc. -
Powder bed fusion (PBF)	Powder	- Continuous gels - Particulates - Ceramics	- Design Flexibility - Material Efficiency - High Mtl. Quality	- Cost Considerations - Limitations in Material Variety - Technical Limitations

Applications of Additive Manufacturing Processes for Composites Plastics

Applications of 3D Printed Composite Plastics in various fields are as discussed below;

- o M\c Tool Industry: Special Machine elements, Tool inserts, Jig & Fixtures.
- o Aerospace: Lightweight and high-strength components and thermal management systems.
- o Automotive: Structural components, impact-resistant parts, and functional prototypes.
- o Medical: Biocompatible implants, prosthetics, and customized medical devices.
- o Consumer Goods: Wearable technology, sports equipment, and household items.
- o Academics & Research: Training, Design & Development, Research & innovation of Additive Mfg., Composite Mtl & its Applications.

Figure 8. Applications scenario of Additive Manufacturing Processes for Composite Plastics

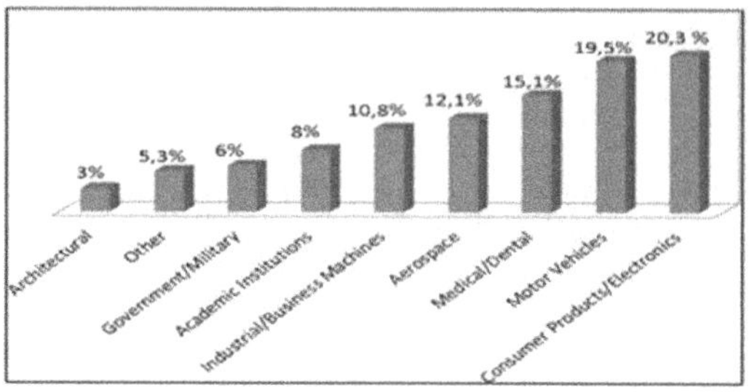

3D PRINTING PROCEDURE: -

There are several typical steps involved in additive manufacturing of composite Plastic parts. Those steps are explained below as shown in fig 8.

1. **Computer aided design (CAD)** of the geometry to be printed. The geometry of the component to be printed is firstly designed through CAD software such as PTC Creo, Solid works, Autodesk Inventor, etc. Object scanners can also be used to obtain the geometry of an object to be printed. For a structural component to be printed, finite element analysis is often carried out for optimizing dimensions and/or fiber orientations to meet design criterion before printing is carried out.
2. **Conversion of files.** The CAD file from the previous step is firstly converted to STL file, the most common format for AM. STL is also known as Stereolithographic, Surface Tessellation Language, Standard Tessellation Language, or Standard Triangle Language. During the file conversion, the surface of the object is tessellated into connected triangles. The STL file carries the surface information of the object. A higher number of triangles denotes a higher resolution for geometric features such as arcs, circles, fillets, etc. This conversion can be done through the CAD software as previously mentioned.
3. **Slicing of the geometry**. Slicing of the geometry in an STL format is a critical step in AM. It demonstrates the working principle of AM. The STL file is imported to a printer equipped with algorithm that can carry out slicing of the geometry. Different printer manufacturers may develop different algorithms and allow a different amount of control on slicing parameters, such as the thickness of one sliced layer.

4. **Setting of printing parameters**. Various printing parameters including nozzle temperature, printing speed, printing path and orientation, printing bed temperature, material selection, and support design (if needed), have to be established before printing. When continuous fiber thermoplastic composites are involved in printing, the printing path is important as it determines the fiber orientation and the material anisotropy. If multiple materials are involved in the printing, the material type has to be assigned for certain layers specifically. For open-source printers, those settings can often be adjusted. However, closed-source printers do not allow much freedom in adjusting certain printing parameters.
5. **Printing of the geometry**. The geometry is printed layer by layer according to the layer thickness, printing path, and material setting pre-determined from the previous steps. When the printing is finished, the component is removed from the printing bed for post- processing if needed.
6. **Post-processing**. Post-processing refers to the operation conducted on the printed component to achieve desired performance, surface finish, and so on. The post- processing operation includes removal of support structures, trimming, sanding, polishing or painting for better surface appearance, or annealing for a higher degree of crystallinity.

Figure 9. 3D Printing Procedure for Composite Plastics

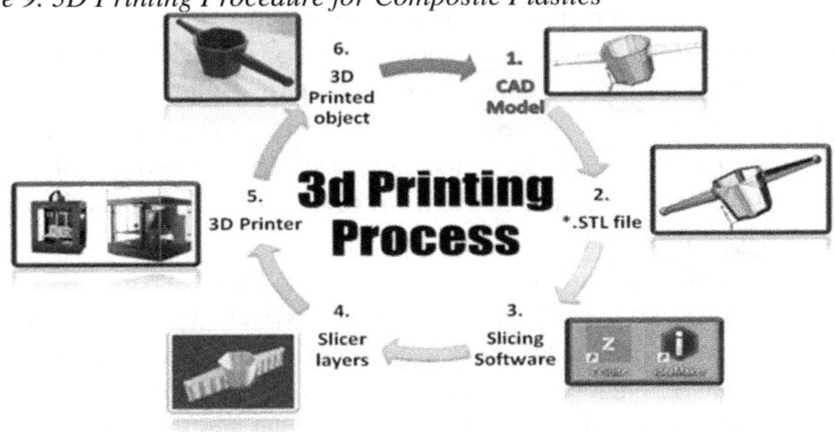

NUMERICAL SIMULATION OF PARTS

To effectively forecast the performance of 3D-printed components, integrated computational material engineering (ICME) approaches are crucial (Schmitz and Prahl, 2016). This is because the parameters and techniques used in the printing process significantly influence the material properties and ensuing structural behaviors. Specifically, the AM process inherently creates spatial variances in material properties. This makes the challenging of multiscale endeavor of modeling these 3D-printed structures. For instance, fused filament fabrication (FFF) may involve filaments reinforced with fillers, while SLS can entail partially sintered powders. In general, there are three primary scales of interest in modeling 3D-printed structures using continuum mechanics: (1) At the microscale, dimensions ranging from 1 to 50 mm typically represent the reinforcing fillers or sintered powder, (2) the mesoscale, featuring dimensions between 0.1 and 1 mm, generally captures the space between filaments or sintered powder, including interlayer structures, and (3) lastly, the macroscale ranges from several millimeters to meters and is concerned with representing the overall geometry, loading conditions, and external boundaries of the structure.

Microscale modeling & Simulation

In microscale modeling, the focus is to understand the behavior of filaments when dealing with FFF or the characteristics of sintered powders in the case of SLS. Using finite element (FE) computational methods, a specific computational unit, known as a representative volume element (RVE), is established to encapsulate the microscopic traits of the material, as shown in Fig. 9. This computational unit serves a dual purpose: it allows for evaluating performance improvements, such as increased stiffness and resilience due to short fiber reinforcement and identifying performance limitations introduced by porosity.

The role of ICME is pivotal at the microscopic level. It provides a bridge between the manufacturing process and the engineering of the resultant structure by focusing on material properties. Microstructural aspects such as filler material orientation and porous regions' existence are typically influenced by the manufacturing process. Despite this, achieving accurate simulations that predict these micro-level features remains an unresolved issue in the field, limiting its current application in an industrial context. As a result, current practice often relies on empirical measurements or assumptions on such microstructural elements rather than computational predictions.

Mesoscale modeling & Simulation

In the case of FFF, the goal at the mesoscale is to capture the behaviors of an assembly of filaments with various orientations and spacings, culminating in a material that can be either densely or loosely arranged. To achieve this, computational methods primarily focus on identifying key features such as the geometric form of filaments, the separation between them, and how they are oriented layer by layer, as indicated in Fig. 7. The attributes of a single filament are derived from microscale models. Often, especially in fiber-reinforced materials, these filaments exhibit anisotropic behaviors. Therefore, their orientations, which are influenced by the microscale attributes, need local adjustments to align with the routes taken by the filaments.

Figure 10. RVE showcasing the internal microstructure of the filament material. Fibers are color-coded in blue and red and are oriented in two dominant directions relative to the horizontal plane. Pockets of porosity are represented by green spheres, while the remaining space is filled with a transparent polymer matrix (Digimat., 2019).

Figure 11. Filament's cross-sectional dimensions and form are considered on one side (left), while the recurring patterns in which these filaments are organized make up the other aspect (right) (Digimat., 2019).

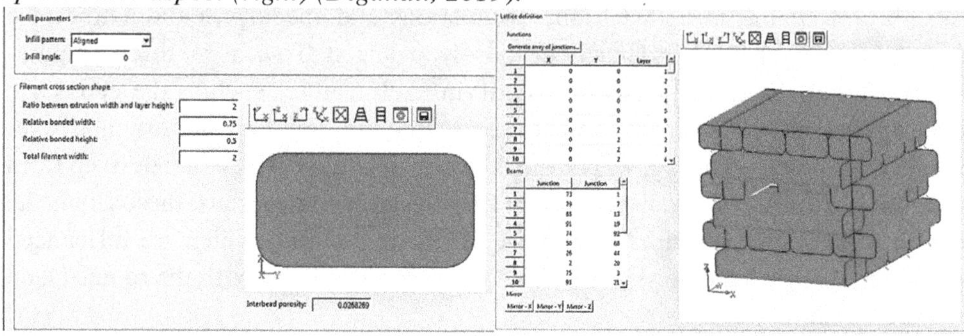

Figure 12. Three meshed RVEs depict varying filament arrangements: one with uniform orientation (left), another with alternating 90 rotations between layers (middle), and a third with a hexagonal lattice structure (right). Each provides unique insights into how filament configurations impact material properties (Digimat., 2019).

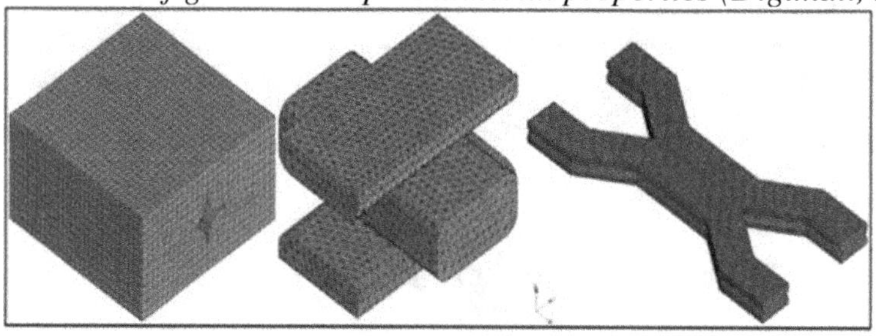

Fig 12 illustrates mesh-ready RVEs, each portraying unique potential FFF print layouts. It is important to highlight that even in scenarios where filaments are consistently oriented, and the material appears fully compacted, voids or spaces between filaments are still present due to their irregular cross-sectional shapes. These RVE models are typically based on artificial or synthesized microstructures. There is a growing trend to utilize simulation methods at the filament level to forecast filament shapes and their interactions with adjacent filaments, using process and printer specifications as the basis.

In addition to examining how the material behaves within a single filament, RVEs can also be employed to study the interactions between adjoining filaments. Utilizing specialized components known as cohesive elements or similar types of interface elements, these models are equipped to identify and analyze instances where failure occurs between adjacent filaments. This allows for a more comprehensive understanding of both internal and interfacial behaviors of filament-based materials.

Macroscale modeling & Simulation

The direction of filaments in the mesoscale models, as detailed in Section 4.2, is governed by the trajectory of the printing nozzle. This trajectory delineates regions where a uniform printing pattern is likely, allowing for reliable material behavior predictions within those zones. However, due to the wide array of geometries and print designs, attaining a completely uniform pattern across an entire object is highly challenging. For example, Fig. 9 shows the development of significant gaps or porosities between filaments resulting from inconsistent arrangements of the contour and infill filaments. It is crucial to explicitly include these porosities in the model, as they locally compromise the material's integrity and create areas vulnerable to the onset of cracks.

FACTORS AFFECTING THE CHARACTERISTICS OF 3D PRINTED COMPOSITE MATERIALS

Various Factors affecting the characteristics of 3D printed composite Materials are as discussed below;

Fiber content: Adjusting the fiber content in continuous FRPCs has unlocked new possibilities for customizing 3D-printed composites. Researchers have explored the manipulation of fiber volume fraction to establish models that lay the groundwork for enhancing the mechanical properties and designing functionally graded composites achievable through 3D printing (Dou et al., 2020; Li et al., 2019). Some studies have focused on assessing the robustness and rigidity of 3D-printed FRPCs by altering the fiber volume percentage, revealing a complex relationship between fiber content and material properties through systematic experimentation. These investigations into the failure mechanisms of 3D-printed composites under various loads have considered composite formation mechanisms, microstructure, and interfacial behavior. The derived material characteristics have been employed to create FE models for additively manufactured composites with changing fiber volume fractions. Furthermore, the analysis of mechanical properties has allowed for predicting, optimizing, and evaluating various material combinations for additively manufactured FRPCs with different fiber volume percentages (Li et al., 2020; Valvez et al., 2020; Yin et al., 2019). Fig. 10 illustrates the failure modes of continuous FRPCs in machinability studies while altering the fiber volume percentage.

Furthermore, other experiments have examined the shape and dispersion of pores in additively manufactured FRPCs, offering insights into the strength and rigidity of these composites. The findings have demonstrated a direct influence of fiber content on the porosity of 3D-printed composite materials. Microstructural analysis revealed that where pores existed along the specimen's length, cracks propagated across the specimen's width under load, reducing the toughness of additively manufactured porous FRPCs. Similar methods have arisen from these experiments, serving as valuable tools for predicting, optimizing, and analyzing the performance of 3D-printed composites. These insights hold relevance across a wide range of applications, including aerospace, aviation, and various engineering fields.

Interfacial bonding: The characteristics of 3D-printed polymer composites are influenced by several factors, including the formation of voids during printing, the bonding between fibers and the matrix at the interface, and the bonding between individually printed beads. Researchers have studied how fiber oxidation affects the fiber-matrix interface in composites produced using the FDM technique (Ramesh, 2016). This study also investigated the effects of vacuum annealing and nozzle geometry on inter-bead bonding, mechanical properties, and microstructure.

Figure 13. Irregular gaps between filaments in FFF/FDM printing, with filament directions indicated by green lines. Red ellipsoids highlight three significant macro-scale porosities, which are crucial to note as they represent weak points in the material (Digimat., 2019).

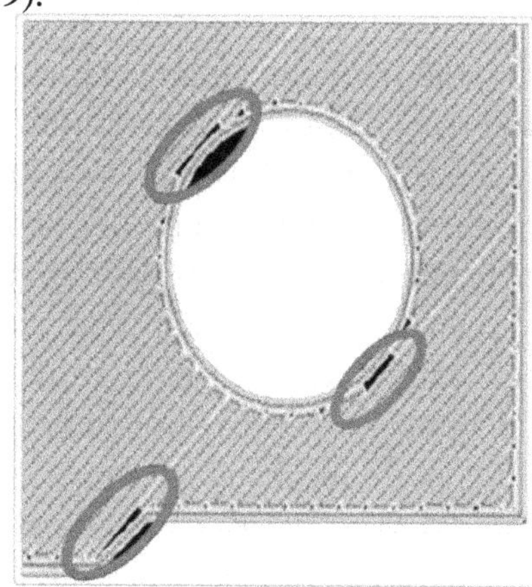

Figure 14. Double shear failure zones for drilled (top) and tailor-woven (bottom) samples. The failed specimen exhibits both compressive and tensile impairment. In the tailor-woven samples, the tensile failure occurs outside the hole area, resulting in less hole distortion. (Dickson and Dowling, 2019).

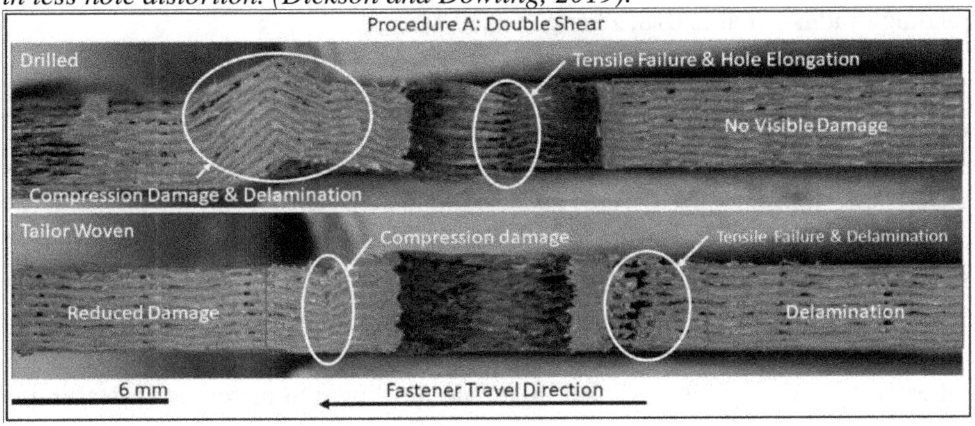

During the oxidation treatment, carbon fibers underwent chemical modification, resulting in an increase in the concentration of -COOH groups on the fiber surface. This increase was confirmed through energy-dispersive X-ray spectroscopy, which showed higher oxygen content. The treated fibers exhibited improved short beam shear strength, tensile behavior, and fracture toughness. Vacuum annealing induced a rearrangement of polymer chains in the crystalline phase, doubling the crystallinity of the 3D-printed composites. This rise resulted in a 7% enhancement in shear strength, a 75% increment in tensile modulus, and a 22% boost in tensile strength as opposed to unreinforced composites. Additionally, vacuum-annealed composite samples with treated functional fibers demonstrated a 24% increase in fracture toughness compared to untreated 3D-printed composites. The structural improvements were evident in the cross-sectional morphology, where factors such as fiber alignment, fiber surface roughness, matrix crystallinity, interfacial debonding, and fiber pull-out hindered crack propagation (Al Abadi et al., 2018; Dutta et al., 2019; Hou et al., 2020; Melenka et al., 2016; Prabhu et al., 2019b).

In another experiment, functional carbon fibers support a PLA matrix in additively manufactured composites using a square-shaped nozzle instead of the conventional circular one. Vacuum annealing was applied to enhance bead bonding and matrix crystallinity. The findings indicated that using a square-shaped nozzle increased the contact surface area between consecutive beads, thereby improving the bonding strength. Annealing further improved this strength while reducing void content from 16% to 3%. The fiber surface treatment increased the presence of oxygen groups and -COOH functional groups on the fiber surface. The bond between the fiber and the matrix was enhanced due to the greater roughness on the fiber surface, which was achieved through surface modification. The formation of microbial 2-pores at the fiber-matrix interface decreased because of the reaction between oxidized carbon fibers and the PLA matrix during 3D printing, further enhancing fiber-matrix bonding (Sathish et al., 2018; Zhang et al., 2018).

Vacuum annealing also altered the thermal response of the 3D-printed composite, significantly increasing crystallinity to 64%, which initially was 30% due to the 3D printing process. This enhancement in crystallinity led to a 21% increase in tensile strength and a 74% increase in tensile modulus compared to unreinforced PLA. Nonetheless, the composites' short beam strength experienced only an 8% increase, as it is primarily affected by the properties of the bead interfaces, and there was no alteration in the voids between the annealed and unannealed composites. Additionally, varying bead orientation at 45 and 90 showed that the composite with a 90 bead orientation exhibited the highest fracture toughness, about 24% greater than the untreated specimen. This rise in strength was attributed to increased crystallinity, improved bonding between the fiber and matrix, and enhanced bonding between adjacent beads. Overall, these findings highlight the significant role of interfacial bonding in the 3D printing of composites (Caminero et al., 2018; Kabir et al., 2020; Prabhu et al., 2019a; van de Werken et al., 2019). Fig. 13 depicts the process of 3D printing fiber-reinforced polymer composites (FRPCs) using a continuous filament and presents a property comparison between 3D printed and conventional composites.

Loads and environmental conditions: AM has become a cost-efficient approach for manufacturing sandwich composite structures widely utilized in industrial applications. Recent advancements in AM technologies have enabled the creation of intricate and complex sandwich polymer structures, garnering significant attention (Akhoundi et al., 2019; Bettini et al., 2017; Peng et al., 2019). Experimental endeavors have focused on designing, manufacturing, and analyzing additively manufactured sandwich composite materials under different environmental and loading conditions. The strength of sandwich composites, particularly in the in-plane lattice layers, is primarily dictated by the core material. To investigate this, two distinct core materials, ABS and acrylonitrile styrene acrylate (ASA), were manufactured and then subjected to experimental strength analysis. Since sandwich structures often operate in varying temperature environments, the thermal behavior of these composites was examined. Accelerated thermal imaging tests were conducted over a temperature range of 22 C to 60 C, simulating potential operational conditions. The impact of thermal aging on the flexural properties of 3D printed sandwich components was studied, and comparisons were made between unaged and aged samples (Gokulkumar et al., 2020; Hashin and Rotem, 1973; Sathish et al., 2017; Thomas and Renaud, 2006). Fig. 14 illustrates diverse cross-cut configurations of carbon fiber/PLA composites, providing insight into the failure mechanisms of additively manufactured composites.

Figure 15. Fabrication of continuous filaments and their associated characteristics (Kabir et al., 2020).

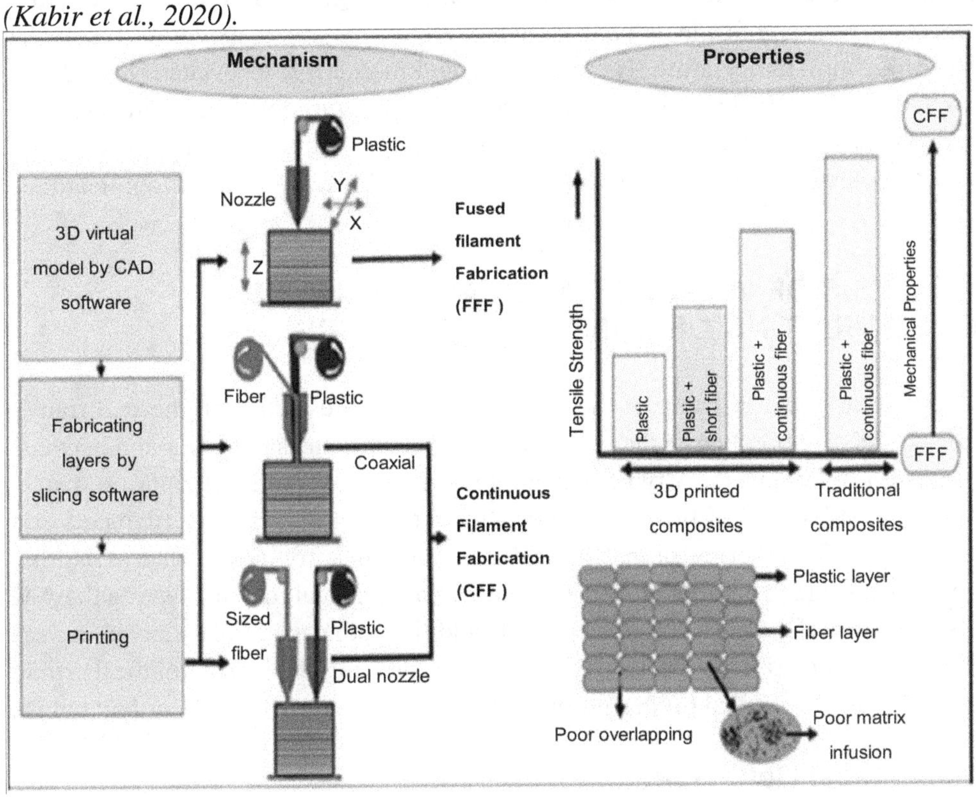

The outcomes indicated that the thermally aged ASA core material with a honeycomb structure displayed the highest flexural strength among the tested materials. Durability studies were also conducted to assess the thermal stability of 3D-printed sandwich structures, providing useful insights for developing robust sandwich materials (Akhoundi et al., 2019; Bettini et al., 2017).

Experiments further explored the influence of core structure geometry on the strength of sandwich composites. ABS and ASA cores and two different core shapes (honeycomb and triangular) underwent thermal aging tests to determine their glass transition temperatures. These cores were additively manufactured and linked to continuous carbon fiber face sheets using adhesives. The flexural failure behavior of the additively manufactured structures was assessed through three-point bending tests at aging temperatures up to 60 C. The results indicated that the ASA sandwich material with a honeycomb structure, which experienced thermal aging, demonstrated the most substantial flexural properties. Additionally, in both aged and unaged composites, sandwich cores featuring a hexagonal lattice pattern exhibited superior load-bearing characteristics compared to those with triangular structures. Both patterned sandwich structures showed high strength, and subjecting these materials to annealing further enhanced their strength at aging temperatures below the glass transition temperature (Hambach and Volkmer, 2017; Kousiatza et al., 2019; Mohammadizadeh et al., 2019). As a result, the configuration of the core structure and the operational temperature of additively manufactured sandwich polymer structures should be tailored to extend their service life across various applications.

ANALYSIS OF VARIOUS MATERIAL VARIABLES OF 3D PRINTING OF COMPOSITE PLASTICS

Additive manufacturing of thermoplastic composites relies on thermal transitions to achieve bonding of the thermoplastic matrix in the same layer and between adjacent printed layers. A typical example is melting of the thermoplastic matrix in extrusion-based AM processes for realizing the adhesion among filaments and layers. Therefore, thermal properties of the thermoplastic composite, such as melting temperature and glass transition temperature, play a critical role in affecting the AM process and performance of printed parts. In addition, other material variables, such as rheological and optical properties, also affect the printing. The effects of the thermal, rheological, and optical properties on additive manufacturing are described below.

Mechanical Behaviors

The properties of a material or a component can significantly vary with its processing methods because different structures can be resulted from those different processing methods. Typical processing parameters, including temperature, pressure, cooling rate, etc., determine the final structure of manufactured components, and therefore, their mechanical properties. AM processes including FDM differ from conventional processes in various aspects, such as the way a component is produced, the amount of consolidation pressure involved, the cooling rate, etc. AM components are built layer by layer while commonly bulk materials are formed into components in one molding step. Additionally, the consolidation pressure in AM processes is significantly less than that in the conventional molding processes and the flow in the AM processes is very limited. Those distinctly different processing conditions in AM processes result in a composite component that is different from the one processed via conventional processes. Therefore, the material properties and behaviors of the AM thermoplastic composite can be distinctly different.

Tensile Behavior

Tensile test is used to study the relationship between the tensile stress and tensile strain of FDM composites and their failure mechanisms. Elastic modulus (also called tensile modulus), tensile strength, and strain to failure are the basic material properties obtained from the test. There is a significant difference in the tensile properties of printed discontinuous reinforcement thermoplastic composites and printed continuous fiber thermoplastic composites. Overall, the discontinuous reinforcement thermoplastic composite, including discontinuous fiber, particulate, or nano-material reinforced thermoplastic composites, shows relatively low tensile properties. Fig.15 compares the tensile properties among neat thermoplastics and their discontinuous fiber reinforced composites that are printed using the FDM process. Elastic modulus can be considerably improved by adding discontinuous reinforcements. However, the increase of tensile strength is marginal compared to the modulus increase. One main reason for the limited strength increase is the presence of voids introduced from the printing process. Those defects cause high stress concentration and weaken the composite.

Figure 16. (a) Elastic modulus and (b) tensile strength of neat thermoplastics and their composites printed by FDM process (adapted from References (3DXTECH, 2019; Abedin et al., 2023; Agarwala et al., 1995, 1996; Ahmed et al., 2021)

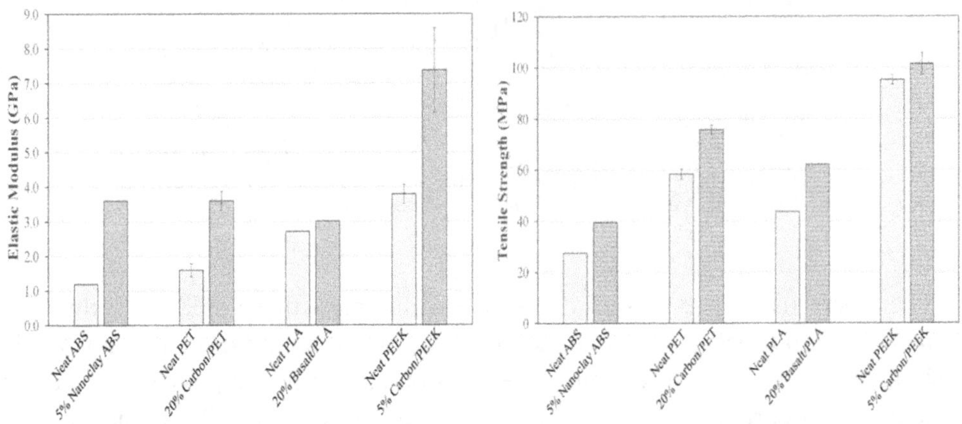

Fiber Matrix

Short fiber reinforced thermoplastic composites printed via the FDM process show various failure modes such as matrix fracture, fiber pullout, and fiber fracture. Those failures modes are also observed in short fiber reinforced composites that are molded using conventional processes, such as injection molding and compression molding. Fig. 16 shows the fracture surface of a printed composite consisting of short carbon fibers and ABS matrix (Al Abadi et al., 2019). The fiber in the composite has an average length of 150 μm and a diameter of about 7 μm. There are obvious fiber pullout, matrix fracture, and fiber fracture. A large number of voids are also present. Those voids are inter-voids, the voids that exist between filaments, and inner-void, the voids that are inside the filament.

Figure 17. Fracture surface showing fiber pullout and matrix fracture as well as the voids in a carbon fiber ABS composite printed by FDM process (①- inner voids; ②- inter voids; ③- cavities from fiber pullout; - fibers that are pulled out) (adapted from Reference (Ahmed et al., 2021; Al Abadi et al., 2019; Alam et al., 2023a).

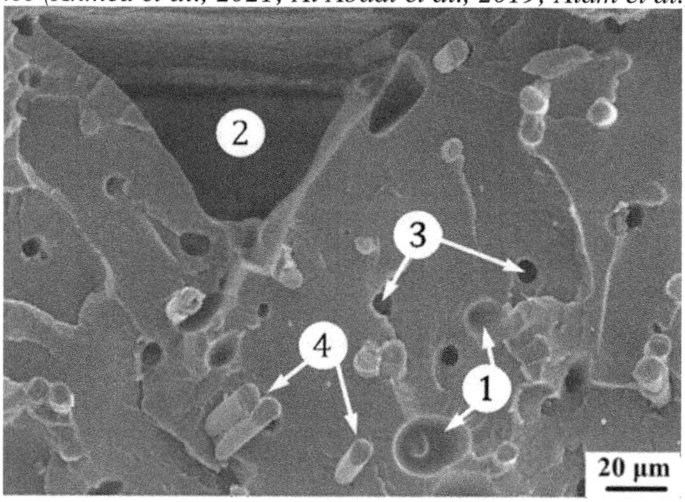

The fibers used in the continuous fiber reinforced thermoplastic composite filament are unidirectional and aligned in the filament length direction. As previously mentioned, the unidirectional fiber reinforced composite possesses superior mechanical properties in the fiber direction (see Section Error! Reference source not found.). Fig. 0.6 shows the tensile properties of some printed continuous fiber thermoplastic composites. The tensile strength and elastic modulus of printed unidirectional carbon fiber reinforced Nylon 6 composite can reach around 800 MPa and 68 GPa, respectively (note: those properties are still lower than the properties of uni-tapes as listed in Error! Reference source not found. because of the limited fiber content in the filament). Therefore, integration of the unidirectional fiber reinforced thermoplastic composite in a printed component can result in significantly improved mechanical properties compared to the one made of only neat thermoplastics or discontinuous fiber reinforced thermoplastic composites.

The properties of the printed unidirectional fiber reinforced thermoplastic composite can vary significantly even when the same type of fiber and matrix are used. The difference in properties is attributed to factors such as percentage of the continuous fiber reinforced composite filament, fiber content in the filament, build orientation, infill percentage, number of walls, layer thickness, printing speed, nozzle temperature, printing chamber temperature, printing bed temperature, and so on. Because of the difference in the abovementioned factors, it is common to see a large variation in the mechanical properties of printed continuous fiber thermoplastic composites even with the same type of fiber and matrix. Nevertheless, the part printed using the continuous fiber reinforced thermoplastic composite can possess properties several times higher than the ones printed using only neat thermoplastics or discontinuous reinforcement thermoplastic composites.

Printed continuous fiber thermoplastic composites normally have a rough surface (Fig. 17), therefore, discontinuous fiber thermoplastic composites are generally printed onto the surface of the continuous fiber composite through dual nozzles. The printing path is designed such that the discontinuous fiber composite surrounds the continuous fiber thermoplastic composite to provide a better surface finish (Fig. 17).

Figure 18. (a) Tensile strength and (b) elastic modulus of printed continuous fiber reinforced thermoplastic composites (adapted from References **(Al Abadi et al., 2018;Alam et al., 2023a,2023b**).

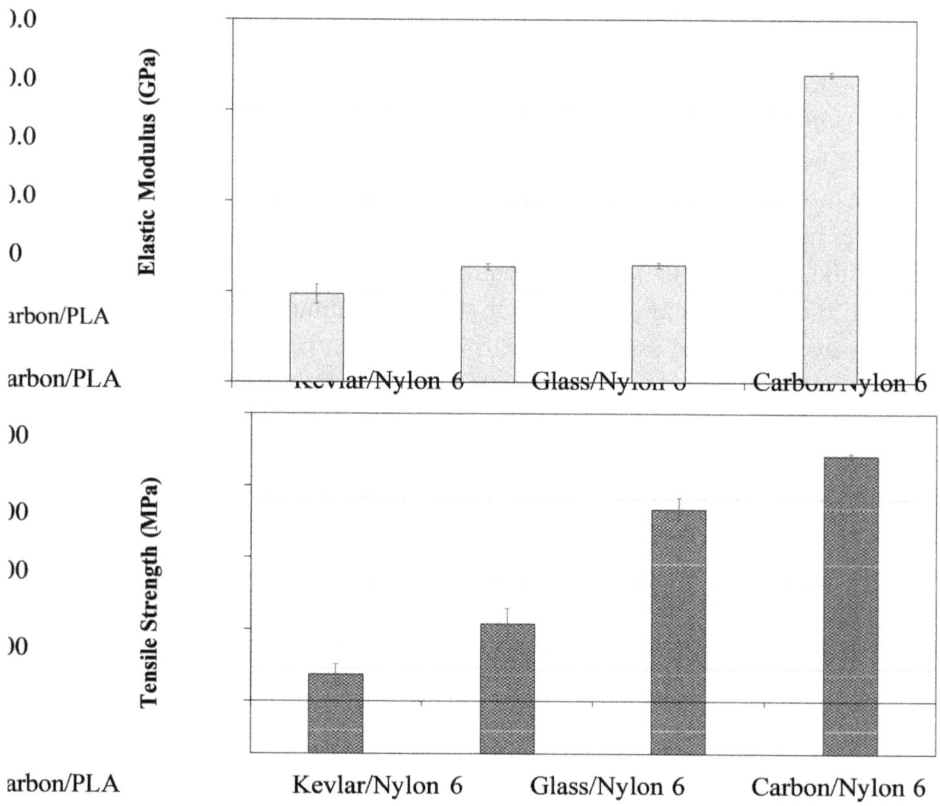

(b)

Impact Behavior

The impact strength of the FDM thermoplastic composite is an indication of its capability of absorbing impact energy. There are several factors that determine its impact properties. Fiber content is one of most important material variables. As the fiber content increases, its impact strength increases. Other processing related variables, such as build orientation, printing speed, layer thickness, nozzle temperature, printing chamber temperature, and printing bed temperature, can also influence the impact performance of the FDM thermoplastic composite.

Fig. 18 compares the impact strength among various thermoplastics and continuous fiber thermoplastic composites printed via FDM. Neat thermoplastics generally have low impact strength. With continuous fibers added, their impact strengths are significantly improved. Glass fiber has more advantage in enhancing the impact strength of thermoplastics than the other fibers, such as carbon fiber and Kevlar® fiber. It is noted that the degradation of Kevlar® fibers during printing can happen when the Kevlar® fiber composite filament goes through the nozzle at a temperature that is normally around 40°C higher than the melting temperature of the thermoplastic matrix. Error! Reference source not found. shows the temperature effect on the tensile strength of Kevlar® fibers. Their tensile strength suffers a drastic decrease when the Kevlar® fiber is exposed to a temperature 250°C or above. In addition, the melt impregnation process that is used for producing Kevlar® fiber thermoplastic composite filament can also cause the degradation of the Kevlar® fiber properties. Therefore, it is necessary to select a thermoplastic matrix with a low melting temperature to combine with Kevlar® fiber to minimize the degradation in its mechanical properties, including impact properties.

Figure 19. Impact strength among neat thermoplastics and continuous fiber thermoplastic composites printed through the fusion deposition modeling process (adapted from Reference (Hashmi, 2015)).

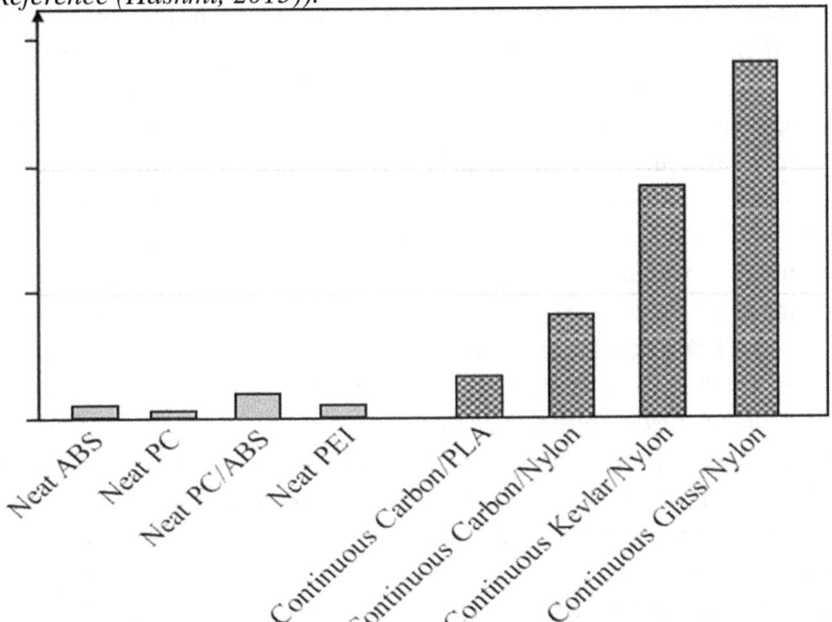

The printed discontinuous and continuous fiber reinforced thermoplastic composites show different failure mechanism during impact. The failure modes are generally matrix fracture and fiber pullout for discontinuous fiber thermoplastic composites as the fiber length is lower than the critical fiber length (see Section Error! Reference source not found.). On the other hand, the printed continuous fiber thermoplastic composite mainly shows fiber fracture, delamination, fiber splitting, and matrix fracture during impact. The delamination occurs between the printed layers while the fiber splitting happens between filaments in the same layer.

Thermal Behaviors

The thermal property of a thermoplastic composite refers to the property that describes its response to heat. The thermal property includes melting temperature, recrystallization temperature, coefficient of thermal expansion (or shrinkage), thermal conductivity, heat capacity, etc. The following sections describe the effects of those major thermal properties on additive manufacturing.

Melting temperature: - It is one of the most important thermal properties that affect the AM process. Most of the AM processes for thermoplastic composites, such as FDM and FPF, require the material to be in a melted state. The melting temperature of the thermoplastic matrix basically determines the temperature for printing the thermoplastic composite, or printing temperature. Printing temperature is referred to as the hot end nozzle temperature at which thermoplastic composites are extruded and deposited for extrusion-based AM processes. The printing temperature results in a thermoplastic composite melt with an adequately low viscosity while still keeps the material from degradation. For the semi-crystalline thermoplastic matrix composite, the printing temperature is normally about 40°C higher than the melting temperature of the thermoplastic matrix. On the other hand, amorphous thermoplastics and their composites do not possess a melting temperature and the printing temperature is often determined to be considerably higher than their glass transition temperature. For example, ABS, an amorphous thermoplastic, has a glass transition temperature of 100°C and its composite is normally printed at a temperature from 220 to 255°C.The printing temperature of the thermoplastic composite influences the geometry stability of the printed part. Thermal shrinkage occurs when the deposited thermoplastic composite is cooled down from the printing temperature. The thermal shrinkage is proportional to the temperature difference ΔT (Eq. (0.2)), that is, the difference between the printing temperature and room temperature. A higher printing temperature decreases the viscosity of the composite, which benefits the printing by enhancing adhesion among deposited layers and reducing the porosities. However, a higher printing temperature can result in larger thermal shrinkage and higher thermal stress in the printed part, which can lead to severe warpage. Therefore, a tradeoff has to be considered between the thermal shrinkage and the printability of a thermoplastic composite when selecting the printing temperature for printing a thermoplastic composite.

Recrystallization temperature: - Recrystallization takes place in the semi-crystalline thermoplastic matrix when its composite is cooled down from the melting state at an adequately low cooling rate. Some disordered polymer chains are transitioned into an ordered status during recrystallization. Because it is an exothermal process, heat is released, which can benefit bonding between deposited layers. However, the disorder- to-order transition induces a relatively large volume change (shrinkage) The shrinkage can cause thermal stress in the composite. For amorphous thermoplastics and their composites, no crystallization occurs and the volume change is from thermal shrinkage only. Recrystallization temperature is also used as a reference to determine the sintering temperature for the selective laser sintering process. Sinter temperature should be above the onset temperature of recrystallization but below the onset temperature of melting of the thermoplastic matrix. When thermoplastic composite particles are heated above the recrystallization temperature of the thermoplastic matrix, diffusion, penetration, and entanglement of the polymer chains in the thermoplastic are enhanced for better adhesion among the particles.

Coefficient of thermal expansion or shrinkage: - (CTE) is sometimes known as coefficient of linear thermal expansion. It defines the change in dimension of a material with temperature (Eq. (1)). When a thermoplastic composite cools from the printing temperature, its dimension decrease and thermal stress is induced. It is another thermal property of the thermoplastic composite that considerably influences the AM process and the quality of the printed part.

$$\Delta L = \alpha L \Delta T \tag{1}$$

where ΔL is change in dimension, i.e. length, width, or thickness, α is the coefficient of thermal expansion or shrinkage (unit: µm/m·K), L is the original dimension, i.e. length, width, or thickness, and ΔT is change in temperature.

The CTE of the thermoplastic matrix can vary at its glass transition temperature and melting temperature (Error! Reference source not found.). The CTE becomes greater when the temperature is raised to glass transition temperature and melting temperature. Addition of reinforcements generally lowers the CTE of thermoplastics, and therefore reduces the stress caused by thermal shrinkage. For example, the addition of short carbon fibers to thermoplastics can effectively reduce their thermal shrinkage and therefore residual stresses. However, continuous fiber thermoplastic composite filaments have significantly different coefficients of thermal expansion or shrinkage in the fiber direction and the transverse direction, which can result in severe geometry instability.

Thermal conductivity: - It is defined as the intrinsic ability of a material to transfer heat. The thermal conductivity of thermoplastics is relatively low. Reinforcements in the thermoplastic composite used for AM have a wide range of thermal conductivity due to the variety of materials used in the reinforcement. Carbon based reinforcements including carbon fibers, graphite, graphene, and nanotubes, have a very high thermal conductivity and therefore can enhance the thermal conductivity of thermoplastic composite feedstock materials and their printed parts. For example, carbon fibers are added to thermoplastics to increase their coefficient of thermal conductivity, which promotes the heating of the filament and enhances the printing rate. Other reinforcements, such as glass fiber and cellulose fiber, have a relatively low thermal conductivity. Some other reinforcements such as glass microspheres have an extremely low thermal conductivity and can lead to a significantly low thermal conductivity of their thermoplastic composites.

The thermoplastic composite with a higher thermal conductivity heats up faster because heat can be conducted from the nozzle to the filament at a faster rate (Eq. (2)). Faster heating benefits the printing by shortening the heat time on the feedstock and improving the printing rate. On the other hand, the composite with a higher thermal conductivity undergoes faster cooling after it leaves the nozzle and makes contact with the printing bed or the previously deposited material.

$$Q = kA\Delta T\, td \tag{2}$$

where Q is heat energy, t is time, k is coefficient of thermal conductivity of the thermoplastic composite (unit: W/m·K), ΔT is change in temperature, A is the contact area, and d is the size of the filament.

6.2.5Specific heat capacity ( ): - 'C' of a material is the amount of heat required to raise one degree or the amount of heat released to reduce one degree in one-gram material. The thermoplastic composite material with a higher specific heat capacity implies that more heat (Q) is required to raise the composite to the same temperature compared to the composite with a less specific heat capacity according to Eq. (3). Therefore, the composite with a high specific heat capacity needs a longer heating time to reach the printing temperature.

$$Q = mc\Delta T \tag{3}$$

where Q is heat energy (unit: joule), m is mass, c is specific heat capacity (unit: joule/g°C), and ΔT is change in temperature.

The heat capacity of the composite also affects cooling of the composite after printing. When the composite has a higher specific heat capacity, there is more heat retained in the composite. As a result, the heat retained in a material with a great heat capacity can induce improved adhesion for the material in the same layer and between adjacent layers.

Printing temperature (Δ) It is one of the simplest but most effective methods to enhance the bonding strength at the interface between adjacent layers. A higher printing temperature reduces the viscosity of thermoplastic composites. Provided that the temperature is below degradation temperature of the thermoplastic matrix. The decrease in viscosity can improve the flow of the composite during printing, which can increase the contact area with the previous layer and therefore enhance the bonding. Higher printing temperature also allows for more material exchange across the interface, including diffusion and penetration of the polymer chains and fiber migration that can provide a bridging effect at the interface, to improve the bonding. A higher printing temperature, however, can have adverse effects on the quality of the printed composites. It can induce greater temperature gradients in the composite, which can cause more residual stress. As a result, severe warpage can occur to the printed composite. In addition, higher printing temperatures can lead to degradation of the thermoplastic matrix and decreased mechanical properties of the printed composite.

Printing Nozzle temperature (Δ n) It is one of the simplest but most effective methods to enhance the bonding strength at the interface between adjacent layers. A higher printing nozzle temperature reduces the viscosity of thermoplastic composites provided that the temperature is below degradation temperature of the thermoplastic matrix. The decrease in viscosity can improve the flow of the composite during printing, which can increase the contact area with the previous layer and therefore enhance the bonding. Higher printing temperature also allows for more material exchange across the interface, including diffusion and penetration of the polymer chains and fiber migration that can provide a bridging effect at the interface, to improve the bonding. A higher nozzle printing temperature, however, can have adverse effects on the quality of the printed composites. It can induce greater temperature gradients in the composite, which can cause more residual stress. As a result, severe warpage can occur to the printed composite. In addition, higher printing temperatures can lead to degradation of the thermoplastic matrix and decreased mechanical properties of the printed composite

Heating Temperature of printing bed(Δ) b: - It is another method to enhance the bonding among printed layers. The heat conducted from the printing bed is able to maintain the previously deposited material at an elevated temperature for an extended time, which enhances its interaction with the newly deposited material, such as diffusion and penetration of the polymer chains between the materials at the interface for better bonding. **Fig. 19** compares the fractured surface of printed short carbon fiber reinforced Nylon 6 composite after tensile testing. **Fig. 19a** shows an SEM image of the fracture surface of a composite printed without a heated bed. Obvious individual filaments and print layers are present. In contrast, the composite printed with a heated printing bed shows better bonding between adjacent layers (**Fig. 19 b**). The filament contour and the layer structure are not obvious anymore. As a result, the bonding strength at the interface of the printed composite is enhanced.

Figure 20. The fracture surface of short carbon fiber reinforced nylon composites printed without a heated printing bed and (b) with a heated printing bed.

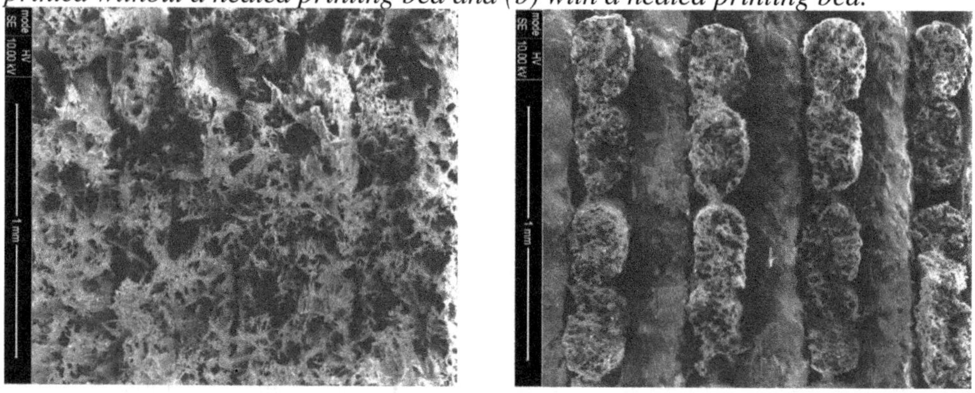

Other Mechanical Property

The other mechanical properties of additively manufactured thermoplastic composites are determined by several major factors such as fiber content, void content, and bonding among print layers. Some of their mechanical properties are shown in Fig. 0.4, Fig. 0.7, Fig. 0.16, and Fig. 0.19. In general, additively manufactured thermoplastic composites possess limited mechanical properties. The major contributing factors are limited fiber content, high void content (see Section 5.8.1 High Void Content), and inadequate bonding among the layers (see Section 5.8.2 Inadequate Bonding).

Fiber content is a major variable that determines the mechanical properties for AM thermoplastic composites. The mechanical properties, such as modulus and strength, increase with fiber content (to a certain extent). Conventional discontinuous and continuous fiber thermoplastic composites have fiber volume percentages up to around 45 vol% and 60 vol%, respectively. However, the fiber content in the thermoplastic composite feedstock (filament, pellets, and powder) for AM processes is still considerably less than those fiber contents. The main reason for that is that a balance between the fiber content and the printability needs to be considered when the amount of fibers is added to the thermoplastic during manufacture of the feedstock material for AM processes. With increasing fiber content, the viscosity of the thermoplastic composite increases (Error! Reference source not found.). The high viscosity can cause issues such as b lockage in the nozzle, poor bonding among printed layers, high void content, etc. For the selective laser sintering process, the zero shear viscosity of the powder feedstock increases with fiber content, which can result in ineffective sintering (Eel (0.1)).

Low Production Rate

Additive manufacturing in general has a low production rate due to its layer-by-layer construction nature. The production rate for printing thermoplastic composites is limited to the material deposition rate in extrusion-based AM processes such as the FDM process or the sintering rate in the SLS process. In comparison to the high production rate in conventional molding processes such as injection molding (seconds per part), the production rate in printing thermoplastic composites is low (generally hours per part dependent on the part volume), which is one of the major hurdles for any of the AM processes to become a mass production process.

The production rate in the extrusion based printing process, such as FDM and FPF, is closely related to printing resolution. Printing resolution, or the resolution of a FDM printer, is defined by its accuracy along three axes, namely, x, y, and z, especially z axis. A smaller thickness of the deposited layer indicates a higher resolution of the printer. The production rate can be increased by increasing flow volume. However, the printing resolution is compromised because of the larger thickness of each deposition layer.

Since the thermoplastic composite melt from the nozzle is a flow process, its volumetric flow rate, q, can be calculated as the product of the cross section of the nozzle, A, and printing speed, v.

$$q = Av$$

For a nozzle aperture with a rounded cross section that has a diameter of d, the volumetric flow rate is expressed in the equation below.

$$\boldsymbol{ql = \pi(d/2)2v}$$

where q_l is the deposition rate at a low resolution.

If the resolution of the printer needs to be improved by *n* times, that is, the diameter of the nozzle has to be reduced to (d/n), the volumetric flow rate is expressed as the equation below assuming the print speed remains the same.

$$qh = \pi[(d/n)/2]2v$$

where q_h is the deposition rate at a high resolution.

The ratio between the volumetric flow rates at different resolutions is defined by **Eq. (4)** below.

$$ql = \pi(d/2)\ 2V * 1$$

$$qh\ \pi(d\backslash n)2\ V\ n\backslash 2$$

(4)

indicates that the deposition rate (printing rate) is inversely proportional to the square of increase in printing resolution. When the resolution is increase by *n*, the deposition rate decreases by n^2. The same relationship stands for nozzle apertures with different cross sections as long as the aperture shape changes with the same proportion. In order to balance the production rate and printing resolution, thermoplastic composites can be deposited at a high flow rate to the areas that do not require a high resolution. For the area that requires a high resolution, for example outer surfaces, the material can be deposited through a smaller nozzle aperture. This approach is able to maintain a good balance between a high production rate and a good printing resolution (Bellini and Bertoldi, 2013).

Residual Stress and Warpage

Residual stress is the stress that remains in manufactured parts, such as AM parts, while there is no externally applied force. It is normally caused by thermal cycles such as heating and cooling during manufacturing and, therefore, it is also called thermal stress or residual thermal stress. Additive manufacturing of thermoplastic composites utilizes heat to achieve adhesion between adjacent layers and the composite undergoes a rapid transition from one state to another, for example, from a solid state to a melted state and back to a solid state within seconds during the fusion deposition modeling process. Residual stress is generally developed when the thermoplastic composite solidifies. The residual stress can cause undesired geometry change such as warpage. The warpage in a printed part can be so severe that it can deboned from the printing bed, which normally results in lifting of the unfinished part and eventually unsuccessful printing.

Residual stress exists in the printed thermoplastic composite at different levels, i.e. constituent level, filament level, print layer level, and part level. The residual stress at the constituent level is mainly caused by the difference in the coefficient of thermal shrinkage between the fiber and the thermoplastic matrix. The CTE between the fiber and the matrix can be significantly different. When the composite filament is heated and extruded through the nozzle, its temperature can generally reach more than 200°C or even 300°C for advanced engineering thermoplastics. When the fiber and the matrix cool down after the composite is deposited, both of them undergo thermal shrinkage. The thermoplastic matrix normally has a significantly higher coefficient of thermal shrinkage than the fiber and therefore much higher volume decrease. In addition, more shrinkage can be induced by the rearrangement of polymer chains into an order structure (or recrystallization) in semi-crystalline thermoplastics. Due to the difference in volume reduction between the fiber and the thermoplastic matrix, residual compressive stress is developed in the fiber and residual tensile stress in the matrix.

The residual stress at the filament level is resulted from the intended shrinkage of a deposited filament and constraints applied by previously printed filaments that are already solidified. When a filament is extruded out of the nozzle and deposited, it makes contact with the previously printed filaments and bonds with those filaments at the contact area. When it cools down, the filament undergoes change from a melted state to a solid state and intends to shrink. However, the shrinkage is constrained at the contact area by the previously printed filaments. Therefore, residual stress is developed in the filament. A higher temperature difference between the deposited filament and the previously printed filaments induces a higher residual stress.

The residual stress at the print layer level is caused by the difference in the shrinkage between adjacent layers. After the top layer is deposited and bonded with the previous layer, the top layer intends to shrink during cooling. Although the previous layer also undergoes thermal shrinkage due to cooling, the amount of the shrinkage of the previous layer is less than that of the top layer due to the temperature difference. Therefore, residual stresses are developed in both layers. The top layer develops tension stress and the previous layer compressive stress. The residual stress can also be developed at the part level. There exist temperature gradients and difference in cooling rate not only through the thickness of the part but also from its center to edge. The bottom layer generally has the lowest temperature and the top layer has the highest temperature. The center of the part has a higher temperature than the edge, or the outer section, of the part. The cooling rate can also vary at different sections of the part due to difference in contact condition, air circulation, etc. Both the temperature gradient and difference in cooling rate induce residual stresses at the part level. The residual stress at the part level can be minimized by reducing the temperature gradient in the printed part through approaches such as reducing printing temperature, heating printing bed, heating print chambers, etc. The same approaches can be used to reduce the residual stress at the other levels as well as enhance the bonding between adjacent layers and improve the fracture toughness of the printed part. Continuous fiber thermoplastic composite filaments have significantly pronounced material anisotropy. It is essentially a unidirectional fiber thermoplastic composite that has all the fibers in the filament direction. The fiber direction has minimal shrinkage because of the low coefficient of thermal shrinkage of the fibers in that direction. On the other hand, there is significant shrinkage in the transverse direction. When one layer is printed using the unidirectional fiber thermoplastic composite filament, the shrinkage in the transverse direction is constrained by the printing bed (if it is the first layer) or the previous layer and residual stress is developed. The residual stress accumulated from printing multiple layers in the same printing direction can overcome the bonding between the part and the printing bed. It can cause debonding between the part and the printing bed and eventually severe bending of the unidirectional fiber composite. When debonding happens, printing has to be stopped and unsuccessful printing is resulted. **Fig. 20a** shows a continuous Kevlar® fiber reinforced nylon composite (fibers aligned in the width direction) debonding from the printing bed during printing (top) and the composite after being removed from the printing bed (bottom). **Fig. 20 b** shows another continuous Kevlar® fiber reinforced nylon composite (fibers aligned in the length direction) debonding from the printing bed during printing (top) and the composite after being removed from the printing bed (bottom). The wall material is a short carbon fiber nylon composite (in black color) that provides a good surface finish to the printed continuous fiber composite.

Figure 21. (a) A continuous Kevlar® fiber nylon composite with all fibers in its width direction debonding from the printing bed (top) and the composite after being removed from the printing bed (bottom); (b) a continuous Kevlar® fiber nylon composite with all fibers in its length direction debonding from the printing bed (top) and the composite after being removed from the printing bed (bottom).

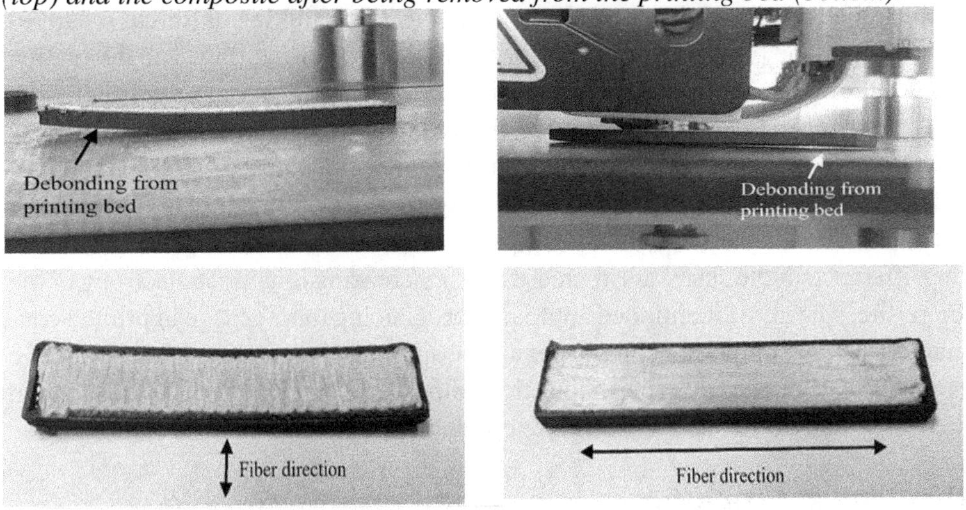

When printing continuous fiber thermoplastic composites, it is necessary to reduce the shrinkage and warpage by properly designing the printing direction (the fiber direction) for each layer to avoid unsuccessful printing. Although it is not possible to eliminate the shrinkage at a specific layer caused by the high material anisotropy, the shrinkage of the layer can be constrained by printing the next layer with an orientation perpendicular to that layer. For example, the layers can be printed in alternate 0 and 90° directions, identical to the cross-ply layup in conventionally processed thermoplastic composites (see Section Error! Reference source not found.). The shrinkage of the 0° fiber layer in the 90° direction is constrained by the 90° fiber layer, and vice versa.

Discontinuous fiber thermoplastic composites tend to have preferred fiber orientation resulted from processing when material flow is involved. Those processes include injection molding, compression molding, extrusion, etc. AM processes such as FDM and FPF are extrusion based processes and deposition of a thermoplastic composite melt through the nozzle is essentially an extrusion process. Flow of the thermoplastic composite melt through the printer nozzle induces fibers oriented in the printing direction. In spite of small fiber aspect ratios, the oriented fibers can result in certain material anisotropy in the printed composite. Fibers can also develop an orientation in powder-bed based AM processes. The feedstock material for selective laser sintering or selective laser melting processes initially has random fiber orientation. There is limited flow involved in the selective laser melting process even for the feedstock that is heated above melting temperature. However, the sintered thermoplastic composite part may still develop oriented fibers from the raking or rolling motion that supplies the feedstock during printing. After sintering is completed on one layer, the part is lowered down and the raking or rolling is carried out to deposit the powder on the part for laser sintering or melting. The interaction between the rake/roller and the powder can align fibers in the raking/rolling direction.

The developed fiber orientation in the discontinuous fiber thermoplastic composite can induce material anisotropy in each layer, which can cause different amount of shrinkage in different directions. The shrinkage in the printing direction is the least while the direction perpendicular to the printing direction has the highest shrinkage. The difference in the shrinkage increases with increasing material anisotropy of the composite. The abovementioned approach for reducing the warpage in printed continuous fiber thermoplastic composites can be applicable to minimizing the warpage for the discontinuous fiber thermoplastic composite by printing the composite with alternate 0 and 90° directions, or more commonly, alternate 45° and −45°.

Rheological Properties

In the AM processes, the matrix in the thermoplastic composite feedstock normally goes through a liquid state (except for the selective laser sintering process in which materials fuse through diffusion) and material flow or deformation is involved during printing. The properties that determine how the thermoplastic composite melt flows or deforms under an applied stress are called rheological properties. The rheological property of a thermoplastic composite material at the liquid state plays a critical role in its printing and mechanical properties. Melt viscosity is one of the most important rheological properties in affecting the printing of thermoplastic composites. It determines the printability of a thermoplastic composite, the bonding strength between adjacent layers, the void content in the printed composite, etc. A common issue during printing thermoplastic composites is their high viscosity due to the high viscosity of the thermoplastic matrix even when it is melted. When reinforcements, such as fibers, Nano-materials, and particulates, are added, the viscosity can be further increased A high viscosity may cause nozzle blockage, poor bonding between the adjacent layers, increased intra-void, etc. Different approaches have been effectively used to reduce the viscosity of thermoplastic composite melts. Those approaches include increasing nozzle temperature, heating deposited layers, heating printing chamber, etc.

Optical Properties

Optical properties are important in printing processes that involve lights, for example, the selective laser sintering or melting process that a light source, such as laser, is used for heating. When the photons in the laser interact with the thermoplastic composite particles, reflection, refraction, absorption, and transmission can occur. It is desired to have the laser sintering feedstock with the least transmission and reflection but highest absorption. The addition of highly absorbing reinforcements, such as carbon fibers and carbon black, in the thermoplastic matrix can significantly enhance the absorption of the laser energy for more efficient heating and enhanced additive manufacturing. Compared to printed neat thermoplastics, thermoplastic composites show higher interlinear shear strength. For example, short carbon fiber reinforced composites have higher shear strength than printed neat nylon (**Fig. 0.16**). Typically, an unreinforced thermoplastic has less viscosity than its composites which enhances bonding b between adjacent layers. However, the fibers in the composite can strengthen the interface by providing a bridging effect at the interface.

Figure 22. Interlinear shear strength of printed thermoplastic composites & compression molded Plastic composites (partially adapted from Reference (Bagaria et al., 2018)).

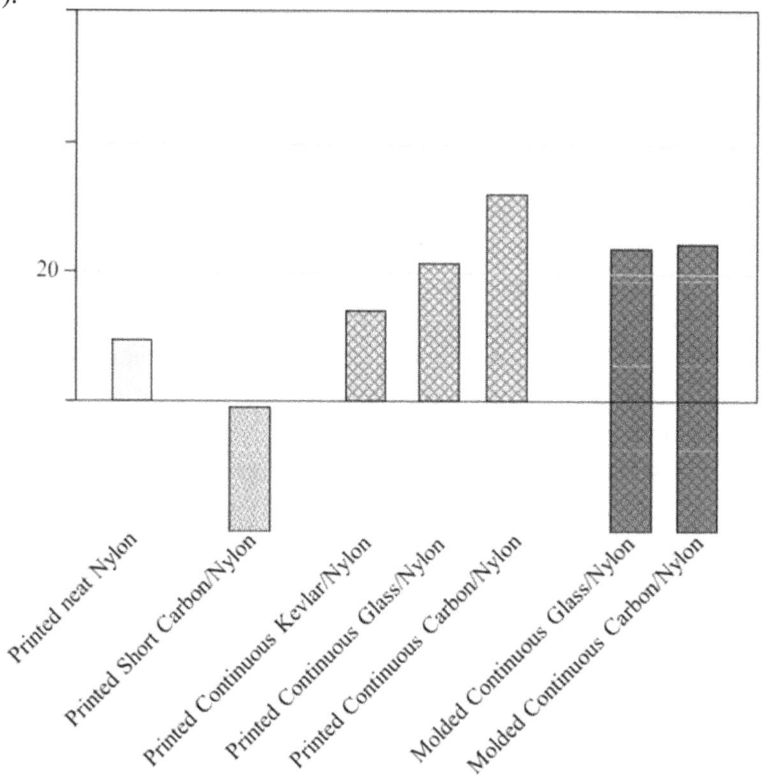

Figure 23. (a) Continuous carbon fiber reinforced Nylon 6 composite and (b) continuous glass fiber reinforced Nylon 6 composite printed through FDM showing delamination resulted from short beam testing (reprinted from (Bagaria et al., 2018)).

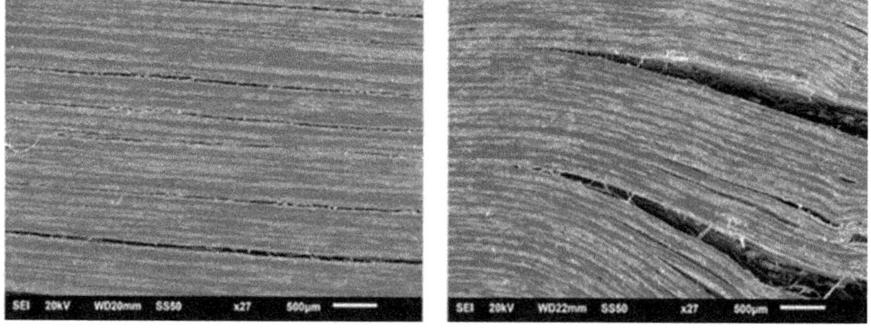

PRESENT CHALLENGES AND FUTURE SCOPES

3D printing has emerged as a transformative force in materials science research and manufacturing. Its unique capability to fabricate complex objects with customizable material properties has opened new horizons in various industries. However, despite the remarkable progress made in this field, several significant challenges still need to be addressed to fully harness the potential of 3D printable composite materials.

One of the primary challenges lies in the compatibility of 3D printers with diverse materials. Many commercial 3D printers are designed for specific resins and may encounter issues when working with modified resins. These issues can manifest as agglomerate formation, the formation of heterogeneous composites, printer head blockages, non-adhesion, and increased curing times. These challenges require careful optimization to ensure the successful 3D printing of composite materials. Furthermore, the resolution of 3D printers remains a critical concern. The quality and precision of the printed objects are influenced by the printer's resolution, making it crucial to address this issue to meet performance requirements. Achieving high-resolution 3D printing using composite materials adds a layer of complexity.

Figure 24. (a) a top view, (b) a cross-sectional view, (c) a printed layer showing fiber breakage at the curvature, and (d-f) magnified cross-sections. Panels (a) and (b) are reproduced from Juan A, Lozada N, Ahuett-garza H, Verbeeten WMH, and Daniel S (2018). Tensile properties and failure behavior of chopped and continuous carbon fiber composites produced by Additive Manufacturing. Additive Manufacturing https://doi.org/10.1016/j.addma.2018.12.020, while (c) and (d) are adapted from Chabaud G, Castro M, Denoual C, and Le Duigou A (2019). Hygromechanical properties of 3D printed continuous carbon and glass fibre reinforced polyamide composite for outdoor structural applications. Additive Manufacturing, 26(November 2018): 94–105. https://doi.org/10.1016/j.addma.2019. 01.005. Panels (e) and (f) are taken from van de Werken N, Hurley J, Khanbolouki P, Sarvestani AN, Tamijani AY, and Tehrani M (2019). Design considerations and modeling of fiber reinforced 3D printed parts. Composites Part B: Engineering, 160: 684–692. https://doi.org/ 10.1016/j.compositesb.2018.12.094.

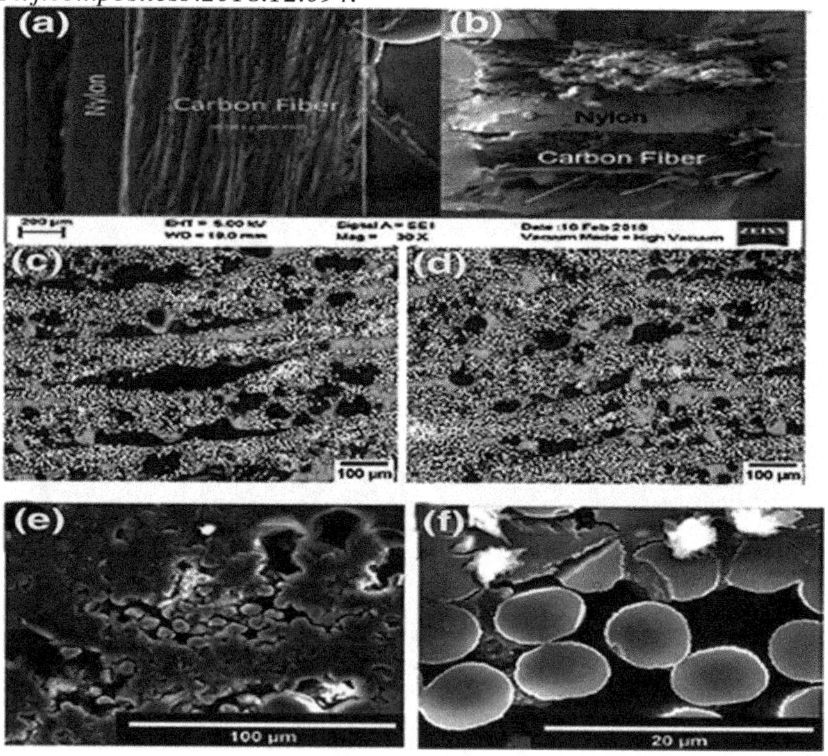

In terms of materials, the versatility of 3D printing is a double-edged sword. While it offers the possibility of using a wide range of materials, including biocompatible ones, for applications such as bioprinting, developing novel materials is an ongoing challenge. Many 3D printers and processes are limited to a single material, limiting their industrial applicability. To fully realize the potential of 3D printing, developing new materials with tunable mechanical, chemical, and physical properties is imperative.

Looking ahead, several directions can shape the future of AM processes. First, hybrid AM processes, which combine additive and subtractive technologies, is gaining traction. These hybrid processes can enhance dimensional accuracy, improve surface finish, and accommodate multi-material printing within the same layer. The demand for higher dimensional accuracy and the elimination of separate post-processing steps drive the development of hybrid AM systems. Second, there is a pressing need for significant breakthroughs in 3D printing materials and processes. Achieving higher dimensional accuracy, especially at smaller scales, and faster printing speeds are critical objectives. Innovations in materials science are essential to develop novel materials that meet the stringent requirements of various applications. Additionally, direct digital manufacturing (DDM) is emerging as a powerful application of AM processes. DDM involves using AM to produce end-use parts and products rather than just prototypes. This approach is particularly attractive for producing low-volume or unique products using 3D printing, opening up new possibilities in customization and small-batch manufacturing.

While these developments hold great promise, challenges persist in applying 3D printing to specific industries. In the aerospace sector, for example, limited resources, high costs, and concerns about the inconsistent quality of 3D-printed components pose barriers to wider adoption. These issues must be addressed to realize the full potential of 3D printing in this field. In parallel, achieving a high-quality surface finish for 3D printed products, regardless of scale, is critical. Surface quality directly impacts the performance and aesthetics of printed objects. While progress has been made in improving surface finish through preprocessing and post-processing techniques, there is a need to explore nanoscale surface finish to achieve superior qualities in 3D printed components. To move forward, future research should focus on making 3D printing techniques more efficient and user-friendly while reducing costs. The influence of various process parameters on the mechanical characteristics of manufactured components should be thoroughly evaluated to optimize performance. By addressing these challenges and exploring new frontiers in materials and processes, 3D printing can become a versatile and cost-effective manufacturing method applicable to a wide range of industries.

Discussion and Summary

This study discussed the AM techniques, modeling, Analysis and simulation of additively manufactured composites its effects on properties of Plastics. The FDM is the most commonly used additive composite manufacturing technique compared to SLA, SLS, and PBF. In addition to making intricate structures, AM offers cost-effective, precise, and efficient solutions to manufacture composites. The use of numerical simulation in the analysis of printed components involves examining micro-, macro-, and meso-scales modeling, hence providing significant insights into the intricate realm of 3D printing. Various observations pointed out are as summarized below;

- Additive manufacturing processes for thermoplastic composites include fusion deposition modeling, fused pellet fabrication, selective laser sintering, laminated object manufacturing, etc. The fusion deposition modeling process is dominantly used to print thermoplastics and thermoplastic composites. • Continuous fiber thermoplastic composites possess excellent mechanical properties but low printability. Limited consolidation pressure, short consolidation time, and high viscosity of thermoplastic composite melts lead to the difficulty in printing the continuous fiber thermoplastic composite. • Discontinues fiber thermoplastic composite filaments are produced through extrusion process. Continues fiber thermoplastic composite filaments are produced through melt impregnation process and have fully impregnated fiber. • Selective laser sintering is a powder-bed based AM method. It uses laser as the energy source to sinter thermoplastic composite powders layer by layer into a part. Sintering occurs between onset temperatures of melting and recrystallization of the thermoplastic matrix and does not involve any liquidation of the material. • Several major processing-related challenges exist in additive manufacturing processes. It includes high void content, inadequate bonding, limited mechanical properties, low production rate, residual stress & warpage. • Voids in fusion deposition modeling are categorized into inner-voids and inter-voids. The inner-void is originated from the porosities inside each feedstock filament while the inter- void comes from the gaps between filaments during printing. Both types of voids, especially the inter-void, contribute to the high void content in the printed thermoplastic composite. • There exist many interfaces due to the layered structure of AM thermoplastic composites. Inadequate bonding can be resulted at the interface. Increasing printing temperature, heating printing bed, heating printing chamber, locally heating previously deposited material, and applying low chamber pressure can be used to improve the bonding. • The production rate in AM is not comparable to the rate achieved in conventional processing methods. The low production rate of AM is considered the largest hurdle for it to become a mass production process. There is a tradeoff between printing rate and printing resolution. The printing rate is inversely proportional to the square of increase in printing resolution. • Continuous fiber thermoplastic composite filaments have significantly pronounced material anisotropy. It is necessary to reduce the shrinkage by properly designing the printing direction (the fiber direction) for each layer to avoid unsuccessful printing. • Oriented fibers in the printing direction caused by the directional flow in the nozzle induce material anisotropy in discontinuous fiber thermoplastic composites. The shrinkage in the printing direction is less than the other directions while the direction perpendicular to the printing direction has the most shrinkage. The difference in the shrinkage increases with increasing material anisotropy of the composite. • Printing thermoplastic composites in alternate printing directions that are perpendicular to each other, for example, 0 and 90°, or 45° and −45°, can minimize the warpage of the printed part. • In materials and processes, 3D printing can become a versatile and cost-effective manufacturing method applicable to a wide range of industries.

CONCLUSIONS

This study discussed the AM techniques, modeling methods, property analysis and simulation of additively manufactured composites. The FDM is the most commonly used additive composite manufacturing technique compared to SLA, SLS, and PBF. In addition to making intricate structures, AM offers cost-effective, precise, and efficient solutions to manufacture composites. The use of numerical simulation in the analysis of printed components involves examining micro-, macro-, and meno-scales modeling, hence providing significant insights into the intricate realm of 3D printing. Factors such as fiber content, interfacial bonding, loads and environmental conditions contribute significantly to the printed composite materials' overall properties and performance. However, to unlock the full potential of the 3D printing technology, challenges related to materials compatibility, resolution, and materials development must be overcome. Effect of varying material properties viz. mechanical, Thermal, Optical occurs due to variation in printing process parameters. This affects the property of finished plastic product. The future of 3D printing lies in hybrid AM techniques, breakthroughs in materials and novel processes, and development in the application. Addressing industry-specific challenges and achieving high-quality product at optimum cost are critical factors for broader adoption of 3D Printing. Through ongoing research and innovation, 3D printing is poised to revolutionize various industries and offer new possibilities in manufacturing and new materials science to evolve as best upcoming process of producing variant products of different types of composite plastics.

REFERENCES

Abedin, M. J., Rahman, M. Z., Rahman, S., Sarker, M., Saha, B., Sazal, Y. I., & Salsabil, Z. (2023). Recent advances in applications of hybrid composites. In *Reference Module in Materials Science and Materials Engineering*. Elsevier., DOI: 10.1016/B978-0-323-96020-5.00134-5

Agarwala, M., Bourell, D., Beaman, J., Marcus, H., & Barlow, J. (1995). Direct selective laser sintering of metals. *Rapid Prototyping Journal, 1*(1), 26–36. DOI: 10.1108/13552549510078113

Agarwala, M. K., Jamalabad, V. R., Langrana, N. A., Safari, A., Whalen, P. J., & Danforth, S. C. (1996). Structural quality of parts processed by fused deposition. *Rapid Prototyping Journal, 2*(4), 4–19. DOI: 10.1108/13552549610732034

Ahmed, A., Zillur Rahman, M., Ou, Y., Liu, S., Mobasher, B., Guo, S., & Zhu, D. (2021). A review on the tensile behavior of fiber-reinforced polymer composites under varying strain rates and temperatures. *Construction & Building Materials, 294*, 123565. DOI: 10.1016/j.conbuildmat.2021.123565

Akhoundi, B., Behravesh, A. H., & Bagheri Saed, A. (2019). Improving mechanical properties of continuous fiber-reinforced thermoplastic composites produced by FDM 3D printer. *Journal of Reinforced Plastics and Composites, 38*(3), 99–116. DOI: 10.1177/0731684418807300

Al Abadi H, Thai HT, Paton-Cole V, and Patel VI (2018) Elastic properties of 3D printed fibre-reinforced structures. Composite Structures 193(March): 8–18. https://doi.org/. compstruct.2018.03.051.DOI: 10.1016/j

Alam, M. J., Abedin, M. J., & Rahman, M. Z. (2023a). Nanoparticles, nanocomposites, green/eco-composites, and hybrid composites and their applications in energy sectors. In *Reference Module in Materials Science and Materials Engineering*. Elsevier., DOI: 10.1016/B978-0-323-96020-5.00121-7

. Alam MR, Dev B, Rahman MA, Islam MR, Mahmud RU, and Rahman MZ (2023b) Advances in bioactive nanoparticles for wound healing, tissue engineering and drug delivery.

Anitha, R., Arunachalam, S., & Radhakrishnan, P. (2001). Critical parameters influencing the quality of prototypes in fused deposition modelling. *Journal of Materials Processing Technology, 118*(1-3), 2–5. DOI: 10.1016/S0924-0136(01)00980-3

ASTM Standards. (2012). *Standard Terminology for Additive Manufacturing Technologies* (Vol. 10).

Azad, H. K. M., & Rahman, M. Z. (2023a). A critical review of developments in the characterization of metal matrix composites with particulate reinforcements. In *Reference Module in Materials Science and Materials Engineering*. Elsevier.

Azad, H. K. M., & Rahman, M. Z. (2023b). Ceramic matrix composites with particulate reinforcements—Progress over the past 15 years. In *Reference Module in Materials Science and Materials Engineering*. Elsevier.

Bagaria, V., Bhansali, R., & Pawar, P. (2018). 3D printing- creating a blueprint for the future of orthopedics: Current concept review and the road ahead! *Journal of Clinical Orthopaedics and Trauma*, *9*(3), 207–212. DOI: 10.1016/j.jcot.2018.07.007 PMID: 30202150

Bellini A and Bertoldi M (2013) Liquefier dynamics in fused deposition. Journal of Manufacturing Science and Engineering 126(May 2004): 237–246. https://doi .org/. 1688377.DOI: 10.1115/1

Bertrand, P., Bayle, F., Combe, C., Goeuriot, P., & Smurov, I. (2007). Ceramic components manufacturing by selective laser sintering. *Applied Surface Science*, *254*(4), 989–992. DOI: 10.1016/j.apsusc.2007.08.085

Bettini, P., Alitta, G., Sala, G., & Di Landro, L. (2017). Fused deposition technique for continuous fiber reinforced thermoplastic. *Journal of Materials Engineering and Performance*, *26*(2), 843–848. DOI: 10.1007/s11665-016-2459-8

Bikas, H., Stavropoulos, P., & Chryssolouris, G. (2016). Additive manufacturing methods and modeling approaches: A critical review. *International Journal of Advanced Manufacturing Technology*, *83*(1–4), 389–405. DOI: 10.1007/s00170-015-7576-2

Caminero, M. A., Chacón, J. M., García-Moreno, I., & Rodríguez, G. P. (2018). Impact damage resistance of 3D printed continuous fibre reinforced thermoplastic composites using fused deposition modelling. *Composites. Part B, Engineering*, *148*, 93–103. DOI: 10.1016/j.compositesb.2018.04.054

Carneiro, O. S., Silva, A. F., & Gomes, R. (2015). Fused deposition modeling with polypropylene. *Materials & Design*, *83*, 768–776. DOI: 10.1016/j.matdes.2015.06.053

Chatterjee, A. N., Kumar, S., Saha, P., Mishra, P. K., & Roy Choudhury, A. (2003). An experimental design approach to selective laser sintering of low carbon steel. *Journal of Materials Processing Technology*, *136*(1–3), 151–157. DOI: 10.1016/S0924-0136(03)00132-8

Chen, C., & Sullivan, P. A. (1996). Predicting total build-time and the resultant cure depth of the 3D stereolithography process. *Rapid Prototyping Journal, 2*(4), 27–40. DOI: 10.1108/13552549610153389

Chen, T., & Zhang, Y. (2004). Numerical simulation of two-dimensional melting and resolidification of a two-component metal powder layer in selective laser sintering process. *Numerical Heat Transfer Part A, 46*(7), 633–649. DOI: 10.1080/104077890504177

Chen, T., & Zhang, Y. (2006). A partial shrinkage model for selective laser sintering of a two-component metal powder layer. *International Journal of Heat and Mass Transfer, 49*(7-8), 1489–1492. DOI: 10.1016/j.ijheatmasstransfer.2005.09.027

Childs, T. H. C., Berzins, M., Ryder, G. R., & Tontowi, A. (1999). Selective laser sintering of an amorphous polymer—Simulations and experiments. *Proceedings of the Institution of Mechanical Engineers. Part B, Journal of Engineering Manufacture, 213*(4), 333–349. DOI: 10.1243/0954405991516822

Chohan, J. S., Singh, R., Boparai, K. S., Penna, R., & Fraternali, F. (2017). Dimensional accuracy analysis of coupled fused deposition modeling and vapour smoothing operations for biomedical applications. *Composites. Part B, Engineering, 117*, 138–149. DOI: 10.1016/j.compositesb.2017.02.045

Choi, B. W., & Leu, M. C. (2000). Determining optimal parameters for stereolithography processes via genetic algorithm. *Journal of Manufacturing Systems, 19*(I), 18–27. DOI: 10.1016/S0278-6125(00)88887-1

Costa, L., Reti, T., Deus, M. A., & Vilar, R. (2002) Simulation of layer overlap tempering kinetics in steel parts deposited by laser cladding. In: *Proceedings of 2002 International Conference on Metal Powder Deposition for Rapid Manufacturing*, May 2014, pp. 172–179.

Crespo, A., Deus, A., & Vilar, R. (2006) Finite element analysis of laser powder deposition of titanium. In: ICALEO 2006 - 25th International Congress on Applications of Laser and Electro-Optics, Congress Proceedings, January. DOI: 10.2351/1.5060805

Crockett, R. S., & Calvert, P. D. (1996) The liquid-to-solid transition in stereodeposition techniques. In: Solid freeform fabrication proceedings, pp. 257–264.

Dickson AN and Dowling DP (2019) Enhancing the bearing strength of woven carbon fibre thermoplastic composites through additive manufacturing. Composite Structures 212(December 2018): 381–388. .DOI: 10.1016/j.compstruct.2019.01.050

Digimat. (2019) The Multi-scale Nonlinear Material Modeling Platform. Part of Hexagon AB: E-Xstream Engineering an MSC Software Company.

Dong, L., Makradi, A., Ahzi, S., & Remond, Y. (2008). Three-dimensional transient finite element analysis of the selective laser sintering process. *Journal of Materials Processing Technology*, *9*(2), 700–706. DOI: 10.1016/j.jmatprotec.2008.02.040

Dou, H., Cheng, Y., Ye, W., Zhang, D., Li, J., Miao, Z., & Rudykh, S. (2020). Effect of process parameters on tensile mechanical properties of 3D printing continuous carbon fiber-reinforced PLA composites. *Materials (Basel)*, *13*(17), 3850. DOI: 10.3390/ma13173850 PMID: 32878337

Dutta, S., Chanda, A., Rahman, M. Z., Das, R., & Bhattacharyya, D. (2019) A numerical model to simulate the impact response of Flax-PP composites. In: *ICCM International Conference on Composite Materials*.

Faisal, M. F., Rahman, S., Rahman, M. Z., Rahman, M. A., Tasnim, J., & Sazal, Y. I. (2023). Modeling approaches for bio-based composites. In *Reference Module in Materials Science and Materials Engineering*. Elsevier., DOI: 10.1016/B978-0-323-96020-5.00181-3

Fakirov, S., Rahman, M. Z., Pötschke, P., & Bhattacharyya, D. (2014). Single polymer composites of poly (B utylene Terephthalate) microfibrils loaded with carbon nanotubes exhibiting electrical conductivity and improved mechanical properties. *Macromolecular Materials and Engineering*, *299*(7), 799–806. DOI: 10.1002/mame.201300322

Fan, Z., & Liou, F. (2012) Numerical modeling of the additive manufacturing (am) processes of titanium alloy. In: Titanium Alloys—Towards Achieving Enhanced Properties for Diversified Applications, pp. 3–29. DOI: 10.5772/34848

Feng, G., Ou, Y., Rahman, M. Z., Zhou, L., Zhao, H., Chen, Q., Mao, D., & Zhu, D. (2023). Improving corrosion resistance of BFRP bars by coating CNTs modified resin in simulated pore solution of seawater sea sand concrete. *Construction & Building Materials*, *392*, 131945. DOI: 10.1016/j.conbuildmat.2023.131945

Gao, X., Fan, S., Pang, J., Rahman, M. Z., Zhu, D., Guo, S., Ma, M., & Li, Z. (2021). Preparation of nano-xylan and its influences on the anti-fungi performance of straw fiber/HDPE composite. *Industrial Crops and Products*, *171*, 113954. DOI: 10.1016/j.indcrop.2021.113954

Gao, X., Zhu, D., Fan, S., Rahman, M. Z., Guo, S., & Chen, F. (2022). Structural and mechanical properties of bamboo fiber bundle and fiber/bundle reinforced composites: A review. *Journal of Materials Research and Technology*, *19*, 1162–1190. DOI: 10.1016/j.jmrt.2022.05.077

Gartner AM report (2014). Information on http://www.gartner.com/ document/2598122

Gokulkumar, S., Thyla, P. R., Prabhu, L., & Sathish, S. (2020). Measuring methods of acoustic properties and influence of physical parameters on natural fibers: A review. *Journal of Natural Fibers*, *17*(12), 1719–1738. DOI: 10.1080/15440478.2019.1598913

Goodridge, R. D., Shofner, M. L., Hague, R. J. M., McClelland, M., Schlea, M. R., Johnson, R. B., & Tuck, C. J. (2011). Processing of a polyamide-12/carbon nanofibre composite by laser sintering. *Polymer Testing*, *30*(1), 94–100. DOI: 10.1016/j.polymertesting.2010.10.011

Gray, R. W.IV, Baird, D. G., & Bohn, J. H. (1998). Effects of processing conditions on short TLCP fiber reinforced FDM parts. *Rapid Prototyping Journal*, *4*(1), 14–25. DOI: 10.1108/13552549810197514

Gu, D. D., Meiners, W., Wissenbach, K., Poprawe, R., Gu, D. D., Meiners, W., Wissenbach, K., & Poprawe, R. (2013). Laser additive manufacturing of metallic components: Materials, processes and mechanisms. *International Materials Reviews*, *57*(3), 133–164. DOI: 10.1179/1743280411Y.0000000014

Hambach M and Volkmer D (2017) Properties of 3D-printed fiber-reinforced Portland cement paste. Cement and Concrete Composites 79: 62–70. https://doi.org/.cemconcomp.2017.02.001.DOI: 10.1016/j

Hao, L., Savalani, M. M., Zhang, Y., Tanner, K. E., & Harris, R. A. (2006). Selective laser sintering of hydroxyapatite reinforced polyethylene composites for bioactive implants and tissue scaffold development. *Proceedings of the Institution of Mechanical Engineers. Part H, Journal of Engineering in Medicine*, *220*(4), 521–531. DOI: 10.1243/09544119JEIM67 PMID: 16808068

Hasan, Z., Rahman, M. Z., Saha, B., & Ador, S. H. (2023). Composite materials from synthetic and natural sources: Fabrication techniques and applications. In *Reference Module in Materials Science and Materials Engineering*. Elsevier., DOI: 10.1016/B978-0-323-96020-5.00171-0

Hashin, Z., & Rotem, A. (1973). A fatigue failure criterion for fiber reinforced materials. *Journal of Composite Materials*, *7*(4), 448–464. DOI: 10.1177/002199837300700404

Hashmi, S. (Ed.). (2015). *Reference module in materials science and materials engineering*. Elsevier BV., DOI: 10.1016/B978-0-323-96020-5.00120-5

Hofmann, M. (2014). 3D printing gets a boost and opportunities with polymer materials. *ACS Macro Letters*, *3*(4), 382–386. DOI: 10.1021/mz4006556 PMID: 35590734

Hosen, M. S. S., Abedin, M. J., Hoque, M. E., & Rahman, M. Z. (2022a) Influence of moisture absorption on physico-chemical properties of natural fiber-based hybrid composites. In: Aging Effects on Natural Fiber-Reinforced Polymer Composites: Durability and Life Prediction, pp. 297–316. Springer. DOI: 10.1007/978-981-16-8360-2_16

Hosen, M. S. S., Hoque, M. E., Rahman, M. Z., & Sagadevan, S. (2022b) Aging effects on mechanical properties of biocomposites with recycled polymers. In: Aging Effects on Natural Fiber-Reinforced Polymer Composites: Durability and Life Prediction, pp. 317–333. Springer. DOI: 10.1007/978-981-16-8360-2_17

Hou, Z., Tian, X., Zheng, Z., Zhang, J., Zhe, L., Li, D., Malakhov, A. V., & Polilov, A. N. (2020). A constitutive model for 3D printed continuous fiber reinforced composite structures with variable fiber content. *Composites. Part B, Engineering*, *189*, 107893. DOI: 10.1016/j.compositesb.2020.107893

Islam, A., & Rahman, M. Z. (2023) A review on eco-friendly basalt fibers, their composites and applications. In: Reference Module in Materials Science and Materials Engineering.

Jacobs PF (1992) Fundamentals of Stereolithography. July.

Jandyal A, Chaturvedi I, Wazir I, Raina A, and Ul Haq MI (2022) 3D printing—A review of processes, materials and applications in industry 4.0. Sustainable Operations and Computers 3(September 2021): 33–42. .DOI: 10.1016/j.susoc.2021.09.004

Kabir, S. M. F., Mathur, K., & Seyam, A. F. M. (2020). A critical review on 3D printed continuous fiber-reinforced composites: History, mechanism, materials and properties. *Composite Structures*, *232*, 111476. DOI: 10.1016/j.compstruct.2019.111476

. Kalsoom U, Nesterenko P, and Paull B (2016) Recent developments in 3D printable composite materials. RSC Advances 6. . https://doi.org/.DOI: 10.1039/C6RA11334F

Karalekas, D., & Aggelopoulos, A. (2003). Study of shrinkage strains in a stereolithography cured acrylic photopolymer resin. *Journal of Materials Processing Technology*, *136*(1-3), 146–150. DOI: 10.1016/S0924-0136(03)00028-1

Karalekas, D. E. (2003). Study of the mechanical properties of nonwoven fibre mat reinforced photopolymers used in rapid prototyping. *Materials & Design, 24*(8), 665–670. DOI: 10.1016/S0261-3069(03)00153-5

Kechagias, J. D., & Chryssolouris, G. (1997) Estimation of build times in rapid prototyping processes. In: Proceedings of the 6th European Conference on Rapid Prototyping and Manufacturing -EuRP&M1997 At: University of Nottingham, UK. January 1997.

Khan, S., Rayhan, S. B., Ibn Salam, S. M. I., Khan, H. A., & Rahman, M. Z. (2023b). Recent advances in sustainable building materials for the construction industry: Mechanical performance and applications. In *Reference Module in Materials Science and Materials Engineering*. Elsevier., DOI: 10.1016/B978-0-323-96020-5.00191-6

Khan, S. H., Rahman, M. Z., Haque, M. R., & Hoque, M. E. (2023a). Characterization and comparative evaluation of structural, chemical, thermal, mechanical, and morphological properties of plant fibers. In Khiari, R., Jawaid, M., & Belgacem, M. N. (Eds.), *Annual Plant: Sources of Fibres, Nanocellulose and Cellulosic Derivatives: Processing, Properties and Applications* (pp. 1–45). Springer Nature Singapore., DOI: 10.1007/978-981-99-2473-8_1

Kolossov, S., Boillat, E., Glardon, R., Fischer, P., & Locher, M. (2004). 3D FE simulation for temperature evolution in the selective laser sintering process. *International Journal of Machine Tools & Manufacture, 44*(2-3), 117–123. DOI: 10.1016/j.ijmachtools.2003.10.019

Kousiatza, C., Tzetzis, D., & Karalekas, D. (2019). In-situ characterization of 3D printed continuous fiber reinforced composites: A methodological study using fiber Bragg grating sensors. *Composites Science and Technology, 174*, 134–141. DOI: 10.1016/j.compscitech.2019.02.008

Kruth, J., Levy, G., Klocke, F., & Childs, T. H. C. (2007). Consolidation phenomena in laser and powder-bed based layered manufacturing. *CIRP Annals, 56*(1), 730–759. DOI: 10.1016/j.cirp.2007.10.004

Kruth, J., Mercelis, P., Van Vaerenbergh, J., Froyen, L., & Rombouts, M. (2005). Binding mechanisms in selective laser sintering and selective laser melting. *Rapid Prototyping Journal, 11*(1), 26–36. DOI: 10.1108/13552540510573365

Kumar, S., & Kruth, J. P. (2010). Composites by rapid prototyping technology. *Materials & Design, 31*(2), 850–856. DOI: 10.1016/j.matdes.2009.07.045

Lan, P., Chou, S., Chent, L., & Gemmill, D. (1997). Determining fabrication orientations for rapid prototyping with stereolithography apparatus. Computer-Aided Design 29(1): 53–62. Li H, Wang T, Joshi S, and Yu Z (2019) The quantitative analysis of tensile strength of additively manufactured continuous carbon fiber reinforced polylactic acid (PLA). *Rapid Prototyping Journal*, *25*(10), 1624–1636. DOI: 10.1108/RPJ-01-2018-0005

Li, N., Link, G., & Jelonnek, J. (2020). 3D microwave printing temperature control of continuous carbon fiber reinforced composites. *Composites Science and Technology*, *187*, 107939. DOI: 10.1016/j.compscitech.2019.107939

Liu, F. R., Zhang, Q., Zhou, W. P., Zhao, J. J., & Chen, J. M. (2012). Micro scale 3D FEM simulation on thermal evolution within the porous structure in selective laser sintering. *Journal of Materials Processing Technology*, *212*(10), 2058–2065. DOI: 10.1016/j.jmatprotec.2012.05.010

Liu, Z. H., Nolte, J. J., Packard, J. I., Hilmas, G., Dogan, F., & Leu, M. C. (2007) Selective Laser Sintering of high-density alumina ceramic parts. In: *Proceedings of the 35th International MATADOR 2007 Conference*, 3, pp. 351–354. DOI: 10.1007/978-1-84628-988-0_79

Lynn, C. M., West, A., & Rosen, D. W. (1998). *A Process Planning Method and Data Format for Achieving Tolerances in Stereolithography*. The University of Texas.

Lynn-charney, C., & Rosen, D. W. (2000). Usage of accuracy models in stereolithography process planning. *Rapid Prototyping Journal*, *6*(2), 77–87. DOI: 10.1108/13552540010323600

Martínez, J., Diéguez, J. L., Ares, E., Pereira, A., Hernández, P., & Pérez, J. A. (2013). Comparative between FEM models for FDM parts and their approach to a real mechanical behaviour. *Procedia Engineering*, *63*(December), 878–884. DOI: 10.1016/j.proeng.2013.08.230

Matsumoto, M., Shiomi, M., Osakada, K., & Abe, F. (2002). Finite element analysis of single layer forming on metallic powder bed in rapid prototyping by selective laser processing. *International Journal of Machine Tools & Manufacture*, *42*(1), 61–67. DOI: 10.1016/S0890-6955(01)00093-1

Melenka, G. W., Cheung, B. K. O., Schofield, J. S., Dawson, M. R., & Carey, J. P. (2016). Evaluation and prediction of the tensile properties of continuous fiber-reinforced 3D printed structures. *Composite Structures*, *153*, 866–875. DOI: 10.1016/j.compstruct.2016.07.018

Michaleris, P. (2014). Modeling metal deposition in heat transfer analyses of additive manufacturing processes. *Finite Elements in Analysis and Design*, *86*, 51–60. DOI: 10.1016/j.finel.2014.04.003

Mohammadizadeh, M., Imeri, A., Fidan, I., & Elkelany, M. (2019). 3D printed fiber reinforced polymer composites—Structural analysis. *Composites. Part B, Engineering*, *175*, 107112. DOI: 10.1016/j.compositesb.2019.107112

Muller, P., Mognol, P., & Hascoet, J. Y. (2013). Modeling and control of a direct laser powder deposition process for Functionally Graded Materials (FGM) parts manufacturing. *Journal of Materials Processing Technology*, *213*(5), 685–692. DOI: 10.1016/j.jmatprotec.2012.11.020

Murr, L. E. (2015). Metallurgy of additive manufacturing: Examples from electron beam melting. *Additive Manufacturing*, *5*, 40–53. DOI: 10.1016/j.addma.2014.12.002

Murr, L. E., Gaytan, S. M., Martinez, E., Medina, F., & Wicker, R. B. (2012). Next generation orthopaedic implants by additive manufacturing using electron beam melting. *International Journal of Biomaterials*, *2012*, 1–14. Advance online publication. DOI: 10.1155/2012/245727 PMID: 22956957

Murugan, B., Rahman, M. Z., Fatimah, I., Anita Lett, J., Annaraj, J., Kaus, N. H. M., Al-Anber, M. A., & Sagadevan, S. (2023). Green synthesis of CuO nanoparticles for biological applications. *Inorganic Chemistry Communications*, *155*, 111088. DOI: 10.1016/j.inoche.2023.111088

Narahara, H., Tanaka, F., Kishinami, T., Igarashi, S., & Saito, K. (1999). Reaction heat effects on initial linear shrinkage and deformation in stereolithography. *Rapid Prototyping Journal*, *5*(3), 120–128. DOI: 10.1108/13552549910278946

Narahara, H., Tanaka, F., Kishinami, T., Igarasm, S., & Saito, K.Department of Mechanical Systems Engineering, and Gunma Polytechnic. (1997). *Reaction Heat Effect on Initial Linear Shrinkage of Stereolithography Resins.*

Ning, F., Cong, W., Jia, Z., Wang, F., & Zhang, M. (2016) Additive manufacturing of CFRP composites using fused deposition modeling: Effects of process parameters. In: ASME 2016 11th International Manufacturing Science and Engineering Conference, MSEC 2016, 3, pp. 1–7. https://doi.org/DOI: 10.1115/MSEC20168561

Onuh, S. O., & Hon, K. K. B. (1998). Optimising build parameters for improved surface finish in stereolithography. *International Journal of Machine Tools & Manufacture*, *38*(4), 329–342. DOI: 10.1016/S0890-6955(97)00068-0

Peng, Y., Wu, Y., Wang, K., Gao, G., & Ahzi, S. (2019). Synergistic reinforcement of polyamide-based composites by combination of short and continuous carbon fibers via fused filament fabrication. *Composite Structures*, *207*, 232–239. DOI: 10.1016/j.compstruct.2018.09.014

Podshivalov, L., Gomes, C. M., Zocca, A., Guenster, J., Bar-yoseph, P., & Fischer, A. (2013). Design, analysis and additive manufacturing of porous structures for biocompatible micro-scale scaffolds. *Procedia CIRP*, *5*, 247–252. DOI: 10.1016/j.procir.2013.01.049

Prabhu, L., Krishnaraj, V., Gokulkumar, S., Sathish, S., & Ramesh, M. (2019a). Mechanical, chemical and acoustical behavior of sisal—Tea waste—Glass fiber reinforced epoxy based hybrid polymer composites. *Materials Today: Proceedings*, *16*, 653–660. DOI: 10.1016/j.matpr.2019.05.142

Prabhu L, Krishnaraj V, Sathish S, Gokulkumar S, and Karthi N (2019b) Study of mechanical and morphological properties of jute-tea leaf fiber reinforced hybrid composites: Effect of glass fiber hybridization. Materials Today Proceedings 27(xxxx): 2372–2375. .DOI: 10.1016/j.matpr.2019.09.132

Raghunath, N., & Pandey, P. M. (2007). Improving accuracy through shrinkage modelling by using Taguchi method in selective laser sintering. *International Journal of Machine Tools & Manufacture*, *47*(6), 985–995. DOI: 10.1016/j.ijmachtools.2006.07.001

Rahman, M. M., Rayhan, S. B., Sultana, J., & Rahman, M. Z. (2023a). Additive manufacturing techniques, their challenges, and additively manufactured composites for advanced engineering applications. In *Reference Module in Materials Science and Materials Engineering*. Elsevier., DOI: 10.1016/B978-0-323-96020-5.00118-7

Rahman, M. Z. (2017). *Static and Dynamic Characterisation of Flax Fibre-reinforced Polypropylene Composites*. ResearchSpace.

Rahman MZ (2021) Mechanical and damping performances of flax fibre composites–A review. Composites Part C: Open Access: 100081.

Rahman, M. Z. (2022). Influence of Fiber treatment on the damping performance of plant Fiber composites. In *Vibration and Damping Behavior of Biocomposites* (pp. 113–135). CRC Press. DOI: 10.1201/9781003173625-8

Rahman, M. Z., Jayaraman, K., & Mace, B. R. (2017). Vibration damping of flax fibre-reinforced polypropylene composites. *Fibers and Polymers*, *18*(11), 2187–2195. DOI: 10.1007/s12221-017-7418-y

Rahman, M. Z., Jayaraman, K., & Mace, B. R. (2018a). Impact energy absorption of flax fiber-reinforced polypropylene composites. *Polymer Composites, 39*(11), 4165–4175. DOI: 10.1002/pc.24486

Rahman, M. Z., Jayaraman, K., & Mace, B. R. (2018b). Influence of damping on the bending and twisting modes of flax fibre-reinforced polypropylene composite. *Fibers and Polymers, 19*(2), 375–382. DOI: 10.1007/s12221-018-7588-7

Rahman, M. Z., Mace, B. R., & Jayaraman, K. (2016) Vibration damping of natural fibre-reinforced composite materials. In: *17th European Conference on Composite Material.*

Rahman, M. Z., & Xu, H. (2023). Damping under varying frequencies, mechanical properties, and failure modes of flax/polypropylene composites. *Polymers, 15*(4), 1042. DOI: 10.3390/polym15041042 PMID: 36850325

Rahman, M. Z. M. M. Z., Rahman, M. Z. M. M. Z., Mahbub, T., Ashiquzzaman, M., Sagadevan, S., & Hoque, M. E. (2023b). Advanced biopolymers for automobile and aviation engineering applications. *Journal of Polymer Research, 30*(3), 106. DOI: 10.1007/s10965-023-03440-z

Rahman, S., Rahman, M. Z., Tasnim, J., Saha, B., & Abedin, M. J. (2023c). Biobased composites reinforced with annual plants—Design, manufacturing techniques, and parameters influencing the overall properties. In *Reference Module in Materials Science and Materials Engineering*. Elsevier., DOI: 10.1016/B978-0-323-96020-5.00142-4

Rahmati, S., & Dickens, P. M. (1995) Stereolithography process improvement. In: *Proceedings of the First National Conference on Rapid Prototyping and Tooling Research*, pp. 111–126.

Ramanath, H. S., Chua, C. K., Leong, K. F., & Shah, K. D. (2008). Melt flow behaviour of poly-e-caprolactone in fused deposition modelling. *Journal of Materials Science. Materials in Medicine, 19*(7), 2541–2550. DOI: 10.1007/s10856-007-3203-6 PMID: 17619957

Ramesh, M. (2016). Kenaf (Hibiscus cannabinus L.) fibre based bio-materials: A review on processing and properties. *Progress in Materials Science, 78–79*, 1–92. DOI: 10.1016/j.pmatsci.2015.11.001

Rawat, P., Liu, S., Guo, S., Rahman, M. Z., Yang, T., Bai, X., Yao, Y., Mobasher, B., & Zhu, D. (2022). A state-of-the-art review on mechanical performance characterization and modelling of high-performance textile reinforced concretes. *Construction & Building Materials, 347*, 128521. DOI: 10.1016/j.conbuildmat.2022.128521

Rayhan, S. B., Rahman, M. M., Rahman, M. Z., & Ibn Salam, S. M. I. (2023). Advances in ceramic composites: Manufacture, performances, and applications. In *Reference Module in Materials Science and Materials Engineering*. Elsevier., DOI: 10.1016/B978-0-323-96020-5.00141-2

Reeves, P. E., & Cobb, R. C. (1997). Reducing the surface deviation of streolithography using an alternative build stratergy. *Rapid Prototyping Journal, 3*(1), 20–31. DOI: 10.1108/13552549710169255

Sagadevan, S., Fatimah, I., Anita Lett, J., Rahman, M. Z., Leonard, E., & Oh, W.-C. (2023a). Eco-friendly green approach of nickel oxide nanoparticles for biomedical applications. *Open Chemistry, 21*(1), 20230141. Advance online publication. DOI: 10.1515/chem-2023-0141

Sagadevan, S., Rahman, M. Z., Léonard, E., Losic, D., & Hessel, V. (2023b). Sensor to electronics applications of graphene oxide through AZO grafting. *Nanomaterials (Basel, Switzerland), 13*(5), 846. DOI: 10.3390/nano13050846 PMID: 36903724

Sagadevan, S., Schirhagl, R., Rahman, M. Z., Ismail, M. F. B., Anita Lett, J., Fatimah, I., Kaus, N. H. M., & Oh, W.-C. (2023c). Recent advancements in polymer matrix nanocomposites for bone tissue engineering applications. *Journal of Drug Delivery Science and Technology, 82*, 104313. DOI: 10.1016/j.jddst.2023.104313

Saha, B., Hoque, M. A., Tahsin, S. H., Rimi, K. N., Prova, A. T., Dey, D., Rahman, S., & Rahman, M. Z. (2023). Biobased composites for advanced applications: Possibilities and difficulties on the path to circularity. In *Reference Module in Materials Science and Materials Engineering*. Elsevier., DOI: 10.1016/B978-0-323-96020-5.00188-6

Salmoria, G. V., Ahrens, C. H., Klauss, P., Paggi, R. A., Oliveira, R. G., & Lago, A. (2007). Rapid manufacturing of polyethylene parts with controlled pore size gradients using selective laser sintering. *Materials Research, 10*(2), 211–214. DOI: 10.1590/S1516-14392007000200019

Sathish, S., Kumaresan, K., Prabhu, L., & Gokulkumar, S. (2018). Experimental investigation of mechanical and ftir analysis of flax fiber/epoxy composites incorporating sic, Al2O3 and graphite. Revista Romana de Materiale/Romanian. *Journal of Materials, 48*(4), 476–482.

Sathish, S., Kumaresan, K., Prabhu, L., & Vigneshkumar, N. (2017). Experimental investigation on volume fraction of mechanical and physical properties of flax and bamboo fibers reinforced hybrid epoxy composites. *Polymers & Polymer Composites, 25*(3), 229–236. DOI: 10.1177/096739111702500309

Schaub, D. A., Chu, K.-R., & Montgomery, D. C. (1997). Optimizing stereolithography throughput. *Journal of Manufacturing Systems, 16*(4), 290–303. DOI: 10.1016/S0278-6125(97)89099-1

Schmidt, M., Pohle, D., & Rechtenwald, T. (2007) Selective laser sintering of PEEK. CIRP Annals - Manufacturing Technology 56(1): 205–208. . Schmitz GJ and Prahl U (eds.) (2016) Handbook of Software Solutions for ICME. Wiley-VCH Verlag GmbH & Co.DOI: 10.1016/j.cirp.2007.05.097

Shishkovsky, I., Yadroitsev, I., Bertrand, P., & Smurov, I. (2007). Alumina-zirconium ceramics synthesis by selective laser sintering/melting. *Applied Surface Science, 254*(4), 966–970. DOI: 10.1016/j.apsusc.2007.09.001

Shofner, M. L., Lozano, K., & Rodrı, F. J. (2002). Nanofiber-reinforced polymers prepared by fused deposition modeling. *Journal of Applied Polymer Science, 89*(11), 3081–3090. DOI: 10.1002/app.12496

Shofner, M. L., Rodríguez-Macías, F. J., Vaidyanathan, R., & Barrera, E. V. (2003). Single wall nanotube and vapor grown carbon fiber reinforced polymers processed by extrusion freeform fabrication. *Composites. Part A, Applied Science and Manufacturing, 34*(12), 1207–1217. DOI: 10.1016/j.compositesa.2003.07.002

Singh, R., Gupta, A., Tripathi, O., Srivastava, S., Singh, B., Awasthi, A., Rajput, S. K., Sonia, P., Singhal, P., & Saxena, K. K. (2020). Powder bed fusion process in additive manufacturing: An overview. *Materials Today: Proceedings, 26*, 3058–3070. Advance online publication. DOI: 10.1016/j.matpr.2020.02.635

Siraj MT, Alshybani I, Payel SB, Shahadat M, Bin R, and Rahman MZ (2023a) Advances in biocomposite fabrication: Emerging technologies and their potential applications.

Siraj, M. T., Jamil, S. N. S., Arka, A. M., Tasnim, S., Ghosh, M., Shahadat, M. R. B., & Rahman, M. Z. (2023). Eco-friendly food packaging innovations: A review of recent progress on recyclable polymers. In *Reference Module in Materials Science and Materials Engineering*. Elsevier., DOI: 10.1016/B978-0-323-96020-5.00077-7

Sood, A. K., Ohdar, R. K., & Mahapatra, S. S. (2012). Experimental investigation and empirical modelling of FDM process for compressive strength improvement. *Journal of Advanced Research, 3*(1), 81–90. DOI: 10.1016/j.jare.2011.05.001

Strano, A. G., Hao, L., Everson, R. M., & Evans, K. E. (2012). Surface roughness analysis, modelling and prediction in selective laser melting. *Journal of Materials Processing Technology, 2010*. Advance online publication. DOI: 10.1016/j.jmatprotec.2012.11.011

Tan, K. H., Chua, C. K., Leong, K. F., Cheah, C. M., Cheang, P., Abu Bakar, M. S., & Cha, S. W. (2003). Scaffold development using selective laser sintering of polyetheretherketone-hydroxyapatite biocomposite blends. *Biomaterials*, *24*(18), 3115–3123. DOI: 10.1016/S0142-9612(03)00131-5 PMID: 12895584

Thomas, J. P., & Renaud, J. E. (2006). Mechanical behavior of acrylonitrile butadiene styrene (ABS) fused deposition materials. Experimental investigation. *Rapid Prototyping Journal*, *7*(3), 148–158.

Toyserkani, E., Khajepour, A., & Corbin, S. (2004). 3-D finite element modeling of laser cladding by powder injection: Effects of laser pulse shaping on the process. *Optics and Lasers in Engineering*, *41*(6), 849–867. DOI: 10.1016/S0143-8166(03)00063-0

Valvez, S., Santos, P., Parente, J. M., Silva, M. P., & Reis, P. N. B. (2020). 3D printed continuous carbon fiber reinforced PLA composites: A short review. *Procedia Structural Integrity*, *25*(May), 394–399. DOI: 10.1016/j.prostr.2020.04.056

van de Werken, N., Hurley, J., Khanbolouki, P., Sarvestani, A. N., Tamijani, A. Y., & Tehrani, M. (2019). Design considerations and modeling of fiber reinforced 3D printed parts. *Composites. Part B, Engineering*, *160*, 684–692. DOI: 10.1016/j.compositesb.2018.12.094

van de Werken, N., Tekinalp, H., Khanbolouki, P., Ozcan, S., Williams, A., & Tehrani, M. (2020). Additively manufactured carbon fiber-reinforced composites: State of the art and perspective. *Additive Manufacturing*, *31*, 100962. DOI: 10.1016/j.addma.2019.100962

Vasinonta, A., & Beuth, J. (2000) Process maps for controlling residual stress and melt pool size in laser-based sff processes. In: *Proceedings 2000 Solid Freeform Fabrication Symposium Solid Freeform Fabrication Symposium*, pp. 200–208.

Venkataraman, N., Rangarajan, S., Matthewson, M. J., Harper, B., Safari, A., Danforth, S. C., Wu, G., Langrana, N., Guceri, S., & Yardimci, A. (2002). Feedstock material property [REMOVED HYPERLINK FIELD] process relationships in fused deposition of ceramics (FDC). *Rapid Prototyping Journal*, *6*, 244–252. DOI: 10.1108/13552540010373344

Venkataraman, N., Rangarajan, S., Matthewson, M. J., Safari, A., Danforth, S. C., Yardimci, A., & Guceri, S. I. (1999) Mechanical and Rheological Properties of Feedstock Material for Fused Deposition of Ceramics and Metals (FDC and FDMet) and their Relationship to Process Performance, pp. 351–360.

Wang, W. L., Cheah, C. M., Fuh, J. Y. H., & Lu, L. (1997). Influence of process parameters on stereolithography part shrinkage. *Materials & Design*, *17*(4), 205–213. DOI: 10.1016/S0261-3069(97)00008-3

Wang, X., Jiang, M., Zhou, Z., Gou, J., & Hui, D. (2017). 3D printing of polymer matrix composites: A review and prospective. *Composites. Part B, Engineering*, *110*, 442–458. DOI: 10.1016/j.compositesb.2016.11.034

West, A. P., Sambu, S. P., & Rosen, D. W. (2001). A process planning method for improving build performance in stereolithography. *Computer Aided Design*, *33*(1), 65–79. DOI: 10.1016/S0010-4485(00)00064-6

Williams, J. M., Adewunmi, A., Schek, R. M., Flanagan, C. L., Krebsbach, P. H., Feinberg, S. E., Hollister, S. J., & Das, S. (2005). Bone tissue engineering using polycaprolactone scaffolds fabricated via selective laser sintering. *Biomaterials*, *26*(23), 4817–4827. DOI: 10.1016/j.biomaterials.2004.11.057 PMID: 15763261

Wohlers, T., & Caffrey, T.Ian Campbell of Wohlers Associates, Inc. (2012). Wohlers Report 2012 [Printing State of the Industry: Annual Worldwide Progress Report. Wohlers Associates.]. *Additive Manufacturing*, •••, 3d.

Wohlers, T. T., & Caffrey, T. (2013). Wohlers Report 2013 [Printing State of the Industry: Annual Worldwide Progress Report. Wohlers Associates.]. *Additive Manufacturing*, •••, 3D.

Yardimci, M. A., Hattori, T., Guceri, S. I., & Danforth, S. C. (1997) Thermal analysis of fused deposition. In: Solid Freeform Fabrication Proceedings, pp. 689–698.

Yin, L., Tian, X., Shang, Z., Wang, X., & Hou, Z. (2019). Characterizations of continuous carbon fiber-reinforced composites for electromagnetic interference shielding fabricated by 3D printing. *Applied Physics. A, Materials Science & Processing*, *125*(4), 266. Advance online publication. DOI: 10.1007/s00339-019-2566-0

Zhang, H., Yang, D., & Sheng, Y. (2018). Performance-driven 3D printing of continuous curved carbon fibre reinforced polymer composites: A preliminary numerical study. *Composites. Part B, Engineering*, *151*, 256–264. DOI: 10.1016/j.compositesb.2018.06.017

Zhang, Y., & Chou, Y. K. (2006). Three-dimensional finite element analysis simulations of the fused deposition modelling process. *Proceedings of the Institution of Mechanical Engineers. Part B, Journal of Engineering Manufacture*, *220*(10), 1663–1671. DOI: 10.1243/09544054JEM572

Zhang, Y., & Chou, Y. K. (2008) A Parametric Study of Part Distortions in FDM Using 3D FEA. Tuscaloosa, AL: Mechanical Engineering Department, The University of Alabama. 35487 Reviewed, accepted September 14, 2006.

Zhong, W., Li, F., Zhang, Z., Song, L., & Li, Z. (2001). Short fiber reinforced composites for fused deposition modeling. *Materials Science and Engineering A*, *301*(2), 125–130. DOI: 10.1016/S0921-5093(00)01810-4

Chapter 11
Techniques and Methods for Optimizing 3D Printing Parameters

Romdhane Ben Khalifa
https://orcid.org/0000-0002-9171-9584
National Engineering School of Gabes (ENIG), Tunisia

Ghalia Nhassa
National Engineering School of Gabes (ENIG), Tunisia

Ali Snoussi
National Engineering School of Gabes (ENIG), Tunisia

ABSTRACT

This paper provides a detailed evaluation of existing methods and techniques for optimizing 3D printing parameters across different additive manufacturing technologies. The optimization of printing parameters is essential to obtain superior mechanical properties, dimensional accuracy, surface quality, and production efficiency in 3D printed components. The review systematically examines methodological approaches, including statistical design of experiments (DOE), artificial intelligence and machine learning (AI/ML) applications, simulation-based optimization, and in-process monitoring and feedback control. The review focuses on material-specific optimization methods for polymers, metals, composites, and ceramics by analyzing process-specific parameters and material-dependent variables. The paper demonstrates real-world applications of optimization methodologies through case studies from aerospace, automotive, medical, and consumer goods manufacturing industries, which show substantial improvements in product performance, production speed, and cost-effectiveness.

DOI: 10.4018/979-8-3373-0533-2.ch011

INTRODUCTION

The technology of additive manufacturing or 3D printing has transformed multiple industries through its ability to create complex designs and customized products with simplified procedures (Wu et al., 2017).

The technology stands apart from conventional subtractive manufacturing processes because it constructs objects through sequential layer addition, which provides designers with extensive design flexibility and material options (Spencer, 2018). The growing requirement for complex components has driven manufacturers to adopt new methods, and additive manufacturing has proven to be a promising solution for creating multi-functional and multi-material designs (Jin et al., 2020).

3D printing demonstrates strong potential for materials, mechanical, and biomedical engineering through its precise material deposition capabilities (Rojek et al., 2025). The ability to precisely position materials in three dimensions has established 3D printing as a revolutionary manufacturing technology (Garcia et al., 2013). The rapid prototyping features of 3D printing technology enable fast design and model production. In contrast, its multiple printing techniques and materials support functional items and scale models, making it suitable for scientific and industrial applications (Królikowski et al., 2019).

The customization capabilities of 3D printing prove essential in healthcare, aerospace, and automotive industries because they enable the creation of customized solutions and optimized designs. The aerospace and defense sectors started using 3D printing for quick prototyping and design verification. This technology creates complex designs at reduced weights, enabling it to make lightweight components such as hinges, brackets, and air ducts (Pathak et al., 2023). The development of new materials and manufacturing techniques enables the 3D printing of essential turbine blades and engine parts, leading to better performance and fuel efficiency, shorter production times, and reduced costs and material waste for sustainable operations.

Over the last ten years, 3D printing has transformed our manufacturing processes through the simplified creation of unusual and complex shapes. The aerospace industry and healthcare practices witnessed significant growth in market demand for complex, detailed designs. Researchers currently explore various product parameters, including temperature layer thickness speed and material properties, to define product strength appearance and price points. Researchers combine physical testing with digital modeling techniques to discover optimal printer parameters for various materials and printing equipment (Natarajan et al., 2024). Researchers use machine learning algorithms to explore automatic printing operations while simultaneously minimizing waste output (Olsson NO et al., 2019). When scientists modify support scaffolds and infill patterns, they discover unexpected results as variables develop connections that create intellectual puzzles (Cristea et al., 2019). Maintaining production efficiency while enhancing product quality remains a delicate challenge, according (Du et al., 2019). The development is substantial, yet particular regions continue to demand further exploration. Most studies examine one experimental factor while neglecting multiple-factor combinations (Oberoi et al., 2018). The behavior of polymers and metals during 3D printing remains a complex problem that researchers have not yet fully solved. Research into product toughness modifications caused by printing behaviors represents an attractive field for upcoming scientific investigation (Morettini et al., 2022). People usually study individual machines or printing methods yet fail to analyze how different techniques might work together to advance production outcomes (Ahmed et al., 2020). The optimization of print settings creates benefits for both creating attractive prints and sustainable, eco-friendly production at reduced costs. The world's growing environmental concerns make it essential to establish precise parameters that minimize waste production while maximizing operational efficiency (Rojek et al., 2025). The research is essential for contemporary makers beyond theoretical concepts (Mladenovska et al., 2021). The review explores widely used methods while discussing present findings and highlighting significant knowledge gaps in the field. The paper examines present methods alongside fresh knowledge and proposes a plan to close knowledge gaps to stimulate innovative additive manufacturing (Zhou et al., 2024).

KEY 3D PRINTING PARAMETERS AND THEIR INFLUENCE

The quality and functionality of 3D-printed objects are highly dependent on a complex interplay of printing parameters, including layer thickness, printing speed, infill density, nozzle temperature, and build orientation (Yin et al., 2022). The resolution and surface finish of the printed part are significantly influenced by layer thickness, with thinner layers resulting in smoother surfaces but requiring longer printing times (Tian et al., 2021). The rate of material deposition and the overall build time are affected by printing speed. However, excessive speeds can compromise print quality due to inadequate layer adhesion and increased material deformation (Zhou et al., 2024). The printed object's internal structure and mechanical properties are determined by infill density, which offers a trade-off between weight, strength, and material consumption (Maurya et al., 2021). The nozzle temperature is crucial for controlling the viscosity and flow behavior of the printing material, ensuring proper melting and bonding of layers. At the same time, the build orientation affects support structure requirements, surface finish, and mechanical strength, necessitating careful consideration to minimize material usage and maximize structural integrity (Kishore et al., 2021) (Chua & Leong, 2013; Wang et al., 2023).

Material Properties and Selection

The selection of materials is a fundamental element in 3D printing because the properties of selected materials must match the requirements of the final printed object (Zhou et al., 2024). The mechanical strength, thermal stability, chemical resistance, and biocompatibility of polymers, metals, ceramics, and composites differ (Mladenovska et al., 2021). The optimization of printing parameters needs to consider material-specific characteristics because temperature, pressure, and printing speed variations produce substantial effects on the final product properties. The production of food products demands precise attention to material attributes and process variables, as well as post-treatment protocols, to reach exact printing outcomes (Liu et al., 2017). The additive manufacturing process produces parts with strength properties that differ from the original material, so designers must evaluate anisotropic material properties when creating structural designs. Researchers continue to develop new materials for 3D printing through research that aims to improve mechanical performance, decrease material costs, and increase printable material diversity (Ahmed et al., 2020). The printing process becomes complex because of multiple factors, including dispensing orifice dimension, solvent and fluid viscosity surface tension and fluid-surface interactions, polymer concentration and humidity, and temperature interaction (Chang et al., 2011). All printing-related aspects require precise control to generate parts with accurate and precise results.

Process Parameters

Build time can vary a lot (Almy & Tontowi, 2018). Going too fast can mess up the print quality because the layers will not stick together well, and the material can change shape. The two most significant factors that affect print time are layer height and print speed. Other important things include how much fill you use, how many walls there are, and the layer height since these affect how much material you use. The surface's roughness and the accuracy of the dimensions mainly depend on layer height (Mendřický & Fris, 2020). It is also key to consider how process settings impact cell accuracy, with printing speed being the most significant influence on accuracy in the printing plane. For height accuracy, the thickness of the layer and what is known as 'bridging' are critical (Filippidis et al., 2020). The quality and performance of parts produced by ME are highly dependent on the build parameters used. These parameters can be grouped into temperature-related parameters, infill-related parameters, and speed-related parameters (Loh et al., 2020). The morphological characteristics of printed parts, such as the degree of crystallinity and the coalescence between deposited lines, are determined by the extrusion temperature and speed (Cabreira & Santana, 2020). The infill percentage affects material consumption, printing time, and mechanical properties, which enables the production of lightweight yet structurally sound parts (Buj-Corral et al., 2019). The selection of process parameters is of extreme importance for the success of printing, especially if the aim is to obtain components that possess load capacity (Morettini et al., 2022). The mechanical characteristics of specimen's decrease when the raster angle is increased (Cristea et al., 2019).

OPTIMIZATION TECHNIQUES FOR 3D PRINTING PARAMETERS

Several methods of optimizing your printing include experimental design, simulation, and feedback control. Experimental design techniques, such as Taguchi and response surface methodology, can systematically explore the parameter space and determine optimal parameters for a given material and printing process (Babagowda et al., 2018). Numerical simulation provides information on material behavior during printing and can be useful to virtually optimize parameters prior to physical printing, e.g., finite element analysis and computational fluid dynamics. Feedback control systems take into account real-time feedback input on temperature, pressure, and material flow, and such systems can change printing parameters continuously to stabilize quality and consider environmental conditions and material specifications. An increasing number of machine learning techniques using artificial neural networks are being used to study how the printing parameters can be used to control part properties, thus paving the way for machine learning techniques to be suggested for parameter-optimized printing quality prediction, quality checking, and printing method (Mahmood et al., 2020).

The optimization methods are particularly relevant where accuracy, print duration, and material usage must be carefully balanced, as producing removable partial dentures using digital light processing (Hussein & Hussein, 2022). Furthermore, part orientation can be optimized to adjust printing parameters such as support structures, printing time, layering, shrinkage, and surface roughness. It shows the importance of optimizing printing processes to improve mechanical properties and print accuracy.

Different methodological approaches can be employed to optimize 3D printing parameters, each with different strengths and limitations regarding efficiency, accuracy, and flexibility.

Experimental design approaches, such as Taguchi methods, response surface methodology, and factorial designs, enable systematic parameter space exploration to identify optimal settings for specific materials and printing processes (Kam et al., 2020). Some of these methods involve conducting experiments wherein printing parameters are varied per a predefined schedule, and printed part characteristics are measured and examined to determine optimal parameter settings (Ren et al., 2017). Numerical simulation techniques, such as finite element analysis and computational fluid dynamics, can predict material behavior during printing, allowing preprinting virtual optimization of parameters (Lee et al., 2019). Heat transfer, fluid flow, and stress distribution in the printed part can be simulated by these techniques and utilized to understand the effect of different printing parameters on part functionality and quality. Machine learning techniques can predict optimal printing parameters by learning from experimental data and iteratively optimizing the parameter values to achieve target part properties (Equbal et al., 2024). Machine learning algorithms can be trained on a database of printing parameters and corresponding part properties. They can then be used to make predictions for the optimal parameter values for new materials or parts (Chari et al., 2018). Real-time monitoring and control systems can change printing parameters during the printing process to balance variations in the environment or material characteristics. Such systems usually consist of sensors that quantify temperature, pressure, and flow of material and control algorithms that modify the printing parameters accordingly. Process parameter optimization is critical to realizing desired results in additive manufacturing, often including incremental tuning of laser power or feed rate for each new layer (Li et al., 2021). As a case in point, process parameter optimization, such as incremental tuning of laser power or feed rate for each new layer, is crucial to desired results in additive manufacturing (Kim et al., 2015).

Statistical Design of Experiments (DOE)

The systematic approach of statistical experimental design optimizes 3D printing parameters by conducting experiments to determine how input parameters affect output responses, according to Wang et al. (2020). Response surface methodology evaluates how parameters affect dimensional accuracy (Almy & Tontowi, 2018). The analysis uses statistical models and regression analysis to identify important parameters and their best values (Fernández & Martín, 2023). Machine learning is a computer program for artificial intelligence to predict optimal process parameters that optimize production processes, according to Hooda et al. (2021). The process has been applied to model the printing process and forecast optimal parameters. Machine learning algorithms in manufacturing operations decrease the time required to configure machines and process parameters for new products. Process engineers perform this task through their experience with product similarity assessments, physical testing, and parameter correlation analysis of desired product quality outputs. The artificial intelligence system uses input characteristics to predict new product process parameters, which it receives as output data. Machine learning models acquire the ability to detect variations in machine parameters that result from different product categories. The system uses learned patterns to forecast new parameter values for upcoming products with specific characteristics (Khdoudi & Masrour, 2018). Physics-based modeling applies mathematical calculations and computational models to predict how processes and materials will act during the 3D printing operation so that parameters can be optimized virtually. These models apply fundamental physics principles such as heat transport, fluid dynamics, and solid mechanics to simulate complex interactions while printing. Multi-scale process modeling, improved in-situ monitoring, and post-fabrication processes are crucial in maintaining consistency for defect avoidance and quality assurance, especially in mission-critical applications (Panwisawas et al., 2020). Physics-based modeling can help optimize the printing parameters by predicting the effect of different parameter values on part performance and quality. For example, finite element analysis can simulate the thermal stresses and deformations during printing, allowing engineers to identify parameter settings that minimize these effects.

Artificial Intelligence and Machine Learning (AI/ML)

3D printing parameter optimization will experience a revolutionary transformation through emerging technologies of machine learning and artificial intelligence, which regulate and optimize print operations. These technologies utilize algorithms and data analysis techniques to extract knowledge from experimental results, leading to continuous parameter setting adjustments to achieve desired part characteristics (Omairi & Ismail, 2021). New additive manufacturing algorithms are being developed, and future uses incorporating state-of-the-art artificial intelligence methods are being explored (Jin et al., 2020). Machine learning integration with 3D printing enables better process control and optimization (Goh et al., 2020). The role of machine learning in additive manufacturing will become increasingly vital because it enables predictive maintenance while improving material characterization and automating process control (Ghilan et al., 2020). The most significant potential of AI and machine learning in 3D printing exists in creating predictive models that use given parameters to forecast printing outcomes.

Machine learning models acquire complex input-output relationships through training with extensive data collections of printing parameters and their associated part characteristics (Rojek et al., 2025). Machine learning in additive manufacturing allows real-time compensation analysis of extensive datasets to produce enhanced results while creating necessary data environments for effective learning and optimization (Anantrasirichai & Bull, 2021). Unsupervised machine learning enables automatic material phase detection through data observation, providing superior material characterization methods, including topological superconductors (Dwivedi et al., 2019). AI techniques for real-time error correction in additive manufacturing eliminate prolonged processing times and real-time loss of control, resulting in unintended goals (Omairi & Ismail, 2021). AI software monitors process parameters and sensor readings online, enabling it to detect printing path deviations and modify parameters for mistake correction.

AI plays a vital role in automation because it performs real-time data analysis and control functions, maximizing manufacturing process optimization. The real-time monitoring capabilities of AI systems provide better performance than human operators because they can handle processes automatically (Dwivedi et al., 2019). Laser processing operations become more precise and efficient when artificial intelligence streamlines them for applications in the automotive, aerospace, and biomedical industries (Murzin, 2024). Machine learning techniques added to metal-based additive manufacturing optimization methods will decrease human intuition requirements and the effort needed (Silbernagel et al., 2019). Modern machine-learning techniques enable algorithms to develop experience-based learning abilities, leading specialists to obtain answers that improve quality (Grünberger, 2020). Implementing machine learning systems in 3D printing operations enables businesses to automate their processes while decreasing human involvement to achieve higher printing efficiency and dependability (Plathottam et al., 2023). Through sensor data analysis, AI optimizes equipment maintenance for laser machining by identifying failure precursors, which reduces downtime and maintenance expenses (Murzin, 2024). AI systems display various advantages, yet researchers need to detect biases that develop during training since these biases affect decision-making processes (Anantrasirichai & Bull, 2021). AI modeling depends heavily on data quality because this factor ensures both reliable and precise results (Cascón et al., 2024).

Simulation-Based Optimization

Simulation-based optimization uses computer simulations to model 3D printing processes to predict how different parameter configurations affect component quality and performance (Murzin, 2024). The simulations enable the optimization of printing parameters by identifying configurations that reduce defects, maximize strength, and realize other target objectives. Simulation methods are increasingly used in manufacturing industries to evaluate and optimize performance metrics of advanced manufacturing systems. The simulation of additive manufacturing processes accurately predicts temperature distribution, residual stress, and deformation of the part being manufactured (Shin et al., 2020). We can observe the effects of modifications or interventions on finished products through simulation, which helps optimize the manufacturing process (D'Emilia et al., 2020). The physics-based models in additive manufacturing use heat transfer fluid flow and structural mechanics to predict the final characteristics of the part (Kumar et al., 2022). The 3D printing process requires numerical models, including finite element analysis comp, mutational fluid dynamics, and discrete element modeling to simulate its various components. The simulation models analyze material movement, thermal heat distribution, and stress development (Druzgalski et al., 2020). Simulation optimization is vital for material property prediction, defect minimization, design, and process parameter enhancement. The determination of end-product quality and performance depends heavily on these factors. The method enables a thorough investigation of design possibilities while making it easier to detect important factors that affect product quality. Implementing multi-physics simulation in production aims to improve product quality and process efficiency while decreasing time requirements and expenses. A complete understanding of the process becomes possible through simulation, which enables continuous improvement by predicting virtual outcomes (Zhang et al., 2024).

Simulation and Modeling Techniques

Simulation and modeling methods play a key role in 3D printing parameter optimization, where manufacturers can predict and improve the printing process before implementation. These methods leverage computational platforms to simulate the printing process, thus making it possible to find the best parametric settings without the need for time-consuming trial-and-error experiments (Alharbi et al., 2020). Finite element analysis is a valuable tool for predicting the mechanical behavior of printed parts under different loading conditions. Computational fluid dynamics simulations can model the material flow through the printer nozzle and predict the resulting layer formation.

Topology optimization serves to create designs that maximize stiffness with, at the same time, weight reduction (Scapin & Peroni, 2021). Models like these may include material behavior, printing conditions, and ambient conditions to project the result of the printing operation.

Forecasting Models

Predictive modeling is the development of mathematical models forecasting the 3D printing process's behavior under different conditions. The models employ machine learning algorithms to process historical data and determine patterns to forecast future outcomes (Balani et al., 2018). The model incorporates input variables like layer thickness, print speed, and material properties to predict the ensuing part quality and mechanical properties. With continuous monitoring and optimizing of print parameters, real-time optimization can translate to substantial improvement in the quality and consistency of the parts produced. Finite Element Analysis Finite Element Analysis (FEA) is a computational method applied to simulate the behavior of a physical system when subjected to different conditions. FEA is utilized to forecast stress patterns, thermal behavior, and deformation of 3D printed parts. FEA recognizes potential failure points and optimizes printing and design parameters, thereby improving structural integrity (Ficzere et al., 2021).

Computational Fluid Dynamics

Computational Fluid Dynamics (CFD) simulates material flow in the 3D printing process. CFD simulations can be utilized to optimize printing parameters and nozzle design, providing even material distribution and avoiding defects (Du et al., 2019).

In-Process Monitoring and Feedback Control

In-process monitoring combined with feedback control methods is used to supervise the 3D printing process, making adjustments to the printing parameters where necessary. This process ensures that the printing process is maintained within specified parameters, guaranteeing that the final component conforms to the specifications. The real-time monitoring provision is achieved by incorporating sensors within the 3D printing machine to take measurements of temperature, pressure, and layer thickness. Feedback control systems interpret sensor information and automatically regulate print parameters like laser power, print speed, and material feed rate to preserve ideal operating conditions.

Enabled by in-process monitoring, this method is a powerful tool for detecting anomalies, reducing defects, and ensuring consistent quality in the parts produced. Real-time feedback loops allow instant adjustment while printing, ensuring accurate control of material deposition and minimizing the chance of error or inconsistency. In-situ monitoring systems with advanced capabilities offer extensive observation of the additive manufacturing process, allowing real-time adjustment and improved process control. This level of regulation significantly enhances the quality of the parts produced and optimizes resource usage efficiency by minimizing waste and reducing the need for subsequent processing. By observing material properties and printing conditions in real-time, manufacturers can institute immediate adjustments to prevent defects and ensure consistent quality. In-situ monitoring is crucial in process control improvement, material loss minimization, and additive manufacturing process optimization. It provides excellent visibility into the additive manufacturing process, allowing real-time anomaly detection and correction to ensure optimum print conditions. Manufacturers can achieve greater precision and efficiency by continuously monitoring critical parameters and providing feedback for real-time correction.

Gathering information about manufacturing operations makes real-time quality control possible, creating a platform for analyzing relationships between quality characteristics and assets.

This approach overlaps process temperature data to determine anomalies (Bernhard et al., 2020). Machine learning software can be trained to recognize patterns and anomalies in the data, enabling the prediction and prevention of defects. This is a crucial aspect of the real-time quality control system. Integrating in-process monitoring and control systems improves printed parts' quality and contributes to waste minimization and optimization of resource usage (Ompusunggu et al., 2020; Wolter et al., 2011). Industries that require assured component quality and reproducible processes, such as aerospace and healthcare sectors, stand to gain significantly from adopting in-situ process monitoring and closed-loop control systems (Scime & Beuth, 2017). Advances in sensor technology and the Internet of Things have facilitated real-time system monitoring, which is necessary to maintain the integrity of the manufacturing environment and prevent unplanned repairs and operational disruptions (Sundaram & Zeid, 2023; Vernezi et al., 2021). Smart manufacturing implements real-time production process monitoring, from material selection to the final product, via Internet of Things technology (Sari et al., 2020). Random failures may result in production interruption and produce waste of raw materials or system malfunctioning (Yuan et al., 2019). Therefore, the ability to monitor in real-time is essential to ensuring safety in the realm of smart manufacturing (Ding et al., 2020).

Material-Specific Optimization Strategies

Material-specific optimization strategies are critical in 3D printing to attain the optimum level of results. Polymers, metals, ceramics, and composites have distinct properties that influence their performance in the printing process (Budzik et al., 2021). Optimization strategies guarantee that the printing process is closely adjusted to get the most out of the material, leading to improved part quality and function. Knowing the thermal behavior of the material is important in managing heat distribution while printing to avert warping or cracking. The mechanical behavior of the printed part can be significantly affected by changing printing settings, such as layer height, printing speed, and infill density. Controlling the printing environment temperature and the cooling rate for polymers is essential in preventing warping and facilitating adequate layer adhesion (Abdulhameed et al., 2019). Also, the infill's configuration and density greatly influence the final component's mechanical strength and weight, providing a flexible means of satisfying individual performance needs. In the case of metals, optimizing laser energy and scanning speed is crucial to attaining the desired density and microstructural characteristics (Yuan et al., 2020). The mechanical properties of metal components can also be improved through post-processing procedures like heat treatment. The support structures aim to give enough support while printing and minimizing the material use and post-processing effort. In the case of ceramics, it is essential to manage the slurry characteristics and binder content to obtain high density and dimensional accuracy. Printing with composite materials involves optimizing the print parameters by carefully balancing the matrix and reinforcement phase characteristics.

It is vital to tune printing parameters to specific material properties to attain the desired mechanical and thermal behavior for the final product (Picard et al., 2020). Acrylonitrile Butadiene Styrene is one of the most common polymers used for Fused Deposition Modeling (Zárybnická et al., 2019). Printing with ABS needs cautious temperature control to evade warping and facilitate layer adhesion (Dul et al., 2021).

CASE STUDIES AND INDUSTRIAL APPLICATIONS

3D printing parameter optimization is essential across multiple industries because case studies demonstrate how it enhances quality productivity and financial performance. Aerospace engineering benefits from optimized print parameters for titanium and nickel-based superalloys to create lightweight aircraft and engine assembly components with enhanced corrosion resistance (Tetsuka & Shin, 2020). The optimization of 3D printing parameters for biocompatible matrices leads to customized medical implants and prosthetics that demonstrate better functionality and biocompatibility. The automobile industry depends on optimized 3D printing parameters for prototyping tooling and component tailoring to produce lightweight components with shorter production times and greater design flexibility (Jankovics & Barari, 2019). The three-dimensional printing technology now operates across multiple sectors, including aircraft manufacturing, geoscience, education, fashion, medical, and pharmaceutical industries (Fan et al., 2020). The extensive utilization of 3D printing across different industries demonstrates its transformative power to transform production methods (Trifol et al., 2021). The technology of 3D printing provides cost-effective solutions and facilitates better collaboration while making design and production more accessible to everyone (Nadagouda et al., 2020; Trenfield et al., 2019). Research and science have experienced breakthroughs through the revolutionary production techniques enabled by 3D printing technology (Coakley & Hurt, 2016). The manufacturing process generates customized structures with special arrangements focusing on individual anatomical needs (Mathews et al., 2020).

CHALLENGES AND FUTURE DIRECTIONS

Notwithstanding the great strides that have been made in optimizing 3D printing parameters, there are still a few challenges that remain, thwarting the realization of the full potential of this technology. Among them is the narrow range of materials that we have available at our disposal, especially when we want to move beyond classical polymers (Oberoi et al., 2018). The internal complexities of such materials during printing make it challenging to predict suitable printing conditions for novel substances. More research is needed to develop better models that more effectively predict material behavior over a range of printing parameters.

The pharmaceutical industry will benefit significantly from how 3D printing is used in personalized medicine, effectively addressing the unique needs of heterogeneous populations by creating customized products with ease and excellent quality (Dhyani et al., 2020) (Wang et al., 2023). The growing need for innovative assembly and production technologies opens a new era for unmatched accuracy, improved product quality, high-speed production rates, and lower costs, all in efforts to maximize efficiency and independence in production, eventually leading to shorter times to market (Spencer, 2018). By successfully combining 3D printing with other Industry 4.0 technologies, production operations get considerable improvements, with consequent economic advantages like lower warehouse and repair expenses, improved quality due to better personalization, and increased employee safety (Molino et al., 2020). Moreover, integrating 3D printing technology into the food supply chain makes food processing personalized to clients' precise requirements, optimizing operations where possible (Zhang et al., 2024). Besides, 3D printing technology provides quick prototyping, low costs, and the ability to create complicated geometries (Ali et al., 2020; Rojek et al., 2025; Tang, 2018). This innovative method has become a revolutionary material, mechanical, and biomedical engineering frontier. There is no question that 3D printing is best suited for producing low-cost customized goods.

While its development has been remarkable, 3D printing is still hampered by limitations like low speeds, a narrow range of materials available, and the need for considerable post-processing to achieve standards within industries (Zhou et al., 2024). These limitations will have to be addressed for broad application across industries. The long-term vision for 3D printing is to create complicated and fully functional parts with unmatched levels of personalization while unlocking opportunities for SMEs to participate in the manufacturing industry (Garcia et al., 2013). Machine learning and artificial intelligence integration can improve real-time printing parameters and part quality without generating waste (Królikowski et al., 2019). As production relies on automation more and more, future factories will most likely use a lot of AI tech (Dwivedi et al., 2019).

With the broader acceptance of AI technologies, its application in the construction industry is expected to expand (Na et al., 2022). AI can potentially revolutionize operations and processes in various industries (Dwivedi et al., 2019). The construction industry has been behind other industries in adopting new technologies (Na et al., 2022). In addition, computer-based systems have the potential to understand human biofield technology, monitor health parameters, and provide remote diagnostic services, eventually making healthcare services delivery better (Dwivedi et al., 2019).

CONCLUSION

Parameter optimization in 3D printing requires a comprehensive approach that involves material science, process management, and advanced monitoring. The ability for researchers and engineers to effectively control these parameters is vital in realizing the full potential of 3D printing for the high-value added personalized parts useful in diverse sectors. Technology-based construction will continue to evolve as research and development move forward to tackle existing problems and expand its application scope. Addressing these challenges with a simultaneous opening of new opportunities will lead to widespread adoption of 3D printing across various industries, making production more efficient, sustainable, and innovative (Jin et al., 2020; Mahmood et al., 2020). New materials and processes with unique design versatility will especially enhance the design freedom when creating parts with the best functional performance; such materials are particularly suitable for additive manufacturing. Moreover, it would provide real-time control over complex production systems through combined intelligence ANS (Anantrasirichai & Bull, 2021). However, construction offers huge potential areas where 3D printing can be used more extensively due to increased levels of automation within processes, greater efficiency at reduced waste rates along with improved geometrical accuracy (Filip et al., 2023; Tahmasebinia et al.,2018). In the future also his application shall revolutionize reduced costs and time savings while reducing environmental pollution as well as onsite accidents injuries Research and development are going to continue playing a vital role in addressing existing constraints as well as enlarging applications of three-dimensional printing under innovative frameworks. Furthermore, machine learning would also contribute significantly by streamlining processes for better efficiency as well as reliability in three-dimensional printing.

REFERENCES

Abdulhameed, O., Al-Ahmari, A., Ameen, W., & Mian, S. H. (2019). Additive manufacturing: Challenges, trends, and applications. *Advances in Mechanical Engineering, 11*(2), 1687814018822880. Advance online publication. DOI: 10.1177/1687814018822880

Ahmed, W., Alnajjar, F., Zaneldin, E., Al-Marzouqi, A. H., Gochoo, M., & Khalid, S. (2020). Implementing FDM 3D Printing Strategies Using Natural Fibers to Produce Biomass Composite [Review of Implementing FDM 3D Printing Strategies Using Natural Fibers to Produce Biomass Composite]. Materials, 13(18), 4065. Multidisciplinary Digital Publishing Institute. DOI: 10.3390/ma13184065

Alharbi, M., Kong, I., & Patel, V. I. (2020). Simulation of uniaxial stress–strain response of 3D-printed polylactic acid by nonlinear finite element analysis. *Applied Adhesion Science, 8*(1), 5. Advance online publication. DOI: 10.1186/s40563-020-00128-1

Ali, A., Ahmad, U., & Akhtar, J. (2020). 3D Printing in Pharmaceutical Sector: An Overview. In *IntechOpen eBooks*. IntechOpen., DOI: 10.5772/intechopen.90738

Almy, R. D., & Tontowi, A. E. (2018). THE EFFECT OF 3D PRINTING MACHINE PARAMETERS IN EXTRUSION PROCESS OF BIOCOMPOSITE MATERIALS (PMMA AND HA) ON DIMENSIONAL ACCURACY. *SINERGI, 22*(2), 83. DOI: 10.22441/sinergi.2018.2.003

Anantrasirichai, N., & Bull, D. (2021). Artificial intelligence in the creative industries: a review [Review of Artificial intelligence in the creative industries: a review]. Artificial Intelligence Review, 55(1), 589. Springer Science+Business Media. DOI: 10.1007/s10462-021-10039-7

Babagowda, M., Kadadevara Math, R. S., Goutham, R., & Srinivas Prasad, K. R. (2018). Study of Effects on Mechanical Properties of PLA Filament which is blended with Recycled PLA Materials. *IOP Conference Series. Materials Science and Engineering, 310*, 12103. DOI: 10.1088/1757-899X/310/1/012103

Balani, S. B., Cantarel, A., Chabert, F., & Nassiet, V. (2018). Influence of parameters controlling the extrusion step in fused filament fabrication (FFF) process applied to polymers using numerical simulation. *AIP Conference Proceedings, 1960*, 140003. DOI: 10.1063/1.5034995

Bernhard, R. J., Neef, P., Wiche, H., Hoff, C., Hermsdorf, J., Kaierle, S., & Wesling, V. (2020). Defect detection in additive manufacturing via a toolpath overlaid melt-pool-temperature tomography. *Journal of Laser Applications*, *32*(2), 022055. Advance online publication. DOI: 10.2351/7.0000055

Brüggemann, H., Stempin, S., & Meier, J.-M. (2020). Consideration of digitalization for the purpose of resource efficiency in a learning factory. *Procedia Manufacturing*, *45*, 140–145. DOI: 10.1016/j.promfg.2020.04.085

Budzik, G., Woźniak, J., Paszkiewicz, A., Przeszłowski, Ł., Dziubek, T., & Dębski, M. (2021). Methodology for the Quality Control Process of Additive Manufacturing Products Made of Polymer Materials. *Materials (Basel)*, *14*(9), 2202. DOI: 10.3390/ma14092202 PMID: 33923048

Buj-Corral, I., Domínguez-Fernández, A., & Durán-Llucià, R. (2019). Influence of Print Orientation on Surface Roughness in Fused Deposition Modeling (FDM) Processes. *Materials (Basel)*, *12*(23), 3834. DOI: 10.3390/ma12233834 PMID: 31766409

Cabreira, V., & Santana, R. M. C. (2020). Effect of infill pattern in Fused Filament Fabrication (FFF) 3D Printing on materials performance. *Matéria (Rio de Janeiro)*, *25*(3), e-12826. Advance online publication. DOI: 10.1590/s1517-707620200003.1126

Cascón, I., Gómez-Omella, M., Fernández, D., Gil, A., Alberdi, N., & González, H. (2024). Towards Zero-Defect Manufacturing Based on Artificial Intelligence through the Correlation of Forces in 5-Axis Milling Process. *Machines (Basel)*, *12*(4), 226. DOI: 10.3390/machines12040226

Chang, C. C., Boland, E. D., Williams, S. K., & Hoying, J. B. (2011). Direct-write bioprinting three-dimensional biohybrid systems for future regenerative therapies [Review of Direct-write bioprinting three-dimensional biohybrid systems for future regenerative therapies]. Journal of Biomedical Materials Research Part B Applied Biomaterials, 1, 160. Wiley. DOI: 10.1002/jbm.b.31831

Chari, V. S., Venkatesh, P. R., Krupashankar, , & Dinesh, V. (1943). Krupashankar, & Dinesh, V. (2018). Effect of processing parameters on FDM process. *AIP Conference Proceedings*, *20061*, 020061. Advance online publication. DOI: 10.1063/1.5029637

Chua, C. K., & Leong, K. F. (2013). 3D Printing and Additive Manufacturing. DOI: 10.1142/9008

Coakley, M. F., & Hurt, D. E. (2016). 3D Printing in the Laboratory: Maximize Time and Funds with Customized and Open-Source Labware. *SLAS Technology*, *21*(4), 489–495. DOI: 10.1177/2211068216649578 PMID: 27197798

Cristea, D., Pop, M. A., Faraian, C., & Munteanu, D. (2019). The influence of additive manufacturing parameters on the structural and mechanical properties of acrylonitrile butadiene styrene (ABS) parts produced by fused filament fabrication. *IOP Conference Series. Materials Science and Engineering*, *682*(1), 12013. DOI: 10.1088/1757-899X/682/1/012013

D'Emilia, G., Ilio, A. D., Gaspari, A., Natale, E., & Stamopoulos, A. G. (2020). Uncertainty assessment for measurement and simulation in selective laser melting: A case study of an aerospace part. *Acta IMEKO*, *9*(4), 96. DOI: 10.21014/acta_imeko.v9i4.720

da Silva, C. E. S., Salgado, E. G., Mello, C. H. P., Oliveira, E., & Leal, F. (2013). Integration of computer simulation in design for manufacturing and assembly. *International Journal of Production Research*, *52*(10), 2851–2866. DOI: 10.1080/00207543.2013.853887

Dhyani, A., Singh, N., Kumar, V., & Dhyani, A. (2020). Applications of 3 Dimensional (3D) Printing in Biomedical Field. *International Journal of Current Research and Review*, *12*(19), 71–75. DOI: 10.31782/IJCRR.2020.121918

Ding, H., Gao, R. X., Isaksson, A., Landers, R. G., Parisini, T., & Yuan, Y. (2020). State of AI-Based Monitoring in Smart Manufacturing and Introduction to Focused Section. *IEEE/ASME Transactions on Mechatronics*, *25*(5), 2143–2154. DOI: 10.1109/TMECH.2020.3022983

Druzgalski, C., Ashby, A. S., Guss, G., King, W. E., Roehling, T. T., & Matthews, M. J. (2020). Process optimization of complex geometries using feed forward control for laser powder bed fusion additive manufacturing. *Additive Manufacturing*, *34*, 101169. DOI: 10.1016/j.addma.2020.101169

Du, Z., Liu, J., & He, Y. (2019). Intelligent Analysis Method of Assembly Process Based on Semantic Elements. *IOP Conference Series. Materials Science and Engineering*, *563*(4), 42055. DOI: 10.1088/1757-899X/563/4/042055

Duan, G., & Yan, X. (2020). A Real-Time Quality Control System Based on Manufacturing Process Data. *IEEE Access : Practical Innovations, Open Solutions*, *8*, 208506–208517. DOI: 10.1109/ACCESS.2020.3038394

Dul, S., Fambri, L., & Pegoretti, A. (2021). High-Performance Polyamide/Carbon Fiber Composites for Fused Filament Fabrication: Mechanical and Functional Performances. *Journal of Materials Engineering and Performance*, *30*(7), 5066–5085. DOI: 10.1007/s11665-021-05635-1

Dwivedi, Y. K., Hughes, L., Ismagilova, E., Aarts, G., Coombs, C., Crick, T., Duan, Y., Dwivedi, R., Edwards, J. S., Eirug, A., Galanos, V., Ilavarasan, P. V., Janssen, M., Jones, P., Kar, A. K., Kizgin, H., Kronemann, B., Lal, B., Lucini, B., & Williams, M. D. (2019). Artificial Intelligence (AI): Multidisciplinary perspectives on emerging challenges, opportunities, and agenda for research, practice and policy. *International Journal of Information Management*, *57*, 101994. DOI: 10.1016/j.ijinfomgt.2019.08.002

Equbal, M. A., Equbal, A., Khan, Z. A., & Badruddin, I. A. (2025). Machine learning in additive manufacturing: A comprehensive insight. *International Journal of Lightweight Materials and Manufacture*, *8*(2), 264–284.

Fan, D., Li, Y., Wang, X., Zhu, T., Wang, Q., Cai, H., Li, W., Tian, Y., & Liu, Z. (2020). Progressive 3D Printing Technology and Its Application in Medical Materials [Review of Progressive 3D Printing Technology and Its Application in Medical Materials]. Frontiers in Pharmacology, 11. Frontiers Media. DOI: 10.3389/fphar.2020.00122

Fernández, F. M., & Martín, M. J. (2023). Analysis of the Effect of the Surface Inclination Angle on the Roughness of Polymeric Parts Obtained with Fused Filament Fabrication Technology. *Polymers*, *15*(3), 585. DOI: 10.3390/polym15030585 PMID: 36771886

Ficzere, P., Lukács, N. L., & Borbás, L. (2021). The Investigation of Interlaminar Failures Caused by Production Parameters in Case of Additive Manufactured Polymers. *Polymers*, *13*(4), 556. DOI: 10.3390/polym13040556 PMID: 33668572

Filip, G. A., Abanda, F. H., & Fru, F. A. (2023). Construction 3D-printing in reducing the incidence of long latency respiratory diseases among construction workers in the UK. *Safety in Extreme Environments*, *5*(3), 177–197. DOI: 10.1007/s42797-023-00078-4

Filippidis, P., Papazetis, G., & Vosniakos, G.-C. (2020). Process parameter investigation for 3D printing of cellular structured parts. *Procedia Manufacturing*, *51*, 717–724. DOI: 10.1016/j.promfg.2020.10.101

Garcia, C. R., Rumpf, R. C., Tsang, H., & Barton, J. H. (2013). Effects of extreme surface roughness on 3D printed horn antenna. *Electronics Letters*, *49*(12), 734–736. DOI: 10.1049/el.2013.1528

Garcia, F. L., Moris, V. A. da S., Nunes, A. O., & Silva, D. A. L. (2018). Environmental performance of additive manufacturing process – an overview. *Rapid Prototyping Journal*, *24*(7), 1166–1177. DOI: 10.1108/RPJ-05-2017-0108

Gargalis, L., Madonna, V., Giangrande, P., Rocca, R., Ashcroft, I., Hague, R., & Galea, M. (2020)... *Development and Testing of Soft Magnetic Rotor for a Switched Reluctance Motor Built Through Additive Manufacturing Technology.*, *263*, 263–268. Advance online publication. DOI: 10.23919/ICEMS50442.2020.9290893

Ghilan, A., Chiriac, A. P., Niţă, L. E., Rusu, A. G., Neamţu, I., & Chiriac, V. M. (2020). Trends in 3D Printing Processes for Biomedical Field: Opportunities and Challenges [Review of Trends in 3D Printing Processes for Biomedical Field: Opportunities and Challenges]. Journal of Polymers and the Environment, 28(5), 1345. Springer Science+Business Media. DOI: 10.1007/s10924-020-01722-x

Ginestra, P. S., Ceretti, E., & Fiorentino, A. (2016). Potential of modeling and simulations of bioengineered devices: Endoprostheses, prostheses and orthoses [Review of Potential of modeling and simulations of bioengineered devices: Endoprostheses, prostheses and orthoses]. Proceedings of the Institution of Mechanical Engineers Part H Journal of Engineering in Medicine, 230(7), 607. SAGE Publishing. DOI: 10.1177/0954411916643343

Goh, G. D., Sing, S. L., & Yeong, W. Y. (2020). A review on machine learning in 3D printing: applications, potential, and challenges [Review of A review on machine learning in 3D printing: applications, potential, and challenges]. Artificial Intelligence Review, 54(1), 63. Springer Science+Business Media. DOI: 10.1007/s10462-020-09876-9

Grünberger, T. (2020). Machine Learning for Process Monitoring Systems. *PhotonicsViews*, *17*(3), 56–59. DOI: 10.1002/phvs.202000026

Hooda, N., Chohan, J. S., Gupta, R., & Kumar, R. (2021). Deposition angle prediction of Fused Deposition Modeling process using ensemble machine learning. *ISA Transactions*, *116*, 121–128. DOI: 10.1016/j.isatra.2021.01.035 PMID: 33518221

Huang, Z., Shi, T., & Yue, X. (2019). The Optimization of FDM 3D Printer's Structure Based on Finite Element Analysis. MATEC Web of Conferences, 257, 2004. DOI: 10.1051/matecconf/201925702004

Hussein, M. O., & Hussein, L. A. (2022). Optimization of Digital Light Processing Three-Dimensional Printing of the Removable Partial Denture Frameworks; The Role of Build Angle and Support Structure Diameter. *Materials (Basel)*, *15*(6), 2316. DOI: 10.3390/ma15062316 PMID: 35329767

Jankovics, D., & Barari, A. (2019). Customization of Automotive Structural Components using Additive Manufacturing and Topology Optimization. *IFAC-PapersOnLine*, *52*(10), 212–217. DOI: 10.1016/j.ifacol.2019.10.066

Jin, Z., Zhang, Z., Demir, K., & Gu, G. X. (2020). Machine Learning for Advanced Additive Manufacturing. *Matter*, *3*(5), 1541–1556. DOI: 10.1016/j.matt.2020.08.023

Kam, M., İpekçi, A., & Şengül, Ö. (2020). Effect of FDM Process Parameters on the Mechanical Properties and Production Costs of 3D Printed PowerABS Samples. [IJAEFEA]. *International Journal of Analytical Experimental and Finite Element Analysis*, *7*(3). Advance online publication. DOI: 10.26706/ijaefea.3.7.20200806

Khdoudi, A., & Masrour, T. (2018). Prediction of Industrial Process Parameters using Artificial Intelligence Algorithms. arXiv (Cornell University). https://doi.org//arxiv.1812.10537DOI: 10.48550

Kim, D. B., Denno, P., & Jones, A. (2015). A model-based approach to refine process parameters in smart manufacturing. *Concurrent Engineering, Research and Applications*, *23*(4), 365–376. DOI: 10.1177/1063293X15591038

Kiran, A. R., Li, Y., Hodek, J., Brázda, M., Urbánek, M., & Džugan, J. (2022). Heat Source Modeling and Residual Stress Analysis for Metal Directed Energy Deposition Additive Manufacturing. *Materials (Basel)*, *15*(7), 2545. DOI: 10.3390/ma15072545 PMID: 35407876

Kishore, R. A., Sivaraj, S., Dhinakaran, V., Prabhagar, V. S. P. P., & Srinivas, S. (2021). Strength of 3D prints with variable print orientation. *Journal of Physics: Conference Series*, *2027*(1), 12021. DOI: 10.1088/1742-6596/2027/1/012021

Królikowski, T., Knitter, R., & Stachnik, M. (2019). Thermo-mechanic tests using 3d printed elements. *Procedia Computer Science*, *159*, 2551–2559. DOI: 10.1016/j.procs.2019.09.430

Kumar, S., Gopi, T., Harikeerthana, N., Gupta, M. K., Gaur, V., Królczyk, G., & Wu, C. (2022). Machine learning techniques in additive manufacturing: A state of the art review on design, processes and production control. *Journal of Intelligent Manufacturing*, *34*(1), 21–55. DOI: 10.1007/s10845-022-02029-5

Lee, J., Walker, J., Natarajan, S., & Yi, S. (2019). Prediction of geometric characteristics in polycaprolactone (PCL) scaffolds produced by extrusion-based additive manufacturing technique for tissue engineering. *Rapid Prototyping Journal*, *26*(2), 238–248. DOI: 10.1108/RPJ-08-2018-0219

Li, G., Odum, K., Yau, C., Soshi, M., & Yamazaki, K. (2021). High productivity fluence based control of Directed Energy Deposition (DED) part geometry. *Journal of Manufacturing Processes*, *65*, 407–417. DOI: 10.1016/j.jmapro.2021.03.028

Liu, Z., Zhang, M., Bhandari, B., & Wang, Y. (2017). 3D printing: Printing precision and application in food sector. *Trends in Food Science & Technology*, *69*, 83–94. DOI: 10.1016/j.tifs.2017.08.018

Loh, G. H., Pei, E., González-Gutiérrez, J., & Monzón, M. (2020). An Overview of Material Extrusion Troubleshooting. *Applied Sciences (Basel, Switzerland)*, *10*(14), 4776. DOI: 10.3390/app10144776

Mahmood, M. A., Visan, A. I., Ristoscu, C., & Mihăilescu, I. N. (2020). Artificial Neural Network Algorithms for 3D Printing [Review of Artificial Neural Network Algorithms for 3D Printing]. Materials, 14(1), 163. Multidisciplinary Digital Publishing Institute. DOI: 10.3390/ma14010163

Mathews, D. A. P., Baird, A. D., & Lucky, M. (2020). Innovation in Urology: Three Dimensional Printing and Its Clinical Application [Review of Innovation in Urology: Three Dimensional Printing and Its Clinical Application]. Frontiers in Surgery, 7. Frontiers Media. DOI: 10.3389/fsurg.2020.00029

Maurya, N. K., Maurya, M., Saxena, A., Tyagi, A., Kiran, G. K., Singh, H., Tripathi, H., Kumar, P., & Yadav, S. (2021). Investigation of mechanical properties and form error of the components fabricated by rapid prototyping: A review [Review of Investigation of mechanical properties and form error of the components fabricated by rapid prototyping: A review]. Materials Today Proceedings, 47, 3901. Elsevier BV. DOI: 10.1016/j.matpr.2021.03.553

Mendřický, R., & Fris, D. (2020). Analysis of the Accuracy and the Surface Roughness of FDM/FFF Technology and Optimisation of Process Parameters. Tehnicki Vjesnik - Technical Gazette, 27(4). DOI: 10.17559/TV-20190320142210

Mladenovska, D., Lazarevska, A. M., & Krstanoski, M. (2021). Attributes relevant for sustainable additive manufacturing – material driven approach. *IOP Conference Series. Materials Science and Engineering*, *1190*(1), 12001. DOI: 10.1088/1757-899X/1190/1/012001

Molino, M., Cortese, C. G., & Ghislieri, C. (2020). The Promotion of Technology Acceptance and Work Engagement in Industry 4.0: From Personal Resources to Information and Training.

Morettini, G., Palmieri, M., Capponi, L., & Landi, L. (2022). Comprehensive characterization of mechanical and physical properties of PLA structures printed by FFF-3D-printing process in different directions. *Progress in Additive Manufacturing*, *7*(5), 1111–1122. DOI: 10.1007/s40964-022-00285-8

Murzin, S. P. (2024). Artificial Intelligence-Driven Innovations in Laser Processing of Metallic Materials. *Metals*, *14*(12), 1458. DOI: 10.3390/met14121458

Na, S., Heo, S., Han, S., Shin, Y.-S., & Roh, Y.-S. (2022). Acceptance Model of Artificial Intelligence (AI)-Based Technologies in Construction Firms: Applying the Technology Acceptance Model (TAM) in Combination with the Technology–Organisation–Environment (TOE) Framework. *Buildings (Basel, Switzerland)*, *12*(2), 90. DOI: 10.3390/buildings12020090

Nadagouda, M. N., Ginn, M., & Rastogi, V. (2020). A review of 3D printing techniques for environmental applications [Review of A review of 3D printing techniques for environmental applications]. Current Opinion in Chemical Engineering, 28, 173. Elsevier BV. DOI: 10.1016/j.coche.2020.08.002

Natarajan, M., Pasupuleti, T., & Kiruthika, J. D. P. (2024), "Optimization and Regression Modeling of Additive Manufacturing (Fused Deposition Modeling) of PETG Material for Automobile Applications," SAE Technical Paper 2024-28-0234. DOI: 10.4271/2024-28-0234

Navarro, J., Din, M., Janes, M. E., Swayambunathan, J., Fisher, J. P., & Dreher, M. L. (2019). Effect of print orientation on microstructural features and mechanical properties of 3D porous structures printed with continuous digital light processing. *Rapid Prototyping Journal*, *25*(6), 1017–1029. DOI: 10.1108/RPJ-10-2018-0276

Oberoi, G., Nitsch, S., Edelmayer, M., Janjić, K., Müller, A. S., & Agis, H. (2018). 3D Printing—Encompassing the Facets of Dentistry [Review of 3D Printing—Encompassing the Facets of Dentistry]. Frontiers in Bioengineering and Biotechnology, 6. Frontiers Media. DOI: 10.3389/fbioe.2018.00172

Olsson, N. O. E., Shafqat, A., Arica, E., & Økland, A. (2019), "3D-Printing Technology in Construction: Results from a Survey", Lill, I. and Witt, E. (Ed.) 10th Nordic Conference on Construction Economics and Organization (Emerald Reach Proceedings Series, Vol. 2), Emerald Publishing Limited, Leeds, pp. 349-356. DOI: 10.1108/S2516-285320190000002044

Omairi, A., & Ismail, Z. H. (2021). Towards Machine Learning for Error Compensation in Additive Manufacturing. *Applied Sciences (Basel, Switzerland)*, *11*(5), 2375. DOI: 10.3390/app11052375

Ompusunggu, A. P., Vonderscher, Y., & Motl, D. (2020). Physics-based vibration feature for detecting eccentric workpiece/runout faults during continuous generating gear grinding processes. *Mechanical Systems and Signal Processing*, *153*, 107536. DOI: 10.1016/j.ymssp.2020.107536

Panwisawas, C., Tang, Y. T., & Reed, R. C. (2020). Metal 3D printing as a disruptive technology for superalloys. *Nature Communications*, *11*(1), 2327. Advance online publication. DOI: 10.1038/s41467-020-16188-7 PMID: 32393778

Pathak, K., Saikia, R., Das, A., Das, D., Islam, M. A., Pramanik, P., Parasar, A., Borthakur, P. P., Sarmah, P., Saikia, M., & Borthakur, B. (2023). 3D printing in biomedicine: Advancing personalized care through additive manufacturing. *Exploration of Medicine*, *1135*, 1135–1167. Advance online publication. DOI: 10.37349/emed.2023.00200

Picard, M., Mohanty, A. K., & Misra, M. (2020). Recent advances in additive manufacturing of engineering thermoplastics: challenges and opportunities [Review of Recent advances in additive manufacturing of engineering thermoplastics: challenges and opportunities]. RSC Advances, 10(59), 36058. Royal Society of Chemistry. DOI: 10.1039/D0RA04857G

Plathottam, S. J., Rzonca, A., Lakhnori, R., & Iloeje, C. O. (2023). A review of artificial intelligence applications in manufacturing operations [Review of A review of artificial intelligence applications in manufacturing operations]. Journal of Advanced Manufacturing and Processing, 5(3). Wiley. DOI: 10.1002/amp2.10159

Ren, L., Zhou, X., Song, Z., Zhao, C., Liu, Q., Xue, J., & Li, X. (2017). Process Parameter Optimization of Extrusion-Based 3D Metal Printing Utilizing PW–LDPE–SA Binder System. *Materials (Basel)*, *10*(3), 305. DOI: 10.3390/ma10030305 PMID: 28772665

Rojek, I., Mikołajewski, D., Kempiński, M., Galas, K., & Piszcz, A. (2025). Emerging Applications of Machine Learning in 3D Printing. *Applied Sciences (Basel, Switzerland)*, *15*(4), 1781. DOI: 10.3390/app15041781

Sari, M. W., Herianto, H., Dharma, I. G. B. B., & Tontowi, A. E. (2020). Design of Product Monitoring System Using Internet of Things Technology for Smart Manufacturing. *IOP Conference Series. Materials Science and Engineering*, *835*(1), 12048. DOI: 10.1088/1757-899X/835/1/012048

Scapin, M., & Peroni, L. (2021). Numerical Simulations of Components Produced by Fused Deposition 3D Printing. *Materials (Basel)*, *14*(16), 4625. DOI: 10.3390/ma14164625 PMID: 34443146

Scime, L., & Beuth, J. (2017). Anomaly detection and classification in a laser powder bed additive manufacturing process using a trained computer vision algorithm. *Additive Manufacturing*, *19*, 114–126. DOI: 10.1016/j.addma.2017.11.009

Shin, D. S., Lee, C. H., Kühn, U., Lee, S., Park, S. J., Schwab, H., Scudino, S., & Kosiba, K. (2020). Optimizing laser powder bed fusion of Ti-5Al-5V-5Mo-3Cr by artificial intelligence. *Journal of Alloys and Compounds*, *862*, 158018. DOI: 10.1016/j.jallcom.2020.158018

Silbernagel, C., Aremu, A., & Ashcroft, I. (2019). Using machine learning to aid in the parameter optimisation process for metal-based additive manufacturing. *Rapid Prototyping Journal*, *26*(4), 625–637. DOI: 10.1108/RPJ-08-2019-0213

Spencer, O. O. (2018). Additive Manufacturing Technology Development: A Trajectory Towards Industrial Revolution. *American Journal of Mechanical and Industrial Engineering*, *3*(5), 80. DOI: 10.11648/j.ajmie.20180305.12

Sundaram, S., & Zeid, A. (2023). Artificial Intelligence-Based Smart Quality Inspection for Manufacturing. *Micromachines*, *14*(3), 570. DOI: 10.3390/mi14030570 PMID: 36984977

Tahmasebinia, F., Niemelä, M., Sepasgozar, S. M. E., Lai, T., Su, W., Reddy, K. R., Shirowzhan, S., Sepasgozar, S. M. E., & Marroquín, F. A. (2018). Three-Dimensional Printing Using Recycled High-Density Polyethylene: Technological Challenges and Future Directions for Construction. *Buildings (Basel, Switzerland)*, *8*(11), 165. DOI: 10.3390/buildings8110165

Tang, Y. (2018). Development Trend and Application Prospect of 3D Printing Technology in the Intelligent Manufacturing. *International Robotics & Automation Journal*, *4*(1). Advance online publication. DOI: 10.15406/iratj.2018.04.00089

Tetsuka, H., & Shin, S. R. (2020). Materials and technical innovations in 3D printing in biomedical applications [Review of Materials and technical innovations in 3D printing in biomedical applications]. Journal of Materials Chemistry B, 8(15), 2930. Royal Society of Chemistry. DOI: 10.1039/D0TB00034E

Tian, Y., Chen, C., Xu, X., Wang, J., Hou, X., Li, K., Lu, X., Shi, H., Lee, E., & Jiang, H. B. (2021). A Review of 3D Printing in Dentistry: Technologies, Affecting Factors, and Applications [Review of A Review of 3D Printing in Dentistry: Technologies, Affecting Factors, and Applications]. Scanning, 2021, 1. Wiley. DOI: 10.1155/2021/9950131

Trenfield, S. J., Awad, A., Madla, C. M., Hatton, G. B., Firth, J., Goyanes, Á., Gaisford, S., & Basit, A. W. (2019). Shaping the future: recent advances of 3D printing in drug delivery and healthcare [Review of Shaping the future: recent advances of 3D printing in drug delivery and healthcare]. Expert Opinion on Drug Delivery, 16(10), 1081. Taylor & Francis. DOI: 10.1080/17425247.2019.1660318

Trifol, J., Jayaprakash, S., Baniasadi, H., Ajdary, R., Kretzschmar, N., Rojas, O. J., Partanen, J., & Seppälä, J. (2021). 3D-Printed Thermoset Biocomposites Based on Forest Residues by Delayed Extrusion of Cold Masterbatch (DECMA). *ACS Sustainable Chemistry & Engineering*, *9*(41), 13979–13987. DOI: 10.1021/acs-suschemeng.1c05587

Tyflopoulos, E., Hofset, T. A., Olsen, A., & Steinert, M. (2021). Simulation-based design: A case study in combining optimization methodologies for angle-ply composite laminates. *Procedia CIRP*, *100*, 607–612. DOI: 10.1016/j.procir.2021.05.131

Vernezi, M., Lukyanov, A., Dolgov, V., & Zotov, V. V. (2021). About diagnostics of technical condition of metal-cutting machines. *IOP Conference Series. Materials Science and Engineering*, *1029*, 12124. DOI: 10.1088/1757-899X/1029/1/012124

Wang, C., Tan, X., Tor, S. B., & Lim, C. S. (2020). Machine learning in additive manufacturing: State-of-the-art and perspectives. *Additive Manufacturing*, *36*, 101538. DOI: 10.1016/j.addma.2020.101538

Wang, S., Chen, X., Han, X., Hong, X., Li, X., Zhang, H., Li, M., Wang, Z., & Zheng, A. (2023). A Review of 3D Printing Technology in Pharmaceutics: Technology and Applications, Now and Future [Review of A Review of 3D Printing Technology in Pharmaceutics: Technology and Applications, Now and Future]. Pharmaceutics, 15(2), 416. Multidisciplinary Digital Publishing Institute. DOI: 10.3390/pharmaceutics15020416

Wang, T., Kaur, L., Beniwal, A. S., Furuhata, Y., Aoyama, H., & Singh, J. (2023). Physico-chemical and Textural Properties of 3D Printed Plant-based and Hybrid Soft Meat Analogs. *Plant Foods for Human Nutrition (Dordrecht, Netherlands)*, *78*(2), 375–382. DOI: 10.1007/s11130-023-01068-4 PMID: 37199825

Wolter, B., Dobmann, G., & Boller, C. (2011). NDT Based Process Monitoring and Control. Strojniški Vestnik –. *Jixie Gongcheng Xuebao*, *2011*(3), 218–226. DOI: 10.5545/sv-jme.2010.172

Wu, J., Aage, N., Lefèbvre, S., & Wang, C. C. L. (2017). Topology Optimization for Computational Fabrication. Eurographics. DOI: 10.2312/egt.20171030

Yin, J., Suo, Y., Lv, T., Ma, K., Wang, X., & Zhang, Z. (2022). *Application of 3D Printing Technology in the Construction Industry and Its Development Prospects*. Atlantis Highlights in Intelligent Systems/Atlantis Highlights in Intelligent Systems., DOI: 10.2991/ahis.k.220601.002

Yuan, S., Li, J., Yao, X., Zhu, J., Gu, X., Gao, T., Xu, Y., & Zhang, W. (2020). Intelligent optimization system for powder bed fusion of processable thermoplastics. *Additive Manufacturing*, *34*, 101182. DOI: 10.1016/j.addma.2020.101182

Yuan, Y., Ma, G., Cheng, C., Zhou, B., Zhao, H., Zhang, H., & Ding, H. (2019). *Artificial Intelligent Diagnosis and Monitoring in Manufacturing. arXiv*. Cornell University., DOI: 10.1093/nsr/nwz190

Zárybnická, L., Dvořák, K., Dostalova, Z., & Vojáčková, H. (2019). Study of Different Printing Design Type Polymer Samples Prepared by Additive Manufacturing. *Periodica Polytechnica. Chemical Engineering*, *64*(2), 255–264. DOI: 10.3311/PPch.13991

Zhang, H., Zhou, X., Fang, Y., Xiong, Z., & Zhang, T. (2024). AI-driven 3D bioprinting for regenerative medicine: From bench to bedside [Review of AI-driven 3D bioprinting for regenerative medicine: From bench to bedside]. Bioactive Materials, 45, 201. Elsevier BV. DOI: 10.1016/j.bioactmat.2024.11.021

Zhang, S., Yu, Q., Wan, S., Cao, H., & Huang, Y. (2024). Digital supply chain: Literature review of seven related technologies. *Manufacturing Review*, *11*, 8. DOI: 10.1051/mfreview/2024006

Zhou, L., Miller, J., Vezza, J., Mayster, M., Raffay, M., Justice, Q., Tamimi, Z. A., Hansotte, G., Sunkara, L. D., & Bernat, J. (2024). Additive Manufacturing: A Comprehensive Review [Review of Additive Manufacturing: A Comprehensive Review]. Sensors, 24(9), 2668. Multidisciplinary Digital Publishing Institute. DOI: 10.3390/s24092668

Chapter 12
The Foundational Principles of FDM:
An Acronym Pivotal t Its Operation and Its Wide Range of Applications

Ruchira Srivastava
https://orcid.org/0000-0002-5957-9534
Amity University, Greater Noida, India

Ayushi Thakur
https://orcid.org/0000-0002-4172-6751
Amity University, Greater Noida, India

ABSTRACT

The use of additive manufacturing is becoming more common in the contemporary industrial sector. This method uses the layer-by-layer deposition concept to create objects. As a result, the current study is critically important and needs to focus on the elements, applications, and materials of the fused-based modeling technique, such as nylon, metals, carbon fiber, polycaprolactone (PCL), polypropylene, polyethylene, and polylactic acid (PLA). Additionally, it has been determined that the FDM approach is frequently employed in the development of items that are necessary in the current environment. Nozzle swabs and face mask supports are made using a variety of additive manufacturing processes. The application and development of fused deposition modeling (FDM), sometimes referred to as three-dimensional (3D) printing, have grown dramatically due to the ongoing demand for increasingly complex designs and efficient material utilization. The fabrication of complicated designs is made possible by AM's ability to create near-net-shaped components with a lower buy-to-fly ratio.

DOI: 10.4018/979-8-3373-0533-2.ch012

1. INTRODUCTION

Additive manufacturing, often known as 3D printing, is a digital manufacturing technique that builds 3D items directly from computer-aided design (CAD) models by layering on materials (Baechle-Clayton et al., 2022). The ability of additive manufacturing to construct complicated geometries that would be challenging or impossible to make using conventional manufacturing processes is one of its key advantages. This allows for greater design freedom and innovation in product development (Tsioukas et al., 2020).

Due to its enticing benefits, including the capacity to create inexpensive, multipurpose things with high levels of detail and unrestricted creative flexibility, 3D printing has been more popular over the past 20 years intricate/delicate constructions in a brief amount of time. Concrete components, for instance, have the ability to be 3D printed, which could save construction waste labor expenses by 50–80%, building time by 50–70%, and by 30–60% (Gordelier et al., 2019). Consequently, 3D printing has emerged as a viable manufacturing method for quick prototyping as well as a number of technical domains, including electronics, mechanical, civil, aerospace, and biomedical. A wide range of materials, including metals, polymers, polymer composites, ceramics, and cement, may be 3D printed using a number of AM techniques (Palanisamy et al., 2022). Using seven broad categories, the ASTM (ISO/ASTM 52900:2015) has categorized the variety of AM techniques (Chaudhari et al., 2018). This grouping of processes, which include sheet lamination, material extrusion, direct energy deposition, powder bed fusion, vat photo polymerization, binder jetting, and material jetting, is based on the fundamental concept of operation (Gong et al., 2019).

Moreover, AM may be divided into three categories—solid-based, powder-based, and liquid-based—based on the kind of base material that is utilized, as seen in fig. 1 Electron beam free form fabrication (EBF3), wire and arc additive manufacturing (WAAM), fused deposition modeling (FDM), and laminated object manufacturing (LOM) are further categories for solid-based AM (Di Angelo, Di Stefano, & Marzola, 2017). There are four types of powder-based additive manufacturing: laser metal deposition (LMD), electron beam melting (EBM), selective laser melting (SLM), and selective laser sintering (SLS). The majority of liquid-based techniques include digital light processing (DLP), stereolithography (SLA), and material jetting (MJ). To have a thorough understanding of the aforementioned AM methods, LOM, FDM, WAAM, EBF, SLS, EBM, SLM, LMD, SLA, DLP, and MJ, we consult these outstanding review papers (Ramazani & Kami, 2022).

Figure 1. Classification Of Additive Manufacturing

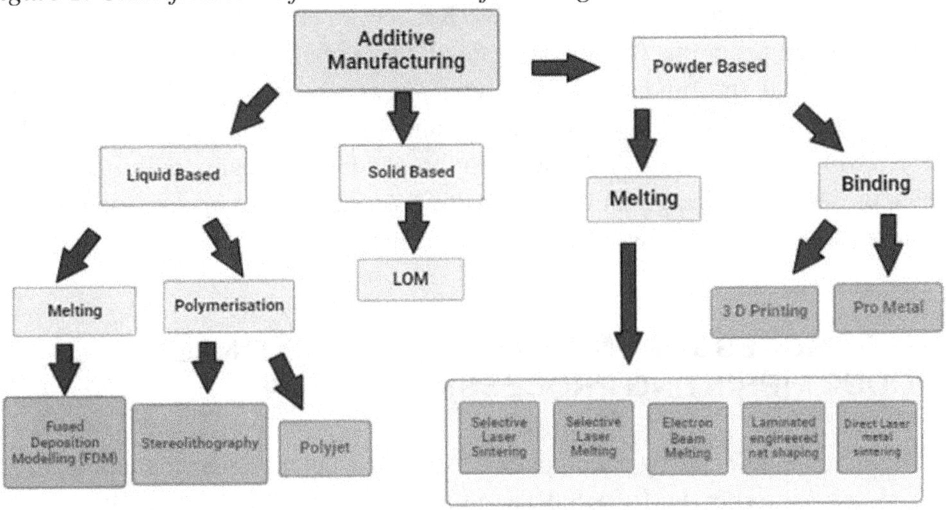

A polymer is a material comprised of numerous repeating subunits that combine to form very huge molecules, often known as macromolecules. Polymers are among the well-known

materials' extensive range of mechanical, thermal, electrical, fire-resistant, and biocompatible qualities make them useful in a variety of applications (Mazurchevici, Nedelcu, & Popa, 2020a). In line with the Web of Science (accessed August 2, 2021), polymers account for more than 60% of AM research. All three of the polymers—that is, solid-based, powder-based, and liquid-centered) are AM methods (Solomon et al., 2021). The most popular and traditional solid-based 3D printing method for producing polymer components is FDM. Conversely, SLS is a well-known AM method that uses polymeric powders as a basis to create polymer components, but another popular early-adopted AM method for producing polymer components is vat-based SLA, which processes the polymeric liquid as a base material (Di Angelo et al., 2020).

To the best of the knowledge, there exist several review articles in the literature that address different facets of polymer printing based on FDM, SLS, and SLA; however, a thorough investigation that focuses on the correlation between material-process-properties for these techniques is lacking for both traditional 3D printing and newly developed 4D printing methods (Tagliaferri et al., 2019). Our goal in writing this article is to provide the link between material-process-properties for the three most popular polymer-based 3D printing processes: stereolithography (SLA), selective laser sintering (SLS), and fused deposition modeling (FDM). We also touch on how these techniques are used to the 4D printing of polymer-based materials

(3D printing of smart materials) in brief, providing an example of 4D printing of polymers that respond to magnetic fields (Kristiawan et al., 2021).

Educational programs that incorporate additive manufacturing's (AM) revolutionary potential and fundamental concepts are necessary given the technology's significance for the future of product design and manufacturing infrastructure (Hiemenz, 2011). Additionally, AM's depth and breadth cross many traditional fields, which presents a challenge for teachers as well as a chance to integrate knowledge through challenging and innovative projects (Patel et al., 2022).

2. THE PROCESS FROM DEVELOPMENT TO FINAL PRODUCT (PRINTING PROCESS) USING FDM

Additive manufacturing encompasses a range of technologies, including Direct Metal Laser Sintering (DMLS), Selective Laser Sintering (SLS), Stereolithography (SLA), and Fused Deposition Modeling (FDM), and others. Every technology has its own strengths, materials, and applications (Cuiffo et al., 2017).

The creation of pieces through the printing of consecutive layers of material layered on top of one another is the basic process of 3D printing. Figure 2 shows the workflow for the 3D printing process (Turner et al., 2014). The standard tessellation language (.stl) file of the CAD is generated once the CAD model of the object to be developed is constructed. The CAD file's continuous geometry is mostly transformed into tiny triangles during the STL file production procedure (Wickramasinghe et al., 2020).

After that, a model slicing program exports the .stl file and generates a tool path for the 3D printer. Here, the cross-sectional data is extracted from the 3D model and converted into 2D slices (Mazurchevici, Nedelcu, & Popa, 2020b). After that, the 3D printer begins the stacking and material processing steps. After then, the finished item is removed from the printer. The overall quality of the printed pieces is determined by several aspects involved in the printing process. For instance, the overall quality of the printed items might be affected by the .stl file (Qattawi & Ablat, 2017). The more faithfulness to the shape of the printed product or part, the smaller the triangles in the.stl file. While environmental elements like humidity and temperature also affect the final product's overall quality, component orientation during the printing process determines the mechanical qualities (Mwema & Akinlabi, 2020).

Layer-by-layer deposition of thermoplastic filaments on a build platform is the method known as fused filament fabrication (FFF) or fused deposition modeling (FDM).

After being heated to a semi-solid state, the polymer filament is placed on the heated platform or print bed (Sbriglia et al., 2016). The nozzle travels in the same direction as the last item in the designated layer. The material is extruded and the nozzle once more follows the object's path in the specified layer as the platform descends one step or the nozzle rises one step for the subsequent layer (Nurhudan et al., 2021).

A slicer cuts the model layer by layer to create a G-code (computer numerical control programming language), which the nozzle follows in each layer to create the route the nozzle takes. The height at which the nozzle descends following each layer is the model's layer thickness (Rahman et al., 2018). The factors that determine the 3D printed part's fractional behaviour are the nozzle temperature, bed temperature, and layer height.

Figure 2. 3D Printing: Workflow

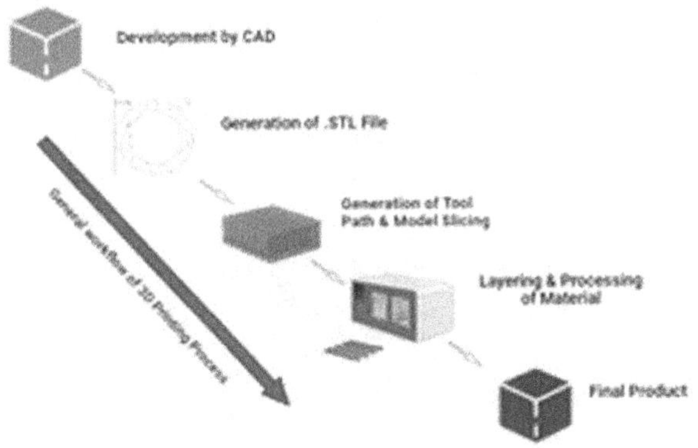

For FDM printing, the bed calibration and nozzle diameter are the machine parameters, while the nozzle temperature, bed temperature, extrusion width, and raster angle are the process parameters (Messimer et al., 2018). The schematic depiction is shown in Figure 3. One of the most crucial factors to take into account is the bed calibration. Since the material is directly deposited by the nozzle in FDM, a consistent spacing must be maintained between the nozzle and the bed across the whole bed (Sehhat, Mahdianikhotbesara, & Yadegari, 2022a). This makes FDM a contact print method. Inadequate bed calibration results in war page and the printer striking the bed and prints at two or more separate locations on the print platform/ bed due to an unequal distance between the nozzle and the bed (Beniak et al., 2019). Although the 3D printing nozzle's diameter may be altered or replaced, doing so affects the part's quality and manufacturing time. The manufacturing time of the part is accelerated when a nozzle with a large hole diameter is used (Karwasz &

Osinski, 2020). Additionally, it has been shown that in FDM 3D printing, increasing the nozzle diameter improves the mechanical qualities and product quality. The surrounding temperature, which contributes to component warpage, is another crucial factor (Samykano et al., 2019). For instance, PLA component warpage of around 50%, 30%, and 10% at ambient print environment temperatures of 10 oC, 15 oC, and 20 oC, respectively, is recorded. The following process variables have an impact on component qualities in FDM: bed temperature, nozzle speed, nozzle temperature, raster angle, extrusion width, and extrusion rate (Thakur et al., 2024). The angle formed between the nozzle's direction and the printing platform's x- (or y-, depending on nomenclature) axis is known as the raster angle.

Figure 3. Overview of the process parameters of 3D printing: FDM (recreated from (Thakur et al., 2024))

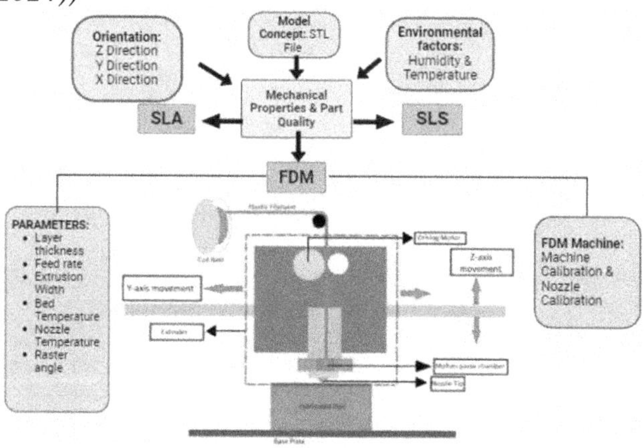

The speed at which the filament is extruded onto the build platform from the nozzle is known as the extrusion rate (Pérez et al., 2018). The construction platform's temperature is referred to as the "bed temperature." In order to prevent warpage and preserve the adhesion between the print portion and the build platform, the bed temperature is necessary (Kumar et al., 2021). The temperature at which a substance melts and extrudes from a nozzle is known as the nozzle temperature. It significantly affects the microstructure and mechanical characteristics of the 3D printed items (Rinaldi et al., 2018). For example, it is claimed that when the nozzle temperature is raised from 370 oC to 390 oC, the relative density increases (from 89.9% to 92.8% for PEEK) (Pérez et al., 2020). The speed at which the nozzle travels when depositing or printing melted filament onto the build platform is known as the nozzle speed (Doshi et al., 2022). Despite a shorter print time, it has a significant impact on the printed items' dimensional accuracy (Singh, Ramakrishna, & Singh, 2017).

For instance, it is stated that when the nozzle speed is increased from 30 mm/s to 90 mm/s, the wall thickness of the ring-shaped design increases from 2.00 mm to 2.17 mm (Quiñonez et al., 2021). Additive manufacturing enables the cost-effective production of customized or personalized products, as each item can be individually tailored without the need for expensive tooling or moulds (Moldovan et al., 2018). This is particularly valuable in fields like healthcare, where patient-specific medical devices and implants can be created (Fazzini et al., 2019).

Table 1. Information is gathered from the supplier's website to create a database of industrial-grade FDM 3D printers that are commercially available, along with their specs. According to the provider, the print material that is most appropriate for the specified printer is indicated by the yellow highlight.

Name	Print Volume in cm^3	Dimension of Printer in mm	Layer Thickness in mm	Available material types					Ref.
Create Bot PEEK- 300	**36,000**	**300*300*400**	**0.04–0.4**	**PEEK**	**ULTEM**	**ABS**	**PLA**	**Nylon**	**(Doshi et al., 2022)**
AON-M2+	129,600	450*450*640	0.05–0.5	PEEK	ULTEM	ABS	PLA	Nylon	(Singh, Ramakrishna, & Singh, 2017)
ULTIMAKER S5	33,660	330*340*300	0.02–.25	PEEK	ULTEM	ABS	PLA	Nylon	(Quiñonez et al., 2021)
3D Gence F420	60,648	380*380*400	0.05	PEEK	ULTEM	ABS	PLA	Nylon	(Moldovan et al., 2018)
Stratasys F900	508756.16	914*609*914	.127–0.5	PEEK	ULTEM	ABS	PLA	Nylon	(Fazzini et al., 2019)

3. 3D Printing materials used in FDM

Additive manufacturing can utilize various materials, includes plastics, metals, ceramics, composites, and even bio-materials. The choice of material depends on factors such as desired properties, application requirements, and the specific additive manufacturing technology being used (Sehhat, Mahdianikhotbesara, & Yadegari, 2022b). The melting and bonding of thermoplastic material—where the polymer is employed as the printing material—is the foundation of the FDM production process (Sun et al., 2023). The features of the thermoplastic material used in the process—such as its low cost, long lifespan, high toughness, ease of finding, ability to be recycled, low shaping temperature, and ability to reshape when heated—allow the approach to become more and more popular (Ning et al., 2017). With this technology, home users may now make a variety of parts. The cost of devices produced using the FDM method is less than that of devices produced using other AM technologies because of the straightforward operating principle and little equipment needed (Mohamed, 2017). The FDM process created desktop-sized 3D printers because of these characteristics. Individual users have also adopted the FDM technology because of its low cost and other benefits associated with both equipment and thermoplastic materials (Abouzaid et al., 2019). Thermoplastics are polymer-class materials that are employed in FDM manufacturing; at low temperatures, they exist in a solid form but may be moulded and softened by heating. The product with the required shape is produced by cooling and solidifying the thermoplastic once the forming process is complete (Salem Bala et al., 2016). The mechanical characteristics of the filament, such as its elastic modulus and strain at yield, are also taken into account in FDM. Fig 3 lists the typical print materials that may be used with FDM 3D printing. Thermoplastic polymers are used as the material for FDM (Kamaal et al., 2021). The material is in filament form for FDM because to the process requirements. There are two types of thermoplastic polymers: amorphous and semi-crystalline. The glass transition temperature (Tg) of amorphous thermoplastic polymers is the point at which they soften and become glassy (Black et al., 2016). Semi-crystalline polymers have a set glass transition temperature (Tg) and melting temperature (Tm), but they lack a constant melting temperature. When the temperature rises above the melting point (Tm), the semi-crystalline thermoplastics' melt viscosity reduces, enabling flow ability (Cerneels et al., 2013).

Figure 4. 3D printing materials used in FDM Technology

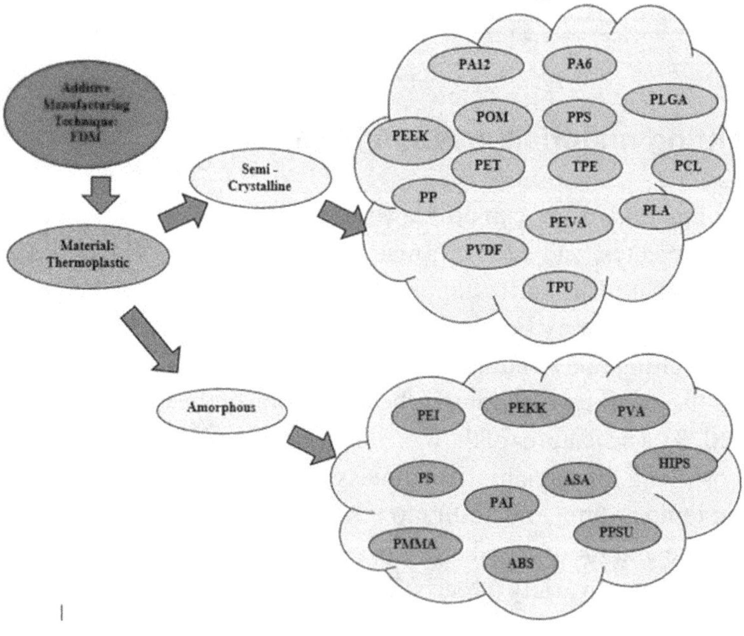

4. GENERAL DEFECTS OF FDM

Because additively created materials are being used widely and growing quickly, it is critical to consider the safety and integrity of the components as well as the testing procedures (Ramesh & Panneerselvam, 2021). In contrast to components that are conventionally created, typical stress/strain testing or flaw inspection tests would cause the part to be damaged or destroyed because of the nature of the polymer-based FDM materials. Porosity and variations in the material's density are now the most prevalent flaws (Agrawal, 2021). Many factors contribute to this, but the most frequent one is the manufacturing and environmental management of the application of temperature and humidity fluctuations throughout the fabrication process. A filament is heated and cooled during the fabrication process of a product using FDM technology. The suggested temperature range for PLA is 401 °F +/- 27 °F. The diameter of the filament and the extrusion nozzle of the printer head, however, will modify the recommended temperature (Ghaffar et al., 2018). In FDM printing, the extrusion temperature is very important. The viscosity and stickiness of the filament, which determine whether the print will perform, are directly impacted by temperature (Hodzic & Pandzic, 2019). The problem is not considered for elements that are intended to be aesthetically pleasing. This problem especially formidable for parts that are manufactured for the functionality they will ultimately provide within a fully integrated system (Valerga et al., 2018).

When a physical item is printed, there's a chance that certain problems will disrupt the flow of material, leading to the possibility of both minor and large flaws. Small flaws might not be taken into account if the layers above and/or below make up for a missing or flawed layer (Sargini et al., 2021). On the other hand, a 3D printed item cannot meet the required output if significant flaws arise. Additionally, to avoid these kinds of errors, it's critical to understand the 3D printing process and take preventative measures. If not, misprinting costs money and causes material and time loss (Shanmugam et al., 2020).

Fused Deposition Modeling (FDM) 3D printing can encounter several defects during the printing process. Here are some common defects:

a. **Stringing:** Stringing occurs when filament oozes out from the nozzle and forms thin strands between different parts of the print. This often happens during travel moves between non-adjacent parts of the print or when retracting filament from the nozzle (Shanmugam et al., 2021). Stringing can result in unwanted connections between features of the print or rough surface finishes.
b. **Warping:** Warping is the lifting or curling of the edges or corners of a print due to uneven cooling or adhesion to the print bed. It typically occurs with materials like ABS, which have high shrinkage rates as they cool (Garzon-Hernandez et

al., 2020). Warping can lead to detachment from the print bed, layer misalignment, and ultimately, print failure.

c. **Layer Misalignment:** Layer misalignment occurs when successive layers are not properly aligned, resulting in a staircase effect on angled or curved surfaces. This defect can be caused by issues such as inconsistent extrusion, mechanical issues with the printer, or inadequate cooling between layers (Frascio et al., 2018). If the 3D printer is not calibrated or adjusted properly, misalignment of the print platform might result in serious issues, including deformation of the print platform or nozzle. The print platform can only move in the Z direction, and its movement dictates the layer thickness (Beniak et al., 2020). Occasionally, if the duration of, steps are not well-tuned, and the subsequent layer does not cling to the preceding layer materials that are extruded or bowed out. The print platform may not align properly, which can lead to a variety of issues including the print not adhering to the platform, the nozzle being too near to the platform, the print bowing out at the bottom—a phenomenon known as "elephant's foot"—the print edges bending (warping syndrome), and the failure to capture fine detail (Kantaros et al., 2023). Therefore, you should inspect the threaded rods or lead screws that are utilized with your printer to see if there is any bending or printing debris that is preventing the print bed from moving freely. It is crucial to clean the rods before putting on a new layer of oil (Tekinalp et al., 2014).

d. **Under-Extrusion:** Under-extrusion happens when the printer does not deposit enough material, leading to weak and brittle parts with gaps or voids between layers. This defect can occur due to clogged nozzles, insufficient filament tension, or incorrect slicing settings (Singh et al., 2019).

e. **Over-Extrusion**: Over-extrusion occurs when the printer deposits too much material, resulting in blobbing, rough surfaces, and loss of fine detail. It can be caused by factors such as incorrect nozzle diameter settings, excessive print speed, or high extrusion temperatures (Hu et al., 2021).

f. **Z-Wobble**: Z-wobble refers to periodic variations in the height of successive layers, resulting in visible ridges or waves on vertical surfaces. This defect is often caused by mechanical issues in the printer's Z-axis, such as bent lead screws or misaligned linear rails (Yasa & Ersoy, 2018). Numerous parameters, including layer thickness, support angle, temperature of the extrusion, temperature of the platform, print speed, flow ratio of the extruder, nozzle distance, surface layers, supports, type of seam, and fan speed, may be controlled by adjusting the printer's settings (Rajaguru et al., 2020). They are all crucial factors to take into account when evaluating print quality. The most frequent issue pertaining to printer settings is stringing, which may be avoided by cleaning the nozzle and turning up the fan speed. Adjusting the filament's melting temperature may be the answer if the problem persists after these measures (Sivamani et al., 2020).

Additionally, holes may be excessively tiny in comparison to indentation due to contradictions; in this instance, enlarging the hole diameters that are around 0.3 mm less than the ideal one can avoid these issues. Another issue with printer settings is hanging strands, which may be avoided by reducing layer thickness, raising the support angle, and speeding up the fan (Hu, Jin, & Wang, 2015).

g. **Layer Adhesion Issues:** Poor layer adhesion can lead to delamination or splitting between layers, resulting in weak and structurally compromised prints. This defect can be caused by insufficient bonding between heated layers, inadequate bed temperature, or improper cooling (Budzik et al., 2021). There are instances where the print fails to adhere to the print platform, resulting in warping syndrome that can eventually destroy the actual object. This issue can be attributed to a number of factors, including an uneven platform, improper calibration, a poorly heated platform, a low-heat filament, and an unclean platform (Adapa & Jagadish, 2023). The platform can be made more textured, the print platform distance can be adjusted, adhesion can be applied, cleaning can be done, and supports can be added. Perforated print platforms are being used by Zortrax to avoid this circumstance, and it's a smart move (García-Dominguez et al., 2020). Applying a small coating of adhesive on other types of print platforms (flat platforms) allows the glue to be removed with hot water after printing. Additionally, one alternative for PLA filament printing is decorator tape. As previously stated, the nozzle's distance from the platform has a significant role in the first layer's adherence. It has to be finely adjusted (Srivastava et al., 2022).

h. **Bridging Issues:** Bridging occurs when the printer attempts to span gaps or unsupported areas between features of the print. If not properly calibrated or supported, bridging can result in sagging, drooping, or collapsed sections of the print (Li et al., 2018). Vibration is one of the important factors that determines print quality. The moving components of 3D printers, particularly the gantry robot's acceleration and jerk, are the cause of vibration or shock. On the print surface, shock or vibrations can produce visible waves (Gutierrez et al., 2011). In order to stop this from happening, the 3D printer has to be placed on a stable, level surface, such as a table, to support the device. Poor maintenance and damaged linear bearings may possibly be the cause of the vibrations (Hosseinzadeh et al., 2023). In order to avoid the gantry robot from tripping, all of the bolts, grub screws, bits, and pulleys must be sufficiently tight, the rails must be clear of debris, dust, and fluff, and the rods must be properly greased. The print speed can be lowered if all technical precautions have been followed but vibration-related visual waves are still present on the print (Tiwary et al., 2021). Due to the fact that Bowden type extruders are substantially lighter than direct type extruders, this issue is not particularly effective for Bowden type extruders employing printers. Regarding the Bowden type, the extruder is stationary and does not move like the direct type; it is situated behind the printer or near the spool. in order to lessen both inertia and the moving part's weight (Parulski et al., 2021).

i. **Raft or Support Structure Residue**: Prints that require rafts or support structures may leave behind residual material after removal, leading to rough or uneven surfaces on the final print. Excessive support material can also be difficult to remove, requiring additional post-processing (Quero et al., 2021). Checking that there is adequate material on the spool is important before beginning to print the physical model. If not, a shortage of material may cause the model to be destroyed and require reprinting if the extrusion stops in the middle of the print. Out of filament is not, however, the only issue that might arise occasionally (Zhao et al., 2019). Other typical issues include low diameter filament, stripped filament, snapped filament, tangled filament, and blocked nozzle. High-quality filaments should be used, and the spool should be examined, to avoid filament-related issues (Li, Cheng, Ahzi et al., 2023). The most common issue caused by the filament is blocked nozzles, which may be cleaned in a various way. The most popular techniques involve heating the nozzle and cleaning it with a smaller diameter than the nozzle (Bacciaglia et al., 2019). But security is significant in this instance because of the nozzle's high temperature. In addition, there are other techniques that can be employed, like heating the nozzle to extremely high temperatures in an oven to turn the filament inside into a carbonized material, attempting to force the filament through with another piece of filament, and cleaning the nozzle by soaking it in a solvent like acetone (Awasthi & Banerjee, 2021).

j. **Nozzle Clogs:** Nozzle clogs can occur due to various reasons, such as debris in the filament, insufficient nozzle temperature, or prolonged printing sessions. Clogs can lead to interruptions in material flow, resulting in incomplete or failed prints (Dey & Yodo, 2019). The gantry-robot, which consists of a two-stepper motor, belts, and pulleys, can move the nozzle. However, some issues affecting the operation of the motors, belts, and pulleys also impact the location of the nozzle, making it impossible to achieve the physical model as modelled using CAD software. Some common problems with these printer parts include loose grub screws, tight belts, and printing debris that prevents the nozzle from moving steadily (Serajuddin, 2023). The aforementioned circumstances may result in the following issues: print head missing the bed; layers not aligning properly; some missing layers; print leaning; and print offset in some areas (shifted layers) (Okafor-Muo et al., 2020). There are steps that may be done to prevent these issues, such tightening the grub screws on the X and Y axis motor pulleys and rolling on a level surface to verify if the rods are straight. Cleaning print dirt, cleaning rods, and oiling necessary parts are also crucial. Missing the print platform issues can also result from end switch/stop issues (Garland & Fadel, 2015). The end switch is not functional on the X or Y axis if the nozzle leans toward one of its edges. Addressing these defects often involves troubleshooting specific issues with printer settings, filament selection, print bed preparation, and maintenance. Regular calibration, proper filament storage, and adherence to recommended printing parameters can help minimize defects and improve print quality in FDM 3D printing (Rahim et al., 2019).

5. APPLICATION OF FDM

Fused Deposition Modeling (FDM) is widely used across various industries for a range of applications. Here are some common uses:

a. **Prototyping:** FDM is extensively used for rapid prototyping in product development. It allows engineers and designers to quickly create physical prototypes for testing form, fit, and function before moving to mass production (Syrlybayev et al., 2021). Thermoplastic material is extruded layer by layer from a nozzle using a temperature-controlled process known as FDM, or fast prototyping. A thermoplastic polymer filament is suctioned into the temperature-controlled FDM extrusion nozzle dome using this method (Cano-Vicent et al., 2021). After that, it is heated to a semi-liquid state that flows freely. A CPU controls the nozzle head's motion, which transfers the material onto a secondary platform in very thin layers. The head precisely guides the material into position. A section of the topic is constructed layer by layer, with the material bonding to the layer below and solidifying in 0.1 seconds after being expelled from the nozzle (Saroia et al., 2020). The supporting structures are later removed by cutting them out of the item; they are designed to accommodate overhanging geometries.
b. **Manufacturing Tooling:** FDM is employed to produce custom manufacturing tools, jigs, fixtures, and molds. These tools can be quickly iterated upon and customized to suit specific production needs, reducing lead times and costs associated with traditional tooling methods (Ngo et al., 2018).
c. **End-Use Parts**: FDM is increasingly used for producing end-use parts, especially in low-volume manufacturing scenarios. While it may not be as fast as injection molding, FDM offers flexibility and cost-effectiveness for producing custom or low-volume parts on demand (Khalid & Billa, 2022).
d. **Education and Research:** FDM is widely used in educational institutions for teaching purposes and in research settings for prototyping and experimentation. Its accessibility and relatively low cost make it an ideal tool for introducing

students and researchers to the principles of 3D printing (Roger & Krawczak, 2015).

e. **Customization**: FDM is utilized for producing personalized and customized products, such as prosthetics, orthotics, hearing aids, and dental aligners. Each item can be tailored to the individual's unique requirements without the need for expensive tooling or setup (Mazzanti et al., 2019).
f. **Architectural Models:** FDM is employed to create detailed architectural models and prototypes, allowing architects and designers to visualize and communicate their designs effectively. It enables the rapid iteration of designs and the exploration of complex geometries (Srivastava & Srivastava, 2020).
g. **Automotive Prototyping:** FDM is used in the automotive industry for rapid prototyping of vehicle components, such as interior panels, dashboard prototypes, and functional prototypes for testing aerodynamics and vehicle performance (Srivastava, Thakur, & Rana, 2024).
h. **Aerospace Applications:** FDM is utilized in the aerospace industry for producing lightweight components, tooling, and prototypes. It enables the production of complex geometries and helps reduce the weight of aircraft parts, leading to fuel savings and improved efficiency (Bahadur, Srivastava, & Srivastava, 2023).
i. **Medical Devices**: FDM is employed in the production of medical devices, such as surgical guides, prosthetics, orthopedic implants, and anatomical models for surgical planning. It allows for the customization of devices to match the patient's anatomy, improving outcomes and patient satisfaction (Srivastava, Arun, & Patel, 2019).

6. CONCLUSION

Overall, additive manufacturing is a transformative technology that is revolutionizing the way objects are designed, prototyped, and manufactured across industries, offering benefits such as design flexibility, customization, and reduced waste. While additive manufacturing offers numerous advantages, it also presents challenges such as limited material selection for certain processes, slower production speeds compared to traditional manufacturing methods for large-scale production, and concerns related to part quality, accuracy, and surface finish.

Additive manufacturing is utilized in various industries, including aerospace, automotive, healthcare, consumer goods, architecture, and more. It is used for prototyping, tooling, production parts, and even in the creation of art and jewellery. Additive manufacturing is often more material-efficient than traditional manufacturing methods since it only uses the material needed to build the object, reducing waste. Additionally, some additive manufacturing processes allow for the recycling or reuse of excess material. : In contrast to conventional subtractive manufacturing techniques, which include removing material from a solid block, additive manufacturing constructs products from digital three-dimensional models layer by layer.

Additive manufacturing is widely used for rapid prototyping and iterative design processes. It allows engineers and designers to quickly produce physical prototypes for testing, evaluation, and validation, accelerating the product development cycle. Fused Deposition Modeling (FDM) finds extensive use in various industries for prototyping, manufacturing tooling, end-use parts production, education, customization, and architectural modeling, among other applications. Its versatility, affordability, and accessibility make it a valuable technology for a wide range of applications

REFERENCES

Abouzaid, K., Guessasma, S., Belhabib, S., Bassir, D., & Chouaf, A. (2019). Thermal mechanical characterization of copolyester for additive manufacturing using FDM. *International Journal for Simulation and Multidisciplinary Design Optimization, 10*, A9. DOI: 10.1051/smdo/2019011

Adapa, S. K., & Jagadish, . (2023). Prospects of natural fiber-reinforced polymer composites for additive manufacturing applications: A review. *Journal of the Minerals Metals & Materials Society*, *75*(3), 920–940. DOI: 10.1007/s11837-022-05670-w

Agrawal, R. (2021). Sustainable material selection for additive manufacturing technologies: A critical analysis of rank reversal approach. *Journal of Cleaner Production, 296*, 126500. DOI: 10.1016/j.jclepro.2021.126500

Awasthi, P., & Banerjee, S. S. (2021). Fused deposition modeling of thermoplastic elastomeric materials: Challenges and opportunities. *Additive Manufacturing, 46*, 102177. DOI: 10.1016/j.addma.2021.102177

Bacciaglia, A., Ceruti, A., & Liverani, A. (2019, September). Additive manufacturing challenges and future developments in the next ten years. In *International Conference of the Italian Association of Design Methods and Tools for Industrial Engineering* (pp. 891-902). Cham: Springer International Publishing.

Baechle-Clayton, M., Loos, E., Taheri, M., & Taheri, H. (2022). Failures and flaws in fused deposition modeling (FDM) additively manufactured polymers and composites. *Journal of Composites Science, 6*(7), 202. DOI: 10.3390/jcs6070202

. Bahadur, P. S., Srivastava, R., & Srivastava, V. (2023). Review on the Development of Smart Materials, Including Shape Memory Alloys, Characterization, and Recent Applications. *Modeling, Characterization, and Processing of Smart Materials*, 107-139.

Beniak, J., Holdy, M., Križan, P., & Matúš, M. (2019). Research on parameters optimization for the Additive Manufacturing process. *Transportation Research Procedia, 40*, 144–149. DOI: 10.1016/j.trpro.2019.07.024

Beniak, J. U. R. A. J., Krizan, P. E. T. E. R., & Matus, M. I. L. O. S. (2020). Conductive material properties for FDM additive manufacturing. *MM Science Journal, 2020*(1), 3846–3851. DOI: 10.17973/MMSJ.2020_03_2019135

. Black, H. T., Celina, M. C., & McElhanon, J. R. (2016). Additive manufacturing of polymers: Materials opportunities and emerging applications.

Budzik, G., Woźniak, J., Paszkiewicz, A., Przeszłowski, Ł., Dziubek, T., & Dębski, M. (2021). Methodology for the quality control process of additive manufacturing products made of polymer materials. *Materials (Basel)*, *14*(9), 2202. DOI: 10.3390/ma14092202 PMID: 33923048

Cano-Vicent, A., Tambuwala, M. M., Hassan, S. S., Barh, D., Aljabali, A. A., Birkett, M., Arjunan, A., & Serrano-Aroca, Á. (2021). Fused deposition modelling: Current status, methodology, applications and future prospects. *Additive Manufacturing*, *47*, 102378. DOI: 10.1016/j.addma.2021.102378

. Cerneels, J., Voet, A., Ivens, J., & Kruth, J. P. (2013). Additive manufacturing of thermoplastic composites. *Composites Week@ Leuven*, 1-7.

Chaudhari, M., Jogi, B. F., & Pawade, R. S. (2018). Comparative study of part characteristics built using additive manufacturing (FDM). *Procedia Manufacturing*, *20*, 73–78. DOI: 10.1016/j.promfg.2018.02.010

Cuiffo, M. A., Snyder, J., Elliott, A. M., Romero, N., Kannan, S., & Halada, G. P. (2017). Impact of the fused deposition (FDM) printing process on polylactic acid (PLA) chemistry and structure. *Applied Sciences (Basel, Switzerland)*, *7*(6), 579. DOI: 10.3390/app7060579

Dey, A., & Yodo, N. (2019). A systematic survey of FDM process parameter optimization and their influence on part characteristics. *Journal of Manufacturing and Materials Processing*, *3*(3), 64. DOI: 10.3390/jmmp3030064

Di Angelo, L., Di Stefano, P., Dolatnezhadsomarin, A., Guardiani, E., & Khorram, E. (2020). A reliable build orientation optimization method in additive manufacturing: The application to FDM technology. *International Journal of Advanced Manufacturing Technology*, *108*(1-2), 263–276. DOI: 10.1007/s00170-020-05359-x

Di Angelo, L., Di Stefano, P., & Marzola, A. (2017). Surface quality prediction in FDM additive manufacturing. *International Journal of Advanced Manufacturing Technology*, *93*(9-12), 3655–3662. DOI: 10.1007/s00170-017-0763-6

Doshi, M., Mahale, A., Singh, S. K., & Deshmukh, S. (2022). Printing parameters and materials affecting mechanical properties of FDM-3D printed Parts: Perspective and prospects. *Materials Today: Proceedings*, *50*, 2269–2275. DOI: 10.1016/j.matpr.2021.10.003

Fazzini, G., Paolini, P., Paolucci, R., Chiulli, D., Barile, G., Leoni, A., & Ferri, G. (2019, June). Print on air: Fdm 3d printing without supports. In *2019 II Workshop on Metrology for Industry 4.0 and IoT (MetroInd4. 0&IoT)* (pp. 350-354). IEEE. DOI: 10.1109/METROI4.2019.8792846

Frascio, M., Avalle, M., & Monti, M. (2018). Fatigue strength of plastics components made in additive manufacturing: First experimental results. *Procedia Structural Integrity*, *12*, 32–43. DOI: 10.1016/j.prostr.2018.11.109

García-Dominguez, A., Claver, J., & Sebastián, M. A. (2020). Optimization methodology for additive manufacturing of customized parts by fused deposition modeling (FDM). Application to a shoe heel. *Polymers*, *12*(9), 2119. DOI: 10.3390/polym12092119 PMID: 32957582

Garland, A., & Fadel, G. (2015). Design and manufacturing functionally gradient material objects with an off the shelf three-dimensional printer: Challenges and solutions. *Journal of Mechanical Design*, *137*(11), 111407. DOI: 10.1115/1.4031097

Garzon-Hernandez, S., Garcia-Gonzalez, D., Jérusalem, A., & Arias, A. (2020). Design of FDM 3D printed polymers: An experimental-modelling methodology for the prediction of mechanical properties. *Materials & Design*, *188*, 108414. DOI: 10.1016/j.matdes.2019.108414

Ghaffar, S. H., Corker, J., & Fan, M. (2018). Additive manufacturing technology and its implementation in construction as an eco-innovative solution. *Automation in Construction*, *93*, 1–11. DOI: 10.1016/j.autcon.2018.05.005

Gong, H., Snelling, D., Kardel, K., & Carrano, A. (2019). Comparison of stainless steel 316L parts made by FDM-and SLM-based additive manufacturing processes. *Journal of the Minerals Metals & Materials Society*, *71*(3), 880–885. DOI: 10.1007/s11837-018-3207-3

Gordelier, T. J., Thies, P. R., Turner, L., & Johanning, L. (2019). Optimising the FDM additive manufacturing process to achieve maximum tensile strength: A state-of-the-art review. *Rapid Prototyping Journal*, *25*(6), 953–971. DOI: 10.1108/RPJ-07-2018-0183

Gutierrez, C., Salas, R., Hernandez, G., Muse, D., Olivas, R., MacDonald, E., & Zufelt, B. (2011, January). Cubesat fabrication through additive manufacturing and micro-dispensing. In *International Symposium on Microelectronics* (Vol. 2011, No. 1, pp. 001021-001027). International Microelectronics Assembly and Packaging Society. DOI: 10.4071/isom-2011-THA4-Paper3

Hiemenz, J. (2011). 3D printing with FDM: How it Works. *Stratasys Inc*, *1*, 1–5.

Hodzic, D., & Pandzic, A. (2019, October). Influence of carbon fibers on mechanical properties of materials in FDM technology. In *Proceedings of the 30th DAAAM International Symposium on Intelligent Manufacturing and Automation, Zadar, ISSN* (pp. 1726-9679). DOI: 10.2507/30th.daaam.proceedings.044

Hosseinzadeh, M., Ghoreishi, M., & Narooei, K. (2023). 4D printing of shape memory polylactic acid beams: An experimental investigation into FDM additive manufacturing process parameters, mathematical modeling, and optimization. *Journal of Manufacturing Processes*, *85*, 774–782. DOI: 10.1016/j.jmapro.2022.12.006

Hu, K., Jin, S., & Wang, C. C. (2015). Support slimming for single material based additive manufacturing. *Computer Aided Design*, *65*, 1–10. DOI: 10.1016/j.cad.2015.03.001

Hu, Y., Ladani, R. B., Brandt, M., Li, Y., & Mouritz, A. P. (2021). Carbon fibre damage during 3D printing of polymer matrix laminates using the FDM process. *Materials & Design*, *205*, 109679. DOI: 10.1016/j.matdes.2021.109679

Kamaal, M., Anas, M., Rastogi, H., Bhardwaj, N., & Rahaman, A. (2021). Effect of FDM process parameters on mechanical properties of 3D-printed carbon fibre–PLA composite. *Progress in Additive Manufacturing*, *6*(1), 63–69. DOI: 10.1007/s40964-020-00145-3

Kantaros, A., Soulis, E., Petrescu, F. I. T., & Ganetsos, T. (2023). Advanced Composite Materials Utilized in FDM/FFF 3D Printing Manufacturing Processes: The Case of Filled Filaments. *Materials (Basel)*, *16*(18), 6210. DOI: 10.3390/ma16186210 PMID: 37763488

Karwasz, A., & Osinski, F. (2020). *Literature review on emissions from additive manufacturing by FDM method and their impact on human health*. Management and Production Engineering Review. DOI: 10.24425/mper.2020.134933

Khalid, G. M., & Billa, N. (2022). Solid dispersion formulations by FDM 3D printing—A review. *Pharmaceutics*, *14*(4), 690. DOI: 10.3390/pharmaceutics14040690 PMID: 35456524

Kristiawan, R. B., Imaduddin, F., Ariawan, D., Ubaidillah, , & Arifin, Z. (2021). A review on the fused deposition modeling (FDM) 3D printing: Filament processing, materials, and printing parameters. *Open Engineering*, *11*(1), 639–649. DOI: 10.1515/eng-2021-0063

Kumar, S. R., Sridhar, S., Venkatraman, R., & Venkatesan, M. (2021). Polymer additive manufacturing of ASA structure: Influence of printing parameters on mechanical properties. *Materials Today: Proceedings*, *39*, 1316–1319. DOI: 10.1016/j.matpr.2020.04.500

Li, S., Cheng, P., Ahzi, S., Peng, Y., Wang, K., Chinesta, F., & Correia, J. P. M. (2023). Advances in hybrid fibers reinforced polymer-based composites prepared by FDM: A review on mechanical properties and prospects. *Composites Communications*, *40*, 101592. DOI: 10.1016/j.coco.2023.101592

Li, Y., Gao, S., Dong, R., Ding, X., & Duan, X. (2018). Additive manufacturing of PLA and CF/PLA binding layer specimens via fused deposition modeling. *Journal of Materials Engineering and Performance*, *27*(2), 492–500. DOI: 10.1007/s11665-017-3065-0

Mazurchevici, A. D., Nedelcu, D., & Popa, R. (2020). Additive manufacturing of composite materials by FDM technology. *RE:view*.

Mazzanti, V., Malagutti, L., & Mollica, F. (2019). FDM 3D printing of polymers containing natural fillers: A review of their mechanical properties. *Polymers*, *11*(7), 1094. DOI: 10.3390/polym11071094 PMID: 31261607

Messimer, S. L., Patterson, A. E., Muna, N., Deshpande, A. P., & Rocha Pereira, T. (2018). Characterization and processing behavior of heated aluminum-polycarbonate composite build plates for the FDM additive manufacturing process. *Journal of Manufacturing and Materials Processing*, *2*(1), 12. DOI: 10.3390/jmmp2010012

Mohamed, O. A. (2017). Analytical modeling and experimental investigation of product quality and mechanical properties in FDM additive manufacturing. *Swinburne University of Technology, Diss.*

Moldovan, C., Cosma, C., Berce, P., & Balc, N. (2018). Theoretical analysis and practical case studies of SLA, Polyjet and FDM manufacturing techniques. Acta Technica Napocensis-Series: Applied Mathematics. *Mechanical Engineering (New York, N.Y.)*, *61*(3).

. Mwema, F. M., Akinlabi, E. T., Mwema, F. M., & Akinlabi, E. T. (2020). Basics of fused deposition modelling (FDM). *Fused deposition modeling: strategies for quality enhancement*, 1-15.

Ngo, T. D., Kashani, A., Imbalzano, G., Nguyen, K. T., & Hui, D. (2018). Additive manufacturing (3D printing): A review of materials, methods, applications and challenges. *Composites. Part B, Engineering*, *143*, 172–196. DOI: 10.1016/j.compositesb.2018.02.012

Ning, F., Cong, W., Hu, Z., & Huang, K. (2017). Additive manufacturing of thermoplastic matrix composites using fused deposition modeling: A comparison of two reinforcements. *Journal of Composite Materials*, *51*(27), 3733–3742. DOI: 10.1177/0021998317692659

Nurhudan, A. I., Supriadi, S., Whulanza, Y., & Saragih, A. S. (2021). Additive manufacturing of metallic based on extrusion process: A review. *Journal of Manufacturing Processes*, *66*, 228–237. DOI: 10.1016/j.jmapro.2021.04.018

Okafor-Muo, O. L., Hassanin, H., Kayyali, R., & ElShaer, A. (2020). 3D printing of solid oral dosage forms: Numerous challenges with unique opportunities. *Journal of Pharmaceutical Sciences*, *109*(12), 3535–3550. DOI: 10.1016/j.xphs.2020.08.029 PMID: 32976900

Palanisamy, C., Raman, R., & Dhanraj, P. K. (2022). Additive manufacturing: A review on mechanical properties of polyjet and FDM printed parts. *Polymer Bulletin*, *79*(9), 7065–7116. DOI: 10.1007/s00289-021-03899-0

Parulski, C., Jennotte, O., Lechanteur, A., & Evrard, B. (2021). Challenges of fused deposition modeling 3D printing in pharmaceutical applications: Where are we now? *Advanced Drug Delivery Reviews*, *175*, 113810. DOI: 10.1016/j.addr.2021.05.020 PMID: 34029646

Patel, R., Desai, C., Kushwah, S., & Mangrola, M. H. (2022). A review article on FDM process parameters in 3D printing for composite materials. *Materials Today: Proceedings*, *60*, 2162–2166. DOI: 10.1016/j.matpr.2022.02.385

Pérez, M., Carou, D., Rubio, E. M., & Teti, R. (2020). Current advances in additive manufacturing. *Procedia CIRP*, *88*, 439–444. DOI: 10.1016/j.procir.2020.05.076

Pérez, M., Medina-Sánchez, G., García-Collado, A., Gupta, M., & Carou, D. (2018). Surface quality enhancement of fused deposition modeling (FDM) printed samples based on the selection of critical printing parameters. *Materials (Basel)*, *11*(8), 1382. DOI: 10.3390/ma11081382 PMID: 30096826

Qattawi, A., & Ablat, M. A. (2017). Design consideration for additive manufacturing: Fused deposition modelling. *Open Journal of Applied Sciences*, *7*(6), 291–318. DOI: 10.4236/ojapps.2017.76024

Quero, R. F., da Silveira, G. D., da Silva, J. A. F., & de Jesus, D. P. (2021). Understanding and improving FDM 3D printing to fabricate high-resolution and optically transparent microfluidic devices. *Lab on a Chip*, *21*(19), 3715–3729. DOI: 10.1039/D1LC00518A PMID: 34355724

Quiñonez, P. A., Ugarte-Sanchez, L., Bermudez, D., Chinolla, P., Dueck, R., Cavender-Word, T. J., & Roberson, D. A. (2021). Design of shape memory thermoplastic material systems for FDM-type Additive manufacturing. *Materials (Basel)*, *14*(15), 4254. DOI: 10.3390/ma14154254 PMID: 34361448

Rahim, T. N. A. T., Abdullah, A. M., & Md Akil, H. (2019). Recent developments in fused deposition modeling-based 3D printing of polymers and their composites. *Polymer Reviews (Philadelphia, Pa.)*, *59*(4), 589–624. DOI: 10.1080/15583724.2019.1597883

Rahman, H., John, T. D., Sivadasan, M., & Singh, N. K. (2018). Investigation on the Scale Factor applicable to ABS based FDM Additive Manufacturing. *Materials Today: Proceedings*, *5*(1), 1640–1648. DOI: 10.1016/j.matpr.2017.11.258

Rajaguru, K., Karthikeyan, T., & Vijayan, V. (2020). Additive manufacturing–State of art. *Materials Today: Proceedings*, *21*, 628–633. DOI: 10.1016/j.matpr.2019.06.728

Ramazani, H., & Kami, A. (2022). Metal FDM, a new extrusion-based additive manufacturing technology for manufacturing of metallic parts: A review. *Progress in Additive Manufacturing*, *7*(4), 609–626. DOI: 10.1007/s40964-021-00250-x

Ramesh, M., & Panneerselvam, K. (2021). Mechanical investigation and optimization of parameter selection for Nylon material processed by FDM. *Materials Today: Proceedings*, *46*, 9303–9307. DOI: 10.1016/j.matpr.2020.02.697

Rinaldi, M., Ghidini, T., Cecchini, F., Brandao, A., & Nanni, F. (2018). Additive layer manufacturing of poly (ether ether ketone) via FDM. *Composites. Part B, Engineering*, *145*, 162–172. DOI: 10.1016/j.compositesb.2018.03.029

Roger, F., & Krawczak, P. (2015, August). 3D-printing of thermoplastic structures by FDM using heterogeneous infill and multi-materials: An integrated design-advanced manufacturing approach for factories of the future. In *CFM 2015-22ème Congrès Français de Mécanique*. AFM, Maison de la Mécanique, 39/41 rue Louis Blanc-92400 Courbevoie.

Salem Bala, A. bin Wahab, M. S., & binti Ahmad, M. (2016, May). Elements and materials improve the FDM products: A review. In Advanced Engineering Forum (Vol. 16, pp. 33-51). Trans Tech Publications Ltd.

Samykano, M., Selvamani, S. K., Kadirgama, K., Ngui, W. K., Kanagaraj, G., & Sudhakar, K. J. T. I. J. O. A. M. T. (2019). Mechanical property of FDM printed ABS: Influence of printing parameters. *International Journal of Advanced Manufacturing Technology*, *102*(9-12), 2779–2796. DOI: 10.1007/s00170-019-03313-0

Sargini, M. I. M., Masood, S. H., Palanisamy, S., Jayamani, E., & Kapoor, A. (2021). Additive manufacturing of an automotive brake pedal by metal fused deposition modelling. *Materials Today: Proceedings*, *45*, 4601–4605. DOI: 10.1016/j.matpr.2021.01.010

Saroia, J., Wang, Y., Wei, Q., Lei, M., Li, X., Guo, Y., & Zhang, K. (2020). A review on 3D printed matrix polymer composites: Its potential and future challenges. *International Journal of Advanced Manufacturing Technology*, *106*(5-6), 1695–1721. DOI: 10.1007/s00170-019-04534-z

Sbriglia, L. R., Baker, A. M., Thompson, J. M., Morgan, R. V., Wachtor, A. J., & Bernardin, J. D. (2016). Embedding sensors in FDM plastic parts during additive manufacturing. In *Topics in Modal Analysis & Testing, Volume 10: Proceedings of the 34th IMAC, A Conference and Exposition on Structural Dynamics 2016* (pp. 205-214). Springer International Publishing. DOI: 10.1007/978-3-319-30249-2_17

Sehhat, M. H., Mahdianikhotbesara, A., & Yadegari, F. (2022). Impact of temperature and material variation on mechanical properties of parts fabricated with fused deposition modeling (FDM) additive manufacturing. *International Journal of Advanced Manufacturing Technology*, *120*(7), 4791–4801. DOI: 10.1007/s00170-022-09043-0

Sehhat, M. H., Mahdianikhotbesara, A., & Yadegari, F. (2022). Verification of stress transformation in anisotropic material additively manufactured by fused deposition modeling (FDM). *International Journal of Advanced Manufacturing Technology*, *123*(5), 1777–1783. DOI: 10.1007/s00170-022-10321-0

Serajuddin, A. (2023). Challenges, current status and emerging strategies in the development of rapidly dissolving FDM 3D-printed tablets: An overview and commentary. *ADMET & DMPK*, *11*(1), 33–55. DOI: 10.5599/admet.1622 PMID: 36778904

Shanmugam, V., Das, O., Babu, K., Marimuthu, U., Veerasimman, A., Johnson, D. J., Neisiany, R. E., Hedenqvist, M. S., Ramakrishna, S., & Berto, F. (2021). Fatigue behaviour of FDM-3D printed polymers, polymeric composites and architected cellular materials. *International Journal of Fatigue*, *143*, 106007. DOI: 10.1016/j.ijfatigue.2020.106007

Shanmugam, V., Das, O., Neisiany, R. E., Babu, K., Singh, S., Hedenqvist, M. S., Berto, F., & Ramakrishna, S. (2020). Polymer recycling in additive manufacturing: An opportunity for the circular economy. *Materials Circular Economy*, *2*(1), 1–11. DOI: 10.1007/s42824-020-00012-0

Singh, R., Singh, H., Farina, I., Colangelo, F., & Fraternali, F. (2019). On the additive manufacturing of an energy storage device from recycled material. *Composites. Part B, Engineering*, *156*, 259–265. DOI: 10.1016/j.compositesb.2018.08.080

Singh, S., Ramakrishna, S., & Singh, R. (2017). Material issues in additive manufacturing: A review. *Journal of Manufacturing Processes*, *25*, 185–200. DOI: 10.1016/j.jmapro.2016.11.006

Sivamani, S., Nadarajan, M., Kameshwaran, R., Bhatt, C. D., Premkumar, M. T., & Hariram, V. (2020). Analysis of cross axis wind turbine blades designed and manufactured by FDM based additive manufacturing. *Materials Today: Proceedings*, *33*, 3504–3509. DOI: 10.1016/j.matpr.2020.05.438

Solomon, I. J., Sevvel, P., & Gunasekaran, J. J. M. T. P. (2021). A review on the various processing parameters in FDM. *Materials Today: Proceedings*, *37*, 509–514. DOI: 10.1016/j.matpr.2020.05.484

Srivastava, M., Rathee, S., Patel, V., Kumar, A., & Koppad, P. G. (2022). A review of various materials for additive manufacturing: Recent trends and processing issues. *Journal of Materials Research and Technology*, *21*, 2612–2641. DOI: 10.1016/j.jmrt.2022.10.015

Srivastava, R., Arun, J., & Patel, D. K. (2019, April). Amalgamating the Service Quality Aspect in Supply Chain Management. In *2019 International Conference on Automation, Computational and Technology Management (ICACTM)* (pp. 63-67). IEEE. DOI: 10.1109/ICACTM.2019.8776839

Srivastava, R., & Srivastava, V. K.Ruchira Srivastava and Vijay Kumar Srivastava. (2020). Changing Scenario of Manufacturing Industries Due to Covid 19. *International Journal for Modern Trends in Science and Technology*, *6*(10), 143–146. DOI: 10.46501/IJMTST061026

. Srivastava, R., Thakur, A., & Rana, A. (2024). The Basic Understanding of Nano Materials: A Review. *Innovations in Nanomaterials-Based Corrosion Inhibitors*, 1-20.

Sun, X., Mazur, M., & Cheng, C. T. (2023). A review of void reduction strategies in material extrusion-based additive manufacturing. *Additive Manufacturing*, *67*, 103463. DOI: 10.1016/j.addma.2023.103463

Syrlybayev, D., Zharylkassyn, B., Seisekulova, A., Akhmetov, M., Perveen, A., & Talamona, D. (2021). Optimisation of strength properties of FDM printed parts—A critical review. *Polymers*, *13*(10), 1587. DOI: 10.3390/polym13101587 PMID: 34069144

Tagliaferri, V., Trovalusci, F., Guarino, S., & Venettacci, S. (2019). Environmental and economic analysis of FDM, SLS and MJF additive manufacturing technologies. *Materials (Basel)*, *12*(24), 4161. DOI: 10.3390/ma12244161 PMID: 31835783

Tekinalp, H. L., Kunc, V., Velez-Garcia, G. M., Duty, C. E., Love, L. J., Naskar, A. K., Blue, C. A., & Ozcan, S. (2014). Highly oriented carbon fiber–polymer composites via additive manufacturing. *Composites Science and Technology*, *105*, 144–150. DOI: 10.1016/j.compscitech.2014.10.009

Thakur, A., Vates, U. K., & Mishra, S. (2024). Parametric optimization of 3D-printed PLA part using response surface methodology for mechanical properties. *Smart Science, 12*(1), 91–104. DOI: 10.1080/23080477.2023.2270819

Tiwary, V. K., Arunkumar, P., & Malik, V. R. (2021). An overview on joining/welding as post-processing technique to circumvent the build volume limitation of an FDM-3D printer. *Rapid Prototyping Journal, 27*(4), 808–821. DOI: 10.1108/RPJ-10-2020-0265

Tsioukas, V., Pikridas, C., & Karolos, I. A. (2020). Challenges, opportunities, and limitations in 3D printing. 3D Printing: Applications in Medicine and Surgery, 151-155.

Turner, N., B., Strong, R., & A. Gold, S. (2014). A review of melt extrusion additive manufacturing processes: I. Process design and modeling. *Rapid Prototyping Journal, 20*(3), 192–204.

Valerga, A. P., Batista, M., Salguero, J., & Girot, F. (2018). Influence of PLA filament conditions on characteristics of FDM parts. *Materials (Basel), 11*(8), 1322. DOI: 10.3390/ma11081322 PMID: 30065153

Wickramasinghe, S., Do, T., & Tran, P. (2020). FDM-based 3D printing of polymer and associated composite: A review on mechanical properties, defects and treatments. *Polymers, 12*(7), 1529. DOI: 10.3390/polym12071529 PMID: 32664374

Yasa, E., & Ersoy, K. (2018). A review of the additive manufacturing of fiber reinforced polymer matrix composites.

Zhao, H., Liu, X., Zhao, W., Wang, G., & Liu, B. (2019, June). An overview of research on FDM 3D printing process of continuous fiber reinforced composites. [). IOP Publishing.]. *Journal of Physics: Conference Series, 1213*(5), 052037. DOI: 10.1088/1742-6596/1213/5/052037

Chapter 13
Rapid Prototyping Bridging Research and Industry in Silicon Photonics

Jyoti Rani
https://orcid.org/0000-0002-5577-7808
Lovely Professional University, India

ABSTRACT

Over the past decade, silicon photonics has grown rapidly in the datacom industry, with applications in LIDAR, sensing, quantum photonics, and AI. This expansion has led commercial CMOS foundries to establish open-access silicon photonics process lines for large-scale production. Meanwhile, university research labs use e-beam lithography for small-scale prototyping, fostering innovation. However, transitioning from research to industry poses challenges due to the incompatibility of research tools with industry-standard fabrication. The CORNERSTONE rapid-prototyping platform addresses this by offering deep-UV lithography-based fabrication, enabling scalable production while maintaining design flexibility. It serves as a bridge between academia and industry, facilitating the commercialization of novel designs. CORNERSTONE's capabilities support the growing demand for photonic solutions and drive collaboration between stakeholders. This review explores its current strengths and future potential, highlighting its critical role in advancing silicon photonics.

DOI: 10.4018/979-8-3373-0533-2.ch013

INTRODUCTION

CMOS (Complementary Metal-Oxide-Semiconductor) and LIDAR (Light Detection and Ranging) are transformative technologies that serve a supporting yet essential role in many contemporary rapid prototyping (RP) and additive manufacturing systems. Although neither of these are direct RP methods, CMOS and LIDAR improve prototyping capabilities by promoting high-precision volume scanning to be used for real time quality control and smart manufacture. Complementary Metal Oxide Semiconductor (CMOS) - a low power, high efficiency semiconductor based circuits technology - is the most common technology type used to develop imaging sensor and embedded electronics in RP systems. For example, CMOS-based image sensors are crucial components of structured-light 3D scanners that acquire complete object 3D geometry and are used for reverse engineering and digital replicas. Finally, CMOS chips are used in DLP (Digital Light Processing) 3D printers to accurately define UV light patterns to harden the photopolymer solution layer by layer. For example, CMOS sensors enable machine vision to perform automated quality inspections of 3D-printed parts, to make sure they meet tight dimensional tolerances.

In contrast, LIDAR uses beams that create detailed three-dimensional maps so large-scale scanning and digitization for rapid prototyping becomes incredible powerful. LIDAR scanners generate point cloud data for complex geometries such as automotive or aerospace parts which can then be converted into CAD models to be printed in 3D. Within the world of robotic additive manufacturing, LIDAR is used to help control robotic arms to move during large-scale printing by ensuring accurate deposition of materials. In addition, LIDAR is deployed in post-processing quality control, comparing the 3D-printed parts with their original digital designs and identifying divergences.

The confluence of CMOS and LIDAR serves to supercharge rapid prototyping by speeding up design validation, increasing accuracy, and facilitating more intelligent and automated workflows. These technologies connect the digital modeling and physical fabrication worlds, making them a vital component of next-gen manufacturing. Used in conjunction with traditional RP techniques for example, FDM, SLA, or SLS they can meaningfully enhance prototyping speed, accuracy, and scalability across verticals including aerospace engineering, automotive engineering, and medical device development.

Silicon photonics has quickly grown into one of the most significant datacom and multiple emerging technologies, offering high integration density and low power consumption by leveraging compatibility with CMOS processes. Due to this technology exhibiting a strong integration of photonic and electronic devices on one chip, it is widely gaining traction in fields such as data communication, LIDAR sensing, quantum photonics and AI (Artificial Intelligence). And in the following sections, we will present examples of silicon photonics in these domains, showcasing existing use cases and future opportunities.

Silicon Photonics in Data Communication

- Si-photonics is a ready-for-use platform for photonic integrated circuits for optical communication; especially for communication transceiver for data-center applications. From thousands to millions, high speed data transmission with low power consumption play a significant role in the proliferation of these devices (Shekhar *et al.*, 2024).
- The union of photonics and electronics has pushed the creation of transceivers to capacities of 1.6 Tbps and beyond, as electronic systems have I/O and power constraints.

Silicon Photonics in LIDAR

- Silicon-based LIDAR systems tend to be smaller, cheaper, and do not have moving parts like traditional LIDAR systems. These solid-state beam steering technologies can make LIDAR systems faster and more efficient (Hu, Pang and Gao, 2023).
- Silicon Photonics Applied in LIDAR Monolithic integration enabled by silicon photonics plays an important role in LIDAR for applications in autonomous driving and sensing technologies (Hu, Pang and Gao, 2023).

Silicon Photonics in Sensing

- Silicon photonic elements are used for integration of multiple optical elements in a single chip, thus being applied in many types of sensing applications. This integration facilitates the design of small and energy-efficient sensors suitable for various applications ranging from environmental monitoring to biomedical diagnostics (Xiao *et al.*, 2023).
- High process speed and low power consumption of the technology make it suitable for processing data in real-time sensing applications (Xiao *et al.*, 2023).

Silicon Photonics in Quantum Photonics

- A promising platform for quantum technologies is silicon photonics, which allows on-chip integration of photon generation and detection. This is the prerequisite for scalable quantum information processing and building quantum computers (Feng *et al.*, 2024).

- However, it yielded interfaces between silicon photonics and CMOS microelectronics, enabling the development of miniaturized quantum devices for advanced quantum communication and computation (Gupta et al., 2023).
- Recent developments have even shown that universal quantum information processing may be implemented with silicon photonics with high fidelity and in a scalable way (Qiang *et al.*, 2018).

Silicon Photonics in Artificial Intelligence

- Silicon photonics is in the center of investigations for applications to AI with photonic neural networks. The high-speed data processing ability of photonics is harnessed by these networks to improve machine learning algorithms and AI Systems (Xiao *et al.*, 2023).
- Furthermore, the incorporation of photonic elements within AI architectures has the potential to deliver substantial enhancements in processing speed and energy efficiency, both of which are vital for the advancement of modern AI UPS (Xiao *et al.*, 2023).

Silicon photonics has made strides in these domains, yet integration and fabrication bottlenecks remain. Addressing these barriers will be key for silicon photonics to achieve widespread deployment in next-generation use cases. Furthermore, robust design approaches and optimization algorithms will be critical to coping with process variation and to maximize yield of microfluidic integrated photonic circuits (Shekhar *et al.*, 2024). With the advancement in technology, it may change the game by providing scalable and efficient solutions across industries.

Role of commercial CMOS foundries in enabling mass production

Commercial CMOS foundries are essential for coordinating the large-scale production of semiconductor devices due to their provision of the infrastructure and technology needed for standard mass-manufacturing. These foundries provide standardization and economies of scale, which are essential for the commercialization processes for a large number of technologies, such as silicon photonics and integrated circuits. CMOS foundries are characterized by high throughput, short cycle times, and high equipment utilization, providing the potential for rapid transition from research to mass production in these areas. The transition is necessary to meet the increasing demands of the semiconductor industry and other sectors. Following are a few important points about the role of commercial CMOS foundries in enabling mass production:

High Throughput and Efficiency

- To accommodate mass production they are coded with throughput driven mission of efficiency by design. To maintain high equipment utilization and short cycle times, which are necessary for on-time delivery and production target, they adopt several types of control polices including input dispatching rules and optimization based rules (Kang and Lee, 2007).
- The production of silicon photonics is an essential component for the commercialization of this technology, and the standardized process technology platforms used by CMOS foundries enable this, providing an efficient channel for its creation (Lim *et al.*, 2014).

Economies of Scale

- You might also note that it is only in CMOS foundries where economies of scale can drive down manufacturing costs sufficiently to make large scale production of semiconductor devices commercially viable. This especially matters for emerging technologies like silicon photonics and integrated circuits which are very cost-driven (Graham *et al.*, 2011).
- By being able to manufacture devices on larger wafers (for example, 6-inch or bigger), the economies of scale are further improved by allowing a greater utilization of all materials and resources per chip produced (Ye *et al.*, 2023).

Standardization and Compatibility

- Standardized CMOS processes promote consistency and compatibility between different CMOS processes and production runs, which is critical for ensuring the quality and reliability of mass-manufactured devices. Moreover, this standardization will help integration of new technologies like silicon spin qubits into existing manufacturing lines (Elsayed *et al.*, 2024).
- CMOS foundries offer a platform where different technologies like optical interconnects and biosensors can be integrated together with standard CMOS processes to enable new applications and products (Changliang *et al.*, 2009).

Environmental Considerations

- CMOS foundries enable mass production but face challenges in terms of sustainability, including power, water consumption, and systemic waste. Investments into renewable energy and waste management are some of the initiatives being explored today to heighten the environmental performance of foundries (Ruberti, 2024).

Challenges, restrictions, and limitations must overcome against the advantages provided by commercial CMOS foundries. As an example, some technologies, e.g. silicon nitride photonic integrated circuits, can face technical challenges when moving from small to larger wafer sizes in CMOS foundries (Ye *et al.*, 2023). Furthermore, although hundreds of companies can take advantage of the economies of scale of CMOS foundries, setting up manufacturing lines for new technologies is equally expensive and painfully slow to custom-fit. However, the contribution of commercial CMOS foundries is still an essential part of bringing semiconductor devices to volume and driving the improvements in the technology.

Challenges faced by academic researchers in scaling their designs for commercial fabrication

When simply trying to scale their designs for commercial fabrication, academic researchers face many struggles. These challenges are technical, non-technical, and logistical in nature, and often require a multidisciplinary approach to find solutions. Mass-production, from prototype, is an extremely complex process that can ultimately prevent the commercialisation of innovative research. Here were identified key challenges from the given papers along with some thoughts about potential research solutions to these challenges.

Technical Challenges

- Design and Manufacturing Knowledge Gaps: Researchers do not understand the technical aspects of large-scale manufacturing, which entails knowing the details of the processing steps and what is possible or not in regards to the current manufacturing technologies. Such a disjointed development process can cause inefficiencies and higher costs during the transition from prototypes to mass productio (Khurana and Hodges, 2020).
- Material and Process Limitations: For example, scaling up additive manufacturing involves pushing through material limitations and processing shutdown restrictions. However, they pose obstacle in large-scale 3D printing

where aspects of print quality and structure integrity play less role in small-scale (Shah *et al.*, 2019) (Zuo *et al.*, 2024).

- Nanomanufacturing Scalability: Nanomanufacturing processes suffer from issues of repeatability and process control that limit scalability. Facilitation of consistent and scalable processes is yet to be achieved to convert nanomaterials from experimental towards commercial viability (Vanli, Rivero and Wang, 2013).

Non-Technical Challenges

- Business and Market Understanding: Researchers may not have the required understanding of the commercial environment, such as market assessments, intellectual property, and regulatory policies. Not talking of these unknowns which can act as barriers to successful commercialization of research outputs (Khurana and Hodges, 2020).
- Network and Relationship Building:
- Establishing a network of business partners and stakeholders is essential for scaling up production. These relationships, which are critical to gaining access to resources, expertise, and market opportunities (Khurana and Hodges, 2020), are often out of reach for researchers.

Manufacturing and Process Optimization

- Economies of Scale in Additive Manufacturing: In additive manufacturing, which operates at a much smaller scale compared to traditional methods, achieving economies of scale necessitates new strategies for optimizing processes and managing resources. This involves factors such as designing for volume and optimizing material flow to save costs and increase time efficiency in production (Huang *et al.*, 2021).
- Micromixer and Coating System Scale-Up: Certain technologies such as micromixers and coating systems are scaled-up specifically with the consideration of process parameters and equipment design. Optimizing the channel width in micromixers for better performance or ensuring consistent coating thickness in processes such as CVD upon industrial applications are just a few examples of the relevant fields (Lee, Krumdieck and Talwar, 2013).

Broader Perspectives

Explain the challenges of scaling academic research for commercial fabrication. Researchers addressing these challenges will help produce new tools and processes that bridge the gap between prototype and production. Additionally, fostering an interdisciplinary knowledge base and building strong industry-academia collaborations can further commercialize research from academia. Understandably, not all research is a proper candidate for commercialization, and there are some projects that just won't leave academia for reasons that include fundamental limitations or niche applications.

INNOVATION HUBS: ACADEMIC VS. INDUSTRIAL APPROACHES

Differences in fabrication tools: e-beam lithography in academia vs. deep-UV lithography in industry.

The nanofabrication techniques that have gained most prominence are electron beam lithography (EBL), and deep-UV lithography (DUV), both of which have unique advantages that lend themselves to different applications. EBL is well suited for academic research as it provides high resolution and flexibility, while DUV is preferred for industrial applications due to its high throughput and low cost. In the next sections, we will summarize what we consider the main benefits of EBL in academic research in relation to what DUV gives you in industrial conditions.

Advantages of E-Beam Lithography in Academic Research

- High Resolution and Precision: EBL excels in providing high-resolution, with the ability to define structures down to 5 nm, which is critical for research applications that demand precise nanoscale patterning (Xomalis *et al.*, 2023) (Gorelick *et al.*, 2010). This large resolution proves to be an advantage when creating advanced materials and devices, such as quantum bits and photonic structures (Kato *et al.*, 2021) (Li *et al.*, 2016).
- Flexibility and Customization: EBL is a maskless lithography method, meaning it can write patterns directly without the use of masks. Such flexibility is excellent for rapid prototyping and iterative design procedures in research settings, allowing fast changes and customization (Hammood *et al.*, 2023).

- Material Versatility: EBL is capable of patterning a variety of materials like metals, polymers and organic crystals, therefore can be applied to a broad spectrum of research applications including biomimetic surfaces and photomechanical devices (Xomalis *et al.*, 2023) (Li *et al.*, 2022). The versatility is further aided by its ability to enable resist-free patterning, thus resulting in cost reduction and increased material compatibility.
- Advanced Patterning Techniques: EBL accommodates sophisticated patterning methods, including lift-off techniques and high aspect ratio structures that facilitate the fabrication of complex nanoelements, and devices having exotic materials characteristics (Sahin, Albayrak and Yapici, 2024).

Advantages of Deep-UV Lithography in Industrial Applications

- High Throughput: DUV lithography developed for high-volume manufacturing is capable of processing large areas rapidly, which is required for industrial applications where time and efficiency are of critical importance. This enables DUV to become the lithography of choice for high volume manufacturing of semiconductor devices.
- Cost-Effectiveness for Large Scale Production: Although DUV lithography may have high upfront costs, with large scale production, the per unit cost is substantially lower, providing an economic advantage to industries manufacturing millions of units (Hammood *et al.*, 2023).
- Established Process and Infrastructure: DUV lithography is a well-established technology with a mature infrastructure, including standardized processes and equipment, which facilitates integration into existing industrial workflows (Hammood *et al.*, 2023).
- Compatibility with Existing Semiconductor Processes: DUV lithography is in line with current semiconductor manufacturing processes, allowing easy integration into the production of integrated circuits and other electronic components (Hammood *et al.*, 2023).

Although EBL has the highest resolution and flexibility, making it well-suited for academic research and prototyping applications, DUV lithography is better suited for industrial applications where high throughput and cost are critical considerations for larger volume production. Every technique has its own benefits designed to meet the distinct requirements of research and industry.

Limitations of e-beam lithography for large-scale production

Nanotechnology advances through the manufacturing of nanoscale devices with electron beam lithography (EBL), which is a high-resolution patterning technique. Although accurate, there are a number of shortcomings that prevent its use for mass production. Primarily, these constraints relate to the throughput, cost, and the technical complexity of the process. The sections that follow elaborate on these main limitations.

Low Throughput

- EBL is by nature a serial process, compromising it to lower throughput hysterically compared to parallel methods (photolithography). This renders EBL not suitable for HVM where speed is paramount (Hager *et al.*, 2023).
- To enhance throughput, different techniques—like character projection and a massively parallel electron beam direct write systems—have been investigated. But these known methods are not yet fast enough in large scales.

High Costs

- EBL demands expensive equipment and high operating costs. The sophistication of the electron beam systems required and the upkeep of the systems render it cost-prohibitive for widespread use (Hager *et al.*, 2023).
- Development and maintenance of the needed infrastructure for EBL (cleanroom facilities specialized resists etc) represents another cost factor (Yu *et al.*, 2012).

Technical Challenges

- EBL suffers technical limitations like proximity effects, where scattered electrons influence nearby areas causing pattern distortion. This involves complicated corrections, which constrain the patterns to a given resolution and precision (Camino *et al.*, 2024) (Manfrinato *et al.*, 2014).
- EBL has also a limitation to encounter with its resolution which is due to the interaction of the electrons with the resist material, causing the energy spread and decrease the precision. However, such advancements are currently not keeping up with systems and tools developments (Saifullah *et al.*, 2022).
- The process is substrate condition-dependent, and obtaining reproducible results on non-planar or uneven substrates is challenging and often necessitates further adjustment or techniques.

Alternative Perspectives

In addition, although EBL is not suitable for large-scale production, it is highly desired in research and development where its high resolution is a prerequisite for prototyping and fabricating complex nanostructures. Additionally, continued research to enhance throughput and decrease costs, like the explorations into new resist materials and parallel processing methods, show potential for future use cases. Moreover, EBL allows us to create features on the atomic scale; replacing optical lithographic methods will be crucial for pushing semiconductor technology to its limits.

Importance of process compatibility for bridging the research-industry gap

Process compatibility is a multifaceted relationship between research and industry bridging gap issues which aims to align the objectives, approaches, and outputs of both sectors for more effective collaboration and successful technology transfer. This concept won in that research through academic methods should flow into industrial practices more efficiently. Aligning efforts is essential to bridge the gap between academia and industry, which often struggle with divergent objectives, communication issues, and resource constraints. The remainder of this article discusses process compatibility and its ability to cross the chasm between research and industry.

Alignment of Objectives

- **Differing Goals**: Academia is more concerned with theoretical breakthroughs and new methods than is industry which is focused on commercial applications and profit. A alignment of these objectives is key to successful collaboration (Han *et al.*, 2022).
- **Shared Vision**: Governance bodies jointly established can set shared long-term vision for the industry (Lavikka *et al.*, 2020), aiding both in the selection of relevant research topics within a specific domain of interest and on the topic addressing both scientific as well as practical relevance .

Communication and Collaboration

- **Effective Communication**: The CODE framework highlights the need for clear communication regarding objectives, deliverables, and evaluations to prevent common pitfalls in technology transfer (Jin, Wanvarie and Le, 2019).

- **Dialogue and Interaction**Frequent engagement between researchers and industry practitioners can help both groups better understand and appreciate the problems and requirements of each.

Resource Sharing and Access

- **Access to Data and Resources**: Academic research can be hampered by limited access to non-public resources (e.g., operational data). With the support of collaborative practices, academia is able to have all required resources to run impactful research (Han *et al.*, 2022).
- **Knowledge Transfer Schemes**: The existing research frameworks could be adapted into practical tools for Knowledge Transfer Schemes where the knowledge gained from science can more readily be applied in practice.

Process Models and Frameworks

- **Process Models**: Creating integrative frameworks and process models for managing research collaborations can increase the efficiency of academic-industry partnerships.
- **Innovation and Implementation**: Cross-sector partnerships between universities and industries have been successful in crossing the chasm of "dissemination" to "implementation" (Lavikka *et al.*, 2020).

Challenges and Opportunities

- **Transparency and Trust**: A transparent process regarding funding, research, and collaboration between academia and industry can help to build trust between the two sides. This must be accompanied at a minimum by clear disclosure of industry payments and contributions.
- **Practice-Based Research**: Involving practitioners in research can help ensure that research findings are relevant to real-world contexts, thus increasing the applicability and impact of academic research (Gordan, 2017).

Even though compatibility of processes is necessary for narrowing the gap between research and industry sectors, it is indispensable to have in mind the inherent difference in the nature of knowledge and purpose of the two sectors. Academics prioritize theoretical interrogations and the creation of long-lasting knowledge, while industry often processes based on the immediacy of applicability and market-driven innovation. Engaging in continuous dialogue, showing mutual respect, and being open to modifying processes for the benefit of both sides are essential in balancing these differences. In this way, everyone comes closer and the distance between research and industry is reduced.

CORNERSTONE: A RAPID-PROTOTYPING SOLUTION

The CORNERSTONE platform is a comprehensive and multifaceted system that aims to improve healthcare delivery through innovative care models and novel population segmentation for better access to these models. The core of the program is not based on a financial mechanism but rather redesign of care models for high-risk populations high to reduce health care costs and improve patient outcomes. Its deployment has seen noted savings in medical costs while optimizing healthcare delivery. Features and functionalities of the CORNERSTONE platform are listed below:

Key Features and Functionalities

Population Segmentation and Care Model Redesign:

The platform is being aimed at high-risk populations, such as patients with late-stage congestive heart failure, oncology patients, Medicare-Medicaid dual eligibles, individuals with multiple chronic conditions, and complex patients with late-stage chronic conditions (Green *et al.*, 2017).

Overall costs for the patients in the program decreased by 12.7%, with hospitalization rates decreasing by 30% across all programs, by implementing changes in care models (Green *et al.*, 2017).

Cost Savings and Efficiency:

Through a fully implemented program, the platform's implementation found that cost savings of between 10% and 16% were yielded.

Decreases in total costs were primarily related to a reduction in hospitalization, although a category called "other" which includes hospice was increased in cost.

Focus on Implementation:

Realized cost savings and efficiency gains depended on the complete rollout of the redesigned care models. Sites that returned to traditional models, did not receive the same benefits.

Additional Perspectives

Again, the CORNERSTONE platform is a health technology platform that explicates various cost- and resource-effective interventions and forecasts future cost-savings based on them, but it is essential to consider other health technology platforms that produce similar results. When looking at other platforms, some include digital design integration for rapid prototyping used in manufacturing (Xie *et al.*, 2019), healthcare promotion and disease prevention through continuous monitoring and AI strategies (Gomez-Garcia, Askar-Rodriguez and Velasco-Medina, 2021), improving collaboration between healthcare professionals and biostatisticians, etc. The benefits of combined systems in a specific field are numerous, as evidenced by these platforms that encapsulate specialty applications and as much integration as potential.

Deep-UV lithography: balancing scalability and flexibility.

Demand for deep-ultraviolet lithography technology (including extreme ultraviolet lithography) tends to be driven by a trade-off between scale and flexibility, that is, between the capacity to manufacture a high volume of parts, and the flexibility to manufacture a diverse range of parts. Mass production demands scalability; innovation and customization require flexibility in semiconductor manufacturing. For example, in deep-uv lithography, there are a number of different methods and technologies that have different compromises and are in some sense exploratory to varying degrees.

Scalability in Deep-UV Lithography

- High Throughput Production: Decrease in feature sizes in semiconductor devices is critical and Deep-UV lithography obtained is one of the best methods for this advanced production of high-density integrated circuits. For example, the utilization of DUV lithography with a wavelength of about 248 nm has been essential in producing conductor components of DRAM and logic gates enabling sub-half-micron features in devices.
- Large Area Patterning: Deep-UV microsphere projection lithography allows large area patterning with high precision, with feature sizes as small as 80 nm. This solution utilizes a macroscopic lens setup to project patterns onto a photoresist film surface, allowing it to remain scalable but not sacrifice pattern accuracy.
- Industrial Implementation: The patterning of organic materials for soft electronics demonstrates the scalability of DUV lithography, where a deep-UV (DUV) irradiation exposure-based system can provide high throughput sub-

micron patterning which could be crucial for large-scale application of the method (Kim *et al.*, 2015).

Flexibility in Deep-UV Lithography

- Adaptability to New Materials: The versatility of DUV lithography is exemplified in its use to fabricate high-density bubble memory circuits through conformable-contact photolithography. By providing single-level masking and the compatibility of different materials, such as amorphous and garnet, this approach can adapt to diverse fabrication requirements.
- Advanced Imaging Techniques: The invention of hyper-numerical-aperture objectives with freeform surfaces increases the flexibility of DUV lithography through aberration reduction and improved imaging performance. Such patterning at smaller nodes demonstrates the versatility of DUV systems to changing technology needs (Mao, 2024).
- Tunable Patterning: Exciting developments in techniques such as tunable ultra-deep subwavelength photolithography(USD-SPR) based on using SPR resonant cavity tunable with cavity length allow for tailored patterning with improved optical resolution. This tunability is important for producing complex designs and tailoring them to meet application needs.

Scalability and flexibility are usually perceived as oppositional development forces in deep-UV lithography because they both are important in corresponding to the future development of semiconductor. Scalability, which can be controlled during the design stage, is crucial for the mass production process to be economically feasible, while flexibility enables access to new materials and applications and adaption of the systems to meet emerging challenges. Ongoing research and technology development are essential to address trade-offs and expand the range of lithographic capabilities.

Unique features of CORNERSTONE that cater to academic and industrial needs

CORNERSTONE represents an innovative and cross-linear educational dosage system, combining academia to the industry while still providing its own key parameters which allow obtaining instant feedback, seamless collaboration, and mutual education for both sides. This integration is facilitated by curriculum alignment, hands-on experience, and creative teaching methods. CORNERSTONE leverages theoretical knowledge to ascertain practical implementation and is adaptable to academia and industry. Here are the key features that go on to make this smooth integration possible:

Curriculum-Aligned Work-Integrated Learning

- Industry-Academic Partnerships: CORNERSTONE does minimize the potential for academic partners to reduce the number of both paid and unpaid positions offered to students. This combination maximizes research impact while delivering an industry-relevant curriculum that equips students with the relevant, real-world experience essential for their degreePathway to a degree:experiential/education (Carmichael *et al.*, 2018).
- Integrated Course Delivery: Theoretical course contents combine with analytical thinking to facilitate work-based learning. And this integration helps students connect real-world applications to their academic studies.

Ill-Structured Problem Solving

- Open-Ended Projects: Cornerstone projects like robotics require ill-structured problem solving from students. This method aids in building self-efficacy and gifted students' problem-solving skills, a necessity for their adjustment to complex challenges in industries (Hulls and Rennick, 2020).
- Focus on Design and Problem Solving: Courses are set up to focus on software design and problem-solving instead of teaching syntax alone Such is the nature of real-world industrial projects, therefore this prepares students for the unclear territory that is often involved.

Critical Thinking and Scientific Literacy

- CREATE Strategy: This strategy is used to help students build critical thinking and scientific literacy. Analyzing scientific literature helps students refine their ability to integrate content and to cultivate sophisticated epistemological beliefs critical to innovation in industrial contexts.
- Adaptability Across Disciplines: The CREATE approach is adaptable to various scientific disciplines, making it a versatile tool for fostering critical thinking across multiple industrial contexts (Gottesman and Hoskins, 2013).

Standardization and Local Adaptation

- Socio-Technical Integration: While technical integration, CORNERSTONE is addressing the socio-technical integration aspects, these works ensure that the systems are tailored to local needs while maintaining standardization across settings. In order for both academia and industry collaboration to be effective, this balance is crucial.

- User-Led Customization: The need for customization based on local practices is critical for understanding the specifics of the diverse industrial environments.

Agile Practices and Capstone Projects

- Agile Organizational Models: Similar to industry practices, capstone projects adopt agile practices to give students authentic learning experiences. It increases leadership and personal responsibility and ensures students are well-prepared for increasingly dynamic industrial roles (Schneider *et al.*, 2020).
- Scalable Project Management: Inspired by a standard used in the industry, the ACE model provides a scalable approach for high-performance management of the capstone projects to help all students gain experience in the working environment (Schneider *et al.*, 2020).

Moving forward, CORNERSTONE provides a detailed, comprehensive framework for integrating academic and industrial settings, albeit with some points acknowledged as strength and challenge. The standardization requirement for example, can lead to rigidity that may stifle innovation and flexibility across dynamic industrial landscapes. Also, the effectiveness of these programs largely depend on the dedication and partnership of academic and industry members, which differ significantly from one environment to another. Although these issues exist, the CORNERSTONE model represents a viable pathway for bridging academia and industry to train the next generation..

CAPABILITIES AND APPLICATIONS OF CORNERSTONE

Key technical capabilities of the platform

It essentially needs a scalable platform to cater to ever-increasing demands and workloads. These platforms need to have core technical competencies such as data management, dynamic resource allocation, load balancer and architecture. These features help the system scale and evolve without losing performance or usability. These core skills are described in the next sections utilizing evidence from a range of research settings.

Efficient Data Management

- Distributed Indexing Structures: Platforms should be capable of handling complex queries efficiently. As an example, a scalable P2P platform administrates object localization and queries in a semantic overlay, using a structured P2P network with distributed trie indexes for broadcast queries.
- Data Replication and Migration: In order to manage million small applications, platform can utilize commodity hardware and free software in providing full SQL features and transactional guarantees by using replication approaches and data migration methodologies.

Dynamic Resource Allocation

- Scalable Cloud Architectures: Cloud PaaS solutions have to manage resources dynamically across multiple layers (app, platform, db, infra) to accommodate rapidly changing demand without impacting users. This includes, among others, employing scalable architectures with various solutions to manage session state.
- Middleware for Cluster Computing: Middleware technology can guarantees the smooth distribution of software components and advanced communication techniques which would enable dynamic allocation of resources in cluster computing environments.

Load Balancing

- Decentralized Load Balancing: A scalable framework must have mechanisms to deal with unevenness in loading caused by heterogeneity in loads and ability of nodes. Decentralized load balancing methods can be used to achieve this.
- Dynamic Load Distribution Policies: In multimedia-on-demand systems, platforms redirect the client requests based on dynamic workload information obtained by the platform to optimize resource utilization, reducing blocking probability and accommodating more users.

Robust Architecture

- Scalable Parallel Computing: Platforms should utilize massively parallel processors (MPPs) and distributed memory architectures to enable time-critical applications while coordinating and functioning for efficiency, assuring their scalable performance.

- Sharded Blockchain Technology: Platforms can use sharded blockchain technology for unlimited scalability and linear scaling in the number of shards, supporting high transaction volumes and maintaining low carbon footprints (Buldas *et al.*, 2022).

These capabilities are important for scalability, but also to weigh the trade-off between performance, scalability, and cost. As an example, balancing these is relevant when re-architecting software for the cloud (Xiong *et al.*, 2014). Furthermore, the unit coordination costs and synergies need to be large enough in order to make it possible to model scalability dynamics better. Such considerations ensure scalable yet cost-effective and manageable platforms.

Real-world examples of innovative silicon photonic devices prototyped using cornerstone.

The CORNERSTONE platform is an entirely new design-for-manufacture framework for these new silicon photonic devices, and is helping to pave the way for its widespread adoption," he adds. Silicon rotates photonics has one of the high integration density and compatibility with existing semiconductor produce process are unique advantages, through this method to develop the general purpose chips made some of complex optical systems of the new solution. Here are some practical examples of these innovative devices.

Si Photonic-Electronic Integrated Optical Receiver

- Si photonic-electronic integrated opto receiver including a ring-resonator based optical filter with temperature control. On the chip, receive circuits and a Ge photodetector are monolithically integrated, which helps keep the device operating efficiently regardless of temperature and wavelength variations. Realized using 0.25-μm photonic BiCMOS technology, it demonstrates the potential for photonic/electronic monolithic integration (Kim *et al.*, 2021).

Silicon-Photonic Tunable Laser

- A dual-wavelength actuator that enables the efficient tuning of two wavelengths with high correlation of their phase noise has been developed in silicon-photonics. It employs a semiconductor optical amplifier for common gain and a silicon-photonic chip for a wavelength-selection element. It reaches a total optical power of 29.3 mW, and proves a continuous tunability of frequency spacing, thus, its application in advanced optical communication systems (Huang *et al.*, 2021).

Silicon Photonics FMCW LiDAR Chip

- An FMCW LiDAR chip based on silicon photonics and a slow-light grating beam scanner. The device allows beam scanning at high speed, high resolution, and wide field of view, suitable for real-time operation and velocity imaging. Silicon photonics has shown potential in automotive and remote sensing applications as well, with its ability to detect targets over long distance in this case (Baba *et al.*, 2022).

Computational Lensless Imaging System

- Developed a silicon photonics based lensless imaging system, using grating couplers and computational processing rather than lenses. This compact form factor is suitable for applications in wearable electronics and IoT devices, featuring a field of view of 45 degrees and optics and sensors integrated within a small volume (White *et al.*, 2020).

Dual-Comb Distance Metrology

- Integrated silicon photonic circuit with a heterodyne receiver structure with tunable power splitting and detector integration on chip for dual-comb distance metrology. According to them, this device provides fast and accurate distance measurements and can be used in high precision metrology applications (Weimann *et al.*, 2017).

These examples showcase the versatility in wearing a silicon photonic hat, especially with prototyping using a platform like CORNERSTONE. Integrating photonic and electronic devices on the same chip enables great benefits in size, power performance, and relative complexity.

These advances underscore the potential of silicon photonics, though obstacles remain in fabrication precision, thermal management, and integration with existing systems. These issues need to be addressed and improve the ability of silicon photonic devices through continued research and development. Cement and mortar are safe in terms of the materials used for this technology, but new materials and fabrication technologies are also even more innovative.

Case studies showcasing its role in bridging the research-commercialization divide.

Case studies also play a significant role in bridging the gap between research and commercialization, as they can help address issues such as the translation of academic research into real-world applications, the cultivation of entrepreneurial ecosystems, and the challenges of interdisciplinary collaboration within the commercialization process. Case studies offer important lessons in the pathways and processes that enable research to enter the commercial world in the commercial world in order to enter the commercial world. They emphasize the role of financial values, policy frameworks, and the development of long-term relationships between academia and industry. Here are the important takeaways from several case studies that describe a productive crossing of that divide.

University Entrepreneurial Ecosystems

- University entrepreneurial ecosystems play a key role that eases commercialization of research. For example, the UK Innovation Knowledge Centres (IKCs) act as intermediaries between research and industry, using public funds to export nascent technologies in fields such as cyber security or synthetic biology. For example, these ecosystems offer the infrastructure, training, and leadership that are necessary to bring research to life as commercial products.
- the NSF I-Corps™ program at University of Michigan is an example of a structured program that brings in funding and entrepreneurship training to researchers and strengthens the university's commercialization capacity (Huang-Saad, Fay and Sheridan, 2017).

Bridging Cultural and Knowledge Gaps

- The cultural gulf between academia and industry is a major impediment to commercialization. Australian universities are working to combat this through the establishment of partnerships that match academic research to industry requirements. Such this through engagement and working together this means identify industry motives knows.
- Interdisciplinary research coordination costs can stymie commercialization. These costs can be minimized by effective team experience and strategic cost coordination, increasing the probability of successful technology transfer and licensing.

Embedded Research Models

- embedded research models have been implemented in public health practice, whereby researchers co-locate with practitioners to co-produce research that directly informs practice. Not only does it serve as a bridge between research and implementation but it also guarantees that research results are applicable to real-world challenges (Akintola, Newbury-Birch and Kilinc, 2024).

Policy and Financial Support

- To successfully commercialize their research, they need sufficient funding and good innovation management. Australian universities recognize these as important focus areas to further their commercialization ambition. This means obtaining funding from both government and industry and promoting academic entrepreneurship.
- The tech-product-market triangulated process illustrates the need for insights on market opportunity and consumer perspectives. This model proposes that developers need to customize technologies that fit the market needs for successful commercialization; however, the knowledge that helps in this process(Kim *et al.*, 2019).

Though the study serves as a useful case study in closing the gap between research and commercialization, there are benefits but also challenges and limitations to these processes. Interdisciplinary research, communication between academia and industry, and strong policy frameworks are perennial issues that need to be addressed. Furthermore, the shifting landscape of market needs and technology requires frequent adjustment and evolution of commercialization approaches.

IMPACT ON THE SILICON PHOTONICS ECOSYSTEM

CORNERSTONE facilitates academic research transitioning to industry

CORNERSTONE is an approach that aids this transition from academia to industry by creating the collaboration of both academic and industry partners that seek to blend their respective strengths. Such partnerships are at the forefront of turning academic innovation into products and services responsive to citizen needs. Academia and industry have themselves, received different treatment to conform to the unique aspects of both ends. Mechanisms like intellectual property management, collaborative frameworks, and resource sharing enable this transition by overcoming the disconnect between theoretical research and practical application.

Industry-Academic Partnerships

- Relationship Building: Successful partnerships are built on relationships, which come from mobility and interaction between academics and industry. It can be tech talks, workshops, and internships to facilitate the exchange of knowledge and innovate.
- Hybrid Models: Seen in the serious games sector, these models combine applied research and business efforts. Bringing together academic and industrial experts, these models help sustain the development of intellectual property and extend research infrastructure.

Intellectual Property and Commercialization

- Intellectual Property Rights (IPR): Managing IPR that plays a significant role in commercializing academic research. Out-licensing and risk-sharing partnerships are applied to exploit new products or processes, for instance in pharmaceutical research.
- Spin-Offs and Start-Ups: Academic spin-offs are assisted by facilitators to mentor them through early stages of growth, offering support with business guidance and network access. Such support is essential for becoming credible in a market and translating a research into a business opportunity (Khodaei *et al.*, 2022).

Collaborative Frameworks and Alliances

- Global Alliances: Initiatives such as the Global Alliance of Leading Drug Discovery and Development Centers bring together resources from multiple translational research centers in order to facilitate the commercialization of biomedical research. These partnerships offer financing, project management, and collaboration with experienced Industry players.
- Public-Private Partnerships (PPPs): PPPs are another common approach, providing a formal way for academia to align with the needs of the industry. These partnerships enable resource sharing and innovation but are complicated by the varying goals of academia and industry.

CORNERSTONE-authorized partnerships are not without their challenges and limitations; however, such partnerships serve to facilitate the transition of academic research to industry. For example, the potential for industry partnerships may redirect academics from theory-based research to more speculative, rapid experimentation, and the development of new products, which may "silt up" the flow of fundamental science. And even then, while alliances and partnerships offer platforms for collaboration, they do not themselves transform. We need radical innovations, such as open innovation and precompetitive data sharing, to begin sparking a real wave of innovation in areas such as pharmaceuticals (Allison, 2013).

Influence on fostering collaborations between universities and commercial foundries

The realisation of successful partnerships between universities and commercial foundries is facilitated by a matrix of organisational, relational and strategic factors. These partnerships engender activity played at the strengths of academic institutions and industry partners alike, propelling them to innovation and commercial success. Three of the important ingredients are the building of strong organizational routines, the cultivation of social capital, and agreement on goals and incentives. These components are integral to overcoming obstacles and increasing the productivity and distribution of teamwork.

Organizational Factors

- Infrastructure and Resources: Universities offer the necessary infrastructure and resources that facilitate research and development processes. This includes state-of-the-art laboratories, equipment, and a highly trained cadre of researchers (Murga Pinillos, 2024).
- Organizational Routines: The characteristics of unabstraining knowledge and mutual learnings, and the absorptive capacity of firms is influenced by the extent to which organizational routines are established (such as with university collaboration). This, subsequently, increases the innovation process efficiency (Lazzarotti *et al.*, 2025).

Relational and Social Capital

- Trust and Communication: These include all elements of trust and proper communication: To address fears of knowledge leakage and intellectual property disputes, common barriers in the early stages of collaboration (O'Dwyer, Filieri and O'Malley, 2023), trust acts as a mitigating factor.

- Cognitive and Relational Social Capital: Building cognitive and relational social capital on an individual, organizational, and alliance level enables the individual and the alliance to identify shared goals and common beliefs that are critical for collaboration (Steinmo, 2015).

Strategic Alignment and Incentives

- Goal Alignment: Goals Alignment: Aligning the strategic goals of universities and industry partners ensures that both parties are working towards common objectives, which improves the success of collaboration (Rybnicek and Königsgruber, 2019).
- Incentives and Policies: It is important to break the expectations of policies and incentives that facilitate innovation and knowledge transfer. Project management methods, educational programs, and relational techniques that motivate both sides to engage actively fall into this category (Murga Pinillos, 2024).

Models and Frameworks

- Collaboration Models Collaboration models including research and development partnerships, internships, and knowledge transfer initiatives were explored as effective mechanisms for universities and industry to work together. These frameworks help guide the collaboration process (Tereshchenko *et al.*, 2024).
- Public Funding and Intellectual Property Management: Public funding is of great importance in successful collaborations, whereas effective management of intellectual property can ensure that both parties are able to benefit from the results of the collaboration (O'Dwyer, Filieri and O'Malley, 2023) (Kneller *et al.*, 2014).

Although these factors are key to successful collaborations, they can lead to problems including conflicts of interest and disputes over intellectual property. Of course, addressing these issues involves flexibility and the need to adjust strategies to the context of the collaboration. In the UK, the Lambert Agreements provide a means for managing publication freedom and intellectual property rights that could be applied in other environments in order to serve both universities and those private sector organisations with which they are increasingly partnering.

Potential for accelerating innovation in emerging applications.

This methodology speeds the innovation process either in part or in whole depending on the best way to implement methods suitable for emerging applications. Using regulatory reforms, advanced data-driven techniques, and emerging technologies can make this possible. There can be different strategies to achieve this goal, each having their own dynamics and hallmarks, which would need to be supported through tailored measures and synergies. The steps involved in accelerating innovation in emerging applications are outlined in the following papers.

Regulatory and Policy Interventions

- Expedited Approval Pathways: Regulatory agencies may establish pilot programs (for example, for drugs that target narrow spectra of pathogens while providing an important weapon against resistant organisms, needed commonly known as antibiotics) to facilitate speedy licensed access. Post-approval studies and active surveillance, which are elements of these programs, both reduce time-to-market and ensure safety.
- Legislative Support: The GAIN Act and other legislative measures can further encourage innovation within targeted therapeutics and diagnostics by creating a pathway for expedited approval of these products tailored towards patients.

Data-Driven and AI/ML Innovations

- Analogy Mining: Developing new solutions by tapping into large idea repositories, such as patent databases, and using analogy mining to find structural similarities between problems. Augmented by machine learning techniques, this method can greatly increase the speed of ideation and creativity (Hope *et al.*, 2018).
- AI/ML in Scientific Discoveries: The role of AI and machine learning models are at the forefront of scientific discoveries by allowing fast data analysis and prediction. These models overcome obstacles such as data scarcity and high system complexity to optimize complex systems like, for example, materials science and molecular design (Alexander *et al.*, 2023).
- Integration of Emerging Technologies
- Digital Health Technologies: The use of digital health technologies — like smartphone applications, wearable devices, etc. in clinical environments can aid telemedicine and personalized health monitoring. The combination of

these technologies provides new opportunities for disease diagnosis and management therefore speeding up innovation in healthcare.

- Population Health Technologies: Implementation of AI and other emerging technologies like robotic, genomics can provide real-time data for monitoring health status, leading to enhanced population health paradigms through optimally priced IOT integrated population health platforms. This means a multipronged approach with resources in transdisciplinary programs and supportive infrastructure.

Organizational and Strategic Approaches

- Intermediary Agencies: In the UK, an intermediary agency like the Cell and Gene Therapy Catapult plays an important role in driving innovation, infrastructure, and regenerative medicine. These agencies assist in navigating the complexities of innovation ecosystems and in aligning the interests of diverse stakeholders (Gardner and Webster, 2017).
- Investment and Funding Models: Global investment strategies, particularly in information and communication technology for health, encourage scaling innovative health solutions to new, larger markets. Sustaining innovation requires social entrepreneurship and specialized financing techniques.

Although these initiatives pose great promise for quickening the pace of innovation, they come with a set of challenges of their own that must be dealt with. For example, fast-track regulatory routes get you there quickly, but they need to be safety-oriented as well or things may go wrong. Data has a crucial role in any machine learning-based solution and machine learning-based solutions in scientific research bring similar considerations as mentioned before related to data quality and ethics. If these challenges are addressed, stakeholders will be able to grasp the full potential of growing applications and propel innovation through the next stage.

FUTURE PERSPECTIVES AND CHALLENGES

Future directions in silicon photonics are shaped by improved fabrication technologies, new applications, and collaboration between academia and industries. Advances in deep-UV lithography and the incorporation of novel materials, including III-Vs and 2D materials, will allow for improved device performance as well as hybrid photonic integration. This development will create new applications beyond datacom, including healthcare, autonomous vehicles and environmental monitoring. As photonic solutions expand into biosensing, LIDAR, and AI accelerators, the need for scalable and cost-effective fabrication methods increases. To make adoption by industry more likely, it will also be essential that research prototypes are compatible with commercial foundry processes. Innovation and a healthy ecosystem need collaboration among the stakeholders. These efforts will be bolstered through partnerships across academia, industry, and government, including open innovation platforms and funding initiatives. Targeted training and educational programs will prepare researchers and engineers, bolstering the human capital underlying this area of technological innovation.

While these developments are promising, several challenges still exist, including fabrication complexities, high production costs, and scalability issues. The barriers should be tertiary concerns, so they can go purely toward technological and less technical minded solutions. This platform will act as a deep nexus between research and industry, paving the way to silicon photonics being truly realized..

CONCLUSION

Today, silicon photonics is raking up its own transformation in industries ranging from datacom to healthcare to autonomous systems. Advances in fabrication technologies, particularly deep-UV lithography and the use of new materials, are enabling innovative solutions to be scalable and efficient. Flexible and scalable prototyping platforms like CORNERSTONE can help bridge the gap between academic research and commercial production. Such advances are helping push new photonic designs from research labs into practice. But it's not all smooth sailing for silicon photonics adoption. Overcoming problems like process standardization, production costs, and photonics-electronics integration (becoming more complex conflict) have to be resolved. Overcoming these barriers will require collaboration among academia, industry, and policymakers to create a sustainable ecosystem. Aditional investment in research, workforce development and openaccess platforms will advance progress even more. In conclusion, silicon photonics is likely to be a game-changing technology for various industries, leading to faster, more efficient, and smaller solutions. Silicon photonics will be crucial to the future of technology and innovation as research continues to stretch the limits of ability.

REFERENCES

Akintola, A., Newbury-Birch, D., & Kilinc, S. (2024). Bridging the gap between research evidence and its implementation in public health practice: Case studies of embedded research model. *BMC Public Health*, *24*(1), 1299. DOI: 10.1186/s12889-024-18727-z PMID: 38741039

Alexander, F. J., Lin, M., Qian, X., & Yoon, B.-J. (2023). Accelerating scientific discoveries through data-driven innovations. *Patterns (New York, N.Y.)*, *4*(11), 100876. DOI: 10.1016/j.patter.2023.100876 PMID: 38035189

Allison, M. (2013). Alliance to bolster translation of academic R&D assets. *Nature Biotechnology*, *31*(4), 275–275. DOI: 10.1038/nbt0413-275 PMID: 23563411

Baba, T. *et al.* (2022) 'Silicon Photonics FMCW LiDAR Chip With a Slow-Light Grating Beam Scanner', *IEEE Journal of Selected Topics in Quantum Electronics*, 28(5: Lidars and Photonic Radars), pp. 1–8. Available at: .DOI: 10.1109/JSTQE.2022.3157824

Buldas, A., Draheim, D., Gault, M., Laanoja, R., Nagumo, T., Saarepera, M., Shah, S. A., Simm, J., Steiner, J., Tammet, T., & Truu, A. (2022). An Ultra-Scalable Blockchain Platform for Universal Asset Tokenization: Design and Implementation. *IEEE Access : Practical Innovations, Open Solutions*, *10*, 77284–77322. DOI: 10.1109/ACCESS.2022.3192837

Camino, F. E., Tiwale, N., Hwang, S., Du, X., & Yang, J. C. (2024). Mitigating challenges in aberration-corrected electron-beam lithography on electron-opaque substrates. *Nanotechnology*, *35*(6), 065301. DOI: 10.1088/1361-6528/ad0908 PMID: 37918028

Carmichael, G.. (2018) 'Curriculum-Aligned Work-Integrated Learning: A New Kind of Industry-Academic Degree Partnership', in *Proceedings of the 49th ACM Technical Symposium on Computer Science Education. SIGCSE '18: The 49th ACM Technical Symposium on Computer Science Education*, Baltimore Maryland USA: ACM, pp. 586–591. Available at: DOI: 10.1145/3159450.3159543

Changliang, Y., Luhong, M., Xindong, X., Sheng, X., & Shilin, Z. (2009). An approach to the optical interconnect made in standard CMOS process. *Journal of Semiconductors*, *30*(5), 055012. DOI: 10.1088/1674-4926/30/5/055012

Elsayed, A. *et al.* (2024) 'Low charge noise quantum dots with industrial CMOS manufacturing', *npj Quantum Information*, 10(1), p. 70. Available at: .DOI: 10.1038/s41534-024-00864-3

Feng, T., Qamruzzaman, M., Sharmin, S. S., & Karim, S. (2024). Bridging Environmental Sustainability and Organizational Performance: The Role of Green Supply Chain Management in the Manufacturing Industry. *Sustainability (Basel), 16*(14), 5918. DOI: 10.3390/su16145918

Gardner, J., & Webster, A. (2017). Accelerating Innovation in the Creation of Biovalue: The Cell and Gene Therapy Catapult. *Science, Technology & Human Values, 42*(5), 925–946. DOI: 10.1177/0162243917702720 PMID: 28845068

Gomez-Garcia, C. A., Askar-Rodriguez, M., & Velasco-Medina, J. (2021). Platform for Healthcare Promotion and Cardiovascular Disease Prevention. *IEEE Journal of Biomedical and Health Informatics, 25*(7), 2758–2767. DOI: 10.1109/JBHI.2021.3051967 PMID: 33449888

Gordan, V. (2017). How to Bridge Research Results to Everyday Clinical Care? *Operative Dentistry, 42*(1), 1–9. DOI: 10.2341/16-154-B PMID: 28002694

Gorelick, S., Guzenko, V. A., Vila-Comamala, J., & David, C. (2010). Direct e-beam writing of dense and high aspect ratio nanostructures in thick layers of PMMA for electroplating. *Nanotechnology, 21*(29), 295303. DOI: 10.1088/0957-4484/21/29/295303 PMID: 20601756

Gottesman, A. J., & Hoskins, S. G. (2013) 'CREATE Cornerstone: Introduction to Scientific Thinking, a New Course for STEM-Interested Freshmen, Demystifies Scientific Thinking through Analysis of Scientific Literature', *CBE—Life Sciences Education*. Edited by J. Vazquez, 12(1), pp. 59–72. Available at: DOI: 10.1187/cbe.12-11-0201

Graham, A. H. D., Robbins, J., Bowen, C. R., & Taylor, J. (2011). Commercialisation of CMOS Integrated Circuit Technology in Multi-Electrode Arrays for Neuroscience and Cell-Based Biosensors. *Sensors (Basel), 11*(5), 4943–4971. DOI: 10.3390/s110504943 PMID: 22163884

Green, D. E., Hamory, B. H., Terrell, G. E., & O'Connell, J. (2017). A Case Report: Cornerstone Health Care Reduced the Total Cost of Care Through Population Segmentation and Care Model Redesign. *Population Health Management, 20*(4), 309–317. DOI: 10.1089/pop.2016.0105 PMID: 28106518

Hager, A., Güniat, L., Morgan, N., Ramanandan, S. P., Rudra, A., Piazza, V., Fontcuberta i Morral, A., & Dede, D. (2023). The implementation of thermal and UV nanoimprint lithography for selective area epitaxy. *Nanotechnology, 34*(44), 445301. DOI: 10.1088/1361-6528/acea87 PMID: 37494897

Hammood, M., Lin, S., Yun, H., Luan, E., Chrostowski, L., & Jaeger, N. A. F. (2023). Emulation of deep-ultraviolet lithography using rapid-prototyping, electron-beam lithography for silicon photonics design. *Optics Letters*, *48*(3), 582. DOI: 10.1364/OL.478212 PMID: 36723536

Han, B., Chen, J., Guo, T., Lee, S.-J., Swaminathan, V., & Venieris, S. I. (2022). Guest Editorial: Bridging the Gap Between Industry and Academia for Networking Research. *IEEE Network*, *36*(1), 8–9. DOI: 10.1109/MNET.2022.9740650

Hope, T.. (2018) 'Accelerating Innovation Through Analogy Mining', in *Proceedings of the Twenty-Seventh International Joint Conference on Artificial Intelligence. Twenty-Seventh International Joint Conference on Artificial Intelligence {IJCAI-18}*, Stockholm, Sweden: International Joint Conferences on Artificial Intelligence Organization, pp. 5274–5278. Available at: DOI: 10.24963/ijcai.2018/736

Hu, M., Pang, Y., & Gao, L. (2023). Advances in Silicon-Based Integrated Lidar. *Sensors (Basel)*, *23*(13), 5920. DOI: 10.3390/s23135920 PMID: 37447770

Huang, X., Doerr, C. R., Qin, C., Heanue, J., Zhu, N., Ton, D., Guan, B., Zhang, S., & Zhao, Y. (2021). Silicon-photonic laser emitting tunable dual wavelengths with highly correlated phase noise. *Optics Letters*, *46*(1), 142. DOI: 10.1364/OL.409729 PMID: 33362036

Huang-Saad, A., Fay, J., & Sheridan, L. (2017). Closing the divide: Accelerating technology commercialization by catalyzing the university entrepreneurial ecosystem with I-Corps™. *The Journal of Technology Transfer*, *42*(6), 1466–1486. DOI: 10.1007/s10961-016-9531-2

Hulls, C. C. W., & Rennick, C. (2020). Use of a Cornerstone Project to Teach Ill-Structured Software Design in First Year. *IEEE Transactions on Education*, *63*(2), 98–107. DOI: 10.1109/TE.2019.2959591

Jin, Y., Wanvarie, D., & Le, P. T. V. (2019) 'Bridging the Gap Between Research and Production with CODE', in Q. Yang et al. (eds) *Advances in Knowledge Discovery and Data Mining*. Cham: Springer International Publishing (Lecture Notes in Computer Science), pp. 277–288. Available at: DOI: 10.1007/978-3-030-16142-2_22

Kang, K. H., & Lee, Y. H. (2007). Make-to-order scheduling in foundry semiconductor fabrication. *International Journal of Production Research*, *45*(3), 615–630. DOI: 10.1080/00207540600792416

Kato, K., Liu, Y., Murakami, S., Morita, Y., & Mori, T. (2021). Electron beam lithography with negative tone resist for highly integrated silicon quantum bits. *Nanotechnology*, *32*(48), 485301. DOI: 10.1088/1361-6528/ac201b PMID: 34425562

Khodaei, H., Scholten, V. E., Wubben, E. F. M., & Omta, S. W. F. O. (2022). The Role of Academic Spin-Offs Facilitators in Navigation of the Early Growth Stage Critical Junctures. *IEEE Transactions on Engineering Management*, *69*(4), 1769–1780. DOI: 10.1109/TEM.2020.2995361

Khurana, R., & Hodges, S. (2020) 'Beyond the Prototype: Understanding the Challenge of Scaling Hardware Device Production', in *Proceedings of the 2020 CHI Conference on Human Factors in Computing Systems. CHI '20: CHI Conference on Human Factors in Computing Systems*, Honolulu HI USA: ACM, pp. 1–11. Available at: DOI: 10.1145/3313831.3376761

Kim, H.-K., Kim, M., Kim, M., Jo, Y., Lischke, S., Mai, C., Zimmermann, L., & Choi, W.-Y. (2021). Si photonic-electronic monolithically integrated optical receiver with a built-in temperature-controlled wavelength filter. *Optics Express*, *29*(6), 9565. DOI: 10.1364/OE.418222 PMID: 33820381

Kim, J., Kim, M.-G., Kim, J., Jo, S., Kang, J., Jo, J.-W., Lee, W., Hwang, C., Moon, J., Yang, L., Kim, Y.-H., Noh, Y.-Y., Yun Jaung, J., Kim, Y.-H., & Kyu Park, S. (2015). Scalable Sub-micron Patterning of Organic Materials Toward High Density Soft Electronics. *Scientific Reports*, *5*(1), 14520. DOI: 10.1038/srep14520 PMID: 26411932

Kim, Y. E., Morton, B. G., Gregorio, J., Rosen, D. S., Edouard, K., & Vallett, R. (2019). Enabling creative collaboration for all levels of learning. *Proceedings of the National Academy of Sciences of the United States of America*, *116*(6), 1878–1885. DOI: 10.1073/pnas.1808678115 PMID: 30718419

Kneller, R., ... (2014) 'Industry-University Collaborations in Canada, Japan, the UK and USA – With Emphasis on Publication Freedom and Managing the Intellectual Property Lock-Up Problem', *PLoS ONE*. Edited by A.R.H. Montoya, 9(3), p. e90302. Available at: DOI: 10.1371/journal.pone.0090302

Lavikka, R., Seppänen, O., Peltokorpi, A., & Lehtovaara, J. (2020). Fostering process innovations in construction through industry–university consortium. *Construction Innovation*, *20*(4), 569–586. DOI: 10.1108/CI-08-2019-0081

Lazzarotti, V., Puliga, G., Manzini, R., Tallarico, S., Pellegrini, L., Eslami, M. H., Ismail, M., & Boer, H. (2025). Collaboration with universities and innovation efficiency: Do relationship depth and organizational routines matter? *European Journal of Innovation Management*, *28*(2), 608–630. DOI: 10.1108/EJIM-03-2023-0241

Lee, D., Krumdieck, S., & Talwar, S. D. (2013). Scale-up design for industrial development of a PP-MOCVD coating system. *Surface and Coatings Technology*, *230*, 39–45. DOI: 10.1016/j.surfcoat.2013.06.064

Li, K., Li, J., Reardon, C., Schuster, C. S., Wang, Y., Triggs, G. J., Damnik, N., Müenchenberger, J., Wang, X., Martins, E. R., & Krauss, T. F. (2016). High speed e-beam writing for large area photonic nanostructures — A choice of parameters. *Scientific Reports*, *6*(1), 32945. DOI: 10.1038/srep32945 PMID: 27633902

Li, W., Kitagawa, D., Kobatake, S., Bekyarova, E., & Bardeen, C. J. (2022). Patterning submicron photomechanical features into single diarylethene crystals using electron beam lithography. *Nanoscale Horizons*, *7*(9), 1065–1072. DOI: 10.1039/D2NH00205A PMID: 35788624

Lim, A. E.-J., Junfeng Song, , Qing Fang, , Chao Li, , Xiaoguang Tu, , Ning Duan, , Tern, R. P.-C., & Tsung-Yang Liow, Kok Kiong Chen. (2014). Review of Silicon Photonics Foundry Efforts. *IEEE Journal of Selected Topics in Quantum Electronics*, *20*(4), 405–416. DOI: 10.1109/JSTQE.2013.2293274

Manfrinato, V. R., Wen, J., Zhang, L., Yang, Y., Hobbs, R. G., Baker, B., Su, D., Zakharov, D., Zaluzec, N. J., Miller, D. J., Stach, E. A., & Berggren, K. K. (2014). Determining the Resolution Limits of Electron-Beam Lithography: Direct Measurement of the Point-Spread Function. *Nano Letters*, *14*(8), 4406–4412. DOI: 10.1021/nl5013773 PMID: 24960635

Mao, X. (2024). The Intersection of Art and Technology: Innovative Thinking and Practice in the Perspective of New Media. *Applied Mathematics and Nonlinear Sciences*, *9*(1), 20241367. DOI: 10.2478/amns-2024-1367

Murga Pinillos, A. Y. (2024). University-industry open innovation: Main enablers and practices based on a scoping review [Preprint]. *International Journal of Innovation Science*. Advance online publication. DOI: 10.1108/IJIS-08-2023-0194

O'Dwyer, M., Filieri, R., & O'Malley, L. (2023). Establishing successful university–industry collaborations: Barriers and enablers deconstructed. *The Journal of Technology Transfer*, *48*(3), 900–931. DOI: 10.1007/s10961-022-09932-2

Qiang, X., Zhou, X., Wang, J., Wilkes, C. M., Loke, T., O'Gara, S., Kling, L., Marshall, G. D., Santagati, R., Ralph, T. C., Wang, J. B., O'Brien, J. L., Thompson, M. G., & Matthews, J. C. F. (2018). Large-scale silicon quantum photonics implementing arbitrary two-qubit processing. *Nature Photonics*, *12*(9), 534–539. DOI: 10.1038/s41566-018-0236-y

Ruberti, M. (2024). Environmental performance and trends of the world's semiconductor foundry industry. *Journal of Industrial Ecology*, *28*(5), 1183–1197. DOI: 10.1111/jiec.13529

Rybnicek, R., & Königsgruber, R. (2019). What makes industry–university collaboration succeed? A systematic review of the literature. *Journal of Business Economics*, *89*(2), 221–250. DOI: 10.1007/s11573-018-0916-6

Sahin, O., Albayrak, O. M., & Yapici, M. K. (2024). Optimization of e-beam lithography parameters for nanofabrication of sub-50 nm gold nanowires and nanogaps based on a bilayer lift-off process. *Nanotechnology*, *35*(39), 395301. DOI: 10.1088/1361-6528/ad5e89 PMID: 38959870

Saifullah, M. S. M., Asbahi, M., Neo, D. C. J., Mahfoud, Z., Tan, H. R., Ha, S. T., Dwivedi, N., Dutta, T., bin Dolmanan, S., Aabdin, Z., Bosman, M., Ganesan, R., Tripathy, S., Hasko, D. G., & Valiyaveettil, S. (2022). Patterning at the Resolution Limit of Commercial Electron Beam Lithography. *Nano Letters*, *22*(18), 7432–7440. DOI: 10.1021/acs.nanolett.2c02339 PMID: 36069429

Schneider, J.-G.. (2020) 'Industry Agile practices in large-scale capstone projects', in *Proceedings of the ACM/IEEE 42nd International Conference on Software Engineering: Companion Proceedings. ICSE '20: 42nd International Conference on Software Engineering*, Seoul South Korea: ACM, pp. 326–327. Available at: DOI: 10.1145/3377812.3390794

Shah, J., Snider, B., Clarke, T., Kozutsky, S., Lacki, M., & Hosseini, A. (2019). Large-scale 3D printers for additive manufacturing: Design considerations and challenges. *International Journal of Advanced Manufacturing Technology*, *104*(9–12), 3679–3693. DOI: 10.1007/s00170-019-04074-6

Shekhar, S., Bogaerts, W., Chrostowski, L., Bowers, J. E., Hochberg, M., Soref, R., & Shastri, B. J. (2024). Roadmapping the next generation of silicon photonics. *Nature Communications*, *15*(1), 751. DOI: 10.1038/s41467-024-44750-0 PMID: 38272873

Steinmo, M. (2015). Collaboration for Innovation: A Case Study on How Social Capital Mitigates Collaborative Challenges in University–Industry Research Alliances. *Industry and Innovation*, *22*(7), 597–624. DOI: 10.1080/13662716.2015.1105127

Tereshchenko, E., Salmela, E., Melkko, E., Phang, S. K., & Happonen, A. (2024). Emerging best strategies and capabilities for university–industry cooperation: Opportunities for MSMEs and universities to improve collaboration. A literature review 2000–2023. *Journal of Innovation and Entrepreneurship*, *13*(1), 28. DOI: 10.1186/s13731-024-00386-4

Vanli, O. A., Rivero, I. V., & Wang, B. (2013). Editorial—advances in nanomanufacturing: Process repeatability, scalability and affordability for nanomaterials and manufacturing. *International Journal of Advanced Manufacturing Technology*, *64*(1–4), 513–514. DOI: 10.1007/s00170-012-4646-6

Weimann, C., Lauermann, M., Hoeller, F., Freude, W., & Koos, C. (2017). Silicon photonic integrated circuit for fast and precise dual-comb distance metrology. *Optics Express*, *25*(24), 30091. DOI: 10.1364/OE.25.030091 PMID: 29221043

White, A., Khial, P., Salehi, F., Hassibi, B., & Hajimiri, A. (2020). A Silicon Photonics Computational Lensless Active-Flat-Optics Imaging System. *Scientific Reports*, *10*(1), 1689. DOI: 10.1038/s41598-020-58027-1 PMID: 32015358

Xiao, Z., Liu, W., Xu, S., Zhou, J., Ren, Z., & Lee, C. (2023). Recent Progress in Silicon-Based Photonic Integrated Circuits and Emerging Applications. *Advanced Optical Materials*, *11*(20), 2301028. DOI: 10.1002/adom.202301028

Xie, J., Yang, Z., Wang, X., & Lai, X. (2019). A cloud service platform for the seamless integration of digital design and rapid prototyping manufacturing. *International Journal of Advanced Manufacturing Technology*, *100*(5–8), 1475–1490. DOI: 10.1007/s00170-018-2809-9

Xiong, H., ... (2014) 'Scalable Architectures for Platform-as-a-Service Clouds: Performance and Cost Analysis', in P. Avgeriou and U. Zdun (eds) *Software Architecture*. Cham: Springer International Publishing (Lecture Notes in Computer Science), pp. 226–233. Available at: DOI: 10.1007/978-3-319-09970-5_21

Xomalis, A., Hain, C., Groetsch, A., Klimashin, F. F., Nelis, T., Michler, J., & Schwiedrzik, J. (2023). Resist-Free E-beam Lithography for Patterning Nanoscale Thick Films on Flexible Substrates. *ACS Applied Nano Materials*, *6*(5), 3388–3394. DOI: 10.1021/acsanm.2c05161

Ye, Z., Jia, H., Huang, Z., Shen, C., Long, J., Shi, B., Luo, Y.-H., Gao, L., Sun, W., Guo, H., He, J., & Liu, J. (2023). Foundry manufacturing of tight-confinement, dispersion-engineered, ultralow-loss silicon nitride photonic integrated circuits. *Photonics Research*, *11*(4), 558. DOI: 10.1364/PRJ.486379

Yu, B.. (2012) 'Dealing with IC manufacturability in extreme scaling', in *Proceedings of the International Conference on Computer-Aided Design. ICCAD '12: The International Conference on Computer-Aided Design*, San Jose California: ACM, pp. 240–242. Available at: DOI: 10.1145/2429384.2429430

Zuo, Z., De Corte, W., Huang, Y., Chen, X., Zhang, Y., Li, J., Zhang, L., Xiao, J., Yuan, Y., Zhang, K., Zhang, L., & Mechtcherine, V. (2024). Strategies towards large-scale 3D printing without size constraints. *Virtual and Physical Prototyping*, *19*(1), e2346821. DOI: 10.1080/17452759.2024.2346821

Chapter 14
Smart Printing: Sensor Integration and Real-Time Monitoring in Additive Manufacturing

M. Maniraj
https://orcid.org/0009-0000-5343-3895
Sardar Raja College of Engineering, Alangulam, India

S. M. Raj Kumar
https://orcid.org/0000-0002-8978-0108
Francis Xavier Engineering College, Tirunelveli, India

ABSTRACT

Additive manufacturing (AM) has emerged as a transformative technology for producing complex, high-performance components across industries such as aerospace, healthcare, and automotive. Central to enhancing the precision, reliability, and scalability of AM is the integration of advanced sensors and real-time monitoring systems. These technologies enable in-situ process control, early defect detection, and adaptive feedback mechanisms that elevate both production quality and efficiency. This chapter delves into the diverse roles of sensor technologies—including thermal, optical, force, gas, and acoustic sensors—and how their deployment facilitates intelligent decision-making through machine learning and artificial intelligence. In addition, we discuss how real time data analytics and computational modeling are being used to predict and control process outcomes and move manufacturing away from reactive to predictive. The growing relevance of hybrid manufacturing, which combines additive and subtractive processes is also examined for its relevance to high precision applications.

DOI: 10.4018/979-8-3373-0533-2.ch014

1. INTRODUCTION

Additive manufacturing (AM) or 3D printing has completely changed modern production by offering the possibility to build complex geometries, customized components with overcomes in materials utilization. Unlike other subtractive manufacturing methods, subtracting material from a piece of stock material to pose the final product, AM constructs objects one layer at a time, which reduces waste and increases design flexibility (Ma et al. 2024). Due to such a paradigm shift new possibilities are opening up in the sectors such as aerospace, healthcare, automotive, and consumer electronics where growing demand for lightweight, high performance and customized designs exists. While it has its benefits, however, the process is not without its own problems of stability, product assurance and scalability. AM processes, however, are susceptible to defects like porosity, warping, or delamination due to the fact that they rely on numerous parameters, which include temperature, material deposition rates and environment conditions, that cannot be controlled properly. These issues must be addressed to ensure that AM produced components can achieve the required standards of industry, and be implemented reliably into intended applications (Zhu et al. 2024). Figure 1 shows the overview of Additive Manufacturing with Sensor Integration (Chen et al. 2024).

Real time monitoring and sensor towards integration become the right solutions to improve consistency, efficiency and reliability of additive manufacturing. Advanced sensors are also incorporated for continuous tracking of some important relevant process parameters for early defect detection and adaptive control mechanisms (Jhun et al. 2024). Through real time data from the printer manufacturers can dynamically optimize printing conditions, reducing material waste and eliminating costly rework. Moreover, it has become possible to predict failures and take corrective actions before defects turn critical because of the innovations in data analytics, machine learning, artificial intelligence, and so on to interpret sensor data. These capabilities are needed for AM to transition from a primarily prototyping technology to a robust manufacturing solution that can be used for large scale manufacturing.

Figure 1. Overview of additive manufacturing with sensor integration (Chen et al. 2024)

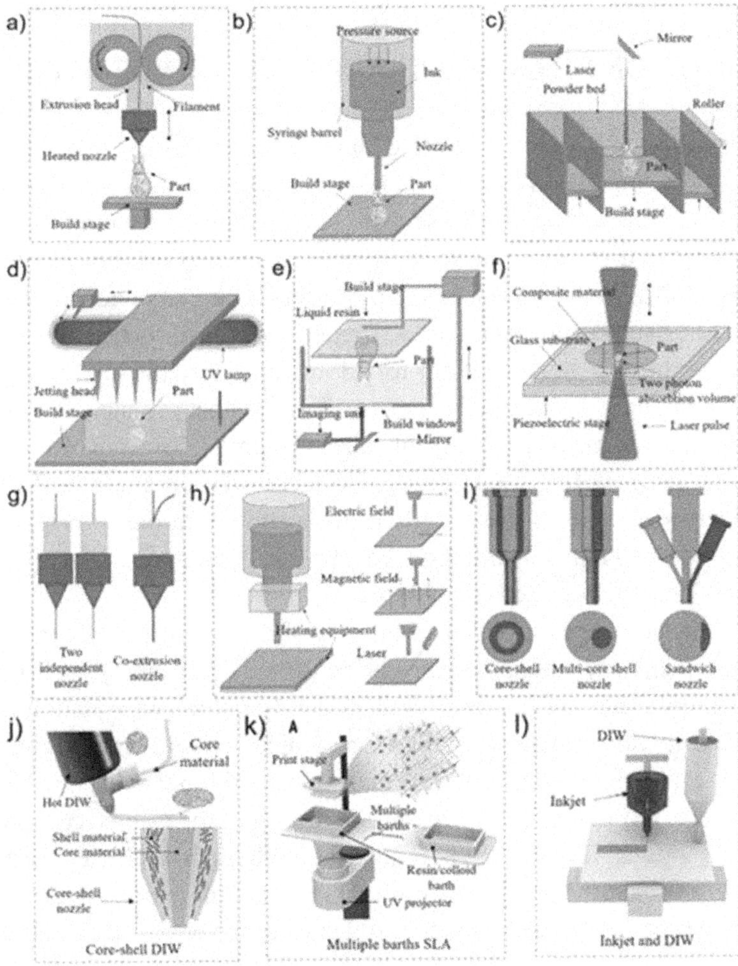

The role of sensor integration and real-time monitoring in additive manufacturing is explored in this chapter as a method of transformative change. It discusses the different kinds of sensors used in AM, their particular functions and the role they play in quality assurance and process optimization (Kanthimathi et al. 2024). The discussion is further extended to challenges related to the implementation of sensor such as calibration, data management and system integration, and the use of artificial intelligence in processing large amount of real time data. The chapter also discusses how these technologies will so broaden as to determine sustainability, cybersecurity, and industrial scalability. This chapter aims to provide a thorough analysis of these developments, in order to shed light on the future of smart printing and how it may revolutionize the additive manufacturing domain.

2. THE ROLE OF SENSORS IN ADDITIVE MANUFACTURING

The integration of sensors in additive manufacturing has key roles in controlling processes, improving product quality, and repeat current process. Essentially, 3D printing systems are equipped with sensors that are constantly monitoring important parameters such as temperature, pressure, material flow and the environment (Hossain et al. 2025). These sensors capture real time data and provide realtime insights into the printing process state, so immediate adjustments and correction can be happened to tuned if any deviations found. In contrast to the traditional manufacturing, in which quality control is usually performed post production by inspection and testing, additive manufacturing substantially profits from in-situ monitoring, which allows for proactive intervention. To tackle the common AM challenges, including inconsistencies in material deposition and variations in print conditions, this research aims to make a shift from reactive to predictive and adaptive control of the pulsed laser powder bed fusion process.

Because the process is layer by layer, one of the biggest challenges to ensure consistent print quality remains. Conventional methods employ well known machining tolerances and are not subjected to the dynamic changes in temperature, material viscosity and environmental factors, which create defects such as warping and porosity; and delamination and residual stress buildup. Defects in these printed components not only degrade the structural integrity of the printed components, but also limit the applicability of AM in industrial applications (Guo et al. 2024). Historically, lack of real time process visibility had meant increased material waste and extensive post processing, both of which add to time and cost associated production. Sensor integration directly responds to these challenges by continuously feeding critical parameters, so that the manufacturer could detect anomalies in process as they occur and modulate the process accordingly.

Beyond defect prevention, real time monitoring in additive manufacturing has benefits related to the optimization of the material and process automation. With sensor data on hand, manufacturers will be able to optimise material deposition rates, energy input for laser based or extrusion based process and make sure layer adhesion is uniform. Such precision leads to minimization of waste and improve the efficiency of AM systems and, in turn, making production more sustainable and cost effective (Li et al. 2024). Real time monitoring also allows process automation through closed loop control whereby sensors communicate with the printing system in real time and make adjustments in real time with no human intervention.

Particularly in high volume, or mission critical, environments such as aerospace or medical device manufacturing where it is critical for the part to return to the plate and be reclaimable, this capability is very valuable. Also, due to the integration of artificial intelligence, and machine learning algorithms with sensor data, predictive maintenance and intelligent optimization of the process have been opened that makes AM systems better with time. Sensor integration, which is acting as technology catalyst, is laying the groundwork for a new age of intelligent printing, with precision, efficiency and adaptability as its future of the additive manufacturing.

3. TYPES OF SENSORS USED IN ADDITIVE MANUFACTURING

It is shown that additive manufacturing uses a wide range of sensors to achieve process stability, improve quality control, and optimize material performance. Of these, thermal sensors are among the most critical, and they are key to monitoring temperature changes as the print is being made. Control of the temperature is critical to avoid the defects like warp, incompleteness of the fusing, and too much residual stress that can impair the mechanical integrity of the final product (Ng et al. 2024). AM Systems can maintain precise thermal conditions utilizing thermocouples, infrared sensors and pyrometers to ensure a complete and correct thermal profile is taken place, for instance in the processes such as selective laser sintering (SLS) or fused deposition modeling (FDM) where heat profile affect layer stacking and the strength of the part. Thermal monitoring in real time enables dynamic change in the heating element and laser power to maintain the optimal processing condition.

Another indispensable category is optical and vision sensors, which give high resolution imaging for the detailed layer by layer quality assessment. The sensors, such as cameras, laser profilometers and structured light scanners, obtain real time visual and geometric information to sense when the surface is defective, is not aligned or other inconsistencies in the material deposition (Wang et al. 2024). The movement of the print head can be tracked by high speed cameras and anomalies in deposition paths can be identified while structured light scanners, can generate 3D reconstructions of the surface to evaluate uniformity of the layer. In metal additive manufacturing, the precise formation of each layer is of great importance for the production of defect free components and optical sensors are very suitable in such situations. Utilizing machine vision and AI based image analysis, these sensors facilitate automated defect detection, and can take action to correct it instantly without the need for extensive post processing.

The print head and the material being deposited are put under tension through use of force and pressure sensors to allow insights into the mechanical interactions. Automatically checking extruder motor current FDM printers as it is a much more economical solution for measuring extrusion force, and keeping consistent material flow to prevent under extrusion or clogging (Waheed & Bernadin, 2024). In powder bed fusion processes, they can be used to assess the compaction force applied during recoating which directly affects the layer density and the final part quality. Manufacturers can fine tune their material delivery mechanisms and also detect malfunctions in progress in real time by monitoring pressure variations. Furthermore, force sensors are also used to regulate the application of binding agents and metal powders in the binder jetting process and in the directed energy deposition (DED) process to ensure uniform distribution (Chen & Moon, 2024).

Monitoring of the printing environment is an essential part of processes based on reactive materials or requiring controlled atmosphere, where gas sensors play an important role. During metal additive manufacturing, argon or nitrogen are used as inert gases to prevent oxidation and contamination during printing. These environments are stable because of gas sensors that detect changes in gas composition or the presence of impurities that may impact print quality (Herzog et al. 2024). On the other hand, gas sensors are also involved with safety aspects, when they are used to identify hazardous emissions, for example, volatile organic compounds or toxic byproducts produced in polymer printing. These sensors are used to monitor air quality, and do so continuously contributing to environmental control, and to workplace safety.

This is where acoustic and ultrasonic sensors come in by providing yet another layer of process monitoring by detecting internal defects of which there is no optical signature. In the case of these sensors, they analyze sound waves and vibrations produced during the printing process to identify irregularities such as voids, delamination, lack of fusion, etc (Khanafer et al. 2024). Due to their ability to penetrate through layers and subsurface integrity can be assessed, ultrasonic sensors offer great value in applications where nondestructive evaluation is needed. Furthermore, acoustic sensors can also be used in laser based AM processes to sense instabilities in the melt pool and respond immediately to ensure a high quality print. Modern AM systems achieve a more complete monitoring by integrating multiple sensing technologies to enable a comprehensive monitoring framework which enables better reliability, fewer defects, fully autonomous AM processes.

Figure 2. Types of sensors used in additive manufacturing (Bas et al. 2024)

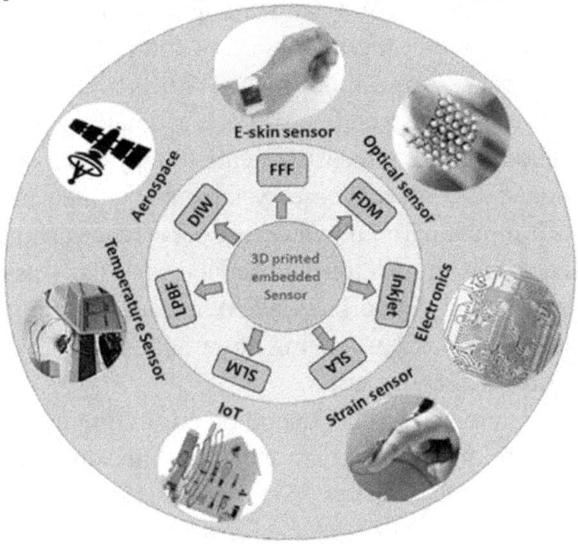

4. DATA PROCESSING AND ANALYSIS IN REAL-TIME MONITORING

With the inclusion of real time monitoring systems in additive manufacturing, tremendous volumes of sensor data are produced, containing data points of the critical parameters like temperature fluctuation, material deposition rate, mechanical stress and environmental condition. Since AM is layer-by-layer, high resolution data is continuously generated and the print cycle, therefore, data processing and analysis is an important part of ensuring precision and quality. With this data being generated at such a velocity, it is only through these sophisticated computational approaches that can one hope to extract meaningful insights in real time. In high precision or high throughput applications, traditional monitoring methods which rely on post process inspection and offline analysis are not satisfactory any more. However, these modern AM systems utilize advanced computational techniques to control these large datasets dynamically so as to make an immediate decision and thereby mitigate defects.

Previously, simple threshold based systems were used for AM monitoring, these systems sent an alert and would halt printing operations when they detected deviations from predefined limits. These conventional methods were effective to some extent, but they were not flexible and did not consider complex interaction among the values of multiple process variables (Franke et al. 2025). Whereas modern AM monitoring relies on machine learning and artificial intelligence to make sense of the patterns in the data, predict the likelihood of future failures, and, at the right time, dynamically improve print conditions. With artificial intelligence driven models, trained on massive data set, one can detect small irregularities that might not show up in standard statistical measurements so they can be used for predictive maintenance and adaptive control. These systems refine their accuracy over time by correlating sensor inputs with known defect occurrences and thus improve the effectiveness of the defect prevention and reduce material waste (Mattera et al. 2024). Convolutional neural networks (CNNs) are deep learning techniques that are applied to optical monitoring systems more and more, for automated image recognition in real time defect detection.

AM monitoring systems typically depend on edge computing for real time decision making and therefore process data locally at the source instead of sending it to centralized servers. In high speed printing processes, any delay could lead to material defects, therefore, it is essential that sensor feedback is acted upon instantaneously, and edge computing helps reduce latency to achieve that. Along with that, edge processing helps 'lean' out bandwidth constraints by filtering and analyzing only what are the most important pieces of information before transferring to higher level systems to evaluate further. Nevertheless, in the speed of response, edge computing has its benefits, compared to cloud based monitoring solutions, as they offer good scalability and data storage capabilities (Ukwaththa et al. 2024). The cloud platforms enable manufacturers to aggregate data from several machines from different locations, which allows manufacturers to optimize large scale processes and fleet management. Using cloud computing, AM companies can apply predictive analytics at an industrial scale to improve the entire process continuously and the entire production network globally.

The addition of real-time sensor data, machine learning, and edge cloud hybrid architectures is transforming the method of controlling and optimizing additive manufacturing processes. In addition to assuring quality, these advances enable autonomous, self correcting AM systems with little or no input from the human operator. The degree of intelligence in future AM systems will be increased through greater computational power, such that the machines will learn from their work and be able to make real time responses to achieve unprecedented precision in their manufacturing.

5. CHALLENGES IN SENSOR INTEGRATION

Challenges to integrating sensors into additive manufacturing and controlling process and monitoring are presented. The hardware and software modifications that need to be performed for deploying the sensor seamlessly is one of the major difficulties. However, many existing 3D printing systems were not originally designed with built in sensing capabilities and hence it required the retrofitting of or redesigning of the hardware components to ensure placement of additional sensors (Rahman et al. 2024). This sometimes means altering print heads, sensor installation in build platforms, or adding external monitoring systems that have to interface with the control unit of the printer. In terms of software, sensor data needs to be properly synchronized with machine operation and requires specialized algorithms and firmware updates to process the incoming data in real time. A further layer of complexity stems from ensuring compatibility between different sensors, control software and data analytics platforms, as the manufacturers typically use a range of AM technologies whose communication protocols are not standardized.

Next to integration, the accurate calibration and positioning of sensors is of great importance to get reliable data. To quantify the process without disturbing the actual printing process, sensors must be situated at specific locations in the AM system. For instance, thermal sensors need to be placed to accurately monitor where heat is distributed and optical sensors have to be carefully set to capture high resolution images of each printed layer. Inaccurate readings, delivering false alarms or failing to detect defects are due to any misalignment or improper calibration (Shen & Li 2024). The positioning of sensors for additive manufacturing is complicated by the continuous changing of build conditions. Automated realignment mechanisms are needed to keep some sensors oriented to the right focus or measurement angle as the build progresses.

Another major challenge is on the durability and reliability of sensors in high temperature and/or reactive environments. The harsh thermal conditions used in many AM processes (especially metal based including selective laser melting and directed energy deposition) can quickly deteriorate sensor performance over the process time. Sensor components can be damaged by exposure to high temperature, molten metal, and reactive gas, leading to erroneous sensor data and the shortening of the useful life of a sensor. For example, optical sensors can be affected by the contamination of lens with smoke or metal spatter and need regular cleaning and maintenance. Just like so, gas sensors used in controlled atmosphere printing environments should be able to tolerate long time exposure to inert gases without sacrificing their sensitivity. The durability issues must be overcome for long term reliability in these AM applications, and for this to be achieved heat resistant, contamination resistant, and self cleaning sensor technologies must be developed (Piran et al. 2025).

Cost implications also play a critical role in deciding the feasibility of sensor enhanced AM systems. Although advanced monitoring technologies provide great benefits in defect reduction and process optimization, they increase the total cost of 3D printing equipment. AM initial investment, especially involving high precision sensors and data acquisition etc., may be prohibitive for smaller manufacturers and research facilities. Moreover, the maintenance costs of sensors, storage of data, and computational resources cost of operations should be taken into account. Companies must make tradeoffs carefully in terms of how much cost should be put into an extra sensor and how much performance improvement warrants the extra expense, specifically associated with reducing material waste, rework, and production downtime. The challenge for researchers is to find a balance between one that offers affordability, but also effectiveness. This means further research in cost effective sensor technologies that provides superior monitoring quality without significant increases in the cost of production (Kong et al. 2024).

However, sensor miniaturization, wireless communication, and AI driven data processing has made it possible for integrating sensors in AM. In order to achieve fully autonomous, self optimising 3D printing systems capable of continuously producing high quality parts for various industries, these obstacles will need to be addressed through innovative engineering solutions.

6. REAL-TIME CONTROL AND ADAPTIVE PRINTING

In order to move beyond these challenges and make additive manufacturing more precise, efficient, and reliable, control of additive manufacturing systems in real time as well as the use of real time adaption are essential. Closed loop control systems have been introduced which utilize sensor feedback adaptively monitored until dynamic process changes are implemented, to improve the ability to maintain print quality, as well as prevent defects. Traditional open loop systems set printing parameters at the beginning of the process and do not change them based on the variations that occur during fabrication (Sousa et al. 2025). Unfortunately, this approach cannot readily account for inconsistencies in the deposition of material, heat fluctuation, or an ill-fated spontaneous environmental change. No such printer exists today, as all of these constraints both limit the resolution that can be achieved, and adversely affect the quality of the print, however closedloop control overcomes these limitations by allowing the printer to adjust its parameters in real time to ensure perfect layer deposition under ideal conditions year round. For instance, temperature sensors can find out how a laser based system fluctuates in terms of heat distribution and can adjust its laser power to maintain uniform fusion. The set of high resolution optical monitoring systems captures images of each printed layer, which is used by the printer to fix layer misalignment or surface irregularities so as not to propagate and compromise the structural integrity of the finished product (Aravind Raj, 2025).

With these sensors, real-time control in additive manufacturing is demonstrated to be effective in a number of case studies. Optical coherence tomography (OCT) sensors have employed in additive manufacturing of metals to study the real time melt pool characteristics with the potential to correct promptly deviations in material fusion. Integrating OCT data with AI driven algorithms has allowed manufacturers to dynamically optimize the laser parameters reducing the likelyhood of defects such as porosity or incomplete fusion. Likewise, thermal cameras and infrared sensors have been implemented in polymer based printing to control extrusion temperatures to avoid problems such as warping or layer delamination. Real time monitoring of powder bed fusion industrial processes within the aerospace industry enabled the detection of anomalies in powder spreading which was then revised with a process modification that produced a component with significant improvements in consistency, and a reduction in material waste. While most of these examples are preventing defects in the first place, they also increase overall a manufacturing efficiency via reduced rework and increased material utilization (McKinney et al. 2025).

Real time control is also playing a leading role in the multi material and hybrid 3D printing where different materials or manufacturing approaches are combined in one build process, well beyond traditional AM. With multi material printing, there are additional complexities as the each material may have different processing needs such as different melting points, deposition rates or adhesion between materials. Adaptive printing as sensor driven guarantees that these variations are taken into account dynamically by altering either the energy input, extrusion speeds or curing times if required in order to smoothly integrate between materials. Real time monitoring in hybrid additive manufacturing, where 3D printing is combined with use of traditional machining or treatment at surfaces, helps coordinate operations during different processes with precision. For instance, hybrid systems may have force sensors that regulate the engagement of the tool during in situ machining of the printed component, so as to ensure that the printed component is being milled accurately without inducing excessive stress or distortion. Adaptive control mechanisms integrated into the additive manufacturing process will be increasingly necessary as AM evolves to provide greater material complexity, improved part customization, and higher levels of automation for manufacturing applications (Makanda et al. 2025).

7. SCALABILITY AND INDUSTRIAL APPLICATIONS

Limited scalability of additive manufacturing for the mass production has been due to process variation, lack of quality control and post processing. To alleviate the challenges above, real time monitoring is a key tool to achieving more predictable and repeatable production outcomes (Rolchigo et al. 2025). Manufacturers can integrate sensors into AM systems to have greater consistency across multiple print runs so that each component is defect free and meets specifications. A capability such as: Real time monitoring improves the capability to automate quality assurance and reduces the dependence for manual inspections, thereby making the production lines run with minimum human intervention. It is critical that such process control exists for the kind of high precision that aerospace and medical device manufacturing requires–any deviation of material properties or geometry just slightly past acceptable control parameters can cause a failure of a critical application. Furthermore, AM systems have adaptive control mechanisms that allow deviations to be corrected allowing mass production to be feasible, but also more cost effective and efficient (Petousis et al. 2025).

Centralized monitoring of multiple 3D printers in a production environment is one of the main advancements which allows additive manufacturing to scale. Several printers are often operated in parallel at large scale manufacturing facilities to make components that are different or work in conjunction to complete a batch production cycle. These printers can provide real time sensor data, and these data can be aggregated into a centralized system so that operators can monitor the whole manufacturing process from a single interface. Remote access to live data streams is possible from a cloud based monitoring platform, offering the ability to see print progress, error detection and performance metrics for the whole fleet of machines. By centralizing that approach, manufacturers are able to provide better productivity without the bottlenecks, optimize print scheduling and dynamically assign resources as demand occurs. Additionally, the application of machine learning algorithms to such data would allow for the identification of patterns in printer performance and predict when maintenance is needed and when it can be minimized, thus helping improve overall operational efficiency (Mahdi et al. 2025).

Already, sensor based additive manufacturing is transforming key industries like aerospace, healthcare, automotive, and most commonly used these days is electronics. Real time monitoring in the aerospace industry is necessary for the structural integrity of printed components to be used in aircraft and spacecraft where lightweight and high performance materials are critical. Sensor driven AM has been integrated into companies such as Boeing and NASA that create complex geometries not possible or too costly for production using traditional manufacturing methods (Fan et al. 2025). The another field which was benefited with the use of real time monitoring is healthcare industry where the recent application of this technology ranges from produce of patient specific implant to the prosthetics. Sensorbased monitoring facilitates the achievement of required level of precision while keeping with the stringent levels of regulatory compliance, and so ensuring biocompatibility and precision fit is crucial. Additive manufacturing is used in the automotive sector for rapid prototyping and production of bespoke and lightweight parts, where real time monitoring enables repeatability of quality and reducing the material waste. AM has been widely adopted by the electronics industry for the fabrication of intricate circuit components and heat exchangers, and in situ monitoring is used to detect microstructural defects and improve electrical performance. With the development of sensor technologies, they will be embedded in additive manufacturing, leading to more industrial applications, more automation, greater reliability, and the scale adoption of additive manufacturing in various manufacturing sectors (Granelli et al. 2025).

8. SUSTAINABILITY AND ENVIRONMENTAL CONSIDERATIONS

Sensors are highly important to utilize in additive manufacturing, to improve sustainability by decreasing material waste and better utilizing resources. Unlike subtractive manufacturing, which removes excess material to form the final product, additive manufacturing does not remove excess material and thereby inherently avoids waste. Though, there are still process inefficiencies like failed prints, too much support structures, and bad material deposition that result in material loss. These issues are mitigated by real – time monitoring systems that sense process parameters in real – time and detecting anomalies before they lead to major defects (Kumar & Kumar, 2025). Manufacturers can save raw materials by preventing errors early on in the printing cycle and thus reducing the need to reprint and post process. This reduction of waste directly reduces cost and is also an environmentally sustainable benefit for industries that use high value materials for titanium or composite polymers. Moreover, sensor based monitoring also allows one to have better control of the support material usage, using the least amount of support material possible and removing it in the most efficient way to reduce overall material consumption (Yadegari et al. 2025).

Along with improving material efficiency, real time monitoring helps in being sustainable by reducing energy usage throughout the additive manufacturing life cycle. AM processes, especially with high temperature operations as in the case of laser based metal printing can have high energy intensity. Adaptative energy management is enabled by sensors that track thermal distribution, power usage and cooling efficiency to ensure that only the minimum required power is deployed while still within a certain quality required for print quality (Ahmad et al. 2025). For instance, thermal imaging sensors can detect application of excessive heat to reach certain points and the system can correct for the laser intensity or deposition rate where it is not necessary and hence unnecessary energy would be spent. Moreover, real time control mechanisms facilitate faster print cycle by guaranteeing that each layer is being deposited with great precision thus reducing idle time and increasing cycle throughput. Predictions can be further made utilizing energy consumption patterns integrated with sensor data, machine learning algorithms can learn from previous prints to suggest energy efficient configurations for future production runs. Collectively, these improvements contribute to a reduced carbon footprint of AM operations thereby, making the technology a more environmentally friendly option for manufacturing (Resendes et al. 2025).

A series of case studies reflect the influence on sustainable AM practices through sensor integration. The aerospace industry benefits from real time monitoring, in particular to improve production and reduced material waste with high structural integrity while producing lightweight metal components. A leading aerospace manufacturer, for instance, achieved the reduction of defect rates and elimination of rework by utilizing optical and thermal sensors in their powder-bed fusion process to detect inconsistencies in material distribution. Real time monitoring of bioresorbable implants in the medical field has enabled manufacturers to accurately control the material deposition such that excessive material is not used and biocompatibility is maintained (Orzari et al. 2025). In the automotive industry, AM has also been used, driven by sensors, to produce lightweight, fuel efficient components and, in turn, adaptive control systems have reduced both the waste of material and energy. The problem is that real time monitoring (RTM) by itself is not all we need to get additive manufacturing to the paradigm where we can have transparent pricing in accordance with Autodesk's mission to help people share ideas, make great new things, and improve the world. There is no doubt that as more industries embrace green production strategies, the use of sensors and real time data analysis will be an integral component to progress AM's sustainability.

9. SECURITY AND ETHICAL CONSIDERATIONS

With the proliferation of additive manufacturing as increasingly connected and reliant systems with real time monitoring, cybersecurity is becoming a growing concern. As sensors, cloud based data processing and networked manufacturing environments are integrated into AM systems, AM systems are exposed to potential cyber threats such as unauthorized access, data breach, and malicious interference. Instead, 3D printing systems can potentially be also subject to cyberattacks, which target the production integrity by altering design files, disrupting print parameters, or introducing defects to critical components that cannot be detected. This is especially concerning in aerospace, defense and healthcare industries where reliability of printed components is critical. Manufacturers must take measures to ensure the security of data and authentication protocols in order to minimize these risks. The examination of the security profiles of the applications is critical to the protection of sensitive data and to the ability of AM processes to remain resilient in the face of cyber threats, and such examination should be an ongoing process of regular security audits and real time threat monitoring (Mattera et al. 2025).

Data privacy and integrity in monitoring of additive manufacturing is equally important as real–time sensor data could be proprietary or sensitive in nature describing manufacturing techniques, material compositions, and design specifications. This data is susceptible to intellectual property theft, industrial espionage or counterfeit production when accessed by unauthorized people. Because of these risks, AM companies must have strict access controls and data encryption methods that prevent unauthorized copying or modification of digital manufacturing files (Singh et al. 2025). As a potential solution, blockchain technology has been explored to provide secure storage and storage of AM data to maintain immutable lists of design revision, sensor feedback and process verification to establish traceability and authenticity. Furthermore, cloud based monitoring technologies need to be compliant with standard data protection rules, such as GDPR, or NIST in the case of cybersecurity frameworks, so that manufacturer still owns the right to their manufacturing data, but can take advantage of the benefits offered by monitoring and predictive analytics in the cloud (Sonavane & Aylani, 2025).

However, ethics in terms of data governance beyond security concerns are critical for regulated industries employing additive manufacturing, such as healthcare. AM handling process used in medical application of prosthetic, implant, and bioprinting also should be strictly regulate by patient privacy regulation and ethical guidelines. Sensitive patient data is being generated by real time monitoring of personalized medical devices that are of concern to data ownership, consent, and use (Sagar, 2025). AM monitoring by third party service providers poses ethical challenges in guaranteeing that patient specific designs and sensor derived performance data are not shared or misused inappropriately. Data governance policies must be transparent with respect to who has access to AM data, how the data is stored, in what context it could be shared, etc. Further, manufacturers have to deal with the bias of AI driven defect detection and process optimization, with sensor based decisions being fair and with no unintended consequences in safety critical applications. Because AM is advancing, its technology, security, and ethical responsibility need to be balanced to instill trust in AM systems for safe and responsible deployment in industry (Leghemo et al. 2025).

10. FUTURE TRENDS AND INNOVATIONS

Continuous advances in sensor technologies, artificial intelligence, and interdisciplinary research are shaping the future of additive manufacturing, driving the industry and shaping the future of additive manufacturing, ever leaner, ever more automated, precise, and scalable. Sensor technology evolution is a tremendous factor that has enabled real time monitoring to be tremendously increased, and new sensor technology like quantum sensors and 5G can revolutionize process control. Ultra precise quantum sensors have the capability of measuring the properties of the environment and materials with exquisite accuracy to nanoteperature, nanostoress and nanoelectromagnetiuc field (Musa & Singh, 2025). This would integrate them into AM systems that would also allow more controlled layer deposition compared to the same plasmas used in conventional CVD systems, resulting in fewer defects and more consistent materials. At the same time, 5G connectivity is expected to bring about a speedier data transmission speed in smart factories, meaning that several printers and monitoring systems can talk to each other immediately. With 5G powered additive manufacturing, real time feedback loops will be taken even further making them faster and more scalable, remote diagnostics and instant adjustment to parameters will be possible, while collaboration between manufacturing facilities spread across the world will be a breeze (Christ et al. 2024).

The applications of artificial intelligence to AM are also producing predictive maintenance and self optimizing printers that can independently modify their operations using information provided by sensors. Machine learning algorithms are utilized in the application of predictive maintenance via AI to analyze historical sensor data to identify patterns that suggest that it may fail or its components degrading. With AI-based monitoring systems, knowing the issues early on and pre – predicting the need for maintenance before any malfunction occurs, printer downtime and lifespan is reduced, and production is unaffected (Jin et al. 2024). Moreover, AI driven process optimization enables 3D printers to become increasingly self-aware by making use of previous builds, and consequently, automatically optimizing printing parameters like speed, temperature etc. to improve the overall efficiency and quality of printing. As the amount of real time data is getting larger, neural networks and deep learning models are being more and more utilized for analyzing such large volumes of data and it allows the AM systems to recognize and correct defects as they appear. This represents a major step towards autodidactic production whereby the amount of human intervention is reduced to a minimum and production reliability is considerably improved (Nasrin et al. 2024).

One theme of the work in intermdisciplinary research is smart printing, bringing together the material scientists, the engineers, the data scientists, the industrial designer as a vital vehicle for collaboration that has been enabling this to happen.

With AM maturation, real time monitoring systems for AM require expertise across different areas of domains such as advanced materials, computational modeling or machine learning. New sensor embedded materials that would provide real time feedback regarding the mechanical properties during the printing and adapt the material deposition in accordance to the structural needs are the subject of research. The development of hybrid manufacturing techniques that utilize additive and subtractive processes with intelligent monitoring systems is being driven by collaborative efforts between universities, research institutions, and industrial leaders (Kahhal et al. 2024). AM is converging with nanotechnology, bioprinting, and smart materials to expand medical applications, aerospace, and sustainable manufacturing. In the future, additive manufacturing will become intelligent, self regulating systems that can perform to a precision, efficiency, and adaptability that was thought unobtainable.

11. CONCLUSION

Additive manufacturing is emerging as a one of the critical enablers of process reliability, efficiency and quality control to integrate sensors and feedback in these processes in real time. As discussed through the chapter, sensor driven systems provide real time feedback to manufacturers of key printing parameters and allow manufacturers to detect and correct defect prior to compromising the printed component integrity. Diverse sensing technologies are used to continuously optimize material deposition, temperature regulation and environmental conditions using thermal and optical sensors, force, gas and acoustic monitoring systems. Such a level of precision is necessary for the industries requiring high performance, safety critical components, for example, aerospace, healthcare, and automotive manufacturing. In addition, the adoption of artificial intelligence with real time monitoring leads to predictive maintenance, adaptive control and self optimizing printers that takes the additive manufacturing technologies very near to the fully autonomous production.

Sensor integration is not limited to quality assurance and defect prevention. Real time monitoring has enabled one of the significant barriers to additive manufacturing being widely adopted; to allow more consistent and scalable production. With sensors integrated in AM systems, material waste and energy consumption are minimized, automation of the process is facilitated, and large scale manufacturing is made possible by reducing the need for manual intervention. Centralizing and quantity analysis of data on multiple printers in a networked production facility increases efficiency for manufacturers to streamline workflows and optimize resource allocation. Also, additive manufacturing is converging with new technologies like 5G, quantum sensing and machine learning, paving the way to smart printing, in which intelligent systems are constantly getting better and smarter at how they do printing tasks.

Yet, the future of additive manufacturing will depend on continued development of real time monitoring as well as of AI based optimization, and interdisciplinary research. With more sophisticated sensor technology and embedded AM system seamlessly, new opportunities for greater precision, reliability and scalability still exist. Smart printing solutions will become more and more the technology used in the industry to manufacture customized, high performance components, stringently reducing cost and environmental impact. This shift towards autonomous and adaptive manufacturing will not only increase production efficiency, but will also open new innovation opportunities for the manufacture in many sectors. Additive manufacturing is headed towards a future where smarter, more responsive systems advance the limits of what is possible in digital fabrication through embracing sensor integration, real time monitoring.

REFERENCES

Ahmad, I., Zulfiqar, A., Shabbir, M., Khan, J. S., Çalışkan, H., & Hong, H. (2025). Advanced sensors, monitoring, and control systems for environmental sustainability. In *Smart Cities and Sustainable Manufacturing* (pp. 47–55). Elsevier. DOI: 10.1016/B978-0-443-26474-0.00007-5

Aravind Raj, S. (2025). Quality control in additive manufacturing: A review of traditional and advanced techniques. *Engineering Research Express*.

Bas, J., Dutta, T., Llamas Garro, I., Velázquez-González, J. S., Dubey, R., & Mishra, S. K. (2024). RETRACTED: Embedded sensors with 3D printing technology. *Sensors (Basel)*, *24*(6), 1955. DOI: 10.3390/s24061955 PMID: 38544218

Chen, D., Han, Z., Zhang, J., Xue, L., & Liu, S. (2024). Additive Manufacturing Provides Infinite Possibilities for Self-Sensing Technology. *Advancement of Science*, *11*(28), 2400816. DOI: 10.1002/advs.202400816 PMID: 38767180

Chen, L., & Moon, S. K. (2024). In-situ defect detection in laser-directed energy deposition with machine learning and multi-sensor fusion. *Journal of Mechanical Science and Technology*, *38*(9), 4477–4484. DOI: 10.1007/s12206-024-2401-1

Christ, M., Zimmermann, C., Neinert, S., Leykauf, B., Döringshoff, K., & Krutzik, M. (2024). Additively manufactured ceramics for compact quantum technologies. *Advanced Quantum Technologies*, *7*(12), 2400076. DOI: 10.1002/qute.202400076

Fan, H., Liu, C., Bian, S., Ma, C., Huang, J., Liu, X., Doyle, M., Lu, T., Chow, E., Chen, L., Fuh, J. Y. H., Lu, W. F., & Li, B. (2025). New era towards autonomous additive manufacturing: A review of recent trends and future perspectives. *International Journal of Extreme Manufacturing*, *7*(3), 032006. DOI: 10.1088/2631-7990/ada8e4

Franke, J., Heinrich, F., & Reisch, R. T. (2025). Vision based process monitoring in wire arc additive manufacturing (WAAM). *Journal of Intelligent Manufacturing*, *36*(3), 1711–1721. DOI: 10.1007/s10845-023-02287-x

Granelli, R., Kovács-Vajna, Z. M., & Torricelli, F. (2025). Additive Manufacturing of Organic Electrochemical Transistors: Methods, Device Architectures, and Emerging Applications. *Small*, *21*(11), 2410499. DOI: 10.1002/smll.202410499 PMID: 39945058

Guo, Y., Zhang, Y., Pan, Z., & Zhou, W. (2024). Recent progress of sensing and machine learning technologies for process monitoring and defects detection in wire arc additive manufacturing. *Journal of Manufacturing Processes*, *125*, 489–511. DOI: 10.1016/j.jmapro.2024.07.060

Herzog, T., Brandt, M., Trinchi, A., Sola, A., & Molotnikov, A. (2024). Process monitoring and machine learning for defect detection in laser-based metal additive manufacturing. *Journal of Intelligent Manufacturing*, *35*(4), 1407–1437. DOI: 10.1007/s10845-023-02119-y

Hossain, M. J., Tabatabaei, B. T., Kiki, M., & Choi, J. W. (2025). Additive manufacturing of sensors: A comprehensive review. *International Journal of Precision Engineering and Manufacturing-Green Technology*, *12*(1), 277–300. DOI: 10.1007/s40684-024-00629-5

Jhun, J., Lee, D. H., Rehman, A. U., Kang, S., & Kim, J. H. (2024). Development of a real-time geometric quality monitoring system for extruded filaments of 3D concrete printing construction. *IEEE Access : Practical Innovations, Open Solutions*, *12*, 69981–69999. DOI: 10.1109/ACCESS.2024.3401472

Jin, L., Zhai, X., Wang, K., Zhang, K., Wu, D., Nazir, A., Jiang, J., & Liao, W. H. (2024). Big data, machine learning, and digital twin assisted additive manufacturing: A review. *Materials & Design*, *244*, 113086. DOI: 10.1016/j.matdes.2024.113086

Kahhal, P., Jo, Y. K., & Park, S. H. (2024). Recent progress in remanufacturing technologies using metal additive manufacturing processes and surface treatment. *International Journal of Precision Engineering and Manufacturing-Green Technology*, *11*(2), 625–658. DOI: 10.1007/s40684-023-00551-2

Kanthimathi, T., Rathika, N., Fathima, A. J., Rajesh, K. S., Srinivasan, S., & Thamizhamuthu, R. (2024, January). Robotic 3D Printing for Customized Industrial Components: IoT and AI-Enabled Innovation. In 2024 14th International Conference on Cloud Computing, Data Science & Engineering (Confluence) (pp. 509-513). IEEE.

Khanafer, K., Cao, J., & Kokash, H. (2024). Condition monitoring in additive manufacturing: A critical review of different approaches. *Journal of Manufacturing and Materials Processing*, *8*(3), 95. DOI: 10.3390/jmmp8030095

Kong, M., Yang, M., Li, R., Long, Y. Z., Zhang, J., Huang, X., Cui, X., Zhang, Y., Said, Z., & Li, C. (2024). Graphene-based flexible wearable sensors: Mechanisms, challenges, and future directions. *International Journal of Advanced Manufacturing Technology*, *131*(5), 3205–3237. DOI: 10.1007/s00170-023-12007-7

Kumar, S., & Kumar, R. (2025). A Comprehensive Study on Additive Manufacturing Techniques, Machine Learning Integration, and Internet of Things-Driven Sustainability Opportunities. *Journal of Materials Engineering and Performance*, •••, 1–68. DOI: 10.1007/s11665-025-10757-x

Leghemo, I. M., Azubuike, C., Segun-Falade, O. D., & Odionu, C. S. (2025). Data governance for emerging technologies: A conceptual framework for managing blockchain, IoT, and AI. *Journal of Engineering Research and Reports*, *27*(1), 247–267. DOI: 10.9734/jerr/2025/v27i11385

Li, B., Zhang, Y., Lei, Y., Wei, H., Chen, C., Liu, F., Zhao, P., & Wang, K. (2024). A single-sensor multi-scale quality monitoring methodology for laser-directed energy deposition: Example with height instability and porosity monitoring in additive manufacturing of ceramic thin-walled parts. *Additive Manufacturing*, *79*, 103923. DOI: 10.1016/j.addma.2023.103923

Ma, T., Zhang, Y., Ruan, K., Guo, H., He, M., Shi, X., Guo, Y., Kong, J., & Gu, J. (2024). Advances in 3D printing for polymer composites: A review. *InfoMat*, *6*(6), e12568. DOI: 10.1002/inf2.12568

Mahdi, M. M., Bajestani, M. S., Do Noh, S., & Kim, D. B. (2025). Digital twin-based architecture for wire arc additive manufacturing using OPC UA. *Robotics and Computer-integrated Manufacturing*, *94*, 102944. DOI: 10.1016/j.rcim.2024.102944

Makanda, I. L. D., Jiang, P., & Yang, M. (2025). Personalized federated unsupervised learning for nozzle condition monitoring using vibration sensors in additive manufacturing. *Robotics and Computer-integrated Manufacturing*, *93*, 102940. DOI: 10.1016/j.rcim.2024.102940

Mattera, G., Nele, L., & Paolella, D. (2024). Monitoring and control the Wire Arc Additive Manufacturing process using artificial intelligence techniques: A review. *Journal of Intelligent Manufacturing*, *35*(2), 467–497. DOI: 10.1007/s10845-023-02085-5

Mattera, G., Polden, J., & Luigi, N. E. L. E. (2025). Monitoring Wire Arc Additive Manufacturing process of Inconel 718 thin-walled structure using wavelet decomposition and clustering analysis of welding signal. *Journal of Advanced Manufacturing Science and Technology*, *5*(1), 2025006. DOI: 10.51393/j.jamst.2025006

McKinney, M., Garland, A., Cillessen, D., Adamczyk, J., Bolintineanu, D., Heiden, M., Fowler, E., & Boyce, B. L. (2025). Unsupervised multimodal fusion of in-process sensor data for advanced manufacturing process monitoring. *Journal of Manufacturing Systems*, *78*, 271–282. DOI: 10.1016/j.jmsy.2024.12.003

Musa, N., & Singh, N. B. (2025). Advanced sensors: classification, fabrication, and applications. In *Advanced Sensors for Smart Healthcare* (pp. 47–67). Elsevier. DOI: 10.1016/B978-0-443-24790-3.00004-1

Nasrin, T., Pourkamali-Anaraki, F., & Peterson, A. M. (2024). Application of machine learning in polymer additive manufacturing: A review. *Journal of Polymer Science*, *62*(12), 2639–2669. DOI: 10.1002/pol.20230649

Ng, W. L., Goh, G. L., Goh, G. D., Ten, J. S. J., & Yeong, W. Y. (2024). Progress and opportunities for machine learning in materials and processes of additive manufacturing. *Advanced Materials*, *36*(34), 2310006. DOI: 10.1002/adma.202310006 PMID: 38456831

Orzari, L. O., Kalinke, C., Silva-Neto, H. A., Rocha, D. S., Camargo, J. R., Coltro, W. K., & Janegitz, B. C. (2025). Screen-Printing vs Additive Manufacturing Approaches: Recent Aspects and Trends Involving the Fabrication of Electrochemical Sensors. Analytical Chemistry.

Petousis, M., Michailidis, N., Papadakis, V., Argyros, A., Spiridaki, M., Mountakis, N., Valsamos, J., Nasikas, N. K., Moutsopoulou, A., & Vidakis, N. (2025). Enhanced engineering and biocidal polypropylene filaments enabling melt reduction of AgNO3 through PVP agent: A scalable process for the defense industry with MEX additive manufacturing. *Defence Technology*, *44*, 52–66. DOI: 10.1016/j.dt.2024.09.007

Piran, F. J., Poduval, P. P., Barkam, H. E., Imani, M., & Imani, F. (2025). Explainable Differential Privacy-Hyperdimensional Computing for Balancing Privacy and Transparency in Additive Manufacturing Monitoring. *Engineering Applications of Artificial Intelligence*, *147*, 110282. DOI: 10.1016/j.engappai.2025.110282

Rahman, M., Islam, K. S., Dip, T. M., Chowdhury, M. F. M., Debnath, S. R., Hasan, S. M. M., Sakib, M. S., Saha, T., Padhye, R., & Houshyar, S. (2024). A review on nanomaterial-based additive manufacturing: Dynamics in properties, prospects, and challenges. *Progress in Additive Manufacturing*, *9*(4), 1197–1224. DOI: 10.1007/s40964-023-00514-8

Resendes, T., Freitas Rodrigues, P., Cruz, F., Gatões, D., Santos, V. M., Ramos, A. S., & Vieira, M. T. (2025). Advanced medical monitoring: 3D printed prosthetics with integrated strain sensor. *Progress in Additive Manufacturing*, *10*(1), 219–229. DOI: 10.1007/s40964-024-00615-y

Rolchigo, M., Reeve, S. T., Stump, B., Coleman, J., & Plotkowski, A. (2025). ExaCA v2. 0: A versatile, scalable, and performance portable cellular automata application for additive manufacturing solidification. *Computational Materials Science*, *251*, 113734. DOI: 10.1016/j.commatsci.2025.113734

Sagar, S. (2025). Data-Driven Excellence: Integrating Analytics into Medical Technology Product Lifecycle Management. *International Journal of Trend in Scientific Research and Development*, *9*(1), 127–136.

Shen, T., & Li, B. (2024). Digital twins in additive manufacturing: A state-of-the-art review. *International Journal of Advanced Manufacturing Technology*, *131*(1), 63–92. DOI: 10.1007/s00170-024-13092-y

Singh, J., Singh, A., Singh, H., & Doyon-Poulin, P. (2025). Implementation and evaluation of a smart machine monitoring system under industry 4.0 concept. *Journal of Industrial Information Integration*, *43*, 100746. DOI: 10.1016/j.jii.2024.100746

Sonavane, A. S., & Aylani, A. (2025). Integration of IoT and Quantum Computing: Revolutionizing Manufacturing. In Machine Learning for Environmental Monitoring in Wireless Sensor Networks (pp. 331-358). IGI Global.

Sousa, J., Brandau, B., Darabi, R., Sousa, A., Brueckner, F., Reis, A., & Reis, L. P. (2025). Artificial Intelligence for Control in Laser-Based Additive Manufacturing: A Systematic Review. *IEEE Access : Practical Innovations, Open Solutions*, *13*, 30845–30860. DOI: 10.1109/ACCESS.2025.3537859

Ukwaththa, J., Herath, S., & Meddage, D. P. P. (2024). A review of machine learning (ML) and explainable artificial intelligence (XAI) methods in additive manufacturing (3D printing). *Materials Today. Communications*, *41*, 110294. DOI: 10.1016/j.mtcomm.2024.110294

Waheed, M. F., & Bernadin, S. (2024, March). In-Situ Analysis of Vibration and Acoustic Data in Additive Manufacturing. In SoutheastCon 2024 (pp. 812-817). IEEE. DOI: 10.1109/SoutheastCon52093.2024.10500055

Wang, R., Wang, R., Dou, C., Yang, S., Gnanasambandam, R., Wang, A., & Kong, Z. (2024). Sub-surface thermal measurement in additive manufacturing via machine learning-enabled high-resolution fiber optic sensing. *Nature Communications*, *15*(1), 7568. DOI: 10.1038/s41467-024-51235-7 PMID: 39217158

Yadegari, M. J., Martucci, A., Biamino, S., Ugues, D., Montanaro, L., Fino, P., & Lombardi, M. (2025). Aluminum Laser Additive Manufacturing: A Review on Challenges and Opportunities Through the Lens of Sustainability. *Applied Sciences (Basel, Switzerland)*, *15*(4), 2221. DOI: 10.3390/app15042221

Zhu, C., Gemeda, H. B., Duoss, E. B., & Spadaccini, C. M. (2024). Toward multiscale, multimaterial 3D printing. *Advanced Materials*, *36*(34), 2314204. DOI: 10.1002/adma.202314204 PMID: 38775924

Chapter 15
An In-Depth Examination of 4D Food Printing Technology

Ruchira Srivastava
https://orcid.org/0000-0002-5957-9534
Amity University, Greater Noida, India

Ayushi Thakur
https://orcid.org/0000-0002-4172-6751
Amity University, Greater Noida, India

ABSTRACT

A relatively new field of 4D printing that is still in its infancy is 4D food printing. The food printing industry and academia have taken a keen interest in 4D printing. An overview of 4D printing is provided in the present article, with a focus on 4D food printing. This article discusses how different stimuli affect the color, flavor, texture, and form of food samples that are 4D printed. It discusses the creation of 4D designs, food printing ink, and several techniques for 4D food printing. Smart materials or multimaterials with shape- or function-changing capabilities may now be printed thanks to recent, quick advancements in 3D printing technology and material research. First, we examine the parallels and divergences between 4D and 3D printing in this review. the key components of 4D printing technology, including clever designs and materials.An overview of the uses for 4D printing at the moment.A succinct overview of two exciting applications that show where and how food printing technology is now developing: the production of programmable food textures and 4D food printing.

DOI: 10.4018/979-8-3373-0533-2.ch015

INTRODUCTION

4D food printing technology depicts an exciting enhancement in the culinary and food science fields, building on the principles of 3D printing. This innovative technology not only allows for the creation of intricate and customized food structures but also introduces the dimension of time, enabling printed foods to change shape, colour, texture, or flavour after printing in response to external stimuli (Ahmed et al., 2021). This dynamic capability opens up new area for culinary creativity, personalized nutrition, and sustainable food production. The majority of 4D food printing research focuses on microwave processing or how pH affects printed food items as a stimulating agent. Food qualities such as colour, flavour, texture, scent, and form were altered by stimuli in 4D printed food (Momeni & Ni, 2020). Food ingredients that have not yet been investigated for 4D printing include starch, soy protein isolate, potato purees, buckwheat dough, etc. The stimulus-responsive substance utilised in 4D printing ink causes spontaneous variations in the structure of 3D printed objects. The term "4D printing" describes the single- or multi-material printing of an apparatus or item that has the ability to morph between several dimensions and change from a 1D strand into a pre-programmed 3D shape or from a 2D surface into a pre-programmed 3D shape (Imam et al., 2021). These changes are made possible, for example, by heating, light, or swelling in a liquid. They may also be produced electrochemically and by programming varying levels of sensitivity to, for example, swelling into different areas of the intended geometry. These methods provide dynamic reaction and flexibility for systems and structures of varying sizes. They also hold up new possibilities for integrating decision-making simplicity and programmability into non-electronic materials (Huang et al., 2021). Applications might include clothing, items, or mechanisms that adapt to changing conditions and human needs, as well as robotics-like activity without the need for intricate electro-mechanical-chemical systems. The main distinction between 3D and 4D printing is that the former method is static (Quanjin et al., 2020). Unlike 3D printing, 4D printing allows the base material to gradually change shape in response to external stimuli (Chu et al., 2020). The deforestation ability of 4D printing technology is the major topic of recent research. Conventional techniques for producing the printed parts actuators, processors, and sensors are frequently needed in order to react to environmental excitations. The 4D printing component itself can reduce structural complexity and weight by realizing the reaction to external stimuli without the need for servo drive equipment (Zafar & Zhao, 2020). The next phase in manufacturing is acknowledged as 4D printing. When a material's typical behaviour is stimulated by the environment, 4D printing can create flexible structures that can change geometrically. The capacity to gradually change the shape of 3D-printed components is provided by 4D printing. The new method integrates 3D printing with sophisticated physics, material science, and applications (Mallakpour et al., 2021). Due to its capacity to produce items that are tailored to each customer's specifications in terms of size, shape, texture, colour, and nutrition, 3D food printing has recently drawn a lot of attention from academic and industry researchers. The field of 4D printing, which enables the gradual development of different features and attributes of 3D printed things by external stimulation, has also been attracting increasing interest (Kuang, 2021). The process known as 4D printing allows a 3D-printed item to transform into a new structure in response to external stimuli like light, temperature, or other variables. 4D printing will bring new opportunities that are practical for large applications, robust in harsh environments, and will aid in the creation of a structure that can be transformed (Subeshan et al., 2021).

2. FACTORS THAT GIVE RISE TO 4D PRINTING:

The five elements are primarily responsible for 4D printing, and while utilizing 4D printing, each of the five criteria has to be considered. These five elements include stimulus, the AM process, the printing material, interaction mechanism and modelling (Mohol & Sharma, 2021). The AM printing method is the first factor. Without the need for any intermediaries, the AM method makes it possible to produce printing material from the digital information supplied by the computer instrument. Many additive manufacturing (AM) technologies are available, and nearly all of them can print. These procedures include stereo lithography (SLA), selective laser sintering (SLS), jet 3D printing (3DP), fused deposition modelling (FDM), electron beam melting (EBM), selective laser melting (SLM), direct ink writing (DIW)etc (Biswas et al., 2021).

Figure 1. Representation of Factors affecting 4D Printing

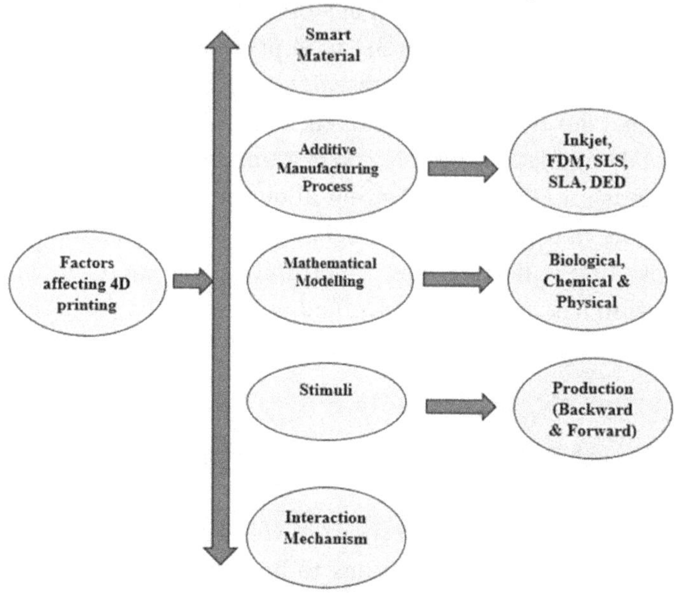

The stimuli that will be employed during 4D printing make up the third element or component. Physical, chemical, and biological stimuli can all be employed. The physical stimuli consist of magnetic, electric, and moisture energy, UV rays, temperature, etc (Tamay et al., 2019). Chemicals, pH levels, and the employment of oxidants and reductants are examples of chemical stimuli. Enzymes and glucose also comprise the biological stimulus. Upon the introduction of a stimulus, the structure experiences physical or chemical alterations that lead to deformation, such as molecules moving, stress relaxing, or phase shifts (Khoo et al., 2015).

The mathematical modeling and the interaction mechanism make up the fourth and fifth factors. Not every smart material can change in the way that is needed when a stimulus is applied. We must provide the interaction mechanisms that will design the sequence of form change, such as mechanical loading or physical manipulation mechanisms (Maniruzzaman, 2018). Once the interaction mechanism has been provided, mathematical modelling is needed to predict how long the stimulus will act on the smart material. A diagrammatical depiction of every element involved in 4D printing is presented in Fig.1. When considering the contributing aspects, 4D printing may be summarised as follows: "The stimulation is put on the smart material utilising mathematical modelling and an appropriate interaction mechanism." in an AM technique that yields a structure that is printed in four dimensions (Oral et al., 2021).

3. MATERIAL TYPES UTILIZED IN 4D PRINTING:

Latest enhancements in 3D printing have made it sure to position materials more accurately and flexibly, which has greatly aided the cause of 4D printing (Ghazal et al., 2023). The substances that go into 4D printing are often referred to as "Smart Materials," these materials can alter over time in terms of their characteristics. These materials exhibit characteristics such as self-assembly, shape memory, self-healing, and self-capability in answer to external stimuli (Deshmukh et al., 2020). In addition to using materials that can change shape, 4D printing also employs materials that can alter hue in answer to visible and ultraviolet light. The kind of smart materials utilized in 4D printing will be covered in this part depending on how they react to stimuli (Mehrpouya et al., 2021). Figure 2 depicts the various types of 4D printing materials utilized in food Industry.

The Table 1 below shows in detail the type of 4D printing method, mechanism, characteristics, material used & application All the six major types of material are covered and described in this part.

Figure 2. Types of 4D printing materials

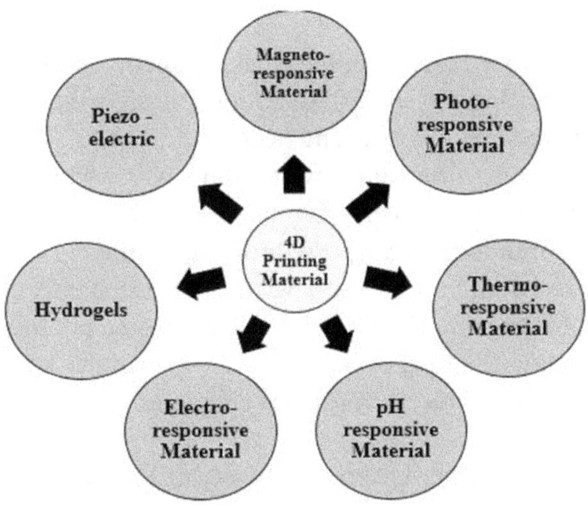

3.1. Hydrogels (Moisture responsive materials)

Hydrogels are a class of smart materials that are particularly suited for 4D printing because of their capacity to undergo significant changes in shape, size, and properties in answer to moisture (Javaid & Haleem, 2019). This responsiveness makes them ideal for creating dynamic structures in food, biomedical, and other applications. Hydrogels consist of networks of cross-linked polymers that have a high water absorption capacity, resulting in swelling or shrinking based on the environmental humidity (Li et al., 2017). There is a huge amount of interest in materials that respond to moisture or water since they have so many uses. These materials are also known as hydrogels because of their remarkable capacity to react to the moisture or water (Rafiee et al., 2020). In reality, they belong to a type of three-dimensional networks of cross linked polymer chains that, when coming into contact with moisture, can swell to a volume of up to 200% of what it was before, because hydrogels is geometrically capable of being folded, bent, stretched, and expanded, a wide variety of forms have been created with them, demonstrating the hydrogels' excellent printing capacity (Momeni et al., 2017). Using direct ink writing, these are very biocompatible and simple to print.

3.1.1. Properties of Hydrogels for 4D Printing

- Hydrogels can contain up to 90% water, making them highly hydrophilic and capable of significant volume changes. The ability to hold water allows for soft, pliable textures suitable for various applications, especially in food and biomedical fields.
- Many hydrogels are biocompatible, making them safe for use in food and medical devices (Joshi et al., 2020). Common biocompatible hydrogels include gelatin, agarose, alginate, and pectin.
- Hydrogels respond to moisture by swelling when hydrated and shrinking when dehydrated. This property can be harnessed to design dynamic structures that evolve in form or purpose throughout time.
- Hydrogels can be engineered to have varying mechanical properties, from soft and flexible to more rigid, relying on the degree of cross-linking and the type of polymer used (Teng et al., 2021). This tenability allows for the design of structures with specific performance characteristics.

3.1.2. Types of Hydrogels Used in 4D Food Printing

Natural Hydrogels: Derived from natural sources such as agar, alginate, gelatin, and pectin. Biocompatible and widely accepted in food applications.

Synthetic Hydrogels: Engineered materials like polyacrylamide and polyvinyl alcohol (PVA) that can be tailored for specific properties. Used in specialized food applications where precise control over properties is needed (Chen et al., 2022a).

Hybrid Hydrogels: Combination of natural and synthetic components to achieve desired characteristics. Balances biocompatibility and functionality.

Smart Hydrogels: Engineered to react to many stimuli, including pH, temperature, and moisture, for sophisticated food uses. Can be used to create multifunctional food items (Yang et al., 2021).

3.1.3. Mechanisms of Moisture Responsiveness

- Hydrogels expand when they absorb water, increasing in volume. Conversely, they contract when water is removed, decreasing in volume. The degree of swelling can be controlled by the polymer composition and cross-linking density (Chen et al., 2022b).
- Water molecules diffuse into the hydrogel network, causing it to swell. Osmotic pressure can drive water movement in and out of the hydrogel, leading to controlled shape changes.
- In porous hydrogels, capillary action can facilitate water absorption and distribution throughout the material, enhancing responsiveness (Navaf et al., 2022).

3.1.4. Applications of Hydrogels in 4D Food Printing

- Hydrogels can create foods that change texture upon hydration, enhancing sensory experiences.
- Edible hydrogels can be designed to transform into specific shapes when exposed to moisture during cooking or consumption (Phuhongsung et al., 2022).
- Hydrogels can encapsulate nutrients that are released upon hydration, providing timed nutrition delivery.

3.2. Thermo-responsive (Temperature responsive material)

Thermo-responsive materials, which change their properties in solution to temperature, are pivotal in the advancement of 4D food printing (Ghazal, Zhang, & Liu, 2019). These materials enable the creation of food items that transform in texture, shape, or even flavour when subjected to specific temperature conditions. This capability opens new avenues for culinary innovation, personalized nutrition, and functional food design (Paunović et al., 2023). These are the intelligent materials that react to changes in temperature or heat. Two basic processes account for these materials' shape alters in response to thermal stimuli: the shape memory effect (SME) and the shape change effect (SCE). SME is the process of applying outside stimuli to return a distorted (plastic) substance to its original shape (Li et al., 2023).

3.2.1. Properties of Thermo-Responsive Materials

- These materials experience chemical or physical changes after being subjected to temperature variations. Common changes include swelling, shrinking, gelation, or melting.
- Many thermo-responsive materials exhibit reversible behaviour, allowing them to back to their original state when the temperature stimulus is removed.
- Materials used in food applications must be safe for consumption and non-toxic (Shun, 2021). Examples include gelatine, starches, and certain proteins.
- The mechanical properties can vary significantly, from soft and flexible to firm and rigid, depending on the temperature.

3.2.2. Types of Thermo-Responsive Materials Used in 4D Food Printing.

Gelatine: A protein derived from collagen, gelatine forms gels at low temperatures and melts at body temperature .Commonly used for creating flexible, melt-in-the-mouth textures (Suryavanshi et al., 2023).

Agar: A polysaccharide obtained from red algae, agar forms strong gels that melt at high temperatures and solidify at lower temperatures. Useful for creating heat-stable structures that can dissolve or change texture when heated.

Methylcellulose: A cellulose derivative that gels upon heating and reverts to a sol state upon cooling. Ideal for creating hot gels and thermal reversibility.

Starches: Various starches exhibit gelatinization and retro gradation properties when heated and cooled. Used for creating textures that change upon heating, such as from firm to soft (Wang et al., 2022).

3.2.3. Mechanisms of Temperature-Responsive behaviour

Gelation and Melting

- Certain materials like gelatine and agar solidify into gels at lower temperatures and melt when heated. This property is exploited to create foods that transition from solid to liquid states (Wang et al., 2023).

Swelling and Shrinking

- Some materials expand when heated and contract upon cooling, or vice versa. These changes can be harnessed for dynamic shape transformations in food.

Phase Transitions

- Materials can switch between different physical states (e.g., solid, liquid, gel) in response to temperature changes. Phase transitions are utilized for creating multi-textured food items (Enferadi et al., 2024).

3.2.4. Applications in 4D Food Printing

- Foods that change shape when heated or cooled can create dynamic culinary presentations. Examples include pasta that curls when boiled or desserts that unfold when warmed (Chen, Liu, Abdiryim, & Liu, 2023).
- Thermo-responsive materials can alter the texture of food in response to temperature, providing unique sensory experiences. Examples include hot gels that become liquid upon cooling or crunchy snacks that soften when heated (Bánhegyi, n.d.).
- Ingredients can be encapsulated in thermo-responsive materials and released at specific temperatures. This technique is used for flavour bursts or nutrient delivery in response to temperature changes during consumption (Enferadi et al., 2024).
- Foods that transform in response to temperature can enhance the dining experience by adding an element of surprise and engagement. Examples include self-heating soups or beverages that change colour and flavour when heated (Carrillo et al., 2024).

3.3. Photo-responsive (Light responsive material)

Indirect light stimulation also contributes to the way intelligent materials distort. A portion of a photo-responsive material, also known as smart material. It is light-responsive is exposed to light, and it will finally cause the heating phenomena in it as a result of absorbing the light (Lui et al., 2019). Furthermore, as was previously said, heat encourages smart material deformation, which is why the photo-responsive material changes shape. Because light does not create a alter immediately like heat and moisture do, it works as an indirect stimulus. A sequential self-folding construction in which light is absorbed by the joints, heating them up and changing their form (Morouço et al., 2020).

3.3.1. Properties of Photo-Responsive Materials

- Photo-responsive materials react to specific wavelengths of light, such as ultraviolet (UV), visible, or infrared (IR) light. These reactions can trigger physical or chemical changes in the material.
- Many photo-responsive materials exhibit reversible behaviour, meaning they can revert to their original state when the light stimulus is removed or altered.
- Materials used in food applications must be safe for consumption and non-toxic (Kuang et al., 2019).

Food-grade photo-responsive materials must meet stringent safety standards.

- The mechanical properties can vary, with some materials becoming softer, harder, or changing shape upon light exposure.

3.3.2. Types of Photo-Responsive Materials Used in 4D Food Printing

Photochromic Compounds: variation in colour when exposed to light of a certain wavelength. Examples include food-safe dyes that change colour under UV light (Rastogi & Kandasubramanian, 2019).

Photo cross-linkable Polymers: Form or break crosslinks in their polymer structure when opens up to light, leading to changes in mechanical properties. Used to create gels or solid structures that can soften or harden with light exposure (Champeau et al., 2020).

Photo-degradable Materials: Break down or degrade when exposed to light, enabling controlled release of encapsulated ingredients. Useful for flavour or nutrient release in response to light (Lee et al., 2017).

Photo-responsive Hydrogels: Swell or shrink when exposed to light, changing their volume and texture. Incorporate light-sensitive molecules that induce these changes (Patdiya & Kandasubramanian, 2021).

3.3.3. Mechanisms of Light-Responsive Behaviour

Colour Change

- Photochromic materials change color when exposed to light, which can be used for aesthetic or functional purposes in food design (Shinde et al., 2023).

Structural Transformation

- Light can induce swelling, shrinking, or deformation in photo-responsive hydrogels and polymers, altering the shape and texture of the food.

Crosslinking and De-crosslinking

- Photo cross linkable materials form or break crosslinks in their structure when exposed to light, changing their mechanical properties from soft to firm or vice versa.

Degradation and Release

Photo-degradable materials break down under light, releasing encapsulated flavours, nutrients, or other ingredients at specific moments (Spiegel et al., 2020).

3.3.4. Applications in 4D Food Printing

- Foods that change colour, shape, or texture when exposed to light during a meal can create engaging and interactive dining experiences.
  Examples include desserts that change colour under UV light or drinks that reveal hidden patterns when illuminated (Shie et al., 2019).
- Light-responsive foods can enhance visual appeal, making meals more visually dynamic and appealing.
  Examples include edible decorations that transform under restaurant lighting or special events.
- Photo-degradable materials can be used to release flavours, aromas, or nutrients at precise times, enhancing the sensory experience and nutritional value of the food.
  Examples include salads that release dressings when exposed to sunlight (Khalid et al., 2022).
- Foods can be customized for individual preferences, changing flavour, texture, or appearance based on light exposure.
  Examples include personalized cakes that reveal messages or images when illuminated (Farid et al., 2021).

3.4. Electro-responsive (Electric energy responsive material)

Electro-responsive materials, which respond to electrical stimuli, represent a ground breaking development in the realm of 4D food printing. These materials can undergo changes in shape, texture, colour, or other properties when exposed to electrical fields or currents. The integration of electro-responsive materials into food printing technologies can lead to innovative culinary techniques, personalized nutrition, and dynamic food experiences (Antezana et al., 2023).

3.4.1. Properties of Electro-Responsive Materials

- These materials alter their physical or chemical properties in response to electrical stimuli, such as fields or currents. Typical changes include deformation, conductivity alterations, or phase transitions.
- Many electro-responsive materials exhibit reversible behaviour, allowing them to return to their actual position when the electrical stimulus is removed.
- For food applications, these materials must be safe for consumption, non-toxic, and comply with food safety standards (Carayon et al., 2020).
- The mechanical properties can range from flexible to rigid, relying on the material composition and the nature of the electrical response.

3.4.2. Types of Electro-Responsive Materials Used in 4D Food Printing

Electro active Polymers (EAPs)

Polymers that change shape or size when an electric field is applied.

Used to create flexible and dynamic structures in food items (Leigh et al., 2012).

Conductive Hydrogels

Hydrogels that can conduct electricity and change properties, such as swelling or shrinking, in response to electrical stimuli.

Suitable for creating textures that can transform when electrified.

Piezoelectric Material

Materials that, when subjected to mechanical stress, produce an electric charge and vice versa.

Useful for creating interactive and responsive food items (Fontana-Escartín et al., 2022).

Ionic Polymers

Polymers that change shape or mechanical properties in response to ionic currents.

Can be used to achieve fine control over food texture and shape transformations.

3.4.3. Mechanisms of Electrical Responsiveness

Deformation

Application of an electric field causes materials to change shape, enabling dynamic structural transformations.

Example: EAPs bending or contracting under electrical influence (Ryan et al., 2021).

Swelling and Shrinking

Conductive hydrogels can swell or shrink in response to electrical stimuli, altering their volume and texture.
Used to create food items that change texture when an electric current is applied.

Phase Transitions

Some materials change their phase (e.g., from solid to gel) when exposed to electric fields.
Useful for creating food with variable textures and consistencies (Jagadeesh et al., 2022).

Conductivity Changes

Materials can alter their electrical conductivity, which can be used to trigger other responses, such as heating or flavour release.
Example: conductive gels that heat up and change texture.

3.4.4. Applications in 4D Food Printing

- Foods that change shape or texture in response to electrical stimuli can create unique dining experiences.
  Example: a dessert that contracts or expands when an electric current is applied (Joharji et al., 2022).
- Foods that respond to touch or proximity, providing interactive elements to dining.
  Example: edible displays or smart packaging that changes upon contact.
- Electro-responsive materials can encapsulate flavours, nutrients, or bioactive compounds, releasing them in response to electrical signals (Zolfagharian et al., 2018).
  Example: a drink that releases vitamins or flavours when activated by an electric field.
- Foods can be personalized to individual preferences by adjusting electrical inputs to modify texture, flavour, or appearance (Zhang et al., 2019).
  Example: personalized snacks that alter their crunchiness or softness based on user preference.

3.5. Piezoelectric material

Mechanical stress can cause piezoelectric materials to produce a charge. Piezoelectricity is the method by which a mechanical force produces electric charges. Materials with piezoelectric properties can also be utilized in 4D printing applications since they can flex when put to mechanical strain. The phenomena is straightforward: when stress is put, charges are created, and because charges may produce deformation, these charges ultimately result in changes to the structure (Li et al., 2016).

3.5.1. Properties of Piezoelectric Materials

- Generates electric charge when mechanically deformed and deforms when an electric field is applied. Useful for creating interactive and responsive food items.
- Changes shape, texture, or structure in response to mechanical or electrical stimuli.
  Provides dynamic and adaptable properties for food design.
- Must be safe for consumption and non-toxic when used in food applications.
  Food-grade piezoelectric materials must comply with stringent safety standards.
- Capable of undergoing multiple cycles of transformation without degrading (Bai & Bu, 2022).
  Ensures consistent performance over time.

3.5.2. Types of Piezoelectric Materials Used in 4D Food Printing

Piezoelectric Polymers: Polymers that exhibit piezoelectric properties, such as poly-vinylidene fluoride (PVDF).Flexible and suitable for creating dynamic textures and shapes in food items (Grinberg et al., 2019).

Piezoelectric Ceramics: Materials like lead zirconate titanate (PZT) that have strong piezoelectric effects. Generally less used directly in food but can be incorporated in food packaging or processing tools (Chen et al., 2023).

Piezoelectric Composites: Combinations of polymers and ceramics to enhance piezoelectric properties while maintaining flexibility. Potentially used for creating specialized food structures or interactive elements (Cataldi et al., 2022).

Natural Piezoelectric Materials: Certain natural materials, like proteins and biomolecules, exhibit piezoelectric properties. Safe and biocompatible for food applications (Sossou et al., 2019).

3.5.3. Mechanisms of Piezoelectric Responsiveness

Mechanical Deformation: Application of mechanical stress generates an electric charge, which can trigger further transformations in food structure or texture.

Example: Chewing action generating electrical signals that cause a food item to change texture (Bodkhe & Ermanni, 2019).

Electromechanical Coupling: Applying an electric field induces mechanical deformation in the material, altering the shape or texture of the food (Chen et al., 2020).

Example: Electrically stimulated food items that change shape or become tenderer.

Interactive Responses: Piezoelectric materials can interact with external stimuli, such as touch or pressure, providing dynamic sensory experiences [(Zeng et al., 2021).

Example: Edible sensors in food that respond to touch, enhancing interactivity.

3.5.4. Applications in 4D Food Printing

- Foods that respond to touch, pressure, or chewing, creating engaging and interactive dining experiences (Mondal & Tripathy, 2021).

  Example: A snack that changes texture when bitten, providing a surprising sensory experience.
- Piezoelectric materials can create foods with textures that change in response to mechanical or electrical stimuli.

  Example: Desserts that become softer or firmer based on the application of an electric field.
- Incorporating piezoelectric materials in packaging or food processing tools to monitor and respond to environmental conditions (Singholi & Sharma, 2021).

  Example: Packaging that changes color or releases preservatives when exposed to mechanical stress.
- Utilizing piezoelectric materials to encapsulate and release flavours or nutrients in response to mechanical or electrical triggers (Megdich et al., 2023).

  Example: A drink that releases vitamins when shaken or stirred.
- Developing edible sensors that can monitor consumption and provide feedback based on mechanical or electrical stimuli.

  Example: A piezoelectric-based edible strip that tracks chewing and sends data to a connected device for dietary monitoring.

3.6. pH responsive material

pH-responsive materials, which alter their properties in response to changes in pH levels, hold significant potential in the field of 4D food printing. These materials can change their shape, texture, color, or other properties when exposed to varying pH environments. Integrating pH-responsive materials into food printing technologies can lead to innovative culinary techniques, personalized nutrition, and dynamic food experiences (Muzaffar et al., 2020).

3.6.1. Properties of pH-Responsive Materials

- These materials change their physical or chemical properties in response to pH variations. Typical changes include swelling, shrinking, sol-gel transitions, or color changes.
- Many pH-responsive materials exhibit reversible behaviour, allowing them to return to their real position when the pH stimulus is removed.
- For food applications, these materials must be safe for consumption, non-toxic, and comply with food safety standards.
- The mechanical properties can have variation from flexible to rigid, depending on the material composition and the nature of the pH response (Shanthamma et al., 2021).

3.6.2. Types of pH-Responsive Materials Used in 4D Food Printing

Hydrogels: hydrogels that react to pH variations by swelling or contracting. Examples include polyacrylic acid and chitosan-based hydrogels.

Polymers: pH-sensitive polymers that undergo sol-gel transitions or change their mechanical properties with pH.

Examples include poly (methacrylic acid) and poly (lactic-co-glycolic acid).

Biopolymers: Natural polymers like proteins and polysaccharides that respond to pH changes.

Examples include gelatine, agar, and alginate.

Dyes and Indicators: Food-safe dyes that change colour in response to pH variations.

Examples include antho- cyanins and turmeric (Teng et al., 2024).

3.6.3. Mechanisms of pH Responsiveness

Swelling and Shrinking: Hydrogels and polymers can swell or shrink in response to changes in pH, altering their volume and texture.

Example: A hydrogel-based dessert that becomes softer in an acidic environment.
Sol-Gel Transitions: Certain polymers can transition from a sol (liquid) to a gel (solid) state or vice versa based on pH changes (Huang et al., 2022).
Example: A sauce that thickens when exposed to the acidic environment of the stomach.
Colour Change: pH-sensitive dyes and indicators change color in response to pH variations, adding visual appeal and functionality.
Example: A frosting that changes color when exposed to acidic fruits like lemon.
Degradation and Release: pH-responsive materials can degrade or release encapsulated ingredients when exposed to specific pH levels.
Example: A capsule that releases flavor or nutrients when it reaches the stomach's acidic environment.

3.6.4. Applications in 4D Food Printing

- Foods that change shape or texture in response to pH variations can create unique dining experiences.
  Example: A pasta that becomes softer and more pliable in a sauce with a specific pH.
- Foods that respond to the pH environment during consumption, providing dynamic sensory experiences.
  Example: Candy that changes texture and flavour when chewed.
- pH- responsive materials can encapsulate flavours, nutrients, or bioactive compounds, releasing them in response to pH changes.
  Example: A protein shake that releases vitamins or supplements when it reaches the stomach.
- pH- sensitive dyes can be used to create visually appealing foods that change color based on the pH of their environment.
  Example: A cake that reveals hidden patterns when topped with an acidic glaze.

3.7. Magneto responsive material (Magnet responsive material)

Magneto-responsive materials, which respond to magnetic fields, hold unique potential in the field of 4D food printing. These materials can change their shape, texture, or other properties in response to magnetic stimuli. Integrating magneto-responsive materials into food printing technologies can lead to innovative culinary techniques, personalized nutrition, and dynamic food experiences (He et al., 2020).

3.7.1. Properties of Magneto-Responsive Materials

- These materials alter their physical properties in answer to magnetic fields.
  Typical changes include deformation, alignment, or phase transitions.
- Many magneto-responsive materials exhibit reversible behaviour, allowing them to revert to their original state when the magnetic stimulus is removed.
- For food applications, these materials must be safe for consumption, non-toxic, and comply with food safety standards.
- The mechanical properties can range from flexible to rigid, depending on the material composition and the nature of the magnetic response (Liu et al., 2023).

3.7.2. Types of Magneto-Responsive Materials Used in 4D Food Printing

Magneto-Rheological Fluids: Fluids that change their viscosity in response to magnetic fields. Can be used to create textures that can be dynamically adjusted.

Ferro-fluids: Colloidal liquids containing nano scale ferromagnetic particles (Ma et al., 2023).

Suitable for creating dynamic and interactive food structures.

Magneto-Responsive Polymers: Polymers embedded with magnetic particles that change shape or mechanical properties when exposed to magnetic fields. Useful for creating flexible and dynamic food items.

Magneto-Responsive Hydrogels: Hydrogels that incorporate magnetic nanoparticles and change their volume or shape in response to magnetic fields. Ideal for creating responsive textures in food.

3.7.3. Mechanisms of Magnetic Responsiveness

- Application of a magnetic field causes materials to change shape, enabling dynamic structural transformations.
  Example: Magneto-responsive gels bending or contracting under magnetic influence.
- Magnetic particles within the material align with the magnetic field, changing the material's overall structure or properties.
  Example: Magnetic alignment creating distinct textures or patterns in a food item.
- Magneto-rheological fluids can alter their viscosity in response to magnetic fields, providing dynamic control over texture.
  Example: A sauce that thickens or thins when exposed to a magnetic field.

- Some materials can transition between different phases (e.g., from liquid to solid) under magnetic influence (Zolfagharian et al., 2023).
  Example: A liquid dressing that solidifies when a magnetic field is applied.

3.7.4. Applications in 4D Food Printing

- Foods that change shape or texture in response to magnetic fields can create unique dining experiences. Example: A dessert that transforms its texture when a magnet is brought close.
- Foods that respond to magnetic stimuli, providing dynamic and interactive dining experiences. Example: A candy that changes shape or releases flavours when exposed to a magnetic field.
- Magneto-responsive materials can encapsulate flavours, nutrients, or bioactive compounds, releasing them in response to magnetic fields.
  Example: A drink that releases vitamins when a magnetic stirrer is used.
- Magnetic fields can create patterns or structures in food items, enhancing visual appeal. Example: A cake with decorative elements that align in patterns when exposed to a magnetic field.
- Incorporating magneto-responsive materials in packaging or food processing tools to monitor and respond to environmental conditions (Vatanparast et al., 2023).
  Example: Packaging that changes properties or indicates freshness when exposed to a magnetic field

Table 1. Illustration of 4D printing materials

Types of 4D printing	Mechanism	Characteristics	Material used	Application
Hydrogels	- Swelling and Shrinking - Diffusion and Osmosis - Capillary Action	-Biocompatibility - Mechanical Properties - Responsive Behaviour - High Water content	- Natural Hydrogels -Synthetic Hydrogels - Hybrid Hydrogels - Smart Hydrogels	- Nutrient Release - Texture Modulation - Shape transformation
Thermo-responsive (Temperature responsive material)	- Gelation and Melting - Swelling and Shrinking - Phase Transition	- Thermal Sensitivity - Reversibility - Biocompatibility - Mechanical Properties	- Gelatine -Methylcellulose - Agar - Starches	- Shape Transformation - Texture Modification - Controlled Release - Interactive Dining Experiences
Photo-responsive (Light responsive material)	-Degradation and Release - Colour Change -Crosslinking & De-crosslinking - Structural Transformation	- Light Sensitivity - Reversibility - Biocompatibility -Mechanical Properties	- Photochromic Compounds - Photo cross-linkable Polymers -Photo responsive hydrogels - Photo-degradable Materials	-Interactive Dining Experiences - Dynamic Food Presentations - Controlled Release of Flavours and Nutrients -Customization and Personalization
Electro-responsive (Electric energy responsive material)	- Swelling and Shrinking -Deformation -Phase transition -Conductivity Changes	-Electrical Sensitivity - Reversibility - Biocompatibility - Mechanical Properties	- Electro active Polymers (EAPs) - Conductive Hydrogels -Ionic Polymers - Piezoelectric Materials	-Shape and Texture Transformation - Interactive and Functional Foods -Customization and Personalization - Controlled Release of Flavours and Nutrients
Piezoelectric material	- Electromechanical Coupling - Mechanical Deformation - Interactive responses	- Piezoelectric Effect - Mechanical Responsiveness - Biocompatibility - Durability and Reversibility	-Piezoelectric Polymers -Piezoelectric Ceramics - Piezoelectric Composites - Natural Piezoelectric Materials	- Smart Packaging and Processing - Interactive Dining Experiences - Dynamic Food Texture - Controlled Flavour and Nutrient Release - Edible Sensors

continued on following page

Table 1. Continued

Types of 4D printing	Mechanism	Characteristics	Material used	Application
pH responsive material	- Swelling and Shrinking - Sol-Gel Transitions - Degradation and Release - Colour Change	-pH Sensitivity - Reversibility - Biocompatibility - Mechanical Properties	- Dyes and Indicators - Polymers -Bio polymers -hydrogels	- Shape and Texture Transformation - Interactive and Functional Foods - Controlled Release Systems - Aesthetic Enhancements
Magneto responsive material (Magnet responsive material)	-Deformation - Alignment - Viscosity Changes - Phase Transitions	-Magnetic Sensitivity - Reversibility - Biocompatibility - Mechanical Properties	- Magneto-Rheological Fluids - Ferro Fluids - Magneto-Responsive Polymers - Magneto-Responsive Hydrogels	- Shape and Texture Transformation - Interactive and Functional Foods - Controlled Release Systems - Smart Packaging and Processing

Advantages of 4D food printing: 4D food printing offers a multitude of advantages that extend beyond traditional culinary practices. From creating dynamic and interactive food experiences to improving nutritional delivery, enhancing sustainability, and offering economic benefits, 4D food printing technology is poised to revolutionize the food industry (Qi et al., 2022). As research and development continue, and as the technology becomes more accessible, the potential applications and benefits of 4D food printing are likely to expand, paving the way for a new era in food production and consumption. Some advantages are highlighted below:

- Economic Benefits
- Accessibility and Convenience
- Enhanced Consumer Engagement
- Educational and Research Opportunities
- Efficient Production Processes
- Dynamic and Interactive Food Experiences
- Innovation in Culinary Techniques
- Improved Nutritional Delivery
- Sustainability and Efficiency
- Enhanced Food Safety and Shelf Life

Limitations Of 4D food printing: While 4D food printing holds immense potential for revolutionizing the culinary world, it faces several limitations and challenges that requires to be addressed for it to become a mainstream technology.

Overcoming these obstacles will require advancements in technology, materials science, consumer education, and regulatory frameworks (Merazzo et al., 2021). As research and development continue, and as the technology matures, many of these limitations may be mitigated, paving the way for broader adoption and more innovative applications in the food industry.

- Disposal Of materials
- Limited Culinary Applications
- Environmental Impact
- Regulatory and Safety Issues
- Scalability and Production Volume
- Processing Time and Efficiency
- Material Constraints
- Technical Challenges

CONCLUSION

4D food printing technology stands at the intersection of culinary art, food science, and material engineering. By incorporating the dimension of time, it offers unprecedented opportunities for innovation in food design, personalized nutrition, and sustainable food production. While challenges remain, ongoing research and development promise to overcome these hurdles, paving the way for this transformative technology to reshape the future of food. Photo-responsive materials bring a new dimension to 4D food printing, enabling the creation of dynamic, interactive, and customizable food experiences. Their ability to change in response to light exposure offers exciting possibilities for culinary innovation, interactive dining, and functional food design. While challenges related to material safety, stability, and consumer acceptance remain, ongoing research and technological advancements are likely to overcome these hurdles, opening its way for wider adoption and new culinary frontiers.

Hydrogels play a important role in 4D food printing, offering unique moisture-responsive properties that enable dynamic transformations in food structures. Their ability to swell, Deswell, and degrade in response to moisture allows for innovative applications in shape transformation, texture modification, and controlled release. While challenges remain in ensuring stability, controlling transformations, and meeting regulatory standards, the potential of hydrogels in 4D food printing is vast, paving the way for novel culinary experiences and advanced food functionalities. Thermo-responsive materials are integral to the development of 4D food printing, enabling the creation of dynamic, interactive, and customizable food experiences.

Their capacity to transform in response to temperature changes offers exciting possibilities for culinary innovation, personalized nutrition, and functional food design. While challenges remain in terms of stability, control, and consumer acceptance, ongoing research and technological advancements promise to overcome these hurdles, openingits area for a new era of food production and consumption. Electro-responsive materials bring a revolutionary dimension to 4D food printing, enabling the creation of dynamic, interactive, and customizable food experiences. Their capacity to alter in response to electrical stimuli offers exciting possibilities for culinary innovation, personalized nutrition, and functional food design. While challenges related to material safety, control, and consumer acceptance remain, ongoing research and technological advancements promise to overcome these hurdles, paving the way for widespread adoption and new culinary frontiers. Magneto-responsive materials offer a revolutionary potential in 4D food printing, enabling the creation of dynamic, interactive, and customizable food experiences. Their ability to respond to magnetic fields provides unique opportunities for culinary innovation, personalized nutrition, and functional food design. While challenges related to material safety, control, and consumer acceptance remain, current research and technological advancements promise to overcome these hurdles, paving the way for widespread adoption and new culinary frontiers.

REFERENCES

Ahmed, A., Arya, S., Gupta, V., Furukawa, H., & Khosla, A. (2021). 4D printing: Fundamentals, materials, applications and challenges. *Polymer, 228*, 123926. DOI: 10.1016/j.polymer.2021.123926

Antezana, P. E., Municoy, S., Ostapchuk, G., Catalano, P. N., Hardy, J. G., Evelson, P. A., Orive, G., & Desimone, M. F. (2023). 4D Printing: The Development of Responsive Materials Using 3D-Printing Technology. *Pharmaceutics, 15*(12), 2743. DOI: 10.3390/pharmaceutics15122743 PMID: 38140084

Bai, J., & Bu, G. (2022). Progress in 4D printing technology. *J. Adv. Manuf. Sci. Technol, 2*(1), 2022001–2022001.

Bánhegyi, G. Advanced Industrial and Engineering Polymer Research.

Biswas, M. C., Chakraborty, S., Bhattacharjee, A., & Mohammed, Z. (2021). 4D printing of shape memory materials for textiles: Mechanism, mathematical modeling, and challenges. *Advanced Functional Materials, 31*(19), 2100257. DOI: 10.1002/adfm.202100257

Bodkhe, S., & Ermanni, P. (2019). Challenges in 3D printing of piezoelectric materials. *Multifunctional Materials, 2*(2), 022001. DOI: 10.1088/2399-7532/ab0c41

Carayon, I., Gaubert, A., Mousli, Y., & Philippe, B. (2020). Electro-responsive hydrogels: Macromolecular and supramolecular approaches in the biomedical field. *Biomaterials Science, 8*(20), 5589–5600. DOI: 10.1039/D0BM01268H PMID: 32996479

Carrillo, L. E. L., Gonzalez, Y. O., Parga, M., Ramos, K. L. D., Neparko, N., & Roberson, D. A. (2024). Development of binary and ternary polyester shape memory blends for additive manufacturing. *Journal of Materials Science, 59*(18), 1–18. DOI: 10.1007/s10853-024-09657-7

Cataldi, P., Liu, M., Bissett, M., & Kinloch, I. A. (2022). A review on printing of responsive smart and 4D structures using 2D materials. *Advanced Materials Technologies, 7*(11), 2200025. DOI: 10.1002/admt.202200025

Champeau, M., Heinze, D. A., Viana, T. N., de Souza, E. R., Chinellato, A. C., & Titotto, S. (2020). 4D printing of hydrogels: A review. *Advanced Functional Materials, 30*(31), 1910606. DOI: 10.1002/adfm.201910606

Chen, A., Su, J., Li, Y., Zhang, H., Shi, Y., Yan, C., & Lu, J. (2023). 3D/4D printed bio-piezoelectric smart scaffolds for next-generation bone tissue engineering. *International Journal of Extreme Manufacturing*, *5*(3), 032007. DOI: 10.1088/2631-7990/acd88f

Chen, C., Wang, X., Wang, Y., Yang, D., Yao, F., Zhang, W., Wang, B., Sewvandi, G. A., Yang, D., & Hu, D. (2020). Additive manufacturing of piezoelectric materials. *Advanced Functional Materials*, *30*(52), 2005141. DOI: 10.1002/adfm.202005141

Chen, L., Liu, F., Abdiryim, T., & Liu, X. (2023). Stimuli-responsive hydrogels as promising platforms for soft actuators. *Materials Today Physics*, •••, 101281.

Chen, X., Zhang, M., Teng, X., & Mujumdar, A. S. (2022). Recent progress in modeling 3D/4D printing of foods. *Food Engineering Reviews*, *14*(1), 1–14. DOI: 10.1007/s12393-021-09297-6

Chen, Y., Zhang, M., Sun, Y., & Phuhongsung, P. (2022). Improving 3D/4D printing characteristics of natural food gels by novel additives: A review. *Food Hydrocolloids*, *123*, 107160. DOI: 10.1016/j.foodhyd.2021.107160

Chu, H., Yang, W., Sun, L., Cai, S., Yang, R., Liang, W., Yu, H., & Liu, L. (2020). 4D printing: A review on recent progresses. *Micromachines*, *11*(9), 796. DOI: 10.3390/mi11090796 PMID: 32842588

Deshmukh, K., Muzaffar, A., Kovářík, T., Křenek, T., Ahamed, M. B., & Pasha, S. K. (2020). *Fundamentals and applications of 3D and 4D printing of polymers: challenges in polymer processing and prospects of future research. In 3D and 4D Printing of Polymer Nanocomposite Materials*. Elsevier.

Enferadi, A., Baniassadi, M., & Baghani, M. (2024). Innovative multiphysics approach for designing high-performance thermo-responsive shape memory polymer microvalve. *European Journal of Mechanics. A, Solids*, *103*, 105174. DOI: 10.1016/j.euromechsol.2023.105174

Enferadi, A., Ostadrahimi, A., Li, G., Baniassadi, M., & Baghani, M. (2024). Insight into single-helix intelligent shape memory polymer cables: Modeling and optimization under finite sliding contact. *Smart Materials and Structures*, *33*(5), 055033. DOI: 10.1088/1361-665X/ad3bf9

Farid, M. I., Wu, W., Liu, X., & Wang, P. (2021). Additive manufacturing landscape and materials perspective in 4D printing. *International Journal of Advanced Manufacturing Technology*, *115*(9-10), 2973–2988. DOI: 10.1007/s00170-021-07233-w PMID: 34092882

Fontana-Escartín, A., Lanzalaco, S., Bertran, O., & Alemán, C. (2022). Electrochemical multi-sensors obtained by applying an electric discharge treatment to 3D-printed poly (lactic acid). *Applied Surface Science*, *597*, 153623. DOI: 10.1016/j.apsusc.2022.153623

Ghazal, A. F., Zhang, M., & Liu, Z. (2019). Spontaneous color change of 3D printed healthy food product over time after printing as a novel application for 4D food printing. *Food and Bioprocess Technology*, *12*(10), 1627–1645. DOI: 10.1007/s11947-019-02327-6

Ghazal, A. F., Zhang, M., Mujumdar, A. S., & Ghamry, M. (2023). Progress in 4D/5D/6D printing of foods: Applications and R&D opportunities. *Critical Reviews in Food Science and Nutrition*, *63*(25), 7399–7422. DOI: 10.1080/10408398.2022.2045896 PMID: 35225117

Grinberg, D., Siddique, S., Le, M. Q., Liang, R., Capsal, J. F., & Cottinet, P. J. (2019). 4D Printing based piezoelectric composite for medical applications. *Journal of Polymer Science. Part B, Polymer Physics*, *57*(2), 109–115. DOI: 10.1002/polb.24763

He, C., Zhang, M., & Guo, C. (2020). 4D printing of mashed potato/purple sweet potato puree with spontaneous color change. *Innovative Food Science & Emerging Technologies*, *59*, 102250. DOI: 10.1016/j.ifset.2019.102250

Huang, J., Xia, S., Li, Z., Wu, X., & Ren, J. (2021). Applications of four-dimensional printing in emerging directions: Review and prospects. *Journal of Materials Science and Technology*, *91*, 105–120. DOI: 10.1016/j.jmst.2021.02.040

Huang, Y., Zhang, M., Pattarapon, P., & Mujumdar, A. S. (2022). 4D printing of mixed vegetable gel based on deformation and discoloration induced by acidification and dehydration. *Journal of Food Process Engineering*, *45*(12), e14172. DOI: 10.1111/jfpe.14172

Imam, S. S., Hussain, A., Altamimi, M. A., & Alshehri, S. (2021). Four-dimensional printing for hydrogel: Theoretical concept, 4d materials, shape-morphing way, and future perspectives. *Polymers*, *13*(21), 3858. DOI: 10.3390/polym13213858 PMID: 34771414

Jagadeesh, P., Puttegowda, M., Rangappa, S. M., Alexey, K., Gorbatyuk, S., Khan, A., Doddamani, M., & Siengchin, S. (2022). A comprehensive review on 3D printing advancements in polymer composites: Technologies, materials, and applications. *International Journal of Advanced Manufacturing Technology*, *121*(1), 127–169. DOI: 10.1007/s00170-022-09406-7

Javaid, M., & Haleem, A. (2019). 4D printing applications in medical field: A brief review. *Clinical Epidemiology and Global Health*, *7*(3), 317–321. DOI: 10.1016/j.cegh.2018.09.007

Joharji, L., Mishra, R. B., Alam, F., Tytov, S., Al-Modaf, F., & El-Atab, N. (2022). 4D printing: A detailed review of materials, techniques, and applications. *Microelectronic Engineering*, *265*, 111874. DOI: 10.1016/j.mee.2022.111874

Joshi, S., Rawat, K., Karunakaran, C., Rajamohan, V., Mathew, A. T., Koziol, K., & Balan, A. S. S. (2020). 4D printing of materials for the future: Opportunities and challenges. *Applied Materials Today*, *18*, 100490. DOI: 10.1016/j.apmt.2019.100490

Khalid, M. Y., Arif, Z. U., Noroozi, R., Zolfagharian, A., & Bodaghi, M. (2022). 4D printing of shape memory polymer composites: A review on fabrication techniques, applications, and future perspectives. *Journal of Manufacturing Processes*, *81*, 759–797. DOI: 10.1016/j.jmapro.2022.07.035

Khoo, Z. X., Teoh, J. E. M., Liu, Y., Chua, C. K., Yang, S., An, J., Leong, K. F., & Yeong, W. Y. (2015). 3D printing of smart materials: A review on recent progresses in 4D printing. *Virtual and Physical Prototyping*, *10*(3), 103–122. DOI: 10.1080/17452759.2015.1097054

Kuang, X. (2021). Introduction to 4D printing: methodologies and materials. In *Additive manufacturing* (pp. 303-342). Elsevier.

Kuang, X., Roach, D. J., Wu, J., Hamel, C. M., Ding, Z., Wang, T., Dunn, M. L., & Qi, H. J. (2019). Advances in 4D printing: Materials and applications. *Advanced Functional Materials*, *29*(2), 1805290. DOI: 10.1002/adfm.201805290

Lee, A. Y., An, J., & Chua, C. K. (2017). Two-way 4D printing: A review on the reversibility of 3D-printed shape memory materials. *Engineering (Beijing)*, *3*(5), 663–674. DOI: 10.1016/J.ENG.2017.05.014

Leigh, S. J., Bradley, R. J., Purssell, C. P., Billson, D. R., & Hutchins, D. A. (2012). A simple, low-cost conductive composite material for 3D printing of electronic sensors. *PLoS One*, *7*(11), e49365. DOI: 10.1371/journal.pone.0049365 PMID: 23185319

Li, S., Yang, H., Chen, G., Zheng, J., Wang, W., Ren, J., Zhu, C., Yang, Y., Cong, Y., & Fu, J. (2023). 4D printing of biomimetic anisotropic self-sensing hydrogel actuators. *Chemical Engineering Journal*, *473*, 145444. DOI: 10.1016/j.cej.2023.145444

Li, X., Shang, J., & Wang, Z. (2017). Intelligent materials: A review of applications in 4D printing. *Assembly Automation*, *37*(2), 170–185. DOI: 10.1108/AA-11-2015-093

Li, Y. C., Zhang, Y. S., Akpek, A., Shin, S. R., & Khademhosseini, A. (2016). 4D bioprinting: The next-generation technology for biofabrication enabled by stimuli-responsive materials. *Biofabrication*, *9*(1), 012001. DOI: 10.1088/1758-5090/9/1/012001 PMID: 27910820

Liu, H., Wang, F., Wu, W., Dong, X., & Sang, L. (2023). 4D printing of mechanically robust PLA/TPU/Fe3O4 magneto-responsive shape memory polymers for smart structures. *Composites. Part B, Engineering*, *248*, 110382. DOI: 10.1016/j.compositesb.2022.110382

Lui, Y. S., Sow, W. T., Tan, L. P., Wu, Y., Lai, Y., & Li, H. (2019). 4D printing and stimuli-responsive materials in biomedical aspects. *Acta Biomaterialia*, *92*, 19–36. DOI: 10.1016/j.actbio.2019.05.005 PMID: 31071476

Ma, B., Zhang, Y., Li, J., Chen, D., Liang, R., Fu, S., & Li, D. (2023). 4D printing of multi-stimuli responsive rigid smart composite materials with self-healing ability. *Chemical Engineering Journal*, *466*, 143420. DOI: 10.1016/j.cej.2023.143420

Mallakpour, S., Tabesh, F., & Hussain, C. M. (2021). 3D and 4D printing: From innovation to evolution. *Advances in Colloid and Interface Science*, *294*, 102482. DOI: 10.1016/j.cis.2021.102482 PMID: 34274721

Maniruzzaman, M. (Ed.). (2018). *3D and 4D printing in biomedical applications: process engineering and additive manufacturing*. John Wiley & Sons.

Megdich, A., Habibi, M., & Laperrière, L. (2023). A review on 3D printed piezoelectric energy harvesters: Materials, 3D printing techniques, and applications. *Materials Today. Communications*, *35*, 105541. DOI: 10.1016/j.mtcomm.2023.105541

Mehrpouya, M., Vahabi, H., Janbaz, S., Darafsheh, A., Mazur, T. R., & Ramakrishna, S. (2021). 4D printing of shape memory polylactic acid (PLA). *Polymer*, *230*, 124080. DOI: 10.1016/j.polymer.2021.124080

Merazzo, K. J., Lima, A. C., Rincón-Iglesias, M., Fernandes, L. C., Pereira, N., Lanceros-Mendez, S., & Martins, P. (2021). Magnetic materials: A journey from finding north to an exciting printed future. *Materials Horizons*, *8*(10), 2654–2684. DOI: 10.1039/D1MH00641J PMID: 34617551

Mohol, S. S., & Sharma, V. (2021). Functional applications of 4D printing: A review. *Rapid Prototyping Journal*, *27*(8), 1501–1522. DOI: 10.1108/RPJ-10-2020-0240

Momeni, F., Liu, X., & Ni, J. (2017). A review of 4D printing. *Materials & Design*, *122*, 42–79. DOI: 10.1016/j.matdes.2017.02.068

Momeni, F., & Ni, J. (2020). Laws of 4D printing. *Engineering (Beijing), 6*(9), 1035–1055. DOI: 10.1016/j.eng.2020.01.015

Mondal, K., & Tripathy, P. K. (2021). Preparation of smart materials by additive manufacturing technologies: A review. *Materials (Basel), 14*(21), 6442. DOI: 10.3390/ma14216442 PMID: 34771968

Morouço, P., Azimi, B., Milazzo, M., Mokhtari, F., Fernandes, C., Reis, D., & Danti, S. (2020). Four-dimensional (bio-) printing: A review on stimuli-responsive mechanisms and their biomedical suitability. *Applied Sciences (Basel, Switzerland), 10*(24), 9143. DOI: 10.3390/app10249143

Muzaffar, A., Ahamed, M. B., Deshmukh, K., Kovářík, T., Křenek, T., & Pasha, S. K. (2020). *3D and 4D printing of pH-responsive and functional polymers and their composites. In 3D and 4D printing of polymer nanocomposite materials.* Elsevier.

Navaf, M., Sunooj, K. V., Aaliya, B., Akhila, P. P., Sudheesh, C., Mir, S. A., & George, J. (2022). 4D printing: A new approach for food printing; effect of various stimuli on 4D printed food properties. A comprehensive review. *Applied Food Research, 2*(2), 100150. DOI: 10.1016/j.afres.2022.100150

Oral, M. O., Derossi, A., Caporizzi, R., & Severini, C. (2021). Analyzing the most promising innovations in food printing. Programmable food texture and 4D foods. *Future Foods : a Dedicated Journal for Sustainability in Food Science, 4*, 100093. DOI: 10.1016/j.fufo.2021.100093

Patdiya, J., & Kandasubramanian, B. (2021). Progress in 4D printing of stimuli responsive materials. *Polymer-Plastics Technology and Materials, 60*(17), 1845–1883. DOI: 10.1080/25740881.2021.1934016

Paunović, N., Meyer, D., Krivitsky, A., Studart, A. R., Bao, Y., & Leroux, J. C. (2023). 4D printing of biodegradable elastomers with tailorable thermal response at physiological temperature. *Journal of Controlled Release, 361*, 417–426. DOI: 10.1016/j.jconrel.2023.07.053 PMID: 37532144

Phuhongsung, P., Zhang, M., Devahastin, S., & Mujumdar, A. S. (2022). Defects in 3D/4D food printing and their possible solutions: A comprehensive review. *Comprehensive Reviews in Food Science and Food Safety, 21*(4), 3455–3479. DOI: 10.1111/1541-4337.12984 PMID: 35678036

Qi, S., Yao, H., Fu, J., Xie, Y., Li, Y., Tian, R., Yu, M., & Guo, H. (2022). Magneto-active soft matter with reprogrammable shape-morphing and self-sensing capabilities. *Composites Science and Technology, 230*, 109789. DOI: 10.1016/j.compscitech.2022.109789

Quanjin, M., Rejab, M. R. M., Idris, M. S., Kumar, N. M., Abdullah, M. H., & Reddy, G. R. (2020). Recent 3D and 4D intelligent printing technologies: A comparative review and future perspective. *Procedia Computer Science*, *167*, 1210–1219. DOI: 10.1016/j.procs.2020.03.434

Rafiee, M., Farahani, R. D., & Therriault, D. (2020). Multi-material 3D and 4D printing: A survey. *Advancement of Science*, *7*(12), 1902307. DOI: 10.1002/advs.201902307 PMID: 32596102

Rastogi, P., & Kandasubramanian, B. (2019). Breakthrough in the printing tactics for stimuli-responsive materials: 4D printing. *Chemical Engineering Journal*, *366*, 264–304. DOI: 10.1016/j.cej.2019.02.085

Ryan, K. R., Down, M. P., & Banks, C. E. (2021). Future of additive manufacturing: Overview of 4D and 3D printed smart and advanced materials and their applications. *Chemical Engineering Journal*, *403*, 126162. DOI: 10.1016/j.cej.2020.126162

Shanthamma, S., Preethi, R., Moses, J. A., & Anandharamakrishnan, C. (2021). 4D printing of sago starch with turmeric blends: A study on pH-triggered spontaneous color transformation. *ACS Food Science & Technology*, *1*(4), 669–679. DOI: 10.1021/acsfoodscitech.0c00151

Shie, M. Y., Shen, Y. F., Astuti, S. D., Lee, A. K. X., Lin, S. H., Dwijaksara, N. L. B., & Chen, Y. W. (2019). Review of polymeric materials in 4D printing biomedical applications. *Polymers*, *11*(11), 1864. DOI: 10.3390/polym11111864 PMID: 31726652

Shinde, S., Mane, R., Vardikar, A., Dhumal, A., & Rajput, A. (2023). 4D printing: From emergence to innovation over 3D printing. *European Polymer Journal*, *197*, 112356. DOI: 10.1016/j.eurpolymj.2023.112356

Shun, L. (2021). *Studies on 4D printing Thermo-responsive PNIPAM-based materials* (Doctoral dissertation, The University of Akron).

Singholi, A. K., & Sharma, A. (2021). Review of Materials and Processes Used in 4D Printing. In *Advances in Engineering Materials: Select Proceedings of FLAME 2020* (pp. 677-684). Springer Singapore.

Sossou, G., Demoly, F., Belkebir, H., Qi, H. J., Gomes, S., & Montavon, G. (2019). Design for 4D printing: Modeling and computation of smart materials distributions. *Materials & Design*, *181*, 108074. DOI: 10.1016/j.matdes.2019.108074

Spiegel, C. A., Hippler, M., Münchinger, A., Bastmeyer, M., Barner-Kowollik, C., Wegener, M., & Blasco, E. (2020). 4D printing at the microscale. *Advanced Functional Materials*, *30*(26), 1907615. DOI: 10.1002/adfm.201907615

Subeshan, B., Baddam, Y., & Asmatulu, E. (2021). Current progress of 4D-printing technology. *Progress in Additive Manufacturing, 6*(3), 495–516. DOI: 10.1007/s40964-021-00182-6

Suryavanshi, P., Kawre, S., Maniruzzaman, M., Seth, K., & Banerjee, S. (2023). Thermo-responsive self-folding feedstock with excellent shape memory programming. *Chemicke Zvesti, 77*(6), 3145–3154. DOI: 10.1007/s11696-023-02693-8

Tamay, D. G., Dursun Usal, T., Alagoz, A. S., Yucel, D., Hasirci, N., & Hasirci, V. (2019). 3D and 4D printing of polymers for tissue engineering applications. *Frontiers in Bioengineering and Biotechnology, 7*, 164. DOI: 10.3389/fbioe.2019.00164 PMID: 31338366

Teng, X., Zhang, M., Adhikari, B., & Ma, Y. (2024). 4D printed pH-responsive labels of methacrylic anhydride grafted konjac glucomannan for detecting quality changes in respiring climacteric fruits. *Food Hydrocolloids, 149*, 109543. DOI: 10.1016/j.foodhyd.2023.109543

Teng, X., Zhang, M., & Mujumdar, A. S. (2021). 4D printing: Recent advances and proposals in the food sector. *Trends in Food Science & Technology, 110*, 349–363. DOI: 10.1016/j.tifs.2021.01.076

Vatanparast, S., Boschetto, A., Bottini, L., & Gaudenzi, P. (2023). New trends in 4D printing: A critical review. *Applied Sciences (Basel, Switzerland), 13*(13), 7744. DOI: 10.3390/app13137744

Wang, Y., Jia, B., Liu, S., Yao, X., & Sun, C. (2022). 3D Printing of Smart Materials and Actuators. In *Proceedings of the 1st International Conference on New Materials, Machinery and Vehicle Engineering* (pp. 128-138). IOS Press. DOI: 10.3233/ATDE220220

Wang, Y., Wang, Y., Liu, M., Wei, Q., & Du, B. (2023). 4D printing light-/thermo-responsive shape memory composites based on thermoplastic polyurethane/polylactic acid/polyaniline blends. *High Performance Polymers, 35*(4), 366–378. DOI: 10.1177/09540083221135499

Yang, W., Tu, A., Ma, Y., Li, Z., Xu, J., Lin, M., Zhang, K., Jing, L., Fu, C., Jiao, Y., & Huang, L. (2021). Chitosan and whey protein bio-inks for 3D and 4D printing applications with particular focus on food industry. *Molecules (Basel, Switzerland), 27*(1), 173. DOI: 10.3390/molecules27010173 PMID: 35011406

Zafar, M. Q., & Zhao, H. (2020). 4D printing: Future insight in additive manufacturing. *Metals and Materials International, 26*(5), 564–585. DOI: 10.1007/s12540-019-00441-w

Zeng, Y., Jiang, L., He, Q., Wodnicki, R., Yang, Y., Chen, Y., & Zhou, Q. (2021). Recent progress in 3D printing piezoelectric materials for biomedical applications. *Journal of Physics. D, Applied Physics*, *55*(1), 013002. DOI: 10.1088/1361-6463/ac27d2

Zhang, Z., Demir, K. G., & Gu, G. X. (2019). Developments in 4D-printing: A review on current smart materials, technologies, and applications. *International Journal of Smart and Nano Materials*, *10*(3), 205–224. DOI: 10.1080/19475411.2019.1591541

Zolfagharian, A., Haq, M. I. U., Nafea, M., & Bodaghi, M. (2023). *4D Printing of Smart Magnetic-Based Robotic Materials. In 3D Printing and Sustainable Product Development*. CRC Press.

. Zolfagharian, A., Kaynak, A., Khoo, S. Y., & Kouzani, A. Z. (2018). Polyelectrolyte soft actuators: 3D printed chitosan and cast gelatin. *3D Printing and Additive Manufacturing, 5*(2), 138-150.

Chapter 16

Effect of Ageing Temperature on Microstructure and Microhardness of L-PBFed 18Ni(300) Maraging Steel

Ashish Kumar Sahu
https://orcid.org/0009-0006-8195-806X
National Institute of Technology, Rourkela, India

Sudipta Swain
https://orcid.org/0009-0003-6042-9108
National Institute of Technology, Rourkela, India

Saurav Datta
National Institute of Technology, Rourkela, India

Kaushik Kumar
https://orcid.org/0000-0002-4237-2836
Birla Institute of Technology, India

ABSTRACT

This study investigates the aging response of Maraging Steel 300 fabricated via Laser-Powder Bed Fusion (L-PBF). The as-built specimens undergo a post-heat treatment sequence comprising Solution Treatment (ST) at 840 °C for 2 hours fol-

DOI: 10.4018/979-8-3373-0533-2.ch016

lowed by water quenching and subsequent Aging Treatment (AT) at 450 °C, 490 °C, and 550 °C for 2 hours each. The effect of post-heat treatment on microstructure and microhardness is analyzed, revealing a solidification sub-structure with semi-elliptical overlapped Melt Pool Boundaries (MPBs) and fine columnar dendritic/ cellular grain morphology due to thermal gradient variations. Solute segregation occurs at grain boundaries, while intense undercooling during L-PBF results in a fine martensitic microstructure. ST dissolves solute elements, preparing the matrix for precipitation during aging, which enhances microhardness. The specimen aged at 450 °C achieves the highest microhardness (~607.5 ± 22.2 HV0.5), influenced by the balance of martensite and austenite phases.

1. BACKGROUND

Laser-Powder Bed Fusion (L-PBF) is a metal 3D printing technology which is attempted to fabricate metallic parts through fusion and melting of powder particulates by means of the laser energy. When compared to other metal-additive manufacturing processes such as Direct Energy Deposition (DED) based Wire Arc Additive Manufacturing (WAAM), a higher relative density of the fabricated part can be ascertained in L-PBF. L-PBF also consumes a lower time while fabricating intricate designed parts. The widespread consumption of maraging steels in aeronautical and aircraft applications, tooling, die and mould making applications is due to its impressive mechanical properties including ultra-high tensile strength (without compromising ductility), outstanding strength-to-weight proportion, high corrosion resistance, low cost, availability and recyclability (Raghuraman and Kumar, 2024). Instead of conventional casting, recently, additively manufactured maraging steel parts are enormously used in aerospace applications (Rao and Rao, 2022). As compared to conventional cast and wrought alloys, microstructural features of the L-PBFed maraging steels are completely distinct. This is due to the combined influence of the large number of process parameters (including powder related, laser related and scan related parameters) and fast cooling/ solidification rate as experienced during the L-PBF process. Appropriate settings of the L-PBF parameters are indeed necessary to accomplish quality part fabrication; else the part may suffer from various defects such as porosity, crack, balling, etc. Depending on the interaction between laser and feedstock powdery material, aforesaid defects adversely affect microstructure, relative density, mechanical properties and surface characteristics of the built part. Post-processing heat treatment is attempted to supress/ eliminate such defects (generally present within the *as fabricated* part) thereby obtaining a more refined microstructure and improved part mechanical properties.

Shamsdini et al. (2020) studied the effect of layer thickness and feedstock powder characteristics on microstructure and mechanical properties of the L-PBFed MS1 cubic parts. They studied the porosity and hardness through the depth of the fabricated specimens. In their study, homogenous properties are evidenced in the core of the material whilst subsurface layers exhibited heterogeneous properties. For the specimens, fabricated at a lower layer thickness, a correlation was established between the volumetric fraction of the austenite and the strength of the material. Allam et al. (2020) studied the influence of heterogeneous structures (dendritic/ cellular regions along with traces of micro-segregations) developed during the L-PBF of maraging steel on the microstructure evolution and formation of nano-precipitates. They applied direct ageing and solution-ageing heat treatment schedules to promote precipitation of nano-particulates in presence and after elimination of heterogeneous microstructural morphology. It was witnessed that such hetero-structures stimulated retention of the austenite phase in the *as built* condition whilst they favoured marternsite-to-austenite reversion in the aged condition. In contrast, the austenite retention and reversion processes got supressed upon solution treatment-ageing. As compared to the direct-aged state, smaller sized nano-particulates (with high number density) were detected in the solution-aged condition. Kannan et al. (2021) attempted to optimize the direct ageing schedule for Ti-free grade 300 maraging steel fabricated through the L-PBF process. They experienced that the direct ageing at 440 °C for 6 h caused the best compatibility between strength and ductility. A lower ageing temperature or shorter ageing duration caused a lower fraction of reverted austenite. On the other hand, ageing at higher temperatures evidenced excessive martensite recrystallization, coarsening of precipitates (increased size of the precipitates) and huge austenite reversion thereby softening the material's matrix and reducing strength.

Huang et al. (2022) reported that dense specimens (with > 99.5% relative density) of 18Ni(300) could be fabricated through the L-PBF when the Volumetric Energy Density (VED) was higher than 80 J/mm^3. The *as built* microstructure was characterized by numerous columnar prior-austenite grains (PAGs) with dendritic structural morphology inside the grains. A few ordered Ni(Al, Fe) intermetallic precipitates were also detected within the *as built* microstructure due to the effect of intrinsic heat treatment. The VED ~ 111 J/mm^3 exhibited the best trade-off between the build rate and part mechanical properties. They found the Ultimate Tensile Strength (UTS) ~1100 ± 11 MPa and elongation at break ~9.8 ± 0.7%; the reported values were comparable to that of the wrought counterpart. Upon solution-aging treatment, an improved strength (~1945 ± 12 MPa) but a lower elongation was witnessed (~5.4 ± 0.5%). It was explained that the improved strength of the solution-aging treated specimens was due to precipitation strengthening, dislocation strengthening and grain boundary strengthening. Wei et al. (2022) studied the effect of post-heat treatment on meso/ micro-structure and mechanical properties of 18Ni(350) maraging steel fabricated through the L-PBF. They reported that the martensitic morphology in the *as printed* as well as direct-aged specimens was influenced by the dendritic growth induced cellular microstructure evolved during the L-PBF process. The martensite transformation in aforesaid conditions occurred within the cell matrix whilst cell boundaries acted as the preferential locations for austenite retention/ reversal. Solution treatment completely dissolved such cellular structural morphology. Presence of the austenite phase in the direct aged specimen lowered strength but increased fracture toughness. In contrast, superior strength but a lower ductility within the solution-aged specimen was due to the absence of austenite. Karlapudy et al. (2024) studied the mechanical properties of post-heat treated L-PBFed maraging steel 300 as a function of the build orientation and laser hatch pattern. They applied a direct ageing treatment (490 °C/ 4 h) and a solution treatment (940 °C/ 2 h) followed by the double ageing treatments (490 °C/ 1 h + 490 °C/ 3 h). The improved tensile strength and increased microhardness were experienced upon heat treatments. It was reported that dual ageing treatments caused significant elimination of orientation-dependent property anisotropy. Raghuraman and Kumar (2025) studied the microstructure and microhardness of the L-PBFed maraging steel specimens both the *as built* as well as the post-heat treated conditions. Different heat treatment schedules such as ageing, double ageing, and solution treatment were attempted. Microhardness value of the *as built*, age treated, double-age treated and solution treated specimens obtained as 450 HV, 650 HV, 760 HV, and 520 HV, respectively. Corresponding specimen's austenite content was 0.4%, 12.6%, 1%, and 0.6%, respectively.

From literature survey, it is understood that in the *as built* condition, the L-BFed maraging steel 300 suffers from the lack of precipitates combined with occurrence of micro-segregation and thus experiences evolution of the non-equilibrium microstructure. Hence, the alloy is further strengthened through post-heat treatments (Kannan et al., 2021). In general, post-heat treatments in additively manufactured maraging steel are crucial for eliminating or suppressing the effects of the fabrication process and for improving material's microstructure as well as mechanical properties. Careful selection and optimization of the post-heat treatment schedule are indeed essential to enhance efficiency and reliability of the additively manufactured maraging steel parts (Podgornik et al., 2021). Although some authors suggested direct ageing of the L-PBFed maraging steel towards achieving desired strength and fracture toughness (Suryawanshi et al., 2017). They reported that a very high rate of undercooling during the L-PBF process doesn't provide sufficient time for severe micro-segregation to occur. However, for maraging steel 300, micro-segregation cannot be totally avoided during the L-PBF process. Therefore, the present work intends to perform solution annealing prior to the ageing treatment. The solution annealing is planned to homogenize alloy's microstructure conducive for precipitation hardening during subsequent ageing treatments. The microstructural morphology of the *as built* specimen is compared with the post-heat treated counterparts. The effect of ageing temperature on the post-heat treated alloy's microhardness is also studied herein.

2. MATERIAL AND METHODS

A Direct Metal Laser Sintering (DMLS) machine (EOSINT M280 EOS, Germany) of 400 W is utilized to fabricate maraging steel 300 cubic specimen of dimension (1×1×1) cm^3. The machine uses Ytterbium fiber laser as the power source. The parametric values are set as follow – 250 W laser power, 500 mm/s scanning speed, 40 μm layer thickness, 80 μm laser spot diameter, 0.05 mm hatch spacing, and chessboard scanning pattern. In between two successive layers deposition, scan rotation angle is set 67°. Gas-atomized maraging steel powder (with 4 - 40 μm particle size) is used as feedstock material. The inert atmosphere within the build chamber is created by providing continuous supply of nitrogen. The temperature of the build platform is maintained at 40 °C. After fabrication, the built part is separated from the substrate by Wire-EDM. Outer surfaces of the *as built* part are smoothened by glass bead blasting. From the *as built* cubic part, small test specimens (dimension ~5×5×5 mm^3) are Wire-EDMed cut for metallographic observations and micro-indentation hardness tests (the *as fabricated* and the post-heat treated conditions both). The post-heat treatment cycle is designed with solution treatment followed by ageing treatment schedules. The required numbers of specimens are solution annealed first at 840 °C for 2 h followed by water quenching. The solution annealed specimens are aged next at different ageing temperatures such as 450 °C, 490 °C and 550 °C for constant holding duration of 2 h followed by water quenching.

In order to study the microstructural morphology, two different etchants are prepared – 3% Nital solution is used to preferentially etch alloy's microstructure thus revealing solidification sub-structure whilst Fry's reagent is used to expose the martensitic morphology. Microstructural observations are carried out through optical microscopy, scanning electron microscopy and field emission scanning electron microscopy. Phase analysis is done by X-Ray diffractometer (Bruker d8 Advance) using Co-kα (λ = 1.78897 Å) radiation. The peak patterns are obtained at scan rate of 10°/min, step size 0.02 with diffraction angle 2θ ranging from 30° to 120°. X'Pert HighScore Plus software is utilized to identify the peaks. Vickers hardness tester (Leco LM810, USA) is used for microhardness testing. Test conditions are 500 gf load and 8 s dwell timing.

3. RESULTS AND DISCUSSION

In the *as built* state, the microstructure exhibits heterogeneous structure. In the *as built* state, a layer wise structure is observed onto the transverse plane (*XZ* plane, parallel to the building direction, *Z*) and laser scan tracks onto the longitudinal plane (*XY* plane, perpendicular to the building direction, *Z*) as shown in Figure 1a and Figure 1b, respectively. Upon etching with Fry's reagent, the *as built* microstructure exposes martensitic morphology (Figure 1c).

Figure 1. The as built microstructure

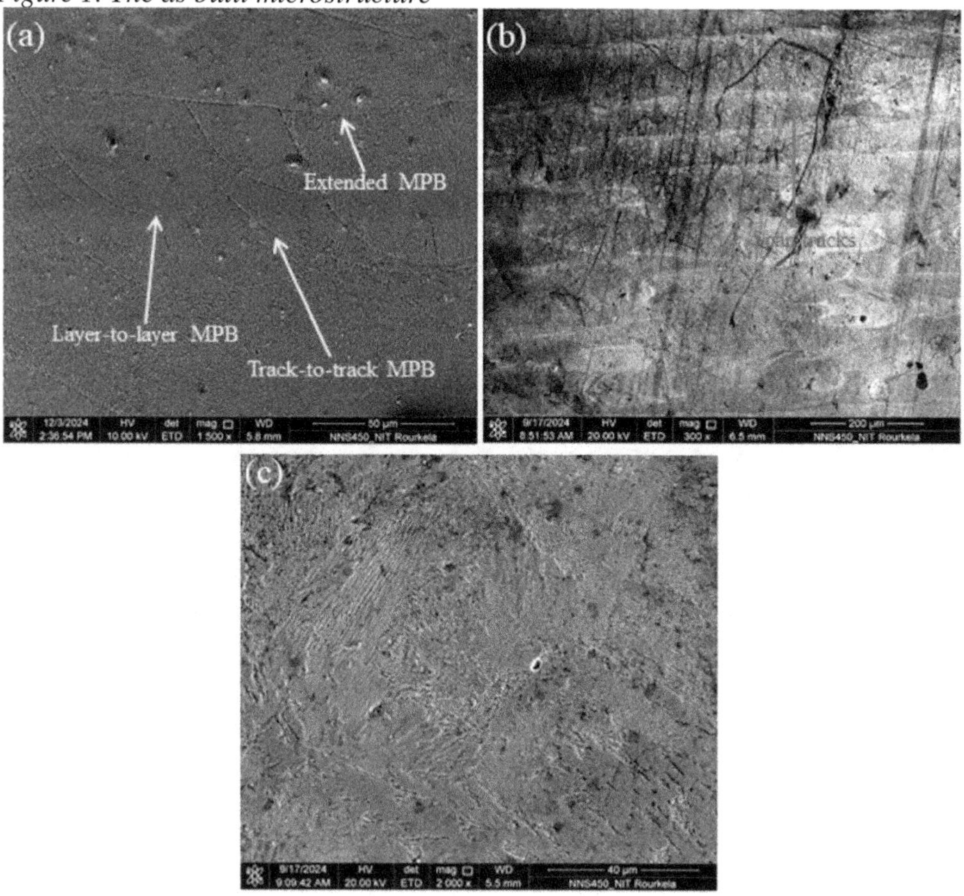

(a) Solidification sub-structure exhibiting MPBs onto the *XZ* plane, (b) Scan tracks onto the *XY* plane, and (c) Martensitic morphology

The solidification sub-structure (when viewed onto the *XZ* plane) exhibits numerous fish-scale marks which clearly indicate track-after-track and layer-upon-layer solidification patterns of melt pools during laser irradiation (Figure 1a). Due to 67° scan rotation, the shape and morphology of melt pools are witnessed non-uniform (Wei et al., 2022). The parabolic (also described as semi-elliptical) Melt Pool Boundaries (MPBs) are developed due to solidification and re-heating of melt pools during deposition of successive tracks and layers (Casalino et al., 2015). When melt pools from two adjacent scan tracks overlap, this creates distinct *track-to-track* MPB. On the other hand, overlapping of melt pools from two different layers create *layer-to-layer* MPB as shown in Figure 1a. Extended MPBs are formed due to intense thermal exposure at the melt pools overlapped regions. These are schematically presented in Figure 2. The dendritic columnar grains tend to grow normal to the MPBs, although the actual orientation is influenced by the building direction and scan direction; which in turn, influences the direction of the local thermal gradient. Hence, dendritic grain growth is evidenced preferentially inclined from the perpendicular line (Wang et al., 2016).

Figure 2. Schematic representation of melt pool solidification during L-PBF process

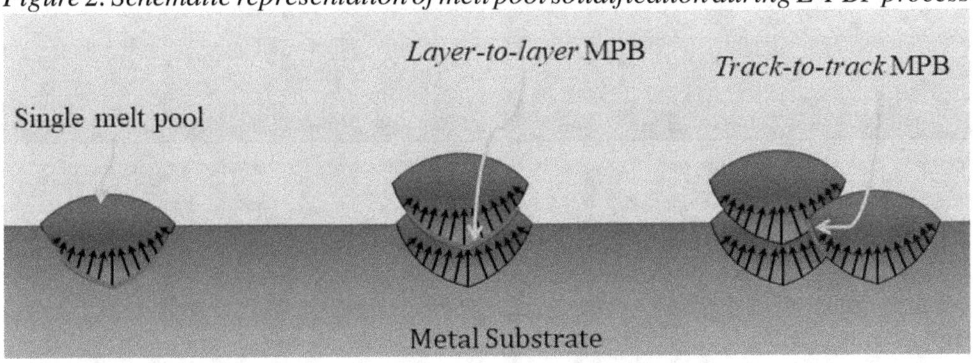

(a) single melt pool, (b) *Layer-to-layer* MPB, and (c) *Track-to-track* MPB (back arrows represent grain orientations)

Upon etching with the Nital solution, the solidification sub-structure reveals very fine columnar dendritic/ cellular grain morphologies (Figure 3). Columnar dendritic bundles are mostly orientated along the build direction, perpendicular to the scan tracks towards the central region of the melt pool which should be the direction of the maximum thermal gradient. The fine microstructure is developed due to constitutional super cooling attained by the L-PBF process as reported by Thijs et al. (2013). Moreover, it is postulated that the heat flux is directed towards the substrate plate irrespective of the complicacy of local thermal conditions of the melt pool (Yuan and Gu, 2015; Collins et al., 2016). Another inherent feature in the L-PBFed *as built* hetero-structure is the traces of micro-segregation of solute elements at locations of dendritic/ cellular grain boundaries which was previously confirmed by Kürnsteiner et al. (2017). As reported by Glicksman (2011), depending on the local thermal gradients (*G*) and growth rates (*R*), the grain morphologies may vary from location to location evolving planner, cellular, columnar dendritic or equiaxed cellular architecture.

Figure 3. The as built microstructure onto the XZ plane: Evolution of columnar / Equiaxed grains (cellular grain morphology)

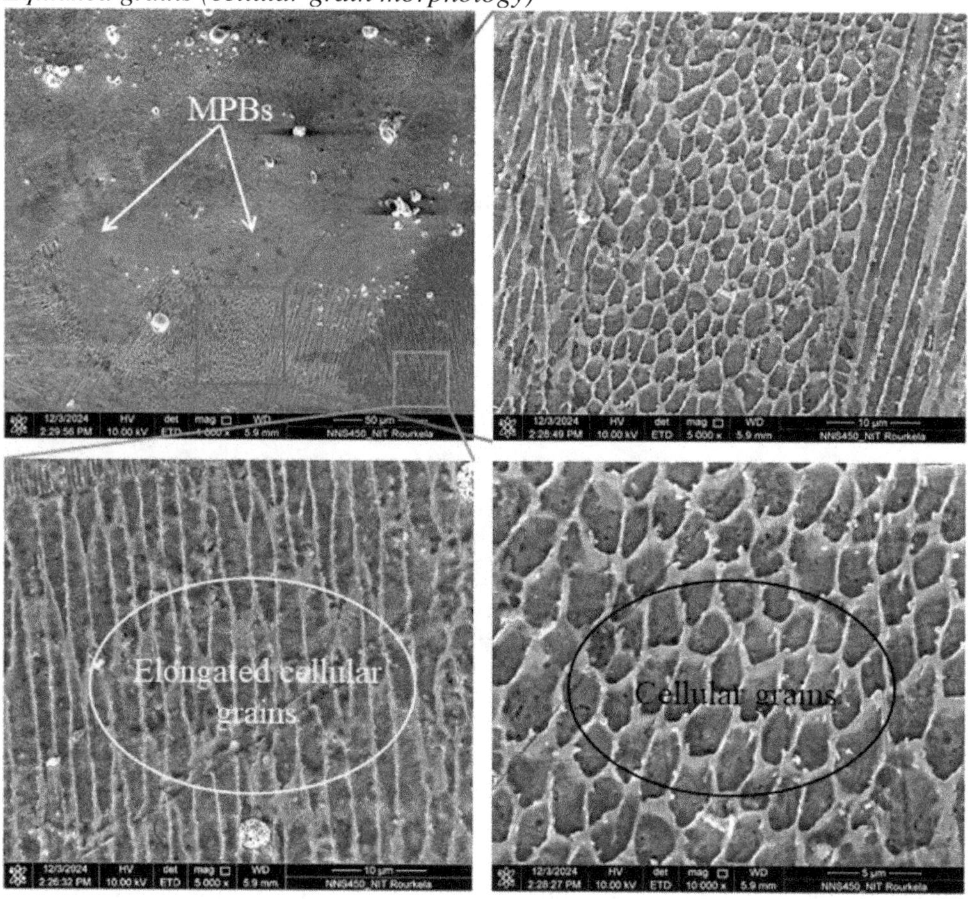

The optical micrographs of the post-heat-treated specimens are collaged in Figure 4. As compared to the *as built* specimen exhibiting lathy martensitic morphology, the solution annealed-aged specimens exhibit approximately needle-shaped (acicular) martensite structure.

Figure 4. Optical micrographs revealing effect of ageing temperature on micro-structure: First STed (840 °C, 2h + water cool) and then ATed at

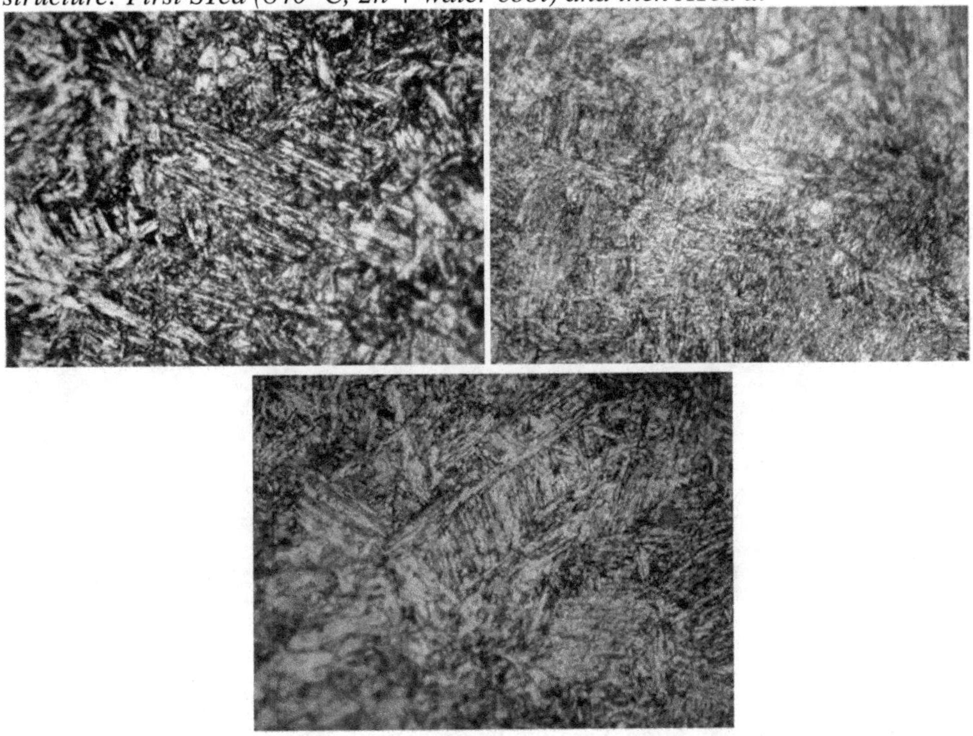

(a) 450 °C, 2h + water cool, (b) 490 °C, 2h + water cool and (c) 550 °C, 2h + water cool

It is surprising to see that the heterogeneous solidification sub-structure is completely eliminated upon solution treatment-ageing condition. It is reported that the solution treatment dissolves segregated alloying elements back into the matrix and prepares a homogeneous microstructure appropriate for precipitation strengthening during subsequent ageing schedule. The elimination of micro-segregation (especially, Ni micro-segregation) leads to high supersaturation and associated increment in the number density of $(Ni, Fe, Co)_3(Ti, Mo, Al)$ precipitates. The resultant Ni-depleted matrix may also inhibit austenite reversion thus forming reversed/ reverted austenite (Allam et al., 2020). Published literature reported different types of precipitates such as Ni_3Ti, Ni_3Mo and Fe_7Mo_6 with varied morphologies are present in the solution-aged maraging steel 300 (Jägle et al., 2014; Paul et al., 2022). However, due their nano-sized morphology, these particulates cannot be traced through optical/ scanning electron microscopy. These precipitates can be identified through Transmission

Electron Microscopy (TEM) as revealed in Figure 5 where red arrows indicate tiny sized nano-precipitates.

Figure 5. TEM micrographs revealing precipitate particulates in the solution-aged specimen

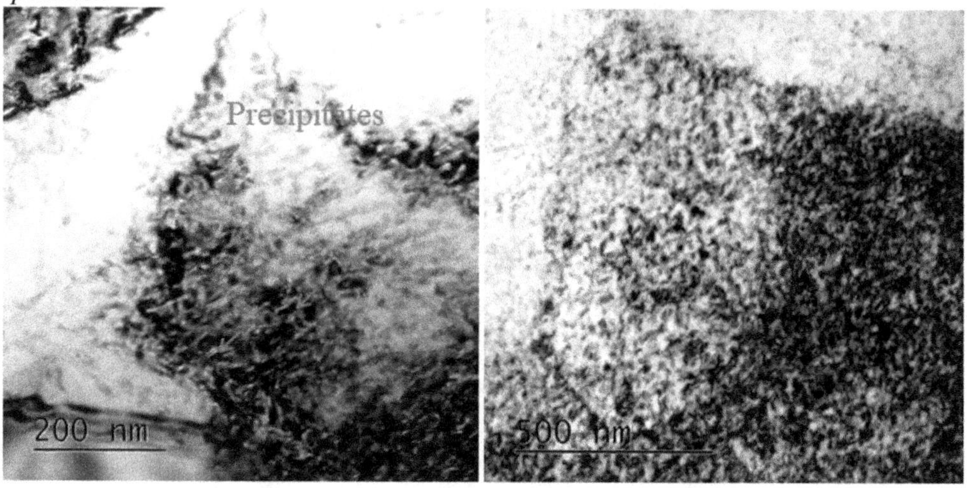

XRD spectra of the *as received* powder and L-PBFed specimens (the *as built* and the post-heat treated conditions) are shown in Figure 6. The *as built* microstructure of maraging steel 300 contains martensite(α) as major proportion with small proportion of austenite(β). Austenite in this case appears as retained austenite. During the L-PBF process, a high degree of undercooling causes non-equilibrium phase transformation hence, martensite transformation is evidenced. However, within the *as built* microstructure, the dendritic/ cellular grain boundaries that are enriched with segregated solute elements act as austenite stabilizer and therefore, cause austenite retention. Wei et al. (2022) also evidenced preserved/ retained austenite within the cellular structures of the *as built* specimen. When the solution annealed specimen is aged at (450 °C/ 2 h), 100% martensite transformation is witnessed. But upon ageing at (490 °C/ 2 h) and at (550 °C/ 2 h), respectively, reverted austenite is detected along with the dominant martensite phase. The corresponding austenite fraction is 9.12% and 8.82% as computed through XRD peak intensity method. Casati et al. (2016) reported that for maraging steel 300, ageing treatment may cause reversion of martensite to softer austenite. They further added that in the solution-aged condition, elimination of the *as built* solidification hetero-structure as well as dissolution of segregated solute elements (during solution annealing) supresses not only austenite retention but also retards martensite-to-austenite reversal on further ageing treatments. This clearly reflects the dominant role of the features of the L-PBFed solidification hetero-structure which can be tuned through post-heat treatments in order to tailor desired microstructures in maraging steels. As reported by Rack and Kalish (1971), the total proportion of austenite is directly related to the ageing temperature and holding duration which impact alloy's strength and toughness. Therefore, it is indeed a necessity to derive ideal ageing conditions to ensure the alloy's peak mechanical properties including strength and toughness.

Figure 6. XRD analysis of the as received powder and L-PBFed specimens (the as built and the post heat-treated conditions)

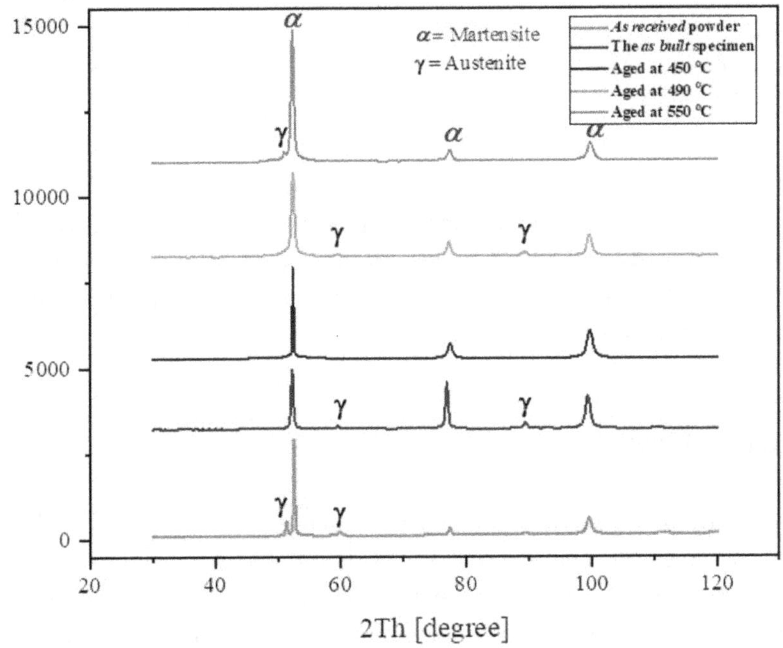

Microhardness test data are provided in Table 1. It is witnessed that microhardness in the *as built* state appears to be the lowest (~393.5 ± 12.7 $HV_{0.05}$). Same observation was made by Kannan et al. (2021). This is due to the lath martensitic microstructure without any matrix-reinforcing precipitates. At the solution-aged state, an increased hardness is evidenced due to the precipitation of intermetallic particulates. The uniform dispersion of precipitate particulates reinforces the matrix. These precipitates act as hindrance to the movement of dislocations; therefore, the alloy gains an increased strength. However, depending on the ageing temperature, microhardness values tend to vary. This is because of the fact that the ageing temperature influences martensite-to-austenite transformation and hence alters the proportion of the reverted austenite within the martensitic matrix.

Table 1. Microhardness test results: Effect of ageing temperature on microhardness

Specimen/ fabrication parameters	Load [gf]	Dwell time [s]	$HV_{0.5}$
The *as built* specimen	500	8	393.5 ± 12.7
The *post heat-treated* specimen (ST + AT ~ STA) ST ~ 840 °C, 2h + water cool AT ~ **450** °C, 2h + water cool	500	8	607.5 ± 22.2
The *post-heat treated* specimen (ST + AT ~ STA) ST ~ 840 °C, 2h + water cool AT ~ **490** °C, 2h + water cool	500	8	529 ± 13
The *post-heat treated* specimen (ST + AT ~ STA) ST ~ 840 °C, 2h + water cool AT ~ **550** °C, 2h + water cool	500	8	584.7 ± 8.1

In the present work, the highest microhardness value (~607.5 ± 22.2 $HV_{0.05}$) is obtained for the solution-treated specimen when aged subsequently at (450 °C/ 2 h). Ageing at 490 °C and 550 °C causes austenite reversal. Presence of reversed austenite imparts ductility to the specimen, and hence a lower microhardness value is obtained (as understood from Table 1).

4. CONCLUSIONS

The L-PBFed *as built* microstructure of maraging steel 300 is found decorated with very fine columnar dendrites as well as cellular lattice structure. Distinct MPBs are clearly visible onto the *XZ* plane (parallel to the build direction, Z). The interior of the solidified melt pools strongly reveals evidences of directional solidification followed by columnar dendritic grain growth. In contrast, the microstructure onto the *XY* plane (perpendicular to the build direction, *Z*) uncovers traces of laser scan tracks.

The overlapped locations between two melt pools (*track-to-track* and *layer-to-layer* overlapping) experience intense thermal exposure when compared with the nearby regions. Therefore, coarser microstructure is witnessed therein. Depending on the direction and magnitude of thermal gradient as well as cooling rate, microstructural morphologies vary location wise. The boundaries of columnar dendrites and cellular grains are enriched with segregated alloying elements. Upon etching with Fry's reagent, the *as built* martensitic morphology is exposed.

The solution annealing treatment (followed by ageing) completely eliminates solidification sub-structure (characterized by melt pools with clear MPBs, scan tracks, etc.) and thus the post-heat treated specimen exhibits a dominant martensitic microstructure. However, depending on the ageing temperature, the proportion of martensite as well as reversed austenite phase within the solution-aged specimens varies. Hence, it is concluded that the ageing temperature influences martensite-to-austenite transformation, known as austenite reversal.

The *as built* specimen exhibits the lowest microhardness value (~393.5 ± 12.7 $HV_{0.5}$). The highest microhardness (~607.5 ± 22.2 $HV_{0.5}$) is evidenced for the specimen aged at 450 °C/ 2 h. This is due to the absence of reversed austenite within the alloy (i.e. 100% martensite). In contrast, an increase in ageing temperature stimulates austenite reversion and hence, the specimens while aged at 490 °C/ 2 h and 550 °C/ 2 h exhibit relatively lower microhardness values (than the specimen, aged at 450 °C/ 2 h) as they contain a significant proportion of reverted austenite (~9.12% and 6.82%, respectively). The aforesaid reduction in the microhardness value is attributed to the presence of reverted austenite (within martensitic matrix) which imparts ductility to the alloy.

The present work examines the effect of the ageing temperature on alloy's microhardness. Apart from microhardness, the tensile properties including strength and toughness need to be studied. This particular aspect is intended to the uncovered in the future work.

ACKNOWLEDGEMENT

The authors appreciate the encouragement provided and sincerely acknowledges the comments and suggestions of the editors and reviewers that have been instrumental for improving and upgrading the paper in its present form.

REFERENCES

Allam, T., Pradeep, K. G., Köhnen, P., Marshal, A., Schleifenbaum, J. H., & Haase, C. (2020). Tailoring the nanostructure of laser powder bed fusion additively manufactured maraging steel. *Additive Manufacturing*, *36*, 101561. DOI: 10.1016/j.addma.2020.101561

Casalino, G., Campanelli, S. L., Contuzzi, N., & Ludovico, A. D. (2015). Experimental investigation and statistical optimisation of the selective laser melting process of a maraging steel. *Optics & Laser Technology*, *65*, 151–158. DOI: 10.1016/j.optlastec.2014.07.021

Casati, R., Lemke, J., Tuissi, A., & Vedani, M. (2016). Aging behaviour and mechanical performance of 18-Ni 300 steel processed by selective laser melting. *Metals*, *6*(9), 218. DOI: 10.3390/met6090218

Collins, P. C., Brice, D. A., Samimi, P., Ghamarian, I., & Fraser, H. L. (2016). Microstructural control of additively manufactured metallic materials. *Annual Review of Materials Research*, *46*(1), 63–91. DOI: 10.1146/annurev-matsci-070115-031816

Glicksman, M. E. (2011). *Principles of solidification*. Springer. DOI: 10.1007/978-1-4419-7344-3

Huang, G., Wei, K., Deng, J., & Zeng, X. (2022). High power laser powder bed fusion of 18Ni300 maraging steel: Processing optimization, microstructure and mechanical properties. *Materials Science and Engineering A*, *856*, 143983. DOI: 10.1016/j.msea.2022.143983

Jägle, E. A., Choi, P.-P., van Humbeeck, J., & Raabe, D. (2014). Precipitation and austenite reversion behavior of a maraging steel produced by selective laser melting. *Journal of Materials Research*, *29*(17), 2072–2079. DOI: 10.1557/jmr.2014.204

Kannan, R., Leonard, D. N., & Nandwana, P. (2021). Optimization of direct aging temperature of Ti free grade 300 maraging steel manufactured using laser powder bed fusion (LPBF). *Materials Science and Engineering A*, *817*, 141266. DOI: 10.1016/j.msea.2021.141266

Karlapudy, S. P., Nancharaiah, T., & Subba Rao, V. V. (2024). Influence of post-heat treatment on microstructure, texture, and mechanical properties of 18Ni-300 maraging steel fabricated by using LPBF technique. *Progress in Additive Manufacturing, 9*(6), 1667–1682. DOI: 10.1007/s40964-023-00530-8

Kürnsteiner, P., Wilms, M. B., Weisheit, A., Barriobero-Vila, P., Jägle, E. A., & Raabe, D. (2017). Massive nanoprecipitation in an Fe-19Ni- x Al maraging steel triggered by the intrinsic heat treatment during laser metal deposition. *Acta Materialia, 129*, 52–60. DOI: 10.1016/j.actamat.2017.02.069

Paul, M. J., Muniandy, Y., Kruzic, J. J., Ramamurty, U., & Gludovatz, B. (2022). Effect of heat treatment on the strength and fracture resistance of a laser powder bed fusion processed 18Ni-300 maraging steel. *Materials Science and Engineering A, 844*, 143167. DOI: 10.1016/j.msea.2022.143167

Podgornik, B., Šinko, M., & Godec, M. (2021). Dependence of the wear resistance of additive-manufactured maraging steel on the build direction and heat treatment. *Additive Manufacturing, 46*, 102123. DOI: 10.1016/j.addma.2021.102123

Rack, H. J., & Kalish, D. (1971). The strength and fracture toughness of 18 Ni (350) maraging steel. *Metallurgical and Materials Transactions. B, Process Metallurgy and Materials Processing Science, 2*, 3011–3020.

Raghuraman, V., & Kumar, T. S. (2024). Influence of in-situ process parameters, post heat treatment effects on microstructure and defects of additively manufactured maraging steel by laser powder bed fusion-A comprehensive review. *Physica Scripta. Physica Scripta, 99*(5), 052001. DOI: 10.1088/1402-4896/ad3681

Raghuraman, V., & Kumar, T. S. (2025). Different post-heat treatment responses on microstructure and mechanical characteristics of maraging steel 1.2709 samples produced by LPBF. *Physica Scripta, 100*(2), 025929. DOI: 10.1088/1402-4896/ada328

Rao, B. S., & Rao, T. B. (2022). Effect of process parameters on powder bed fusion maraging steel 300: A review. *Lasers in Manufacturing and Materials Processing, 9*(3), 338–375. DOI: 10.1007/s40516-022-00182-6

Shamsdini, S., Shakerin, S., Hadadzadeh, A., Amirkhiz, B. S., & Mohammadi, M. (2020). A trade-off between powder layer thickness and mechanical properties in additively manufactured maraging steels. *Materials Science and Engineering A, 776*, 139041. DOI: 10.1016/j.msea.2020.139041

Suryawanshi, J., Prashanth, K. G., & Ramamurty, U. (2017). Tensile, fracture, and fatigue crack growth properties of a 3D printed maraging steel through selective laser melting. *Journal of Alloys and Compounds*, *725*, 355–364. DOI: 10.1016/j.jallcom.2017.07.177

Thijs, L., Kempen, K., Kruth, J.-P., & van Humbeeck, J. (2013). Fine-structured aluminium products with controllable texture by selective laser melting of pre-alloyed AlSi10Mg powder. *Acta Materialia*, *61*(5), 1809–1819. DOI: 10.1016/j.actamat.2012.11.052

Wang, D., Song, C., Yang, Y., & Bai, Y. (2016). Investigation of crystal growth mechanism during selective laser melting and mechanical property characterization of 316L stainless steel parts. *Materials & Design*, *100*, 291–299. DOI: 10.1016/j.matdes.2016.03.111

Wei, S., Kumar, P., Lau, K. B., Wuu, D., Liew, L. L., Wei, F., & Ramamurty, U. (2022). Effect of heat treatment on the microstructure and mechanical properties of 2.4 GPa grade maraging steel fabricated by laser powder bed fusion. *Additive Manufacturing*, *59*, 103190. DOI: 10.1016/j.addma.2022.103190

Yuan, P., & Gu, D. (2015). Molten pool behaviour and its physical mechanism during selective laser melting of TiC/AlSi10Mg nanocomposites: Simulation and experiments. *Journal of Physics. D, Applied Physics*, *48*(3), 35303. DOI: 10.1088/0022-3727/48/3/035303

Chapter 17
Trends and Developments in Additive Manufacturing Technologies

G. Boopathy
https://orcid.org/0000-0003-2515-7277
Vel Tech Rangarajan Dr. Sagunthala R&D Institute of Science and Technology, Chennai, India

S. Sathish
Aalim Muhammed Salegh College of Engineering, India

M. Vignesh Kumar
https://orcid.org/0000-0002-5606-106X
SRM Institute of Science and Technology, India

C. Suresh
https://orcid.org/0000-0001-7549-3506
Global Academy of Technology, India

ABSTRACT

Additive Manufacturing (AM), or 3D printing, is enabling material efficient, customizable, and adaptable design. It examines the progress from prototyping pioneering AM, its current multifaceted status as an industrial technology and its implications in multiple sectors such as aerospace, healthcare, automotive, and consumer goods. We study the principal AM techniques, including Direct Metal Laser Sintering (DMLS), Stereolithography (SLA), Fused Deposition Modeling (FDM) and Selective Laser Sintering (SLS), and materials comprising metals, polymers and

DOI: 10.4018/979-8-3373-0533-2.ch017

composites. End-to-end, the entire scope of the persistent challenges are covered, including cost, quality assurance, regulatory complexities, and intellectual property. In addition, the chapter imagines the future of AM, contemplating the incorporation of artificial intelligence (AI), the Internet of Things (IoT), bioprinting, and its potential to democratize manufacturing, for instance through localized production and disruptive practices.

INTRODUCTION

Additive manufacturing (AM) also known as three dimensional (3D) printing has vastly changed the manufacturing field by allowing the generation of elaborate geometries and modified products in a manner never thought possible before. In this chapter, we study the ancient history of AM, specifically highlighting how it is a key to many industries and disrupts the conventional manufacturing methods. In addition, it attempts to define major breakthroughs, technological innovations, and possible directions in additive manufacturing to offer a thorough summary for enthusiasts and decision makers.

Background and Significance

AM comes from stereolithography in the 1980s, which produced digital diagrams first, then assembled iteratively objects produced by the digital schematics. Through subsequent decades, AM technologies have continued to evolve into variety of methods such as fused deposition modelling (FDM), selective laser sintering (SLS), and more recently binder jetting and material jetting. The nature of AM is so significant within the manufacturing sector due to its outstanding design versatility, reduced material waste, and accelerated production time lines that AM can be an attractive alternative to traditional subtractive manufacturing techniques.

AM is far more relevant than mere productivity improvements. It has provided opportunity to create highly personalized, tailored products that meet that which the consumers demand and desire. AM technologies have been exploited in various industries such as the consumer goods, healthcare, automotive, aerospace and more to facilitate innovation as well as to improve and streamline manufacturing workflows in those industries. As a consequence, AM is not only changing the manufacturing paradigms, but is also enabling the emergence of new supply chain structures and business models.

Objective of the Chapter

The purpose of this chapter is to conduct a full assessment of the current landscape for AM, with a view to the future. Key objectives include:

- **Identifying Key Trends**: An historic and complete analysis of the current state of AM – including the impact of the emerging bio printing industry; new advancements in material science to better accommodate functionalization requirements; and the integration of Artificial Intelligence (AI) as a design and production methodology.
- **Technological Advancements**: It illuminates key technological breakthroughs that have dramatically improved AM capability such as accelerated printing speeds, multi material printing capability and new post processing techniques.
- **Future Developments**: logically discussing the future of additive manufacturing, emphasizing potential uses in emerging industries, the role of AM in sustainable manufacturing models, and the challenges that still exist with regard to standards, regulations, and market acceptance.

The chapter will offer important insights into how additive manufacturing is positioned to influence the manufacturing industry and beyond through this investigation.

Definition and Principles

In contrast to traditional manufacturing processes, which primarily entail the removal of material from a solid mass (such as machining and milling), additive manufacturing is a technique that creates objects by gradually adding material in layers. The fundamental principle of additive manufacturing involves building components from a digital model, usually created using computer-aided design (CAD) software. The development of complex geometries and designs that would be extremely difficult or impossible to achieve with conventional methods is made possible by this gradual layering technique.

TYPES OF ADDITIVE MANUFACTURING

Figure 1 lists the several kinds of additive manufacturing technology, each having distinct applications and guiding principles:

Figure 1. Types of Additive Manufacturing Processes

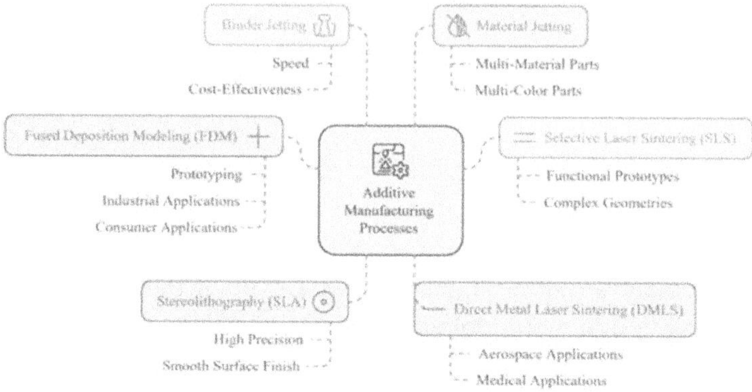

Fused Deposition Modeling (FDM):

FDM constitutes a predominant technique in additive manufacturing that fabricates three-dimensional entities by systematically layering thermoplastic substrates. The FDM process is initiated by the extrusion of a thermoplastic filament through a thermally regulated nozzle, which facilitates the melting of the material. The nozzle traverses a pre-defined trajectory, methodically depositing the liquefied filament in successive layers upon a build platform. Upon cooling, each layer undergoes solidification and establishes a bond with the preceding layer, thereby incrementally constructing the final product. As shown in Figure 2, the process typically consists of the following successive steps.

Figure 2. FDM process sequence

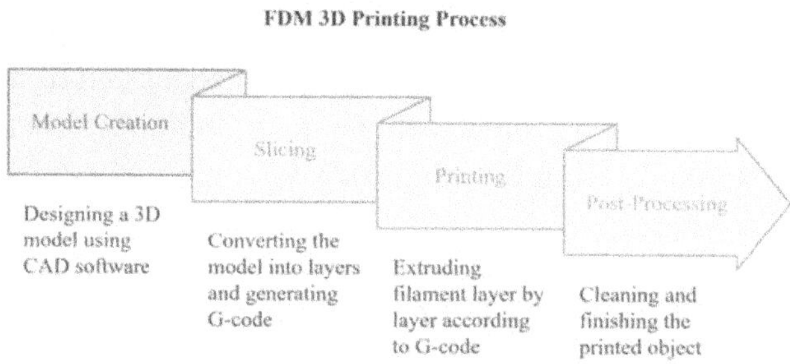

- **Model Creation**: CAD software is used to create a 3D model.

- **Slicing**: G-code instructions for the printer are created by utilizing slicing software to cut the model into tiny horizontal layers.
- **Printing**: The printer builds the object layer by layer by extruding the filament according to the G-code.
- **Post-Processing**: After printing is finished, the item could need to be cleaned, have its supports removed, or have its surface finished.

The geometrical accuracy and surface finish of FDM components are significantly influenced by variables such as print layer height, printing speed, and nozzle diameter. For instance, smaller nozzle sizes and lower printing speeds improve surface texture and flatness. Changes in characteristics like perpendicularity and parallelism depend on the materials chosen and the layer height, so particular adjustments catered to each application are required (Petruse et al.,2024). Technological developments could enhance impact energy by 28% and tensile strength by up to 40%, for example, by adding titanium dioxide-infused resin to printed structures (Eryildiz et al.,2024). By incorporating continuous fibers into the FDM process, the inherent flaws in traditional FDM components may be addressed and the mechanical integrity of printed elements can be significantly improved (Karimi et al., 2024). The development of digital material filaments makes it possible to create functionally graded materials, allowing for differences in mechanical and electrical characteristics inside a single printed object (Ahn et al., 2024).

FDM Process has the following advantages:

- **Cost-Effective**: When compared to other 3D printing technologies, FDM machines and materials are often less expensive.
- **Material Variety**: A large number of thermoplastic materials, such as PLA, ABS, and PETG, can be utilized; each has unique qualities for a range of applications.
- **Ease of Use**: Beginners and laypeople may easily use and maintain FDM printers due to their relative ease of use.
- **Scalability**: FDM may be used for low-volume manufacturing runs as well as prototyping.

Despite its advantages, FDM has some limitations:

- **Surface Finish**: A rough surface finish and noticeable layer lines may be the consequence of the layer-by-layer process.
- **Mechanical Properties**: Anisotropic qualities refer to the possibility of parts becoming weaker along specific axes.

- **Print Speed**: FDM could be slower than other 3D printing methods for complicated geometry.

Selective Laser Sintering (SLS):

SLS, a sophisticated additive manufacturing technique, combines powdered material into solidified constructs by laser. This technique is widely adopted by numerous sectors ranging from aerospace to automotive, to medical, due to its capability to fabricate complex geometries with high precision. SLS in particular stands out as having the capability to process a wide variety of material from metals, ceramics and polymers. Figure 3 shows the procedural phases that are part of SLS.

Figure 3. SLS Process Sequence

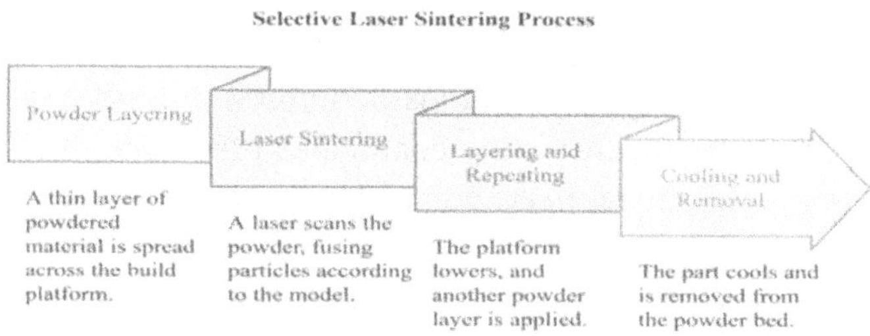

Selective sintering of powder particles is enabled by the application of laser technology to generate solid objects through thermal fusion of materials. This methodology distinguishes itself through the ability to manufacture components with intricate, substantial internal architectures and mechanical integrity, making it suitable for high demand applications. Understanding of material properties as well as processing parameters are reflected by current research endeavors to refine SLS techniques using computational modeling (Castro et al., 2024). Across industries as diverse as biomedical, automotive, aerospace, SLS plays a large role in the manufacture of complex tooling systems and patient specific implants (Park et al., 2024). Additionally, SLS leads to the formulation of zolpidem tartrate tablets with unique attributes (Adamov et al, 2023). New developments in printer design, such as open-source SLS systems, aim to make additive manufacturing more accessible and adaptable (Pierson et al., 2024).

- **Powder Layering**: The SLS process begins with a thin layer of powdered material that is carefully applied across the building platform.
- **Laser Sintering**: In compliance with the requirements outlined by the three-dimensional model, a strong laser beam carefully searches the powder layer's surface, melting and combining the particles as needed

- **Layering and Repeating**: The construction platform falls once a layer has been sintered, and a second layer of powder is placed. Until the thing is completely fabricated, this cycle is repeated.
- **Cooling and Removal**: After the printing process is finished, the part is allowed to cool before being removed from the powder bed. It is often possible to utilize unfused powder in future projects.

SLS Process has the following advantages:

- **Complex Geometries**: SLS may create intricate designs that are difficult or impossible to achieve using traditional manufacturing methods.
- **Material Efficiency**: Reusing and recycling unfused powder reduces waste production and material costs.
- **Functional Parts**: SLS-produced components have the potential to be significantly stronger and more durable, making them suitable for end-use applications.

The following are some applications of SLS:

- **Prototyping**: To effectively iterate on their design concepts, designers commonly utilize SLS for rapid prototyping.
- **Production Parts**: SLS is used by several sectors to fabricate low-volume production parts that require extraordinary performance and precision.
- **Medical Devices**: This technology is also used to create customized prosthetic devices and medical implants.

SLA is a major game changer in the world of three dimensional printing technologies, utilizing a laser to selectively harden drop by drop liquid resin into solidified layer by layer. Photo polymerization, on the other hand, is how SLA works, whereby a photo sensitive resin is thoroughly and selectively solidified by an ultraviolet (UV) laser. A reservoir of liquid resin is filled with liquid resin at the beginning of this process. The geometry of the object is established by the position of the laser beam in regards to the surface of the resin, defining its requisite cross sectional outline. The laser causes a phase transition upon contact with the resin which solidifies the liquid. After each layer is finished, the build platform is lowered just enough so that the layer above the layer is now free to wrap around the layer just under, continuing the cycle. Figure 4 shows the entirety of this methodology.

Figure 4. SLA Process Sequence

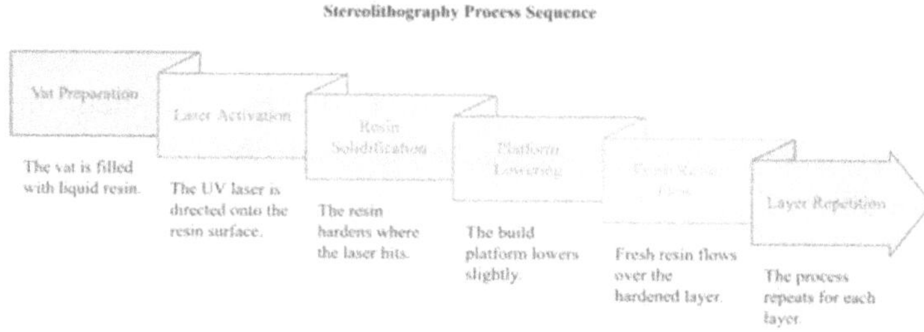

SLA is widely used within sectors including aerospace, medical instrumentation and automotive production due to its ability to produce complex geometries with high precision. However, it is also found that components fabricated through SLA may be successfully metallized in order to improve both its electrical conductivity and mechanical strength, and make it an appropriate candidate for structures in hostile environments, such as outer space (Miller et al., 2024). Their versatility for the fabrication of medical components, such as a blood pump, may allow SLA usage to mitigate risk for thrombus formation (Major et al., 2023). Augmenting mechanical properties of SLA manufactured items: post processing techniques such as UV curing and machining are critical UV treatment has been proven to raise the hardness level, for instance, by more than 200%. In addition, pre-curing prior to machining has been demonstrated to increase the surface quality but simultaneously increase brittleness, potentially compromising the durability of the final product (Kim et al., 2024). When added to SLA resins, nanomaterials like graphene oxide have been shown to increase mechanical performance under a variety of stress scenarios (Garcia et al., 2024).

SLA process the following advantages:

- **High Precision**: The widespread acceptance of the SLA capability to make incredibly intricate and detailed patterns with fine surface finish is understood.
- **Complex Geometries**: The technology itself has the ability to create elaborate geometrical configurations that, if not impossible, would be a challenge to produce using conventional fabrication techniques.
- **Material Variety**: Among these materials are a wide variety of resins with specific properties, such as flexibility, toughness, and thermal resistance.

SLA is widely used in various industries, including:

- **Prototyping**: This methodology provides designers with rapid prototyping for product development providing the ability to generate and evaluate models efficiently.

- **Dental and Medical**: Used in the fabrication of dental molds, surgical guides and custom prosthetics.
- **Jewelry**: This is a technology necessary to prepare intricate jewellery designs and mold that will cast on.
- **Aerospace and Automotive**: The material is used in the fabrication of lightweight and complex components for both test and production.

Despite its advantages, SLA has some limitations:

- **Material Cost**: Resins used in SLA more expensive than other 3D printing materials.
- **Post-Processing**: Finished components often require additional post processing such as cleaning and curing and need to optimise both strength and surface finish to get the best performance.
- **Build Size**: It is different from other 3D printing techniques since SLA printers typically operate with a reduced build volume limitating product size.

Direct Metal Laser Sintering (DMLS):

The DMLS is a very advanced kind of AM process that uses a very high energy that the laser fuses metallic specialized powder to the solid cohesive structures. It is most famous for being able to create very intricate forms, very complex geometries, which are usually not imaginable using traditional approaches to fabrication. The sequential processes taken in order in Figure 5 shown the DMLS had taken place in. DMLS applications are extensive in precision and performance demanding industries such as aerospace, automotive and medical.

Figure 5. SLA Process Sequence

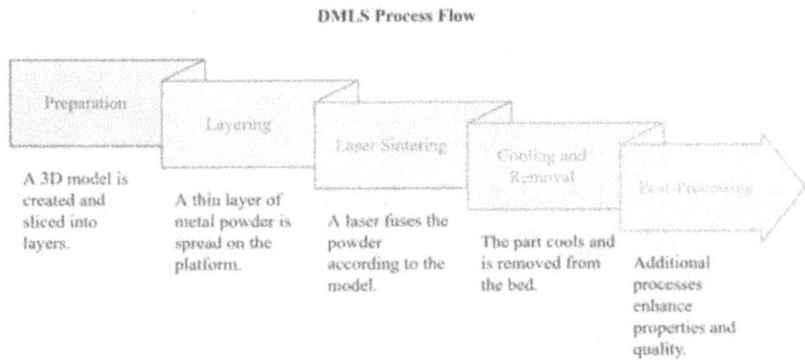

DMLS is a method of constructing solid components from metallic powders, layer by layer, incrementally, through the process of selective amalgamation of the metallic powders with a high intensity laser. This intricate process is governed by the parameter of laser power, scanning velocity, hatch spacing and layer thickness. Due to its application in fabricating components for automotive, medical implant and aerospace sectors where precision and material efficiency is crucial, DMLS has found extensive application (Nancharaiah et al., 2023) (Boopathy et al., 2024). The methodology enables creation of complex geometries that are rarely made use of by conventional manufacturing techniques However, negative consequences of these defects, such as residual stresses and micro porosity, can lead to poor mechanical properties of the resultant product from DMLS (Tebianian et al., 2023). Here, DMLS has a lower sustainability index than alternative methods such as Additive Friction Stir Deposition (AFSD), and there are increasing apprehensions about sustainability (Ahmed et al., 2024).

The key features of DMLS are as follows:

- **Material Versatility**: DMLS's capacity to process a range of metal powders, including titanium, aluminum, stainless steel, and cobalt-chrome alloys, enables a broad range of applications.
- **Complex Geometries**: DMLS's layering technique makes it possible to create intricate, lightweight structures with lattice patterns and interior channels.
- **Rapid Prototyping**: DMLS significantly reduces lead times for both production and prototyping, allowing for quicker iterations and a shorter time to market.
- **Customization**: Since this cutting-edge technology allows for component customisation, it is especially well-suited for applications requiring for personalized solutions, such medical implants.

Some Advantages of DMLS Process are:

- **Design Freedom**: Complex geometries that are impossible to fabricate using traditional manufacturing methods can be made possible by the DMLS technology.
- **Material Efficiency**: Since any extra powder can be used for other industrial projects, the production process produces very little waste.
- **High Strength-to-Weight Ratio**: When compared to components made using conventional techniques, DMLS-fabricated components usually exhibit improved mechanical properties.

Here are the some challenges of DMLS process:

- **Cost**: The initial investment required for DMLS equipment and related supplies can be high, making it less accessible to smaller businesses.

- **Surface Finish**: In order to achieve the necessary surface integrity, components may require additional finishing techniques.
- **Thermal Distortion**: The rapid heat cycling involved in the process may cause residual strains or distortion in the finished part.

Binder Jetting (BJ):

Along with many other 3D Additive Manufacturing techniques, Binder Jetting is a state of the art technique for fabricating industry parts on an individual basis. This process, however, requires the precise deposition of a liquid binder onto a substrate consisting of powdered materials in order to create the intricate geometrical configurations and assemblies. Figure 6 shows the comprehensive stages of the binder jetting procedure.

Figure 6. Process Sequence of Binder Jetting

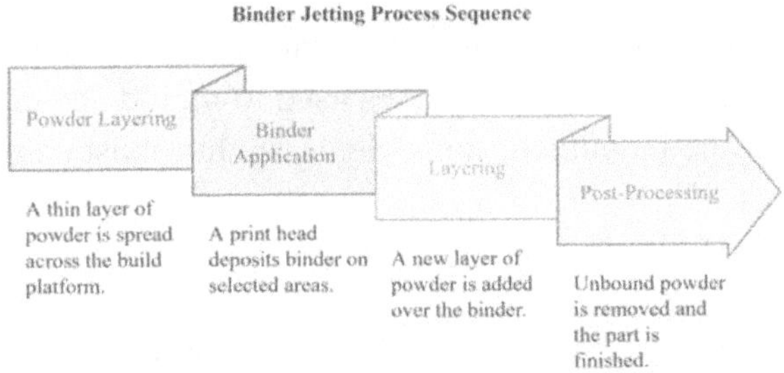

High speed imaging investigation reveals the interaction between powder particles and binder droplets as fundamental for the quality of the manufactured part. Power ejection and powder bed density downstream are critical in justifying production defects and are significantly affected by factors like moisture content. BJ components may be operating inefficiently because of defects (Lawrence et al., 2024). Research emphasizes the importance of real time fault diagnosis and subsequently the corrective measures to improve part quality. Investigations of microstructures and defects especially in metallic materials of AISI M2 tool steel have employed advanced imaging methodologies (Choudhari et al., 2024). Utilizing bimodal powder compositions for silicon carbide ceramics has recently been the topic of much investigation. The density and strength achieved from this strategy hold promise for the use of tailored material compositions in BJ processes (Feng et al., 2024).

The procedural overview of BJ is outlined as follows:

- **Powder Layering**: To begin, the build platform is accurately coated by layer and layer of powder material.

- **Binder Application**: The print head traverses the powder bed with a selected liquid binder dispensed to targeted regions of the powder bed, for which the component is being fabricated. In this binder the adhesive holds the powder particles together.
- **Layering**: After the binder, another layer of powder is superposed on top of the one just applied and the whole procedure is repeated. It keeps repeating until the whole of the component is constructed.
- **Post-Processing**: After printing, the powder is collected from the unbound part, which may then be sintered or infiltrated to increase its mechanical properties.

BJ can employ an extensive range of materials, covering:

- **Metals**: Stainless steel, aluminum and bronze can be included but not limited to.
- **Ceramics**: Used for applications that demand elevated temperature resilience.
- **Sand**: Often used in casting processes used to mould and core.
- **Polymers**: For the manufacturing of lightweight robust components.

Some Advantages of binder jetting technique are:

- **Complex Geometries**: The Binder Jetting process makes it possible to fabricate intricate designs that are often impossible to achieve with other manufacturing techniques.
- **Material Efficiency**: Since only the necessary amount of binder is used, this method produces very little waste.
- **Speed**: Compared to other additive manufacturing techniques, binder jetting may produce components relatively quickly.
- **Cost-Effective**: In general, this method is more economically feasible for producing components in large quantities, especially for applications involving metal and ceramics.

Binder Jetting is widely used across various industries, including:

- **Aerospace**: For producing lightweight components with complex geometries.
- **Automotive**: In the creation of prototypes and production parts.
- **Medical**: For custom implants and surgical tools.
- **Foundry**: To create molds and cores for casting processes.

Material Jetting:

Material Jetting (MJ) is an advanced additive manufacturing process that creates three-dimensional objects by sequentially depositing materials. The ability to produce high-resolution components with complex geometries and precise characteristics sets this technology apart. MJ, which involves ejecting material droplets onto a substrate and then curing or solidifying them to produce the finished product, is commonly compared to inkjet printing. Figure 7 shows the steps that are part of the Material Jetting process.

Figure 7. Process Sequence of MJ Technique

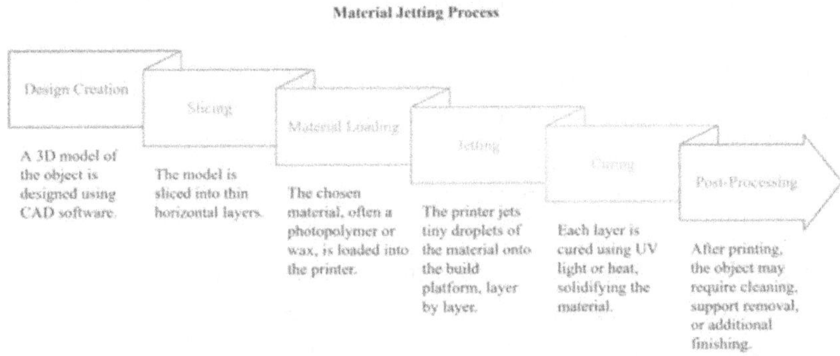

The pre-processing workflow, which includes the choice of CAD software and slicing methods, has a significant impact on the accuracy of MJ. While rasterization is essential to the authenticity of smaller parts, empirical research indicates that the repetition technique used in CAD systems affects the accuracy of larger elements (Karin et al., 2024). The mechanical qualities of manufactured components can be improved by MJ. For example, the use of Metal Material Jetting (MMJ) with sub-micron powders has produced SS316L samples with higher tensile strength than those made using traditional techniques, which is due to the microstructures being refined (Ibrahim et al., 2024). The Slurry-based Photoinitiator Jetting Process (SPJ), which makes it easier to produce high-density ceramic components with low porosity, is proof that MJ is equally useful in the ceramic manufacturing industry (Jia et al., 2024). Moreover, the integrated production of low-temperature co-fired ceramics demonstrates MJ's capacity to manufacture reliable electronic components (Liang et al., 2023).

The primary materials utilized in MJ process are photopolymers, or substances that change their properties in response to light. Typical materials consist of:

- **Photopolymers**: These are used to make intricate models and prototypes.
- **Waxes**: Frequently utilized in patterns for investment casting.

- **Composites**: For improved qualities, certain MJ systems can jet composite materials.

Material Jetting offers several advantages, including:

- **High Resolution**: The capacity to produce fine details and produce smooth surface treatments.
- **Multi-Material Printing**: By combining various materials into a single print, it makes possible to create intricate assemblies and color variations.
- **Rapid Prototyping**: ncourages quicker turnaround times for the creation of models and prototypes.
- **Minimal Waste**: Compared to traditional subtractive manufacturing methods, the layer-by-layer approach greatly reduces material waste.

Material Jetting is utilized across various industries, including:

- **Aerospace**: For producing lightweight components and prototypes.
- **Automotive**: Used for rapid prototyping and tooling.
- **Healthcare**: Creating custom implants, dental models, and surgical guides.
- **Consumer Products**: creating complex designs for home goods and electronics.

MATERIALS USED IN AM TECHNIQUES

A diverse array of materials is employed in additive manufacturing, each meticulously engineered for a specific application:

- **Plastics:** ABS, PLA and nylon are broadly used as thermoplastic materials in the framework of SLA and FDM methodologies. These substances are cheap, adaptable and user friendly which makes them esteemed.
- **Metals:** DMLS and SLS techniques utilize alloys includes titanium, aluminum, and stainless steel. As applications requiring improved strength and resilience are moving into the automotive and aerospace sectors, these materials are critical.
- **Ceramics:** Ceramic materials can be added in the additive manufacturing methodology to produce those components with exceptional mechanical characteristics and superior thermal resistance. High performance components are down for notable applications in many industries and dental prosthetics.

- **Composites:** Additively manufactured composite materials, which combine one or more dissimilar substances, are being employed to further improve attributes such as tensile strength, mass efficiency and thermal stability. Consequently, the automotive and aerospace domains are of great interest to these materials.

Polymers are commonly employed because of their adaptability and processing simplicity. They are frequently employed in FDM and SLA procedures. Ceramics are popular because of their remarkable resistance to wear and stability at high temperatures, especially in situations where durability is required. The development of smart materials that can react to environmental factors has been made possible by advancements in AM (Choudhary et al., 2024). Particularly in cryogenic and aeronautical applications, metals like copper and Inconel 718 are valued for their mechanical qualities and resistance to heat (Rehme et al., 2024). Copper and polyethylene, for instance, combine to generate a composite material that enhances performance and allows for tailored features in functionally graded constructions (Prakash et al., 2024).

CURRENT TRENDS IN ADDITIVE MANUFACTURING

The current state of AM is examined in detail in this section, with particular attention to the integration of AM in different industries and its wide range of uses; the shift to mass customization, environmentally friendly methods, and the combination of AM with other advanced technologies are all thoroughly explained in the chart as shown in Figure 8.

Figure 8. Indusrty Adoptation of AM

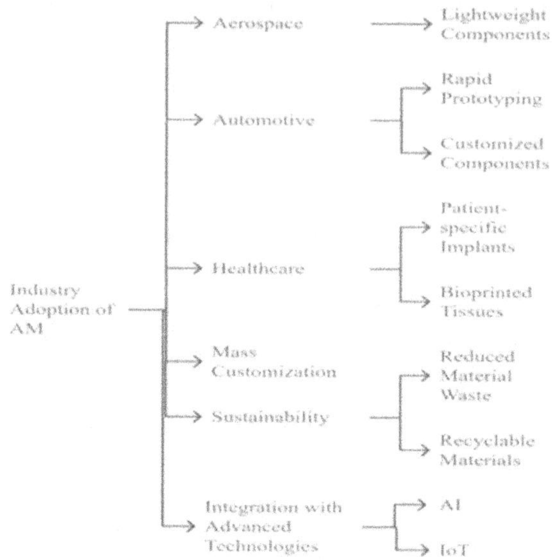

Industry Adoption and Applications

Rapid integration of AM is occurring in a variety of industries, including healthcare, automotive, and aerospace. In the aerospace industry, businesses are using AM techniques to create lightweight parts that increase fuel efficiency and reduce emissions at the same time. For instance, 3D printing is being used by Boeing and Airbus to create complex parts that would be difficult or impossible to fabricate using traditional methods.

Manufacturers in the automotive industry are using AM to fabricate end-use components and speed prototypes. Companies like Ford and BMW are looking into AM techniques for the creation of bespoke parts since it speeds up design iterations and reduces time-to-market. Significant advancements are also being seen in the healthcare industry, where AM is being used to create prosthetics, implants, and even bio printed tissues tailored to each patient, proving the technology's potential to improve patient outcomes.

Customized implants made of materials like titanium and cobalt chromium can be made through AM, which improves their mechanical properties and biocompatibility. Significant barriers to the broad use in medical applications still exist, including issues like high Young's modulus and corrosion resistance (Aufa et al., 2024). AM is gradually being used in the automobile sector to improve design flexibility, reduce waste, and streamline production procedures. To meet certain performance requirements in automotive components, a wide range of materials, including metals and polymers, are being researched for AM processes (Yang et al., 2024).

Customization and Mass Production

One of the most important AM methodologies is the transition towards mass customization. While conventional manufacturing relies heavily on a standard set of products, additive manufacturing allows products to be made more to fit the customer specifications. Given its capability, this is particularly valuable in sectors where demand for custom offers is also on the rise, for example, in consumer goods and fashion.

In addition, AM is developing to allow for broader production. Innovative advanced 3D printing technologies that produce significant amounts of complex components are being developed by the enterprises. This mass production driven shift not only increases the economic feasibility for AM but also opens up new frontiers for manufacturers to serve a multitude of consumer demands at a speed and with quality not feasible in prior technologies.

AM greatly improves customer satisfaction and entire journey experience through ease of making customized items like orthotic devices and prostheses. With advancing technological innovations, clients are integrated in the design process, which improves loyalty and can also heighten profit margins (Munoz et al., 2013). AM can reduce material consumption by 35 to 80%, and given that, it is a feasible, on demand, localized production process that can mitigate waste and transport needs (Jung et al., 2023). By combining AM with conventional mass customisation methods, production strategies can be improved, allowing firms to manage capacity constraints and satisfy a range of client demands (Lacroix et al., 2023).

Sustainability and Eco-friendly Practices

In the field of manufacturing, sustainability has become a key concern, with additive manufacturing leading the way in the development of environmentally friendly practices. By building things layer by layer, AM significantly reduces material waste when compared to traditional subtractive manufacturing methods. Additionally, several AM techniques use a recyclable material, which lessens the environmental impact even more.

In keeping with the goals of global sustainability, organizations are looking more and more into using bio-based and biodegradable materials in their AM processes. Manufacturers who use AM have the potential to reduce their carbon emissions and contribute to a circular economy, which makes it a desirable option for businesses, which value environmental responsibility.

Compared to traditional methods, AM techniques like FDM and SLS have shown a decrease in material waste. Enhancing sustainability in AM requires the application of circular economy ideas. This includes both material recycling and disassembly design. The importance of creating biodegradable materials and improving recycling techniques to lessen environmental effects is highlighted by research (Suresh et al., 2024). Cost savings and competitive benefits can be obtained through sustainable practices in metal additive manufacturing, underscoring the need for both environmental and economic assessments. Integrating social responsibility into AM procedures encourages moral production methods and industry inclusion (Baskaran, 2024).

Integration with Other Technologies

Combining Additive Manufacturing with other innovative technologies is a significant trend that is shaping the direction of manufacturing techniques. Manufacturing operations' efficiency and functional capabilities are being greatly increased by the convergence of AM with artificial intelligence (AI), the Internet of Things (IoT), and other Industry 4.0 technologies.

While IoT devices allow for real-time observation and data collection throughout the AM process, AI algorithms are being used more and more to improve the optimization of design and production workflows. Better supply chain management, strengthened quality control systems, and predictive maintenance are made possible by this connection. The mutually beneficial interaction between AM and Industry 4.0 will promote innovation and create new business models in the manufacturing industry as both technologies advance.

FDM is applied for device fabrication and the robotic integration of preassembled components is employed as the discourse of 3D printing synthesis with automated assembly. It addresses contemporary trends seeking to enhance additive manufacturing capability with automation and multi material integration (Munguia Galeano et al 2023). 3D modeling software boosts accuracy, efficiency through AI, it greatly boosts the accuracy as well as the efficiency of the design process in enhancing 3D modeling software (Tarasov et al. 2024). Further, with IoT technologies the advent of real time monitoring of AM processes increases efficiency and control. IOT applications in the AM are enabled through the utilization of smart materials that can change to accommodate environmental changes (Chigilipalli et al., 2023). IoT also, facilitates production data collection and analysis, and thus aids in informed operational decision making (Rath et al., 2024).

TECHNOLOGICAL DEVELOPMENTS IN ADDITIVE MANUFACTURING

The most recent developments in AM are examined in this section, with particular attention paid to automation, enhanced material qualities, and the incorporation of AI into software and machines. These developments are essential for improving manufacturing processes' performance, effectiveness, and affordability as AM develops.

Advancements in Hardware and Software

New developments in AM equipment and software are significantly advancing the field. Currently, new 3D printers with increased accuracy, speed, and versatility are being developed to enable a wider range of applications. For example, developments in multi-material printing enable the creation of complex geometries with a variety of material qualities in a single build. Furthermore, software solutions are becoming more complicated, using advanced algorithms for improved simulation and design optimization. These tools enable engineers to predict how materials will behave in different scenarios, leading to more reliable and effective designs.

The discussion focuses on the on-going developments in AM technology, namely the increased use of laser systems as the main energy source, which greatly increases fabrication accuracy and streamlines automated procedures. It highlights the development of software and hardware innovations that have improved design capabilities in a variety of industrial sectors (Arunprasath et al., 2022). Energy deposition and powder-bed fusion are two of the many processing approaches that have helped AM to develop. Important developments include the creation of sophisticated parts, more accessible materials, and reasonably priced machinery that enable complicated designs and optimize production efficiency without the limitations of conventional tooling methods (Agrawal et al., 2021).

Improvement in Material Properties

The focus of current scholarly effort is on material property enhancement for use in AM. Investigators are looking at novel alloys and composite materials with an improved strength, durability, and thermal resistance. Metal powders and high performance polymers enable the manufacture of parts that can withstand the most extreme operating conditions, for example. In addition, the integration of bio based materials is catalyzing the promotion of sustainable manufacturing methodologies.

These improvements make it applicable to a range of applications, spanning from medical devices to aerospace applications, and increase AM product performance.

Recent progress in functional materials, such as magnetic and piezoelectric substances, makes it possible to fabricate multifunctional parts. These materials are merged to develop more functional products and facilitate the manufacturing process of products with complex geometrical configurations (Kim et al., 2024) (Boopathy et al.,2024). Al-Ce-Mg alloys have shown significant improvements in mechanical properties caused by solid-state AM techniques including additive friction stir deposition, which have been shown to increase yield strength by 22% and ductility by 181%. The processes of particle fragmentation and grain refinement that take place during manufacture are responsible for the increased mechanical performance (Gumaste et al., 2023).

Automation and AI in Additive Manufacturing

As automation and AI are being introduced into AM methodologies, these methods are evolving. Automated technologies are integrating with production workflows, removing the errors or eliminating human intervention. To make the manufacturing process consistent quality and prevent waste, the manufacturing process itself is supervised in real time by AI algorithms that adjust print settings to make changes. The same technology helps with the boosting of productivity and predictive maintenance of machinery, which reduces the cost. Therefore, AM becomes a more viable option for large scale production since manufacturers can obtain greater volumes and better resource management.

Design for Additive Manufacturing (DfAM) has been automated using AI networks such as convolutional auto encoders and generative adversarial networks. This includes impressive accuracy at predicting optimal building orientations 96% that concurrently avoids the need for support structures and decreases stresses by 50% (Mohseni et al., 2024). This AI enhanced three dimensional modeling software offers a platform for raise in design capabilities and turn manufacturing processes more efficient and productive (Tarasov et al, 2024). In addition, the power of machine learning allows engineers to have a more detailed knowledge of complicated AM processes, thus enabling engineers to develop new ways and perfect model designs (Zhou et al., 2024). AI also helps maintain real time monitoring and adjusting of new generations of printing techniques, such as 4D and 5D, with potential to largely shorten manufacturing time. (Yadav et al., 2024).

CHALLENGES AND LIMITATIONS OF ADDITIVE MANUFACTURING

Although AM offers unmatched flexibility and personalization, it has completely changed the methods used in product design and production. However, despite all of its benefits, AM faces a number of obstacles and restrictions that prevent its adoption. In the context of additive manufacturing, this section explores the technical difficulties, legal and certification barriers, and security and intellectual property issues were shown in Figure 9 in the flow chart format.

Figure 9. Challenges in AM Processes

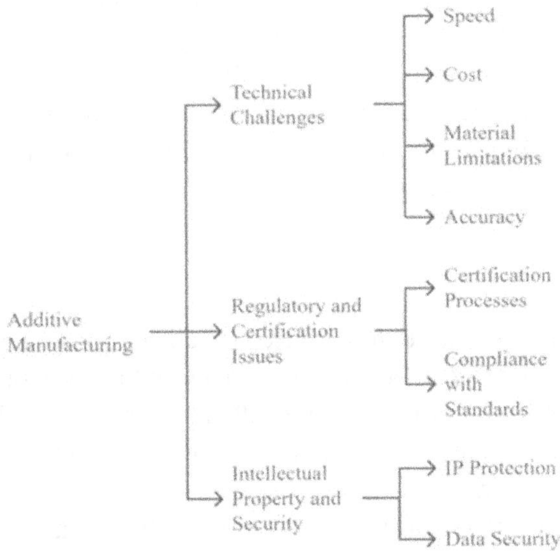

Technical Challenges

Speed

The issue of speed is one of the main technical challenges in the area of AM. It demonstrates the ability to create complex geometries beyond the scope of what other manufacturing processes have traditionally achieved, but at an often extended printing duration. One of the particularly important aspects of this challenge is that we are talking about sectors that require quick turnaround cycles and rapid prototyping.

Cost

The financial aspects seem to be a big barrier to widespread adoption of AM. It has the potential to reduce material waste and enable construction of complex geometries, but the prohibitive initial costs to purchase high fidelity printers and necessary materials can be significant. Additionally, the amount of ongoing operational costs associated to maintaining and utilizing the AM equipment can be accumulated to the point that it becomes less financially viable for some applications.

Material Limitations

One more obstacle in this domain is material constraints. Though the spectrum of materials suitable for AM is expanding, many of the AM techniques remain constrained to one class of materials that, while useful for some applications, may not meet their performance requirements. However, such limitations could complicate the taking of AM to sectors such as aerospace and automotive, for which high performance materials are required.

Accuracy

The precise levels of AM systems vary greatly depending on the specific processes and materials employed. It can be difficult to achieve the necessary tolerances and surface finishes, particularly when dealing with intricate designs. For industries that require accuracy, this discrepancy could lead to differences in the finished product, which is a serious concern.

In addition to their high melting points, materials like tungsten cause challenges for AM because of increased heat stress and cracking sensitivity (Zarinejad et al., 2024). Both the printing process and the end product's quality are greatly impacted by the properties and quality of the powders used, especially the distribution of particle sizes. The range of possible uses is limited by the restricted compatibility of particular materials with current AM technology (Choma et al., 2024).

Regulatory and Certification Issues

In industries like aircraft and healthcare, it might be difficult to certify parts made with AM. Rules often place strict restrictions on the materials and manufacturing processes, and the speed at which AM technologies are developing could exceed the ability to develop appropriate rules.

Certification Processes

Certification processes for AM components can be drawn out and complex, requiring extensive testing and validation to ensure reliability and safety. Such procedures have the potential to increase manufacturing costs and delay the release of new items onto the market. Furthermore, the certification process is made more difficult by the lack of standardized testing procedures for AM components.

Compliance with Standards

Maintaining industry standards is crucial, especially for applications that are essential. However, producers may become concerned as a result of the absence of widely accepted standards for additive manufacturing. This circumstance could encourage reluctance to adopt AM technology since businesses may be concerned with possible non-compliance with legal requirements.

To standardize AM procedures for parts designed for aircraft use, the aviation sector requires the creation of a certification roadmap. To ensure the safety and dependability of parts made by AM, national and international airworthiness authorities have implemented particular standards (Saracyakupoglu,2022). Within the parameters of current regulatory pathways, the Food and Drug Administration (FDA) has approved the use of 3D printed medications, such as Spritam®, highlighting that manufacturing processes comply with identification, strength, and quality criteria. Innovative manufacturing technologies should be incorporated, according to the FDA's Emerging technologies Program (Rahman et al., 2023). A strong regulatory framework must ensure safety and effectiveness as the healthcare industry has adopted AM for the creation of customized medical equipment. For certain compliance to established safety protocols, standards are being developed in various geographical locations to fill in the regulatory gaps related to AM in medical equipment (Rafi et al., 2022).

Intellectual Property and Security

Concerns regarding data security and intellectual property (IP) protection have also been raised by the growth of additive manufacturing.

IP Protection

The ease with which designs can be copied is the AM difficulty. It may be hard to stop businesses from illegally copying their patented ideas. This problem arises in industries where maintaining a competitive edge primarily rely on constant innovation.

Data Security

Data security becomes still another important concern since digital files used in AM processes are subject to attacks. Unauthorized access to these data could result in the theft of valuable designs and trade secrets. For businesses hoping to use AM technology without compromising their intellectual property, protecting digital assets is essential.

There is evidence that security flaws in AM systems can intercept and edit 3D models while they are being transmitted, leading to illegal changes and possible intellectual property theft. Attackers could alter printed models' internal configurations, lowering their quality and operation and putting safety-critical components at serious risk (Alkofahi et al., 2024). By using separated monitoring systems, components' quality can be verified without exposing them to network-related vulnerabilities, safeguarding both intellectual property and manufacturing integrity (Raeker-Jordan et al., 2024).

CASE STUDIES AND REAL-WORLD APPLICATIONS OF ADDITIVE MANUFACTURING

This section analyses the significant revolutionary implications of AM in a variety of industries. We focus on the application of AM technologies as mentioned in Figure 10 fields like as aerospace, healthcare, automotive, and consumer goods by analysing specific case studies. Furthermore, we look at exemplary success stories that explain the benefits and challenges that organisations face, providing important insights that can be used to influence future AM initiatives.

Figure 10. Application of AM Technologies

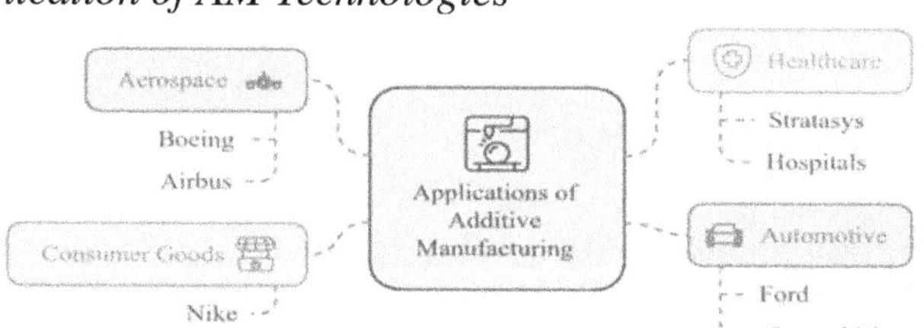

Industry-specific Case Studies

Aerospace

AM has been used by aerospace corporations such as Boeing and Airbus to produce lightweight parts that great reduce fuel consumption. Boeing, for instance, has used 3D printing for making components in the 787 Dreamliner, thereby drastically reducing money and time costs. Innovative designs to augment overall performance have been enabled through the ability to produce complex geometries beyond the capabilities of conventional manufacturing.

Healthcare

AM has made a lot of progress in the healthcare industry, especially for developing tailor made prosthetics and implants. Medical institutions use 3D printing to create patient specific models for surgery. Not only does this practice help pre-operatively plan, it also lends to a more precise procedural outcome. Partnering medical enterprises, organizations such as Stratasys have created biocompatible materials for 3D printed implants to reduce the chances of infections and assist in patient recovery.

Automotive

AM is used in the automotive sector by manufacturers like Ford and General Motors to improve component variability and reduce prototyping time, as well as to replace damaged parts. Through the use of 3D printing technology, Ford is using 3D printing to fabricate tools and fixtures in order to decrease lead time and cost related to traditional manufacturing practices. Besides, the capability to generate low volume components on demand further increased flexibility in production lines, thus providing a way to fast adapt the production to the market variation.

Consumer Goods

AM is increasingly being recognised as a production process that has the potential to support the production of customized consumer goods. For instance, Nike has taken 3D printing to personalize its footwear to meet the millimetres of consumer's feet. In addition to making customer happy, this approach mitigates waste by manufacturing items as demand is required. AM integration in this field is an excellent example of enterprises can pursue innovation simultaneously as they put sustainability as a priority.

Success Stories and Lessons Learned

Implementation of AM in GE Aviation jet engine components has been revolutionary. To make fuel nozzles, GE turned to 3D printing, combining 20 individual component parts into one and reducing weight and inefficiency in the engine. This is an example of how AM technology can have a positive impact and how an organization has to invest in them to foster organizational culture that encourages innovation.

- **Collaboration is Essential:** Implementing a successful AM process typically requires the cooperation of a broad range of stakeholders, including engineers, designers, and suppliers. In all likelihood more efficacious solutions can be created through engaging all reasonable relevant parties from the outset.
- **Material Selection is Crucial:** Above all it depends on the right choice of the material for a particular purpose. To make sure that materials meet performance and regulatory standards, organizations must spend resources on research and development.
- **Iterative Design Methodology:** The advantage here is that AM allows enterprises to rapidly prototype their design concepts. An iterative methodology, if adopted, may result in better products and shorter time to market.
- **Scalability Considerations:** AM appreciates specific advantages in respect of low volume production, however, its conversion to mass production may also require some considerations. Firms must carefully consider their production requirements, and consider hybrid manufacturing methods.

FUTURE TRENDS AND DEVELOPMENTS IN ADDITIVE MANUFACTURING

This section analyses the future trajectories and breakthroughs in AM, with a focus on next-generation materials, potentially transformational technologies, and AM's global implications. As the industry evolves, the introduction of innovative materials and technologies has the potential to radically alter manufacturing techniques, enhance product functions, and reshape global marketplaces. This study aims to provide insights into how these advances may influence the future configuration of AM.

Next-Generation Materials

The growth of additive manufacturing requires the development of novel materials. Figure 11 illustrates some possible future developments.

Figure 11. Novel Materials in AM

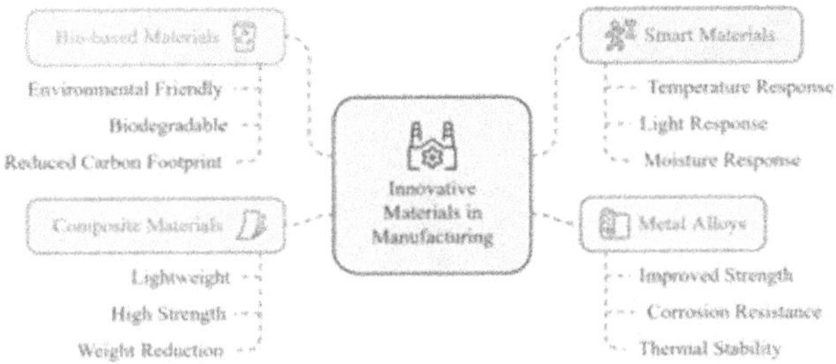

- **Bio-based Materials**: The rise of sustainable practices is driving the development of bio-based materials that are both ecologically beneficial and biodegradable. These materials may drastically lower the carbon footprint of industrial operations.
- **Smart Materials**: Advances in smart materials that can respond to environmental stimuli (such as temperature, light, or moisture) could pave the way for applications in fields such as healthcare, where implants can adjust to the physiological needs of the body.
- **Metal Alloys**: The development of innovative metal alloys designed specifically for AM could improve mechanical properties and overall performance. These alloys may have better strength, corrosion resistance, and thermal stability, making them suitable for aerospace and automotive applications.
- **Composite Materials**: Lightweight and robust components may be produced by combining complex composites, such as fibers or nanoparticles with polymers. This would be particularly helpful in sectors where weight reduction is essential.

Potential Disruptive Technologies

The flow chart in Figure 12 highlights a number of innovative technologies that have the potential to completely transform the AM sector.

Figure 12. Modern Technologies in AM

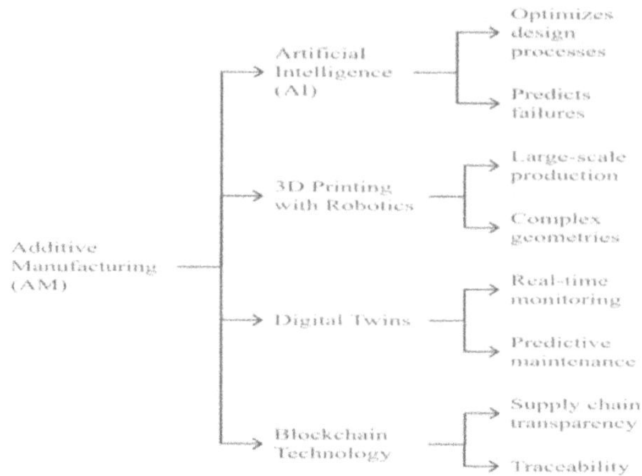

- **Artificial Intelligence (AI)**: AM can benefit not only in quality assurance, but also design refinement and material selection optimization using AI. Machine learning algorithms have the capability of predicting problems using machine learning algorithms and recommending changes in real time and ways that could lead to more efficient manufacturing methodologies.
- **3D Printing with Robotics**: The combination of robotics and 3D printing offers a way to realize intricate geometries and large scale manufacture, which are currently difficult to achieve. If materials were to be handled robotically, they could be arranged in a manner that is not possible with conventional printers.
- **Digital Twins**: Using digital twins in AM could allow for real-time manufacturing process simulation and monitoring. This technological development has the potential to improve overall operating efficiency and predictive maintenance.
- **Blockchain Technology**: Supply chain traceability and transparency may be strengthened by integrating blockchain technology into AM. This would be particularly important for industries where quality assurance is crucial, like aerospace and healthcare.

Global Impact and Future Outlook

The global consequences of additive manufacturing are expected to be enormous, with numerous key trends and opportunities arising:

- **Decentralized Manufacturing**: AM enables localised production, reducing the need for vast supply chains. This could result in more resilient manufacturing systems, especially during times of global instability.
- **Customization and Personalization**: The ability to produce customised products on demand is likely to shift consumer expectations. Fashion, healthcare, and consumer goods industries may shift to more personalised offers.
- **Sustainability**: As industries increasingly focus on sustainability, AM's ability to reduce waste and energy consumption will provide a significant benefit. The development of environmentally friendly materials and methods will match with global sustainability goals.
- **Job Creation and Skills Development**: While additive manufacturing may threaten traditional manufacturing jobs, it will also provide new opportunities in design, engineering, and materials research. To meet the demands of this rapidly changing business, the staff's skill set must be expanded.

SUMMARY

Throughout the chapter, we examined several major trends and advancements in AM techniques. Key points include:

- **Technological Advancements**: The rapid advancement of 3D printing technology has resulted in improved materials, faster production speeds, and greater precision, making additive manufacturing more practicable across a wide range of industries.
- **Industry Integration:** AM is increasingly integrated into production workflows of sectors that recognize their potential, not only in customization but also in operational efficiency, such as aerospace, automotive and healthcare.
- **Sustainability:** AM also encourages a more sustainable approach as it eliminates waste allowing for the use of recycled materials and promotes a more environmentally sustainable manufacturing effort in accordance with global efforts focused on greener manufacturing.
- **Regulatory Advancements:** Quality and safety standards and guidelines within additive manufacturing applications are starting to be developed by regulatory agencies as technological progress continues.
- **Economic Implications:** The common thread that has arisen from industries realizing significant material cost savings and cutting production timelines is formalizing AM's economic viability leading to a reorganization of traditional manufacturing processes.

AM is dignified for a highly promising future in the long term. As the technology advances, one can be sure of even greater integration into conventional production processes. The ability to fabricate complex geometries and customized products as required will probably restructure supply chain dynamics and stock management, facilitating more flexible manufacturing systems.

Additionally, the continued focus on sustainability will lead to further development of materials and processes, elevating AM as a critical enabler to manufacturing practices of more environmentally friendly processes. As industries adapt to these transformations, AM is expected to play a role expanding to the point that it becomes capable of fundamentally altering both the manufacturing process as well as design and distribution practices.

REFERENCES

Adamov, I., Stanojević, G., Kočović, D., Mugoša, S., Grujić, B., & Ibrić, S. (2023). Selective laser sintering (SLS) 3D printing process: Influence of model design on the properties of zolpidem tartrate tablets. *Macedonian Pharmaceutical Bulletin*, *69*(3, Suppl 1), 133–134. DOI: 10.33320/maced.pharm.bull.2023.69.03.065

Agrawal, M. K. (2021). Additive manufacturing: Advances in trends and technology. *Turkish Journal of Computer and Mathematics Education*, *12*(1S), 452–458. DOI: 10.17762/turcomat.v12i1S.1903

Ahmed, A. A., Nazzal, M. A., Darras, B. M., Eltaggaz, A., & Deiab, I. M. (2024). Comparative sustainability assessment of powder bed fusion and solid-state additive manufacturing processes: The case of direct metal laser sintering versus additive friction stir deposition. *Sustainable Materials and Technologies*, *39*, e00858. DOI: 10.1016/j.susmat.2024.e00858

Ahn, S. J., Lee, H., & Cho, K. J. (2024). 3D printing with a 3D printed digital material filament for programming functional gradients. *Nature Communications*, *15*(1), 3605. Advance online publication. DOI: 10.1038/s41467-024-47480-5 PMID: 38714684

Alkofahi, H., Alawneh, H., & Skjellum, A. (2024). MitM attacks on intellectual property and integrity of additive manufacturing systems: A security analysis. *Computers & Security*, *140*, 103810. DOI: 10.1016/j.cose.2024.103810

Arunprasath, K., Arumugaprabu, V., Amuthakkannan, P., Deepak Joel Johnson, R., & Vigneshwaran, S. (2022). Development in additive manufacturing techniques. In *Innovations in Additive Manufacturing* (pp. 33–53). Springer International Publishing., DOI: 10.1007/978-3-030-89401-6_2

Aufa, A. N., Hassan, M. Z., Ismail, Z., Ramlie, F., Jamaludin, K. R., Daud, M. Y. M., & Ren, J. (2024). Current trends in additive manufacturing of selective laser melting for biomedical implant applications. *Journal of Materials Research and Technology*, *26*, 1–15. DOI: 10.1016/j.jmrt.2024.06.041

Baskaran, J. (2024). Roadmap to achieve sustainable and eco-friendly metal additive manufacturing processes. In *Futuristic Technology for Sustainable Manufacturing* (pp. 185–215). IGI Global., DOI: 10.4018/979-8-3693-1862-1.ch011

Boopathy, G., Palani, S., Ganesan, B., & Kumaar, K. (2024). Advancements in design, fabrication, and applications of functionally graded composite materials. In *Engineering materials for efficient energy storage and conversion* (pp. 25–50). IGI Global., DOI: 10.4018/979-8-3693-2798-2.ch002

Boopathy, G., Sathish, S., & Kumar, M. H. (2024). Introduction to precision manufacturing for micro- and nanofabrication. In *Microfabrication and Nanofabrication: Precision Manufacturing* (Vol. 11, p. 1). Walter de Gruyter GmbH. https://doi.org/DOI: 10.1515/9783110795369-001

Castro, J., Nóbrega, J. M., & Costa, R. (2024). Computational framework to model the selective laser sintering process. *Materials (Basel), 17*(8), 1845. DOI: 10.3390/ma17081845 PMID: 38673203

Chigilipalli, B. K., Karri, T., Chetti, S. N., Bhiogade, G., Kottala, R. K., & Cheepu, M. (2023). A review on recent trends and applications of IoT in additive manufacturing. *Applied System Innovation*, *6*(2), 50. DOI: 10.3390/asi6020050

Choma, T., Żrodowski, Ł., Ciftci, J., Morończyk, B., & Kalicki, B. (2024, February). Overcoming challenges in custom powder manufacturing—From low-melting and reactive materials to refractories. In *TMS Annual Meeting & Exhibition* (pp. 257–269). Springer Nature Switzerland. https://doi.org/DOI: 10.1007/978-3-031-50349-8_24

Choudhari, A., Elder, J., Mugale, M., Karki, S., Digole, S., Omeike, S., & Borkar, T. (2024). Enhancing quality control: Image-based quantification of carbides and defect remediation in binder jetting additive manufacturing. *Materials (Basel), 17*(10), 2174. DOI: 10.3390/ma17102174 PMID: 38793241

Choudhary, N., Singh, A. K., & Sharma, V. (2024). Materials for additive manufacturing. In Sharma, P., & Badheka, V. (Eds.), *Additive manufacturing for advance applications: Technologies, challenges and case studies* (pp. 1–20). CRC Press., DOI: 10.1201/9781003484325-1

Eryildiz, M. (2024). Tailoring mechanical properties of FDM-3D-printed parts through titanium dioxide-reinforced resin filling technique. *Proceedings of the Institution of Mechanical Engineers, Part B: Journal of Engineering Manufacture*. Advance online publication. https://doi.org/DOI: 10.1177/09544054241260467

Feng, K., Hu, S., Zhao, W., Sun, J., Mao, Y., Cai, D., Wu, J., & Wei, Q. (2024). Bimodal powder optimization in SiC binder jetting for mechanical performance. *International Journal of Mechanical Sciences*, *274*, 109278. DOI: 10.1016/j.ijmecsci.2024.109278

Garcia, G. E. S., de Sousa, R. R.Junior, Gouveia, J. R., & Dos Santos, D. J. (2024). Graphene oxide-based nanocomposites for stereolithography (SLA) 3D printing: Comprehensive mechanical characterization under combined loading modes. *Polymers, 16*(9), 1261. DOI: 10.3390/polym16091261 PMID: 38732730

Gumaste, A., Dhal, A., Agrawal, P., Haridas, R. S., Vasudevan, V. K., Weiss, D., & Mishra, R. S. (2023). A novel approach for enhanced mechanical properties in solid-state additive manufacturing by additive friction stir deposition using thermally stable Al-Ce-Mg alloy. *Journal of the Minerals Metals & Materials Society*, *75*(10), 4185–4198. DOI: 10.1007/s11837-023-06044-6

Ibrahim, B., Lopez, L., Kulkarni, S., Jobes, D., Forgiarini, M., Barber, J. R., & Gordon, J. V. (2024). Increasing strength properties in sinter-based additive manufacturing of SS316L via metal material jetting of sub-micron powders. *Additive Manufacturing*, *65*, 104268. DOI: 10.1016/j.addma.2024.104268

Jia, Q., Liu, B., Li, S., Hu, C., Yang, Y., & Wang, G. (2024). Slurry-based photoinitiator jetting process for ceramic additive manufacturing. *Additive Manufacturing*, *86*, 104185. DOI: 10.1016/j.addma.2024.104185

Jung, S., Kara, L. B., Nie, Z., Simpson, T. W., & Whitefoot, K. S. (2023). Is additive manufacturing an environmentally and economically preferred alternative for mass production? *Environmental Science & Technology*, *57*(16), 6373–6386. DOI: 10.1021/acs.est.2c04927 PMID: 37066969

Karimi, A., Rahmatabadi, D., & Baghani, M. (2024). Various FDM mechanisms used in the fabrication of continuous-fiber reinforced composites: A review. *Polymers*, *16*(6), 831. DOI: 10.3390/polym16060831 PMID: 38543436

Karin, J., Ahmed, E., Scholz, S., & Hagenmeyer, V. (2024). On the correlation between pre-processing workflow and dimensional accuracy of 3D printed parts in high-precision material jetting. *Additive Manufacturing*, *85*, 104335. DOI: 10.1016/j.addma.2024.104335

Kim, H., Kim, K. H., Jeong, J., Lee, Y., & Jung, I. D. (2024). Recent progress on materials for functional additive manufacturing. *Materials Science in Additive Manufacturing*, *3*(2), 3323. DOI: 10.36922/msam.3323

Kim, Y. J., Kim, H. N., & Kim, D. Y. (2024). A study on effects of curing and machining conditions in post-processing of SLA additive manufactured polymer. *Journal of Manufacturing Processes*, *119*, 511–519. DOI: 10.1016/j.jmapro.2024.03.112

Lacroix, R., Timonina-Farkas, A., & Seifert, R. W. (2023). Utilizing additive manufacturing and mass customization under capacity constraints. *Journal of Intelligent Manufacturing*, *34*(1), 281–301. DOI: 10.1007/s10845-022-02007-x PMID: 36618339

Lawrence, J. E., Lawrence, M. P., Fezzaa, K., Clark, S. J., & Crane, N. B. (2024). High-speed X-ray imaging of droplet-powder interaction in binder jet additive manufacturing. *Additive Manufacturing*, *65*, 104269. DOI: 10.1016/j.addma.2024.104269

Liang, C., Huang, J., Wang, J., Gong, H., Bai, D., & Zhao, P. (2023). Additive manufacturing of low-temperature co-fired ceramic substrates and surface conductors based on material jetting. *Additive Manufacturing*, *78*, 103856. DOI: 10.1016/j.addma.2023.103856

Major, R., Gawlikowski, M., Surmiak, M., Janiczak, K., Więcek, J., Kurtyka, P., Schwentenwein, M., Jasek-Gajda, E., Kopernik, M., & Lackner, J. M. (2023). In vitro and in vivo testing of stereolithography (SLA)-manufactured haemocompatible photopolymers for blood pump. *Applied Sciences (Basel, Switzerland)*, *14*(1), 383. DOI: 10.3390/app14010383

Miller, Z., Fernandes, P., Hunter, B., Maldonado, C., Patterson, B., & Skoug, R. (2024, March). *The use of metallized stereolithography additively manufactured components in space. In 2024 IEEE Aerospace Conference*. IEEE., DOI: 10.1109/AERO60161.2024.10509144

Mohseni, M., & Khodaygan, S. (2024). Design for additive manufacturing of topology-optimized structures based on deep learning and transfer learning. *Rapid Prototyping Journal*, *30*(7), 1411–1433. DOI: 10.1108/RPJ-02-2024-0102

Munguia-Galeano, F., Ortega-Arroyo, L., Villarreal-Cervantes, M. G., & Ji, Z. (2023, August). Towards integrating 3D printing and automated assembly. In *2023 IEEE 19th International Conference on Automation Science and Engineering (CASE)* (pp. 1–7). IEEE. https://doi.org/DOI: 10.1109/CASE56687.2023.10259847

Munoz, C., Kim, C., & Armstrong, L. (2013). Layer-by-layer: Opportunities in 3D printing—Technology trends, growth drivers and the emergence of innovative applications in 3D printing. *Lux Research*. (No DOI found – gray literature)

Nancharaiah, T., Reddy, V. S. K., Chakravarthi, T., Chowdary, G. T. S., & Teja, Y. B. (2023). Experimental study on the effect of process parameters on the build time and part accuracy of direct metal laser sintering components. *Applied Mechanics and Materials*, *917*, 49–56. DOI: 10.4028/p-fzsqM6

Park, J. H., Tucker, S. J., Yoon, J. K., Kim, Y., & Hollister, S. J. (2024). 3D printing modality effect: Distinct printing outcomes dependent on selective laser sintering (SLS) and melt extrusion. *Journal of Biomedical Materials Research. Part A*, *112*(7), 1015–1024. DOI: 10.1002/jbm.a.37682 PMID: 38348580

Petruse, R. E., Simion, C., & Bondrea, I. (2024). Geometrical and dimensional deviations of fused deposition modelling (FDM) additive-manufactured parts. *Metrology*, *4*(3), 411–429. DOI: 10.3390/metrology4030025

Pierson, S., Marshall, L., Lemmons, L., Efrid, L., Shou, W., & Hu, H. (2024, July). Work-in-progress: Open-source selective laser sintering (SLS) test bed. In *2023 ASEE Midwest Section Conference*. https://peer.asee.org/45312

Prakash, B. V., Koushik, N., Jha, S. K., & Ramasamy, K. (2024). Additive manufacturing of composite materials and functionally graded structures using archerfish hunting technique. *Lubrication Science*, *36*(8), 595–609. Advance online publication. DOI: 10.1002/ls.1716

Raeker-Jordan, N., Chung, J., Kong, Z. J., & Williams, C. (2024). Ensuring additive manufacturing quality and cyber–physical security via side-channel measurements and transmissions. *Journal of Manufacturing Systems*, *73*, 275–286. DOI: 10.1016/j.jmsy.2024.02.005

Rafi, K., Liu, A. Z., Di Prima, M., Bates, P., & Seifi, M. (2022). Regulatory and standards development in medical additive manufacturing. *MRS Bulletin*, *47*(1), 98–105. DOI: 10.1557/s43577-022-00266-w

Rahman, Z., Charoo, N. A., Mohamed, E. M., Kuttolamadom, M., & Khan, M. A. (2023). Regulatory perspective of additive manufacturing in the field of pharmaceuticals. In *Additive Manufacturing in Pharmaceuticals* (pp. 327–348). DOI: 10.1007/978-981-99-2404-2_10

Rath, K. C., Khang, A., Mishra, S. K., Patnaik, P. K., Mohanty, G. K., & Dash, T. (2024). Integration of artificial intelligence and Internet of Things technology solutions in smart manufacturing. In *Machine Vision and Industrial Robotics in Manufacturing* (pp. 155–177). CRC Press., DOI: 10.1201/9781003438137-9

Rehme, O., & Weidermann, C. (2024, May). Additive manufacturing of cryogenic materials. *IOP Conference Series. Materials Science and Engineering*, *1301*(1), 012047. DOI: 10.1088/1757-899X/1301/1/012047

Saracyakupoglu, T. (2022). The certification steps for the additively manufactured aviation-grade parts. *The European Journal of Research and Development*, 2(4), 33–42. DOI: 10.56038/ejrnd.v2i4.133

Suresh, S., Natarajan, E., Sudhagar, S., Boopathi, S., & Elayaraja, R. (2024). Additive manufacturing for a sustainable production: Materials, processes, and impact. In *Futuristic Technology for Sustainable Manufacturing* (pp. 1–16). IGI Global., DOI: 10.4018/978-1-7998-8451-0.ch001

Tarasov, N., & Khamula, O. (2024). Integrating artificial intelligence into 3D modeling software for optimized additive manufacturing. *Grail of Science*, (39), 348–352. DOI: 10.36074/grail-of-science.10.05.2024.052

Tebianian, M., Aghaie, S., Razavi Jafari, N. S., Elmi Hosseini, S. R., Pereira, A. B., Fernandes, F. A., & Huo, Y. (2023). A review of the metal additive manufacturing processes. *Materials (Basel)*, *16*(24), 7514. DOI: 10.3390/ma16247514 PMID: 38138655

Yadav, S., & Mishra, M. K. (2024). Advancements in artificial intelligence-mediated fabrication of 3D, 4D, and 5D printed drug delivery formulations. *Drug Delivery Letters*, *14*(4), 273–289. DOI: 10.2174/0122103031309831240531084125

Yang, J., Li, B., Liu, J., Tu, Z., & Wu, X. (2024). Application of additive manufacturing in the automobile industry: A mini review. *Processes (Basel, Switzerland)*, *12*(6), 1101. DOI: 10.3390/pr12061101

Zarinejad, M., Tong, Y., Salehi, M., Mu, C., Wang, N., Xu, Y., & Chen, X. (2024). Advancements and perspectives in additive manufacturing of tungsten alloys and composites: Challenges and solutions. *Additive Manufacturing*, *85*, 104333. DOI: 10.1016/j.addma.2024.104333

Zhou, H. R., Yang, H., Li, H. Q., Ma, Y. C., Yu, S., Shi, J., Cheng, J., Gao, P., Yu, B., Miao, Z., & Wei, Y. P. (2024). Advancements in machine learning for material design and process optimization in the field of additive manufacturing. *China Foundry*, *21*(2), 101–115. DOI: 10.1007/s41230-024-3145-3

Compilation of References

Abazari, R., Sanati, S., Bajaber, M. A., Javed, M. S., Junk, P. C., Nanjundan, A. K., Qian, J., & Dubal, D. P. (2024). Design and advanced manufacturing of NU-1000 metal–organic frameworks with future perspectives for environmental and renewable energy applications. *Small, 20*(15), 2306353. DOI: 10.1002/smll.202306353 PMID: 37997226

Abdulhameed, O., Al-Ahmari, A., Ameen, W., & Mian, S. H. (2019). Additive manufacturing: Challenges, trends, and applications. *Advances in Mechanical Engineering, 11*(2), 1687814018822880. Advance online publication. DOI: 10.1177/1687814018822880

Abedin, M. J., Rahman, M. Z., Rahman, S., Sarker, M., Saha, B., Sazal, Y. I., & Salsabil, Z. (2023). Recent advances in applications of hybrid composites. In *Reference Module in Materials Science and Materials Engineering*. Elsevier., DOI: 10.1016/B978-0-323-96020-5.00134-5

Abouzaid, K., Guessasma, S., Belhabib, S., Bassir, D., & Chouaf, A. (2019). Thermal mechanical characterization of copolyester for additive manufacturing using FDM. *International Journal for Simulation and Multidisciplinary Design Optimization, 10*, A9. DOI: 10.1051/smdo/2019011

Adamov, I., Stanojević, G., Kočović, D., Mugoša, S., Grujić, B., & Ibrić, S. (2023). Selective laser sintering (SLS) 3D printing process: Influence of model design on the properties of zolpidem tartrate tablets. *Macedonian Pharmaceutical Bulletin, 69*(3, Suppl 1), 133–134. DOI: 10.33320/maced.pharm.bull.2023.69.03.065

Adapa, S. K., & Jagadish, . (2023). Prospects of natural fiber-reinforced polymer composites for additive manufacturing applications: A review. *Journal of the Minerals Metals & Materials Society, 75*(3), 920–940. DOI: 10.1007/s11837-022-05670-w

Adarsh, S. H., & Nagamadhu, M. (2025). Effect of Printing Parameters on Mechanical Properties and Warpage of 3D-Printed PEEK/CF-PEEK Composites Using Multi-Objective Optimization Technique. *Journal of Composites Science*, *9*(5), 208. DOI: 10.3390/jcs9050208

Agarwal, A., & Ojha, R. (2024). Prioritizing implications of Industry-4.0 on the sustainable development goals: A perspective from the analytic hierarchy process in manufacturing operations. *Journal of Cleaner Production*, *444*, 141189. DOI: 10.1016/j.jclepro.2024.141189

Agarwala, M. K., Jamalabad, V. R., Langrana, N. A., Safari, A., Whalen, P. J., & Danforth, S. C. (1996). Structural quality of parts processed by fused deposition. *Rapid Prototyping Journal*, *2*(4), 4–19. DOI: 10.1108/13552549610732034

Agarwala, M., Bourell, D., Beaman, J., Marcus, H., & Barlow, J. (1995). Direct selective laser sintering of metals. *Rapid Prototyping Journal*, *1*(1), 26–36. DOI: 10.1108/13552549510078113

Agrawal, M. K. (2021). Additive manufacturing: Advances in trends and technology. *Turkish Journal of Computer and Mathematics Education*, *12*(1S), 452–458. DOI: 10.17762/turcomat.v12i1S.1903

Agrawal, R. (2021). Sustainable material selection for additive manufacturing technologies: A critical analysis of rank reversal approach. *Journal of Cleaner Production*, *296*, 126500. DOI: 10.1016/j.jclepro.2021.126500

Ahmad, I., Zulfiqar, A., Shabbir, M., Khan, J. S., Çalışkan, H., & Hong, H. (2025). Advanced sensors, monitoring, and control systems for environmental sustainability. In *Smart Cities and Sustainable Manufacturing* (pp. 47–55). Elsevier. DOI: 10.1016/B978-0-443-26474-0.00007-5

Ahmad, M., Javaid, M., & Haleem, A. (2024). A study on fused deposition modeling (FDM) and laser-based additive manufacturing (LBAM) in the medical field. *Intelligent Pharmacy*, *2*(3), 381–391. DOI: 10.1016/j.ipha.2024.02.010

Ahmed, W., Alnajjar, F., Zaneldin, E., Al-Marzouqi, A. H., Gochoo, M., & Khalid, S. (2020). Implementing FDM 3D Printing Strategies Using Natural Fibers to Produce Biomass Composite [Review of Implementing FDM 3D Printing Strategies Using Natural Fibers to Produce Biomass Composite]. Materials, 13(18), 4065. Multidisciplinary Digital Publishing Institute. DOI: 10.3390/ma13184065

Ahmed, A. A., Nazzal, M. A., Darras, B. M., Eltaggaz, A., & Deiab, I. M. (2024). Comparative sustainability assessment of powder bed fusion and solid-state additive manufacturing processes: The case of direct metal laser sintering versus additive friction stir deposition. *Sustainable Materials and Technologies*, *39*, e00858. DOI: 10.1016/j.susmat.2024.e00858

Ahmed, A., Arya, S., Gupta, V., Furukawa, H., & Khosla, A. (2021). 4D printing: Fundamentals, materials, applications and challenges. *Polymer*, *228*, 123926. DOI: 10.1016/j.polymer.2021.123926

Ahmed, A., Zillur Rahman, M., Ou, Y., Liu, S., Mobasher, B., Guo, S., & Zhu, D. (2021). A review on the tensile behavior of fiber-reinforced polymer composites under varying strain rates and temperatures. *Construction & Building Materials*, *294*, 123565. DOI: 10.1016/j.conbuildmat.2021.123565

Ahmed, W., & Zaneldin, E. (2020). The Innovation of Using 3D Printing Technology in Mechanical and Manufacturing Engineering. *IOP Conference Series. Materials Science and Engineering*, *938*(1), 12002. DOI: 10.1088/1757-899X/938/1/012002

Ahn, S. J., Lee, H., & Cho, K. J. (2024). 3D printing with a 3D printed digital material filament for programming functional gradients. *Nature Communications*, *15*(1), 3605. Advance online publication. DOI: 10.1038/s41467-024-47480-5 PMID: 38714684

Aimar, A., Palermo, A., & Innocenti, B. (2019). The Role of 3D Printing in Medical Applications: A State of the Art [Review of The Role of 3D Printing in Medical Applications: A State of the Art]. Journal of Healthcare Engineering, 2019, 1. Hindawi Publishing Corporation. DOI: 10.1155/2019/5340616

Akhoundi, B., Behravesh, A. H., & Bagheri Saed, A. (2019). Improving mechanical properties of continuous fiber-reinforced thermoplastic composites produced by FDM 3D printer. *Journal of Reinforced Plastics and Composites*, *38*(3), 99–116. DOI: 10.1177/0731684418807300

Akintola, A., Newbury-Birch, D., & Kilinc, S. (2024). Bridging the gap between research evidence and its implementation in public health practice: Case studies of embedded research model. *BMC Public Health*, *24*(1), 1299. DOI: 10.1186/s12889-024-18727-z PMID: 38741039

Al Abadi H, Thai HT, Paton-Cole V, and Patel VI (2018) Elastic properties of 3D printed fibre-reinforced structures. Composite Structures 193(March): 8–18. https://doi.org/. compstruct.2018.03.051.DOI: 10.1016/j

Al Jassmi, H., Al Najjar, F., & Mourad, A. H. I. (2018, March). Large-Scale 3D printing: The way forward. []. IOP Publishing.]. *IOP Conference Series. Materials Science and Engineering*, *324*, 012088. DOI: 10.1088/1757-899X/324/1/012088

Al Rashid, A., & Koç, M. (2023). Additive manufacturing for sustainability and circular economy: Needs, challenges, and opportunities for 3D printing of recycled polymeric waste. *Materials Today Sustainability*, *24*, 100529. DOI: 10.1016/j.mtsust.2023.100529

Alabi, M. O. (2020). Big data, 3D printing technology, and industry of the future. In *Additive Manufacturing: Breakthroughs in Research and Practice* (pp. 503-520). IGI Global.

Alexander, F. J., Lin, M., Qian, X., & Yoon, B.-J. (2023). Accelerating scientific discoveries through data-driven innovations. *Patterns (New York, N.Y.)*, *4*(11), 100876. DOI: 10.1016/j.patter.2023.100876 PMID: 38035189

Alfaify, A., Saleh, M., Abdullah, F. M., & Al-Ahmari, A. M. (2020). Design for additive manufacturing: A systematic review. *Sustainability (Basel)*, *12*(19), 7936. DOI: 10.3390/su12197936

Alghamdi, S. S., John, S., Choudhury, N. R., & Dutta, N. K. (2021). Additive Manufacturing of Polymer Materials: Progress, Promise and Challenges [Review of Additive Manufacturing of Polymer Materials: Progress, Promise and Challenges]. Polymers, 13(5), 753. Multidisciplinary Digital Publishing Institute. DOI: 10.3390/polym13050753

Alharbi, M., Kong, I., & Patel, V. I. (2020). Simulation of uniaxial stress–strain response of 3D-printed polylactic acid by nonlinear finite element analysis. *Applied Adhesion Science*, *8*(1), 5. Advance online publication. DOI: 10.1186/s40563-020-00128-1

Alhnan, M. A., Okwuosa, T. C., Sadia, M., Wan, K.-W., Ahmed, W., & Arafat, B. (2016). Emergence of 3D Printed Dosage Forms : Opportunities and Challenges. *Pharmaceutical Research*, *33*(8), 1817–1832. DOI: 10.1007/s11095-016-1933-1 PMID: 27194002

Ali, A., Ahmad, U., & Akhtar, J. (2020). 3D Printing in Pharmaceutical Sector: An Overview. In *IntechOpen eBooks*. IntechOpen., DOI: 10.5772/intechopen.90738

Alibe, A. M., Janga, A., Girgiri, A., & Alibe, I. (2023). Residual stress characterization of an additively manufactured titanium alloy. *Interantional Journal of Scientific Research in Engineering and Management*, *07*(09), 1–11. DOI: 10.55041/IJSREM25885

Aljabali, B. A., Parupelli, S. K., & Desai, S. (2025). Generalized Design for Additive Manufacturing (DfAM) Expert System Using Compliance and Design Rules. *Machines (Basel), 13*(1), 29. DOI: 10.3390/machines13010029

Alkofahi, H., Alawneh, H., & Skjellum, A. (2024). MitM attacks on intellectual property and integrity of additive manufacturing systems: A security analysis. *Computers & Security, 140*, 103810. DOI: 10.1016/j.cose.2024.103810

Allam, T., Pradeep, K. G., Köhnen, P., Marshal, A., Schleifenbaum, J. H., & Haase, C. (2020). Tailoring the nanostructure of laser powder bed fusion additively manufactured maraging steel. *Additive Manufacturing, 36*, 101561. DOI: 10.1016/j.addma.2020.101561

Allison, M. (2013). Alliance to bolster translation of academic R&D assets. *Nature Biotechnology, 31*(4), 275–275. DOI: 10.1038/nbt0413-275 PMID: 23563411

Almeida, H., Pei, E., & Vitorino, L. (2021). Sustainablity for 3D Printing. In *Sustainability for 3D Printing* (pp. 1–13). Springer International Publishing.

Almy, R. D., & Tontowi, A. E. (2018). The effect of 3d printing machine parameters in the extrusion process of biocomposite materials (PMMA and HA) on dimensional accuracy. *Sinergi, 22*(2), 83. DOI: 10.22441/sinergi.2018.2.003

Alqutaibi, A. Y., Alghauli, M. A., Aljohani, M. H. A., & Zafar, M. S. (2024). Advanced additive manufacturing in implant dentistry: 3D printing technologies, printable materials, current applications and future requirements. *Bioprinting (Amsterdam, Netherlands), 42*, 00356. DOI: 10.1016/j.bprint.2024.e00356

Alzyod, H., & Ficzere, P. (2023). Material-Dependent Effect of Common Printing Parameters on Residual Stress and Warpage Deformation in 3D Printing: A Comprehensive Finite Element Analysis Study. *Polymers, 15*(13), 2893. DOI: 10.3390/polym15132893 PMID: 37447538

Amaya, J., Perero, B. S., Helguero, C. G., Hurel, J., Peralta, J., Flores, F. A., & Alvarado, J. D. (2024). Future trends of additive manufacturing in medical applications: An overview [Review of Future trends of additive manufacturing in medical applications: An overview]. Heliyon, 10(5). Elsevier BV. DOI: 10.1016/j.heliyon.2024.e26641

Amores, I. D., González-Gutiérrez, J., García, I. M., Franco, J. M., & Gallegos, C. (2022). 3D printing–Present and future–A Chemical Engineering perspective. *Chemical Engineering Research & Design, 187*, 598–610. DOI: 10.1016/j.cherd.2022.08.049

Anantrasirichai, N., & Bull, D. (2021). Artificial intelligence in the creative industries: a review [Review of Artificial intelligence in the creative industries: a review]. Artificial Intelligence Review, 55(1), 589. Springer Science+Business Media. DOI: 10.1007/s10462-021-10039-7

Andronov, V. N., Šimota, J., Pelikán, L., Beránek, L., & Novák, V. (2021). Analysis of strength properties of 3D printed lightweight structures made of AlSi10Mg material. In IOP Conference Series Materials Science and Engineering (Vol. 1178, Issue 1, p. 12002). IOP Publishing. DOI: 10.1088/1757-899X/1178/1/012002

Anitha, R., Arunachalam, S., & Radhakrishnan, P. (2001). Critical parameters influencing the quality of prototypes in fused deposition modelling. *Journal of Materials Processing Technology*, *118*(1-3), 2–5. DOI: 10.1016/S0924-0136(01)00980-3

An, J., Teoh, J. E. M., Suntornnond, R., & Chua, C. K. (2015). Design and 3D Printing of Scaffolds and Tissues. *Engineering (Beijing)*, *1*(2), 261–268. DOI: 10.15302/J-ENG-2015061

Antezana, P. E., Municoy, S., Ostapchuk, G., Catalano, P. N., Hardy, J. G., Evelson, P. A., Orive, G., & Desimone, M. F. (2023). 4D printing: The development of responsive materials using 3D-printing technology. *Pharmaceutics*, *15*(12), 2743. DOI: 10.3390/pharmaceutics15122743 PMID: 38140084

Anwajler, B., Zdybel, E., & Tomaszewska-Ciosk, E. (2023). Innovative Polymer Composites with Natural Fillers Produced by Additive Manufacturing (3D Printing)—A Literature Review [Review of Innovative Polymer Composites with Natural Fillers Produced by Additive Manufacturing (3D Printing)—A Literature Review]. Polymers, 15(17), 3534. Multidisciplinary Digital Publishing Institute. DOI: 10.3390/polym15173534

Aravind Raj, S. (2025). Quality control in additive manufacturing: A review of traditional and advanced techniques. *Engineering Research Express*.

Arıkan, V. (2023). Temperature-dependent debonding resistance of 316 stainless steel, inconel 625, and ti-6al-4v alloys. *Black Sea Journal of Engineering and Science*, *6*(3), 287–295. DOI: 10.34248/bsengineering.1312954

Armstrong, M., Mehrabi, H., & Naveed, N. (2022). An overview of modern metal additive manufacturing technology. *Journal of Manufacturing Processes*, *84*, 1001–1029. DOI: 10.1016/j.jmapro.2022.10.060

Armstrong, M., Mehrabi, H., Naveed, N., & Gregg, C. (2023). Evaluating energy consumption in atomic diffusion additive manufacturing versus sand casting. Journal of Engineering &Amp. *Technological Advances*, *8*(1), 46–65. DOI: 10.35934/segi.v8i1.81

Arróyave, R., & McDowell, D. L. (2019). Systems Approaches to Materials Design: Past, Present, and Future. In Annual Review of Materials Research (Vol. 49, Issue 1, p. 103). Annual Reviews. DOI: 10.1146/annurev-matsci-070218-125955

Arunprasath, K., Arumugaprabu, V., Amuthakkannan, P., Deepak Joel Johnson, R., & Vigneshwaran, S. (2022). Development in additive manufacturing techniques. In *Innovations in Additive Manufacturing* (pp. 33–53). Springer International Publishing., DOI: 10.1007/978-3-030-89401-6_2

Ashby, M. F., & Jones, D. R. (2012). *Engineering materials 1: an introduction to properties, applications and design* (Vol. 1). Elsevier.

Aslam, J., Saleem, A., Lai, K. H., & Kim, Y. B. (2025). Critical successes factors for the adoption of additive manufacturing: Integrated impact for circular economy model. *Technological Forecasting and Social Change*, *213*, 124041. DOI: 10.1016/j.techfore.2025.124041

Astafurova, E., Reunova, K., Zagibalova, E., Astapov, D., Astafurov, S., & Kolubaev, E. (2024). Microstructure, phase composition, and mechanical properties of intermetallic ni-al-cr material produced by dual-wire electron-beam additive manufacturing. *Metals*, *14*(1), 75. DOI: 10.3390/met14010075

ASTM Standards. (2012). *Standard Terminology for Additive Manufacturing Technologies* (Vol. 10).

Aufa, A. N., Hassan, M. Z., Ismail, Z., Ramlie, F., Jamaludin, K. R., Daud, M. Y. M., & Ren, J. (2024). Current trends in additive manufacturing of selective laser melting for biomedical implant applications. *Journal of Materials Research and Technology*, *26*, 1–15. DOI: 10.1016/j.jmrt.2024.06.041

Awad, A., Xu, X., Ong, J. J., Goyanes, A., & Basit, A. W. (2023). Vat Photopolymerisation additive manufacturing for pharmaceutical applications. In *Nano-and Microfabrication Techniques in Drug Delivery : Recent Developments and Future Prospects* (p. 99-124). Springer. DOI: 10.1007/978-3-031-26908-0_5

Awad, A., Fina, F., Goyanes, A., Gaisford, S., & Basit, A. W. (2020). 3D printing: Principles and pharmaceutical applications of selective laser sintering. *International Journal of Pharmaceutics*, *586*, 119594. DOI: 10.1016/j.ijpharm.2020.119594 PMID: 32622811

Awasthi, P., & Banerjee, S. S. (2021). Fused deposition modeling of thermoplastic elastomeric materials: Challenges and opportunities. *Additive Manufacturing, 46*, 102177. DOI: 10.1016/j.addma.2021.102177

Azlin, M. N. M., Ilyas, R. A., Zuhri, M. Y. M., Sapuan, S. M., Harussani, M. M., Sharma, S., Nordin, A. H., Nurazzi, N. M., & Afiqah, A. N. (2022). 3D Printing and Shaping Polymers, Composites, and Nanocomposites : A Review. *Polymers, 14*(1), 180. Advance online publication. DOI: 10.3390/polym14010180 PMID: 35012202

Baba, T. *et al.* (2022) 'Silicon Photonics FMCW LiDAR Chip With a Slow-Light Grating Beam Scanner', *IEEE Journal of Selected Topics in Quantum Electronics*, 28(5: Lidars and Photonic Radars), pp. 1–8. Available at: .DOI: 10.1109/JSTQE.2022.3157824

Babagowda, K. M. R. S., Goutham, R., & Srinivas Prasad, K. R. (2018, February). Study of Effects on Mechanical Properties of PLA Filament which is blended with Recycled PLA Materials. In IOP conference series: materials science and engineering (Vol. 310, p. 012103). IOP Publishing.

Babagowda, M., Kadadevara Math, R. S., Goutham, R., & Srinivas Prasad, K. R. (2018). Study of Effects on Mechanical Properties of PLA Filament, which is blended with Recycled PLA Materials. *IOP Conference Series. Materials Science and Engineering, 310*, 12103. DOI: 10.1088/1757-899X/310/1/012103

Babbar, A., Rai, A., & Sharma, A. (2021). The latest trend in building construction: Three-dimensional printing. *Journal of Physics: Conference Series, 1950*(1), 12007. DOI: 10.1088/1742-6596/1950/1/012007

Bacciaglia, A., Ceruti, A., & Liverani, A. (2019, September). Additive manufacturing challenges and future developments in the next ten years. In *International Conference of the Italian Association of Design Methods and Tools for Industrial Engineering* (pp. 891-902). Cham: Springer International Publishing.

Baechle-Clayton, M., Loos, E., Taheri, M., & Taheri, H. (2022). Failures and flaws in fused deposition modeling (FDM) additively manufactured polymers and composites. *Journal of Composites Science, 6*(7), 202. DOI: 10.3390/jcs6070202

Bagaria, V., Bhansali, R., & Pawar, P. (2018). 3D printing- creating a blueprint for the future of orthopedics: Current concept review and the road ahead! *Journal of Clinical Orthopaedics and Trauma, 9*(3), 207–212. DOI: 10.1016/j.jcot.2018.07.007 PMID: 30202150

Bahadur, P. S., Srivastava, R., & Srivastava, V. (2023). Review on the Development of Smart Materials, Including Shape Memory Alloys, Characterization, and Recent Applications. *Modeling, Characterization, and Processing of Smart Materials*, 107-139.

Bahraminasab, M. (2020). Challenges on optimization of 3D-printed bone scaffolds [Review of Challenges on optimization of 3D-printed bone scaffolds]. BioMedical Engineering OnLine, 19(1). BioMed Central. DOI: 10.1186/s12938-020-00810-2

Baigarina, A., Shehab, E., & Ali, M. H. (2023). Construction 3D printing: A critical review and future research directions. *Progress in Additive Manufacturing*, *8*(6), 1393–1421. DOI: 10.1007/s40964-023-00409-8

Bai, J., & Bu, G. (2022). Progress in 4D printing technology. *J. Adv. Manuf. Sci. Technol*, *2*(1), 2022001–2022001.

Baktheer, A., & Classen, M. (2024). A review of recent trends and challenges in numerical modeling of the anisotropic behavior of hardened 3D printed concrete. *Additive Manufacturing*, *89*, 104309. DOI: 10.1016/j.addma.2024.104309

Balani, S. B., Cantarel, A., Chabert, F., & Nassiet, V. (2018). Influence of parameters controlling the extrusion step in fused filament fabrication (FFF) process applied to polymers using numerical simulation. *AIP Conference Proceedings*, *1960*, 140003. DOI: 10.1063/1.5034995

Baldi, J., Grò, A., Orsini, U., & Favale, L. (2022). *3D-Printed Custom-Made Instruments. In 3D Printing in Bone Surgery*. Springer International Publishing.

Bałon, P., Rejman, E., Świątoniowski, A., Kiełbasa, B., Smusz, R., Szostak, J., Cieślik, J., & Kowalski, Ł. (2020). Thin-walled Integral Constructions in the Aircraft Industry. *Procedia Manufacturing*, *47*, 498–504. DOI: 10.1016/j.promfg.2020.04.153

Bánhegyi, G. Advanced Industrial and Engineering Polymer Research.

Bas, J., Dutta, T., Llamas Garro, I., Velázquez-González, J. S., Dubey, R., & Mishra, S. K. (2024). RETRACTED: Embedded sensors with 3D printing technology. *Sensors (Basel)*, *24*(6), 1955. DOI: 10.3390/s24061955 PMID: 38544218

Baskaran, J. (2024). Roadmap to achieve sustainable and eco-friendly metal additive manufacturing processes. In *Futuristic Technology for Sustainable Manufacturing* (pp. 185–215). IGI Global., DOI: 10.4018/979-8-3693-1862-1.ch011

Bastola, N., Jahan, M. P., Rangasamy, N., & Rakurty, C. S. (2023). A review of the residual stress generation in metal additive manufacturing: Analysis of cause, measurement, effects, and prevention. *Micromachines*, *14*(7), 1480. DOI: 10.3390/mi14071480 PMID: 37512791

Batu, T., Lemu, H. G., & Shimels, H. (2023). Application of artificial intelligence for surface roughness prediction of additively manufactured components. *Materials (Basel)*, *16*(18), 6266. DOI: 10.3390/ma16186266 PMID: 37763543

Baumann, F. W., & Roller, D. (2017). Survey on additive manufacturing, cloud 3D printing and services. arXiv preprint arXiv:1708.04875.

Behseresht, S., Love, A., Valdez Pastrana, O. A., & Park, Y. H. (2024). Enhancing fused deposition modeling precision with serial communication-driven closed-loop control and image analysis for fault diagnosis-correction. *Materials (Basel)*, *17*(7), 1459. DOI: 10.3390/ma17071459 PMID: 38611974

Behseresht, S., & Park, Y. H. (2024). Additive manufacturing of composite polymers: Thermomechanical FEA and experimental study. *Materials (Basel)*, *17*(8), 1912. DOI: 10.3390/ma17081912 PMID: 38673269

Bellini A and Bertoldi M (2013) Liquefier dynamics in fused deposition. Journal of Manufacturing Science and Engineering 126(May 2004): 237–246. https://doi .org/. 1688377.DOI: 10.1115/1

Bendoly, E., Chandrasekaran, A., Lima, M. D. R. F., Handfield, R., Khajavi, S. H., & Roscoe, S. (2024). The role of generative design and additive manufacturing capabilities in developing human–AI symbiosis: Evidence from multiple case studies. *Decision Sciences*, *55*(4), 325–345. DOI: 10.1111/deci.12619

Beniak, J. U. R. A. J., Krizan, P. E. T. E. R., & Matus, M. I. L. O. S. (2020). Conductive material properties for FDM additive manufacturing. *MM Science Journal*, *2020*(1), 3846–3851. DOI: 10.17973/MMSJ.2020_03_2019135

Beniak, J., Holdy, M., Križan, P., & Matúš, M. (2019). Research on parameters optimization for the Additive Manufacturing process. *Transportation Research Procedia*, *40*, 144–149. DOI: 10.1016/j.trpro.2019.07.024

Bernhard, R. J., Neef, P., Wiche, H., Hoff, C., Hermsdorf, J., Kaierle, S., & Wesling, V. (2020). Defect detection in additive manufacturing via a toolpath overlaid melt-pool-temperature tomography. *Journal of Laser Applications*, *32*(2), 022055. Advance online publication. DOI: 10.2351/7.0000055

Bertrand, P., Bayle, F., Combe, C., Goeuriot, P., & Smurov, I. (2007). Ceramic components manufacturing by selective laser sintering. *Applied Surface Science, 254*(4), 989–992. DOI: 10.1016/j.apsusc.2007.08.085

Besklubova, S., Skibniewski, M. J., & Zhang, X. (2021). Factors affecting 3D printing technology adaptation in construction. *Journal of Construction Engineering and Management, 147*(5), 04021026. DOI: 10.1061/(ASCE)CO.1943-7862.0002034

Bettini, P., Alitta, G., Sala, G., & Di Landro, L. (2017). Fused deposition technique for continuous fiber reinforced thermoplastic. *Journal of Materials Engineering and Performance, 26*(2), 843–848. DOI: 10.1007/s11665-016-2459-8

Bhoominathan, R., Boopathy, G., Jacob, S., Pakiyaraj, A., & Karthik, M. K. (2024). Analysis of failure modes and energy absorption in glass fiber and chopped basalt sandwich panels under low-velocity impact. *Materials Today: Proceedings*. Advance online publication. DOI: 10.1016/j.matpr.2024.05.054

Bian, D., Tong, Z., Gong, G., Huang, H., Fang, L., Yang, H., Gu, W., Yu, H., & Zheng, Y. (2024). Additive manufacturing of biodegradable molybdenum–from powder to vascular stent. *Advanced Materials, 36*(32), 2401614. DOI: 10.1002/adma.202401614 PMID: 38837830

Bigliardi, B., Bottani, E., Gianatti, E., Monferdini, L., Pini, B., & Petroni, A. (2024). Sustainable Additive Manufacturing in the context of Industry 4.0: A literature review. *Procedia Computer Science, 232*, 766–774. DOI: 10.1016/j.procs.2024.01.076

Bijalwan, V., Rana, S., Yun, G. J., Singh, K. P., Jamil, M., & Schlögl, S. (2024). 3D Printing of Covalent Adaptable Networks: Overview, Applications and Future Prospects. *Polymer Reviews (Philadelphia, Pa.), 64*(1), 36–79. DOI: 10.1080/15583724.2023.2227692

Bikas, H., Stavropoulos, P., & Chryssolouris, G. (2016). Additive manufacturing methods and modeling approaches: A critical review. *International Journal of Advanced Manufacturing Technology, 83*(1–4), 389–405. DOI: 10.1007/s00170-015-7576-2

Birtchnell, T., & Hoyle, W. (2014). What is 3D Printing? In *Palgrave Macmillan UK eBooks* (p. 36). Palgrave Macmillan., DOI: 10.1057/9781137365668_3

Biswas, M. C., Chakraborty, S., Bhattacharjee, A., & Mohammed, Z. (2021). 4D printing of shape memory materials for textiles: Mechanism, mathematical modeling, and challenges. *Advanced Functional Materials, 31*(19), 2100257. DOI: 10.1002/adfm.202100257

Blanco, I. (2020). The Use of Composite Materials in 3D Printing. *Journal of Composites Science*, *4*(2), 42. DOI: 10.3390/jcs4020042

Bodkhe, S., & Ermanni, P. (2019). Challenges in 3D printing of piezoelectric materials. *Multifunctional Materials*, *2*(2), 022001. DOI: 10.1088/2399-7532/ab0c41

Boopathy, G., Sathish, S., & Kumar, M. H. (2024). Introduction to precision manufacturing for micro- and nanofabrication. In *Microfabrication and Nanofabrication: Precision Manufacturing* (Vol. 11, p. 1). Walter de Gruyter GmbH. https://doi.org/ DOI: 10.1515/9783110795369-001

Boopathy, G., Palani, S., Ganesan, B., & Kumaar, K. (2024). Advancements in design, fabrication, and applications of functionally graded composite materials. In *Engineering materials for efficient energy storage and conversion* (pp. 25–50). IGI Global., DOI: 10.4018/979-8-3693-2798-2.ch002

Boopathy, G., Sathish, S., & Kumar, M. H. (2024). Introduction to precision manufacturing for micro-and nanofabrication. *Microfabrication and Nanofabrication: Precision Manufacturing*, *11*, 1–24. DOI: 10.1515/9783111335537-001

Borgianni, Y., Pradel, P., Berni, A., Obi, M., & Bibb, R. (2022). An investigation into the current state of education in Design for Additive Manufacturing. *Journal of Engineering Design*, *33*(7), 461–490. DOI: 10.1080/09544828.2022.2102893

Boschetto, A., Bottini, L., & Veniali, F. (2016). Integration of FDM surface quality modeling with process design. In *Additive manufacturing* (Vol. 12, p. 334). Elsevier BV., DOI: 10.1016/j.addma.2016.05.008

Bowo, F. P., Suzen, Z. S., & Oktriadi, Y. (2021). Pengaruh Infill Geometry, Printing Speed dan Nozzle Temperature terhadap Kekuatan Impak Menggunakan Filamen ST PLA. *Jurnal Syntax Admiration*, *2*(7), 1257–1268. DOI: 10.46799/jsa.v2i7.265

Bozkurt, Y., & Karayel, E. (2021). 3D printing technology; methods, biomedical applications, future opportunities and trends. *Journal of Materials Research and Technology*, *14*, 1430–1450. DOI: 10.1016/j.jmrt.2021.07.050

Brackett, D., Ashcroft, I., & Hague, R. (2011). Topology Optimization for Additive Manufacturing. DOI: 10.26153/tsw/15300

Brüggemann, H., Stempin, S., & Meier, J.-M. (2020). Consideration of digitalization for the purpose of resource efficiency in a learning factory. *Procedia Manufacturing*, *45*, 140–145. DOI: 10.1016/j.promfg.2020.04.085

Budinoff, H. D., & McMains, S. (2021). Will it print: A manufacturability toolbox for 3D printing. *International Journal on Interactive Design and Manufacturing*, *15*(4), 613–630. DOI: 10.1007/s12008-021-00786-w

Budinski, K. G., & Budinski, M. K. (1983). *Engineering materials*. Pearson Education India.

Budzik, G., Woźniak, J., Paszkiewicz, A., Przeszłowski, Ł., Dziubek, T., & Dębski, M. (2021). Methodology for the Quality Control Process of Additive Manufacturing Products Made of Polymer Materials. *Materials (Basel)*, *14*(9), 2202. DOI: 10.3390/ma14092202 PMID: 33923048

Buffa, G., Costa, A., Palmeri, D., Pollara, G., Barcellona, A., & Fratini, L. (2023). A new control parameter to predict micro-warping-induced job failure in lpbf of ti6al4v titanium alloy. *International Journal of Advanced Manufacturing Technology*, *126*(3-4), 1143–1157. DOI: 10.1007/s00170-023-11179-6

Buj-Corral, I., Domínguez-Fernández, A., & Durán-Llucià, R. (2019). Influence of Print Orientation on Surface Roughness in Fused Deposition Modeling (FDM) Processes. *Materials (Basel)*, *12*(23), 3834. DOI: 10.3390/ma12233834 PMID: 31766409

Buldas, A., Draheim, D., Gault, M., Laanoja, R., Nagumo, T., Saarepera, M., Shah, S. A., Simm, J., Steiner, J., Tammet, T., & Truu, A. (2022). An Ultra-Scalable Blockchain Platform for Universal Asset Tokenization: Design and Implementation. *IEEE Access : Practical Innovations, Open Solutions*, *10*, 77284–77322. DOI: 10.1109/ACCESS.2022.3192837

Burkhalter, D. (2020). *Simulation-Driven Design of a Portable Basketball Hoop System.*, DOI: 10.3390/proceedings2020049131

Cabreira, V., & Santana, R. M. C. (2020). Effect of infill pattern in Fused Filament Fabrication (FFF) 3D Printing on materials performance. *Matéria (Rio de Janeiro)*, *25*(3), e-12826. Advance online publication. DOI: 10.1590/s1517-707620200003.1126

Çakmak, G., Dönmez, M. B., Akay, C., Abou-Ayash, S., Schimmel, M., & Yılmaz, B. (2022). Effect of thermal cycling on the flexural strength and hardness of new-generation denture base materials. *Journal of Prosthodontics*, *32*(S1), 81–86. DOI: 10.1111/jopr.13615 PMID: 36266244

Camacho, D. D., Clayton, P., O'Brien, W. J., Seepersad, C. C., Juenger, M. C. G., Ferron, R. D., & Salamone, S. (2018). Applications of additive manufacturing in the construction industry – A forward-looking review [Review of Applications of additive manufacturing in the construction industry – A forward-looking review]. Automation in Construction, 89, 110. Elsevier BV. DOI: 10.1016/j.autcon.2017.12.031

Caminero, M. A., Chacón, J. M., García-Moreno, I., & Rodríguez, G. P. (2018). Impact damage resistance of 3D printed continuous fibre reinforced thermoplastic composites using fused deposition modelling. *Composites. Part B, Engineering*, *148*, 93–103. DOI: 10.1016/j.compositesb.2018.04.054

Camino, F. E., Tiwale, N., Hwang, S., Du, X., & Yang, J. C. (2024). Mitigating challenges in aberration-corrected electron-beam lithography on electron-opaque substrates. *Nanotechnology*, *35*(6), 065301. DOI: 10.1088/1361-6528/ad0908 PMID: 37918028

Cannella, F., Garinei, A., Marsili, R., & Speranzini, E. (2017). Dynamic mechanical analysis and thermoelasticity investigate composite structural elements made with additive manufacturing. In *Composite Structures* (Vol. 185, p. 466). Elsevier BV., DOI: 10.1016/j.compstruct.2017.11.029

Cano-Vicent, A., Tambuwala, M. M., Hassan, S. S., Barh, D., Aljabali, A. A., Birkett, M., Arjunan, A., & Serrano-Aroca, Á. (2021). Fused deposition modelling: Current status, methodology, applications and future prospects. *Additive Manufacturing*, *47*, 102378. DOI: 10.1016/j.addma.2021.102378

Carayon, I., Gaubert, A., Mousli, Y., & Philippe, B. (2020). Electro-responsive hydrogels: Macromolecular and supramolecular approaches in the biomedical field. *Biomaterials Science*, *8*(20), 5589–5600. DOI: 10.1039/D0BM01268H PMID: 32996479

Carmichael, G.. (2018) 'Curriculum-Aligned Work-Integrated Learning: A New Kind of Industry-Academic Degree Partnership', in *Proceedings of the 49th ACM Technical Symposium on Computer Science Education. SIGCSE '18: The 49th ACM Technical Symposium on Computer Science Education*, Baltimore Maryland USA: ACM, pp. 586–591. Available at: DOI: 10.1145/3159450.3159543

Carneiro, O. S., Silva, A. F., & Gomes, R. (2015). Fused deposition modeling with polypropylene. *Materials & Design*, *83*, 768–776. DOI: 10.1016/j.matdes.2015.06.053

Carrier, A. (1987). Sequential steps in project execution. *International Journal of Project Management*, *5*(1), 25–28. DOI: 10.1016/0263-7863(87)90006-8

Carrillo, L. E. L., Gonzalez, Y. O., Parga, M., Ramos, K. L. D., Neparko, N., & Roberson, D. A. (2024). Development of binary and ternary polyester shape memory blends for additive manufacturing. *Journal of Materials Science*, *59*(18), 1–18. DOI: 10.1007/s10853-024-09657-7

Carrola, M., Asadi, A., Zhang, H., Papageorgiou, D. G., Bilotti, E., & Koerner, H. (2021). Best of Both Worlds: Synergistically Derived Material Properties via Additive Manufacturing of Nanocomposites. In Advanced Functional Materials (Vol. 31, Issue 46). Wiley. DOI: 10.1002/adfm.202103334

Casalino, G., Campanelli, S. L., Contuzzi, N., & Ludovico, A. D. (2015). Experimental investigation and statistical optimisation of the selective laser melting process of a maraging steel. *Optics & Laser Technology*, *65*, 151–158. DOI: 10.1016/j.optlastec.2014.07.021

Casati, R., Lemke, J., Tuissi, A., & Vedani, M. (2016). Aging behaviour and mechanical performance of 18-Ni 300 steel processed by selective laser melting. *Metals*, *6*(9), 218. DOI: 10.3390/met6090218

Cascón, I., Gómez-Omella, M., Fernández, D., Gil, A., Alberdi, N., & González, H. (2024). Towards Zero-Defect Manufacturing Based on Artificial Intelligence through the Correlation of Forces in 5-Axis Milling Process. *Machines (Basel)*, *12*(4), 226. DOI: 10.3390/machines12040226

Castro, J., Nóbrega, J. M., & Costa, R. (2024). Computational framework to model the selective laser sintering process. *Materials (Basel)*, *17*(8), 1845. DOI: 10.3390/ma17081845 PMID: 38673203

Cataldi, P., Liu, M., Bissett, M., & Kinloch, I. A. (2022). A review on printing of responsive smart and 4D structures using 2D materials. *Advanced Materials Technologies*, *7*(11), 2200025. DOI: 10.1002/admt.202200025

Chafik, R., Ebeid, S., & Sayed, M. (2024). Printing parameters and material characteristics affecting mechanical properties of fdm printed parts. *Engineering Research Journal (Shoubra)*, *53*(2), 220–225. DOI: 10.21608/erjsh.2023.251818.1252

Chahin, S., Afify, A., Mohsen, H., & Youssef, M. (2022). ROLE OF 3D PRINTED GREEN WALLS IN HEALING ARCHITECTURE. *BAU Journal - Health and Wellbeing*, *5*(1, Article 1).

Champeau, M., Heinze, D. A., Viana, T. N., de Souza, E. R., Chinellato, A. C., & Titotto, S. (2020). 4D printing of hydrogels: A review. *Advanced Functional Materials*, *30*(31), 1910606. DOI: 10.1002/adfm.201910606

Chang, C. C., Boland, E. D., Williams, S. K., & Hoying, J. B. (2011). Direct-write bioprinting three-dimensional biohybrid systems for future regenerative therapies [Review of Direct-write bioprinting three-dimensional biohybrid systems for future regenerative therapies]. Journal of Biomedical Materials Research Part B Applied Biomaterials, 1, 160. Wiley. DOI: 10.1002/jbm.b.31831

Changliang, Y., Luhong, M., Xindong, X., Sheng, X., & Shilin, Z. (2009). An approach to the optical interconnect made in standard CMOS process. *Journal of Semiconductors*, *30*(5), 055012. DOI: 10.1088/1674-4926/30/5/055012

Chari, V. S., Venkatesh, P. R., Krupashankar, , & Dinesh, V. (1943). Krupashankar, & Dinesh, V. (2018). Effect of processing parameters on the FDM process. *AIP Conference Proceedings*, *20061*, 020061. Advance online publication. DOI: 10.1063/1.5029637

Chatterjee, A. N., Kumar, S., Saha, P., Mishra, P. K., & Roy Choudhury, A. (2003). An experimental design approach to selective laser sintering of low carbon steel. *Journal of Materials Processing Technology*, *136*(1–3), 151–157. DOI: 10.1016/S0924-0136(03)00132-8

Chatzipapas, K., Nika, A., & Krimpenis, A. A. (2024). Introduction of Hybrid Additive Manufacturing for Producing Multi-Material Artificial Organs for Education and In Vitro Testing. *Designs*, *8*(3), 51. DOI: 10.3390/designs8030051

Chaudhari, M., Jogi, B. F., & Pawade, R. S. (2018). Comparative study of part characteristics built using additive manufacturing (FDM). *Procedia Manufacturing*, *20*, 73–78. DOI: 10.1016/j.promfg.2018.02.010

Chaudhary, R. (2023). *Additive Manufacturing by Digital Light Processing*. DLP. DOI: 10.1007/s40964-022-00336-0

Chaudhuri, A., Rogers, H., Søberg, P., Baricz, N., & Pawar, K. (2017, July). Identifying future 3D printing related services: insights from Denmark and Germany. In *22nd International Symposium on Logistics* (p. 73). Centre for Concurrent Enterprise, Nottingham University Business School.

Chavez, L. A., Ibave, P., Wilburn, B. R., Alexander, D., Stewart, C. M., Wicker, R. B., & Lin, Y. (2020). The Influence of Printing Parameters, Post-Processing, and Testing Conditions on the Properties of Binder Jetting Additive Manufactured Functional Ceramics. In Ceramics (Vol. 3, Issue 1, p. 65). Multidisciplinary Digital Publishing Institute. DOI: 10.3390/ceramics3010008

Chen, A., Su, J., Li, Y., Zhang, H., Shi, Y., Yan, C., & Lu, J. (2023). 3D/4D printed bio-piezoelectric smart scaffolds for next-generation bone tissue engineering. *International Journal of Extreme Manufacturing*, *5*(3), 032007. DOI: 10.1088/2631-7990/acd88f

Chen, C., & Sullivan, P. A. (1996). Predicting total build-time and the resultant cure depth of the 3D stereolithography process. *Rapid Prototyping Journal*, *2*(4), 27–40. DOI: 10.1108/13552549610153389

Chen, C., Wang, X., Wang, Y., Yang, D., Yao, F., Zhang, W., Wang, B., Sewvandi, G. A., Yang, D., & Hu, D. (2020). Additive manufacturing of piezoelectric materials. *Advanced Functional Materials*, *30*(52), 2005141. DOI: 10.1002/adfm.202005141

Chen, D., Han, Z., Zhang, J., Xue, L., & Liu, S. (2024). Additive Manufacturing Provides Infinite Possibilities for Self-Sensing Technology. *Advancement of Science*, *11*(28), 2400816. DOI: 10.1002/advs.202400816 PMID: 38767180

Chen, K., Liu, Q., Chen, B., Zhang, S., Ferrara, L., & Li, W. (2024). Effect of raw materials on the performance of 3D printing geopolymer: A review. *Journal of Building Engineering*, *84*, 108501. DOI: 10.1016/j.jobe.2024.108501

Chen, L., Duan, G., Zhang, C., Cheng, P., & Wang, Z. (2022). 3d printed hydrogel for soft thermo-responsive smart window. *International Journal of Extreme Manufacturing*, *4*(2), 025302. DOI: 10.1088/2631-7990/ac5ae3

Chen, L., Liu, F., Abdiryim, T., & Liu, X. (2023). Stimuli-responsive hydrogels as promising platforms for soft actuators. *Materials Today Physics*, •••, 101281.

Chen, L., & Moon, S. K. (2024). In-situ defect detection in laser-directed energy deposition with machine learning and multi-sensor fusion. *Journal of Mechanical Science and Technology*, *38*(9), 4477–4484. DOI: 10.1007/s12206-024-2401-1

Chen, L., Yao, X., Liu, K., Tan, C., & Moon, S. K. (2023). Multisensor fusion-based digital twin in additive manufacturing for in-situ quality monitoring and defect correction. *Proceedings of the Design Society*, *3*, 2755–2764. DOI: 10.1017/pds.2023.276

Chen, M. T., Gong, Z., Zhang, T., Zuo, W., Zhao, Y., Zhao, O., Zhang, G., & Wang, Z. (2024). Mechanical behavior of austenitic stainless steels produced by wire arc additive manufacturing. *Thin-walled Structures*, *196*, 111455. DOI: 10.1016/j.tws.2023.111455

Chen, P., Wang, H., Su, J., Tian, Y., Wen, S., Su, B., Yang, C., Chen, B., Zhou, K., Yan, C., & Shi, Y. (2022). Recent advances on high-performance polyaryletherketone materials for additive manufacturing. *Advanced Materials*, *34*(52), 2200750. Advance online publication. DOI: 10.1002/adma.202200750 PMID: 35385149

Chen, T., & Zhang, Y. (2004). Numerical simulation of two-dimensional melting and resolidification of a two-component metal powder layer in selective laser sintering process. *Numerical Heat Transfer Part A*, *46*(7), 633–649. DOI: 10.1080/104077890504177

Chen, T., & Zhang, Y. (2006). A partial shrinkage model for selective laser sintering of a two-component metal powder layer. *International Journal of Heat and Mass Transfer*, *49*(7-8), 1489–1492. DOI: 10.1016/j.ijheatmasstransfer.2005.09.027

Chen, X. (2016).. . *Making Fabrication Real.*, *17*, 17–20. Advance online publication. DOI: 10.1145/2984751.2984785

Chen, X., Zhang, M., Teng, X., & Mujumdar, A. S. (2022). Recent progress in modeling 3D/4D printing of foods. *Food Engineering Reviews*, *14*(1), 1–14. DOI: 10.1007/s12393-021-09297-6

Chen, Y., Zhang, M., Sun, Y., & Phuhongsung, P. (2022). Improving 3D/4D printing characteristics of natural food gels by novel additives: A review. *Food Hydrocolloids*, *123*, 107160. DOI: 10.1016/j.foodhyd.2021.107160

Chen, Z., Li, Z., Li, J., Liu, C., Lao, C., Fu, Y., Liu, C., Li, Y., Wang, P., & He, Y. (2019). 3D printing of ceramics: A review. *Journal of the European Ceramic Society*, *39*(4), 661–687. DOI: 10.1016/j.jeurceramsoc.2018.11.013

Chigilipalli, B. K., Karri, T., Chetti, S. N., Bhiogade, G., Kottala, R. K., & Cheepu, M. (2023). A review on recent trends and applications of IoT in additive manufacturing. *Applied System Innovation*, *6*(2), 50. DOI: 10.3390/asi6020050

Childs, T. H. C., Berzins, M., Ryder, G. R., & Tontowi, A. (1999). Selective laser sintering of an amorphous polymer—Simulations and experiments. *Proceedings of the Institution of Mechanical Engineers. Part B, Journal of Engineering Manufacture*, *213*(4), 333–349. DOI: 10.1243/0954405991516822

Chmelko, V., Šulko, M., Škriniarová, J., Margetin, M., Gašparík, M., Koščo, T., & Semeš, M. (2023). Strength and cyclic properties of additive vs. conventionally produced material alsi10mg. *Materials (Basel)*, *16*(7), 2598. DOI: 10.3390/ma16072598 PMID: 37048892

Chohan, J. S., & Singh, R. (2017). Pre and post processing techniques to improve surface characteristics of FDM parts: A state of art review and future applications. *Rapid Prototyping Journal*, *23*(3), 495–513. DOI: 10.1108/RPJ-05-2015-0059

Chohan, J. S., Singh, R., Boparai, K. S., Penna, R., & Fraternali, F. (2017). Dimensional accuracy analysis of coupled fused deposition modeling and vapour smoothing operations for biomedical applications. *Composites. Part B, Engineering, 117*, 138–149. DOI: 10.1016/j.compositesb.2017.02.045

Choi, B. W., & Leu, M. C. (2000). Determining optimal parameters for stereolithography processes via genetic algorithm. *Journal of Manufacturing Systems, 19*(I), 18–27. DOI: 10.1016/S0278-6125(00)88887-1

Choi, S.-H., & Kwok, K. T. (2002). Hierarchical slice contours for layered-manufacturing. *Computers in Industry, 48*(3), 219–239. DOI: 10.1016/S0166-3615(02)00040-4

Cho, K. T., Nuñez, L., Shelton, J., & Sciammarella, F. M. (2023). Investigation of effect of processing parameters for direct energy deposition additive manufacturing technologies. *Journal of Manufacturing and Materials Processing, 7*(3), 105. DOI: 10.3390/jmmp7030105

Choma, T., Żrodowski, Ł., Ciftci, J., Morończyk, B., & Kalicki, B. (2024, February). Overcoming challenges in custom powder manufacturing—From low-melting and reactive materials to refractories. In *TMS Annual Meeting & Exhibition* (pp. 257–269). Springer Nature Switzerland. https://doi.org/DOI: 10.1007/978-3-031-50349-8_24

Choonara, Y. E., Toit, L. C. du, Kumar, P., Kondiah, P. P. D., & Pillay, V. (2016). 3D-printing and the effect on medical costs: a new era? [Review of 3D-printing and the effect on medical costs: a new era?]. Expert Review of Pharmacoeconomics & Outcomes Research, 16(1), 23. Taylor & Francis. DOI: 10.1586/14737167.2016.1138860

Choudhari, A., Elder, J., Mugale, M., Karki, S., Digole, S., Omeike, S., & Borkar, T. (2024). Enhancing quality control: Image-based quantification of carbides and defect remediation in binder jetting additive manufacturing. *Materials (Basel), 17*(10), 2174. DOI: 10.3390/ma17102174 PMID: 38793241

Choudhary, N., Singh, A. K., & Sharma, V. (2024). Materials for additive manufacturing. In Sharma, P., & Badheka, V. (Eds.), *Additive manufacturing for advance applications: Technologies, challenges and case studies* (pp. 1–20). CRC Press., DOI: 10.1201/9781003484325-1

Christensen, A., & Rybicki, F. J. (2017). Maintaining safety and efficacy for 3D printing in medicine [Review of Maintaining safety and efficacy for 3D printing in medicine]. 3D Printing in Medicine, 3(1). BioMed Central. DOI: 10.1186/s41205-016-0009-5

Christ, M., Zimmermann, C., Neinert, S., Leykauf, B., Döringshoff, K., & Krutzik, M. (2024). Additively manufactured ceramics for compact quantum technologies. *Advanced Quantum Technologies*, *7*(12), 2400076. DOI: 10.1002/qute.202400076

Chua, C. K., & Leong, K. F. (2013). 3D Printing and Additive Manufacturing. DOI: 10.1142/9008

Chu, H., Yang, W., Sun, L., Cai, S., Yang, R., Liang, W., Yu, H., & Liu, L. (2020). 4D printing: A review on recent progresses. *Micromachines*, *11*(9), 796. DOI: 10.3390/mi11090796 PMID: 32842588

Chung, J. J., Im, H., Kim, S. H., Park, J. W., & Jung, Y. (2020). Toward biomimetic scaffolds for tissue engineering: 3D printing techniques in regenerative medicine. *Frontiers in Bioengineering and Biotechnology*, *8*, 586406. DOI: 10.3389/fbioe.2020.586406 PMID: 33251199

Chun, H. (2021). A study on the impact of 3D printing and artificial intelligence on education and learning process. *Scientific Programming*, *2021*(1), 2247346. DOI: 10.1155/2021/2247346

Coakley, M. F., & Hurt, D. E. (2016). 3D Printing in the Laboratory: Maximize Time and Funds with Customized and Open-Source Labware. *SLAS Technology*, *21*(4), 489–495. DOI: 10.1177/2211068216649578 PMID: 27197798

Cojocaru, V., Frunzăverde, D., & Micloşină, C.-O. (2022). On the Behavior of Honeycomb, Grid and Triangular PLA Structures under Symmetric and Asymmetric Bending. *Micromachines*, *14*(1), 120. DOI: 10.3390/mi14010120 PMID: 36677181

Collins, P. C., Brice, D. A., Samimi, P., Ghamarian, I., & Fraser, H. L. (2016). Microstructural control of additively manufactured metallic materials. *Annual Review of Materials Research*, *46*(1), 63–91. DOI: 10.1146/annurev-matsci-070115-031816

Cornejo, J., Cornejo-Aguilar, J. A., Vargas, M., Helguero, C. G., Milanezi de Andrade, R., Torres-Montoya, S., Asensio-Salazar, J., Rivero Calle, A., Martínez Santos, J., Damon, A., Quiñones-Hinojosa, A., Quintero-Consuegra, M. D., Umaña, J. P., Gallo-Bernal, S., Briceño, M., Tripodi, P., Sebastian, R., Perales-Villarroel, P., De la Cruz-Ku, G., & Russomano, T. (2022). Anatomical Engineering and 3D printing for surgery and medical devices: International review and future exponential innovations. *BioMed Research International*, *2022*(1), 6797745. DOI: 10.1155/2022/6797745 PMID: 35372574

Corral, L. C., Walker, K. J., Hubert, S. K., Smith, K. R., & Cheramie, L. M. (2017). *Exploring the abilities of 3D printing and its viability for consumption in the fashion industry. 74*(1).

Costa, L., Reti, T., Deus, M. A., & Vilar, R. (2002) Simulation of layer overlap tempering kinetics in steel parts deposited by laser cladding. In: *Proceedings of 2002 International Conference on Metal Powder Deposition for Rapid Manufacturing*, May 2014, pp. 172–179.

Cozzolino, E., Franchitti, S., Borrelli, R., Pirozzi, C., & Astarita, A. (2023). Energy consumption assessment in manufacturing ti6al4v electron beam melted parts post-processed by machining. *International Journal of Advanced Manufacturing Technology, 125*(3-4), 1289–1303. DOI: 10.1007/s00170-022-10794-z PMID: 36644782

Crespo, A., Deus, A., & Vilar, R. (2006) Finite element analysis of laser powder deposition of titanium. In: ICALEO 2006 - 25th International Congress on Applications of Laser and Electro-Optics, Congress Proceedings, January. DOI: 10.2351/1.5060805

Cristea, D., Pop, M. A., Faraian, C., & Munteanu, D. (2019). The influence of additive manufacturing parameters on the structural and mechanical properties of acrylonitrile butadiene styrene (ABS) parts produced by fused filament fabrication. *IOP Conference Series. Materials Science and Engineering, 682*(1), 12013. DOI: 10.1088/1757-899X/682/1/012013

Crockett, R. S., & Calvert, P. D. (1996) The liquid-to-solid transition in stereodeposition techniques. In: Solid freeform fabrication proceedings, pp. 257–264.

Cucinotta, F., Raffaele, M., & Salmeri, F. (2019). A stress-based topology optimization method by a Voronoi tessellation, Additive Manufacturing oriented. In The International Journal of Advanced Manufacturing Technology (Vol. 103, p. 1965). Springer Science+Business Media. DOI: 10.1007/s00170-019-03676-4

Cuiffo, M. A., Snyder, J., Elliott, A. M., Romero, N., Kannan, S., & Halada, G. P. (2017). Impact of the fused deposition (FDM) printing process on polylactic acid (PLA) chemistry and structure. *Applied Sciences (Basel, Switzerland), 7*(6), 579. DOI: 10.3390/app7060579

Czaja, A. U., Trukhan, N., & Müller, U. (2009). Industrial applications of metal–organic frameworks. *Chemical Society Reviews, 38*(5), 1284–1293. DOI: 10.1039/b804680h PMID: 19384438

D'Emilia, G., Ilio, A. D., Gaspari, A., Natale, E., & Stamopoulos, A. G. (2020). Uncertainty assessment for measurement and simulation in selective laser melting: A case study of an aerospace part. *Acta IMEKO, 9*(4), 96. DOI: 10.21014/acta_imeko.v9i4.720

da Silva, C. E. S., Salgado, E. G., Mello, C. H. P., Oliveira, E., & Leal, F. (2013). Integration of computer simulation in design for manufacturing and assembly. *International Journal of Production Research*, *52*(10), 2851–2866. DOI: 10.1080/00207543.2013.853887

Daly, M., Tarfaoui, M., Bouali, M., & Bendarma, A. (2024). Effects of infill density and pattern on the tensile mechanical behavior of 3d-printed glycolyzed polyethylene terephthalate reinforced with carbon-fiber composites by the fdm process. *Journal of Composites Science*, *8*(4), 115. DOI: 10.3390/jcs8040115

Daminabo, S. I. C., Goel, S., Grammatikos, S., Nezhad, H. Y., & Thakur, V. K. (2020). Fused deposition modeling-based additive manufacturing (3D printing): Techniques for polymer material systems. *Materials Today. Chemistry*, *16*, 100248. DOI: 10.1016/j.mtchem.2020.100248

Dani, I., Drossel, W., Milaev, N., Korn, H., Hannemann, C., Hohlfeld, J., & Wertheim, R. (2020). Sustainability of industrial components using additive manufacturing and foam materials. *Procedia Manufacturing*, *43*, 10–17. DOI: 10.1016/j.promfg.2020.02.102

Dasappa Ashoka, V. K., & Sidnekoppa, L. (2024). Challenges and Opportunities within Design for Additive Manufacturing.

Dass, A., & Moridi, A. (2019). State of the Art in Directed Energy Deposition: From Additive Manufacturing to Materials Design. *Coatings*, *9*(7), 418. DOI: 10.3390/coatings9070418

Davydov, R., Polyakov, N., Korenchenkova, D. E., Чулкин, С. Г., Rud', V. Yu., & Krasnoscshekov, V. (2021). Testing the methodology for evaluating the accuracy of the 3d printing. *Journal of Physics: Conference Series*, *1942*(1), 12057. DOI: 10.1088/1742-6596/1942/1/012057

Dawood, A., Marti, B. M., Sauret-Jackson, V., & Darwood, A. (2015). 3D printing in dentistry [Review of 3D printing in dentistry]. BDJ, 219(11), 521. Springer Nature. DOI: 10.1038/sj.bdj.2015.914

Deepak Kumar, S., Arun Manohar, G., & Surya Teja, R. (2022). The state of art 3D printing: A case study of Ganesh Idol. *Materials Today: Proceedings*, *56*, 455–461. Advance online publication. DOI: 10.1016/j.matpr.2022.01.418

Deshmukh, K., Muzaffar, A., Kovářík, T., Křenek, T., Ahamed, M. B., & Pasha, S. K. (2020). *Fundamentals and applications of 3D and 4D printing of polymers: challenges in polymer processing and prospects of future research. In 3D and 4D Printing of Polymer Nanocomposite Materials*. Elsevier.

Despeisse, M., & Ford, S. (2015). The Role of Additive Manufacturing in Improving Resource Efficiency and Sustainability. In IFIP, information and communication technology advances are made p. 129). Springer Science+Business Media. DOI: 10.1007/978-3-319-22759-7_15

Dey, A., & Yodo, N. (2019). A systematic survey of FDM process parameter optimization and their influence on part characteristics. *Journal of Manufacturing and Materials Processing*, *3*(3), 64. DOI: 10.3390/jmmp3030064

Dey, D., Srinivas, D., Panda, B., Suraneni, P., & Sitharam, T. G. (2022). Use of industrial waste materials for 3D printing of sustainable concrete: A review. *Journal of Cleaner Production*, *340*, 130749. DOI: 10.1016/j.jclepro.2022.130749

Dezaki, M. L., & Ariffin, M. K. A. M. (2020). The Effects of Combined Infill Patterns on Mechanical Properties in FDM Process. *Polymers*, *12*(12), 2792. DOI: 10.3390/polym12122792 PMID: 33255897

Dezaki, M. L., Ariffin, M. K. A. M., Serjouei, A., Zolfagharian, A., Hatami, S., & Bodaghi, M. (2021). Influence of Infill Patterns Generated by CAD and FDM 3D Printer on Surface Roughness and Tensile Strength Properties. *Applied Sciences (Basel, Switzerland)*, *11*(16), 7272. DOI: 10.3390/app11167272

Dhangar, M., Chaturvedi, K., Mili, M., Patel, S. S., Khan, M. A., Bhargaw, H. N., Srivastava, A. K., & Verma, S. (2023). Emerging 3d printed thermal insulating materials for sustainable approach: A review and a way forward. *Polymers for Advanced Technologies*, *34*(5), 1425–1434. DOI: 10.1002/pat.5989

Dhinakaran, V., Stalin, B., Ravichandran, M., Balasubramanian, M., Chairman, C. A., & Pritima, D. (2020). Wire Arc Additive Manufacturing Perspectives and Recent Developments. In IOP Conference Series Materials Science and Engineering (Vol. 988, Issue 1, p. 12102). IOP Publishing. DOI: 10.1088/1757-899X/988/1/012102

Dhyani, A., Singh, N., Kumar, V., & Dhyani, A. (2020). Applications of 3-Dimensional (3D) Printing in the Biomedical Field. *International Journal of Current Research and Review*, *12*(19), 71–75. DOI: 10.31782/IJCRR.2020.121918

Di Angelo, L., Di Stefano, P., Dolatnezhadsomarin, A., Guardiani, E., & Khorram, E. (2020). A reliable build orientation optimization method in additive manufacturing: The application to FDM technology. *International Journal of Advanced Manufacturing Technology*, *108*(1-2), 263–276. DOI: 10.1007/s00170-020-05359-x

Di Angelo, L., Di Stefano, P., & Marzola, A. (2017). Surface quality prediction in FDM additive manufacturing. *International Journal of Advanced Manufacturing Technology*, *93*(9-12), 3655–3662. DOI: 10.1007/s00170-017-0763-6

Di Salvo, S. (2024). Insights into new trends and contemporary challenges in 3D printing in architectural sector. *Engineering Innovations*, *8*, 43–61. DOI: 10.4028/p-VPN0Le

Dickson AN and Dowling DP (2019) Enhancing the bearing strength of woven carbon fibre thermoplastic composites through additive manufacturing. Composite Structures 212(December 2018): 381–388. .DOI: 10.1016/j.compstruct.2019.01.050

Diegel, O., Nordin, A., & Motte, D. (2020). *A practical guide to design for additive manufacturing*. Springer.

Digimat. (2019) The Multi-scale Nonlinear Material Modeling Platform. Part of Hexagon AB: E-Xstream Engineering an MSC Software Company.

Di, L., Yang, Y., & Wang, S. (2024). Additive manufacturing thermoplastic recycling: Profit-driven planning and optimization. *Journal of Cleaner Production*, *436*, 140598. DOI: 10.1016/j.jclepro.2024.140598

Dilberoglu, U. M., Gharehpapagh, B., Yaman, U., & Dolen, M. (2021). Current trends and research opportunities in hybrid additive manufacturing. *International Journal of Advanced Manufacturing Technology*, *113*(3-4), 623–648. DOI: 10.1007/s00170-021-06688-1

Diment, L., Thompson, M. S., & Bergmann, J. (2017). Clinical efficacy and effectiveness of 3D printing: a systematic review [Review of Clinical efficacy and effectiveness of 3D printing: a systematic review]. BMJ Open, 7(12). *BMJ (Clinical Research Ed.)*, *7*(12), e016891. Advance online publication. DOI: 10.1136/bmjopen-2017-016891 PMID: 29273650

Dimitrov, D., Schreve, K., & de Beer, N. (2006). Advances in three dimensional printing–state of the art and future perspectives. *Rapid Prototyping Journal*, *12*(3), 136–147. DOI: 10.1108/13552540610670717

Ding, H., Gao, R. X., Isaksson, A., Landers, R. G., Parisini, T., & Yuan, Y. (2020). State of AI-Based Monitoring in Smart Manufacturing and Introduction to Focused Section. *IEEE/ASME Transactions on Mechatronics*, *25*(5), 2143–2154. DOI: 10.1109/TMECH.2020.3022983

Di, Y., Zheng, Z., Pang, S., & Li, J. J. (2022). Numerical and experimental investigation of the thermal behavior and microstructure of wire arc additive manufactured 316l stainless steel straight wall part. *IOP Conference Series. Materials Science and Engineering*, *1270*(1), 012084. DOI: 10.1088/1757-899X/1270/1/012084

Djokikj, J., & Kandikjan, T. (2023). DfAM: Application of the design rules in the early design stages. *Procedia CIRP*, *118*, 659–663. DOI: 10.1016/j.procir.2023.06.113

Dominguez-Alfaro, A., Mitoudi-Vagourdi, E., Dimov, I. B., Picchio, M. L., Lopez-Larrea, N., de Lacalle, J. L., Tao, X., Serrano, R. R., Gallastegui, A., Vassardanis, N., Mecerreyes, D., & Malliaras, G. G. (2024). Light-Based 3D Multi-Material Printing of Micro-Structured Bio-Shaped, Conducting and Dry Adhesive Electrodes for Bioelectronics. *Advancement of Science, 11*(27), 2306424. Advance online publication. DOI: 10.1002/advs.202306424 PMID: 38251224

Dong, L., Makradi, A., Ahzi, S., & Remond, Y. (2008). Three-dimensional transient finite element analysis of the selective laser sintering process. *Journal of Materials Processing Technology, 9*(2), 700–706. DOI: 10.1016/j.jmatprotec.2008.02.040

Doshi, M., Mahale, A., Singh, S. K., & Deshmukh, S. (2022). Printing parameters and materials affecting mechanical properties of FDM-3D printed Parts: Perspective and prospects. *Materials Today: Proceedings, 50*, 2269–2275. DOI: 10.1016/j.matpr.2021.10.003

Dou, H., Cheng, Y., Ye, W., Zhang, D., Li, J., Miao, Z., & Rudykh, S. (2020). Effect of process parameters on tensile mechanical properties of 3D printing continuous carbon fiber-reinforced PLA composites. *Materials (Basel), 13*(17), 3850. DOI: 10.3390/ma13173850 PMID: 32878337

Drebezgova, M. Yu., Чернышева, Н., Klimenko, V. A., & Olisov, A. V. (2019). Architectural design of residential modules using additive technologies. In IOP Conference Series Materials Science and Engineering (Vol. 698, Issue 3, p. 33007). IOP Publishing. DOI: 10.1088/1757-899X/698/3/033007

Druzgalski, C., Ashby, A. S., Guss, G., King, W. E., Roehling, T. T., & Matthews, M. J. (2020). Process optimization of complex geometries using feed-forward control for laser powder bed fusion additive manufacturing. *Additive Manufacturing, 34*, 101169. DOI: 10.1016/j.addma.2020.101169

Duan, G., & Yan, X. (2020). A Real-Time Quality Control System Based on Manufacturing Process Data. *IEEE Access : Practical Innovations, Open Solutions, 8*, 208506–208517. DOI: 10.1109/ACCESS.2020.3038394

Duarte, L. C., Figueredo, F., Chagas, C. L., Cortón, E., & Coltro, W. K. (2024). A review of the recent achievements and future trends on 3D printed microfluidic devices for bioanalytical applications. *Analytica Chimica Acta, 1299*, 342429. DOI: 10.1016/j.aca.2024.342429 PMID: 38499426

Dul, S., Fambri, L., & Pegoretti, A. (2021). High-Performance Polyamide/Carbon Fiber Composites for Fused Filament Fabrication: Mechanical and Functional Performances. *Journal of Materials Engineering and Performance, 30*(7), 5066–5085. DOI: 10.1007/s11665-021-05635-1

Dutta, S., Chanda, A., Rahman, M. Z., Das, R., & Bhattacharyya, D. (2019) A numerical model to simulate the impact response of Flax-PP composites. In: *ICCM International Conference on Composite Materials.*

Du, Z., Liu, J., & He, Y. (2019). Intelligent Analysis Method of Assembly Process Based on Semantic Elements. *IOP Conference Series. Materials Science and Engineering, 563*(4), 42055. DOI: 10.1088/1757-899X/563/4/042055

Dvorak, K., Dvorakova, J., Zarybnicka, L., & Horak, Z. (2021). Influence of 3D printing topology by DMLS method on crack propagation. *Materials (Basel), 14*(23), 7483. DOI: 10.3390/ma14237483 PMID: 34885639

Dwivedi, H., Tushar, , Singh, S., Rani, P., Ananya, , & Kumar, S. (2024). Analysis of the microstructural and mechanical properties of 3d-printed removable partial denture base materials. *Journal of Pharmacy & Bioallied Sciences, 16*(Suppl 1), S681–S683. DOI: 10.4103/jpbs.jpbs_939_23 PMID: 38595564

Dwivedi, Y. K., Hughes, L., Ismagilova, E., Aarts, G., Coombs, C., Crick, T., Duan, Y., Dwivedi, R., Edwards, J. S., Eirug, A., Galanos, V., Ilavarasan, P. V., Janssen, M., Jones, P., Kar, A. K., Kizgin, H., Kronemann, B., Lal, B., Lucini, B., & Williams, M. D. (2019). Artificial Intelligence (AI): Multidisciplinary perspectives on emerging challenges, opportunities, and agenda for research, practice, and policy. *International Journal of Information Management, 57*, 101994. DOI: 10.1016/j.ijinfomgt.2019.08.002

Egan, P. (2019). Integrated Design Approaches for 3D Printed Tissue Scaffolds: Review and Outlook [Review of Integrated Design Approaches for 3D Printed Tissue Scaffolds: Review and Outlook]. Materials, 12(15), 2355. Multidisciplinary Digital Publishing Institute. DOI: 10.3390/ma12152355

Eisenberg, M. (2013). 3D printing for children: What to build next? *International Journal of Child-Computer Interaction, 1*(1), 7–13. DOI: 10.1016/j.ijcci.2012.08.004

Ejjami, R., & Boussalham, K. (2024). Industry 5.0 in Manufacturing: Enhancing Resilience and Responsibility through AI-Driven Predictive Maintenance, Quality Control, and Supply Chain Optimization. International Journal For Multidisciplinary Research, 6(4).

El-Eskandarany, M. S. (2013). *Mechanical alloying: For fabrication of advanced engineering materials.* Univ. Press of Mississippi.

Elsayed, A. *et al.* (2024) 'Low charge noise quantum dots with industrial CMOS manufacturing', *npj Quantum Information*, 10(1), p. 70. Available at: .DOI: 10.1038/s41534-024-00864-3

El-Sayegh, S., Romdhane, L., & Manjikian, S. (2020). A critical review of 3D printing in construction: Benefits, challenges, and risks. *Archives of Civil and Mechanical Engineering*, *20*(2), 1–25. DOI: 10.1007/s43452-020-00038-w

Enferadi, A., Baniassadi, M., & Baghani, M. (2024). Innovative multiphysics approach for designing high-performance thermo-responsive shape memory polymer microvalve. *European Journal of Mechanics. A, Solids*, *103*, 105174. DOI: 10.1016/j.euromechsol.2023.105174

Enferadi, A., Ostadrahimi, A., Li, G., Baniassadi, M., & Baghani, M. (2024). Insight into single-helix intelligent shape memory polymer cables: Modeling and optimization under finite sliding contact. *Smart Materials and Structures*, *33*(5), 055033. DOI: 10.1088/1361-665X/ad3bf9

Equbal, A., Equbal, M. A., Sood, A. K., Pranav, R., & Equbal, M. I. (2018, December). A review and reflection on part quality improvement of fused deposition modelled parts. [). IOP Publishing.]. *IOP Conference Series. Materials Science and Engineering*, *455*(1), 012072.

Equbal, M. A., Equbal, A., Khan, Z. A., & Badruddin, I. A. (2025). Machine learning in additive manufacturing: A comprehensive insight. *International Journal of Lightweight Materials and Manufacture*, *8*(2), 264–284.

Erdaş, M. U., Yıldız, B. S., & Yıldız, A. R. (2024). Experimental analysis of the effects of different production directions on the mechanical characteristics of ABS, PLA, and PETG materials produced by FDM. *Materialprüfung (Munchen)*, *66*(2), 198–206. DOI: 10.1515/mt-2023-0206

Eriksson, M., Lervåg, M., Sørensen, C., Robertstad, A., Brønstad, B. M., Nyhus, B., Aune, R. E., Ren, X., & Akselsen, O. M. (2018). Additive manufacture of superduplex stainless steel using WAAM. In MATEC Web of Conferences (Vol. 188, p. 3014). EDP Sciences. DOI: 10.1051/matecconf/201818803014

Eryildiz, M. (2024). Tailoring mechanical properties of FDM-3D-printed parts through titanium dioxide-reinforced resin filling technique. *Proceedings of the Institution of Mechanical Engineers, Part B: Journal of Engineering Manufacture*. Advance online publication. https://doi.org/DOI: 10.1177/09544054241260467

Esene, J. E., Nasman, P. R., Akuoko, Y., Tahir, A., & Woolley, A. T. (2023). Past, current, and future roles of 3D printing in the development of capillary electrophoresis systems. *Trends in Analytical Chemistry*, *162*, 117032. DOI: 10.1016/j.trac.2023.117032 PMID: 37008739

Espalin, D., Muse, D. W., MacDonald, E., & Wicker, R. B. (2014). 3D Printing multifunctionality: Structures with electronics. *International Journal of Advanced Manufacturing Technology, 72*(5-8), 963–978. DOI: 10.1007/s00170-014-5717-7

Ewart, P. (2019). *A comparison of processing techniques for producing prototype injection moulding inserts.*

Faes, M. G. R., Abbeloos, W., Vogeler, F., Valkenaers, H., Coppens, K., Goedemé, T., & Ferraris, E. (2014). Process Monitoring of Extrusion-Based 3D Printing via Laser Scanning. In arXiv (Cornell University) (Vol. 6, p. 363). Cornell University. https://arxiv.org/abs/1612.02219

Faizaan, M., Shenoy, S., & Kini, C. R. (2024). Tensile and flexural performance of hybrid fdm and compression moulded pla/basalt biocomposite. *Materials Science Forum, 1120,* 77–84. DOI: 10.4028/p-duYO7M

Fakirov, S., Rahman, M. Z., Pötschke, P., & Bhattacharyya, D. (2014). Single polymer composites of poly (B utylene Terephthalate) microfibrils loaded with carbon nanotubes exhibiting electrical conductivity and improved mechanical properties. *Macromolecular Materials and Engineering, 299*(7), 799–806. DOI: 10.1002/mame.201300322

Fan, D., Li, Y., Wang, X., Zhu, T., Wang, Q., Cai, H., Li, W., Tian, Y., & Liu, Z. (2020). Progressive 3D Printing Technology and Its Application in Medical Materials [Review of Progressive 3D Printing Technology and Its Application in Medical Materials]. Frontiers in Pharmacology, 11. Frontiers Media. DOI: 10.3389/fphar.2020.00122

Fan, Z., & Liou, F. (2012) Numerical modeling of the additive manufacturing (am) processes of titanium alloy. In: Titanium Alloys—Towards Achieving Enhanced Properties for Diversified Applications, pp. 3–29. DOI: 10.5772/34848

Fan, H., Liu, C., Bian, S., Ma, C., Huang, J., Liu, X., Doyle, M., Lu, T., Chow, E., Chen, L., Fuh, J. Y. H., Lu, W. F., & Li, B. (2025). New era towards autonomous additive manufacturing: A review of recent trends and future perspectives. *International Journal of Extreme Manufacturing, 7*(3), 032006. DOI: 10.1088/2631-7990/ada8e4

Farid, M. I., Wu, W., Liu, X., & Wang, P. (2021). Additive manufacturing landscape and materials perspective in 4D printing. *International Journal of Advanced Manufacturing Technology, 115*(9-10), 2973–2988. DOI: 10.1007/s00170-021-07233-w PMID: 34092882

Fayazbakhsh, K., Movahedi, M., & Kalman, J. (2018). The impact of defects on tensile properties of 3D printed parts manufactured by fused filament fabrication. *Materials Today. Communications, 18*, 140–148. DOI: 10.1016/j.mtcomm.2018.12.003

Fazzini, G., Paolini, P., Paolucci, R., Chiulli, D., Barile, G., Leoni, A., & Ferri, G. (2019, June). Print on air: Fdm 3d printing without supports. In *2019 II Workshop on Metrology for Industry 4.0 and IoT (MetroInd4. 0&IoT)* (pp. 350-354). IEEE. DOI: 10.1109/METROI4.2019.8792846

Fei, G., Parra-Cabrera, C., Zhong, K., Clays, K., & Ameloot, R. (2024). From gray-scale photopolymerization 3D printing to functionally graded materials. *Advanced Functional Materials, 34*(32), 2314635. DOI: 10.1002/adfm.202314635

Feng, G., Ou, Y., Rahman, M. Z., Zhou, L., Zhao, H., Chen, Q., Mao, D., & Zhu, D. (2023). Improving corrosion resistance of BFRP bars by coating CNTs modified resin in simulated pore solution of seawater sea sand concrete. *Construction & Building Materials, 392*, 131945. DOI: 10.1016/j.conbuildmat.2023.131945

Feng, K., Hu, S., Zhao, W., Sun, J., Mao, Y., Cai, D., Wu, J., & Wei, Q. (2024). Bimodal powder optimization in SiC binder jetting for mechanical performance. *International Journal of Mechanical Sciences, 274*, 109278. DOI: 10.1016/j.ijmecsci.2024.109278

Feng, T., Qamruzzaman, M., Sharmin, S. S., & Karim, S. (2024). Bridging Environmental Sustainability and Organizational Performance: The Role of Green Supply Chain Management in the Manufacturing Industry. *Sustainability (Basel), 16*(14), 5918. DOI: 10.3390/su16145918

Fernández, F. M., & Martín, M. J. (2023). Analysis of the Effect of the Surface Inclination Angle on the Roughness of Polymeric Parts Obtained with Fused Filament Fabrication Technology. *Polymers, 15*(3), 585. DOI: 10.3390/polym15030585 PMID: 36771886

Festas, A., Ramos, A., & Davim, J. (2020). Medical devices biomaterials–A review. *Proceedings of the institution of mechanical engineers, part L. Journal of Materials: Design and Applications, 234*(1), 218–228.

Ficzere, P., Lukács, N. L., & Borbás, L. (2021). The Investigation of Interlaminar Failures Caused by Production Parameters in Case of Additive Manufactured Polymers. *Polymers, 13*(4), 556. DOI: 10.3390/polym13040556 PMID: 33668572

Fidan, I., Naikwadi, V., Alkunte, S., Mishra, R., & Tantawi, K. (2024). Energy efficiency in additive manufacturing: Condensed review. *Technologies, 12*(2), 21. DOI: 10.3390/technologies12020021

Filip, G. A., Abanda, F. H., & Fru, F. A. (2023). Construction 3D-printing reduces the incidence of long latency respiratory diseases among construction workers in the UK. *Safety in Extreme Environments*, *5*(3), 177–197. DOI: 10.1007/s42797-023-00078-4

Filippidis, P., Papazetis, G., & Vosniakos, G.-C. (2020). Process parameter investigation for 3D printing of cellular structured parts. *Procedia Manufacturing*, *51*, 717–724. DOI: 10.1016/j.promfg.2020.10.101

Firtikiadis, L., Minaoglou, P., Manavis, A., Aidinli, K., & Efkolidis, N. (2022). Circular economy through customised 3d printed products: a case of souvenir. Proceedings - The Eleventh International Symposium GRID 2022, 467-471. DOI: 10.24867/GRID-2022-p51

Fleck, N. A., Kang, K. J., & Ashby, M. F. (1994). Overview no. 112: The cyclic properties of engineering materials. *Acta Metallurgica et Materialia*, *42*(2), 365–381. DOI: 10.1016/0956-7151(94)90493-6

Fontana-Escartín, A., Lanzalaco, S., Bertran, O., & Alemán, C. (2022). Electrochemical multi-sensors obtained by applying an electric discharge treatment to 3D-printed poly (lactic acid). *Applied Surface Science*, *597*, 153623. DOI: 10.1016/j.apsusc.2022.153623

Ford, S., & Despeisse, M. (2016). Additive manufacturing and sustainability: an exploratory study of the advantages and challenges. In *Journal of Cleaner Production* (Vol. 137, p. 1573). Elsevier BV., DOI: 10.1016/j.jclepro.2016.04.150

Ford, S., & Minshall, T. (2019). Invited review article: Where and how 3D printing is used in teaching and education. *Additive Manufacturing*, *25*, 131–150. DOI: 10.1016/j.addma.2018.10.028

Franke, J., Heinrich, F., & Reisch, R. T. (2025). Vision based process monitoring in wire arc additive manufacturing (WAAM). *Journal of Intelligent Manufacturing*, *36*(3), 1711–1721. DOI: 10.1007/s10845-023-02287-x

Frascio, M., Avalle, M., & Monti, M. (2018). Fatigue strength of plastics components made in additive manufacturing: First experimental results. *Procedia Structural Integrity*, *12*, 32–43. DOI: 10.1016/j.prostr.2018.11.109

Frizziero, L., Liverani, A., Donnici, G., Osti, F., Neri, M., Maredi, E., Trisolino, G., & Stilli, S. (2019). New Methodology for Diagnosis of Orthopedic Diseases through Additive Manufacturing Models. *Symmetry*, *11*(4), 542. DOI: 10.3390/sym11040542

Fu, K., Yao, Y., Dai, J., & Hu, L. (2017). Progress in 3D printing of carbon materials for energy-related applications. *Advanced Materials, 29*(9), 1603486. DOI: 10.1002/adma.201603486 PMID: 27982475

Gao, J., Hu, Z., Yang, Q., Liang, X., & Wu, H. (2022). Fluid flow and heat transfer in microchannel heat sinks: Modelling review and recent progress. *Thermal Science and Engineering Progress, 29*, 101203. DOI: 10.1016/j.tsep.2022.101203

Gao, X., Fan, S., Pang, J., Rahman, M. Z., Zhu, D., Guo, S., Ma, M., & Li, Z. (2021). Preparation of nano-xylan and its influences on the anti-fungi performance of straw fiber/HDPE composite. *Industrial Crops and Products, 171*, 113954. DOI: 10.1016/j.indcrop.2021.113954

Gao, X., Zhu, D., Fan, S., Rahman, M. Z., Guo, S., & Chen, F. (2022). Structural and mechanical properties of bamboo fiber bundle and fiber/bundle reinforced composites: A review. *Journal of Materials Research and Technology, 19*, 1162–1190. DOI: 10.1016/j.jmrt.2022.05.077

Garcia, C. R., Rumpf, R. C., Tsang, H., & Barton, J. H. (2013). Effects of extreme surface roughness on 3D printed horn antenna. *Electronics Letters, 49*(12), 734–736. DOI: 10.1049/el.2013.1528

García-Dominguez, A., Claver, J., & Sebastián, M. A. (2020). Optimization methodology for additive manufacturing of customized parts by fused deposition modeling (FDM). Application to a shoe heel. *Polymers, 12*(9), 2119. DOI: 10.3390/polym12092119 PMID: 32957582

Garcia, F. L., Moris, V. A. da S., Nunes, A. O., & Silva, D. A. L. (2018). Environmental performance of additive manufacturing process – an overview. *Rapid Prototyping Journal, 24*(7), 1166–1177. DOI: 10.1108/RPJ-05-2017-0108

Garcia, G. E. S., de Sousa, R. R.Junior, Gouveia, J. R., & Dos Santos, D. J. (2024). Graphene oxide-based nanocomposites for stereolithography (SLA) 3D printing: Comprehensive mechanical characterization under combined loading modes. *Polymers, 16*(9), 1261. DOI: 10.3390/polym16091261 PMID: 38732730

Garcia-Granada, A. A., Rostro-González, H., Puigoriol-Forcada, J. M., & Reyes-Pozo, G. (2023). Open material database for tensile test properties of additive manufacturing materials. *IOP Conference Series. Materials Science and Engineering, 1294*(1), 012043. DOI: 10.1088/1757-899X/1294/1/012043

Gardan, J. (2020). Additive manufacturing technologies for polymer composites: State-of-the-art and future trends. In *Elsevier eBooks* (p. 3). Elsevier BV., DOI: 10.1016/B978-0-12-819535-2.00001-6

Gardner, J., & Webster, A. (2017). Accelerating Innovation in the Creation of Biovalue: The Cell and Gene Therapy Catapult. *Science, Technology & Human Values, 42*(5), 925–946. DOI: 10.1177/0162243917702720 PMID: 28845068

Gargalis, L., Madonna, V., Giangrande, P., Rocca, R., Ashcroft, I., Hague, R., & Galea, M. (2020)... *Development and Testing of Soft Magnetic Rotor for a Switched Reluctance Motor Built Through Additive Manufacturing Technology., 263*, 263–268. Advance online publication. DOI: 10.23919/ICEMS50442.2020.9290893

Garland, A., & Fadel, G. (2015). Design and manufacturing functionally gradient material objects with an off the shelf three-dimensional printer: Challenges and solutions. *Journal of Mechanical Design, 137*(11), 111407. DOI: 10.1115/1.4031097

Garner, E., Wu, J., & Zadpoor, A. A. (2022). Multi-objective design optimization of 3D micro-architected implants. *Computer Methods in Applied Mechanics and Engineering, 396*, 115102. DOI: 10.1016/j.cma.2022.115102

Gartner AM report (2014). Information on http://www.gartner.com/ document/2598122

Garzon-Hernandez, S., Garcia-Gonzalez, D., Jérusalem, A., & Arias, A. (2020). Design of FDM 3D printed polymers: An experimental-modelling methodology for the prediction of mechanical properties. *Materials & Design, 188*, 108414. DOI: 10.1016/j.matdes.2019.108414

Gebrehiwot, S., Gebrehiwot, A. S. Z., Espinosa-Leal, L., Linderbäck, P., & Remes, H. (2023). Optimising the mechanical properties of additive-manufactured recycled polylactic acid (rpla) using single and multi-response analyses methods... DOI: 10.21203/rs.3.rs-3115087/v1

Geiregat, M., & De Vos, S. (2024). *Leerpad over FFF 3D-printing voor FabLab-gebruikers.*

Gerdes, S., Gaikwad, A., Ramesh, S., Rivero, I. V., Tamayol, A., & Rao, P. (2024). Monitoring and control of biological additive manufacturing using machine learning. *Journal of Intelligent Manufacturing, 35*(3), 1055–1077. DOI: 10.1007/s10845-023-02092-6

Ghaffar, S. H., Corker, J., & Fan, M. (2018). Additive manufacturing technology and its implementation in construction as an eco-innovative solution. *Automation in Construction, 93*, 1–11. DOI: 10.1016/j.autcon.2018.05.005

Ghazal, A. F., Zhang, M., & Liu, Z. (2019). Spontaneous color change of 3D printed healthy food product over time after printing as a novel application for 4D food printing. *Food and Bioprocess Technology, 12*(10), 1627–1645. DOI: 10.1007/s11947-019-02327-6

Ghazal, A. F., Zhang, M., Mujumdar, A. S., & Ghamry, M. (2023). Progress in 4D/5D/6D printing of foods: Applications and R&D opportunities. *Critical Reviews in Food Science and Nutrition, 63*(25), 7399–7422. DOI: 10.1080/10408398.2022.2045896 PMID: 35225117

Ghilan, A., Chiriac, A. P., Niţă, L. E., Rusu, A. G., Neamţu, I., & Chiriac, V. M. (2020). Trends in 3D Printing Processes for Biomedical Field: Opportunities and Challenges [Review of Trends in 3D Printing Processes for Biomedical Field: Opportunities and Challenges]. Journal of Polymers and the Environment, 28(5), 1345. Springer Science+Business Media. DOI: 10.1007/s10924-020-01722-x

Giannopoulos, A. A., Chepelev, L., Sheikh, A., Wang, A., Dang, W., Akyuz, E., Hong, C. J., Wake, N., Pietila, T., Dydynski, P., Mitsouras, D., & Rybicki, F. J. (2015). 3D printed ventricular septal defect patch: a primer for the 2015 Radiological Society of North America (RSNA) hands-on course in 3D printing. 3D Printing in Medicine, 1(1). DOI: 10.1186/s41205-015-0002-4

Gibson, I., Rosen, D., Stucker, B., Khorasani, M., Gibson, I., Rosen, D., ... & Khorasani, M. (2021b). Design for additive manufacturing. Additive manufacturing technologies, 555-607.

Gibson, I., Rosen, D. W., Stucker, B., Khorasani, M., Rosen, D., Stucker, B., & Khorasani, M. (2021a). *Additive manufacturing technologies* (Vol. 17). Springer. DOI: 10.1007/978-3-030-56127-7

Gift, M. M., Senthil, T. S., Hasan, D. S., Alagarraja, K., Jayaseelan, P., & Boopathi, S. (2024). Additive Manufacturing and 3D Printing Innovations: Revolutionizing Industry 5.0. In *Technological Advancements in Data Processing for Next Generation Intelligent Systems* (pp. 255-287). IGI Global.

Ginestra, P. S., Ceretti, E., & Fiorentino, A. (2016). Potential of modeling and simulations of bioengineered devices: Endoprostheses, prostheses and orthoses [Review of Potential of modeling and simulations of bioengineered devices: Endoprostheses, prostheses and orthoses]. Proceedings of the Institution of Mechanical Engineers Part H Journal of Engineering in Medicine, 230(7), 607. SAGE Publishing. DOI: 10.1177/0954411916643343

Glicksman, M. E. (2011). *Principles of solidification.* Springer. DOI: 10.1007/978-1-4419-7344-3

Goala, S., & Sarkar, P. (2024). Selection methodology for additive manufacturing processes considering design, material, and manufacturability. *Rapid Prototyping Journal, 30*(1), 15–32. DOI: 10.1108/RPJ-05-2023-0155

Goh, G. D., Sing, S. L., & Yeong, W. Y. (2020). A review on machine learning in 3D printing: applications, potential, and challenges [Review of A review on machine learning in 3D printing: applications, potential, and challenges]. Artificial Intelligence Review, 54(1), 63. Springer Science+Business Media. DOI: 10.1007/s10462-020-09876-9

Goh, G. D., Wong, K. K., Tan, N., Seet, H. L., & Nai, M. L. S. (2024). Large-format additive manufacturing of polymers: A review of fabrication processes, materials, and design. *Virtual and Physical Prototyping, 19*(1), e2336160. DOI: 10.1080/17452759.2024.2336160

Gokulkumar, S., Thyla, P. R., Prabhu, L., & Sathish, S. (2020). Measuring methods of acoustic properties and influence of physical parameters on natural fibers: A review. *Journal of Natural Fibers, 17*(12), 1719–1738. DOI: 10.1080/15440478.2019.1598913

Golubović, Z., Milovanović, A., & Mitrović, A. (2022). *Characterization of 3D printed parts*. 56-56.

Golvaskar, M., Ojo, S. A., & Kannan, M. (2024). Recent Advancements in Material Waste Recycling: Conventional, Direct Conversion, and Additive Manufacturing Techniques. *Recycling, 9*(3), 43. DOI: 10.3390/recycling9030043

Gomez, S. (2023). *An Experimental Study on the Mechanical Properties and Chemical Composition of LCD 3D Printed Specimens.*

Gomez-Garcia, C. A., Askar-Rodriguez, M., & Velasco-Medina, J. (2021). Platform for Healthcare Promotion and Cardiovascular Disease Prevention. *IEEE Journal of Biomedical and Health Informatics, 25*(7), 2758–2767. DOI: 10.1109/JBHI.2021.3051967 PMID: 33449888

Gong, H., Snelling, D., Kardel, K., & Carrano, A. (2019). Comparison of stainless steel 316L parts made by FDM-and SLM-based additive manufacturing processes. *Journal of the Minerals Metals & Materials Society, 71*(3), 880–885. DOI: 10.1007/s11837-018-3207-3

Goodridge, R. D., Shofner, M. L., Hague, R. J. M., McClelland, M., Schlea, M. R., Johnson, R. B., & Tuck, C. J. (2011). Processing of a polyamide-12/carbon nanofibre composite by laser sintering. *Polymer Testing, 30*(1), 94–100. DOI: 10.1016/j.polymertesting.2010.10.011

Gopinathan, J., & Noh, I. (2018). Recent trends in bioinks for 3D printing. *Biomaterials Research, 22*(1), 11. DOI: 10.1186/s40824-018-0122-1 PMID: 29636985

Gordan, V. (2017). How to Bridge Research Results to Everyday Clinical Care? *Operative Dentistry, 42*(1), 1–9. DOI: 10.2341/16-154-B PMID: 28002694

Gordelier, T. J., Thies, P. R., Turner, L., & Johanning, L. (2019). Optimising the FDM additive manufacturing process to achieve maximum tensile strength: A state-of-the-art review. *Rapid Prototyping Journal, 25*(6), 953–971. DOI: 10.1108/RPJ-07-2018-0183

Gorelick, S., Guzenko, V. A., Vila-Comamala, J., & David, C. (2010). Direct e-beam writing of dense and high aspect ratio nanostructures in thick layers of PMMA for electroplating. *Nanotechnology, 21*(29), 295303. DOI: 10.1088/0957-4484/21/29/295303 PMID: 20601756

Gottesman, A. J., & Hoskins, S. G. (2013) 'CREATE Cornerstone: Introduction to Scientific Thinking, a New Course for STEM-Interested Freshmen, Demystifies Scientific Thinking through Analysis of Scientific Literature', *CBE—Life Sciences Education*. Edited by J. Vazquez, 12(1), pp. 59–72. Available at: DOI: 10.1187/cbe.12-11-0201

Gowda, R. B. S., Udayagiri, C. S., & Narendra, D. D. (2014). Studies on the Process Parameters of Rapid Prototyping Technique (Stereolithography) for the Betterment of Part Quality. *International Journal of Manufacturing Engineering, 2014*(1), 804705. DOI: 10.1155/2014/804705

Graham, A. H. D., Robbins, J., Bowen, C. R., & Taylor, J. (2011). Commercialisation of CMOS Integrated Circuit Technology in Multi-Electrode Arrays for Neuroscience and Cell-Based Biosensors. *Sensors (Basel), 11*(5), 4943–4971. DOI: 10.3390/s110504943 PMID: 22163884

Granelli, R., Kovács-Vajna, Z. M., & Torricelli, F. (2025). Additive Manufacturing of Organic Electrochemical Transistors: Methods, Device Architectures, and Emerging Applications. *Small, 21*(11), 2410499. DOI: 10.1002/smll.202410499 PMID: 39945058

Gray, R. W.IV, Baird, D. G., & Bohn, J. H. (1998). Effects of processing conditions on short TLCP fiber reinforced FDM parts. *Rapid Prototyping Journal, 4*(1), 14–25. DOI: 10.1108/13552549810197514

Graziosi, S., Rosa, F., Casati, R., Solarino, P., Vedani, M., & Bordegoni, M. (2017). Designing for Metal Additive Manufacturing: A Case Study in the Professional Sports Equipment Field. In Procedia Manufacturing (Vol. 11, p. 1544). Elsevier BV. DOI: 10.1016/j.promfg.2017.07.288

Graziosi, S., Faludi, J., Stanković, T., Borgianni, Y., Meisel, N., Hallstedt, S. I., & Rosen, D. W. (2024). A vision for sustainable additive manufacturing. *Nature Sustainability*, *7*(6), 698–705. DOI: 10.1038/s41893-024-01313-x

Green, D. E., Hamory, B. H., Terrell, G. E., & O'Connell, J. (2017). A Case Report: Cornerstone Health Care Reduced the Total Cost of Care Through Population Segmentation and Care Model Redesign. *Population Health Management*, *20*(4), 309–317. DOI: 10.1089/pop.2016.0105 PMID: 28106518

Grinberg, D., Siddique, S., Le, M. Q., Liang, R., Capsal, J. F., & Cottinet, P. J. (2019). 4D Printing based piezoelectric composite for medical applications. *Journal of Polymer Science. Part B, Polymer Physics*, *57*(2), 109–115. DOI: 10.1002/polb.24763

Grozav, S. D., Sterca, A. D., Kočiško, M., Pollák, M., & Ceclan, V. (2023). Artificial neural network-based predictive model for finite element analysis of additive-manufactured components. *Machines (Basel)*, *11*(5), 547. DOI: 10.3390/machines11050547

Grünberger, T. (2020). Machine Learning for Process Monitoring Systems. *PhotonicsViews*, *17*(3), 56–59. DOI: 10.1002/phvs.202000026

Gu, D. (2016). Materials creation adds new dimensions to 3D printing. *Science Bulletin*, *61*(22), 1718–1722. DOI: 10.1007/s11434-016-1191-y

Gu, D. D., Meiners, W., Wissenbach, K., Poprawe, R., Gu, D. D., Meiners, W., Wissenbach, K., & Poprawe, R. (2013). Laser additive manufacturing of metallic components: Materials, processes and mechanisms. *International Materials Reviews*, *57*(3), 133–164. DOI: 10.1179/1743280411Y.0000000014

Guenther, E., Kahlert, M., Vollmer, M., Niendorf, T., & Greiner, C. (2021). Tribological Performance of Additively Manufactured AISI H13 Steel in Different Surface Conditions. In Materials (Vol. 14, Issue 4, p. 928). Multidisciplinary Digital Publishing Institute. DOI: 10.3390/ma14040928

Guessasma, S., Zhang, W., Zhu, J., Belhabib, S., & Nouri, H. (2015). Challenges of additive manufacturing technologies from an optimisation perspective. In *International Journal for Simulation and Multidisciplinary Design Optimization* (Vol. 6). EDP Sciences., DOI: 10.1051/smdo/2016001

Gumaste, A., Dhal, A., Agrawal, P., Haridas, R. S., Vasudevan, V. K., Weiss, D., & Mishra, R. S. (2023). A novel approach for enhanced mechanical properties in solid-state additive manufacturing by additive friction stir deposition using thermally stable Al-Ce-Mg alloy. *Journal of the Minerals Metals & Materials Society*, *75*(10), 4185–4198. DOI: 10.1007/s11837-023-06044-6

Guo, Y., Zhang, Y., Pan, Z., & Zhou, W. (2024). Recent progress of sensing and machine learning technologies for process monitoring and defects detection in wire arc additive manufacturing. *Journal of Manufacturing Processes*, *125*, 489–511. DOI: 10.1016/j.jmapro.2024.07.060

Gutierrez, C., Salas, R., Hernandez, G., Muse, D., Olivas, R., MacDonald, E., & Zufelt, B. (2011, January). Cubesat fabrication through additive manufacturing and micro-dispensing. In *International Symposium on Microelectronics* (Vol. 2011, No. 1, pp. 001021-001027). International Microelectronics Assembly and Packaging Society. DOI: 10.4071/isom-2011-THA4-Paper3

Hager, A., Güniat, L., Morgan, N., Ramanandan, S. P., Rudra, A., Piazza, V., Fontcuberta i Morral, A., & Dede, D. (2023). The implementation of thermal and UV nanoimprint lithography for selective area epitaxy. *Nanotechnology*, *34*(44), 445301. DOI: 10.1088/1361-6528/acea87 PMID: 37494897

Hager, I., Golonka, A., & Putanowicz, R. (2016). 3D Printing of Buildings and Building Components as the Future of Sustainable Construction? In *Procedia Engineering* (Vol. 151, p. 292). Elsevier BV., DOI: 10.1016/j.proeng.2016.07.357

Hahn, M., Gies, S., & Tekkaya, A. E. (2018). Light enough or go lighter? In *Materials & Design* (Vol. 163, p. 107545). Elsevier BV., DOI: 10.1016/j.matdes.2018.107545

Hammood, M., Lin, S., Yun, H., Luan, E., Chrostowski, L., & Jaeger, N. A. F. (2023). Emulation of deep-ultraviolet lithography using rapid-prototyping, electron-beam lithography for silicon photonics design. *Optics Letters*, *48*(3), 582. DOI: 10.1364/OL.478212 PMID: 36723536

Han, B., Chen, J., Guo, T., Lee, S.-J., Swaminathan, V., & Venieris, S. I. (2022). Guest Editorial: Bridging the Gap Between Industry and Academia for Networking Research. *IEEE Network*, *36*(1), 8–9. DOI: 10.1109/MNET.2022.9740650

Hao, B., & Lin, G. (2020). 3D Printing Technology and Its Application in Industrial Manufacturing. *IOP Conference Series. Materials Science and Engineering*, *782*(2), 22065. DOI: 10.1088/1757-899X/782/2/022065

Hao, C. (2024). Artificial intelligence and operational efficiency of chinese manufacturing firms. *International Cognitive Journal*, *1*(1), 28–39. DOI: 10.69659/qcqt7937

Hao, L., Savalani, M. M., Zhang, Y., Tanner, K. E., & Harris, R. A. (2006). Selective laser sintering of hydroxyapatite reinforced polyethylene composites for bioactive implants and tissue scaffold development. *Proceedings of the Institution of Mechanical Engineers. Part H, Journal of Engineering in Medicine*, *220*(4), 521–531. DOI: 10.1243/09544119JEIM67 PMID: 16808068

Hashin, Z., & Rotem, A. (1973). A fatigue failure criterion for fiber reinforced materials. *Journal of Composite Materials*, *7*(4), 448–464. DOI: 10.1177/002199837300700404

Hashmi, S. (Ed.). (2015). *Reference module in materials science and materials engineering*. Elsevier BV., DOI: 10.1016/B978-0-323-96020-5.00120-5

Havenga, S., van Zyl, I., Snaddon, B., & Chisin, A. (2023). Towards the effective development of Design for Additive Manufacturing (DFAM) curricula: an exploration of strategies and solutions in education. In MATEC Web of Conferences (Vol. 388, p. 05009). EDP Sciences.

He, F., & Khan, M. (2021). Effects of Printing Parameters on the Fatigue Behaviour of 3D-Printed ABS under Dynamic Thermo-Mechanical Loads. In Polymers (Vol. 13, Issue 14, p. 2362). Multidisciplinary Digital Publishing Institute. DOI: 10.3390/polym13142362

He, C., Zhang, M., & Guo, C. (2020). 4D printing of mashed potato/purple sweet potato puree with spontaneous color change. *Innovative Food Science & Emerging Technologies*, *59*, 102250. DOI: 10.1016/j.ifset.2019.102250

Heilala, J., Parchegani, S., & Piili, H. (2023). Additive manufacturing systems integration. *IOP Conference Series. Materials Science and Engineering*, *1296*(1), 012024. DOI: 10.1088/1757-899X/1296/1/012024

Hendrixson, S. (2020). Infographic: How additive manufacturing supports the circular economy. *Additive Manufacturing*.

Hermosilla, R., Oñate, A., Castillo, R., Fuente, A. D., Sepúlveda, J., Escudero, B., & Medina, C. (2023). Influence stacking sequence and heat treatments on the out-of-plane mechanical properties of 3d-printed fiberglass-reinforced thermoplastics. *International Journal of Advanced Manufacturing Technology*, *125*(9-10), 4753–4764. DOI: 10.1007/s00170-023-11050-8

Herzog, T., Brandt, M., Trinchi, A., Sola, A., & Molotnikov, A. (2024). Process monitoring and machine learning for defect detection in laser-based metal additive manufacturing. *Journal of Intelligent Manufacturing*, *35*(4), 1407–1437. DOI: 10.1007/s10845-023-02119-y

He, X., Wang, H., Zhu, Z., Wang, L. Z., Liu, J. Q., Haghdadi, N., Nai, S. M. L., Huang, J., Primig, S., Ringer, S. P., & Liao, X. (2022). Texture evolution in a crmnfeconi high-entropy alloy manufactured by laser powder bed fusion. *Journal of Materials Science*, *57*(21), 9714–9725. DOI: 10.1007/s10853-022-07025-x

He, Y., Wei, J., Peng, Y., Wang, F., Wang, Y., & Liu, J. (2023). Deformation prediction and experimental investigation on alternating additive-subtractive hybrid manufacturing of 316l stainless steel thin-walled parts. *International Journal of Advanced Manufacturing Technology*, *129*(11-12), 5271–5285. DOI: 10.1007/s00170-023-12592-7

Hiemenz, J. (2011). 3D printing with FDM: How it Works. *Stratasys Inc*, *1*, 1–5.

Hoang, D., Perrault, D., Stevanović, M., & Ghiassi, A. (2016). Surgical applications of three-dimensional printing: a review of the current literature & how to get started [Review of Surgical applications of three-dimensional printing: a review of the current literature & how to get started]. Annals of Translational Medicine, 4(23), 456. AME Publishing Company. DOI: 10.21037/atm.2016.12.18

Hodzic, D., & Pandzic, A. (2019, October). Influence of carbon fibers on mechanical properties of materials in FDM technology. In *Proceedings of the 30th DAAAM International Symposium on Intelligent Manufacturing and Automation, Zadar, ISSN* (pp. 1726-9679). DOI: 10.2507/30th.daaam.proceedings.044

Hofmann, M. (2014). 3D printing gets a boost and opportunities with polymer materials. *ACS Macro Letters*, *3*(4), 382–386. DOI: 10.1021/mz4006556 PMID: 35590734

Hooda, N., Chohan, J. S., Gupta, R., & Kumar, R. (2021). Deposition angle prediction of Fused Deposition Modeling process using ensemble machine learning. *ISA Transactions*, *116*, 121–128. DOI: 10.1016/j.isatra.2021.01.035 PMID: 33518221

Hope, T.. (2018) 'Accelerating Innovation Through Analogy Mining', in *Proceedings of the Twenty-Seventh International Joint Conference on Artificial Intelligence. Twenty-Seventh International Joint Conference on Artificial Intelligence {IJCAI-18}*, Stockholm, Sweden: International Joint Conferences on Artificial Intelligence Organization, pp. 5274–5278. Available at: DOI: 10.24963/ijcai.2018/736

Hosen, M. S. S., Abedin, M. J., Hoque, M. E., & Rahman, M. Z. (2022a) Influence of moisture absorption on physico-chemical properties of natural fiber-based hybrid composites. In: Aging Effects on Natural Fiber-Reinforced Polymer Composites: Durability and Life Prediction, pp. 297–316. Springer. DOI: 10.1007/978-981-16-8360-2_16

Hosen, M. S. S., Hoque, M. E., Rahman, M. Z., & Sagadevan, S. (2022b) Aging effects on mechanical properties of biocomposites with recycled polymers. In: Aging Effects on Natural Fiber-Reinforced Polymer Composites: Durability and Life Prediction, pp. 317–333. Springer. DOI: 10.1007/978-981-16-8360-2_17

Hossain, M. J., Tabatabaei, B. T., Kiki, M., & Choi, J. W. (2025). Additive manufacturing of sensors: A comprehensive review. *International Journal of Precision Engineering and Manufacturing-Green Technology*, *12*(1), 277–300. DOI: 10.1007/s40684-024-00629-5

Hosseinzadeh, M., Ghoreishi, M., & Narooei, K. (2023). 4D printing of shape memory polylactic acid beams: An experimental investigation into FDM additive manufacturing process parameters, mathematical modeling, and optimization. *Journal of Manufacturing Processes*, *85*, 774–782. DOI: 10.1016/j.jmapro.2022.12.006

Hou, Z., Tian, X., Zheng, Z., Zhang, J., Zhe, L., Li, D., Malakhov, A. V., & Polilov, A. N. (2020). A constitutive model for 3D printed continuous fiber reinforced composite structures with variable fiber content. *Composites. Part B, Engineering*, *189*, 107893. DOI: 10.1016/j.compositesb.2020.107893

Hsieh, C. T. (2017). Investigation of Delta Robot 3D Printer for a Good Quality of Printing. *Applied Mechanics and Materials*, *870*, 164–169. . DOI: 10.4028/www.scientific.net/AMM.870.164

Huang, Z., Shi, T., & Yue, X. (2019). The Optimization of FDM 3D Printer's Structure Based on Finite Element Analysis. MATEC Web of Conferences, 257, 2004. DOI: 10.1051/matecconf/201925702004

Huang, G., Wei, K., Deng, J., & Zeng, X. (2022). High power laser powder bed fusion of 18Ni300 maraging steel: Processing optimization, microstructure and mechanical properties. *Materials Science and Engineering A*, *856*, 143983. DOI: 10.1016/j.msea.2022.143983

Huang, J., Xia, S., Li, Z., Wu, X., & Ren, J. (2021). Applications of four-dimensional printing in emerging directions: Review and prospects. *Journal of Materials Science and Technology*, *91*, 105–120. DOI: 10.1016/j.jmst.2021.02.040

Huang-Saad, A., Fay, J., & Sheridan, L. (2017). Closing the divide: Accelerating technology commercialization by catalyzing the university entrepreneurial ecosystem with I-Corps™. *The Journal of Technology Transfer*, *42*(6), 1466–1486. DOI: 10.1007/s10961-016-9531-2

Huang, X., Doerr, C. R., Qin, C., Heanue, J., Zhu, N., Ton, D., Guan, B., Zhang, S., & Zhao, Y. (2021). Silicon-photonic laser emitting tunable dual wavelengths with highly correlated phase noise. *Optics Letters*, *46*(1), 142. DOI: 10.1364/OL.409729 PMID: 33362036

Huang, Y., Zhang, M., Pattarapon, P., & Mujumdar, A. S. (2022). 4D printing of mixed vegetable gel based on deformation and discoloration induced by acidification and dehydration. *Journal of Food Process Engineering*, *45*(12), e14172. DOI: 10.1111/jfpe.14172

Hu, K., Jin, S., & Wang, C. C. (2015). Support slimming for single material based additive manufacturing. *Computer Aided Design*, *65*, 1–10. DOI: 10.1016/j.cad.2015.03.001

Hulls, C. C. W., & Rennick, C. (2020). Use of a Cornerstone Project to Teach Ill-Structured Software Design in First Year. *IEEE Transactions on Education*, *63*(2), 98–107. DOI: 10.1109/TE.2019.2959591

Hu, M., Pang, Y., & Gao, L. (2023). Advances in Silicon-Based Integrated Lidar. *Sensors (Basel)*, *23*(13), 5920. DOI: 10.3390/s23135920 PMID: 37447770

Humnabad, P. S., Tarun, R., & Das, I. (2022). An overview of direct metal laser sintering (DMLS) technology for metal 3D printing. *Journal of Mines, Metals and Fuels*, 127-133.

Hussein, M. O., & Hussein, L. A. (2022). Optimization of Digital Light Processing Three-Dimensional Printing of the Removable Partial Denture Frameworks; The Role of Build Angle and Support Structure Diameter. *Materials (Basel)*, *15*(6), 2316. DOI: 10.3390/ma15062316 PMID: 35329767

Hu, Y., Ladani, R. B., Brandt, M., Li, Y., & Mouritz, A. P. (2021). Carbon fibre damage during 3D printing of polymer matrix laminates using the FDM process. *Materials & Design*, *205*, 109679. DOI: 10.1016/j.matdes.2021.109679

Hyer, H. (2022). Status of additive manufacturing capabilities for processing refractory alloys under the mo-99 program.. DOI: 10.2172/1909108

Ibhadode, O., Zhang, Z., Sixt, J., Nsiempba, K. M., Orakwe, J., Martinez-Marchese, A., Ero, O., Shahabad, S. I., Bonakdar, A., & Toyserkani, E. (2023). Topology optimization for metal additive manufacturing: Current trends, challenges, and future outlook. *Virtual and Physical Prototyping*, *18*(1), e2181192. Advance online publication. DOI: 10.1080/17452759.2023.2181192

Ibrahim, B., Lopez, L., Kulkarni, S., Jobes, D., Forgiarini, M., Barber, J. R., & Gordon, J. V. (2024). Increasing strength properties in sinter-based additive manufacturing of SS316L via metal material jetting of sub-micron powders. *Additive Manufacturing*, *65*, 104268. DOI: 10.1016/j.addma.2024.104268

Iftekar, S. F., Aabid, A., Amir, A., & Baig, M. (2023). Advancements and limitations in 3D printing materials and technologies: A critical review. *Polymers*, *15*(11), 2519. DOI: 10.3390/polym15112519 PMID: 37299318

Imam, S. S., Hussain, A., Altamimi, M. A., & Alshehri, S. (2021). Four-dimensional printing for hydrogel: Theoretical concept, 4d materials, shape-morphing way, and future perspectives. *Polymers*, *13*(21), 3858. DOI: 10.3390/polym13213858 PMID: 34771414

Immonen, K., Metsä-Kortelainen, S., Nurmio, J., Tribot, A., Turpeinen, T., Mikkelson, A., Kalpio, T., Kaukoniemi, O.-V., & Kangas, H. (2022). Recycling of 3d printable thermoplastic cellulose-composite. *Sustainability (Basel)*, *14*(5), 2734. DOI: 10.3390/su14052734

Iordachescu, D., & Quintino, L. (2008). Steps toward a new classification of metal transfer in gas metal arc welding. *Journal of Materials Processing Technology*, *202*(1-3), 391–397. DOI: 10.1016/j.jmatprotec.2007.08.081

Islam, A., & Rahman, M. Z. (2023) A review on eco-friendly basalt fibers, their composites and applications. In: Reference Module in Materials Science and Materials Engineering.

Ismail, M. I., & Ibrahim, M. (2021). Analysis on Lighweighting Approach Topology Optimization Design for Additive Manufacturing (DfAM). *Research Progress in Mechanical and Manufacturing Engineering*, *2*(2), 860–868.

Issal, A. (2023). Ender 3 (V2/Pro) Calibration: 5 Simple Steps. https://all3dp.com/2/ender-3-calibration-how-to-calibrate-your-ender-3/

Ituarte, I. F., Chekurov, S., Salmi, M., Tuomi, J., & Partanen, J. (2015). Post-processing opportunities of professional and consumer grade 3D printing equipment: A comparative study. *International Journal of Rapid Manufacturing*, *5*(1), 58. DOI: 10.1504/IJRAPIDM.2015.073548

Jacobs PF (1992) Fundamentals of Stereolithography. July.

Jagadeesh, P., Puttegowda, M., Rangappa, S. M., Alexey, K., Gorbatyuk, S., Khan, A., Doddamani, M., & Siengchin, S. (2022). A comprehensive review on 3D printing advancements in polymer composites: Technologies, materials, and applications. *International Journal of Advanced Manufacturing Technology, 121*(1), 127–169. DOI: 10.1007/s00170-022-09406-7

Jägle, E. A., Choi, P.-P., van Humbeeck, J., & Raabe, D. (2014). Precipitation and austenite reversion behavior of a maraging steel produced by selective laser melting. *Journal of Materials Research, 29*(17), 2072–2079. DOI: 10.1557/jmr.2014.204

Jain, K., Shukla, R., Yadav, A., Ujjwal, R. R., & Flora, S. J. S. (2021). 3D printing in development of nanomedicines. *Nanomaterials (Basel, Switzerland), 11*(2), 420. DOI: 10.3390/nano11020420 PMID: 33562310

Jaksic, N. (2015). What to Do When 3D Printers Go Wrong: Laboratory Experiences. DOI: 10.18260/p.25066

Jamal, M. A., Shah, O. R., Ghafoor, U., Qureshi, Y., & Bhutta, M. R. (2024). Additive manufacturing of continuous fiber-reinforced polymer composites via fused deposition modelling: A comprehensive review. *Polymers, 16*(12), 1622. DOI: 10.3390/polym16121622 PMID: 38931971

Jandyal, A., Chaturvedi, I., Wazir, I., Raina, A., & Haq, M. I. U. (2022). 3D printing–A review of processes, materials and applications in industry 4.0. *Sustainable Operations and Computers, 3*, 33–42. DOI: 10.1016/j.susoc.2021.09.004

Jankovics, D., & Barari, A. (2019). Customization of Automotive Structural Components using Additive Manufacturing and Topology Optimization. *IFAC-PapersOnLine, 52*(10), 212–217. DOI: 10.1016/j.ifacol.2019.10.066

Jared, B. H., Aguiló, M. A., Beghini, L. L., Boyce, B., Clark, B. W., Cook, A., Kaehr, B., & Robbins, J. (2017). Additive manufacturing: Toward holistic design. In *Scripta Materialia* (Vol. 135, p. 141). Elsevier BV., DOI: 10.1016/j.scriptamat.2017.02.029

Jasim, M., Abbas, T., & Huayier, A. (2022). The effect of infill pattern on tensile strength of pla material in fused deposition modeling (fdm) process. *Engineering and Technology Journal, 40*(12), 1–8. DOI: 10.30684/etj.2021.131733.1054

Javaid, M., Haleem, A., Singh, R. P., Suman, R., & Rab, S. (2021). Role of additive manufacturing applications towards environmental sustainability. In Advanced Industrial and Engineering Polymer Research (Vol. 4, Issue 4, p. 312). Elsevier BV. DOI: 10.1016/j.aiepr.2021.07.005

Javaid, M., & Haleem, A. (2019). 4D printing applications in medical field: A brief review. *Clinical Epidemiology and Global Health, 7*(3), 317–321. DOI: 10.1016/j.cegh.2018.09.007

Javaid, M., Haleem, A., Singh, R. P., & Suman, R. (2022). 3D printing applications for healthcare research and development. *Global Health Journal (Amsterdam, Netherlands), 6*(4), 217–226. DOI: 10.1016/j.glohj.2022.11.001

Jayakrishna, M., Vijay, M., & Khan, B. (2023). An Overview of Extensive Analysis of 3D Printing Applications in the Manufacturing Sector. *Journal of Engineering, 1*, 1–23. Advance online publication. DOI: 10.1155/2023/7465737

Jeong, M., Radomski, K., López, D., Liu, J. T., Lee, J. D., & Lee, S. J. (2023). Materials and Applications of 3D Printing Technology in Dentistry: An Overview [Review of Materials and Applications of 3D Printing Technology in Dentistry: An Overview]. Dentistry Journal, 12(1), 1. Multidisciplinary Digital Publishing Institute. DOI: 10.3390/dj12010001

Jhun, J., Lee, D. H., Rehman, A. U., Kang, S., & Kim, J. H. (2024). Development of a real-time geometric quality monitoring system for extruded filaments of 3D concrete printing construction. *IEEE Access : Practical Innovations, Open Solutions, 12*, 69981–69999. DOI: 10.1109/ACCESS.2024.3401472

Jia, Q., Liu, B., Li, S., Hu, C., Yang, Y., & Wang, G. (2024). Slurry-based photoinitiator jetting process for ceramic additive manufacturing. *Additive Manufacturing, 86*, 104185. DOI: 10.1016/j.addma.2024.104185

Jihong, Z. H. U., Han, Z. H. O. U., Chuang, W. A. N. G., Lu, Z. H. O. U., Shangqin, Y. U. A. N., & Zhang, W. (2021). A review of topology optimization for additive manufacturing: Status and challenges. *Chinese Journal of Aeronautics, 34*(1), 91–110. DOI: 10.1016/j.cja.2020.09.020

Jiménez, M., Romero, L., Domínguez, I. A., & Escudero, M. (2019). del M. E., & Somonte, M. D. (2019). Additive Manufacturing Technologies: An Overview about 3D Printing Methods and Future Prospects. *Complexity, 2019*(1), 9656938. Advance online publication. DOI: 10.1155/2019/9656938

Jin, Y., Wanvarie, D., & Le, P. T. V. (2019) 'Bridging the Gap Between Research and Production with CODE', in Q. Yang et al. (eds) *Advances in Knowledge Discovery and Data Mining*. Cham: Springer International Publishing (Lecture Notes in Computer Science), pp. 277–288. Available at: DOI: 10.1007/978-3-030-16142-2_22

Jin, L., Zhai, X., Wang, K., Zhang, K., Wu, D., Nazir, A., Jiang, J., & Liao, W. H. (2024). Big data, machine learning, and digital twin assisted additive manufacturing: A review. *Materials & Design*, *244*, 113086. DOI: 10.1016/j.matdes.2024.113086

Jin, Z., Zhang, Z., Demir, K., & Gu, G. X. (2020). Machine Learning for Advanced Additive Manufacturing. *Matter*, *3*(5), 1541–1556. DOI: 10.1016/j.matt.2020.08.023

Joharji, L., Mishra, R. B., Alam, F., Tytov, S., Al-Modaf, F., & El-Atab, N. (2022). 4D printing: A detailed review of materials, techniques, and applications. *Microelectronic Engineering*, *265*, 111874. DOI: 10.1016/j.mee.2022.111874

Joshi, S., Rawat, K., Karunakaran, C., Rajamohan, V., Mathew, A. T., Koziol, K., & Balan, A. S. S. (2020). 4D printing of materials for the future: Opportunities and challenges. *Applied Materials Today*, *18*, 100490. DOI: 10.1016/j.apmt.2019.100490

Jung, S., Kara, L. B., Nie, Z., Simpson, T. W., & Whitefoot, K. S. (2023). Is additive manufacturing an environmentally and economically preferred alternative for mass production? *Environmental Science & Technology*, *57*(16), 6373–6386. DOI: 10.1021/acs.est.2c04927 PMID: 37066969

Kabir, S. M. F., Mathur, K., & Seyam, A. F. M. (2020). A critical review on 3D printed continuous fiber-reinforced composites: History, mechanism, materials and properties. *Composite Structures*, *232*, 111476. DOI: 10.1016/j.compstruct.2019.111476

Kafle, A., Luis, E., Silwal, R., Pan, H. M., Shrestha, P. L., & Bastola, A. K. (2021). 3D/4D printing of polymers : Fused deposition modelling (FDM), selective laser sintering (SLS), and stereolithography (SLA). *Polymers*, *13*(18), 3101. DOI: 10.3390/polym13183101 PMID: 34578002

Kahhal, P., Jo, Y. K., & Park, S. H. (2024). Recent progress in remanufacturing technologies using metal additive manufacturing processes and surface treatment. *International Journal of Precision Engineering and Manufacturing-Green Technology*, *11*(2), 625–658. DOI: 10.1007/s40684-023-00551-2

Kalaskar, D. M. (Ed.). (2022). *3D printing in medicine*. Woodhead Publishing.

Kaliaraj, G. S., Shanmugam, D. K., Dasan, A., & Mosas, K. K. A. (2023). Hydrogels—A promising materials for 3D printing technology. *Gels (Basel, Switzerland)*, *9*(3), 260. DOI: 10.3390/gels9030260 PMID: 36975708

Kamaal, M., Anas, M., Rastogi, H., Bhardwaj, N., & Rahaman, A. (2021). Effect of FDM process parameters on mechanical properties of 3D-printed carbon fibre–PLA composite. *Progress in Additive Manufacturing*, *6*(1), 63–69. DOI: 10.1007/s40964-020-00145-3

Kam, M., İpekçi, A., & Şengül, Ö. (2020). Effect of FDM Process Parameters on the Mechanical Properties and Production Costs of 3D Printed PowerABS Samples. [IJAEFEA]. *International Journal of Analytical Experimental and Finite Element Analysis, 7*(3). Advance online publication. DOI: 10.26706/ijaefea.3.7.20200806

Kang, K. H., & Lee, Y. H. (2007). Make-to-order scheduling in foundry semiconductor fabrication. *International Journal of Production Research, 45*(3), 615–630. DOI: 10.1080/00207540600792416

Kannan, R., Leonard, D. N., & Nandwana, P. (2021). Optimization of direct aging temperature of Ti free grade 300 maraging steel manufactured using laser powder bed fusion (LPBF). *Materials Science and Engineering A, 817*, 141266. DOI: 10.1016/j.msea.2021.141266

Kantaros, A., & Piromalis, D. (2021). Fabricating Lattice Structures via 3D Printing: The Case of Porous Bio-Engineered Scaffolds. *Applied Mechanics, 2*(2), 289–302. DOI: 10.3390/applmech2020018

Kantaros, A., Soulis, E., Petrescu, F. I. T., & Ganetsos, T. (2023). Advanced Composite Materials Utilized in FDM/FFF 3D Printing Manufacturing Processes: The Case of Filled Filaments. *Materials (Basel), 16*(18), 6210. DOI: 10.3390/ma16186210 PMID: 37763488

Kanthimathi, T., Rathika, N., Fathima, A. J., Rajesh, K. S., Srinivasan, S., & Thamizhamuthu, R. (2024, January). Robotic 3D Printing for Customized Industrial Components: IoT and AI-Enabled Innovation. In 2024 14th International Conference on Cloud Computing, Data Science & Engineering (Confluence) (pp. 509-513). IEEE.

Karalekas, D. E. (2003). Study of the mechanical properties of nonwoven fibre mat reinforced photopolymers used in rapid prototyping. *Materials & Design, 24*(8), 665–670. DOI: 10.1016/S0261-3069(03)00153-5

Karalekas, D., & Aggelopoulos, A. (2003). Study of shrinkage strains in a stereolithography cured acrylic photopolymer resin. *Journal of Materials Processing Technology, 136*(1-3), 146–150. DOI: 10.1016/S0924-0136(03)00028-1

Kargar, E., & Ghasemi-Ghalebahman, A. (2023). Experimental investigation on fatigue life and tensile strength of carbon fiber-reinforced pla composites based on fused deposition modeling. *Scientific Reports, 13*(1), 18194. Advance online publication. DOI: 10.1038/s41598-023-45046-x PMID: 37875509

Karimi, A., Rahmatabadi, D., & Baghani, M. (2024). Various FDM mechanisms used in the fabrication of continuous-fiber reinforced composites: A review. *Polymers, 16*(6), 831. DOI: 10.3390/polym16060831 PMID: 38543436

Karin, J., Ahmed, E., Scholz, S., & Hagenmeyer, V. (2024). On the correlation between pre-processing workflow and dimensional accuracy of 3D printed parts in high-precision material jetting. *Additive Manufacturing*, *85*, 104335. DOI: 10.1016/j.addma.2024.104335

Karkun, M. S., & Dharmalingam, S. (2022). 3D printing technology in aerospace industry–a review. *International Journal of Aviation, Aeronautics, and Aerospace*, *9*(2), 4. DOI: 10.15394/ijaaa.2022.1708

Karlapudy, S. P., Nancharaiah, T., & Subba Rao, V. V. (2024). Influence of post-heat treatment on microstructure, texture, and mechanical properties of 18Ni-300 maraging steel fabricated by using LPBF technique. *Progress in Additive Manufacturing*, *9*(6), 1667–1682. DOI: 10.1007/s40964-023-00530-8

Karwasz, A., & Osinski, F. (2020). *Literature review on emissions from additive manufacturing by FDM method and their impact on human health.* Management and Production Engineering Review. DOI: 10.24425/mper.2020.134933

Karyappa, R., Zhang, D., Zhu, Q., Ji, R., Suwardi, A., & Liu, H. (2024). Newtonian liquid-assisted material extrusion 3D printing: Progress, challenges and future perspectives. *Additive Manufacturing*, *79*, 103903. DOI: 10.1016/j.addma.2023.103903

Kaščák, Ľ., Varga, J., Bidulská, J., Bidulský, R., & Grande, M. A. (2023). Simulation tool for material behaviour prediction in additive manufacturing. *Acta Metallurgica Slovaca*, *29*(2), 113–118. DOI: 10.36547/ams.29.2.1846

Kato, K., Liu, Y., Murakami, S., Morita, Y., & Mori, T. (2021). Electron beam lithography with negative tone resist for highly integrated silicon quantum bits. *Nanotechnology*, *32*(48), 485301. DOI: 10.1088/1361-6528/ac201b PMID: 34425562

Kechagias, J. D., & Chryssolouris, G. (1997) Estimation of build times in rapid prototyping processes. In: Proceedings of the 6th European Conference on Rapid Prototyping and Manufacturing -EuRP&M1997 At: University of Nottingham, UK. January 1997.

Kędziora, S., Decker, T., & Museyibov, E. (2023). Application of Functionally Graded Shell Lattice as Infill in Additive Manufacturing. *Materials (Basel)*, *16*(12), 4401. DOI: 10.3390/ma16124401 PMID: 37374591

Kent, N., Jolivet, L., O'Neill, P., & Brabazon, D. (2017). An evaluation of components manufactured from a range of materials, fabricated using PolyJet technology. *Advances in Materials and Processing Technologies*, *3*(3), 318–329. DOI: 10.1080/2374068X.2017.1330856

Khalid, G. M., & Billa, N. (2022). Solid dispersion formulations by FDM 3D printing—A review. *Pharmaceutics, 14*(4), 690. DOI: 10.3390/pharmaceutics14040690 PMID: 35456524

Khalid, M. Y., Arif, Z. U., Noroozi, R., Zolfagharian, A., & Bodaghi, M. (2022). 4D printing of shape memory polymer composites: A review on fabrication techniques, applications, and future perspectives. *Journal of Manufacturing Processes, 81*, 759–797. DOI: 10.1016/j.jmapro.2022.07.035

Khalifa, R. B., Nhassa, G., & Snoussi, A. (2024). An Overview of Additive Manufacturing's Role in industry 4.0. Modeling, Analysis, and Control of Smart Energy Systems, 137-152.

Khalil, Y., Hopkinson, N., Kowalski, A., & Fairclough, J. P. A. (2022). Investigating the feasibility of processing activated carbon/uhmwpe polymer composite using laser powder bed fusion. *Polymers, 14*(16), 3320. DOI: 10.3390/polym14163320 PMID: 36015577

Khaliq, J., Gurrapu, D. R., & Elfakhri, F. (2023). Effects of Infill Line Multiplier and Patterns on Mechanical Properties of Lightweight and Resilient Hollow Section Products Manufactured Using Fused Filament Fabrication. *Polymers, 15*(12), 2585. DOI: 10.3390/polym15122585 PMID: 37376230

Khanafer, K., Cao, J., & Kokash, H. (2024). Condition monitoring in additive manufacturing: A critical review of different approaches. *Journal of Manufacturing and Materials Processing, 8*(3), 95. DOI: 10.3390/jmmp8030095

Khan, S. F., Zakaria, H., Chong, Y. L., Saad, M. A. M., & Basaruddin, K. S. (2018). Effect of infill on tensile and flexural strength of 3D printed PLA parts. *IOP Conference Series. Materials Science and Engineering, 429*, 12101. DOI: 10.1088/1757-899X/429/1/012101

Khan, S. H., Rahman, M. Z., Haque, M. R., & Hoque, M. E. (2023a). Characterization and comparative evaluation of structural, chemical, thermal, mechanical, and morphological properties of plant fibers. In Khiari, R., Jawaid, M., & Belgacem, M. N. (Eds.), *Annual Plant: Sources of Fibres, Nanocellulose and Cellulosic Derivatives: Processing, Properties and Applications* (pp. 1–45). Springer Nature Singapore., DOI: 10.1007/978-981-99-2473-8_1

Khdoudi, A., & Masrour, T. (2018). Prediction of Industrial Process Parameters using Artificial Intelligence Algorithms. arXiv (Cornell University). https://doi .org//arxiv.1812.10537DOI: 10.48550

Khodaei, H., Scholten, V. E., Wubben, E. F. M., & Omta, S. W. F. O. (2022). The Role of Academic Spin-Offs Facilitators in Navigation of the Early Growth Stage Critical Junctures. *IEEE Transactions on Engineering Management*, *69*(4), 1769–1780. DOI: 10.1109/TEM.2020.2995361

Khoo, Z. X., Teoh, J. E. M., Liu, Y., Chua, C. K., Yang, S., An, J., Leong, K. F., & Yeong, W. Y. (2015). 3D printing of smart materials: A review on recent progresses in 4D printing. *Virtual and Physical Prototyping*, *10*(3), 103–122. DOI: 10.1080/17452759.2015.1097054

Khurana, R., & Hodges, S. (2020) 'Beyond the Prototype: Understanding the Challenge of Scaling Hardware Device Production', in *Proceedings of the 2020 CHI Conference on Human Factors in Computing Systems. CHI '20: CHI Conference on Human Factors in Computing Systems*, Honolulu HI USA: ACM, pp. 1–11. Available at: DOI: 10.1145/3313831.3376761

Kim, D. B., Denno, P., & Jones, A. (2015). A model-based approach to refining process parameters in smart manufacturing. *Concurrent Engineering, Research and Applications*, *23*(4), 365–376. DOI: 10.1177/1063293X15591038

Kim, D. W., Han, S. B., Lee, Y. S., Park, D. Y., Lee, H., Park, S. H., & Song, H. (2023). Correlation between cellular structure morphology and anisotropic yield in additively manufactured stainless steel 316l. *Materials (Basel)*, *16*(4), 1666. DOI: 10.3390/ma16041666 PMID: 36837296

Kim, H.-K., Kim, M., Kim, M., Jo, Y., Lischke, S., Mai, C., Zimmermann, L., & Choi, W.-Y. (2021). Si photonic-electronic monolithically integrated optical receiver with a built-in temperature-controlled wavelength filter. *Optics Express*, *29*(6), 9565. DOI: 10.1364/OE.418222 PMID: 33820381

Kim, H., Kim, K. H., Jeong, J., Lee, Y., & Jung, I. D. (2024). Recent progress on materials for functional additive manufacturing. *Materials Science in Additive Manufacturing*, *3*(2), 3323. DOI: 10.36922/msam.3323

Kim, J., Kim, M.-G., Kim, J., Jo, S., Kang, J., Jo, J.-W., Lee, W., Hwang, C., Moon, J., Yang, L., Kim, Y.-H., Noh, Y.-Y., Yun Jaung, J., Kim, Y.-H., & Kyu Park, S. (2015). Scalable Sub-micron Patterning of Organic Materials Toward High Density Soft Electronics. *Scientific Reports*, *5*(1), 14520. DOI: 10.1038/srep14520 PMID: 26411932

Kim, Y. E., Morton, B. G., Gregorio, J., Rosen, D. S., Edouard, K., & Vallett, R. (2019). Enabling creative collaboration for all levels of learning. *Proceedings of the National Academy of Sciences of the United States of America*, *116*(6), 1878–1885. DOI: 10.1073/pnas.1808678115 PMID: 30718419

Kim, Y. J., Kim, H. N., & Kim, D. Y. (2024). A study on effects of curing and machining conditions in post-processing of SLA additive manufactured polymer. *Journal of Manufacturing Processes, 119*, 511–519. DOI: 10.1016/j.jmapro.2024.03.112

Kiran, A. R., Li, Y., Hodek, J., Brázda, M., Urbánek, M., & Džugan, J. (2022). Heat Source Modeling and Residual Stress Analysis for Metal Directed Energy Deposition Additive Manufacturing. *Materials (Basel), 15*(7), 2545. DOI: 10.3390/ma15072545 PMID: 35407876

Kishore, R. A., Sivaraj, S., Dhinakaran, V., Prabhagar, V. S. P. P., & Srinivas, S. (2021). Strength of 3D prints with variable print orientation. *Journal of Physics: Conference Series, 2027*(1), 12021. DOI: 10.1088/1742-6596/2027/1/012021

Klenam, D., Asumadu, T., Bodunrin, M., Obiko, J., Genga, R., Maape, S., McBagonluri, F., & Soboyejo, W. (2025). Toward sustainable industrialization in Africa: The potential of additive manufacturing–an overview. *Frontiers in Manufacturing Technology, 4*, 1410653. DOI: 10.3389/fmtec.2024.1410653

Kneller, R., ... (2014) 'Industry-University Collaborations in Canada, Japan, the UK and USA – With Emphasis on Publication Freedom and Managing the Intellectual Property Lock-Up Problem', *PLoS ONE*. Edited by A.R.H. Montoya, 9(3), p. e90302. Available at: DOI: 10.1371/journal.pone.0090302

Kok, Y., Tan, X. P., Wang, P., Nai, M. L. S., Loh, N. H., Liu, E., & Tor, S. B. (2018). Anisotropy and Heterogeneity of Microstructure and Mechanical Properties in Metal Additive Manufacturing: A Critical Review. *Materials & Design, 139*, 565–586. DOI: 10.1016/j.matdes.2017.11.021

Kolossov, S., Boillat, E., Glardon, R., Fischer, P., & Locher, M. (2004). 3D FE simulation for temperature evolution in the selective laser sintering process. *International Journal of Machine Tools & Manufacture, 44*(2-3), 117–123. DOI: 10.1016/j.ijmachtools.2003.10.019

Kong, M., Yang, M., Li, R., Long, Y. Z., Zhang, J., Huang, X., Cui, X., Zhang, Y., Said, Z., & Li, C. (2024). Graphene-based flexible wearable sensors: Mechanisms, challenges, and future directions. *International Journal of Advanced Manufacturing Technology, 131*(5), 3205–3237. DOI: 10.1007/s00170-023-12007-7

Kousiatza, C., Tzetzis, D., & Karalekas, D. (2019). In-situ characterization of 3D printed continuous fiber reinforced composites: A methodological study using fiber Bragg grating sensors. *Composites Science and Technology, 174*, 134–141. DOI: 10.1016/j.compscitech.2019.02.008

Krishna, R. S., Rehman, A. U., Mishra, J., Saha, S., Korniejenko, K., Rehman, R. U., & Qureshi, T. S. (2024). Additive manufacturing of geopolymer composites for sustainable construction: Critical factors, advancements, challenges, and future directions. *Progress in Additive Manufacturing*, •••, 1–59.

Kristiawan, R. B., Imaduddin, F., Ariawan, D., Ubaidillah, , & Arifin, Z. (2021). A review on the fused deposition modeling (FDM) 3D printing: Filament processing, materials, and printing parameters. *Open Engineering*, *11*(1), 639–649. DOI: 10.1515/eng-2021-0063

Królikowski, T., Knitter, R., & Stachnik, M. (2019). Thermo-mechanical tests using 3d printed elements. *Procedia Computer Science*, *159*, 2551–2559. DOI: 10.1016/j.procs.2019.09.430

Kruth, J., Levy, G., Klocke, F., & Childs, T. H. C. (2007). Consolidation phenomena in laser and powder-bed based layered manufacturing. *CIRP Annals*, *56*(1), 730–759. DOI: 10.1016/j.cirp.2007.10.004

Kruth, J., Mercelis, P., Van Vaerenbergh, J., Froyen, L., & Rombouts, M. (2005). Binding mechanisms in selective laser sintering and selective laser melting. *Rapid Prototyping Journal*, *11*(1), 26–36. DOI: 10.1108/13552540510573365

Kuang, X. (2021). Introduction to 4D printing: methodologies and materials. In *Additive manufacturing* (pp. 303-342). Elsevier.

Kuang, X., Roach, D. J., Wu, J., Hamel, C. M., Ding, Z., Wang, T., Dunn, M. L., & Qi, H. J. (2019). Advances in 4D printing: Materials and applications. *Advanced Functional Materials*, *29*(2), 1805290. DOI: 10.1002/adfm.201805290

Kumaar, K. R., Boopathy, G., Bejaxhin, B. H., & Ramanan, N. (2024). Considerations on fractured edge dislocations of mechanical tested hand-laid hemp flax GFRP hybrid polymer composites with SEM verifications. *Journal of Polymer & Composites*, *11*(13), 1–19. https://journals.stmjournals.com/jopc/article=2024/view=133571/

Kumar, A., Verma, S., & Jeng, J. (2020). Supportless Lattice Structures for Energy Absorption Fabricated by Fused Deposition Modeling. 3D Printing and Additive Manufacturing, 7(2), 85. DOI: 10.1089/3dp.2019.0089

Kumar, A., Nune, K. C., Murr, L. E., & Misra, R. D. K. (2016). Biocompatibility and mechanical behaviour of three-dimensional scaffolds for biomedical devices: Process–structure–property paradigm. *International Materials Reviews*, *61*(1), 20–45. DOI: 10.1080/09506608.2015.1128310

Kumar, N., Ramakrishnan, S. A., Lopez, K. G., Chin, B. Z., Devyapriya, S., Kumar, L., & Anantharajan, S. K. (2022). Current trends and future scope in 3D printing for surgical management of spine pathologies. *Bioprinting (Amsterdam, Netherlands)*, *26*, e00197. DOI: 10.1016/j.bprint.2022.e00197

Kumar, R., Kumar, M., & Chohan, J. S. (2021). Material-specific properties and applications of additive manufacturing techniques: A comprehensive review. *Bulletin of Materials Science*, *44*(3), 181. DOI: 10.1007/s12034-021-02364-y

Kumar, S. R., Sridhar, S., Venkatraman, R., & Venkatesan, M. (2021). Polymer additive manufacturing of ASA structure: Influence of printing parameters on mechanical properties. *Materials Today: Proceedings*, *39*, 1316–1319. DOI: 10.1016/j.matpr.2020.04.500

Kumar, S., Gopi, T., Harikeerthana, N., Gupta, M. K., Gaur, V., Królczyk, G., & Wu, C. (2022). Machine learning techniques in additive manufacturing: A state of the art review on design, processes, and production control. *Journal of Intelligent Manufacturing*, *34*(1), 21–55. DOI: 10.1007/s10845-022-02029-5

Kumar, S., & Kruth, J. P. (2010). Composites by rapid prototyping technology. *Materials & Design*, *31*(2), 850–856. DOI: 10.1016/j.matdes.2009.07.045

Kumar, S., & Kumar, R. (2025). A Comprehensive Study on Additive Manufacturing Techniques, Machine Learning Integration, and Internet of Things-Driven Sustainability Opportunities. *Journal of Materials Engineering and Performance*, •••, 1–68. DOI: 10.1007/s11665-025-10757-x

Kurian, M., & Thankachan, S. (2023). Introduction: Ceramics classification and applications. In *Ceramic catalysts* (pp. 1-17). Elsevier.

Kürnsteiner, P., Wilms, M. B., Weisheit, A., Barriobero-Vila, P., Jägle, E. A., & Raabe, D. (2017). Massive nanoprecipitation in an Fe-19Ni- x Al maraging steel triggered by the intrinsic heat treatment during laser metal deposition. *Acta Materialia*, *129*, 52–60. DOI: 10.1016/j.actamat.2017.02.069

Lacroix, R., Timonina-Farkas, A., & Seifert, R. W. (2023). Utilizing additive manufacturing and mass customization under capacity constraints. *Journal of Intelligent Manufacturing*, *34*(1), 281–301. DOI: 10.1007/s10845-022-02007-x PMID: 36618339

Ladani, L. J. (2021). Applications of artificial intelligence and machine learning in metal additive manufacturing. *JPhys Materials*, *4*(4), 042009. DOI: 10.1088/2515-7639/ac2791

Lan, P., Chou, S., Chent, L., & Gemmill, D. (1997). Determining fabrication orientations for rapid prototyping with stereolithography apparatus. Computer-Aided Design 29(1): 53–62. Li H, Wang T, Joshi S, and Yu Z (2019) The quantitative analysis of tensile strength of additively manufactured continuous carbon fiber reinforced polylactic acid (PLA). *Rapid Prototyping Journal, 25*(10), 1624–1636. DOI: 10.1108/RPJ-01-2018-0005

Lavikka, R., Seppänen, O., Peltokorpi, A., & Lehtovaara, J. (2020). Fostering process innovations in construction through industry–university consortium. *Construction Innovation, 20*(4), 569–586. DOI: 10.1108/CI-08-2019-0081

Lawrence, J. E., Lawrence, M. P., Fezzaa, K., Clark, S. J., & Crane, N. B. (2024). High-speed X-ray imaging of droplet-powder interaction in binder jet additive manufacturing. *Additive Manufacturing, 65*, 104269. DOI: 10.1016/j.addma.2024.104269

Lazzarotti, V., Puliga, G., Manzini, R., Tallarico, S., Pellegrini, L., Eslami, M. H., Ismail, M., & Boer, H. (2025). Collaboration with universities and innovation efficiency: Do relationship depth and organizational routines matter? *European Journal of Innovation Management, 28*(2), 608–630. DOI: 10.1108/EJIM-03-2023-0241

Lee, A. Y., An, J., & Chua, C. K. (2017). Two-way 4D printing: A review on the reversibility of 3D-printed shape memory materials. *Engineering (Beijing), 3*(5), 663–674. DOI: 10.1016/J.ENG.2017.05.014

Lee, D., Krumdieck, S., & Talwar, S. D. (2013). Scale-up design for industrial development of a PP-MOCVD coating system. *Surface and Coatings Technology, 230*, 39–45. DOI: 10.1016/j.surfcoat.2013.06.064

Lee, J., Walker, J., Natarajan, S., & Yi, S. (2019). Prediction of geometric characteristics in polycaprolactone (PCL) scaffolds produced by extrusion-based additive manufacturing technique for tissue engineering. *Rapid Prototyping Journal, 26*(2), 238–248. DOI: 10.1108/RPJ-08-2018-0219

Leghemo, I. M., Azubuike, C., Segun-Falade, O. D., & Odionu, C. S. (2025). Data governance for emerging technologies: A conceptual framework for managing blockchain, IoT, and AI. *Journal of Engineering Research and Reports, 27*(1), 247–267. DOI: 10.9734/jerr/2025/v27i11385

Leigh, S. J., Bradley, R. J., Purssell, C. P., Billson, D. R., & Hutchins, D. A. (2012). A simple, low-cost conductive composite material for 3D printing of electronic sensors. *PLoS One, 7*(11), e49365. DOI: 10.1371/journal.pone.0049365 PMID: 23185319

Leonetti, D., de Munnik, M., Kassing, L., Khan, D., Moulin, J., & Snijder, B. (2023). Rotating bending fatigue behaviour and quasi-static tensile properties of wire arc additively manufactured 308l stainless steel. *ce/papers, 6*(3-4), 732–738. DOI: 10.1002/cepa.2441

Leontiadis, K., Tsioptsias, C., Messaritakis, S., Terzaki, A., Xidas, P., Mystikos, K., Tzimpilis, E., & Tsivintzelis, I. (2022). Optimization of thermal and mechanical properties of polypropylene-wollastonite composite drawn fibers based on surface response analysis. *Polymers, 14*(5), 924. DOI: 10.3390/polym14050924 PMID: 35267749

Le, V. T., Bui, M. C., Pham, T. Q. D., Tran, H. S., & Tran, X. V. (2023). Efficient prediction of thermal history in wire and arc additive manufacturing combining machine learning and numerical simulation. *International Journal of Advanced Manufacturing Technology, 126*(9-10), 4651–4663. DOI: 10.1007/s00170-023-11473-3

Levashkin, D., Ogin, P. A., & Vasilyev, F. V. (2019). Efficiency of Hybrid Cyclic Processing with the Use of Additive Technologies on CNC Machines for the Manufacture of Composite Aviation Parts due to the Reduction of Processing Errors. In *Materials Science Forum* (Vol. 946, p. 959). Trans Tech Publications. https://doi.org/DOI: 10.4028/www.scientific.net/MSF.946.959

Li, X., Chua, J. W., Yu, X., Li, Z., Zhao, M., Wang, Z., & Zhai, W. (2023). 3D-Printed Lattice Structures for Sound Absorption: Current Progress, Mechanisms and Models, Structural-Property Relationships, and Future Outlook [Review of 3D-Printed Lattice Structures for Sound Absorption: Current Progress, Mechanisms and Models, Structural-Property Relationships, and Future Outlook]. Advanced Science, 11(4). Wiley. DOI: 10.1002/advs.202305232

Li, Z., Wang, Q., & Liu, G. (2022). A Review of 3D Printed Bone Implants [Review of A Review of 3D Printed Bone Implants]. Micromachines, 13(4), 528. Multidisciplinary Digital Publishing Institute. DOI: 10.3390/mi13040528

Liang, C., Huang, J., Wang, J., Gong, H., Bai, D., & Zhao, P. (2023). Additive manufacturing of low-temperature co-fired ceramic substrates and surface conductors based on material jetting. *Additive Manufacturing, 78*, 103856. DOI: 10.1016/j.addma.2023.103856

Liaw, C.-Y., & Güvendiren, M. (2017). Current and emerging applications of 3D printing in medicine [Review of Current and emerging applications of 3D printing in medicine]. Biofabrication, 9(2), 24102. IOP Publishing. DOI: 10.1088/1758-5090/aa7279

Li, B., Zhang, Y., Lei, Y., Wei, H., Chen, C., Liu, F., Zhao, P., & Wang, K. (2024). A single-sensor multi-scale quality monitoring methodology for laser-directed energy deposition: Example with height instability and porosity monitoring in additive manufacturing of ceramic thin-walled parts. *Additive Manufacturing*, *79*, 103923. DOI: 10.1016/j.addma.2023.103923

Li, G., Odum, K., Yau, C., Soshi, M., & Yamazaki, K. (2021). High productivity fluence-based control of Directed Energy Deposition (DED) part geometry. *Journal of Manufacturing Processes*, *65*, 407–417. DOI: 10.1016/j.jmapro.2021.03.028

Li, H., Shi, X., Wu, B., Corradi, D. R., Pan, Z., & Li, H. (2024). Wire arc additive manufacturing: A review on digital twinning and visualization process. *Journal of Manufacturing Processes*, *116*, 293–305. DOI: 10.1016/j.jmapro.2024.03.001

Li, K., Li, J., Reardon, C., Schuster, C. S., Wang, Y., Triggs, G. J., Damnik, N., Müenchenberger, J., Wang, X., Martins, E. R., & Krauss, T. F. (2016). High speed e-beam writing for large area photonic nanostructures — A choice of parameters. *Scientific Reports*, *6*(1), 32945. DOI: 10.1038/srep32945 PMID: 27633902

Lim, A. E.-J., Junfeng Song, , Qing Fang, , Chao Li, , Xiaoguang Tu, , Ning Duan, , Tern, R. P.-C., & Tsung-Yang Liow, Kok Kiong Chen. (2014). Review of Silicon Photonics Foundry Efforts. *IEEE Journal of Selected Topics in Quantum Electronics*, *20*(4), 405–416. DOI: 10.1109/JSTQE.2013.2293274

Li, N., Link, G., & Jelonnek, J. (2020). 3D microwave printing temperature control of continuous carbon fiber reinforced composites. *Composites Science and Technology*, *187*, 107939. DOI: 10.1016/j.compscitech.2019.107939

Li, S., Cheng, P., Ahzi, S., Peng, Y., Wang, K., Chinesta, F., & Correia, J. P. M. (2023). Advances in hybrid fibers reinforced polymer-based composites prepared by FDM: A review on mechanical properties and prospects. *Composites Communications*, *40*, 101592. DOI: 10.1016/j.coco.2023.101592

Li, S., Yang, H., Chen, G., Zheng, J., Wang, W., Ren, J., Zhu, C., Yang, Y., Cong, Y., & Fu, J. (2023). 4D printing of biomimetic anisotropic self-sensing hydrogel actuators. *Chemical Engineering Journal*, *473*, 145444. DOI: 10.1016/j.cej.2023.145444

Liu, J.-Y., & Yan, C. (2018). 3D Printing of Scaffolds for Tissue Engineering. In InTech eBooks. DOI: 10.5772/intechopen.78145

Liu, Z., Jiang, Q., Zhang, Y., Li, T., & Zhang, H. (2016). Sustainability of 3D Printing: A Critical Review and Recommendations [Review of Sustainability of 3D Printing: A Critical Review and Recommendations]. Volume 2: Materials; Biomanufacturing; Properties, Applications and Systems; Sustainable Manufacturing. DOI: 10.1115/MSEC2016-8618

Liu, B., Qian, B., Liang, Y., Dai, P., Li, R., & Wei, Q. (2025). Fast and accurate auto-disturbances-rejection temperature control system based on particle swarm optimized fuzzy control: Applied for particle three-dimensional (3D) printing prosthetic orthotic plate. *Additive Manufacturing*, *101*, 104704. DOI: 10.1016/j.addma.2025.104704

Liu, F. R., Zhang, Q., Zhou, W. P., Zhao, J. J., & Chen, J. M. (2012). Micro scale 3D FEM simulation on thermal evolution within the porous structure in selective laser sintering. *Journal of Materials Processing Technology*, *212*(10), 2058–2065. DOI: 10.1016/j.jmatprotec.2012.05.010

Liu, H., Wang, F., Wu, W., Dong, X., & Sang, L. (2023). 4D printing of mechanically robust PLA/TPU/Fe3O4 magneto-responsive shape memory polymers for smart structures. *Composites. Part B, Engineering*, *248*, 110382. DOI: 10.1016/j.compositesb.2022.110382

Liu, J., Su, Z., Wang, C., & Xu, Z. (2022). Effect of an Adaptive-Density Filling Structure on the Mechanical Properties of FDM Parts with a Variable Cross-Section. *Materials (Basel)*, *15*(24), 8746. DOI: 10.3390/ma15248746 PMID: 36556552

Liu, L., Yang, X., Bhandari, B., Meng, Y., & Prakash, S. (2020). Optimization of the Formulation and Properties of 3D-Printed Complex Egg White Protein Objects. *Foods*, *9*(2), 164. DOI: 10.3390/foods9020164 PMID: 32046351

Liu, Y., Hu, X., & Chen, S. (2024). Multi-Material 3D Printing and Computational Design in Pharmaceutical Tablet Manufacturing. *Journal of Computer Science and Artificial Intelligence*, *1*(1), 34–38. DOI: 10.54097/40ftw337

Liu, Z. H., Nolte, J. J., Packard, J. I., Hilmas, G., Dogan, F., & Leu, M. C. (2007) Selective Laser Sintering of high-density alumina ceramic parts. In: *Proceedings of the 35th International MATADOR 2007 Conference*, 3, pp. 351–354. DOI: 10.1007/978-1-84628-988-0_79

Liu, Z., Zhang, M., Bhandari, B., & Wang, Y. (2017). 3D printing: Printing precision and application in the food sector. *Trends in Food Science & Technology*, *69*, 83–94. DOI: 10.1016/j.tifs.2017.08.018

Livesu, M., Ellero, S., Martínez, J., Lefebvre, S., & Attene, M. (2017). *From 3D models to 3D prints : An overview of the processing pipeline*. *36*(2), 537-564.

Li, W., Kitagawa, D., Kobatake, S., Bekyarova, E., & Bardeen, C. J. (2022). Patterning submicron photomechanical features into single diarylethene crystals using electron beam lithography. *Nanoscale Horizons*, *7*(9), 1065–1072. DOI: 10.1039/D2NH00205A PMID: 35788624

Li, X., Shang, J., & Wang, Z. (2017). Intelligent materials: A review of applications in 4D printing. *Assembly Automation*, *37*(2), 170–185. DOI: 10.1108/AA-11-2015-093

Li, Y. C., Zhang, Y. S., Akpek, A., Shin, S. R., & Khademhosseini, A. (2016). 4D bioprinting: The next-generation technology for biofabrication enabled by stimuli-responsive materials. *Biofabrication*, *9*(1), 012001. DOI: 10.1088/1758-5090/9/1/012001 PMID: 27910820

Li, Y., Gao, S., Dong, R., Ding, X., & Duan, X. (2018). Additive manufacturing of PLA and CF/PLA binding layer specimens via fused deposition modeling. *Journal of Materials Engineering and Performance*, *27*(2), 492–500. DOI: 10.1007/s11665-017-3065-0

Li, Y., Wang, S., Qian, S., Liu, Z., Weng, Y., & Zhang, Y. (2024). Depolymerization and Re/upcycling of biodegradable PLA plastics. *ACS Omega*, *9*(12), 13509–13521. DOI: 10.1021/acsomega.3c08674 PMID: 38559974

Loh, G. H., Pei, E., González-Gutiérrez, J., & Monzón, M. (2020). An Overview of Material Extrusion Troubleshooting. *Applied Sciences (Basel, Switzerland)*, *10*(14), 4776. DOI: 10.3390/app10144776

Lolla, R., Srinath, A., Murali, G., Murugan, M., Venkatesan, E. P., & Hasan, N. (2023). Effect of Infill Patterns with Machine Learning Techniques on the Tensile Properties of Polylactic Acid-Based Ceramic Materials with Fused Filament Fabrication. *ACS Omega*, *8*(28), 24786–24796. DOI: 10.1021/acsomega.2c08164 PMID: 37483243

Lopez Taborda, L. L., Maury, H., & Pacheco, J. (2021). Design for additive manufacturing: A comprehensive review of the tendencies and limitations of methodologies. *Rapid Prototyping Journal*, *27*(5), 918–966. DOI: 10.1108/RPJ-11-2019-0296

Lubombo, C., & Huneault, M. A. (2018). Effect of infill patterns on the mechanical performance of lightweight 3D-printed cellular PLA parts. *Materials Today. Communications*, *17*, 214–228. DOI: 10.1016/j.mtcomm.2018.09.017

Lui, Y. S., Sow, W. T., Tan, L. P., Wu, Y., Lai, Y., & Li, H. (2019). 4D printing and stimuli-responsive materials in biomedical aspects. *Acta Biomaterialia*, *92*, 19–36. DOI: 10.1016/j.actbio.2019.05.005 PMID: 31071476

Luo, W., Hui, J., Wang, J., & Yan, Z. (2024). Study on the influence of selective laser melting temperature on the melting bath. *Journal of Physics: Conference Series, 2791*(1), 012024. DOI: 10.1088/1742-6596/2791/1/012024

Luo, X., Wu, H., Wang, C., Jin, Q., Luo, C., Ma, G., Guo, W., & Long, Y. (2024). 3D printing of self-healing and degradable conductive ionoelastomers for customized flexible sensors. *Chemical Engineering Journal, 483*, 149330. DOI: 10.1016/j.cej.2024.149330

Lynn, C. M., West, A., & Rosen, D. W. (1998). *A Process Planning Method and Data Format for Achieving Tolerances in Stereolithography*. The University of Texas.

Lynn-charney, C., & Rosen, D. W. (2000). Usage of accuracy models in stereolithography process planning. *Rapid Prototyping Journal, 6*(2), 77–87. DOI: 10.1108/13552540010323600

Ma, B., Zhang, Y., Li, J., Chen, D., Liang, R., Fu, S., & Li, D. (2023). 4D printing of multi-stimuli responsive rigid smart composite materials with self-healing ability. *Chemical Engineering Journal, 466*, 143420. DOI: 10.1016/j.cej.2023.143420

Magri, A. E., Ouassil, S., & Vaudreuil, S. (2022). Effects of printing parameters on the tensile behavior of 3d-printed acrylonitrile styrene acrylate (asa) material in z direction. Polymer Engineering &Amp. *Polymer Engineering and Science, 62*(3), 848–860. DOI: 10.1002/pen.25891

Mahdi, M. M., Bajestani, M. S., Do Noh, S., & Kim, D. B. (2025). Digital twin-based architecture for wire arc additive manufacturing using OPC UA. *Robotics and Computer-integrated Manufacturing, 94*, 102944. DOI: 10.1016/j.rcim.2024.102944

Mahmood, M. A., Visan, A. I., Ristoscu, C., & Mihăilescu, I. N. (2020). Artificial Neural Network Algorithms for 3D Printing [Review of Artificial Neural Network Algorithms for 3D Printing]. Materials, 14(1), 163. Multidisciplinary Digital Publishing Institute. DOI: 10.3390/ma14010163

Mahmood, M. A., Chioibasu, D., Rehman, A. U., Mihai, S., & Popescu, A. C. (2022). Post-processing techniques to enhance the quality of metallic parts produced by additive manufacturing. *Metals, 12*(1), 77. DOI: 10.3390/met12010077

Maines, E. M., Porwal, M. K., Ellison, C. J., & Reineke, T. M. (2021). Sustainable advances in SLA/DLP 3D printing materials and processes. *Green Chemistry, 23*(18), 6863–6897. DOI: 10.1039/D1GC01489G

Majko, J., Sága, M., Vaško, M., Handrik, M., & Ságová, Z. (2020). FE modeling of continuous fiber reinforced thermoplastic composite structures produced by additive manufacturing. *IOP Conference Series. Materials Science and Engineering*, *776*(1), 12080. DOI: 10.1088/1757-899X/776/1/012080

Major, R., Gawlikowski, M., Surmiak, M., Janiczak, K., Więcek, J., Kurtyka, P., Schwentenwein, M., Jasek-Gajda, E., Kopernik, M., & Lackner, J. M. (2023). In vitro and in vivo testing of stereolithography (SLA)-manufactured haemocompatible photopolymers for blood pump. *Applied Sciences (Basel, Switzerland)*, *14*(1), 383. DOI: 10.3390/app14010383

Makanda, I. L. D., Jiang, P., & Yang, M. (2025). Personalized federated unsupervised learning for nozzle condition monitoring using vibration sensors in additive manufacturing. *Robotics and Computer-integrated Manufacturing*, *93*, 102940. DOI: 10.1016/j.rcim.2024.102940

Mallakpour, S., Tabesh, F., & Hussain, C. M. (2021). 3D and 4D printing: From innovation to evolution. *Advances in Colloid and Interface Science*, *294*, 102482. DOI: 10.1016/j.cis.2021.102482 PMID: 34274721

Mandal, A., Morali, A., Skorobogatiy, M., & Bodkhe, S. (2024). 3D Printing of Polyvinylidene Fluoride-Based Piezoelectric Sensors for Noninvasive Continuous Blood Pressure Monitoring. *Advanced Engineering Materials*, *26*(1), 2301292. DOI: 10.1002/adem.202301292

Manfrinato, V. R., Wen, J., Zhang, L., Yang, Y., Hobbs, R. G., Baker, B., Su, D., Zakharov, D., Zaluzec, N. J., Miller, D. J., Stach, E. A., & Berggren, K. K. (2014). Determining the Resolution Limits of Electron-Beam Lithography: Direct Measurement of the Point-Spread Function. *Nano Letters*, *14*(8), 4406–4412. DOI: 10.1021/nl5013773 PMID: 24960635

Maniruzzaman, M. (Ed.). (2018). *3D and 4D printing in biomedical applications: process engineering and additive manufacturing*. John Wiley & Sons.

Mantecón, R., Rufo-Martín, C., Castellanos, R., & Díaz-Álvarez, J. (2022). Experimental assessment of thermal gradients and layout effects on the mechanical performance of components manufactured by fused deposition modeling. *Rapid Prototyping Journal*, *28*(8), 1598–1608. DOI: 10.1108/RPJ-12-2021-0329

Mao, X. (2024). The Intersection of Art and Technology: Innovative Thinking and Practice in the Perspective of New Media. *Applied Mathematics and Nonlinear Sciences*, *9*(1), 20241367. DOI: 10.2478/amns-2024-1367

Ma, Q., Dong, K., Li, F., Jia, Q., Tian, J., Yu, M., & Xiong, Y. (2025). Additive manufacturing of polymer composite millimeter-wave components: Recent progress, novel applications, and challenges. *Polymer Composites*, *46*(1), 14–37. DOI: 10.1002/pc.28985

Martelli, N., Serrano, C., Brink, H. van den, Pineau, J., Prognon, P., Borget, I., & Batti, S. E. (2016). Advantages and disadvantages of 3-dimensional printing in surgery: A systematic review [Review of Advantages and disadvantages of 3-dimensional printing in surgery: A systematic review]. Surgery, 159(6), 1485. Elsevier BV. DOI: 10.1016/j.surg.2015.12.017

Martin. (2024). 3D Printer Calibration: Step-by-Step to Perfect Results. https://the3dprinterbee.com/3d-printer-calibration/

Martínez, J., Diéguez, J. L., Ares, E., Pereira, A., Hernández, P., & Pérez, J. A. (2013). Comparative between FEM models for FDM parts and their approach to a real mechanical behaviour. *Procedia Engineering*, *63*(December), 878–884. DOI: 10.1016/j.proeng.2013.08.230

Martinez-Marquez, D., Mirnajafizadeh, A., Carty, C. P., & Stewart, R. A. (2018). Application of quality by design for 3D printed bone prostheses and scaffolds. *PLoS One*, *13*(4), e0195291. Advance online publication. DOI: 10.1371/journal.pone.0195291 PMID: 29649231

Martín, M. J., Auñón, J. A., & Fernández, F. M. (2021). Influence of Infill Pattern on Mechanical Behavior of Polymeric and Composites Specimens Manufactured Using Fused Filament Fabrication Technology. *Polymers*, *13*(17), 2934. DOI: 10.3390/polym13172934 PMID: 34502974

Ma, S., Fu, S., Yang, H., He, P., Sun, Z., Duan, X., Jia, D., Colombo, P., & Zhou, Y. (2024). Exploiting bifunctional 3D-Printed geopolymers for efficient cesium removal and immobilization: An approach for hazardous waste management. *Journal of Cleaner Production*, *437*, 140599. DOI: 10.1016/j.jclepro.2024.140599

Ma, T., Zhang, Y., Ruan, K., Guo, H., He, M., Shi, X., Guo, Y., Kong, J., & Gu, J. (2024). Advances in 3D printing for polymer composites: A review. *InfoMat*, *6*(6), e12568. DOI: 10.1002/inf2.12568

Mathews, D. A. P., Baird, A. D., & Lucky, M. (2020). Innovation in Urology: Three-Dimensional Printing and Its Clinical Application [Review of Innovation in Urology: Three-Dimensional Printing and Its Clinical Application]. Frontiers in Surgery, 7. Frontiers Media. DOI: 10.3389/fsurg.2020.00029

Mathur, V., Dsouza, V., Srinivasan, V., & Vasanthan, K. S. (2025). Volumetric Additive Manufacturing for Cell Printing: Bridging Industry Adaptation and Regulatory Frontiers. *ACS Biomaterials Science & Engineering*, *11*(1), 156–181. DOI: 10.1021/acsbiomaterials.4c01837 PMID: 39746181

Matsumoto, M., Shiomi, M., Osakada, K., & Abe, F. (2002). Finite element analysis of single layer forming on metallic powder bed in rapid prototyping by selective laser processing. *International Journal of Machine Tools & Manufacture*, *42*(1), 61–67. DOI: 10.1016/S0890-6955(01)00093-1

Matte, C.-D., Pearson, M., Trottier-Cournoyer, F., Dafoe, A., & Kwok, T. H. (2019). Automated storage and active cleaning for multi-material digital-light-processing printer. *Rapid Prototyping Journal*, *25*(5), 864–874. DOI: 10.1108/RPJ-08-2018-0211

Mattera, G., Caggiano, A., & Nele, L. (2025). Energy Efficiency Optimisation in Wire arc Additive Manufacturing of Invar 36 Alloy via Intelligent Data-Driven Techniques. *International Journal of Precision Engineering and Manufacturing-Green Technology*, 1-13.

Mattera, G., Nele, L., & Paolella, D. (2024). Monitoring and control the Wire Arc Additive Manufacturing process using artificial intelligence techniques: A review. *Journal of Intelligent Manufacturing*, *35*(2), 467–497. DOI: 10.1007/s10845-023-02085-5

Mattera, G., Polden, J., & Luigi, N. E. L. E. (2025). Monitoring Wire Arc Additive Manufacturing process of Inconel 718 thin-walled structure using wavelet decomposition and clustering analysis of welding signal. *Journal of Advanced Manufacturing Science and Technology*, *5*(1), 2025006. DOI: 10.51393/j.jamst.2025006

Maurya, N. K., Maurya, M., Saxena, A., Tyagi, A., Kiran, G. K., Singh, H., Tripathi, H., Kumar, P., & Yadav, S. (2021). Investigation of mechanical properties and form error of the components fabricated by rapid prototyping: A review [Review of Investigation of mechanical properties and form error of the components fabricated by rapid prototyping: A review]. Materials Today Proceedings, 47, 3901. Elsevier BV. DOI: 10.1016/j.matpr.2021.03.553

Mayfield, C. K., Ayad, M., Lechtholz-Zey, E., Chen, Y., & Lieberman, J. R. (2022). 3D-printing for critical sized bone defects: Current concepts and future directions. *Bioengineering (Basel, Switzerland)*, *9*(11), 680. DOI: 10.3390/bioengineering9110680 PMID: 36421080

Mazurchevici, A. D., Nedelcu, D., & Popa, R. (2020). Additive manufacturing of composite materials by FDM technology. *RE:view*.

Mazzanti, V., Malagutti, L., & Mollica, F. (2019). FDM 3D printing of polymers containing natural fillers: A review of their mechanical properties. *Polymers, 11*(7), 1094. DOI: 10.3390/polym11071094 PMID: 31261607

McKinney, M., Garland, A., Cillessen, D., Adamczyk, J., Bolintineanu, D., Heiden, M., Fowler, E., & Boyce, B. L. (2025). Unsupervised multimodal fusion of in-process sensor data for advanced manufacturing process monitoring. *Journal of Manufacturing Systems, 78*, 271–282. DOI: 10.1016/j.jmsy.2024.12.003

Mearns, C. D. (2017). *Analysis of Selective Laser Sintering print parameter modelling methodologies for energy input minimisation : A thesis presented in partial fulfilment of the requirements for the degree of Master of Engineering in Mechatronics at Massey University, Albany, New Zealand.*

Megdich, A., Habibi, M., & Laperrière, L. (2023). A review on 3D printed piezoelectric energy harvesters: Materials, 3D printing techniques, and applications. *Materials Today. Communications, 35*, 105541. DOI: 10.1016/j.mtcomm.2023.105541

Mehrpouya, M., Dehghanghadikolaei, A., Fotovvati, B., Vosooghnia, A., Emamian, S. S., & Gisario, A. (2019). The potential of additive manufacturing in the smart factory industrial 4.0 : A review. *Applied Sciences (Basel, Switzerland), 9*(18), 3865. DOI: 10.3390/app9183865

Mehrpouya, M., Vahabi, H., Janbaz, S., Darafsheh, A., Mazur, T. R., & Ramakrishna, S. (2021). 4D printing of shape memory polylactic acid (PLA). *Polymer, 230*, 124080. DOI: 10.1016/j.polymer.2021.124080

Melenka, G. W., Cheung, B. K. O., Schofield, J. S., Dawson, M. R., & Carey, J. P. (2016). Evaluation and prediction of the tensile properties of continuous fiber-reinforced 3D printed structures. *Composite Structures, 153*, 866–875. DOI: 10.1016/j.compstruct.2016.07.018

Mendřický, R., & Fris, D. (2020). Analysis of the Accuracy and the Surface Roughness of FDM/FFF Technology and Optimisation of Process Parameters. Tehnicki Vjesnik - Technical Gazette, 27(4). DOI: 10.17559/TV-20190320142210

Meng, M., Wang, J., Huang, H., Liu, X., Zhang, J., & Li, Z. (2023). 3D printing metal implants in orthopedic surgery: Methods, applications and future prospects. *Journal of Orthopaedic Translation, 42*, 94–112. DOI: 10.1016/j.jot.2023.08.004 PMID: 37675040

Merazzo, K. J., Lima, A. C., Rincón-Iglesias, M., Fernandes, L. C., Pereira, N., Lanceros-Mendez, S., & Martins, P. (2021). Magnetic materials: A journey from finding north to an exciting printed future. *Materials Horizons*, *8*(10), 2654–2684. DOI: 10.1039/D1MH00641J PMID: 34617551

Messimer, S. L., Patterson, A. E., Muna, N., Deshpande, A. P., & Rocha Pereira, T. (2018). Characterization and processing behavior of heated aluminum-polycarbonate composite build plates for the FDM additive manufacturing process. *Journal of Manufacturing and Materials Processing*, *2*(1), 12. DOI: 10.3390/jmmp2010012

Miao, R. (2024). Development, prospect and challenge for additive manufacturing of polymer materials. Transactions on Environment. *Energy and Earth Sciences*, *1*, 90–96. DOI: 10.62051/s182f353

Michaleris, P. (2014). Modeling metal deposition in heat transfer analyses of additive manufacturing processes. *Finite Elements in Analysis and Design*, *86*, 51–60. DOI: 10.1016/j.finel.2014.04.003

Michiels, S., D'Hollander, A., Lammens, N., Kersemans, M., Zhang, G., Denis, J., Poels, K., Sterpin, E., Nuyts, S., Haustermans, K., & Depuydt, T. (2016). Towards 3D printed multifunctional immobilization for proton therapy : Initial materials characterization. *Medical Physics*, *43*(10), 5392–5402. DOI: 10.1118/1.4962033 PMID: 27782703

Mieloszyk, M., Majewska, K., & Andrearczyk, A. (2022). Embedded optical fibre with fibre bragg grating influence on additive manufactured polymeric structure durability. *Materials (Basel)*, *15*(7), 2653. DOI: 10.3390/ma15072653 PMID: 35407984

Mieloszyk, M., & Muna, I. I. (2023). Numerical modeling of thermal effects on mechanical behavior of additive manufactured carbon fiber reinforced polymer. *Health Monitoring of Structural and Biological Systems*, *XVII*, 35. Advance online publication. DOI: 10.1117/12.2657985

Miller, Z., Fernandes, P., Hunter, B., Maldonado, C., Patterson, B., & Skoug, R. (2024, March). *The use of metallized stereolithography additively manufactured components in space. In 2024 IEEE Aerospace Conference*. IEEE., DOI: 10.1109/AERO60161.2024.10509144

Minaoglou, P., Efkolidis, N., Manavis, A., & Kyratsis, P. (2024). A Review on Wearable Product Design and Applications [Review of A Review on Wearable Product Design and Applications]. Machines, 12(1), 62. Multidisciplinary Digital Publishing Institute. DOI: 10.3390/machines12010062

Miryousefi Ata, S., Kazemian, A., & Jafari, A. (2022). Application of concrete 3D printing for bridge construction: Current challenges and future directions. In *Construction Research Congress 2022* (pp. 869-879). DOI: 10.1061/9780784483961.091

Mirzaali, M. J., Moosabeiki, V., Rajaai, S., Zhou, J., & Zadpoor, A. A. (2022). Additive manufacturing of biomaterials—Design principles and their implementation. *Materials (Basel)*, *15*(15), 5457. DOI: 10.3390/ma15155457 PMID: 35955393

Mladenovska, D., Lazarevska, A. M., & Krstanoski, M. (2021). Attributes relevant for sustainable additive manufacturing – material-driven approach. *IOP Conference Series. Materials Science and Engineering*, *1190*(1), 12001. DOI: 10.1088/1757-899X/1190/1/012001

Mohamed, O. A. (2017). Analytical modeling and experimental investigation of product quality and mechanical properties in FDM additive manufacturing. *Swinburne University of Technology, Diss.*

Mohamed, R. A., & Mohamed, A. F. A. (2025). Exploring the environmental benefits of 3D printing technology in concrete construction; a review. *Progress in Additive Manufacturing*, *10*(1), 279–289. DOI: 10.1007/s40964-024-00622-z

Mohammadizadeh, M., Imeri, A., Fidan, I., & Elkelany, M. (2019). 3D printed fiber reinforced polymer composites—Structural analysis. *Composites. Part B, Engineering*, *175*, 107112. DOI: 10.1016/j.compositesb.2019.107112

Mohammed, R., Shaik, A. S., Mohammed, S., Bunga, K. K., Aggala, C., Babu, B. P., & Badruddin, I. A. (2025). Advancements and Challenges in Additive Manufacturing: Future Directions and Implications for Sustainable Engineering. *Advance Sustainable Science Engineering and Technology*, *7*(1), 0250108–0250108. DOI: 10.26877/asset.v7i1.1079

Mohapatra, S., Kar, R. K., Biswal, P. K., & Bindhani, S. (2022). Approaches of 3D printing in current drug delivery. *Sensors International*, *3*, 100146. DOI: 10.1016/j.sintl.2021.100146

Mohd Yusof, A. A. B., Md Yusop, A. H. B., & Abu Kasim, H. B. (2024). Analysis of crack propagation of pla fabricated by the additive manufacturing technique. *Engineering Research Express*, *6*(3), 035518. DOI: 10.1088/2631-8695/ad62ae

Mohol, S. S., & Sharma, V. (2021). Functional applications of 4D printing: A review. *Rapid Prototyping Journal*, *27*(8), 1501–1522. DOI: 10.1108/RPJ-10-2020-0240

Mohseni, M., & Khodaygan, S. (2024). Design for additive manufacturing of topology-optimized structures based on deep learning and transfer learning. *Rapid Prototyping Journal*, *30*(7), 1411–1433. DOI: 10.1108/RPJ-02-2024-0102

Moldovan, C., Cosma, C., Berce, P., & Balc, N. (2018). Theoretical analysis and practical case studies of SLA, Polyjet and FDM manufacturing techniques. Acta Technica Napocensis-Series: Applied Mathematics. *Mechanical Engineering (New York, N.Y.)*, *61*(3).

Molino, M., Cortese, C. G., & Ghislieri, C. (2020). The Promotion of Technology Acceptance and Work Engagement in Industry 4.0: From Personal Resources to Information and Training.

Momeni, F., Liu, X., & Ni, J. (2017). A review of 4D printing. *Materials & Design*, *122*, 42–79. DOI: 10.1016/j.matdes.2017.02.068

Momeni, F., & Ni, J. (2020). Laws of 4D printing. *Engineering (Beijing)*, *6*(9), 1035–1055. DOI: 10.1016/j.eng.2020.01.015

Mondal, K., & Tripathy, P. K. (2021). Preparation of smart materials by additive manufacturing technologies: A review. *Materials (Basel)*, *14*(21), 6442. DOI: 10.3390/ma14216442 PMID: 34771968

Morales, M. A., Maranon, A., Hernández, C., & Porras, A. (2021). Development and characterization of a 3d printed cocoa bean shell filled recycled polypropylene for sustainable composites. *Polymers*, *13*(18), 3162. DOI: 10.3390/polym13183162 PMID: 34578062

Morampudi, P., Ramana, V. S. N. V., Prabha, K. A., Swetha, S., & Rao, A. (2021). 3D-printing analysis of surface finish. *Materials Today: Proceedings*, *43*, 587–592. DOI: 10.1016/j.matpr.2020.12.085

Moreira, C. A., Caicedo, M., Cervera, M., Chiumenti, M., & Baiges, J. (2022). A multi-criteria h-adaptive finite-element framework for industrial part-scale thermal analysis in additive manufacturing processes. *Engineering with Computers*, *38*(6), 4791–4813. DOI: 10.1007/s00366-022-01655-0

Moreno Nieto, D., & Moreno Sánchez, D. (2021). Design for additive manufacturing: Tool review and a case study. *Applied Sciences (Basel, Switzerland)*, *11*(4), 1571. DOI: 10.3390/app11041571

Morettini, G., Palmieri, M., Capponi, L., & Landi, L. (2022). Comprehensive characterization of the mechanical and physical properties of PLA structures printed by the FFF-3D printing process in different directions. *Progress in Additive Manufacturing*, *7*(5), 1111–1122. DOI: 10.1007/s40964-022-00285-8

Morouço, P., Azimi, B., Milazzo, M., Mokhtari, F., Fernandes, C., Reis, D., & Danti, S. (2020). Four-dimensional (bio-) printing: A review on stimuli-responsive mechanisms and their biomedical suitability. *Applied Sciences (Basel, Switzerland), 10*(24), 9143. DOI: 10.3390/app10249143

Mueller, U., Schubert, M., Teich, F., Puetter, H., Schierle-Arndt, K., & Pastre, J. (2006). Metal–organic frameworks—Prospective industrial applications. *Journal of Materials Chemistry, 16*(7), 626–636. DOI: 10.1039/B511962F

Mu, H., He, F., Yuan, L., Hatamian, H., Commins, P., & Pan, Z. (2024). Online distortion simulation using generative machine learning models: A step toward digital twin of metallic additive manufacturing. *Journal of Industrial Information Integration, 38*, 100563. DOI: 10.1016/j.jii.2024.100563

Mujtaba, A., Islam, F., Kaeding, P., Lindemann, T., & Prusty, B. G. (2022). Machine-learning based process monitoring of automated composites manufacturing to enable digital twin development. DOI: 10.21203/rs.3.rs-2220331/v1

Mukai, S., Mukai, E., Santos-Junior, J. A., Shibli, J. A., Faveri, M., & Giro, G. (2021). Assessment of the reproducibility and precision of milling and 3D printing surgical guides. *BMC Oral Health, 21*(1), 1–7. DOI: 10.1186/s12903-020-01362-6 PMID: 33388028

Muller, P., Mognol, P., & Hascoet, J. Y. (2013). Modeling and control of a direct laser powder deposition process for Functionally Graded Materials (FGM) parts manufacturing. *Journal of Materials Processing Technology, 213*(5), 685–692. DOI: 10.1016/j.jmatprotec.2012.11.020

Muna, I. I., Mieloszyk, M., Rimašauskienė, R., Maqsood, N., & Rimašauskas, M. (2022). Thermal effects on mechanical strength of additive manufactured cfrp composites at stable and cyclic temperature. *Polymers, 14*(21), 4680. DOI: 10.3390/polym14214680 PMID: 36365673

Munguia-Galeano, F., Ortega-Arroyo, L., Villarreal-Cervantes, M. G., & Ji, Z. (2023, August). Towards integrating 3D printing and automated assembly. In *2023 IEEE 19th International Conference on Automation Science and Engineering (CASE)* (pp. 1–7). IEEE. https://doi.org/DOI: 10.1109/CASE56687.2023.10259847

Munoz, C., Kim, C., & Armstrong, L. (2013). Layer-by-layer: Opportunities in 3D printing—Technology trends, growth drivers and the emergence of innovative applications in 3D printing. *Lux Research*. (No DOI found – gray literature)

Murga Pinillos, A. Y. (2024). University-industry open innovation: Main enablers and practices based on a scoping review [Preprint]. *International Journal of Innovation Science*. Advance online publication. DOI: 10.1108/IJIS-08-2023-0194

Murr, L. E. (2015). Metallurgy of additive manufacturing: Examples from electron beam melting. *Additive Manufacturing*, *5*, 40–53. DOI: 10.1016/j.addma.2014.12.002

Murr, L. E., Gaytan, S. M., Martinez, E., Medina, F., & Wicker, R. B. (2012). Next generation orthopaedic implants by additive manufacturing using electron beam melting. *International Journal of Biomaterials*, *2012*, 1–14. Advance online publication. DOI: 10.1155/2012/245727 PMID: 22956957

Murtezani, I., Sharma, N., & Thieringer, F. M. (2022). Medical 3D printing with a focus on Point-of-Care in Cranio-and Maxillofacial Surgery. A systematic review of literature. *Annals of 3D Printed Medicine, 6*, 100059.

Murugan, B., Rahman, M. Z., Fatimah, I., Anita Lett, J., Annaraj, J., Kaus, N. H. M., Al-Anber, M. A., & Sagadevan, S. (2023). Green synthesis of CuO nanoparticles for biological applications. *Inorganic Chemistry Communications*, *155*, 111088. DOI: 10.1016/j.inoche.2023.111088

Murzin, S. P. (2024). Artificial Intelligence-Driven Innovations in Laser Processing of Metallic Materials. *Metals*, *14*(12), 1458. DOI: 10.3390/met14121458

Musa, N., & Singh, N. B. (2025). Advanced sensors: classification, fabrication, and applications. In *Advanced Sensors for Smart Healthcare* (pp. 47–67). Elsevier. DOI: 10.1016/B978-0-443-24790-3.00004-1

Muzaffar, A., Ahamed, M. B., Deshmukh, K., Kovářík, T., Křenek, T., & Pasha, S. K. (2020). *3D and 4D printing of pH-responsive and functional polymers and their composites. In 3D and 4D printing of polymer nanocomposite materials*. Elsevier.

Mwema, F. M., & Akinlabi, E. T. (2020). Basics of Fused Deposition Modelling (FDM). In SpringerBriefs in applied sciences and technology (p. 1). Springer Nature. DOI: 10.1007/978-3-030-48259-6_1

Nadagouda, M. N., Ginn, M., & Rastogi, V. (2020). A review of 3D printing techniques for environmental applications [Review of A review of 3D printing techniques for environmental applications]. Current Opinion in Chemical Engineering, 28, 173. Elsevier BV. DOI: 10.1016/j.coche.2020.08.002

Nagano, H. Umemoto, R., Ueno, A., & Ogushi, T. (2022). Effective thermal conductivity measurement of additively manufactured lattice structures by application of modified temperature profile method. SSRN Electronic Journal. DOI: 10.2139/ssrn.4313766

Nancharaiah, T., Reddy, V. S. K., Chakravarthi, T., Chowdary, G. T. S., & Teja, Y. B. (2023). Experimental study on the effect of process parameters on the build time and part accuracy of direct metal laser sintering components. *Applied Mechanics and Materials, 917*, 49–56. DOI: 10.4028/p-fzsqM6

Narahara, H., Tanaka, F., Kishinami, T., Igarashi, S., & Saito, K. (1999). Reaction heat effects on initial linear shrinkage and deformation in stereolithography. *Rapid Prototyping Journal, 5*(3), 120–128. DOI: 10.1108/13552549910278946

Narahara, H., Tanaka, F., Kishinami, T., Igarasm, S., & Saito, K.Department of Mechanical Systems Engineering, and Gunma Polytechnic. (1997). *Reaction Heat Effect on Initial Linear Shrinkage of Stereolithography Resins.*

Na, S., Heo, S., Han, S., Shin, Y.-S., & Roh, Y.-S. (2022). Acceptance Model of Artificial Intelligence (AI)-Based Technologies in Construction Firms: Applying the Technology Acceptance Model (TAM) in Combination with the Technology–Organisation–Environment (TOE) Framework. *Buildings (Basel, Switzerland), 12*(2), 90. DOI: 10.3390/buildings12020090

Nasrin, T., Pourkamali-Anaraki, F., & Peterson, A. M. (2024). Application of machine learning in polymer additive manufacturing: A review. *Journal of Polymer Science, 62*(12), 2639–2669. DOI: 10.1002/pol.20230649

Natarajan, M., Pasupuleti, T., & Kiruthika, J. D. P. (2024). "Optimization and Regression Modeling of Additive Manufacturing (Fused Deposition Modeling) of PETG Material for Automobile Applications," SAE Technical Paper 2024-28-0234, DOI: 10.4271/2024-28-0234

Navaf, M., Sunooj, K. V., Aaliya, B., Akhila, P. P., Sudheesh, C., Mir, S. A., & George, J. (2022). 4D printing: A new approach for food printing; effect of various stimuli on 4D printed food properties. A comprehensive review. *Applied Food Research, 2*(2), 100150. DOI: 10.1016/j.afres.2022.100150

Navarro, J., Din, M., Janes, M. E., Swayambunathan, J., Fisher, J. P., & Dreher, M. L. (2019). Effect of print orientation on microstructural features and mechanical properties of 3D porous structures printed with continuous digital light processing. *Rapid Prototyping Journal, 25*(6), 1017–1029. DOI: 10.1108/RPJ-10-2018-0276

Nesic, D., Durual, S., Marger, L., Mekki, M., Sailer, I., & Scherrer, S. S. (2020). Could 3D printing be the future for oral soft tissue regeneration? *Bioprinting (Amsterdam, Netherlands), 20*, e00100. DOI: 10.1016/j.bprint.2020.e00100 PMID: 32679657

Ngo, T. D., Kashani, A., Imbalzano, G., Nguyen, K. T., & Hui, D. (2018). Additive manufacturing (3D printing): A review of materials, methods, applications and challenges. *Composites. Part B, Engineering*, *143*, 172–196. DOI: 10.1016/j.compositesb.2018.02.012

Ng, W. L., Goh, G. L., Goh, G. D., Ten, J. S. J., & Yeong, W. Y. (2024). Progress and opportunities for machine learning in materials and processes of additive manufacturing. *Advanced Materials*, *36*(34), 2310006. DOI: 10.1002/adma.202310006 PMID: 38456831

Nhassa, G., Ben khalifa, R., & Snoussi, A. (2023, December). Numerical Analysis of the Mechanical Behavior of 3D-Printed Parts. In International Conference Design and Modeling of Mechanical Systems (pp. 474-486). Cham: Springer Nature Switzerland.

Nhassa, G., Ben Khalifa, R., & Snoussi, A. (2025). Impact of 3D Printing Settings on Polylactic Acid Filament Mechanical Behaviors Based on the Taguchi Method. *Recent Patents on Mechanical Engineering*, *18*, 1–11. DOI: 10.2174/0122127976328835241016094600

Niesler, F., & Hermatschweiler, M. (2015). Two-Photon Polymerization — A Versatile Microfabrication Tool. *Laser-Technik-Journal*, *12*(3), 44–47. DOI: 10.1002/latj.201500019

Ning, F., Cong, W., Jia, Z., Wang, F., & Zhang, M. (2016) Additive manufacturing of CFRP composites using fused deposition modeling: Effects of process parameters. In: ASME 2016 11th International Manufacturing Science and Engineering Conference, MSEC 2016, 3, pp. 1–7. https://doi.org/DOI: 10.1115/MSEC20168561

Ning, F., Cong, W., Hu, Z., & Huang, K. (2017). Additive manufacturing of thermoplastic matrix composites using fused deposition modeling: A comparison of two reinforcements. *Journal of Composite Materials*, *51*(27), 3733–3742. DOI: 10.1177/0021998317692659

Ntousia, M., & Fudos, I. (2019, June). 3D printing technologies & applications: an overview. In *Proceedings of the CAD 2020 Conference,* Singapore (Vol. 243, p. 248).

Nurhudan, A. I., Supriadi, S., Whulanza, Y., & Saragih, A. S. (2021). Additive manufacturing of metallic based on extrusion process: A review. *Journal of Manufacturing Processes*, *66*, 228–237. DOI: 10.1016/j.jmapro.2021.04.018

Nyamekye, P., Lakshmanan, R., Tepponen, V., & Westman, S. (2024). Sustainability aspects of additive manufacturing: Leveraging resource efficiency via product design optimization and laser powder bed fusion. *Heliyon, 10*(1), e23152. DOI: 10.1016/j.heliyon.2023.e23152 PMID: 38163216

O'Dwyer, M., Filieri, R., & O'Malley, L. (2023). Establishing successful university–industry collaborations: Barriers and enablers deconstructed. *The Journal of Technology Transfer*, *48*(3), 900–931. DOI: 10.1007/s10961-022-09932-2

Oberoi, G., Nitsch, S., Edelmayer, M., Janjić, K., Müller, A. S., & Agis, H. (2018). 3D Printing—Encompassing the Facets of Dentistry [Review of 3D Printing—Encompassing the Facets of Dentistry]. Frontiers in Bioengineering and Biotechnology, 6. Frontiers Media. DOI: 10.3389/fbioe.2018.00172

Obi, M. (2022). Investigating an approach for the dissemination of formalised Design for Additive Manufacturing knowledge (Doctoral dissertation, Loughborough University).

Obilanade, D., Törlind, P., & Dordlofva, C. (2022). Surface roughness and design for additive manufacturing: A design artefact investigation. *Proceedings of the Design Society*, *2*, 1421–1430. DOI: 10.1017/pds.2022.144

Ogoh, G. I., & Fairweather, N. B. (2019). The state of the responsible research and innovation programme : A case for its application in additive manufacturing. *Journal of Information. Communication and Ethics in Society*, *17*(2), 145–166. DOI: 10.1108/JICES-12-2018-0093

Okafor-Muo, O. L., Hassanin, H., Kayyali, R., & ElShaer, A. (2020). 3D printing of solid oral dosage forms: Numerous challenges with unique opportunities. *Journal of Pharmaceutical Sciences*, *109*(12), 3535–3550. DOI: 10.1016/j.xphs.2020.08.029 PMID: 32976900

Okoli, N. J., & Kabaso, B. (2024). Building a smart water city: Iot smart water technologies, applications, and future directions. *Water (Basel)*, *16*(4), 557. DOI: 10.3390/w16040557

Okuyelu, O., & Adaji, O. (2024). Ai-driven real-time quality monitoring and process optimization for enhanced manufacturing performance. *Journal of Advances in Mathematics and Computer Science*, *39*(4), 81–89. DOI: 10.9734/jamcs/2024/v39i41883

Olaiya, N. G., Maraveas, C., Salem, M., Raja, S. P., Rashedi, A., El-Bahy, Z. M., & Olaiya, F. G. (2022). Viscoelastic and properties of amphiphilic chitin in plasticised polylactic acid/starch biocomposite. *Polymers*, *14*(11), 2268. DOI: 10.3390/polym14112268 PMID: 35683940

Olasek, K., & Wiklak, P. (2014). Application of 3D printing technology in aerodynamic study. *Journal of Physics: Conference Series*, *530*, 12009. DOI: 10.1088/1742-6596/530/1/012009

Olawumi, M. A., Oladapo, B. I., & Olugbade, T. O. (2024). Evaluating the impact of recycling on polymer of 3D printing for energy and material sustainability. *Resources, Conservation and Recycling*, *209*, 107769. DOI: 10.1016/j.resconrec.2024.107769

Oleksy, M., Dynarowicz, K., & Aebisher, D. (2023). Advances in Biodegradable Polymers and Biomaterials for Medical Applications—A Review [Review of Advances in Biodegradable Polymers and Biomaterials for Medical Applications—A Review]. Molecules, 28(17), 6213. Multidisciplinary Digital Publishing Institute. DOI: 10.3390/molecules28176213

Olsson, N. O. E., Shafqat, A., Arica, E., & Økland, A. (2019). "3D-Printing Technology in Construction: Results from a Survey", Lill, I. and Witt, E. (Ed.) 10th Nordic Conference on Construction Economics and Organization (Emerald Reach Proceedings Series, Vol. 2), Emerald Publishing Limited, Leeds, pp. 349-356. DOI: 10.1108/S2516-285320190000002044

Omairi, A., & Ismail, Z. H. (2021). Towards Machine Learning for Error Compensation in Additive Manufacturing. *Applied Sciences (Basel, Switzerland)*, *11*(5), 2375. DOI: 10.3390/app11052375

Ompusunggu, A. P., Vonderscher, Y., & Motl, D. (2020). Physics-based vibration feature for detecting eccentric workpiece/runout faults during continuous generating gear grinding processes. *Mechanical Systems and Signal Processing*, *153*, 107536. DOI: 10.1016/j.ymssp.2020.107536

Ononiwu, N., Ako, P., Anyaoha, C., Ikwuagwu, C., & Jacobs, I. (2024). Sustainable considerations in additive manufacturing processes: A review. *Journal of Mechanical Engineering Science*, •••, 9853–9871. DOI: 10.15282/jmes.18.1.2024.5.0780

Onuh, S. O., & Hon, K. K. B. (1998). Optimising build parameters for improved surface finish in stereolithography. *International Journal of Machine Tools & Manufacture*, *38*(4), 329–342. DOI: 10.1016/S0890-6955(97)00068-0

Oral, M. O., Derossi, A., Caporizzi, R., & Severini, C. (2021). Analyzing the most promising innovations in food printing. Programmable food texture and 4D foods. *Future Foods : a Dedicated Journal for Sustainability in Food Science*, *4*, 100093. DOI: 10.1016/j.fufo.2021.100093

Oropallo, W., & Piegl, L. A. (2016). Ten challenges in 3D printing. *Engineering with Computers*, *32*(1), 135–148. DOI: 10.1007/s00366-015-0407-0

Orzari, L. O., Kalinke, C., Silva-Neto, H. A., Rocha, D. S., Camargo, J. R., Coltro, W. K., & Janegitz, B. C. (2025). Screen-Printing vs Additive Manufacturing Approaches: Recent Aspects and Trends Involving the Fabrication of Electrochemical Sensors. Analytical Chemistry.

Otitoju, T. A., Okoye, P. U., Chen, G., Li, Y., Okoye, M. O., & Li, S. (2020). Advanced ceramic components: Materials, fabrication, and applications. *Journal of Industrial and Engineering Chemistry*, *85*, 34–65. DOI: 10.1016/j.jiec.2020.02.002

Păcurar, A. (2019). Introductory Chapter: Applications of Design for Manufacturing and Assembly. In IntechOpen eBooks. IntechOpen. DOI: 10.5772/intechopen.82549

Palanisamy, C., Raman, R., & Dhanraj, P. K. (2022). Additive manufacturing: A review on mechanical properties of polyjet and FDM printed parts. *Polymer Bulletin*, *79*(9), 7065–7116. DOI: 10.1007/s00289-021-03899-0

Palanivel, S., Nelaturu, P., Glass, B., & Mishra, R. S. (2014). Friction stir additive manufacturing for high structural performance through microstructural control in an Mg-based WE43 alloy. In *Materials & Design (1980-2015)* (Vol. 65, p. 934). Elsevier BV., DOI: 10.1016/j.matdes.2014.09.082

Palmara, G., Frascella, F., Roppolo, I., Chiappone, A., & Chiadò, A. (2021). Functional 3D printing: Approaches and bio applications. *Biosensors & Bioelectronics*, *175*, 112849. DOI: 10.1016/j.bios.2020.112849 PMID: 33250333

Pandey, S., Srivastava, R., & Narain, R. (2020). Study on the effect of powder feed rate on direct metal deposition layer of precipitation hardened steel. In *Materials Today Proceedings* (Vol. 26, p. 2272). Elsevier BV., DOI: 10.1016/j.matpr.2020.02.492

Panwisawas, C., Tang, Y. T., & Reed, R. C. (2020). Metal 3D printing as a disruptive technology for superalloys. *Nature Communications*, *11*(1), 2327. Advance online publication. DOI: 10.1038/s41467-020-16188-7 PMID: 32393778

Paprocka, I., & Kempa, W. M. (2021). Model of Production System Evaluation with the Influence of FDM Machine Reliability and Process-Dependent Product Quality. *Materials (Basel)*, *14*(19), 5806. DOI: 10.3390/ma14195806 PMID: 34640202

Parisi, F., Sangiorgio, V., Parisi, N., Mangini, A. M., Fanti, M. P., & Adam, J. M. (2024). A new concept for large additive manufacturing in construction: Tower crane-based 3D printing controlled by deep reinforcement learning. *Construction Innovation*, *24*(1), 8–32. DOI: 10.1108/CI-10-2022-0278

Park, J. H., Tucker, S. J., Yoon, J. K., Kim, Y., & Hollister, S. J. (2024). 3D printing modality effect: Distinct printing outcomes dependent on selective laser sintering (SLS) and melt extrusion. *Journal of Biomedical Materials Research. Part A*, *112*(7), 1015–1024. DOI: 10.1002/jbm.a.37682 PMID: 38348580

Park, Y. G., Yun, I., Chung, W. G., Park, W., Lee, D. H., & Park, J. U. (2022). High-resolution 3D printing for electronics. *Advancement of Science*, *9*(8), 2104623. PMID: 35038249

Parulski, C., Jennotte, O., Lechanteur, A., & Evrard, B. (2021). Challenges of fused deposition modeling 3D printing in pharmaceutical applications: Where are we now? *Advanced Drug Delivery Reviews*, *175*, 113810. DOI: 10.1016/j.addr.2021.05.020 PMID: 34029646

Patdiya, J., & Kandasubramanian, B. (2021). Progress in 4D printing of stimuli responsive materials. *Polymer-Plastics Technology and Materials*, *60*(17), 1845–1883. DOI: 10.1080/25740881.2021.1934016

Patel, R., Desai, C., Kushwah, S., & Mangrola, M. H. (2022). A review article on FDM process parameters in 3D printing for composite materials. *Materials Today: Proceedings*, *60*, 2162–2166. DOI: 10.1016/j.matpr.2022.02.385

Pathak, K., Saikia, R., Das, A., Das, D., Islam, M. A., Pramanik, P., Parasar, A., Borthakur, P. P., Sarmah, P., Saikia, M., & Borthakur, B. (2023). 3D printing in biomedicine: Advancing personalized care through additive manufacturing. *Exploration of Medicine*, *1135*, 1135–1167. Advance online publication. DOI: 10.37349/emed.2023.00200

Patpatiya, P., Chaudhary, K., Shastri, A., & Sharma, S. (2022). A review on polyjet 3D printing of polymers and multi-material structures. *Proceedings of the Institution of Mechanical Engineers. Part C, Journal of Mechanical Engineering Science*, *236*(14), 7899–7926. DOI: 10.1177/09544062221079506

Paul, M. J., Muniandy, Y., Kruzic, J. J., Ramamurty, U., & Gludovatz, B. (2022). Effect of heat treatment on the strength and fracture resistance of a laser powder bed fusion processed 18Ni-300 maraging steel. *Materials Science and Engineering A*, *844*, 143167. DOI: 10.1016/j.msea.2022.143167

Paunović, N., Meyer, D., Krivitsky, A., Studart, A. R., Bao, Y., & Leroux, J. C. (2023). 4D printing of biodegradable elastomers with tailorable thermal response at physiological temperature. *Journal of Controlled Release*, *361*, 417–426. DOI: 10.1016/j.jconrel.2023.07.053 PMID: 37532144

Paxton, N. C., Zhao, J., & Sauret, E. (2024). Polymer 3D printing in perspective: Assessing challenges and opportunities in industrial translation against the metal benchmark. *International Journal of Advanced Manufacturing Technology*, *133*(1), 59–80. DOI: 10.1007/s00170-024-13744-z

Peng, B., Xu, H., Song, F., Wen, P., Tian, Y., & Zheng, Y. (2024). Additive manufacturing of porous magnesium alloys for biodegradable orthopedic implants: Process, design, and modification. *Journal of Materials Science and Technology*, *182*, 79–110. DOI: 10.1016/j.jmst.2023.08.072

Peng, Y., Wu, Y., Wang, K., Gao, G., & Ahzi, S. (2019). Synergistic reinforcement of polyamide-based composites by combination of short and continuous carbon fibers via fused filament fabrication. *Composite Structures*, *207*, 232–239. DOI: 10.1016/j.compstruct.2018.09.014

Pereira Morais, A., & Festas, J. (2022). *Medical Applications of Materials Manufactured by the AM Process.*

Pereira, G. R., Gasi, F., & Lourenço, S. R. (2019). Review, analysis, and classification of 3D printing literature: Types of research and technology benefits. *IJAERS*, *6*(6), 167–187. DOI: 10.22161/ijaers.6.6.18

Perez-de-Eulate, N. G., Ares-Elejoste, P., Goenaga, G., Urrutxua, M., Vallejo, F. J. L., Ballestero, J., Allue, A., & Gómez-Alonso, J. L. (2023). On the influence of flame-retardant additives on uv-curable thermosetting glass fiber-reinforced composites. *Polymers*, *15*(1), 240. DOI: 10.3390/polym15010240 PMID: 36616588

Pérez, M., Carou, D., Rubio, E. M., & Teti, R. (2020). Current advances in additive manufacturing. *Procedia CIRP*, *88*, 439–444. DOI: 10.1016/j.procir.2020.05.076

Pérez, M., Medina-Sánchez, G., García-Collado, A., Gupta, M., & Carou, D. (2018). Surface quality enhancement of fused deposition modeling (FDM) printed samples based on the selection of critical printing parameters. *Materials (Basel)*, *11*(8), 1382. DOI: 10.3390/ma11081382 PMID: 30096826

Perkins, I., & Skitmore, M. (2015). Three-dimensional printing in the construction industry: A review [Review of Three-dimensional printing in the construction industry: A review]. International Journal of Construction Management, 15(1), 1. Taylor & Francis. DOI: 10.1080/15623599.2015.1012136

Pernet, B., Nagel, J., & Zhang, H. (2022). Compressive Strength Assessment of 3D Printing Infill Patterns. *Procedia CIRP*, *105*, 682–687. DOI: 10.1016/j.procir.2022.02.114

Petousis, M., Michailidis, N., Papadakis, V., Argyros, A., Spiridaki, M., Mountakis, N., Valsamos, J., Nasikas, N. K., Moutsopoulou, A., & Vidakis, N. (2025). Enhanced engineering and biocidal polypropylene filaments enabling melt reduction of AgNO3 through PVP agent: A scalable process for the defense industry with MEX additive manufacturing. *Defence Technology*, *44*, 52–66. DOI: 10.1016/j.dt.2024.09.007

Petruse, R. E., Simion, C., & Bondrea, I. (2024). Geometrical and dimensional deviations of fused deposition modelling (FDM) additive-manufactured parts. *Metrology*, *4*(3), 411–429. DOI: 10.3390/metrology4030025

Pfaff, A., Jäcklein, M., Hoschke, K., & Wickert, M. (2018). Designed Materials by Additive Manufacturing—Impact of Exposure Strategies and Parameters on Material Characteristics of AlSi10Mg Processed by Laser Beam Melting. *Metals*, *8*(7), 491. DOI: 10.3390/met8070491

Pham, G. N., Lee, S.-H., Kwon, O.-H., & Kwon, K.-R. (2018). A 3D Printing Model Watermarking Algorithm Based on 3D Slicing and Feature Points. *Electronics (Basel)*, *7*(2), 23. DOI: 10.3390/electronics7020023

Phuhongsung, P., Zhang, M., Devahastin, S., & Mujumdar, A. S. (2022). Defects in 3D/4D food printing and their possible solutions: A comprehensive review. *Comprehensive Reviews in Food Science and Food Safety*, *21*(4), 3455–3479. DOI: 10.1111/1541-4337.12984 PMID: 35678036

Picard, M., Mohanty, A. K., & Misra, M. (2020). Recent advances in additive manufacturing of engineering thermoplastics: challenges and opportunities [Review of Recent advances in additive manufacturing of engineering thermoplastics: challenges and opportunities]. RSC Advances, 10(59), 36058. Royal Society of Chemistry. DOI: 10.1039/D0RA04857G

Pierson, S., Marshall, L., Lemmons, L., Efrid, L., Shou, W., & Hu, H. (2024, July). Work-in-progress: Open-source selective laser sintering (SLS) test bed. In *2023 ASEE Midwest Section Conference*. https://peer.asee.org/45312

Pilch, M., Topa-Skwarczyńska, M., Chachaj-Brekiesz, A., Jamróz, P., Kiesiewicz, D., Noworyta, M., & Ortyl, J. (2024). Luminescence labeled surfaces mapping system dedicated for use in quality control of 3D prints produced by stereolithography 3D printing (SLA) and laser engraving. *Sensors and Actuators. A, Physical*, *365*, 114828. DOI: 10.1016/j.sna.2023.114828

Pinkerton, A. J. (2016). Lasers in additive manufacturing. *Optics & Laser Technology, 78*, 25–32. DOI: 10.1016/j.optlastec.2015.09.025

Piran, F. J., Poduval, P. P., Barkam, H. E., Imani, M., & Imani, F. (2025). Explainable Differential Privacy-Hyperdimensional Computing for Balancing Privacy and Transparency in Additive Manufacturing Monitoring. *Engineering Applications of Artificial Intelligence, 147*, 110282. DOI: 10.1016/j.engappai.2025.110282

Placone, J. K., Mahadik, B., & Fisher, J. P. (2020). Addressing present pitfalls in 3D printing for tissue engineering to enhance future potential. *APL Bioengineering, 4*(1), 010901. DOI: 10.1063/1.5127860 PMID: 32072121

Plancher, E., Suard, M., Dendievel, R., Blandin, J., Martin, G., & Lhuissier, P. (2020). Behavior by design made possible by additive manufacturing: The case of a whistle-blower mechanical response. In *Materials Letters* (Vol. 282, p. 128669). Elsevier BV., DOI: 10.1016/j.matlet.2020.128669

Plathottam, S. J., Rzonca, A., Lakhnori, R., & Iloeje, C. O. (2023). A review of artificial intelligence applications in manufacturing operations [Review of A review of artificial intelligence applications in manufacturing operations]. Journal of Advanced Manufacturing and Processing, 5(3). Wiley. DOI: 10.1002/amp2.10159

Podgornik, B., Šinko, M., & Godec, M. (2021). Dependence of the wear resistance of additive-manufactured maraging steel on the build direction and heat treatment. *Additive Manufacturing, 46*, 102123. DOI: 10.1016/j.addma.2021.102123

Podshivalov, L., Gomes, C. M., Zocca, A., Guenster, J., Bar-yoseph, P., & Fischer, A. (2013). Design, analysis and additive manufacturing of porous structures for biocompatible micro-scale scaffolds. *Procedia CIRP, 5*, 247–252. DOI: 10.1016/j.procir.2013.01.049

Prabhakar, M. M., Saravanan, A. K., Lenin, A. H., Mayandi, K., & Ramalingam, P. S. (2021). A short review on 3D printing methods, process parameters and materials. *Materials Today: Proceedings, 45*, 6108–6114.

Prabhu L, Krishnaraj V, Sathish S, Gokulkumar S, and Karthi N (2019b) Study of mechanical and morphological properties of jute-tea leaf fiber reinforced hybrid composites: Effect of glass fiber hybridization. Materials Today Proceedings 27(xxxx): 2372–2375. .DOI: 10.1016/j.matpr.2019.09.132

Prabhu, R., Simpson, T. W., Miller, S. R., Cutler, S. L., & Meisel, N. A. (2023). Teaching designing for additive manufacturing: formulating educational interventions that encourage design creativity. 3D printing and additive manufacturing, 10(2), 356-372.

Prabhu, L., Krishnaraj, V., Gokulkumar, S., Sathish, S., & Ramesh, M. (2019a). Mechanical, chemical and acoustical behavior of sisal—Tea waste—Glass fiber reinforced epoxy based hybrid polymer composites. *Materials Today: Proceedings*, *16*, 653–660. DOI: 10.1016/j.matpr.2019.05.142

Prabhu, R., Bracken, J., Armstrong, C. B., Jablokow, K., Simpson, T. W., & Meisel, N. A. (2020). Additive creativity: Investigating the use of design for additive manufacturing to encourage creativity in the engineering design industry. *International Journal of Design Creativity and Innovation*, *8*(4), 198–222. DOI: 10.1080/21650349.2020.1813633

Pradhan, S. R., Singh, R., & Banwait, S. S. (2023). 3D printing assisted investment casting of dental crowns for recycling of DMLS waste. *Arabian Journal for Science and Engineering*, *48*(3), 3289–3299. DOI: 10.1007/s13369-022-07155-z

Prakash, B. V., Koushik, N., Jha, S. K., & Ramasamy, K. (2024). Additive manufacturing of composite materials and functionally graded structures using archerfish hunting technique. *Lubrication Science*, *36*(8), 595–609. Advance online publication. DOI: 10.1002/ls.1716

Prashar, G., Vasudev, H., & Bhuddhi, D. (2023). Additive manufacturing: Expanding 3D printing horizon in industry 4.0. [IJIDeM]. *International Journal on Interactive Design and Manufacturing*, *17*(5), 2221–2235. DOI: 10.1007/s12008-022-00956-4

Priyadarshini, J., Singh, R. K., Mishra, R., He, Q., & Braganza, A. (2024). Implementation of additive manufacturing in the healthcare supply chain for circular economy goals: Paradoxical tensions and solutions from an Industry 5.0 perspective. *Information Systems Frontiers*, •••, 1–23. DOI: 10.1007/s10796-024-10482-1

Przekop, R. E., Gabriel, E., Pakuła, D., & Sztorch, B. (2023). Liquid for Fused Deposition Modeling Technique (L-FDM)—A Revolution in Application Chemicals to 3D Printing Technology : Color and Elements. *Applied Sciences (Basel, Switzerland)*, *13*(13), 7393. Advance online publication. DOI: 10.3390/app13137393

Puerta, A. P. V., Fernández-Vidal, S., & Girot, F. (2023). Impact of in-soil ageing effect on pla printed parts tensile properties. *Polymers*, *15*(4), 862. DOI: 10.3390/polym15040862 PMID: 36850145

Punzo, D., van der Hulst, J., Roerdink, J. B., Fillion-Robin, J.-C., & Yu, L. (2017). SlicerAstro : A 3-D interactive visual analytics tool for HI data. *Astronomy and Computing*, *19*, 45–59. DOI: 10.1016/j.ascom.2017.03.004

Qattawi, A., & Ablat, M. A. (2017). Design consideration for additive manufacturing: Fused deposition modelling. *Open Journal of Applied Sciences*, *7*(6), 291–318. DOI: 10.4236/ojapps.2017.76024

Qiang, X., Zhou, X., Wang, J., Wilkes, C. M., Loke, T., O'Gara, S., Kling, L., Marshall, G. D., Santagati, R., Ralph, T. C., Wang, J. B., O'Brien, J. L., Thompson, M. G., & Matthews, J. C. F. (2018). Large-scale silicon quantum photonics implementing arbitrary two-qubit processing. *Nature Photonics*, *12*(9), 534–539. DOI: 10.1038/s41566-018-0236-y

Qi, S., Yao, H., Fu, J., Xie, Y., Li, Y., Tian, R., Yu, M., & Guo, H. (2022). Magneto-active soft matter with reprogrammable shape-morphing and self-sensing capabilities. *Composites Science and Technology*, *230*, 109789. DOI: 10.1016/j.compscitech.2022.109789

Quan, H., Zhang, T., Xu, H., Luo, S., Nie, J., & Zhu, X. (2020). Photo-curing 3D printing technique and its challenges. *Bioactive Materials*, *5*(1), 110–115. DOI: 10.1016/j.bioactmat.2019.12.003 PMID: 32021945

Quanjin, M., Rejab, M. R. M., Idris, M. S., Kumar, N. M., Abdullah, M. H., & Reddy, G. R. (2020). Recent 3D and 4D intelligent printing technologies: A comparative review and future perspective. *Procedia Computer Science*, *167*, 1210–1219. DOI: 10.1016/j.procs.2020.03.434

Quero, R. F., da Silveira, G. D., da Silva, J. A. F., & de Jesus, D. P. (2021). Understanding and improving FDM 3D printing to fabricate high-resolution and optically transparent microfluidic devices. *Lab on a Chip*, *21*(19), 3715–3729. DOI: 10.1039/D1LC00518A PMID: 34355724

Quiñonez, P. A., Ugarte-Sanchez, L., Bermudez, D., Chinolla, P., Dueck, R., Cavender-Word, T. J., & Roberson, D. A. (2021). Design of shape memory thermoplastic material systems for FDM-type Additive manufacturing. *Materials (Basel)*, *14*(15), 4254. DOI: 10.3390/ma14154254 PMID: 34361448

Rack, H. J., & Kalish, D. (1971). The strength and fracture toughness of 18 Ni (350) maraging steel. *Metallurgical and Materials Transactions. B, Process Metallurgy and Materials Processing Science*, *2*, 3011–3020.

Raeker-Jordan, N., Chung, J., Kong, Z. J., & Williams, C. (2024). Ensuring additive manufacturing quality and cyber–physical security via side-channel measurements and transmissions. *Journal of Manufacturing Systems*, *73*, 275–286. DOI: 10.1016/j.jmsy.2024.02.005

Rael, R., & San Fratello, V. (2017). Clay bodies: Crafting the future with 3D printing. *Architectural Design*, *87*(6), 92–97. DOI: 10.1002/ad.2243

Rafiee, M., Farahani, R. D., & Therriault, D. (2020). Multi-material 3D and 4D printing: A survey. *Advancement of Science*, *7*(12), 1902307. DOI: 10.1002/advs.201902307 PMID: 32596102

Rafi, K., Liu, A. Z., Di Prima, M., Bates, P., & Seifi, M. (2022). Regulatory and standards development in medical additive manufacturing. *MRS Bulletin*, *47*(1), 98–105. DOI: 10.1557/s43577-022-00266-w

Raghunath, N., & Pandey, P. M. (2007). Improving accuracy through shrinkage modelling by using Taguchi method in selective laser sintering. *International Journal of Machine Tools & Manufacture*, *47*(6), 985–995. DOI: 10.1016/j.ijmachtools.2006.07.001

Raghuraman, V., & Kumar, T. S. (2024). Influence of in-situ process parameters, post heat treatment effects on microstructure and defects of additively manufactured maraging steel by laser powder bed fusion-A comprehensive review. *Physica Scripta. Physica Scripta*, *99*(5), 052001. DOI: 10.1088/1402-4896/ad3681

Raghuraman, V., & Kumar, T. S. (2025). Different post-heat treatment responses on microstructure and mechanical characteristics of maraging steel 1.2709 samples produced by LPBF. *Physica Scripta*, *100*(2), 025929. DOI: 10.1088/1402-4896/ada328

Raheem, A., Hameed, P., Whenish, R., Elsen, R., Aswin, G., Jaiswal, A. K., Prashanth, K. G., & Manivasagam, G. (2021). A Review on Development of Bio-Inspired Implants Using 3D Printing [Review of A Review on Development of Bio-Inspired Implants Using 3D Printing]. Biomimetics, 6(4), 65. Multidisciplinary Digital Publishing Institute. DOI: 10.3390/biomimetics6040065

Rahim, T. N. A. T., Abdullah, A. M., & Md Akil, H. (2019). Recent developments in fused deposition modeling-based 3D printing of polymers and their composites. *Polymer Reviews (Philadelphia, Pa.)*, *59*(4), 589–624. DOI: 10.1080/15583724.2019.1597883

Rahman MZ (2021) Mechanical and damping performances of flax fibre composites–A review. Composites Part C: Open Access: 100081.

Rahman, Z., Charoo, N. A., Mohamed, E. M., Kuttolamadom, M., & Khan, M. A. (2023). Regulatory perspective of additive manufacturing in the field of pharmaceuticals. In *Additive Manufacturing in Pharmaceuticals* (pp. 327–348). DOI: 10.1007/978-981-99-2404-2_10

Rahman, H., John, T. D., Sivadasan, M., & Singh, N. K. (2018). Investigation on the Scale Factor applicable to ABS based FDM Additive Manufacturing. *Materials Today: Proceedings*, *5*(1), 1640–1648. DOI: 10.1016/j.matpr.2017.11.258

Rahman, M. Z. (2017). *Static and Dynamic Characterisation of Flax Fibre-reinforced Polypropylene Composites*. ResearchSpace.

Rahman, M. Z. (2022). Influence of Fiber treatment on the damping performance of plant Fiber composites. In *Vibration and Damping Behavior of Biocomposites* (pp. 113–135). CRC Press. DOI: 10.1201/9781003173625-8

Rahman, M. Z. M. M. Z., Rahman, M. Z. M. M. Z., Mahbub, T., Ashiquzzaman, M., Sagadevan, S., & Hoque, M. E. (2023b). Advanced biopolymers for automobile and aviation engineering applications. *Journal of Polymer Research*, *30*(3), 106. DOI: 10.1007/s10965-023-03440-z

Rahman, M. Z., Jayaraman, K., & Mace, B. R. (2017). Vibration damping of flax fibre-reinforced polypropylene composites. *Fibers and Polymers*, *18*(11), 2187–2195. DOI: 10.1007/s12221-017-7418-y

Rahman, M. Z., Jayaraman, K., & Mace, B. R. (2018a). Impact energy absorption of flax fiber-reinforced polypropylene composites. *Polymer Composites*, *39*(11), 4165–4175. DOI: 10.1002/pc.24486

Rahman, M. Z., Jayaraman, K., & Mace, B. R. (2018b). Influence of damping on the bending and twisting modes of flax fibre-reinforced polypropylene composite. *Fibers and Polymers*, *19*(2), 375–382. DOI: 10.1007/s12221-018-7588-7

Rahman, M. Z., Mace, B. R., & Jayaraman, K. (2016) Vibration damping of natural fibre-reinforced composite materials. In: *17th European Conference on Composite Material*.

Rahman, M. Z., & Xu, H. (2023). Damping under varying frequencies, mechanical properties, and failure modes of flax/polypropylene composites. *Polymers*, *15*(4), 1042. DOI: 10.3390/polym15041042 PMID: 36850325

Rahman, M., Islam, K. S., Dip, T. M., Chowdhury, M. F. M., Debnath, S. R., Hasan, S. M. M., Sakib, M. S., Saha, T., Padhye, R., & Houshyar, S. (2024). A review on nanomaterial-based additive manufacturing: Dynamics in properties, prospects, and challenges. *Progress in Additive Manufacturing*, *9*(4), 1197–1224. DOI: 10.1007/s40964-023-00514-8

Rahmati, S., & Dickens, P. M. (1995) Stereolithography process improvement. In: *Proceedings of the First National Conference on Rapid Prototyping and Tooling Research*, pp. 111–126.

Rajaguru, K., Karthikeyan, T., & Vijayan, V. (2020). Additive manufacturing–State of art. *Materials Today: Proceedings, 21*, 628–633. DOI: 10.1016/j.matpr.2019.06.728

Rajan, K., Samykano, M., Kadirgama, K., Harun, W. S. W., & Rahman, M. M. (2022). Fused deposition modeling : Process, materials, parameters, properties, and applications. *International Journal of Advanced Manufacturing Technology, 120*(3), 1531–1570. DOI: 10.1007/s00170-022-08860-7

Ramanath, H. S., Chua, C. K., Leong, K. F., & Shah, K. D. (2008). Melt flow behaviour of poly-e-caprolactone in fused deposition modelling. *Journal of Materials Science. Materials in Medicine, 19*(7), 2541–2550. DOI: 10.1007/s10856-007-3203-6 PMID: 17619957

Ramazani, H., & Kami, A. (2022). Metal FDM, a new extrusion-based additive manufacturing technology for manufacturing of metallic parts: A review. *Progress in Additive Manufacturing, 7*(4), 609–626. DOI: 10.1007/s40964-021-00250-x

Ramesh, M. (2016). Kenaf (Hibiscus cannabinus L.) fibre based bio-materials: A review on processing and properties. *Progress in Materials Science, 78–79*, 1–92. DOI: 10.1016/j.pmatsci.2015.11.001

Ramesh, M., & Panneerselvam, K. (2021). Mechanical investigation and optimization of parameter selection for Nylon material processed by FDM. *Materials Today: Proceedings, 46*, 9303–9307. DOI: 10.1016/j.matpr.2020.02.697

Randriambololona, A. M., Shaeri, M. R., & Sarabi, S. (2022). Artificial neural network models to predict heat transfer coefficients and pressure drops in cold plates with surface roughness. *Proceedings of the 8th World Congress on Mechanical, Chemical, and Material Engineering*. DOI: 10.11159/htff22.167

Raney, K., Lani, E., & Kalla, D. K. (2017). Experimental characterization of the tensile strength of ABS parts manufactured by fused deposition modeling process. *Materials Today: Proceedings, 4*(8), 7956–7961. DOI: 10.1016/j.matpr.2017.07.132

Ranjan, R., Kumar, D., Kundu, M., & Moi, S. C. (2022). A critical review on Classification of materials used in 3D printing process. *Materials Today: Proceedings, 61*, 43–49. DOI: 10.1016/j.matpr.2022.03.308

Ranjan, R., & Saha, A. (2025). Selection of best 3d printing parameters for fused deposition modeling part using a novel hybrid multi-criteria optimization technique. *Surface Review and Letters, 32*(03), 2450121. DOI: 10.1142/S0218625X2450121X

Ransikarbum, K., Pitakaso, R., & Kim, N. (2020). A Decision-Support Model for Additive Manufacturing Scheduling Using an Integrative Analytic Hierarchy Process and Multi-Objective Optimization. In Applied Sciences (Vol. 10, Issue 15, p. 5159). Multidisciplinary Digital Publishing Institute. DOI: 10.3390/app10155159

Rao, B. S., & Rao, T. B. (2022). Effect of process parameters on powder bed fusion maraging steel 300: A review. *Lasers in Manufacturing and Materials Processing*, *9*(3), 338–375. DOI: 10.1007/s40516-022-00182-6

Rao, C. H., Avinash, K., Varaprasad, B. K. S. V. L., & Goel, S. (2022). A review on printed electronics with digital 3D printing: Fabrication techniques, materials, challenges and future opportunities. *Journal of Electronic Materials*, *51*(6), 2747–2765. DOI: 10.1007/s11664-022-09579-7

Rastogi, P., & Kandasubramanian, B. (2019). Breakthrough in the printing tactics for stimuli-responsive materials: 4D printing. *Chemical Engineering Journal*, *366*, 264–304. DOI: 10.1016/j.cej.2019.02.085

Rath, K. C., Khang, A., Mishra, S. K., Patnaik, P. K., Mohanty, G. K., & Dash, T. (2024). Integration of artificial intelligence and Internet of Things technology solutions in smart manufacturing. In *Machine Vision and Industrial Robotics in Manufacturing* (pp. 155–177). CRC Press., DOI: 10.1201/9781003438137-9

Raut, J. M., Pimpalshende, A., Chandak, M. A., Patil, T. R., Pinjarkar, L., & Nair, S. (2025). A novel multi-method framework for 3D printed fiber-reinforced polymer concrete utilizing advance additive manufacturing techniques. Asian Journal of Civil Engineering, 1-14.

Rautio, S., & Broms, M. (2024). Security scenarios: 3D printed firearms. *Security and Defence Quarterly*, *48*(4), 59–79.

Rawat, P., Liu, S., Guo, S., Rahman, M. Z., Yang, T., Bai, X., Yao, Y., Mobasher, B., & Zhu, D. (2022). A state-of-the-art review on mechanical performance characterization and modelling of high-performance textile reinforced concretes. *Construction & Building Materials*, *347*, 128521. DOI: 10.1016/j.conbuildmat.2022.128521

Rayna, T., Striukova, L., & Darlington, J. (2015). Co-creation and user innovation : The role of online 3D printing platforms. *Journal of Engineering and Technology Management*, *37*, 90–102. DOI: 10.1016/j.jengtecman.2015.07.002

Razzaghi, M., Alexander Ninan, J., & Akbari, M. (2024). Advancements in Materials for 3D-Printed Microneedle Arrays : Enhancing Performance and Biocompatibility. *Micromachines*, *15*(12), 1433. Advance online publication. DOI: 10.3390/mi15121433 PMID: 39770187

Reeves, P. E., & Cobb, R. C. (1997). Reducing the surface deviation of streolithography using an alternative build stratergy. *Rapid Prototyping Journal, 3*(1), 20–31. DOI: 10.1108/13552549710169255

Regehly, M., Garmshausen, Y., Reuter, M., König, N. F., Israel, E., Kelly, D. P., Chou, C.-Y., Koch, K., Asfari, B., & Hecht, S. (2020). Xolography for linear volumetric 3D printing. *Nature, 588*(7839), 620–624. DOI: 10.1038/s41586-020-3029-7 PMID: 33361791

Rehme, O., & Weidermann, C. (2024, May). Additive manufacturing of cryogenic materials. *IOP Conference Series. Materials Science and Engineering, 1301*(1), 012047. DOI: 10.1088/1757-899X/1301/1/012047

Ren, L., Zhou, X., Song, Z., Zhao, C., Liu, Q., Xue, J., & Li, X. (2017). Process Parameter Optimization of Extrusion-Based 3D Metal Printing Utilizing PW–LDPE–SA Binder System. *Materials (Basel), 10*(3), 305. DOI: 10.3390/ma10030305 PMID: 28772665

Resendes, T., Freitas Rodrigues, P., Cruz, F., Gatões, D., Santos, V. M., Ramos, A. S., & Vieira, M. T. (2025). Advanced medical monitoring: 3D printed prosthetics with integrated strain sensor. *Progress in Additive Manufacturing, 10*(1), 219–229. DOI: 10.1007/s40964-024-00615-y

Rezaie, F., Farshbaf, M., Dahri, M., Masjedi, M., Maleki, R., Amini, F., Wirth, J., Moharamzadeh, K., Weber, F. E., & Tayebi, L. (2023). 3D printing of dental prostheses : Current and emerging applications. *Journal of Composites Science, 7*(2), 80. DOI: 10.3390/jcs7020080 PMID: 38645939

Ribeiro, D., Cimino, S. R., Mayo, A. L., Ratto, M., & Hitzig, S. L. (2021). 3D printing and amputation: A scoping review. *Disability and Rehabilitation. Assistive Technology, 16*(2), 221–240. DOI: 10.1080/17483107.2019.1646825 PMID: 31418306

Riensche, A., Severson, J., Yavari, R., Piercy, N., Cole, K. D., & Rao, P. (2022). Thermal modeling of directed energy deposition additive manufacturing using graph theory. *Rapid Prototyping Journal, 29*(2), 324–343. DOI: 10.1108/RPJ-07-2021-0184

Rinaldi, M., Ghidini, T., Cecchini, F., Brandao, A., & Nanni, F. (2018). Additive layer manufacturing of poly (ether ether ketone) via FDM. *Composites. Part B, Engineering, 145*, 162–172. DOI: 10.1016/j.compositesb.2018.03.029

Ritchie, S. M., Kovačević, S., Deshmukh, P., Christodoulides, A. D., Malen, J. A., Mesarovic, S. Dj., & Panat, R. (2023). Shape distortion in sintering results from nonhomogeneous temperature activating a long-range mass transport. In Nature Communications (Vol. 14, Issue 1). Nature Portfolio. DOI: 10.1038/s41467-023-38142-z

Robayo-Salazar, R., de Gutiérrez, R. M., Villaquirán-Caicedo, M. A., & Arjona, S. D. (2023). 3D printing with cementitious materials: Challenges and opportunities for the construction sector. *Automation in Construction*, *146*, 104693. DOI: 10.1016/j.autcon.2022.104693

Robles Poblete, F., Ireland, M., Slattery, L., Davids, W. G., & Lopez-Anido, R. A. (2023). In situ, real-time temperature mapping and thermal fe simulations of large-format 3d printed petg/cf vertical wall. *Materials (Basel)*, *16*(19), 6486. DOI: 10.3390/ma16196486 PMID: 37834624

Rodrigues Filho, C. R. (2023). *Analysing the Impact of Data Quality and Processing Techniques on the Accuracy of Virtual Reconstruction-A Case Study of a Deactivated Grain Milling Factory*.

Roger, F., & Krawczak, P. (2015, August). 3D-printing of thermoplastic structures by FDM using heterogeneous infill and multi-materials: An integrated design-advanced manufacturing approach for factories of the future. In *CFM 2015-22ème Congrès Français de Mécanique*. AFM, Maison de la Mécanique, 39/41 rue Louis Blanc-92400 Courbevoie.

Rojek, I., Mikołajewski, D., Kempiński, M., Galas, K., & Piszcz, A. (2025). Emerging Applications of Machine Learning in 3D Printing. *Applied Sciences (Basel, Switzerland)*, *15*(4), 1781. DOI: 10.3390/app15041781

Rolchigo, M., Reeve, S. T., Stump, B., Coleman, J., & Plotkowski, A. (2025). ExaCA v2. 0: A versatile, scalable, and performance portable cellular automata application for additive manufacturing solidification. *Computational Materials Science*, *251*, 113734. DOI: 10.1016/j.commatsci.2025.113734

Romani, A., & Levi, M. (2024). Large-Format Material Extrusion Additive Manufacturing for Circular Economy Practices: A Focus on Product Applications with Materials from Recycled Plastics and Biomass Waste. *Sustainability (Basel)*, *16*(18), 7966. DOI: 10.3390/su16187966

Romani, A., Perusin, L., Ciurnelli, M., & Levi, M. (2024). Characterization of PLA feedstock after multiple recycling processes for large-format material extrusion additive manufacturing. *Materials Today Sustainability*, *25*, 100636. DOI: 10.1016/j.mtsust.2023.100636

Roudný, P., & Syrový, T. (2022). Thermal conductive composites for FDM 3D printing: A review, opportunities and obstacles, future directions. *Journal of Manufacturing Processes, 83*, 667–677. DOI: 10.1016/j.jmapro.2022.09.026

Rouf, S., Malik, A., Singh, N., Raina, A., Naveed, N., Siddiqui, M. I. H., & Haq, M. I. U. (2022). Additive manufacturing technologies: Industrial and medical applications. *Sustainable Operations and Computers, 3*, 258–274. DOI: 10.1016/j.susoc.2022.05.001

Ruberti, M. (2024). Environmental performance and trends of the world's semiconductor foundry industry. *Journal of Industrial Ecology, 28*(5), 1183–1197. DOI: 10.1111/jiec.13529

Rusmayadi, G., Salawati, U., Haslinah, A., & Judijanto, L. (2023). The effect of investment in green technology and renewable technology adoption on energy efficiency and carbon emissions reduction in indonesian manufacturing companies. *West Science Interdisciplinary Studies, 1*(11), 1175–1183. DOI: 10.58812/wsis.v1i11.347

Ryan, K. R., Down, M. P., & Banks, C. E. (2021). Future of additive manufacturing: Overview of 4D and 3D printed smart and advanced materials and their applications. *Chemical Engineering Journal, 403*, 126162. DOI: 10.1016/j.cej.2020.126162

Rybnicek, R., & Königsgruber, R. (2019). What makes industry–university collaboration succeed? A systematic review of the literature. *Journal of Business Economics, 89*(2), 221–250. DOI: 10.1007/s11573-018-0916-6

Saad, I., Maalej, S., & Zaghdoudi, M. C. (2022). Investigation of the thermal performance of a nanofluid-filled grooved cylindrical heat pipe for electronics cooling. *Journal of Advanced Research in Fluid Mechanics and Thermal Sciences, 99*(2), 135–154. DOI: 10.37934/arfmts.99.2.135154

Sabbatini, B., Cambriani, A., Cespi, M., Palmieri, G. F., Perinelli, D. R., & Bonacucina, G. (2021). An overview of natural polymers as reinforcing agents for 3D printing. *ChemEngineering, 5*(4), 78. DOI: 10.3390/chemengineering5040078

Sæterbø, M., & Solvang, W. D. (2024). Metal additive manufacturing adoption in SMEs: Technical attributes, challenges, and opportunities. *Journal of Manufacturing Processes, 128*, 175–189. DOI: 10.1016/j.jmapro.2024.07.074

Safonov, A., Maltsev, E., Chugunov, S., Tikhonov, A., Konev, S., Evlashin, S., Popov, D., Pasko, A., & Akhatov, I. (2020). Design and fabrication of complex-shaped ceramic bone implants via 3d printing based on laser stereolithography. *Applied Sciences (Basel, Switzerland), 10*(20), 7138. DOI: 10.3390/app10207138

Sagadevan, S., Fatimah, I., Anita Lett, J., Rahman, M. Z., Leonard, E., & Oh, W.-C. (2023a). Eco-friendly green approach of nickel oxide nanoparticles for biomedical applications. *Open Chemistry*, *21*(1), 20230141. Advance online publication. DOI: 10.1515/chem-2023-0141

Sagadevan, S., Rahman, M. Z., Léonard, E., Losic, D., & Hessel, V. (2023b). Sensor to electronics applications of graphene oxide through AZO grafting. *Nanomaterials (Basel, Switzerland)*, *13*(5), 846. DOI: 10.3390/nano13050846 PMID: 36903724

Sagadevan, S., Schirhagl, R., Rahman, M. Z., Ismail, M. F. B., Anita Lett, J., Fatimah, I., Kaus, N. H. M., & Oh, W.-C. (2023c). Recent advancements in polymer matrix nanocomposites for bone tissue engineering applications. *Journal of Drug Delivery Science and Technology*, *82*, 104313. DOI: 10.1016/j.jddst.2023.104313

Sagar, S. (2025). Data-Driven Excellence: Integrating Analytics into Medical Technology Product Lifecycle Management. *International Journal of Trend in Scientific Research and Development*, *9*(1), 127–136.

Sağbaş, B. (2020). Effect of Orientation Angle on Surface Quality and Dimensional Accuracy of Functional Parts Manufactured by Multi Jet Fusion Technology. *European Mechanical Science*, *4*(2), 47–52. DOI: 10.26701/ems.678901

Sahafnejad-Mohammadi, I., Karamimoghadam, M., Zolfagharian, A., Akrami, M., & Bodaghi, M. (2022). 4D printing technology in medical engineering : A narrative review. *Journal of the Brazilian Society of Mechanical Sciences and Engineering*, *44*(6), 233. DOI: 10.1007/s40430-022-03514-x

Sahin, O., Albayrak, O. M., & Yapici, M. K. (2024). Optimization of e-beam lithography parameters for nanofabrication of sub-50 nm gold nanowires and nanogaps based on a bilayer lift-off process. *Nanotechnology*, *35*(39), 395301. DOI: 10.1088/1361-6528/ad5e89 PMID: 38959870

Saifullah, M. S. M., Asbahi, M., Neo, D. C. J., Mahfoud, Z., Tan, H. R., Ha, S. T., Dwivedi, N., Dutta, T., bin Dolmanan, S., Aabdin, Z., Bosman, M., Ganesan, R., Tripathy, S., Hasko, D. G., & Valiyaveettil, S. (2022). Patterning at the Resolution Limit of Commercial Electron Beam Lithography. *Nano Letters*, *22*(18), 7432–7440. DOI: 10.1021/acs.nanolett.2c02339 PMID: 36069429

Sajjad, R., Butt, S. U., Ali, A., & Saeed, H. A. (2019). Investigating the Impacts of Heterogeneous Infills on Structural Strength of 3D Printed Parts. *Key Engineering Materials*, *799*, 276–281. . DOI: 10.4028/www.scientific.net/KEM.799.276

Sakin, M., & Kiroglu, Y. C. (2017). 3D Printing of Buildings: Construction of the Sustainable Houses of the Future by BIM. In *Energy Procedia* (Vol. 134, p. 702). Elsevier BV., DOI: 10.1016/j.egypro.2017.09.562

Salem Bala, A. bin Wahab, M. S., & binti Ahmad, M. (2016, May). Elements and materials improve the FDM products: A review. In Advanced Engineering Forum (Vol. 16, pp. 33-51). Trans Tech Publications Ltd.

Salmoria, G. V., Ahrens, C. H., Klauss, P., Paggi, R. A., Oliveira, R. G., & Lago, A. (2007). Rapid manufacturing of polyethylene parts with controlled pore size gradients using selective laser sintering. *Materials Research*, *10*(2), 211–214. DOI: 10.1590/S1516-14392007000200019

Samykano, M., Selvamani, S. K., Kadirgama, K., Ngui, W. K., Kanagaraj, G., & Sudhakar, K. J. T. I. J. O. A. M. T. (2019). Mechanical property of FDM printed ABS: Influence of printing parameters. *International Journal of Advanced Manufacturing Technology*, *102*(9-12), 2779–2796. DOI: 10.1007/s00170-019-03313-0

Sandeep, B., Kannan, T. T. M., Chandradass, J., Ganesan, M., & Rajan, A. J. (2021). Scope of 3D printing in manufacturing industries-A review. *Materials Today: Proceedings*, *45*, 6941–6945. DOI: 10.1016/j.matpr.2021.01.394

Santhoshkumar, P., Negi, A., & Moses, J. A. (2024). 3D printing for space food applications: Advancements, challenges, and prospects. *Life Sciences in Space Research*, *40*, 158–165. DOI: 10.1016/j.lssr.2023.08.002 PMID: 38245341

Sapkota, A., Ghimire, S. K., & Adanur, S. (2024). A review on fused deposition modeling (FDM)-based additive manufacturing (AM) methods, materials and applications for flexible fabric structures. *Journal of Industrial Textiles*, *54*, 15280837241282110. DOI: 10.1177/15280837241282110

Saracyakupoglu, T. (2022). The certification steps for the additively manufactured aviation-grade parts. *The European Journal of Research and Development*, *2*(4), 33–42. DOI: 10.56038/ejrnd.v2i4.133

Sargini, M. I. M., Masood, S. H., Palanisamy, S., Jayamani, E., & Kapoor, A. (2021). Additive manufacturing of an automotive brake pedal by metal fused deposition modelling. *Materials Today: Proceedings*, *45*, 4601–4605. DOI: 10.1016/j.matpr.2021.01.010

Sari, M. W., Herianto, H., Dharma, I. G. B. B., & Tontowi, A. E. (2020). Design of Product Monitoring System Using Internet of Things Technology for Smart Manufacturing. *IOP Conference Series. Materials Science and Engineering*, *835*(1), 12048. DOI: 10.1088/1757-899X/835/1/012048

Saroia, J., Wang, Y., Wei, Q., Lei, M., Li, X., Guo, Y., & Zhang, K. (2020). A review on 3D printed matrix polymer composites: Its potential and future challenges. *International Journal of Advanced Manufacturing Technology, 106*(5-6), 1695–1721. DOI: 10.1007/s00170-019-04534-z

Sathish, S., Kumaresan, K., Prabhu, L., & Gokulkumar, S. (2018). Experimental investigation of mechanical and ftir analysis of flax fiber/epoxy composites incorporating sic, Al2O3 and graphite. Revista Romana de Materiale/Romanian. *Journal of Materials, 48*(4), 476–482.

Sathish, S., Kumaresan, K., Prabhu, L., & Vigneshkumar, N. (2017). Experimental investigation on volume fraction of mechanical and physical properties of flax and bamboo fibers reinforced hybrid epoxy composites. *Polymers & Polymer Composites, 25*(3), 229–236. DOI: 10.1177/096739111702500309

Satterfield, C. (2020). How to Calibrate your FDM 3D Printer: The Ultimate Guide. https://3dprintknowledge.com/how-to-calibrate-your-fdm-3d-printer-the-ultimate-guide/

Savastano, M., Amendola, C., Bellini, F., & D'Ascenzo, F. (2019). Contextual impacts on industrial processes brought by the digital transformation of manufacturing : A systematic review. *Sustainability (Basel), 11*(3), 891. DOI: 10.3390/su11030891

Sbriglia, L. R., Baker, A. M., Thompson, J. M., Morgan, R. V., Wachtor, A. J., & Bernardin, J. D. (2016). Embedding sensors in FDM plastic parts during additive manufacturing. In *Topics in Modal Analysis & Testing, Volume 10: Proceedings of the 34th IMAC, A Conference and Exposition on Structural Dynamics 2016* (pp. 205-214). Springer International Publishing. DOI: 10.1007/978-3-319-30249-2_17

Sbrugnera Sotomayor, N. A., Caiazzo, F., & Alfieri, V. (2021). Enhancing design for additive manufacturing workflow: Optimization, design and simulation tools. *Applied Sciences (Basel, Switzerland), 11*(14), 6628. DOI: 10.3390/app11146628

Scaffaro, R., Mistretta, M. C., & Balsamo, M. (2024). Innovative 3D-printed devices for water pollutant removal : Comprehensive review on printing parameters, composition, properties and performances of the latest 3D-systems. *Polymer Testing, 140*, 108627. DOI: 10.1016/j.polymertesting.2024.108627

Scapin, M., & Peroni, L. (2021). Numerical Simulations of Components Produced by Fused Deposition 3D Printing. *Materials (Basel), 14*(16), 4625. DOI: 10.3390/ma14164625 PMID: 34443146

Schaub, D. A., Chu, K.-R., & Montgomery, D. C. (1997). Optimizing stereolithography throughput. *Journal of Manufacturing Systems, 16*(4), 290–303. DOI: 10.1016/S0278-6125(97)89099-1

Schmidt, M., Pohle, D., & Rechtenwald, T. (2007) Selective laser sintering of PEEK. CIRP Annals - Manufacturing Technology 56(1): 205–208. . Schmitz GJ and Prahl U (eds.) (2016) Handbook of Software Solutions for ICME. Wiley-VCH Verlag GmbH & Co.DOI: 10.1016/j.cirp.2007.05.097

Schneberger, J.-H., Häfele, T., Kaspar, J., & Vielhaber, M. (2019). Hybrid Additive Manufacturing - Requirements Engineering Framework for Process Chain Considerations. Proceedings of the ... International Conference on Engineering Design, 1(1), 3651. DOI: 10.1017/dsi.2019.372

Schneider, J.-G.. (2020) 'Industry Agile practices in large-scale capstone projects', in *Proceedings of the ACM/IEEE 42nd International Conference on Software Engineering: Companion Proceedings. ICSE '20: 42nd International Conference on Software Engineering*, Seoul South Korea: ACM, pp. 326–327. Available at: DOI: 10.1145/3377812.3390794

Scime, L., & Beuth, J. (2017). Anomaly detection and classification in a laser powder bed additive manufacturing process using a trained computer vision algorithm. *Additive Manufacturing, 19*, 114–126. DOI: 10.1016/j.addma.2017.11.009

Seharing, A., Azman, A. H., & Abdullah, S. (2014). A review on integration of lightweight gradient lattice structures in additive manufacturing parts [Review of A review on integration of lightweight gradient lattice structures in additive manufacturing parts]. Advances in Mechanical Engineering, 12(6). SAGE Publishing. DOI: 10.1177/1687814020916951

Sehhat, M. H., Mahdianikhotbesara, A., & Yadegari, F. (2022). Impact of temperature and material variation on mechanical properties of parts fabricated with fused deposition modeling (FDM) additive manufacturing. *International Journal of Advanced Manufacturing Technology, 120*(7), 4791–4801. DOI: 10.1007/s00170-022-09043-0

Sehhat, M. H., Mahdianikhotbesara, A., & Yadegari, F. (2022). Verification of stress transformation in anisotropic material additively manufactured by fused deposition modeling (FDM). *International Journal of Advanced Manufacturing Technology, 123*(5), 1777–1783. DOI: 10.1007/s00170-022-10321-0

Sekmen, K., Rehbein, T., Johlitz, M., Lion, A., & Constantinescu, A. (2024). Thermal analysis and shrinkage characterization of the photopolymers for DLP additive manufacturing processes. *Continuum Mechanics and Thermodynamics, 36*(2), 351–368. DOI: 10.1007/s00161-022-01137-0

Selvam, P. P., Prabhakaran, S., Vinod, B., & Jishnu, T. (2024). A review of energy efficiency and Machine learning analysis for additive manufacturing of direct laser metal deposition. *Materials Today: Proceedings*. Advance online publication. DOI: 10.1016/j.matpr.2024.03.034

Serajuddin, A. (2023). Challenges, current status and emerging strategies in the development of rapidly dissolving FDM 3D-printed tablets: An overview and commentary. *ADMET & DMPK*, *11*(1), 33–55. DOI: 10.5599/admet.1622 PMID: 36778904

Shaheen, I., Hussain, I., Zahra, T., Javed, M. S., Shah, S. S. A., Khan, K., Hanif, M. B., Assiri, M. A., Said, Z., Arifeen, W. U., Akkinepally, B., & Zhang, K. (2023). Recent advancements in metal oxides for energy storage materials: Design, classification, and electrodes configuration of supercapacitor. *Journal of Energy Storage*, *72*, 108719. DOI: 10.1016/j.est.2023.108719

Shah, J., Snider, B., Clarke, T., Kozutsky, S., Lacki, M., & Hosseini, A. (2019). Large-scale 3D printers for additive manufacturing: Design considerations and challenges. *International Journal of Advanced Manufacturing Technology*, *104*(9–12), 3679–3693. DOI: 10.1007/s00170-019-04074-6

Shahrubudin, N., Lee, T. C., & Ramlan, R. J. P. M. (2019). An overview on 3D printing technology: Technological, materials, and applications. *Procedia Manufacturing*, *35*, 1286–1296. DOI: 10.1016/j.promfg.2019.06.089

Shahzad, Q., Umair, M., & Waqar, S. (2022). Bibliographic analysis on 3D printing in the building and construction industry: Printing systems, material properties, challenges, and future trends. *Journal of Sustainable Construction Materials and Technologies*, *7*(3), 198–220. DOI: 10.47481/jscmt.1143239

Shamsdini, S., Shakerin, S., Hadadzadeh, A., Amirkhiz, B. S., & Mohammadi, M. (2020). A trade-off between powder layer thickness and mechanical properties in additively manufactured maraging steels. *Materials Science and Engineering A*, *776*, 139041. DOI: 10.1016/j.msea.2020.139041

Shanmugam, V., Babu, K., Kannan, G., Mensah, R. A., Samantaray, S. K., & Das, O. (2024). The thermal properties of FDM printed polymeric materials: A review. *Polymer Degradation & Stability*, *228*, 110902. DOI: 10.1016/j.polymdegradstab.2024.110902

Shanmugam, V., Das, O., Babu, K., Marimuthu, U., Veerasimman, A., Johnson, D. J., Neisiany, R. E., Hedenqvist, M. S., Ramakrishna, S., & Berto, F. (2021). Fatigue behaviour of FDM-3D printed polymers, polymeric composites and architected cellular materials. *International Journal of Fatigue, 143*, 106007. DOI: 10.1016/j.ijfatigue.2020.106007

Shanmugam, V., Das, O., Neisiany, R. E., Babu, K., Singh, S., Hedenqvist, M. S., Berto, F., & Ramakrishna, S. (2020). Polymer recycling in additive manufacturing: An opportunity for the circular economy. *Materials Circular Economy, 2*(1), 1–11. DOI: 10.1007/s42824-020-00012-0

Shanthamma, S., Preethi, R., Moses, J. A., & Anandharamakrishnan, C. (2021). 4D printing of sago starch with turmeric blends: A study on pH-triggered spontaneous color transformation. *ACS Food Science & Technology, 1*(4), 669–679. DOI: 10.1021/acsfoodscitech.0c00151

Shao, L., Chang, Y. C., Hao, C., Fei, M. E., Zhao, B., Bliss, B. J., & Zhang, J. (2022). A chemical approach for the future of PLA upcycling: From plastic wastes to new 3D printing materials. *Green Chemistry, 24*(22), 8716–8724. DOI: 10.1039/D2GC01745H

Sheikh, M. F., Padmanabhan, R., Hadid, M., & Kerbache, L. (2025). Towards Additive Manufacturing Enabled Sustainable Spare Parts Supply Chain: A Multi-Objective MILP Model for Cost and Carbon Emissions Minimization. *Procedia Computer Science, 253*, 952–963. DOI: 10.1016/j.procs.2025.01.157

Shekhar, S., Bogaerts, W., Chrostowski, L., Bowers, J. E., Hochberg, M., Soref, R., & Shastri, B. J. (2024). Roadmapping the next generation of silicon photonics. *Nature Communications, 15*(1), 751. DOI: 10.1038/s41467-024-44750-0 PMID: 38272873

Shen, T., & Li, B. (2024). Digital twins in additive manufacturing: A state-of-the-art review. *International Journal of Advanced Manufacturing Technology, 131*(1), 63–92. DOI: 10.1007/s00170-024-13092-y

Shen, Y., Lin, L., Wei, S., Yan, J., & Xu, T. (2022). Research on the Preparation and Mechanical Properties of Solidified 3D Printed Concrete Materials. *Buildings (Basel, Switzerland), 12*(12), 2264. DOI: 10.3390/buildings12122264

Shie, M. Y., Shen, Y. F., Astuti, S. D., Lee, A. K. X., Lin, S. H., Dwijaksara, N. L. B., & Chen, Y. W. (2019). Review of polymeric materials in 4D printing biomedical applications. *Polymers, 11*(11), 1864. DOI: 10.3390/polym11111864 PMID: 31726652

Shin, D. S., Lee, C. H., Kühn, U., Lee, S., Park, S. J., Schwab, H., Scudino, S., & Kosiba, K. (2020). Optimizing laser powder bed fusion of Ti-5Al-5V-5Mo-3Cr by artificial intelligence. *Journal of Alloys and Compounds*, *862*, 158018. DOI: 10.1016/j.jallcom.2020.158018

Shinde, S., Mane, R., Vardikar, A., Dhumal, A., & Rajput, A. (2023). 4D printing: From emergence to innovation over 3D printing. *European Polymer Journal*, *197*, 112356. DOI: 10.1016/j.eurpolymj.2023.112356

Shishkovsky, I., Yadroitsev, I., Bertrand, P., & Smurov, I. (2007). Alumina-zirconium ceramics synthesis by selective laser sintering/melting. *Applied Surface Science*, *254*(4), 966–970. DOI: 10.1016/j.apsusc.2007.09.001

Shofner, M. L., Lozano, K., & Rodrı, F. J. (2002). Nanofiber-reinforced polymers prepared by fused deposition modeling. *Journal of Applied Polymer Science*, *89*(11), 3081–3090. DOI: 10.1002/app.12496

Shofner, M. L., Rodríguez-Macías, F. J., Vaidyanathan, R., & Barrera, E. V. (2003). Single wall nanotube and vapor grown carbon fiber reinforced polymers processed by extrusion freeform fabrication. *Composites. Part A, Applied Science and Manufacturing*, *34*(12), 1207–1217. DOI: 10.1016/j.compositesa.2003.07.002

Shun, L. (2021). *Studies on 4D printing Thermo-responsive PNIPAM-based materials* (Doctoral dissertation, The University of Akron).

Sierra-Vega, N. O., Ashraf, M., O'Connor, T., Kopcha, M., Di Prima, M., Coburn, J., & Zidan, A. (2025). Emerging 3D printing technologies for solid oral dosage forms: Processes, materials and analytical tools for real-time assessment. *International Journal of Pharmaceutics*, *673*, 125415. DOI: 10.1016/j.ijpharm.2025.125415 PMID: 40023346

Silbernagel, C., Aremu, A., & Ashcroft, I. (2019). Using machine learning to aid in the parameter optimisation process for metal-based additive manufacturing. *Rapid Prototyping Journal*, *26*(4), 625–637. DOI: 10.1108/RPJ-08-2019-0213

Singh, J., Singh, A., Singh, H., & Doyon-Poulin, P. (2025). Implementation and evaluation of a smart machine monitoring system under industry 4.0 concept. *Journal of Industrial Information Integration*, *43*, 100746. DOI: 10.1016/j.jii.2024.100746

Singholi, A. K., & Sharma, A. (2021). Review of Materials and Processes Used in 4D Printing. In *Advances in Engineering Materials: Select Proceedings of FLAME 2020* (pp. 677-684). Springer Singapore.

Singh, R., Gupta, A., Tripathi, O., Srivastava, S., Singh, B., Awasthi, A., Rajput, S. K., Sonia, P., Singhal, P., & Saxena, K. K. (2020). Powder bed fusion process in additive manufacturing: An overview. *Materials Today: Proceedings*, *26*, 3058–3070. Advance online publication. DOI: 10.1016/j.matpr.2020.02.635

Singh, R., Singh, H., Farina, I., Colangelo, F., & Fraternali, F. (2019). On the additive manufacturing of an energy storage device from recycled material. *Composites. Part B, Engineering*, *156*, 259–265. DOI: 10.1016/j.compositesb.2018.08.080

Singh, S., Kaur, D., Singh, M., Balu, R., Mehta, A., & Vasudev, H. (2024). Challenges and issues in manufacturing of components using polymer-based selective laser sintering (SLS) : A review. [IJIDeM]. *International Journal on Interactive Design and Manufacturing*. Advance online publication. DOI: 10.1007/s12008-024-02049-w

Singh, S., Ramakrishna, S., & Singh, R. (2017). Material issues in additive manufacturing: A review. *Journal of Manufacturing Processes*, *25*, 185–200. DOI: 10.1016/j.jmapro.2016.11.006

Siraj MT, Alshybani I, Payel SB, Shahadat M, Bin R, and Rahman MZ (2023a) Advances in biocomposite fabrication: Emerging technologies and their potential applications.

Sivamani, S., Nadarajan, M., Kameshwaran, R., Bhatt, C. D., Premkumar, M. T., & Hariram, V. (2020). Analysis of cross axis wind turbine blades designed and manufactured by FDM based additive manufacturing. *Materials Today: Proceedings*, *33*, 3504–3509. DOI: 10.1016/j.matpr.2020.05.438

Solomon, I. J., Sevvel, P., & Gunasekaran, J. J. M. T. P. (2021). A review on the various processing parameters in FDM. *Materials Today: Proceedings*, *37*, 509–514. DOI: 10.1016/j.matpr.2020.05.484

Somers, P., Münchinger, A., Maruo, S., Moser, C., Xu, X., & Wegener, M. (2024). The physics of 3D printing with light. *Nature Reviews. Physics*, *6*(2), 99–113. DOI: 10.1038/s42254-023-00671-3

Sonavane, A. S., & Aylani, A. (2025). Integration of IoT and Quantum Computing: Revolutionizing Manufacturing. In Machine Learning for Environmental Monitoring in Wireless Sensor Networks (pp. 331-358). IGI Global.

Sood, A. K., Ohdar, R. K., & Mahapatra, S. S. (2012). Experimental investigation and empirical modelling of FDM process for compressive strength improvement. *Journal of Advanced Research*, *3*(1), 81–90. DOI: 10.1016/j.jare.2011.05.001

Soori, M., Dastres, R., Arezoo, B., & Jough, F. K. G. (2024). Intelligent robotic systems in Industry 4.0: A review. *Journal of Advanced Manufacturing Science and Technology*, 2024007-0.

Sossou, G., Demoly, F., Belkebir, H., Qi, H. J., Gomes, S., & Montavon, G. (2019). Design for 4D printing: Modeling and computation of smart materials distributions. *Materials & Design*, *181*, 108074. DOI: 10.1016/j.matdes.2019.108074

Sousa, J., Brandau, B., Darabi, R., Sousa, A., Brueckner, F., Reis, A., & Reis, L. P. (2025). Artificial Intelligence for Control in Laser-Based Additive Manufacturing: A Systematic Review. *IEEE Access : Practical Innovations, Open Solutions*, *13*, 30845–30860. DOI: 10.1109/ACCESS.2025.3537859

Sovetova, M., & Calautit, J. K. (2024). Design, calibration and performance evaluation of a small-scale 3D printer for accelerating research in additive manufacturing in construction. *Cleaner Engineering and Technology*, *22*, 100786. DOI: 10.1016/j.clet.2024.100786

Spencer, O. O. (2018). Additive Manufacturing Technology Development: A Trajectory Towards Industrial Revolution. *American Journal of Mechanical and Industrial Engineering*, *3*(5), 80. DOI: 10.11648/j.ajmie.20180305.12

Spiegel, C. A., Hippler, M., Münchinger, A., Bastmeyer, M., Barner-Kowollik, C., Wegener, M., & Blasco, E. (2020). 4D printing at the microscale. *Advanced Functional Materials*, *30*(26), 1907615. DOI: 10.1002/adfm.201907615

Srinivas, M., & Babu, B. S. (2017). A Critical Review on Recent Research Methodologies in Additive Manufacturing [Review of A Critical Review on Recent Research Methodologies in Additive Manufacturing]. Materials Today Proceedings, 4(8), 9049. Elsevier BV. DOI: 10.1016/j.matpr.2017.07.258

Srinivasan, D., Meignanamoorthy, M., Ravichandran, M., Mohanavel, V., Alagarsamy, S. V., Chanakyan, C., Sakthivelu, S., Karthick, A., Prabhu, T. R., & Rajkumar, S. (2021). 3D Printing Manufacturing Techniques, Materials, and Applications: An Overview. Advances in Materials Science and Engineering, 2021(1). DOI: 10.1155/2021/5756563

Srivastava, R., Thakur, A., & Rana, A. (2024). The Basic Understanding of Nano Materials: A Review. *Innovations in Nanomaterials-Based Corrosion Inhibitors*, 1-20.

Srivastava, R., Thakur, A., Bahadur, P. S., & Rana, A. (2024). A State-of-the-Art Review of Recent Developments and Challenges for 2D Implantable Materials. *Engineering Materials for Efficient Energy Storage and Conversion*, 1-24.

Srivastava, M., Rathee, S., Patel, V., Kumar, A., & Koppad, P. G. (2022). A review of various materials for additive manufacturing: Recent trends and processing issues. *Journal of Materials Research and Technology*, *21*, 2612–2641. DOI: 10.1016/j.jmrt.2022.10.015

Srivastava, R., Arun, J., & Patel, D. K. (2019, April). Amalgamating the Service Quality Aspect in Supply Chain Management. In *2019 International Conference on Automation, Computational and Technology Management (ICACTM)* (pp. 63-67). IEEE. DOI: 10.1109/ICACTM.2019.8776839

Srivastava, R., & Srivastava, V. K., (2020). Changing Scenario of Manufacturing Industries Due to Covid 19. *International Journal for Modern Trends in Science and Technology*, *6*(10), 143–146. DOI: 10.46501/IJMTST061026

Starikov, K., Polozov, I., Borisov, E., Kim, A., Voevodenko, D., Gracheva, A., Shamshurin, A., & Popovich, A. (2023). Selective laser melting of non-weldable nickel superalloy: Microstructure, cracks and texture. *Metals*, *13*(11), 1886. DOI: 10.3390/met13111886

Stefańska, A., Cygan, M., Batte, K., & Pietrzak, J. (2021). Application of Timber and Wood-based Materials in Architectural Design using Multi-objective Optimisation Tools. *Construction Economics and Building*, *21*(3). Advance online publication. DOI: 10.5130/AJCEB.v21i3.7642

Steinmo, M. (2015). Collaboration for Innovation: A Case Study on How Social Capital Mitigates Collaborative Challenges in University–Industry Research Alliances. *Industry and Innovation*, *22*(7), 597–624. DOI: 10.1080/13662716.2015.1105127

Stevens, C. V. (2010). *Industrial applications of natural fibres: structure, properties and technical applications*. John Wiley & Sons.

Strano, A. G., Hao, L., Everson, R. M., & Evans, K. E. (2012). Surface roughness analysis, modelling and prediction in selective laser melting. *Journal of Materials Processing Technology*, *2010*. Advance online publication. DOI: 10.1016/j.jmatprotec.2012.11.011

Subeshan, B., Baddam, Y., & Asmatulu, E. (2021). Current progress of 4D-printing technology. *Progress in Additive Manufacturing*, *6*(3), 495–516. DOI: 10.1007/s40964-021-00182-6

Subramani, R., Mustafa, M. A., Ghadir, G. K., Al-Tmimi, H. M., Alani, Z. K., Haridas, D., Rusho, M. A., Rajeswari, N., Rajan, A. J., & Kumar, A. P. (2024). Advancements in 3D printing materials: A comparative analysis of performance and applications. *Applied Chemical Engineering*, *7*(2), ACE-3867. DOI: 10.59429/ace.v7i2.3867

Subramani, R., Mustafa, M. A., Ghadir, G. K., Al-Tmimi, H. M., Alani, Z. K., Rusho, M. A., & Kumar, A. P. (2024). Exploring the use of biodegradable polymer materials in sustainable 3D printing. *Applied Chemical Engineering*, *7*(2), 3870–3870.

Su, C., Chen, Y., Tian, S., Lu, C., & Lv, Q. (2022). Natural materials for 3D printing and their applications. *Gels (Basel, Switzerland)*, *8*(11), 748. DOI: 10.3390/gels8110748 PMID: 36421570

Su, J., Ng, W. L., An, J., Yeong, W. Y., Chua, C. K., & Sing, S. L. (2024). Achieving sustainability by additive manufacturing: A state-of-the-art review and perspectives. *Virtual and Physical Prototyping*, *19*(1), e2438899. DOI: 10.1080/17452759.2024.2438899

Sundaram, S., & Zeid, A. (2023). Artificial Intelligence-Based Smart Quality Inspection for Manufacturing. *Micromachines*, *14*(3), 570. DOI: 10.3390/mi14030570 PMID: 36984977

Sun, X., Mazur, M., & Cheng, C. T. (2023). A review of void reduction strategies in material extrusion-based additive manufacturing. *Additive Manufacturing*, *67*, 103463. DOI: 10.1016/j.addma.2023.103463

Suryavanshi, P., Kawre, S., Maniruzzaman, M., Seth, K., & Banerjee, S. (2023). Thermo-responsive self-folding feedstock with excellent shape memory programming. *Chemicke Zvesti*, *77*(6), 3145–3154. DOI: 10.1007/s11696-023-02693-8

Suryawanshi, J., Prashanth, K. G., & Ramamurty, U. (2017). Tensile, fracture, and fatigue crack growth properties of a 3D printed maraging steel through selective laser melting. *Journal of Alloys and Compounds*, *725*, 355–364. DOI: 10.1016/j.jallcom.2017.07.177

Syrlybayev, D., Zharylkassyn, B., Seisekulova, A., Akhmetov, M., Perveen, A., & Talamona, D. (2021). Optimisation of strength properties of FDM printed parts—A critical review. *Polymers*, *13*(10), 1587. DOI: 10.3390/polym13101587 PMID: 34069144

Tabassum, T., & Mir, A. A. (2023). A review of 3d printing technology-the future of sustainable construction. *Materials Today: Proceedings*, *93*, 408–414. DOI: 10.1016/j.matpr.2023.08.013

Taghizadeh, M., & Zhu, Z. H. (2024). A comprehensive review on metal laser additive manufacturing in space: Modeling and perspectives. *Acta Astronautica*, *222*, 403–421. DOI: 10.1016/j.actaastro.2024.06.027

Tagliaferri, V., Trovalusci, F., Guarino, S., & Venettacci, S. (2019). Environmental and economic analysis of FDM, SLS and MJF additive manufacturing technologies. *Materials (Basel)*, *12*(24), 4161. DOI: 10.3390/ma12244161 PMID: 31835783

Tahmasebinia, F., Niemelä, M., Sepasgozar, S. M. E., Lai, T., Su, W., Reddy, K. R., Shirowzhan, S., Sepasgozar, S. M. E., & Marroquín, F. A. (2018). Three-Dimensional Printing Using Recycled High-Density Polyethylene: Technological Challenges and Future Directions for Construction. *Buildings (Basel, Switzerland)*, *8*(11), 165. DOI: 10.3390/buildings8110165

Tamay, D. G., Dursun Usal, T., Alagoz, A. S., Yucel, D., Hasirci, N., & Hasirci, V. (2019). 3D and 4D printing of polymers for tissue engineering applications. *Frontiers in Bioengineering and Biotechnology*, *7*, 164. DOI: 10.3389/fbioe.2019.00164 PMID: 31338366

Tang, T., Wang, L., Zhu, M., Zhang, H., Dong, J., Yue, W., & Xia, H. (2024). Topology Optimization: A Review for Structural Designs Under Statics Problems. *Materials (Basel)*, *17*(23), 5970. DOI: 10.3390/ma17235970 PMID: 39685406

Tang, Y. (2018). Development Trend and Application Prospect of 3D Printing Technology in Intelligent Manufacturing. *International Robotics & Automation Journal*, *4*(1). Advance online publication. DOI: 10.15406/iratj.2018.04.00089

Tan, K. H., Chua, C. K., Leong, K. F., Cheah, C. M., Cheang, P., Abu Bakar, M. S., & Cha, S. W. (2003). Scaffold development using selective laser sintering of polyetheretherketone-hydroxyapatite biocomposite blends. *Biomaterials*, *24*(18), 3115–3123. DOI: 10.1016/S0142-9612(03)00131-5 PMID: 12895584

Tan, L., Wang, Z., Jiang, H., Han, B., Tang, J., Kang, C., Zhang, N., & Xu, Y. (2022). Full color 3D printing of anatomical models. *Clinical Anatomy (New York, N.Y.)*, *35*(5), 598–608. DOI: 10.1002/ca.23875 PMID: 35384062

Tao, Y., Yu, P., Zhang, X., Li, M., Liu, T., & Zhao, J. (2024). Displacement-reconstruction-realized components by structure-sensing integration via a hybrid 3D printing strategy. *Composites. Part A, Applied Science and Manufacturing*, *177*, 107896. DOI: 10.1016/j.compositesa.2023.107896

Tarasov, N., & Khamula, O. (2024). Integrating artificial intelligence into 3D modeling software for optimized additive manufacturing. *Grail of Science*, (39), 348–352. DOI: 10.36074/grail-of-science.10.05.2024.052

Tareq, M. S., Rahman, T., Hossain, M., & Dorrington, P. (2021). Additive manufacturing and the COVID-19 challenges : An in-depth study. *Journal of Manufacturing Systems*, *60*, 787–798. DOI: 10.1016/j.jmsy.2020.12.021 PMID: 33897085

Ţârtea, D. A., Ionescu, M., Manolea, H. O., Mercu , V., Obădan, E., Amărăscu, M. O., Mără escu, P. C., Dăguci, L., & Popescu, S. M. (2023). Comparative study of dental custom cad-cam implant abutments and dental implant stock abutments. *Journal of Clinical Medicine*, *12*(6), 2128. DOI: 10.3390/jcm12062128 PMID: 36983129

Taufik, M., & Jain, P. K. (2020). Part surface quality improvement studies in fused deposition modelling process: a review [Review of Part surface quality improvement studies in fused deposition modelling process: a review]. Australian Journal of Mechanical Engineering, 20(2), 527. Taylor & Francis. DOI: 10.1080/14484846.2020.1723342

Tay, Y. W. D., Panda, B., Paul, S. C., Tan, M. J., Qian, S. Z., Leong, K. F., & Chua, C. K. (2016, August). Processing and properties of construction materials for 3D printing. []). Trans Tech Publications Ltd.]. *Materials Science Forum*, *861*, 177–181. DOI: 10.4028/www.scientific.net/MSF.861.177

Teacher, M., & Velu, R. (2024). Additive manufacturing of functionally graded materials: A comprehensive review. *International Journal of Precision Engineering and Manufacturing*, *25*(1), 165–197. DOI: 10.1007/s12541-023-00864-x

Tebianian, M., Aghaie, S., Jafari, N. R., Hosseini, S. R. E., Pereira, A. B., Fernandes, F. A. O., Farbakhti, M., Chen, C., & Huo, Y. (2023). A Review of the Metal Additive Manufacturing Processes [Review of A Review of the Metal Additive Manufacturing Processes]. Materials, 16(24), 7514. Multidisciplinary Digital Publishing Institute. DOI: 10.3390/ma16247514

Tekinalp, H. L., Kunc, V., Velez-Garcia, G. M., Duty, C. E., Love, L. J., Naskar, A. K., Blue, C. A., & Ozcan, S. (2014). Highly oriented carbon fiber–polymer composites via additive manufacturing. *Composites Science and Technology*, *105*, 144–150. DOI: 10.1016/j.compscitech.2014.10.009

Teng, X., Zhang, M., Adhikari, B., & Ma, Y. (2024). 4D printed pH-responsive labels of methacrylic anhydride grafted konjac glucomannan for detecting quality changes in respiring climacteric fruits. *Food Hydrocolloids*, *149*, 109543. DOI: 10.1016/j.foodhyd.2023.109543

Teng, X., Zhang, M., & Mujumdar, A. S. (2021). 4D printing: Recent advances and proposals in the food sector. *Trends in Food Science & Technology*, *110*, 349–363. DOI: 10.1016/j.tifs.2021.01.076

Tereshchenko, E., Salmela, E., Melkko, E., Phang, S. K., & Happonen, A. (2024). Emerging best strategies and capabilities for university–industry cooperation: Opportunities for MSMEs and universities to improve collaboration. A literature review 2000–2023. *Journal of Innovation and Entrepreneurship*, *13*(1), 28. DOI: 10.1186/s13731-024-00386-4

Tetsuka, H., & Shin, S. R. (2020). Materials and technical innovations in 3D printing in biomedical applications [Review of Materials and technical innovations in 3D printing in biomedical applications]. Journal of Materials Chemistry B, 8(15), 2930. Royal Society of Chemistry. DOI: 10.1039/D0TB00034E

Thakar, C. M., Parkhe, S. S., Jain, A., Phasinam, K., Murugesan, G., & Ventayen, R. J. M. (2022). 3d Printing: Basic principles and applications. *Materials Today: Proceedings*, *51*, 842–849. DOI: 10.1016/j.matpr.2021.06.272

Thakur, A., Srivastava, R., Bahadur, P. S., & Rana, A. (2024). Exploring the Various Biomedical Applications of Smart Materials. In *Discovery, Disruption, and Future Implications of Nanomaterials* (pp. 249-274). IGI Global. DOI: 10.4018/979-8-3693-4397-5.ch008

Thakur, A., Vates, U. K., & Mishra, S. (2024). Parametric optimization of 3D-printed PLA part using response surface methodology for mechanical properties. *Smart Science*, *12*(1), 91–104. DOI: 10.1080/23080477.2023.2270819

Thijs, L., Kempen, K., Kruth, J.-P., & van Humbeeck, J. (2013). Fine-structured aluminium products with controllable texture by selective laser melting of pre-alloyed AlSi10Mg powder. *Acta Materialia*, *61*(5), 1809–1819. DOI: 10.1016/j.actamat.2012.11.052

Thomas, H. R., Hopkinson, N., & Erasenthiran, P. (2006). *High speed sintering–continuing research into a new rapid manufacturing process.*

Thomas, J. P., & Renaud, J. E. (2006). Mechanical behavior of acrylonitrile butadiene styrene (ABS) fused deposition materials. Experimental investigation. *Rapid Prototyping Journal*, *7*(3), 148–158.

Tian, H., Yi, X., Lu, L., Chen, X., Zhang, G., Cai, Y., & Han, J. (2022). Heat propagation simulation of thin-walled ti6al4v alloy components fabricated by double-wire gtaw additive manufacturing. *Science and Technology of Welding and Joining*, *27*(6), 446–454. DOI: 10.1080/13621718.2022.2063566

Tian, X., Wang, Q., Liu, G., Liu, Y., Xie, Y., & Deng, W. (2019). Topology optimization design for offshore platform jacket structure. *Applied Ocean Research*, *84*, 38–50. DOI: 10.1016/j.apor.2019.01.003

Tian, Y., Chen, C., Xu, X., Wang, J., Hou, X., Li, K., Lu, X., Shi, H. Y., Lee, E.-S., & Jiang, H. B. (2021). A review of 3D printing in dentistry: Technologies, affecting factors, and applications. *Scanning*, *2021*(1), 9950131. DOI: 10.1155/2021/9950131 PMID: 34367410

Tissen, L. N. (2022). 3D printing and the art world: current developments and future perspectives. In *Advances in 3D printing*. IntechOpen.

Tiwary, V. K., Arunkumar, P., & Malik, V. R. (2021). An overview on joining/ welding as post-processing technique to circumvent the build volume limitation of an FDM-3D printer. *Rapid Prototyping Journal*, *27*(4), 808–821. DOI: 10.1108/ RPJ-10-2020-0265

Toyserkani, E., Khajepour, A., & Corbin, S. (2004). 3-D finite element modeling of laser cladding by powder injection: Effects of laser pulse shaping on the process. *Optics and Lasers in Engineering*, *41*(6), 849–867. DOI: 10.1016/S0143-8166(03)00063-0

Trenfield, S. J., Awad, A., Madla, C. M., Hatton, G. B., Firth, J., Goyanes, Á., Gaisford, S., & Basit, A. W. (2019). Shaping the future: recent advances of 3D printing in drug delivery and healthcare [Review of Shaping the future: recent advances of 3D printing in drug delivery and healthcare]. Expert Opinion on Drug Delivery, 16(10), 1081. Taylor & Francis. DOI: 10.1080/17425247.2019.1660318

Treutler, K., Lorenz, S., Hamje, J., & Wesling, V. (2022). Wire and arc additive manufacturing of a cocrfemoniv complex concentrated alloy using metal-cored wire—Process, properties, and wear resistance. *Applied Sciences (Basel, Switzerland)*, *12*(13), 6308. DOI: 10.3390/app12136308

Trifol, J., Jayaprakash, S., Baniasadi, H., Ajdary, R., Kretzschmar, N., Rojas, O. J., Partanen, J., & Seppälä, J. (2021). 3D-Printed Thermoset Biocomposites Based on Forest Residues by Delayed Extrusion of Cold Masterbatch (DECMA). *ACS Sustainable Chemistry & Engineering*, *9*(41), 13979–13987. DOI: 10.1021/acssuschemeng.1c05587

Truxova, V., Safka, J. I. R. I., Seidl, M., Kovalenko, I., Volesky, L. U. K. A. S., & Ackermann, M. I. C. H. A. L. (2020). Ceramic 3D printing: Comparison of SLA and DLP technologies. *MM Sci. J*, *2020*(2), 3905–3911. DOI: 10.17973/ MMSJ.2020_06_2020006

Tsioukas, V., Pikridas, C., & Karolos, I. A. (2020). Challenges, opportunities, and limitations in 3D printing. 3D Printing: Applications in Medicine and Surgery, 151-155.

Tuli, N. T., Khatun, S., & Rashid, A. B. (2024). Unlocking the future of precision manufacturing: A comprehensive exploration of 3D printing with fiber-reinforced composites in aerospace, automotive, medical, and consumer industries. *Heliyon*, *10*(5), e27328. DOI: 10.1016/j.heliyon.2024.e27328 PMID: 38495162

Turk, T., Dominguez, C. E., Sutton, A. T., Bernardin, J. D., Park, J., & Leu, M. C. (2024). Spot pattern welding scanning strategy for sensor embedding and residual stress reduction in laser-foil-printing additive manufacturing. *Rapid Prototyping Journal*, *31*(1), 83–99. DOI: 10.1108/RPJ-01-2024-0042

Turner, N., B., Strong, R., & A. Gold, S. (2014). A review of melt extrusion additive manufacturing processes: I. Process design and modeling. *Rapid Prototyping Journal*, *20*(3), 192–204.

Tyagi, N., Bhardwaj, V., Sharma, D., Tomar, R., Chaudhary, V., Khanuja, M., Singh, M. K., & Sharma, G. (2024). 3D Printing Technology in the Pharmaceutical and Biomedical Applications : A Critical Review. *Biomedical Materials & Devices*, *2*(1), 178–190. DOI: 10.1007/s44174-023-00105-9

Tyflopoulos, E., Hofset, T. A., Olsen, A., & Steinert, M. (2021). Simulation-based design: A case study combining optimization methodologies for angle-ply composite laminates. *Procedia CIRP*, *100*, 607–612. DOI: 10.1016/j.procir.2021.05.131

Ukobitz, D. V. (2021). Organizational adoption of 3D printing technology: A semi-systematic literature review. *Journal of Manufacturing Technology Management*, *32*(9), 48–74. DOI: 10.1108/JMTM-03-2020-0087

Ukwaththa, J., Herath, S., & Meddage, D. P. P. (2024). A review of machine learning (ML) and explainable artificial intelligence (XAI) methods in additive manufacturing (3D printing). *Materials Today. Communications*, *41*, 110294. DOI: 10.1016/j.mtcomm.2024.110294

Vadivelu, V., Nachimuthu, P., & Karthikeyan, P. (2025). Transforming Supply Chains (SCs) with Additive Manufacturing (AM) A Paradigm Shift in Production. *Metal Additive Manufacturing: Principles, Techniques and Applications*, 359-406.

Valerga, A. P., Batista, M., Salguero, J., & Girot, F. (2018). Influence of PLA filament conditions on characteristics of FDM parts. *Materials (Basel)*, *11*(8), 1322. DOI: 10.3390/ma11081322 PMID: 30065153

Valvez, S., Santos, P., Parente, J. M., Silva, M. P., & Reis, P. N. B. (2020). 3D printed continuous carbon fiber reinforced PLA composites: A short review. *Procedia Structural Integrity*, *25*(May), 394–399. DOI: 10.1016/j.prostr.2020.04.056

van Blitterswijk, R. H., Farshidianfar, M. H., & Khajepour, A. (2024). Model predictive control for real-time microstructure control in laser heat treatment. *Journal of Manufacturing Processes*, *118*, 116–127. DOI: 10.1016/j.jmapro.2024.03.038

van de Werken, N., Hurley, J., Khanbolouki, P., Sarvestani, A. N., Tamijani, A. Y., & Tehrani, M. (2019). Design considerations and modeling of fiber reinforced 3D printed parts. *Composites. Part B, Engineering*, *160*, 684–692. DOI: 10.1016/j.compositesb.2018.12.094

van de Werken, N., Tekinalp, H., Khanbolouki, P., Ozcan, S., Williams, A., & Tehrani, M. (2020). Additively manufactured carbon fiber-reinforced composites: State of the art and perspective. *Additive Manufacturing*, *31*, 100962. DOI: 10.1016/j.addma.2019.100962

Vaneker, T., Bernard, A., Moroni, G., Gibson, I., & Zhang, Y. (2020). Design for additive manufacturing: Framework and methodology. *CIRP Annals*, *69*(2), 578–599. DOI: 10.1016/j.cirp.2020.05.006

Vanli, O. A., Rivero, I. V., & Wang, B. (2013). Editorial—advances in nanomanufacturing: Process repeatability, scalability and affordability for nanomaterials and manufacturing. *International Journal of Advanced Manufacturing Technology*, *64*(1–4), 513–514. DOI: 10.1007/s00170-012-4646-6

Vasinonta, A., & Beuth, J. (2000) Process maps for controlling residual stress and melt pool size in laser-based sff processes. In: *Proceedings 2000 Solid Freeform Fabrication Symposium Solid Freeform Fabrication Symposium*, pp. 200–208.

Vatanparast, S., Boschetto, A., Bottini, L., & Gaudenzi, P. (2023). New trends in 4D printing: A critical review. *Applied Sciences (Basel, Switzerland)*, *13*(13), 7744. DOI: 10.3390/app13137744

Vayre, B., Vignat, F., & Villeneuve, F. (2012). Metallic additive manufacturing: state-of-the-art review and prospects. In Mechanics & Industry (Vol. 13, Issue 2, p. 89). EDP Sciences. DOI: 10.1051/meca/2012003

Vemuganti, S., Soliman, E., & Taha, M. M. R. (2024). Manipulating interfacial bond for controlling load transfer in 3D printed fiber reinforced polymer composites. *Journal of Reinforced Plastics and Composites*, •••, 07316844241238502. DOI: 10.1177/07316844241238502

Venkatalakshmi, S. (2024). Role of Cybersecurity to Safeguard 3D Bioprinting in Healthcare: Challenges and Opportunities. Computational Intelligence in Bioprinting, 93-126.

Venkataraman, N., Rangarajan, S., Matthewson, M. J., Safari, A., Danforth, S. C., Yardimci, A., & Guceri, S. I. (1999) Mechanical and Rheological Properties of Feedstock Material for Fused Deposition of Ceramics and Metals (FDC and FDMet) and their Relationship to Process Performance, pp. 351–360.

Venkataraman, N., Rangarajan, S., Matthewson, M. J., Harper, B., Safari, A., Danforth, S. C., Wu, G., Langrana, N., Guceri, S., & Yardimci, A. (2002). Feedstock material property [REMOVED HYPERLINK FIELD] process relationships in fused deposition of ceramics (FDC). *Rapid Prototyping Journal*, *6*, 244–252. DOI: 10.1108/13552540010373344

Vernezi, M., Lukyanov, A., Dolgov, V., & Zotov, V. V. (2021). About diagnostics of the technical condition of metal-cutting machines. *IOP Conference Series. Materials Science and Engineering*, *1029*, 12124. DOI: 10.1088/1757-899X/1029/1/012124

Vidal, L., Kampleitner, C., Brennan, M. Á., Hoornaert, A., & Layrolle, P. (2020). Reconstruction of large skeletal defects: Current clinical therapeutic strategies and future directions using 3D printing. *Frontiers in Bioengineering and Biotechnology*, *8*, 61. DOI: 10.3389/fbioe.2020.00061 PMID: 32117940

Vigliotti, A., & Pasini, D. (2011). *Structural Optimization of Lattice Materials.*, DOI: 10.1115/DETC2011-47390

Villa, A., Gianchandani, P. K., & Baino, F. (2024). Sustainable approaches for the additive manufacturing of ceramic materials. *Ceramics*, *7*(1), 291–309. DOI: 10.3390/ceramics7010019

Volpe, S., Sangiorgio, V., Petrella, A., Notarnicola, M., Varum, H., & Fiorito, F. (2023). 3d printed concrete blocks made with sustainable recycled material. VITRUVIO - International Journal of Architectural Technology and Sustainability, 8, 70-83. DOI: 10.4995/vitruvioijats.2023.18832

Vorkapić, M., Hasan, M. A., Tanović, D., Baltić, M., & Tomić, B. (2020). Implementation of 6R strategy in FDM printing process: Case: Small electronic enclosure box. In Journal of Engineering Management and Competitiveness (Vol. 10, Issue 2, p. 141). University of Novi Sad, Technical faculty Mihajlo Pupin, Zrenjanin. DOI: 10.5937/jemc2002141V

Vysochinskiy, D., & Akhtar, N. (2022). On the influence of cross-section size on measured strength of slm-produced alsi10mg-alloy. *Key Engineering Materials*, *926*, 132–140. DOI: 10.4028/p-g078sc

Výtisk, J., Kočí, V., Honus, S., & Vrtek, M. (2019). Current options in the life cycle assessment of additive manufacturing products. In Open Engineering (Vol. 9, Issue 1, p. 674). De Gruyter. DOI: 10.1515/eng-2019-0073

Wagmare, R., Harshe, R., Pednekar, J., & Patro, T. U. (2024). Additive manufacturing of continuous fiber-reinforced polymer composites: Current trend and future directions. *Progress in Additive Manufacturing*, •••, 1–28.

Wagner, J. J., & Higgs, C. F.III. (2024). Coupled CFD-DEM simulation of interfacial fluid–particle interaction during binder jet 3D printing. *Computer Methods in Applied Mechanics and Engineering*, *421*, 116747. DOI: 10.1016/j.cma.2024.116747

Waheed, M. F., & Bernadin, S. (2024, March). In-Situ Analysis of Vibration and Acoustic Data in Additive Manufacturing. In SoutheastCon 2024 (pp. 812-817). IEEE. DOI: 10.1109/SoutheastCon52093.2024.10500055

Wang, X., Jiang, M., Zhou, Z., Gou, J., & Hui, D. (2016). 3D printing of polymer matrix composites: A review and prospective [Review of 3D printing of polymer matrix composites: A review and prospective]. Composites Part B Engineering, 110, 442. Elsevier BV. DOI: 10.1016/j.compositesb.2016.11.034

Wang, B., Zhang, Z., Pei, Z., Qiu, J., & Wang, S. (2020). Current progress on the 3D printing of thermosets. *Advanced Composites and Hybrid Materials*, *3*(4), 462–472. DOI: 10.1007/s42114-020-00183-z

Wang, C., Sun, Y., Chen, G., Chen, X., Ding, J., Suder, W., Diao, C., & Williams, S. (2022). A simplified modelling approach for thermal behaviour analysis in hybrid plasma arc-laser additive manufacturing. *International Journal of Heat and Mass Transfer*, *195*, 123157. DOI: 10.1016/j.ijheatmasstransfer.2022.123157

Wang, C., Tan, X., Tor, S. B., & Lim, C. S. (2020). Machine learning in additive manufacturing: State-of-the-art and perspectives. *Additive Manufacturing*, *36*, 101538. DOI: 10.1016/j.addma.2020.101538

Wang, D., Song, C., Yang, Y., & Bai, Y. (2016). Investigation of crystal growth mechanism during selective laser melting and mechanical property characterization of 316L stainless steel parts. *Materials & Design*, *100*, 291–299. DOI: 10.1016/j.matdes.2016.03.111

Wang, R., Wang, R., Dou, C., Yang, S., Gnanasambandam, R., Wang, A., & Kong, Z. (2024). Sub-surface thermal measurement in additive manufacturing via machine learning-enabled high-resolution fiber optic sensing. *Nature Communications*, *15*(1), 7568. DOI: 10.1038/s41467-024-51235-7 PMID: 39217158

Wang, S., Chen, X., Han, X., Hong, X., Li, X., Zhang, H., Li, M., Wang, Z., & Zheng, A. (2023). A review of 3D printing technology in pharmaceutics: Technology and applications, now and future. *Pharmaceutics*, *15*(2), 416. DOI: 10.3390/pharmaceutics15020416 PMID: 36839738

Wang, T., Kaur, L., Beniwal, A. S., Furuhata, Y., Aoyama, H., & Singh, J. (2023). Physico-chemical and Textural Properties of 3D Printed Plant-based and Hybrid Soft Meat Analogs. *Plant Foods for Human Nutrition (Dordrecht, Netherlands)*, *78*(2), 375–382. DOI: 10.1007/s11130-023-01068-4 PMID: 37199825

Wang, W. L., Cheah, C. M., Fuh, J. Y. H., & Lu, L. (1997). Influence of process parameters on stereolithography part shrinkage. *Materials & Design*, *17*(4), 205–213. DOI: 10.1016/S0261-3069(97)00008-3

Wang, Y., Jia, B., Liu, S., Yao, X., & Sun, C. (2022). 3D Printing of Smart Materials and Actuators. In *Proceedings of the 1st International Conference on New Materials, Machinery and Vehicle Engineering* (pp. 128-138). IOS Press. DOI: 10.3233/ATDE220220

Wang, Y., Wang, Y., Liu, M., Wei, Q., & Du, B. (2023). 4D printing light-/thermo-responsive shape memory composites based on thermoplastic polyurethane/polylactic acid/polyaniline blends. *High Performance Polymers*, *35*(4), 366–378. DOI: 10.1177/09540083221135499

Wei, X., Zou, N., Zeng, L., & Pei, Z. (2022). PolyJet 3D printing: Predicting color by multilayer perceptron neural network. *Annals of 3D Printed Medicine*, *5*, 100049.

Wei, H., Tang, L., Qin, H., Wang, H., Chen, C., Li, Y., & Wang, C. (2024). Optimizing FDM 3D printing parameters for improved tensile strength using the Takagi–Sugeno fuzzy neural network. *Materials Today. Communications*, *38*, 108268. DOI: 10.1016/j.mtcomm.2024.108268

Weimann, C., Lauermann, M., Hoeller, F., Freude, W., & Koos, C. (2017). Silicon photonic integrated circuit for fast and precise dual-comb distance metrology. *Optics Express*, *25*(24), 30091. DOI: 10.1364/OE.25.030091 PMID: 29221043

Wei, S., Kumar, P., Lau, K. B., Wuu, D., Liew, L. L., Wei, F., & Ramamurty, U. (2022). Effect of heat treatment on the microstructure and mechanical properties of 2.4 GPa grade maraging steel fabricated by laser powder bed fusion. *Additive Manufacturing*, *59*, 103190. DOI: 10.1016/j.addma.2022.103190

Wen, Y., Che, Q. T., Wang, S., Park, H. J., & Kim, H. W. (2024). Elaboration of dimensional quality in 3D-printed food: Key factors in process steps. *Comprehensive Reviews in Food Science and Food Safety*, *23*(1), e13267. DOI: 10.1111/1541-4337.13267 PMID: 38284586

West, A. P., Sambu, S. P., & Rosen, D. W. (2001). A process planning method for improving build performance in stereolithography. *Computer Aided Design*, *33*(1), 65–79. DOI: 10.1016/S0010-4485(00)00064-6

White, A., Khial, P., Salehi, F., Hassibi, B., & Hajimiri, A. (2020). A Silicon Photonics Computational Lensless Active-Flat-Optics Imaging System. *Scientific Reports*, *10*(1), 1689. DOI: 10.1038/s41598-020-58027-1 PMID: 32015358

Wiberg, A., Persson, J. A., & Ölvander, J. (2023). A Design Automation Framework Supporting Design for Additive Manufacturing. In International Design Engineering Technical Conferences and Computers and Information in Engineering Conference (Vol. 87295, p. V002T02A083). American Society of Mechanical Engineers. DOI: 10.1115/DETC2023-116415

Wickramasinghe, S., Do, T., & Tran, P. (2020). FDM-based 3D printing of polymer and associated composite: A review on mechanical properties, defects and treatments. *Polymers*, *12*(7), 1529. DOI: 10.3390/polym12071529 PMID: 32664374

Williams, J. M., Adewunmi, A., Schek, R. M., Flanagan, C. L., Krebsbach, P. H., Feinberg, S. E., Hollister, S. J., & Das, S. (2005). Bone tissue engineering using polycaprolactone scaffolds fabricated via selective laser sintering. *Biomaterials*, *26*(23), 4817–4827. DOI: 10.1016/j.biomaterials.2004.11.057 PMID: 15763261

Wilson, S., Thomas, R. J., Mary, N., Bosco, E. T., & Gopinath, A. (2021). Development and fabrication of fused deposition modelling 3D printer. *IOP Conference Series. Materials Science and Engineering*, *1132*(1), 12019. DOI: 10.1088/1757-899X/1132/1/012019

Wohlers, T. T., & Caffrey, T. (2013). Wohlers Report 2013 [Printing State of the Industry: Annual Worldwide Progress Report. Wohlers Associates.]. *Additive Manufacturing*, •••, 3D.

Wohlers, T., & Caffrey, T.Ian Campbell of Wohlers Associates, Inc. (2012). Wohlers Report 2012 [Printing State of the Industry: Annual Worldwide Progress Report. Wohlers Associates.]. *Additive Manufacturing*, •••, 3d.

Wolter, B., Dobmann, G., & Boller, C. (2011). NDT-Based Process Monitoring and Control. Strojniški Vestnik –. *Jixie Gongcheng Xuebao*, *2011*(3), 218–226. DOI: 10.5545/sv-jme.2010.172

Wong, K. C. (2016). 3D-printed patient-specific applications in orthopedics [Review of 3D-printed patient-specific applications in orthopedics]. Orthopedic Research and Reviews, 57. Dove Medical Press. DOI: 10.2147/ORR.S99614

Wong, K. V., & Hernandez, A. (2012). A Review of Additive Manufacturing [Review of A Review of Additive Manufacturing]. ISRN Mechanical Engineering, 2012, 1. Hindawi Publishing Corporation. DOI: 10.5402/2012/208760

Wooldridge, E. (2020). Phantom holes: Optimised internal structural design for use with additive manufacturing, typical fused filament fabrication systems. *International Journal of Rapid Manufacturing*, *9*(2/3), 137. DOI: 10.1504/IJRAPIDM.2020.107732

Wu, P., Wang, J., & Wang, X. (2016). A critical review of the use of 3-D printing in the construction industry [Review of A critical review of the use of 3-D printing in the construction industry]. Automation in Construction, 68, 21. Elsevier BV. DOI: 10.1016/j.autcon.2016.04.005

Wu, D., Song, J., Zhai, Z., Hua, M., Kim, C., Frenkel, I., Jiang, H., & He, X. (2019). Visualizing morphogenesis through instability formation in 4-D printing. *ACS Applied Materials & Interfaces*, *11*(50), 47468–47475. DOI: 10.1021/acsami.9b19730 PMID: 31741387

Wu, J., Aage, N., Lefèbvre, S., & Wang, C. C. L. (2017). Topology Optimization for Computational Fabrication. Eurographics. DOI: 10.2312/egt.20171030

Wu, Q., Tang, W., Yang, J., & Shi, X. (2023). Comparison of thermomechanical responses of single-arc and dual-arc parallel additive manufacturing. *Science and Technology of Welding and Joining*, *28*(7), 580–588. DOI: 10.1080/13621718.2023.2219909

Xiao, Z., Liu, W., Xu, S., Zhou, J., Ren, Z., & Lee, C. (2023). Recent Progress in Silicon-Based Photonic Integrated Circuits and Emerging Applications. *Advanced Optical Materials*, *11*(20), 2301028. DOI: 10.1002/adom.202301028

Xie, J., Yang, Z., Wang, X., & Lai, X. (2019). A cloud service platform for the seamless integration of digital design and rapid prototyping manufacturing. *International Journal of Advanced Manufacturing Technology*, *100*(5–8), 1475–1490. DOI: 10.1007/s00170-018-2809-9

Xiong, H., ... (2014) 'Scalable Architectures for Platform-as-a-Service Clouds: Performance and Cost Analysis', in P. Avgeriou and U. Zdun (eds) *Software Architecture*. Cham: Springer International Publishing (Lecture Notes in Computer Science), pp. 226–233. Available at: DOI: 10.1007/978-3-319-09970-5_21

Xomalis, A., Hain, C., Groetsch, A., Klimashin, F. F., Nelis, T., Michler, J., & Schwiedrzik, J. (2023). Resist-Free E-beam Lithography for Patterning Nanoscale Thick Films on Flexible Substrates. *ACS Applied Nano Materials*, *6*(5), 3388–3394. DOI: 10.1021/acsanm.2c05161

Xu, K., Liu, S., & Ai, Y. (2015). Application of Shearlet transform to classification of surface defects for metals. *Image and Vision Computing*, *35*, 23–30. DOI: 10.1016/j.imavis.2015.01.001

Yadav, S., & Mishra, M. K. (2024). Advancements in artificial intelligence-mediated fabrication of 3D, 4D, and 5D printed drug delivery formulations. *Drug Delivery Letters*, *14*(4), 273–289. DOI: 10.2174/0122103031309831240531084125

Yadegari, M. J., Martucci, A., Biamino, S., Ugues, D., Montanaro, L., Fino, P., & Lombardi, M. (2025). Aluminum Laser Additive Manufacturing: A Review on Challenges and Opportunities Through the Lens of Sustainability. *Applied Sciences (Basel, Switzerland)*, *15*(4), 2221. DOI: 10.3390/app15042221

Yakubov, V., Ostergaard, H., Bhagavath, S., Leung, C. L. A., Hughes, J., Yasa, E., Khezri, M., Löschke, S. K., Li, Q., & Paradowska, A. M. (2024). Recycled aluminium feedstock in metal additive manufacturing: A state of the art review. *Heliyon*, *10*(5), e27243. DOI: 10.1016/j.heliyon.2024.e27243 PMID: 38463898

Yang, Y., Jiang, R., Han, C., Chen, J., Li, H., Wang, Y., ... & Wang, D. (2024). Frontiers in laser additive manufacturing technology. Additive Manufacturing Frontiers, 200160.

Yang, C., & Li, Q. M. (2020). Advanced lattice material with high energy absorption based on topology optimisation. In *Mechanics of Materials* (Vol. 148, p. 103536). Elsevier BV., DOI: 10.1016/j.mechmat.2020.103536

Yang, J., Li, B., Liu, J., Tu, Z., & Wu, X. (2024). Application of additive manufacturing in the automobile industry: A mini review. *Processes (Basel, Switzerland)*, *12*(6), 1101. DOI: 10.3390/pr12061101

Yang, W., Tu, A., Ma, Y., Li, Z., Xu, J., Lin, M., Zhang, K., Jing, L., Fu, C., Jiao, Y., & Huang, L. (2021). Chitosan and whey protein bio-inks for 3D and 4D printing applications with particular focus on food industry. *Molecules (Basel, Switzerland)*, *27*(1), 173. DOI: 10.3390/molecules27010173 PMID: 35011406

Yankin, A., Alipov, Y., Temirgali, A., Serik, G., Danenova, S., Talamona, D., & Perveen, A. (2023). Optimization of Printing Parameters to Enhance Tensile Properties of ABS and Nylon Produced by Fused Filament Fabrication. *Polymers*, *15*(14), 3043. DOI: 10.3390/polym15143043 PMID: 37514431

Yardimci, M. A., Hattori, T., Guceri, S. I., & Danforth, S. C. (1997) Thermal analysis of fused deposition. In: Solid Freeform Fabrication Proceedings, pp. 689–698.

Yasa, E., & Ersoy, K. (2018). A review of the additive manufacturing of fiber reinforced polymer matrix composites.

Yehia, H. M., Hamada, A., Sebaey, T. A., & Abd-Elaziem, W. (2024). Selective laser sintering of polymers : Process parameters, machine learning approaches, and future directions. *Journal of Manufacturing and Materials Processing*, *8*(5), 197. DOI: 10.3390/jmmp8050197

Yeole, S. N., Madhav, C. H. V., & Kesav, R. S. N. H. (2016). Importance and utilization of 3D printing in various applications. https://www.researchgate.net/profile/Shivraj_Yeole/publication/293487409_Importance_and_utilization_of_3D_printing_in_various_applications/links/5720d8aa08aeaced78907019.pdf

Ye, Z., Jia, H., Huang, Z., Shen, C., Long, J., Shi, B., Luo, Y.-H., Gao, L., Sun, W., Guo, H., He, J., & Liu, J. (2023). Foundry manufacturing of tight-confinement, dispersion-engineered, ultralow-loss silicon nitride photonic integrated circuits. *Photonics Research*, *11*(4), 558. DOI: 10.1364/PRJ.486379

Yi, J., Yang, S., Yue, L., & Lei, I. M. (2025). Digital light processing 3D printing of flexible devices : Actuators, sensors and energy devices. *Microsystems & Nanoengineering*, *11*(1), 51. DOI: 10.1038/s41378-025-00885-8 PMID: 40108126

Yin, H., Qu, M., Zhang, H., & Lim, Y. (2018). 3D Printing and Buildings: A Technology Review and Future Outlook [Review of 3D Printing and Buildings: A Technology Review and Future Outlook]. Technology|Architecture + Design, 2(1), 94. Taylor & Francis. DOI: 10.1080/24751448.2018.1420968

Yin, J., Suo, Y., Lv, T., Ma, K., Wang, X., & Zhang, Z. (2022). *Application of 3D Printing Technology in the Construction Industry and Its Development Prospects*. Atlantis Highlights in Intelligent Systems/Atlantis Highlights in Intelligent Systems., DOI: 10.2991/ahis.k.220601.002

Yin, L., Tian, X., Shang, Z., Wang, X., & Hou, Z. (2019). Characterizations of continuous carbon fiber-reinforced composites for electromagnetic interference shielding fabricated by 3D printing. *Applied Physics. A, Materials Science & Processing*, *125*(4), 266. Advance online publication. DOI: 10.1007/s00339-019-2566-0

Yuan, P., & Gu, D. (2015). Molten pool behaviour and its physical mechanism during selective laser melting of TiC/AlSi10Mg nanocomposites: Simulation and experiments. *Journal of Physics. D, Applied Physics*, *48*(3), 35303. DOI: 10.1088/0022-3727/48/3/035303

Yuan, S., Li, J., Yao, X., Zhu, J., Gu, X., Gao, T., Xu, Y., & Zhang, W. (2020). Intelligent optimization system for powder bed fusion of processable thermoplastics. *Additive Manufacturing, 34*, 101182. DOI: 10.1016/j.addma.2020.101182

Yuan, Y., Ma, G., Cheng, C., Zhou, B., Zhao, H., Zhang, H., & Ding, H. (2019). *Artificial Intelligence Diagnosis and Monitoring in Manufacturing. arXiv*. Cornell University., DOI: 10.1093/nsr/nwz190

Yu, B.. (2012) 'Dealing with IC manufacturability in extreme scaling', in *Proceedings of the International Conference on Computer-Aided Design. ICCAD '12: The International Conference on Computer-Aided Design*, San Jose California: ACM, pp. 240–242. Available at: DOI: 10.1145/2429384.2429430

Yu, K., & Dunn, M. L. (2024). Design and extrusion-based 3D printing of continuous fiber composites: Status, challenges, and opportunities. *Langmuir, 40*(17), 8751–8759. DOI: 10.1021/acs.langmuir.4c00139 PMID: 38644644

Zafar, A., Alruwaili, N. K., Rizwanullah, Md., Bukhari, S. N. A., Amir, M., Ahmed, M. M., & Fazil, M. (2019). 3D Printing Technology in Design of Pharmaceutical Products [Review of 3D Printing Technology in Design of Pharmaceutical Products]. Current Pharmaceutical Design, 24(42), 5009. Bentham Science Publishers. DOI: 10.2174/1381612825666190116104620

Zafar, M. Q., & Zhao, H. (2020). 4D printing: Future insight in additive manufacturing. *Metals and Materials International, 26*(5), 564–585. DOI: 10.1007/s12540-019-00441-w

Zaid, O., & El Ouni, M. H. (2024). Advancements in 3D printing of cementitious materials: A review of mineral additives, properties, and systematic developments. *Construction & Building Materials, 427*, 136254. DOI: 10.1016/j.conbuildmat.2024.136254

Zarinejad, M., Tong, Y., Salehi, M., Mu, C., Wang, N., Xu, Y., & Chen, X. (2024). Advancements and perspectives in additive manufacturing of tungsten alloys and composites: Challenges and solutions. *Additive Manufacturing, 85*, 104333. DOI: 10.1016/j.addma.2024.104333

Zárybnická, L., Dvořák, K., Dostalova, Z., & Vojáčková, H. (2019). Study of Different Printing Design Type Polymer Samples Prepared by Additive Manufacturing. *Periodica Polytechnica. Chemical Engineering, 64*(2), 255–264. DOI: 10.3311/PPch.13991

Zastrow, M. (2020). The new 3D printing. *Nature, 578*(7793), 20–23. DOI: 10.1038/d41586-020-00271-6 PMID: 32025009

Zeng, Y., Jiang, L., He, Q., Wodnicki, R., Yang, Y., Chen, Y., & Zhou, Q. (2021). Recent progress in 3D printing piezoelectric materials for biomedical applications. *Journal of Physics. D, Applied Physics*, *55*(1), 013002. DOI: 10.1088/1361-6463/ac27d2

Zhang, H., Zhou, X., Fang, Y., Xiong, Z., & Zhang, T. (2024). AI-driven 3D bio-printing for regenerative medicine: From bench to bedside [Review of AI-driven 3D bioprinting for regenerative medicine: From bench to bedside]. Bioactive Materials, 45, 201. Elsevier BV. DOI: 10.1016/j.bioactmat.2024.11.021

Zhang, Y., & Chou, Y. K. (2008) A Parametric Study of Part Distortions in FDM Using 3D FEA. Tuscaloosa, AL: Mechanical Engineering Department, The University of Alabama. 35487 Reviewed, accepted September 14, 2006.

Zhang, Y., Gulati, K., Li, Z., Di, P., & Liu, Y. (2021). Dental Implant Nano-Engineering: Advances, Limitations and Future Directions [Review of Dental Implant Nano-Engineering: Advances, Limitations and Future Directions]. Nanomaterials, 11(10), 2489. Multidisciplinary Digital Publishing Institute. DOI: 10.3390/nano11102489

Zhang, H., Yang, D., & Sheng, Y. (2018). Performance-driven 3D printing of continuous curved carbon fibre reinforced polymer composites: A preliminary numerical study. *Composites. Part B, Engineering*, *151*, 256–264. DOI: 10.1016/j.compositesb.2018.06.017

Zhang, J., Hu, Q., Wang, S., Tao, J., & Gou, M. (2019). Digital Light Processing Based Three-dimensional Printing for Medical Applications [Review of Digital Light Processing Based Three-dimensional Printing for Medical Applications]. *International Journal of Bioprinting*, *6*(1), 242. DOI: 10.18063/ijb.v6i1.242 PMID: 32782984

Zhang, M., & Shang, J. (2024). Molecular dynamics simulation of ultrasound-assisted 3d printing of aluminum matrix composites. *Fourth International Conference on Applied Mathematics, Modelling, and Intelligent Computing (CAMMIC 2024)*, 66. DOI: 10.1117/12.3035661

Zhang, S., Yu, Q., Wan, S., Cao, H., & Huang, Y. (2024). Digital supply chain: Literature review of seven related technologies. *Manufacturing Review*, *11*, 8. DOI: 10.1051/mfreview/2024006

Zhang, W., Li, J., Fang, J., Hui, L., Qin, L., Gong, T., Sun, F., & Feng, H. (2023). Atomic layer deposited high quality aln thin films for efficient thermal management. *Journal of Materials Chemistry. A, Materials for Energy and Sustainability*, *11*(40), 21846–21856. DOI: 10.1039/D3TA04618D

Zhang, Y., & Chou, Y. K. (2006). Three-dimensional finite element analysis simulations of the fused deposition modelling process. *Proceedings of the Institution of Mechanical Engineers. Part B, Journal of Engineering Manufacture*, *220*(10), 1663–1671. DOI: 10.1243/09544054JEM572

Zhang, Z., Demir, K. G., & Gu, G. X. (2019). Developments in 4D-printing: A review on current smart materials, technologies, and applications. *International Journal of Smart and Nano Materials*, *10*(3), 205–224. DOI: 10.1080/19475411.2019.1591541

Zhao, H., Wang, Y., Liu, X., Wang, X., Chen, Z., Lei, Z., ... & Singh, A. (2024). Review on solid wastes incorporated cementitious material using 3D concrete printing technology. *Case Studies in Construction Materials*, e03676.

Zhao, C. (2020). Research on Cultural and Creative Product Design System Based on 3DP Technology. *Journal of Physics: Conference Series*, *1648*(4), 42097. DOI: 10.1088/1742-6596/1648/4/042097

Zhao, H., Liu, X., Zhao, W., Wang, G., & Liu, B. (2019, June). An overview of research on FDM 3D printing process of continuous fiber reinforced composites. [). IOP Publishing.]. *Journal of Physics: Conference Series*, *1213*(5), 052037. DOI: 10.1088/1742-6596/1213/5/052037

Zhao, J., Yang, Y., Kobir, M. H., Faludi, J., & Zhao, F. (2024). Driving additive manufacturing towards circular economy: State-of-the-art and future research directions. *Journal of Manufacturing Processes*, *124*, 621–637. DOI: 10.1016/j.jmapro.2024.06.018

Zhao, R., Shmatok, A., Fischer, R., & Prorok, B. C. (2023). Investigation of causal relationships between printing parameters, pore properties and porosity in laser powder bed fusion. *Metals*, *13*(2), 330. DOI: 10.3390/met13020330

Zhong, N., & Zhao, X. (2017). 3D printing for clinical application in otorhinolaryngology. *European Archives of Oto-Rhino-Laryngology*, *274*(12), 4079–4089. DOI: 10.1007/s00405-017-4743-0 PMID: 28929219

Zhong, W., Li, F., Zhang, Z., Song, L., & Li, Z. (2001). Short fiber reinforced composites for fused deposition modeling. *Materials Science and Engineering A*, *301*(2), 125–130. DOI: 10.1016/S0921-5093(00)01810-4

Zhou, H. R., Yang, H., Li, H. Q., Ma, Y. C., Yu, S., Shi, J., Cheng, J., Gao, P., Yu, B., Miao, Z., & Wei, Y. P. (2024). Advancements in machine learning for material design and process optimization in the field of additive manufacturing. *China Foundry*, *21*(2), 101–115. DOI: 10.1007/s41230-024-3145-3

Zhou, L. Y., Fu, J., & He, Y. (2020). A review of 3D printing technologies for soft polymer materials. *Advanced Functional Materials, 30*(28), 2000187. DOI: 10.1002/adfm.202000187

Zhou, L., Miller, J., Vezza, J., Mayster, M., Raffay, M., Justice, Q., Al Tamimi, Z., Hansotte, G., Sunkara, L. D., & Bernat, J. (2024). Additive manufacturing: A comprehensive review. *Sensors (Basel), 24*(9), 2668. DOI: 10.3390/s24092668 PMID: 38732776

Zhu, C., Gemeda, H. B., Duoss, E. B., & Spadaccini, C. M. (2024). Toward multiscale, multimaterial 3D printing. *Advanced Materials, 36*(34), 2314204. DOI: 10.1002/adma.202314204 PMID: 38775924

Zhu, G., Houck, H. A., Spiegel, C. A., Selhuber-Unkel, C., Hou, Y., & Blasco, E. (2023). Introducing dynamic bonds in light-based 3d printing. *Advanced Functional Materials, 34*(20), 2300456. Advance online publication. DOI: 10.1002/adfm.202300456

Živanović, S., Tabaković, S., & Randjelović, S. (2019). Rapid prototyping of art sculptural shapes according to the sample. Advanced Technologies and Materials, 44(1), 27-32. DOI: <ALIGNMENT.qj></ALIGNMENT>10.24867/ATM-2019-1-005

Zohdi, N., & Yang, R. (2021). Material anisotropy in additively manufactured polymers and polymer composites: A review. *Polymers, 13*(19), 3368. DOI: 10.3390/polym13193368 PMID: 34641184

Zolfagharian, A., Haq, M. I. U., Nafea, M., & Bodaghi, M. (2023). *4D Printing of Smart Magnetic-Based Robotic Materials. In 3D Printing and Sustainable Product Development*. CRC Press.

Zou, Y., Song, Y., & Liu, X. (2023). Designing for Strength: Enhancing Mechanical Performance through Structured Patterns in 3D Printed Elastomer. *Materials Research Express, 10*(9), 95302. DOI: 10.1088/2053-1591/acf64f

Zub, K., Hoeppener, S., & Schubert, U. S. (2022). Inkjet printing and 3D printing strategies for biosensing, analytical, and diagnostic applications. *Advanced Materials, 34*(31), 2105015. DOI: 10.1002/adma.202105015 PMID: 35338719

Zuo, Z., De Corte, W., Huang, Y., Chen, X., Zhang, Y., Li, J., Zhang, L., Xiao, J., Yuan, Y., Zhang, K., Zhang, L., & Mechtcherine, V. (2024). Strategies towards large-scale 3D printing without size constraints. *Virtual and Physical Prototyping, 19*(1), e2346821. DOI: 10.1080/17452759.2024.2346821

Дородов, П. В., Kasatkin, V., Lekomcev, P. L., Petrov, V. A., & Niyazov, A. M. (2021). Rigidity, creep and dynamic strength of plastics for three-dimensional printing of machine parts. In IOP Conference Series Earth and Environmental Science (Vol. 677, Issue 5, p. 52045). IOP Publishing. DOI: 10.1088/1755-1315/677/5/052045

Дородов, П. В., Kasatkin, V., Kasatkina, N. Y., Petrov, V. A., & Litvinyuk, A. A. (2021). Investigation of the main mechanical characteristics of plastics for three-dimensional printing of machine parts models. *IOP Conference Series. Earth and Environmental Science*, *677*(5), 52041. DOI: 10.1088/1755-1315/677/5/052041

About The Contributors

Romdhane Ben Khalifa is assistant professor, academic and researcher in the field of mechanical engineering, primarily associated with the University of Gabes. He holds a position at the National engineering school of Gabes (ENIG) where his research and teaching focus on optimization, neural networks, and CAD/CAM technologies. His work is well-regarded for its contributions to the improvement of manufacturing processes and the application of artificial intelligence in engineering. He obtained his PhD Doctoral thesis in Mechanical Engineering at National Engineering School of Tunisia in 2012. In 2005, he obtained the Master degree of Mechanical Engineering from the National Engineering School of Tunisia. He has authored a number of research papers in the field of Mechanical engineering in impacted international journals. Publications

***

Suresh C Associate Professor Global Academy of Technology, Bangalore I am specializes in aeronautical engineering research, focusing on Impact analysis, Fluid structure interactions, aircraft structures, propulsion systems, and additive manufacturing (3D printing) for aerospace applications. I am actively integrates advanced technologies like Digital Twin into my research and academic initiatives. i am dedicated to fostering innovation through experiment design, conference contributions, and industry-focused projects, ensuring impactful outcomes in both academia and aerospace research.

Saurav Datta is presently working as Professor in the Department of Mechanical Engineering, National Institute of Technology, Rourkela (Odisha). After receiving B.E. (Honors) in Mechanical Engineering from Jadavpur University, Kolkata in 2003, he was awarded Ph.D. (Engineering) by Jadavpur University in December 2008. He joined the Department of Mechanical Engineering of NIT Rourkela in August 2008.

His domain of research includes Manufacturing, Quality Optimization, Statistical Modeling/ Optimization of manufacturing Processes, Industrial Decision Making, Traditional and Non-traditional machining, and Metal Additive Manufacturing. He is Life Associate Member of Indian Institute of Welding. He has published more than 200 articles in International Journals, National Journals of repute, and proceedings of different International and National level Conferences. To date, he mentored 44 M. Tech. projects and supervised a total number of 10 PhDs with the capacity of Sole/ Principal Supervisor. He executed two sponsored projects funded by SERB and CSIR, with capacity of Principal Investigator and Co-Investigator, respectively. He is currently executing 02 more sponsored projects on Metal Additive Manufacturing funded by the Science and Technology Department, Govt. of Odisha and CSIR, respectively. Recently, his name has been included within the Top 2% of the most Influential Scientists – 2023, by the Stanford University list.

Boopathy Govindarajan earned his undergraduate degree in mechanical engineering from University of Madras, his master's degree in aeronautical engineering from Madras Institute of Technology, and his doctorate from the Department of Aeronautical Engineering, St. Peter's Institute of Higher Education and Research, Chennai. At Vel Tech University, a professorship is currently held in the department of aeronautical engineering. In addition to guiding 2 Ph.D Scholars, guided 8 PG & 90 UG scholars on their significant projects, he has written in 30 international publications. He is a Life Member of the Society of Automotive Engineers INDIA and the Aeronautical Society of India.

Balaji Ganesan is a distinguished academic and researcher associated with the Hindustan Institute of Technology and Science (HITS) in Chennai, India. Known for his contributions to fields like Aerodynamics, Wind Turbiens, High Speed flow, Nose cone and Composite materials, Dr. Balaji holds a Ph.D. in Aerospace Engineering and has published numerous research papers in international journals and conferences. At HITS, Dr. Balaji has played a significant role in curriculum development and student mentoring, actively collaborating on research projects that enhance academic and practical knowledge. His work often involves interdisciplinary approaches, OBE and SDG to address real-world challenges.In addition to his research and teaching, Dr. Balaji is a sought-after speaker at national and international academic events. He is known for his passion for technology-driven solutions and is dedicated to fostering an innovative learning environment that prepares students for modern technological landscapes.

Kaushik Kumar holds a Ph.D. in Engineering from Jadavpur University, India, an MBA in Marketing Management from Indira Gandhi National Open University,

India and a Bachelor of Technology from Regional Engineering College (Now National Institute of Technology), Warangal, India. For 11 years, he worked in a manufacturing unit of Global repute. He is currently working as a Professor in the Department of Mechanical Engineering, Birla Institute of Technology, Mesra, Ranchi, India. He has 23 Years of Teaching and Research Experience. His research interests include Composites, Optimization, Non-Conventional Machining, CAD / CAM, Rapid Prototyping and Quality Management Systems towards product development for societal and industrial usage and has received 35+ patents for them most of which are donated to Start-ups and Industrial fraternity through Jharkhand Small Industries Association (JSIA). He has published 55+ Books (They are referred as Text Books and Reference Books by 40+ Universities / Institutes in their academic curriculum), 95+ Book Chapters and 200+ Research Papers in peer-reviewed reputed national and international journals. Kaushik has also served as Editor–in–Chief, Series Editor, Guest Editor, Editor, Editorial Board Member and Reviewers for International and National Journals. He has been felicitated with many awards and honours including Distinguished Alumnus Award for Professional Excellence 2023 under Academic and Research from National Institute of Technology, Warangal, India. He has also received Sponsored Research and Consultancy Projects of more than 1 Crore from Govt. of India and abroad. Kaushik has delivered expert lectures as Keynote Speaker at International & National Conferences, Resource Person at Various Workshops, FDP's and Short-Term Courses. He has guided many students of Doctoral, Masters and Under Graduate programmes of his home and other institutions in India and abroad. He also has served as Reviewer and Examiner of Doctoral and Masters dissertation for institutes in India and abroad.

Vignesh Kumar Murugesan was born in Madurai, India, in 1985. He received the B.E. degree in aeronautical engineering from Anna University, Chennai, India, in 2008, and the M.E. degree in CAD/CAM from the Anna University of Technology, Tirunelveli, India, in 2011. He completed Ph.D in open jet studies in the year 2022. Since 2013, he has been an Assistant Professor with the Department of Aerospace Engineering, SRM Institute of Science and Technology, India. His research interests include jet flow dynamics, wind engineering, and wind tunnel testing. He is an Associate Member of the Aeronautical Society of India.

M. Maniraj is an Associate Professor in the Department of Mechanical Engineering at Sardar Raja College of Engineering, Alangulam, Tamil Nadu, India. With 15 years of teaching experience, he is currently pursuing his Ph.D. at Anna University, Chennai. His research interests include additive manufacturing and process parameter optimization.

S. M. Raj Kumar is a distinguished academician and researcher in Mechanical Engineering, born on May 25, 1979. He holds a Ph.D. from Anna University, Chennai, and has received awards for his academic excellence, including the Gold Medal for his M.E. in Computer Aided Design. With 22+ years of experience in teaching, research, and industry, Raj Kumar has worked in various institutions and has published numerous research papers in reputable journals and conferences.

Jyoti Rani, Assistant professor, school of fashion design and technology. Research and Academic experience. 2 year experience in academics and research. Research area: fashion, textile, natural dyeing, consumer behaviour, AI, merchandising. she has published 28 research paper in refered journal and 5 book chapters.

Ashish Kumar Sahu is a Mechanical Engineering student at NIT Rourkela (CGPA: 9.08/10), skilled in Abaqus-CAE, Ansys, SolidWorks, MATLAB, and Python. His research spans material modeling and dynamic stability analysis, with projects in robotics and power plant optimization. A Global Winner of ASME E-Fests Ocean Cleanup Challenge and Finalist at IIT Bombay's eYantra Robotics Competition, he also co-leads the Cyborg Robotics Club and has organized major tech events. With a CSWA certification and multiple accolades, Ashish is passionate about robotics, automation, and mechanical design.

Ali Snoussi is an Associate Professor at the Department of Mechanical Engineering of the National Engineering School of Gabes (Tunisia). He holds a PHD degree in Energetic Engineering from National Engineering School of Monastir (Tunisia) and has 25 years of experience of teachnig and scientific research. He has authored a number of research papers in the field of thermal engineering in impacted international journals. The main research activities involve solar energy, desalination, thermodynamic analysis of processes. Besides, he has been a member of a number of projects in the framework of European programs (Tempus, Erasmus, Horizon).

Sudipta Swain is working as a project research scholar in the Department of Mechanical Engineering under the guidance of Prof Saurav Datta . My research work is on Additive Manufacturing. With a strong research background in manufacturing, I have contributed significantly to this area.

V. Srinivasan earned his undergraduate degree in mechanical engineering from S.K.P Institute of Technology affiliated to Anna University, his master's degree in CAD/CAM from Vel Tech Rangarajan Dr. Sagunthala R & D Institute of Science and Technology and currently pursuing Ph. D in Aeronautical Engineering from Vel Tech Rangarajan Dr. Sagunthala R & D Institute of Science and Technology.

Currently doing research in LiDAR (Light detection and Ranging) and other remote sensing and imaging sensors, its integration with UAVs. Obstacle avoidance and autonomous navigation systems in UAV using sensor fusion are his key areas for research.

Index

Symbols

A

B

C

D

F

G

H

I

L

M

O

P

R

S

T

W

www.ingramcontent.com/pod-product-compliance
Lightning Source LLC
LaVergne TN
LVHW080616300926
849162LV00005B/1179